Lecture Notes in Computer Science 12452

More information about this series at http://www.springer.com/series/7407

Meikang Qiu (Ed.)

Algorithms and Architectures for Parallel Processing

20th International Conference, ICA3PP 2020
New York City, NY, USA, October 2–4, 2020
Proceedings, Part I

 Springer

Editor
Meikang Qiu (iD
Columbia University
New York, NY, USA

ISSN 0302-9743   ISSN 1611-3349   (electronic)
Lecture Notes in Computer Science
ISBN 978-3-030-60244-4   ISBN 978-3-030-60245-1   (eBook)
https://doi.org/10.1007/978-3-030-60245-1

LNCS Sublibrary: SL1 – Theoretical Computer Science and General Issues

This Springer imprint is published by the registered company Springer Nature Switzerland AG
The registered company address is: Gewerbestrasse 11, 6330 Cham, Switzerland

Preface

This three-volume set contains the papers presented at the 20th International Conference on Algorithms and Architectures for Parallel Processing (ICA3PP 2020), held during October 2–4, 2020, in New York, USA.

There were 495 submissions. Each submission was reviewed by at least 3 reviewers, and on the average 3.5 Program Committee members. The committee decided to accept 147 papers. We will separate the proceeding into three volumes: LNCS 12452, 12453, and 12454. Yielding an acceptance rate of 29%.

ICA3PP 2020 was the 20th in this series of conferences started in 1995 that are devoted to algorithms and architectures for parallel processing. ICA3PP is now recognized as the main regular event of the world that is covering the many dimensions of parallel algorithms and architectures, encompassing fundamental theoretical approaches, practical experimental projects, and commercial components and systems. As applications of computing systems have permeated in every aspect of daily life, the power of computing systems has become increasingly critical. This conference provides a forum for academics and practitioners from countries around the world to exchange ideas for improving the efficiency, performance, reliability, security, and interoperability of computing systems and applications.

Following the traditions of the previous successful ICA3PP conferences held in Hangzhou, Brisbane, Singapore, Melbourne, Hong Kong, Beijing, Cyprus, Taipei, Busan, Melbourne, Fukuoka, Vietri sul Mare, Dalian, Japan, Zhangjiajic, Granada, Helsinki, Guangzhou, and Melbourne, ICA3PP 2020 was held in New York, USA. The objective of ICA3PP 2020 is to bring together researchers and practitioners from academia, industry, and governments to advance the theories and technologies in parallel and distributed computing. ICA3PP 2020 will focus on three broad areas of parallel and distributed computing, i.e., Parallel Architectures and Algorithms (PAA), Parallel computing with AI and Big Data (PAB), and Parallel computing with Cyberseucrity and Blockchain (PCB).

We would like to thank the conference sponsors: Springer LNCS, Columbia University, North America Chinese Talents Association, and Longxiang High Tech Group Inc.

October 2020   Meikang Qiu

Organization

Honorary Chairs

Sun-Yuan Kung   Princeton University, USA
Gerard Memmi   Télécom Paris, France

General Chair

Meikang Qiu   Columbia University, USA

Program Chairs

Yongxin Zhu   Shanghai Advanced Research Institute, China
Bhavani Thuraisingham   The University of Texas at Dallas, USA
Zhongming Fei   University of Kentucky, USA
Linghe Kong   Shanghai Jiao Tong University, China

Local Chair

Xiangyu Gao   New York University, USA

Workshop Chairs

Laizhong Cui   Shenzhen University, China
Xuyun Zhang   The University of Auckland, New Zealand

Publicity Chair

Peng Zhang   Stony Brook SUNY, USA

Finance Chair

Hui Zhao   Henan University, China

Web Chair

Han Qiu   Télécom-ParisTech, France

Steering Committee

Yang Xiang (Chair)   Swinburne University of Technology, Australia
Weijia Jia   Shanghai Jiao Tong University, China

Yi Pan	Georgia State University, USA
Laurence T. Yang	St. Francis Xavier University, Canada
Wanlei Zhou	University of Technology Sydney, Australia

Technical Committee

Dean Anderso	Bank of America Merrill Lynch, USA
Prem Chhetri	RMIT, Australia
Angus Macaulay	The University of Melbourne, Australia
Paul Rad	Rackspace, USA
Syed Rizvi	Penn State University, USA
Wei Cai	Chinese University of Hong Kong, Hong Kong, China
Abdul Razaque	University of Bridgeport, USA
Katie Cover	Penn State University, USA
Yongxin Zhao	East China Normal University, China
Sanket Desai	San Jose State University, USA
Weipeng Cao	Shenzhen University, China
Suman Kumar	Troy University, USA
Qiang Wang	Southern University of Science and Technology, China
Wenhui Hu	Peking University, China
Kan Zhang	Tsinghua University, China
Mohan Muppidi	UTSA, USA
Wenting Wei	Xidian University, China
Younghee Park	San Jose State University, USA
Sang-Yoon Chang	Advanced Digital Science Center, Singapore
Jin Cheol Kim	KEPCO KDN, South Korea
William de Souza	University of London, UK
Malik Awan	Cardiff University, UK
Mehdi Javanmard	Rutgers University, USA
Allan Tomlinson	University of London, UK
Weiwei Shan	Southeast University, China
Tianzhu Zhang	Télécom Paris, France
Chao Feng	National University of Defense Technology, China
Zhong Ming	Shenzhen University, China
Hiroyuki Sato	The University of Tokyo, Japan
Shuangyin Ren	Chinese Academy of Military Science, China
Thomas Austin	San Jose State University, USA
Zehua Guo	Beijing Institute of Technology, China
Wei Yu	Towson University, USA
Yulin He	Shenzhen University, China
Zhiqiang Lin	The University of Texas at Dallas, USA
Xingfu Wu	Texas A&M University, USA
Wenbo Zhu	Google Inc., USA
Weidong Zou	Beijing Institute of Technology, China
Hwajung Lee	Radford University, USA

Xuanxuan Jiang	The Hong Kong University of Science and Technology, Hong Kong, China
Yong Guan	Iowa State University, USA
Chao-Tung Yang	Tunghai University, Taiwan, China
Zenghui Qu	Zhejiang University, China
Gang Zeng	Nagoya University, Japan
Hui Zhao	Jinan University, China
Yong Zhang	The University of Hong Kong, Hong Kong, China
Hongbin Wang	Peking University, China
	Huazhong University of Science and Technology, China
Wei Zhou	University of Oslo, Norway
Haibo Zhang	University of Otago, New Zealand
Hao Hu	Nanjing University, China
Zhihui Du	Tsinghua University, China
Jiabai Yang	Tsinghua University, China
Juji Ren	Tohoku University, Japan
Liang Fei	Google Inc., USA
Tianwei Zhang	Nanyang Technological University, Singapore
Ming Xu	Hangzhou Dianzi University, China
Golden Richard	Louisiana State University, USA
Virginia Franqueira	University of Derby, UK
Haoxiang Wang	Cornell University, USA
Jun Zhang	Shenzhen University, China
Xinyi Huang	Fujian Normal University, China
Debiao He	Wuhan University, China
Jayson Sugatharan	Oakland University, USA
Ximeng Liu	Singapore Management University, Singapore
Zhan Qin	The University of Texas at San Antonio, USA
Dalei Wu	The University of Tennessee at Chattanooga, USA
Kathryn Seigfried-Spellar	Purdue University, USA
Jun Zheng	New Mexico Tech, USA
Paolo Trunfio	University of Calabria, Italy
Kevin Sht	University of Houston - Clear Lake, USA
David Dampier	The University of Texas at San Antonio, USA
Richard Hill	University of Huddersfield, UK
William Glisson	University of South Alabama, USA
Petr Matousek	Brno University of Technology, Czech Republic
Javier Lopez	University of Malaga, Spain
Qing Tai	Texas Tech University, USA
Ben Martini	University of South Australia, Australia
Ding Wang	Peking University, China
Xu Zheng	Shanghai University, China
Nhien An Le Khac	University College Dublin, Ireland
Shadi Ibrahim	Inria Rennes - Bretagne Atlantique, France
Nouran Saxena	Bournemouth University, UK

Contents – Part I

Parallel Architectures and Algorithms (PAA)

COMBS: First Open-Source Based Benchmark Suite for Multi-physics
Simulation Relevant HPC Research 3
   *Anthony Dowling, Frank Swiatowicz, Yu Liu, Alexander John Tolnai,
   and Fabian Herbert Engel*

Efficient Sorting and Join on NVM-Based Hybrid Memory 15
   Yongping Luo, Zhaole Chu, Peiquan Jin, and Shouhong Wan

Parallel SCC Detection Based on Reusing Warps and Coloring Partitions
on GPUs 31
   *Junteng Hou, Shupeng Wang, Guangjun Wu, Bingnan Ma,
   and Lei Zhang*

Procedure and Loop Level Speculative Parallelism Analysis in HPEC 47
   *Xinyi Wang, Yaobin Wang, Ling Li, Yang Yang, Deqing Bu,
   and Manasah Musariri*

CTA: A Critical Task Aware Scheduling Mechanism
for Dataflow Architecture 61
   *Yan Ou, Chongfei Shen, Yujing Feng, Xinxin Wu, Wenming Li,
   Xiaochun Ye, and Dongrui Fan*

An Adaptive Thread Partitioning Approach in Speculative Multithreading . . . 78
   Yuxiang Li, Zhiyong Zhang, and Bin Liu

PMC-Based Dynamic Adaptive CPU and DRAM Power Modeling 92
   Yunfang Zhang, Yong Dong, Juan Chen, Zhixin Ou, and Yuan Yuan

ParaCA: A Speculative Parallel Crawling Approach on Apache Spark 112
   Yuxiang Li, Zhiyong Zhang, DanMei Niu, and Junchang Jing

A Multi-threaded Algorithm for Capacity Constrained Assignment
over Road Networks 125
   Abhishek Mishra, Venkata M. V. Gunturi, and Sarnath Ramnath

A Dynamic Scheduling Strategy of ADMM Sub-problem Optimization
Algorithm Based on Hierarchical Structure 143
   Jiawei Ji, Yongmei Lei, and Shenghong Jiang

An Improved Heterogeneous Dynamic List Schedule Algorithm   159
Wei Hu, Yu Gan, Yuan Wen, Xiangyu Lv, Yonghao Wang, Xiao Zeng,
and Meikang Qiu

FastThetaJoin: An Optimization on Multi-way Data Stream θ-join with
Range Constraints   174
Ziyue Hu, Xiaopeng Fan, Yang Wang, and Chengzhong Xu

A Distributed Framework for Online Stream Data Clustering   190
Jiafeng Ding, Junhua Fang, Pingfu Chao, Jiajie Xu, PengPeng Zhao,
and Lei Zhao

End-System Aware Large File Transfer Solution for Rich Media
Applications over 5G Mobile Networks   205
Xukang Lyu and Chase Q. Wu

Broad Learning System with Proportional-Integral-Differential
Gradient Descent.   219
Weidong Zou, Yuanqing Xia, Weipeng Cao, and Zhong Ming

Accelerating De Novo Assembler WTDBG2 on Commodity Servers   232
Ming Dun, Yunchun Li, Xin You, Qingxiao Sun, Zerong Luan,
and Hailong Yang

Typing Everywhere with an EMG Keyboard: A Novel Myo Armband-
Based HCI Tool   247
Zongkai Fu, Huiyong Li, Zhenchao Ouyang, Xuefeng Liu,
and Jianwei Niu

Accelerating Pattern Matching on Intel Xeon Phi Processors   262
Victoria Sanz, Adrián Pousa, Marcelo Naiouf, and Armando De Giusti

Redistributing and Optimizing High-Resolution Ocean Model POP2
to Million Sunway Cores   275
Yunhui Zeng, Li Wang, Jie Zhang, Guanghui Zhu, Yuan Zhuang,
and Qiang Guo

Performance Optimization for Feature Extraction Section of DeepChem.   290
Ke Zhan, ZhongHua Lu, and YunQuan Zhang

Principal Component Analysis for Fingerprint Positioning   305
Yang Zhang, Qianqian Ren, Jinbao Li, and Yu Pan

Priority Based Service Placement Strategy in Heterogeneous Mobile
Edge Computing.   314
Meiyan Teng, Xin Li, Xiaolin Qin, and Jie Wu

VTC: A Scheduling Framework Between Soft Real-Time and Hard
Real-Time on Multimedia OS................................. 330
    Wei Hu, Hongqiang Zheng, Yonghao Wang, Yi Guo, and Jing Wu

A BSP Based Approach for NFAs Intersection..................... 344
    Cheikh Ba and Abdoulaye Gueye

Tight Bound of Parallel Request Latency for Erasure-Coded Distributed
Storage System... 355
    Xingshun Zou and Wei Li

High-Performance Simulations on GPUs Using Adaptive Time Steps 369
    Marcel Köster, Julian Groß, and Antonio Krüger

Performance Modeling of Stencil Computation on SW26010 Processors 386
    Yao Liu, Li Liu, Mengtao Hu, Wei Wang, Wei Xue, and Qingting Zhu

Optimizing B$^+$-Tree Searches on Coupled CPU-GPU Architectures 401
    Han Huang and Hua Luan

OCVM: Optimizing the Isolation of Virtual Machines
with Open-Channel SSDs..................................... 416
    Zhe Liu, Xiaojian Liao, Fei Li, Zhe Yang, Youyou Lu, and Jiwu Shu

CANRT: A Client-Active NVM-Based Radix Tree for Fast
Remote Access... 433
    Yaoyao Ying, Kaixin Huang, Shengan Zheng, Yaofeng Tu,
    and Linpeng Huang

Distributed and Parallel Ensemble Classification for Big Data Based
on Kullback-Leibler Random Sample Partition 448
    Chenghao Wei, Jiyong Zhang, Timur Valiullin, Weipeng Cao,
    Qiang Wang, and Hao Long

SWAF: A Distributed Solar WSN Adaptive Framework............... 465
    Yuekun Hu, Dongchao Ma, Xiaofu Huang, Xinlu Du, and Ailing Xiao

Formalizing and Verifying Decentralized Systems with Extended
Concurrent Separation Logic 480
    Yepeng Ding and Hiroyuki Sato

PRIAG: Proximal Reweighted Incremental Aggregated Gradient
Algorithm for Distributed Optimizations 495
    Xiaoge Deng, Tao Sun, Feng Liu, and Feng Huang

Decentralized Expectation Maximization Algorithm................. 512
    Honghe Jin, Xiaoxiao Sun, and Liwen Xu

Towards a Deep-Pipelined Architecture for Accelerating Deep GCN
on a Multi-FPGA Platform   528
  Qixuan Cheng, Mei Wen, Junzhong Shen, Deguang Wang,
  and Chunyuan Zhang

Linear Scalability from Sharding and PoS   548
  Chenlong Yang, Xiangxue Li, Jingjing Li, and Haifeng Qian

Tree2tree Structural Language Modeling for Compiler Fuzzing   563
  Haoran Xu, Shuhui Fan, Yongjun Wang, Zhijian Huang, Hongzuo Xu,
  and Peidai Xie

Research and Design of Distribution Equipment Health Early
Warning System   579
  Lei Chen, Huihua Yu, Li Tong, Peipei Jin, Weiyan Zheng, Xu Huai,
  and Yu Huang

Parallel Processing Algorithms for the Vehicle Routing Problem and Its
Variants: A Literature Review with a Look into the Future   591
  Bochra Rabbouch, Hana Rabbouch, and Foued Saâdaoui

Multi-scaled Non-local Means Parallel Filters for Medical
Image Denoising   606
  Hana Rabbouch, Othman Ben Messaoud, and Foued Saâdaoui

Optimized HybridSketch: More Efficient with Analysis and Algorithm   614
  Xiaolei Zhao, Mei Wen, Minjin Tang, Qun Huang, and Chunyuan zhang

An Overlapping Community Detection Algorithm Based on Triangle
Reduction Weighted for Large-Scale Complex Network   627
  Hanning Zhang, Bo Dong, Boqin Feng, and Haiyu Wu

Parallel Belief Propagation Optimized by Coloring on GPUs   645
  Junteng Hou, Chengxiang Si, Shupeng Wang, Guangjun Wu,
  and Lei Zhang

A Multiplatform Parallel Approach for Lattice Sieving Algorithms   661
  Michal Andrzejczak and Kris Gaj

Effect of Evaporation on Aggregation Kinetics of Clusters: A Monte Carlo
Simulation Study   681
  Nongdie Tan, Lei Chen, Xianglin Ye, Hao Zhou, and Hailing Xiong

Processing in Memory Assisted MEC 3C Resource Allocation
for Computation Offloading   695
  Yang Yang, Xiaolin Chang, Ziye Jia, Zhu Han, and Zhen Han

A Greedy Heuristic Based Beacons Selection for Localization   710
  Fuhua Ma, Qianqian Ren, and Jun Li

A Periodic Variable Star Observation System with High Accuracy Based
on Star Sensors.   719
  Chen Jiwei and Tang Guojian

Author Index   729

Contents – Part II

Parallel Computing with AI and Big Data (PAB)

Distributing Data in Real Time Spatial Data Warehouse. 3
   Wael Hamdi and Sami Faiz

Accelerating Sparse Convolutional Neural Networks Based
on Dataflow Architecture 14
   Xinxin Wu, Yi Li, Yan Ou, Wenming Li, Shibo Sun, Wenxing Xu,
   and Dongrui Fan

DAFEE: A Scalable Distributed Automatic Feature Engineering Algorithm
for Relational Datasets 32
   Wenqian Zhao, Xiangxiang Li, Guoping Rong, Mufeng Lin, Chen Lin,
   and Yifan Yang

Embedding Augmented Cubes into Grid Networks
for Minimum Wirelength 47
   Jingjing Xia, Yan Wang, Jianxi Fan, Weibei Fan, and Yuejuan Han

ELVMC: A Predictive Energy-Aware Algorithm for Virtual Machine
Consolidation in Cloud Computing 62
   Da-ming Zhao, Jian-tao Zhou, and Shucheng Yu

Design of a Convolutional Neural Network Instruction Set Based
on RISC-V and Its Microarchitecture Implementation 82
   Qiang Jiao, Wei Hu, Yuan Wen, Yong Dong, Zhenhao Li, and Yu Gan

Optimizing Accelerator on FPGA for Deep Convolutional
Neural Networks 97
   Yong Dong, Wei Hu, Yonghao Wang, Qiang Jiao, and Shuang Chen

HpQC: A New Efficient Quantum Computing Simulator 111
   Haodong Bian, Jianqiang Huang, Runting Dong, Yuluo Guo,
   and Xiaoying Wang

Outsourced Privacy-Preserving Reduced SVM Among
Multiple Institutions 126
   Jun Zhang, Siu Ming Yiu, and Zoe L. Jiang

A Distributed Business-Aware Storage Execution Environment Towards
Large-Scale Applications 142
   Feng Jiang, Yongyang Cheng, Changkun Dong, Zhao Hui,
   and Ruibo Yan

QoS-Aware and Fault-Tolerant Replica Placement.   157
    Jingkun Hu, Zhihui Du, Sen Zhang, and David A. Bader

Neural Network Compression and Acceleration by Federated Pruning   173
    Songwen Pei, Yusheng Wu, and Meikang Qiu

Scalable Aggregation Service for Satellite Remote Sensing Data.   184
    Jianwu Wang, Xin Huang, Jianyu Zheng, Chamara Rajapakshe,
    Savio Kay, Lakshmi Kandoor, Thomas Maxwell, and Zhibo Zhang

Edge-Assisted Federated Learning: An Empirical Study from Software
Decomposition Perspective.   200
    Yimin Shi, Haihan Duan, Yuanfang Chi, Keke Gai, and Wei Cai

A Dynamic Partitioning Framework for Edge-Assisted Cloud Computing. . . .   215
    Zhengjia Cao, Bowen Xiao, Haihan Duan, Lei Yang, and Wei Cai

Deep Reinforcement Learning for Intelligent Migration of Fog Services
in Smart Cities   230
    Dapeng Lan, Amir Taherkordi, Frank Eliassen, Zhuang Chen,
    and Lei Liu

A Novel Clustering-Based Filter Pruning Method for Efficient Deep
Neural Networks.   245
    Xiaohui Wei, Xiaoxian Shen, Changbao Zhou, and Hengshan Yue

Fast Segmentation-Based Object Tracking Model
for Autonomous Vehicles.   259
    Xiaoyun Dong, Jianwei Niu, Jiahe Cui, Zongkai Fu,
    and Zhenchao Ouyang

A Data Augmentation-Based Defense Method Against Adversarial Attacks
in Neural Networks.   274
    Yi Zeng, Han Qiu, Gerard Memmi, and Meikang Qiu

User Recruitment with Budget Redistribution in Edge-Aided
Mobile Crowdsensing   290
    Yanlin Zhang, Peng Li, and Tao Zhang

Multi-user Service Migration for Mobile Edge Computing Empowered
Connected and Autonomous Vehicles   306
    Shuxin Ge, Weixu Wang, Chaokun Zhang, Xiaobo Zhou,
    and Qinglin Zhao

A Precise Telecom Customer Tariff Promotion Method Based
on Multi-route Radial Basis Kernel Fuzzy C-means Clustering   321
    Chenghao Wei, Timur Valiullin, and Long Hao

Clustering by Unified Principal Component Analysis and Fuzzy C-Means
with Sparsity Constraint.   337
   Jikui Wang, Quanfu Shi, Zhengguo Yang, and Feiping Nie

A Hierarchical-Tree-Based Method for Generative Zero-Shot Learning   352
   Xizhao Wang, Zhongwu Xie, Weipeng Cao, and Zhong Ming

Fast Computation of the Exact Number of Magic Series with an Improved
Montgomery Multiplication Algorithm.   365
   Yukimasa Sugizaki and Daisuke Takahashi

I am Smartglasses, and I Can Assist Your Reading   383
   *Baojie Yuan, Yetong Han, Jialu Dai, Yongpan Zou, Ye Liu,
   and Kaishun Wu*

CHEAPS2AGA: Bounding Space Usage in Variance-Reduced
Stochastic Gradient Descent over Streaming Data and Its Asynchronous
Parallel Variants   398
   Yaqiong Peng, Haiqiang Fei, Lun Li, Zhenquan Ding, and Zhiyu Hao

A Quantum Computer Operating System   415
   Reid Honan, Trent W. Lewis, Scott Anderson, and Jake Cooke

Dynamic Knowledge Graph Completion with Jointly Structural
and Textual Dependency   432
   Wenhao Xie, Shuxin Wang, Yanzhi Wei, Yonglin Zhao, and Xianghua Fu

MEFE: A Multi-fEature Knowledge Fusion and Evaluation Method Based
on BERT   449
   *Yimu Ji, Lin Hu, Shangdong Liu, Zhengyang Xu, Yanlan Liu,
   Kaihang Liu, Shuning Tang, Qiang Liu, and Wan Xiao*

Comparative Analysis of Three Kinds of Laser SLAM Algorithms   463
   Xin Liu, Yang Lin, Hua Huang, and Meikang Qiu

Aspect-Level Sentiment Difference Feature Interaction Matching Model
Based on Multi-round Decision Mechanism   477
   *Yanzhi Wei, Xianghua Fu, Shuxin Wang, Wenhao Xie, Jianwei He,
   and Yonglin Zhao*

Horus: An Interference-Aware Resource Manager for Deep
Learning Systems   492
   *Gingfung Yeung, Damian Borowiec, Renyu Yang, Adrian Friday,
   Richard Harper, and Peter Garraghan*

Attribute Bagging-Based Extreme Learning Machine.   509
   Xuan Ye, Yulin He, and Joshua Zhexue Huang

A Semi-supervised Joint Entity and Relation Extraction Model Based
on Tagging Scheme and Information Gain   523
Yonglin Zhao, Xudong Sun, Shuxin Wang, Jianwei He, Yanzhi Wei,
and Xianghua Fu

Research Progress of Zero-Shot Learning Beyond Computer Vision   538
Weipeng Cao, Cong Zhou, Yuhao Wu, Zhong Ming, Zhiwu Xu,
and Jiyong Zhang

An Optimization of Deep Sensor Fusion Based on Generalized Intersection
over Union.   552
Lianxiao Meng, Lin Yang, Gaigai Tang, Shuangyin Ren, and Wu Yang

A Hot/Cold Task Partition for Energy-Efficient Neural Network
Deployment on Heterogeneous Edge Device.   563
Jihe Wang, Jiaxiang Zhao, and Danghui Wang

Towards Energy Efficient Architecture for Spaceborne Neural
Networks Computation   575
Shiyu Wang, Shengbing Zhang, Jihe Wang, and Xiaoping Huang

Roda: A Flexible Framework for Real-Time On-demand
Data Aggregation   587
Jiawei Xu, Weidong Zhu, Shiyou Qian, Guangtao Xue, Jian Cao,
Yanmin Zhu, Zongyao Zhu, and Junwei Zhu

Structured Data Encoder for Neural Networks Based on Gradient Boosting
Decision Tree.   603
Wenhui Hu, Xueyang Liu, Yu Huang, Yu Wang, Minghui Zhang,
and Hui Zhao

Stochastic Model-Based Quantitative Analysis of Edge UPF
Service Dependability   619
Haoran Zhu, Jing Bai, Xiaolin Chang, Jelena Mišić, Vojislav Mišić,
and Yang Yang

QoE Estimation of DASH-Based Mobile Video Application Using Deep
Reinforcement Learning.   633
Biao Hou and Junxing Zhang

Modeling and Analyzing for Data Durability Towards Cloud
Storage Services   646
Feng Jiang, Yongyang Cheng, Zhao Hui, and Ruibo Yan

CC-MOEA: A Parallel Multi-objective Evolutionary Algorithm
for Recommendation Systems.   662
Guoshuai Wei and Quanwang Wu

CS-Dict: Accurate Indoor Localization with CSI Selective Amplitude
and Phase Based Regularized Dictionary Learning   677
    *Jian-guo Jiang, Shang Jiang, Bo-bai Zhao, Si-ye Wang, Meng-nan Cai,
    and Yan-fang Zhang*

Recommendation with Temporal Dynamics Based on Sequence
Similarity Search.   690
    Guang Yang, Xiaoguang Hong, and Zhaohui Peng

A Software Stack for Composable Cloud Robotics System.   705
    Yuan Xu, Tianwei Zhang, Sa Wang, and Yungang Bao

Poster Paper

An OpenMP-based Parallel Execution of Neural Networks Specified
in NNEF   723
    Nakhoon Baek and Seung-Jong Park

Author Index   727

Contents – Part III

Parallel Computing with Cybersecurity and Blockchain (PCB)

Understanding Privacy-Preserving Techniques in Digital Cryptocurrencies . . .   3
  Yue Zhang, Keke Gai, Meikang Qiu, and Kai Ding

LNBFSM: A Food Safety Management System Using Blockchain
and Lightning Network   19
  Zhengkang Fang, Keke Gai, Liehuang Zhu, and Lei Xu

Reputation-Based Trustworthy Supply Chain Management
Using Smart Contract   35
  Haochen Li, Keke Gai, Liehuang Zhu, Peng Jiang, and Meikang Qiu

Content-Aware Anomaly Detection with Network
Representation Learning.   50
  Zhong Li, Xiaolong Jin, Chuanzhi Zhuang, and Zhi Sun

Blockchain Based Data Integrity Verification for Cloud Storage
with T-Merkle Tree.   65
  Kai He, Jiaoli Shi, Chunxiao Huang, and Xinrong Hu

IM-ACS: An Access Control Scheme Supporting Informal Non-malleable
Security in Mobile Media Sharing System   81
  Anyuan Deng, Jiaoli Shi, Kai He, and Fang Xu

Blockchain-Based Secure and Privacy-Preserving Clinical Data Sharing
and Integration   93
  Hao Jin, Chen Xu, Yan Luo, and Peilong Li

Blockchain Meets DAG: A BlockDAG Consensus Mechanism. 110
  Keke Gai, Ziyue Hu, Liehuang Zhu, Ruili Wang, and Zijian Zhang

Dynamic Co-located VM Detection and Membership Update for Residency
Aware Inter-VM Communication in Virtualized Clouds 126
  Zhe Wang, Yi Ren, Jianbo Guan, Ziqi You, Saqing Yang,
  and Yusong Tan

An Attack-Immune Trusted Architecture for Supervisory
Intelligent Terminal. 144
  Dongxu Cheng, Jianwei Liu, Zhenyu Guan, and Jiale Hu

A Simulation Study on Block Generation Algorithm Based
on TPS Model   155
  Shubin Cai, Huaifeng Zhou, NingSheng Yang, and Zhong Ming

Improving iForest for Hydrological Time Series Anomaly Detection   170
  Pengpeng Shao, Feng Ye, Zihao Liu, Xiwen Wang, Ming Lu,
  and Yupeng Mao

Machine Learning-Based Attack Detection Method in Hadoop   184
  Ningwei Li, Hang Gao, Liang Liu, and Jianfei Peng

Profiling-Based Big Data Workflow Optimization in a Cross-layer Coupled
Design Framework   197
  Qianwen Ye, Chase Q. Wu, Wuji Liu, Aiqin Hou, and Wei Shen

Detection of Loose Tracking Behavior over Trajectory Data   218
  Jiawei Li, Hua Dai, Yiyang Liu, Jianqiu Xu, Jie Sun, and Geng Yang

Behavioral Fault Modelling and Analysis with BIP: A Wheel Brake System
Case Study   231
  Xudong Tang , Qiang Wang , and Weikai Miao

H2P: A Novel Model to Study the Propagation of Modern Hybrid
Worm in Hierarchical Networks   251
  Tianbo Wang and Chunhe Xia

Design of Six-Rotor Drone Based on Target Detection
for Intelligent Agriculture   270
  Chenyang Liao, Jiahao Huang, Fangkai Zhou, and Yang Lin

Consensus in Lens of Consortium Blockchain: An Empirical Study   282
  Hao Yin, Yihang Wei, Yuwen Li, Liehuang Zhu, Jiakang Shi,
  and Keke Gai

Collaborative Design Service System Based on Ceramic
Cloud Service Platform   297
  Yu Nie, Yu Liu, Chao Li, Hua Huang, Fubao He, and Meikang Qiu

Towards NoC Protection of HT-Greyhole Attack   309
  Soultana Ellinidou, Gaurav Sharma, Olivier Markowitch,
  Jean-Michel Dricot, and Guy Gogniat

Cross-shard Transaction Processing in Sharding Blockchains   324
  Yizhong Liu, Jianwei Liu, Jiayuan Yin, Geng Li, Hui Yu,
  and Qianhong Wu

Lexicon-Enhanced Transformer with Pointing for Domains Specific
Generative Question Answering 340
　　Jingying Yang, Xianghua Fu, Shuxin Wang, and Wenhao Xie

Design of Smart Home System Based on Collaborative Edge Computing
and Cloud Computing 355
　　Qiangfei Ma, Hua Huang, Wentao Zhang, and Meikang Qiu

Classification of Depression Based on Local Binary Pattern and Singular
Spectrum Analysis 367
　　Lijuan Duan, Hongli Liu, Huifeng Duan, Yuanhua Qiao,
　　and Changming Wang

Cloud Allocation and Consolidation Based on a Scalability Metric 381
　　Tarek Menouer, Amina Khedimi, Christophe Cérin, and Congfeng Jiang

Adversarial Attacks on Deep Learning Models of Computer Vision:
A Survey 396
　　Jia Ding and Zhiwu Xu

FleetChain: A Secure Scalable and Responsive Blockchain Achieving
Optimal Sharding 409
　　Yizhong Liu, Jianwei Liu, Dawei Li, Hui Yu, and Qianhong Wu

DSBFT: A Delegation Based Scalable Byzantine False Tolerance
Consensus Mechanism. 426
　　Yuan Liu, Zhengpeng Ai, Mengmeng Tian, Guibing Guo,
　　and Linying Jiang

A Privacy-Preserving Approach for Continuous Data Publication 441
　　Mengjie Zhang, Xingsheng Zhang, Zhijun Chen, and Dunhui Yu

Web Attack Detection Based on User Behaviour Semantics 459
　　Yunyi Zhang, Jintian Lu, and Shuyuan Jin

A Supervised Anonymous Issuance Scheme of Central Bank Digital
Currency Based on Blockchain 475
　　Wenhao Dai, Xiaozhuo Gu, and Yajun Teng

IncreAIBMF: Incremental Learning for Encrypted Mobile
Application Identification 494
　　Yafei Sang, Mao Tian, Yongzheng Zhang, Peng Chang,
　　and Shuyuan Zhao

A Multi-level Features Fusion Network for Detecting Obstructive Sleep
Apnea Hypopnea Syndrome 509
　　Xingfeng Lv and Jinbao Li

BIMP: Blockchain-Based Incentive Mechanism with Privacy Preserving
in Location Proof 520
   Zhen Lin, Yuchuan Luo, Shaojing Fu, and Tao Xie

Indoor Positioning and Prediction in Smart Elderly Care: Model, System
and Applications 537
   Yufei Liu, Xuqi Fang, Fengyuan Lu, Xuxin Chen, and Xinli Huang

Research on Stylization Algorithm of Ceramic Decorative Pattern Based
on Ceramic Cloud Design Service Platform 549
   Xinxin Liu, Hua Huang, Meikang Qiu, and Meiqin Liu

Blockchain Consensus Mechanisms and Their Applications in IoT:
A Literature Survey 564
   Yujuan Wen, Fengyuan Lu, Yufei Liu, Peijin Cong, and Xinli Huang

Towards a Secure Communication of Data in IoT Networks:
A Technical Research Report 580
   Bismark Tei Asare, Kester Quist-Aphetsi, and Laurent Nana

Efficient Thermography Guided Learning for Breast Cancer Detection 592
   *Vishwas Rajashekar, Ishaan Lagwankar, Durga Prasad S N,
   and Rahul Nagpal*

PTangle: A Parallel Detector for Unverified Blockchain Transactions 601
   *Ashish Christopher Victor, Akhilarka Jayanthi,
   Atul Anand Gopalakrishnan, and Rahul Nagpal*

DOS-GAN: A Distributed Over-Sampling Method Based on Generative
Adversarial Networks for Distributed Class-Imbalance Learning 609
   Hongtao Guan, Xingkong Ma, and Siqi Shen

Effective Sentiment Analysis for Multimodal Review Data on the Web 623
   Peiquan Jin, Jianchuan Li, Lin Mu, Jingren Zhou, and Jie Zhao

An Energy-Efficient AES Encryption Algorithm Based
on Memristor Switch 639
   Danghui Wang, Chen Yue, Ze Tian, Ru Han, and Lu Zhang

Digital Currency Investment Strategy Framework Based on Ranking 654
   *Chuangchuang Dai, Xueying Yang, Meikang Qiu, Xiaobing Guo,
   Zhonghua Lu, and Beifang Niu*

Authentication Study for Brain-Based Computer Interfaces
Using Music Stimulations 663
   Sukun Li and Meikang Qiu

An Ensemble Learning Approach to Detect Malwares Based
on Static Information.   676
    Lin Chen, Huahui Lv, Kai Fan, Hang Yang, Xiaoyun Kuang, Aidong Xu,
    and Siliang Suo

Automatic Medical Image Report Generation with Multi-view
and Multi-modal Attention Mechanism   687
    Shaokang Yang, Jianwei Niu, Jiyan Wu, and Xuefeng Liu

Poster Paper

PLRS: Personalized Literature Hybrid Recommendation System
with Paper Influence   703
    Fanghan Liu, Wenzheng Cai, and Kun Ma

Author Index   707

Parallel Architectures and Algorithms
(PAA)

COMBS: First Open-Source Based Benchmark Suite for Multi-physics Simulation Relevant HPC Research

Anthony Dowling[1], Frank Swiatowicz[2], Yu Liu[1]([✉]), Alexander John Tolnai[3], and Fabian Herbert Engel[4]

[1] Clarkson University, Potsdam, NY 13699, USA
{dowlinah,yuliu}@clarkson.edu
[2] Vassar College, Poughkeepsie, NY 12604, USA
fxswiatowicz@vassar.edu
[3] Royal Melbourne Institute of Technology, Melbourne, VIC 3000, Australia
s3540645@student.rmit.edu.au
[4] Hasso Plattner Institute, 14482 Potsdam, Germany
fabian.engel@student.hpi.de

Abstract. Recent scientific computing increasingly relies on multi-scale multi-physics simulations to enhance predictive capabilities by replacing a suite of stand-alone simulation codes that independently simulate various physical phenomena. Inevitably, multi-physics simulation demands high performance computing (HPC) through advanced hardware and software accelerating due to its intensive computing workload and run-time communication needs. Thus, its research has become a hotspot across different disciplines. However, it is observed that most benchmarks used in the evaluation of corresponding work are through commercial or in-house codes. Then, the lack of accessible open-source multi-physics benchmark suites has presented a challenge in uniformly evaluating simulation performance across related disciplines. This work proposes the first open-source based benchmark suite with 12 selected benchmarks for research in multi-physics simulation, the Clarkson Open-Source Multi-physics Benchmark Suite (COMBS). Multiple metrics have been gathered for these benchmarks, such as instructions per second and memory usage. Also provided are build and benchmark scripts to improve usability. Additionally, their source codes and installation guides are available for downloading through a github repository built by the authors. The selected benchmarks are from key applications of multi-physics simulation and highly cited publications. It is believed that this benchmark suite will facilitate to harness the full potential of HPC research in the field of multi-physics simulation.

Keywords: Multi-physics simulation · High performance computing · Benchmark suite · Open-source

© Springer Nature Switzerland AG 2020
M. Qiu (Ed.): ICA3PP 2020, LNCS 12452, pp. 3–14, 2020.
https://doi.org/10.1007/978-3-030-60245-1_1

1 Introduction

Recent trends in scientific computing increasingly rely on multi-scale multi-physics computer simulations to enhance predictive capabilities by replacing conventional methods that are largely empirically based with a more scientifi-cally based methodology. Through this approach, one addresses the issue of tra-ditionally employing a suite of stand-alone codes that independently simulate various physical phenomena that were previously disconnected [1]. For exam-ple, coupled multi-physics codes have been developed worldwide to address the growing concerns of reactor performance and nuclear safety [1–5]. Also, multi-physics simulations are helping researchers make progress in finding effective cancer treatments through simulating and modeling the background dose of tar-geted alpha therapy (TAT) [6–8]. Moreover, multi-physics simulation plays an important role in semiconductor design [9] and aerospace engineering [10].

Multi-physics simulation requires capturing many physical interactions among complex phenomena through rigorous coupling at run-time, facilitating the need for high performance computing (HPC) through advanced hardware and software acceleration due to its massive computing workload and fast run-time communication needs. Thus, prior research has been done to boost the general software and hardware HPC computing environment for multi-physics simulations in recent years, which is briefly described in Sect. 2. However, the authors realize that one of the research obstacles in this domain is the lack of an open-source benchmark suite for research evaluation. The lack of accessible open-source multi-physics benchmark suites has presented a challenge in uni-formly evaluating simulation performance across related disciplines.

To the best of our knowledge, this work proposes the first open-source bench-mark suite for multi-physics simulation relevant HPC research, the Clarkson Open-Source Multi-physics Benchmark Suite (COMBS). These benchmarks have been installed and evaluated on an Ubuntu Linux server, e.g., instructions per second or memory usage. Also provided are build and benchmark scripts to improve usability. Additionally, their source code and installation guides are available for downloading through a Github repository built by the authors. The selected 12 benchmarks are from key applications of multi-physics simula-tions and highly cited publications, which will be detailed in Sect. 3. It is believed that this benchmark suite will facilitate to harness the full potential of the HPC research in the field of multi-physics simulation. The contributions of this paper are three-fold:

- Propose the first benchmark suite for multi-physics simulation relevant HPC research;
- Each benchmark is open-source, allowing for free access and modification as desired;
- All benchmarks have been installed and evaluated on an Ubuntu Linux server.

The rest of the paper is organized as follows. Section 2 describes the related work and motivation of this paper. Section 3 introduces the selected benchmarks in detail, along with the benchmark selection criteria. The benchmark data and website are illustrated in Sect. 4. Lastly, the authors summarize the conclusion of this work and discuss planned future work in Sect. 6.

2   Related Work and Motivation

To satisfy the HPC needs of multi-physics simulation, many research projects have been done to adapt simulation codes, optimize the HPC based software framework, and explore novel computer architectures for HPC systems. One research focus is in exploring software coupling platforms, simulation codes (a.k.a., applications), algorithms with the objective of reaching a performance boost through the HPC systems, exchange run-time data effectively, and cost-effective transition. The other focus is in hardware design, e.g., investigating multi-physics simulation aware computer architectures and HPC systems. The objective of recent research is to improve the general software and hardware HPC computing environment for large-scale multi-physics simulations.

Many HPC technique based code coupling platforms have been developed; the capabilities and intended applications of each code differ, such as the Backbone [1], MOOSE [11], LIME [12], and SALOME [13]. The Backbone, developed at the Canadian Nuclear Laboratories (CNL), has the capability of synchronizing various reactor performance and safety analysis codes, which permits appropriate and efficient coupling at the time step level. Idaho National Laboratory (INL) proposed the Multi-physics Object Oriented Simulation Environment (MOOSE) [11], and a code matching MOOSE standard can "plug and play" into the environment. The Open Source Integration Platform for Numerical Simulation (SALOME) [13], funded by the French Energy Commission (CEA), is based on the model of distributed components as a distributed objects architecture. The Lightweight Integrating Multi-physics Environment (LIME) toolset [12], developed by Sandia National Laboratory (SNL), is used within the Consortium for Advanced Simulation of Light Water Reactors (CASL) hub and also supports coupling of multiple codes in other fields. Besides this platform based work, other research focuses on resource allocation and load balancing to achieve an effective run-time computing environment for multi-physics simulations [14,15].

Compared to general HPC applications, multi-physics simulations have their own unique characteristics. For example, the simulation codes involved in a simulation may request intensive run-time data to be exchanged by them. Also, some applications (e.g., Monte Carlo (MC)) with long consequential computing-intensive codes may run faster on multi-core CPUs, which offer high degrees of instruction level parallelism (ILP). Other codes (e.g., Computational Fluid Dynamics (CFD)) might have a high degree of parallelism, which could make use of the hundreds of simple and effective cores of general purpose GPUs. Thus, much work has been done to satisfy these unique needs through designing advanced computer architectures or HPC systems, such as heterogeneous design, 3D architecture, etc. [16].

However, it is observed that their evaluation is based on either commercial/in-house simulation codes (a.k.a., applications), or limited applications in specific areas. Research on designing a benchmark suite for multi-physics simulations is inspired by the lack of available open-source suites that measure performance evaluation. Some of the available benchmarks are locked behind licenses and use in-house codes that make easy comparisons across different simulations challenging. Benchmark suites have been created to measure the performance of multi-physics simulations but suffer from certain flaws. No open source benchmark suite has ever been proposed to facilitate the relevant evaluation work specific to multi-physics simulation.

For example, COMSOL and other commercial packages provide benchmarks to test simulations but the codes are not open-source, and require a license to use. In [1,6,7], in-house simulation codes (e.g., Element Loss-Of-Coolant Accidents (ELOCA), Canadian Algorithm for Thermalhydraulic Network Analysis (CATHENA)) are used, which are confidential and not accessible by researchers outside CNL. In [14], the authors only mentioned the benchmark is a MC simulation used as a bigJob, while [15] used 7 problems named P1 to P7. Both [14] and [15] work on resource allocation algorithms for multi-physics simulations.

Since very few of these works provide detailed information regarding their benchmarks, **is it possible for us to evaluate their advantages and disadvantages? Is it fair if one of them declares performance improvement over the other?** In addition, general purpose open-source benchmark suites like CORAL-2 [17] are just a collection and still require users to individually install, build, and run each benchmark in the suite. The CORAL-2 benchmarks also include wide variations in lines of code and other attributes like uneven testing of CPU compared to GPU.

Thus, it is very difficult for researchers to duplicate the achievements of previous publications. In addition, evaluating performance between new proposed systems and legacy ones is almost impossible, since the evaluations are based on different benchmarks. Thus, the authors believe that this situation becomes an obstacle for future research regarding multi-physics simulations. Thus, it is necessary to propose an open-source based benchmark suite to address this concern. The authors believe that the open-source nature of such a suite could guarantee the access of benchmarks and build a general foundation for relevant research.

3   Benchmark Suite

Through the compilation of 12 benchmarks that measure performance indicative of multiple physics algorithm designs, research in the field will benefit from an open-source suite that is modifiable and free to use. The benchmark codes have been tested to successfully build on an Ubuntu 18.04 server, allowing for the codes to be ran in a stable, easily configurable environment.

3.1   Selection Criteria

In order to construct a benchmark suite that would provide adequate coverage and measure desired performance, potential benchmarks have to meet certain criteria, as follows:

- open-source;
- able to be built and compiled with few additional external libraries;
- written in C or C++ as to allow for the use of a single compiler;
- measures performance typical of an aspect of multi-physics simulations (e.g., tightly coupled codes).

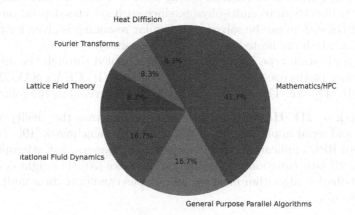

Fig. 1. A pie chart showing the different multi-physics-related benchmarks that are present in the suite. Note the relatively even distribution among selected topics.

The last point was of particular importance, as it is difficult to determine the most salient aspects of a field with limited background experience in multi-physics algorithm design. Upon consulting literature in the field, it was determined that finding algorithms that solve partial differential equations (PDEs) was necessary, as multi-physics simulation solution times depend on how quickly PDEs can be solved. Secondary characteristics used for the selection of benchmarks included their relatedness to the field in terms of the problem they solved or their presence in licensed benchmark suites. In this work, queries were conducted primarily through GitHub.

3.2 Benchmarks

The benchmark suite proposed in this paper is named the Clarkson Open-Source Multi-physics Benchmark Suite (COMBS). A total 12 benchmarks are divided into 2 groups, including 8 key application/framework benchmarks, and 4 supplemental benchmarks. The distribution of the selected 12 benchmarks are shown in Fig. 1.

The 8 Selected Key Application/platform Benchmarks. These 8 benchmarks consist of 8 applications. Each benchmark is open-source, allowing for free access and modification as desired. Each benchmark consists of algorithm(s) that solve problems related to the field. As an example, one of the selected benchmarks is a solver for the Kuramoto-Sivashinsky (KS) equation. The KS equation is one of the principal equations in connecting PDEs and dynamical systems. Since multi-physics simulations rely heavily on the tight coupling between dynamic algorithms, this benchmark appears to be reasonable to test baseline performance, at least on a much simpler, one dimensional problem. The KS equation is a highly cited problem in the field dating back over 30 years and being a PDE problem, fits directly in to multi-physics, since such systems depend on the how quickly a PDE system can be solved [18]. Similar reasoning is given for the other benchmarks, which are hosted on a GitHub repository.

The 8 application type benchmarks can be coupled through the distributed code coupling platforms, such as CNL's Backbone [1], CEA's SALOME [13], SNL's LIME [12], etc. The 8 key benchmarks are described in the following.

Benchmark - 2D Heat: This benchmark compares the utility of MPI, OpenMP, and serial implementations of the 2d-heat benchmark [19]. This work is related to HPC applications as shown in related papers that attempt to parallelize the 2D heat equation to improve performance [20]. The tight coupling of a heat distribution algorithm makes it an obvious candidate for a multi-physics benchmark suite.

Benchmark - Advection-Diffusion Equation: Advection-diffusion equations [21] are used in the field of CFD to measure phenomena like the time evolution of chemical or biological species in a flowing medium such as water or air. Since such an evolution can be modeled with PDEs, an algorithm that models a simpler, one dimensional aspect of advection-diffusion can be useful in measuring simulation performance. This benchmark was adapted from open-source codes at [22].

Benchmark - Fidibench: FiDiBench is a finite difference suite of codes that can be used to benchmark hardware performance on HPC and other systems. The code examples are small enough to be well understood, typically averaging a few hundred lines of code, but are also relevant to scientific computing, which often involves nearest neighbor communication. FiDiBench can be used to compare the execution speed obtained by implementing a given algorithm in different languages (e.g., C++ vs Python vs Julia). This benchmark was adapted from open-source codes at [23].

Benchmark - High Performance Conjugate Gradient (HPCG): HPCG [24] is a software package that performs a fixed number of multigrid preconditioned (using a symmetric Gauss-Seidel smoother) conjugate gradient (PCG) iterations using double precision floating point values. The HPCG rating is a weighted GFLOP/s (billion floating operations per second) value that is composed of the operations performed in the PCG iteration phase over the time taken. The overhead time of problem construction and any modifications to improve performance are divided by 500 iterations (the amortization weight) and added to the run-time.

Benchmark - KS-PDE: The Kuramoto-Sivashinsky (KS) equation [18] is one of the principal equations in connecting partial differential equations (PDEs) and dynamical systems. Since multi-physics simulations rely heavily on the tight coupling between dynamic algorithms, this benchmark appears to be reasonable to test baseline performance, at least on a much simpler, one dimensional problem. The KS equation is a highly cited problem in the field dating back over 30 years and being a PDE problem, fits directly in to multi-physics, since such systems depend on the how quickly a PDE system solved.

The code for this benchmark is adapted from a recent open-source work [25], which is in an attempt to compare run-time performance between different programming languages, with the goal of legitimizing Julia as having the potential to be among the fastest languages if used correctly. Additionally, the benchmark can be adapted to test the performance of just the C code.

Benchmark - 2D Lid-Driven Cavity: This is a concise finite difference method based code for solving the Navier-Stokes equation in a 2D Lid-driven cavity. The problem is to move fluid in a box from one corner to another, measuring the fluid's velocity. The Navier-Stokes equation is commonly used to solve CFD relevant problems and then couple with other codes in many multi-physics simulation scenarios.

Benchmark - Phase Retrieval: The mathematical and algorithmic aspects of the phase retrieval problem have received considerable attention. In crystallography, a principal application in this area, the signal to be recovered is periodic and comprised of atomic distributions arranged homogeneously in the unit cell of the crystal. The crystallographic problem is both the leading application and one of the hardest forms of phase retrieval [26]. This benchmark is used to solve phase retrieval, which uses techniques such as 3D Fourier Transforms that are applicable to multi-physics simulation designs. This benchmark is adapted from a recent open-source implementation at [27].

Benchmark - SOMBRERO: In physics, lattice field theory is the study of lattice models of quantum field theory, i.e., of field theory on a spacetime that has been discretized onto a lattice. Lattice field theory is directly applicable to high performance computing and adjacent multi-physics simulations. This benchmark is adapted from open-source codes at [28]. SOMBRERO runs 6 sub-benchmarks, each representing a theory under active study by the lattice field theory community. For the purposes of benchmarking, the models vary only in the amount of data communicated and the number of floating point operation required.

The 4 Supplement Benchmarks of HPC Mathematics/Algorithms.
Along with the 8 benchmarks selected from the key areas involved in the multi-physics simulation, the authors also chose 4 benchmarks regarding general HPC mathematics and algorithms, e.g., Monte Carlo, MPI matrix multiplication, Gaussian Blur, Radix Sort, etc. These benchmarks are a good complement to the key benchmarks for evaluating the basic performance and optimization of the

proposed new software and hardware systems, such as inter-node communication latency, shared last level cache (LLC) performance, etc.

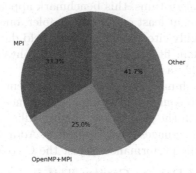

Fig. 2. HPC Techniques Coverage

3.3   HPC Coverage Analysis

Figure 2 shows the coverage of HPC techniques of the 10 key application/framework benchmarks. It is observed that about 70% of them utilize Message Passing Interface (MPI) [29] and OpenMP [30], which are the most popular HPC techniques used in the field of multi-physics simulation. Although 30% of them are based on other techniques, they can be easily coupled through the distribution style software framework like Backbone, SALOME and LIME.

4   Benchmark Repository on Github

A Github repository has been designed to include the source codes of all benchmarks, the installation guide (tested on Ubuntu Linux 18.04). This website is shown in Fig. 3, and the link is at [31]. Users can download the source codes of all benchmarks through the git utility, and then follow the guide to install the dependencies for the benchmarks and run them.

5   Characteristics of the Benchmark Suite

By building and running the benchmark suite using the provided scripts in the github page built by authors, it is possible and convenient to create an overview of all the benchmarks with multiple metrics as the characteristics of this benchmark suite. Table 1 shows this characteristics of COMBS for selected benchmarks when formatted as a table. This run of the benchmarks was gathered on a Dell Precision 7920 Tower server with two Intel Xeon Silver 4110 16 3 GHz processors and 32 GB of RAM. The server also has two GPUs: an NVIDIA NVS 310 and

Table 1. Example Output from the Benchmark Suite showing selected Benchmarks

Benchmark	Time (s)	Instruction count (Millions)	Max memory usage (kB)	MIPS
2d-heat	12.628	62091.213	10400	4917
FourierBenchmarks	85.937	288618.737	53272	3358
advection-diffusion	9.638	170011.780	10360	17640
fidibench	43.565	50.128	178028	1.151
hpcg	99.667	222584.356	971244	2233
ks-pde	24.375	174739.281	21392	7169
lid-driven-cavity	3.553	31345.988	6492	8822
matrix-mpi	8.541	46539.884	535608	5449
monte-carlo	116.524	233543.444	1506492	2004
phase-retrieval-benchmarks	4.141	26685.456	15888	6444
radix_sort	4.968	37136.804	16052	7475
sombrero	59.478	8.747	22124	0.147

an NVIDIA TITAN Xp. The server is running Ubuntu 18.04.3 LTS x86_64 with Linux kernel 5.0.0-32-generic.

These metrics were gathered using tools from the valgrind toolset, and the timing functionality of the Linux time command [32]. First, the `time` command is used to time the wall-clock execution time of the program. This gives a feel for how long the program takes to run. Then, maximum memory usage is gathered using the `massif` [33] tool in the valgrind toolchain. This measurement shows the space complexity of the program as it is running, along with any extra memory that might be used by the program while it is in operation. Lastly, the `callgrind` [34] tool of valgrind was used to gather the instruction count of the program. This, along with the timing information and size of a program gives a way to see if a program may spend a lot of time in a loop, or if it is using more complex instructions.

These metrics were chosen because they are similar to the metrics used in [35], another research-oriented benchmark suite. Also, a comparison of the features of COMBS with CORAL-2 [17] mentioned in Sect. 2 is given in Table 2, which indicates the difference between the two open source benchmark suites. Furthermore, as described above, they give a method for determining certain platform-specific characteristics of a program, such as whether a long-running program simply spends a large amount of time in loops, or if a program is compiled to use complex instructions. Moreover, the memory usage measurement gives a method to

Table 2. Features of COMBS compared to CORAL-2

	COMBS	CORAL-2 [17]
Focus	Multi-physics Simulation	General Purpose
Compilation	Scripts to compile all	Compile each individually
Running	Automatic Run	Run each individually
Benchmark Size	Similar Size	Wide Variations

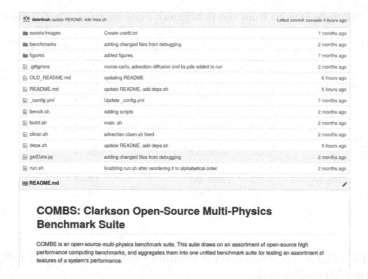

Fig. 3. Github Repository of Clarkson Open-Source Multi-physics Benchmark Suite (COMBS).

determine if a program may be accessing memory to a large degree; as high memory use would typically indicate high memory access. The authors believe these metrics are necessary for researchers to determine and adapt this benchmark suite based on their research purpose in the field of multi-physics simulation.

6 Conclusion and Future Work

The lack of accessible open-source multi-physics benchmark suites has presented a challenge in uniformly evaluating simulation performance across related disciplines. Through the compilation of 12 benchmarks that measure performance indicative of multiple physics algorithm designs, research in the field will benefit from an open-source suite that is modifiable and free to use. A dedicated Github repository is setup to host this benchmark suite, where users will have access to all documentation for each benchmark along with building scripts to facilitate easy use of the suite. Also provided are build and benchmark scripts to automatically perform the aforementioned steps. In the future, the authors would like to have the suite build and compile with a simulated computer architecture. Also, GPU acceleration-aware benchmarks will be selected and evaluated.

Acknowledgments. This work is partially funded by NSF Award OAC-1852102.

1 Artifact Description Appendix

A Github website [31] has been designed to include the source codes of all benchmarks and the installation guide (tested on Ubuntu Linux 18.04). Users can download the source codes of all benchmarks through the command below, and then follow the installation guide of each benchmark to install and test them.

References

1. Liu, Y., Nishimura, M., Seydaliev, M., Piro, M.: Backbone: a multi-physics framework for coupling nuclear codes based on CORBA and MPI. Nucl. Eng. Radiat. Sci. (2017)
2. Gouja, I., Avramova, M., Rubin, A.: Development and optimization of coupling interfaces between reactor core neutronics and thermal-hydraulic codes. In: The International Conference on Advances in Reactor Physics to Power the Nuclear Renaissance (2010)
3. Gomez-Torres, A.M., Sanchez-Espinoza, V., Ivanov, K., Macian-Juan, R.: DYN-SUB: a high fidelity coupled code system for the evaluation of local safety parameters-part I: development, implementation and verification. Ann. Nucl. Energy 48, 108–122 (2012)
4. Sanchez, V., Al-Hamry, A.: Development of a coupling scheme between MCNP and COBRA-TF for the prediction of the pin power of a PWR fuel assembly. In: The International Conference on Mathematics, Computational Methods and Reactor Physics (2009)
5. Chen, Z., Chen, X.-N., Rineiski, A., Zhao, P., Chen, H.: Coupling a CFD code with neutron kinetics and pin thermal models for nuclear reactor safety analyses. Ann. Nucl. Energy 83, 41–49 (2015)
6. Tao, X., Liu, Yu., Liu, T., Li, G., Aydemir, N.: Multiphysics modelling of background dose by systemic targeted alpha therapy. Med. Imaging Radiat. Sci. (2018). https://doi.org/10.1016/j.jmir.2018.06.002
7. Xu, T., Liu, T., Li, G., Dugal, C., Li, Y.: Microdosimetric and biokinetic modelling of alpha-immuno-conjugate transport in endothelial cells. J. Med. Imaging Radiat. Sci. 50, S1–S2 (2019)
8. Xu, T., et al.: Technical note: the development of a multi-physics simulation tool to estimate the background dose by systemic targeted alpha therapy. Med. Phys. (2020)
9. Xiao, H.: A multi-physics approach to the co-design of 3D multi-core processors. Ph.D. dissertation (2018)
10. Errera, M., et al.: Multi-physics coupling approaches for aerospace numerical simulations. J. Aerosp. Lab (2011)
11. Schmidt, R., Hooper, R., Belcourt, N., Pawlowski, R.: MOOSE: a parallel computational framework for coupled systems of nonlinear equations. Nucl. Eng. Des. 239(10), 1768–1778 (2009)
12. Schmidt, R., Belcourt, N., Hooper, R., Pawlowski, R.: An introduction to lime 1.0 and its use in coupling codes for multiphysics simulations. Sandia Report, SAND2011-8524 (2011)
13. SALOME official webpage (2019)
14. Ko, S.-H., Kim, N., Kim, J., Thota, A., Jha, S.: Efficient runtime environment for coupled multi-physics simulations: dynamic resource allocation and load-balancing. In: 10th IEEE/ACM International Conference on Cluster, Cloud and Grid Computing (2010)

15. Sfika, N., Korfiati, A., Alexakos, C., Likothanassis, S., Daloukas, K., Tsompanopoulou, P.: Dynamic cloud resources allocation on multidomain/multiphysics problems. In: 3rd International Conference on Future Internet of Things and Cloud (2015)
16. Hermann, E., Raffin, B., Faure, F., Gautier, T., Allard, J.: Multi-GPU and multi-CPU parallelization for interactive physics simulations. In: D'Ambra, P., Guarracino, M., Talia, D. (eds.) Euro-Par 2010. LNCS, vol. 6272, pp. 235–246. Springer, Heidelberg (2010). https://doi.org/10.1007/978-3-642-15291-7_23
17. CORAL-2 Benchmarks (2019). https://asc.llnl.gov/coral-2-benchmarks/
18. Hyman, J.M., Nicolaenko, B.: The Kuramoto-Sivashinsky equation: a bridge between PDE's and dynamical systems. Physica D: Nonlinear Phenomena **18**, 113–126 (1986)
19. 2D Heat Benchmark Source Codes (2011)
20. Horak, V., Gruber, P.: Multi-physics coupling approaches for aerospace numerical simulations. Parallel Numerics (2005)
21. Hundsdorfer, W.H., Verwer, J.G.: Numerical solution of time-dependent advection-diffusion-reaction equations. Parallel Numerics (2011)
22. Advection-Diffusion Equation Benchmark Source Codes (2017). https://github.com/antoine-levitt/benchmark_heat
23. Fidibench Benchmark Source Codes (2019). https://github.com/pletzer/fidibench
24. HPCG Benchmark Website (2019). http://www.hpcg-benchmark.org/
25. KS-PDE Benchmark Source Codes (2018). https://github.com/johnfgibson/julia-pde-benchmark
26. TAMIR BENDORY VEIT ELSER, TI-YEN LAN. Benchmark problems for phase retrieval (2017)
27. Phase Retrieval Benchmark Source Codes (2019). https://github.com/veitelser/phase-retrieval-benchmarks
28. Sombrero Benchmark Source Codes (2019). https://github.com/sa2c/sombrero
29. OpenMPI (2019). https://www.open-mpi.org/
30. OpenMP (2019). https://www.openmp.org/
31. COMBS Github (2020). https://github.com/dowlinah/COMBS
32. Nethercote, N., Seward, J.: Valgrind: a framework for heavyweight dynamic binary instrumentation. ACM SIGPLAN Not. **42**(6), 89–100 (2007)
33. Massif: a heap profiler (2020). https://valgrind.org/docs/manual/ms-manual.html
34. Callgrind: a call-graph generating cache and branch prediction profiler (2020). http://valgrind.org/docs/manual/cl-manual.html
35. Bienia, C., Kumar, S., Singh, J.P., Li, K.: The parsec benchmark suite: characterization and architectural implications. In: Proceedings of the 17th International Conference on Parallel Architectures and Compilation Techniques, pp. 72–81 (2008)

Efficient Sorting and Join on NVM-Based Hybrid Memory

Yongping Luo[1], Zhaole Chu[1], Peiquan Jin[1,2(✉)], and Shouhong Wan[1,2]

[1] University of Science and Technology of China, Hefei 230027, Anhui, China
jpq@ustc.edu.cn
[2] Key Laboratory of Electromagnetic Space Information, China Academy of Science,
Hefei 230027, Anhui, China

Abstract. Non-volatile memory (NVM) as a new kind of future memories has a number of special properties such as non-volatility, read/write asymmetry, and byte addressability. This makes it difficult to directly replace DRAM with NVM in current memory hierarchy. Thus, a practical way is to construct hybrid memory composed of both NVM and DRAM. Such hybrid memory architecture introduces many new challenges for existing algorithms. In this paper, we focus on improving sorting and join algorithms for DRAM-NVM-based hybrid memory. In particular, we start with a theoretical study on the data placement issue in DRAM-NVM-based hybrid memory systems and propose an optimal data placement model to store data structures in DRAM and NVM during the sorting process. We present the theoretical proof to the optimal data placement model to ensure the correctness of the model. Further, based on the optimal data placement model, we propose a new NVM-aware sorting algorithm named NVMSort that adopts heap structures to accelerate the sorting process. Compared with traditional sorting algorithms, NVMSort is write-friendly and more efficient on DRAM-NVM-based hybrid memory. We further apply NVMSort into the traditional merge-sort join algorithm to optimize merge-sort join on DRAM-NVM-based hybrid memory. We conduct comparative experiments with existing sorting algorithms including HeapSort and QuickSort. The results show that NVMSort is much faster than the classical Heapsort and QuickSort. In addition, NVMSort is more NVM-friendly as it can reduce more NVM writes. When integrated into the traditional merge-sort join algorithm, NVMSort also achieves the best performance.

Keywords: Non-volatile memory · Sorting · Data placement model · Sort join

1 Introduction

The performance of traditional computer systems is highly limited by the high latency between DRAM and disks. This has been widely regarded as the "storage wall" problem [1]. Although building in-memory systems [2] seems to be a possible solution to the storage-wall problem, the volatility of DRAM makes it difficult to tackle data consistency and durability. Recently, the advance in non-volatile memory (NVM) [3] brings new

© Springer Nature Switzerland AG 2020
M. Qiu (Ed.): ICA3PP 2020, LNCS 12452, pp. 15–30, 2020.
https://doi.org/10.1007/978-3-030-60245-1_2

opportunities to build NVM-based in-memory systems that can not only support in-memory data processing but also ensure data consistency and durability. On the other hand, traditional algorithms such as sorting algorithms and database join schemes, which are designed either for disks or for DRAM, need to be revisited to suit for the special properties of NVM.

NVM has some special properties [4–6]. First, differing from DRAM, it is non-volatile, meaning that all data written into NVM will not be lost when the host computer is shut down. Second, differing from magnetic disks or solid-state drives (SSD) [7] that only support block-based data accesses, NVM is byte addressable, which is similar to DRAM. Third, NVM has a high level of density, which is comparable to SSD and higher than NVM. To this end, NVM has the advantages of both disks and DRAM. Moreover, NVM is more like a new kind of memory, but not a new type of disk. However, NVM also has some limitations compared to DRAM and disks. Firstly, the read and write latencies of NVM are not balanced. Particularly, NVM has the similar read latency as DRAM, but its write latency is higher than that of DRAM. In addition, the endurance of NVM is limited, meaning that after a certain number of writes (~10^8 at present), NVM will become unstable. Thus, algorithms running on NVM have to be write-friendly.

Due to the special features of NVM, existing in-memory algorithms such as sorting algorithms must be re-designed to make them efficient on NVM. However, it is not a trivial task. There are a few challenges that need to be carefully considered. First, as NVM has a lower write speed than DRAM, currently it is more reasonable to construct a hybrid memory system composed of DRAM and NVM. In this hybrid-memory environment, how to optimally place data in DRAM and NVM is a new and challenging issue. Second, traditional external sorting algorithms like Merge Sort [8] are designed toward reducing I/O operations on block-based disks or SSDs and are not suitable for byte-addressable NVM. Traditional in-memory sorting algorithms [8] like Quick Sort or Heap Sort do not consider the high-write-latency of NVM as well as the reduction of NVM writes for endurance. Thus, they cannot be applied to NVM. Here, the challenge is that reducing NVM writes may lower the sorting performance of the algorithm. Therefore, we need to devise new sorting schemes that are not only time efficient, but also write friendly.

In this paper, we focus on re-designing sorting algorithms on NVM and further improving the sort join algorithm in database systems. We aim to devise a new sorting algorithm that is not only time-efficient but also NVM friendly. Specially, we start with a theoretical study on the data placement issue in DRAM-NVM-based hybrid memory systems and propose an optimal data placement model to store data structures in DRAM and NVM during the running of an algorithm. We present the theoretical proof to the optimal data placement model to ensure the correctness of the model. Further, based on the optimal data placement model, we propose a new sorting algorithm on NVM as well as a new sort join algorithm. In summary, we make the following contributions in this paper:

(1) We study the data placement issue on DRAM-NVM-based hybrid memory systems and present an optimal data placement model for algorithms running on hybrid memory. We theoretically prove that the proposed model can offer the best data placement during the execution of algorithms.

(2) We present a new sorting algorithm named NVMSort that is optimized for DRAM-NVM-based hybrid memory. NVMSort is based on the optimal data placement model and adopts heap structures to accelerate the sorting process. Compared with traditional sorting algorithms, NVMSort is more efficient on DRAM-NVM-based hybrid memory. In addition, NVMSort can reduce more NVM writes. We further integrate the NVMSort algorithm into the traditional merge-sort join algorithm to accelerate the join process on the hybrid memory.

(3) We conduct extensive experiments to verify the performance of NVMSort and NVMSort-based join algorithm. Compared with QuickSort and HeapSort, the proposed NVMSort has the best sorting performance. In addition, its NVM writes are much fewer than that of QuickSort and HeapSort. Overall, NVMSort reaches a better trade-off between time performance and NVM writes. When integrated into the merge-sort join algorithm, NVMSort also shows better time performance and is more NVM-friendly than its competitors.

2 Related Work

Recently, the big data concept leads to a special focus on the use of main memory. However, the increasing capacity of main memory introduces many problems, such as increasing of total costs and energy consumption [9]. Both academia and industries are looking for new greener memory media. Emerging NVM technologies, such as Phase Change Memory (PCM) and Resistive Memory (ReRAM), can provide faster persistence than traditional disks and flash memory. NVMs can provide similar read latency but higher write latency than DRAM. Like flash memory, the write endurance of NVM is limited. Thus, reducing write operations to NVM is critical for software system design [4–6].

NVM can provide better support for data durability than DRAM does. Further, it differs from other media such as flash memory in that it supports byte addressability. However, NVM has some limitations [3, 4], e.g., high write latency, limited lifecycle, slower access speed than DRAM, etc. Therefore, it is not a feasible design to completely replace DRAM with NVM in current computer architectures. A more exciting idea is to use both NVM and DRAM to construct hybrid memory systems, so that we can utilize the advantages from both media [10, 11]. NVM has the advantages of low energy consumption and high density, and DRAM can afford nearly unlimited writes. Specially, NVM can be used to expand the capacity of main memory, whereas DRAM can be used as a buffer for NVM. Presently, both the architectures are hot topics in academia and industries. Many issues need to be further explored. The biggest challenge for DRAM-NVM-based hybrid memory systems is that we have to cope with heterogeneous memories. In this paper, we also focus on the architecture of hybrid memory systems.

A few prior works [12, 13] have explored algorithms for asymmetric read-write costs in emerging NVMs within the context of databases. Chen et al. [12] presented analytical formulas for PCM latency and energy, as well as algorithms for B-trees and hash joins that are tuned for PCM. For example, their B-tree variant does not sort the keys in a leaf node nor repack a leaf after a deleted key, thereby avoiding the write cost of sorting and repacking, at the expense of additional reads when searching. Similarly, Viglas [13] traded off fewer writes for additional reads by rebalancing a B+ -tree only if the cost of rebalancing has been amortized.

To the best of our knowledge, few studies have been focused on the improvement of fundamental sorting and join algorithms on NVM. In a word, there are only two previous works that are related to this study. The first study [14] presented a write-limited sorting algorithm but it was toward in-memory sorting. On the contrary, this study is toward traditional disk-based sorting and join, i.e., the involved relations are initially stored in disks. The second work [15] proposed a cost model for sorting on storage devices with asymmetric read and write latencies. However, these related works are both towards page-based storage devices, such as flash-memory-based SSDs. Although flash memory also has limited write endurance and low write latency, it is much different from NVM, because NVM can be used as main memory while flash memory can only be used as secondary storage.

3 Data Placement Model on Hybrid Memory

In this section, we study the data placement issue on DRAM-NVM-based hybrid memory. This issue is raised because of the existence of the two heterogeneous memories in hybrid memory systems. Thus, when a sorting algorithm is running, we have to decide where any intermediate data should be placed. Below we first introduce the concepts and problem definitions of the data placement issue. Then, we present the optimal data placement model as well as its proof for hybrid memory.

3.1 Basic Concepts

Logical Data Structures (DS) need to be placed on physical memory space. In a hybrid memory system, the memory space of a data structure can be allocated completely from DRAM, completely from NVM or partially from the two memory devices. A *data placement model* (DPM) cares only about how the physical memory allocation of data structures affects the total memory read and write cost. If not mentioned specially, all the memory units we refer to afterwards represent one cacheline, which is the unit of one memory reference, typically 64 bytes in current computer systems. The read and write time to a unit, as well as other involved symbols, are summarized in Table 1.

If one algorithm uses t data structures, which can be referred as $\Phi = \{DS_i | i = 1, 2, \ldots t\}$. For any data structure DS_i, we refer the read times of each unit of DS_i to be m_i and write times to be n_i. When the unit is allocated on DRAM, the total read and write cost (memory cost) of that unit is $C_i^d = m_i r^d + n_i w^d$; when the unit is allocated on NVM, the memory cost is $C_i^p = m_i r^p + n_i w^p$. If we swap the unit from NVM to DRAM, the memory cost of that unit will decrease by $C_i^p - C_i^d$, we name it the cost gain C_i^g of the unit, represented by (1).

$$C_i^g = C_i^p - C_i^d \tag{1}$$

Moreover, we define the memory allocation scheme of Φ in an algorithm as a data placement scheme. In a data placement scheme, if DS_i occupies B_i^d DRAM units and B_i^p

Table 1. Symbols in DPM

Symbol	Description
r^d	Latency of read a DRAM unit
w^d	Latency of write a DRAM unit
C^d	Memory cost of a unit if placed in DRAM
r^p	Latency of read an NVM unit
w^p	Latency of write an NVM unit
C^p	Memory cost of a unit if placed in NVM
C^g	Memory cost gain when swap unit from NVM to DRAM

NVM units, the memory cost C_i of DS_i is then $C_i^d B_i^d + C_i^p B_i^p$. Hence, the total memory cost C is calculated by (2)

$$C = \sum_{i=1}^{t} \left(C_i^d B_i^d + C_i^p B_i^p \right) \tag{2}$$

3.2 Optimal Data Placement Model

Theorem 1. For $\forall x, y \in 1, 2, \ldots t$ that holds $C_x^g > C_y^g$, if the DRAM space of a hybrid memory system cannot accommodate both DS_x and DS_y, then the DRAM space must be allocated to DS_x preferentially. ∎

Proof. Given that the total memory cost in the data placement scheme is minimal, we assume that although $C_x^g > C_y^g$ and DRAM is not large enough to accommodate both DS_x and DS_y, DRAM is needless to allocate to DS_x preferentially. Thus, there must be some DRAM spaces that are allocated to DS_y rather than DS_x, i.e. $B_x^p > 0$ and $B_y^d > 0$. Without loss of generality, we let $B_x^p > B_y^d$. Then, if we swap B_y^d units DS_y in DRAM with B_y^d units DS_x in NVM, we will get a new data placement scheme. Let C' be the new memory cost under this scheme, we have the following result.

$$C' = \sum_{i=1,\ldots t \text{ and } i \neq x,y} \left(C_i^d B_i^d + C_i^p B_I^p \right) + C_x^d \left(B_x^d + B_y^d \right) + C_x^p \left(B_x^p - B_y^d \right) + C_y^d \left(B_y^d - B_y^d \right)$$

$$+ C_y^p \left(B_y^p + B_y^d \right)$$

$$= \sum_{i=1,\ldots t} \left(C_i^d B_i^d + C_i^p B_I^p \right) + B_y^d \left(C_x^d - C_x^p + C_y^p - C_y^d \right) = C + B_y^d (C_y^g - C_x^g). \tag{3}$$

Since $B_y^d > 0$ and $C_x^g > C_y^g$, we can derive that $B_y^d (C_y^g - C_x^g) < 0$, resulting in $C' < C$. This is in contrast with the above assumption that C is the minimal memory cost. So, we can conclude that DRAM space must be preferentially allocated to DS_x when $C_x^g > C_y^g$. ∎

Theorem 2. Suppose that under a data placement scheme, the maximal memory cost is denoted as C_{max}. If $\forall x, y \in 1, 2, \ldots t$ that holds $C_x^g > C_y^g$ and the DRAM space cannot accommodate both DS_x and DS_y, the DRAM space must be preferentially allocated to DS_y. ∎

The proof of Theorem 2 is similar to that of Theorem 1.

We name the data placement scheme with the minimum memory cost in Theorem 1 as the *optimal data placement scheme*. Accordingly, the scheme defined by Theorem 2 is regarded as the *worst data placement scheme*. According to the two theorems, the *optimal data placement scheme* allocates DRAM space to data structure with larger C^g value with higher priority, while the *worst data placement scheme* exactly does the opposite.

4   Sorting and Join with the Optimal Data Placement Model

The optimal data placement model is helpful to improve sorting and join algorithms. In this section, we present the new sorting and join algorithms that are based on the optimal data placement model. We first propose the NVMSort algorithm in Sect. 4.1, and then explore the NVMSort-based sort join algorithm in Sect. 4.2.

4.1   NVMSort

The sorting problem we consider is the standard comparison-based sorting with n records, each of which contains a key. We assume that the input is an unsorted array, and the output to be a sorted array which is sorted on their keys.

Based on the optimal data placement scheme, we propose an efficient sorting algorithm on hybrid memory systems. This algorithm is motivated by the read/write property of the classical HeapSort algorithm [8]. As Fig. 1 shows, during the execution of Heap-Sort, most writes are focused on a few portions of heap nodes. The following part will explain why heap structure has this read/write property and how we can leverage it to reduce the NVM writes.

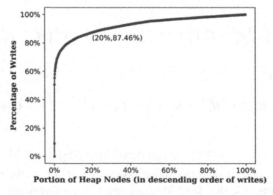

Fig. 1. Relationship between the portion of heap nodes and its percentage of total writes

Heap is physically stored in consecutive memory space like an array, while logically it is a binary tree with several constraints. Taking Min-heap as an example, the root node stores the minimal value and every node contains a smaller value than its child nodes. The HeapSort algorithm builds a heap at first; each time it replaces the root node with the last node in the heap, heapifies the heap, and outputs the original root to the result set until the heap goes empty. Note that every time we replace the root with new node, the heapify process will sift-down the new root until the heap meets Min-heap properties again. Therefore, every time the heap pops a root, the following heapify process does a couple of read/write operations to the heap memory space. Since the heap is a binary tree structure, if a node is close to the root, it is much likely to be read and written frequently. In other words, the nodes near to the root will have a higher C^g value than remote nodes. According to Theorem 1, if we place nodes close to the root into DRAM and nodes close to leaf into NVM, the total rea/write operations occurring on NVM as well as the total memory cost can be reduced.

Next, we want to present a more formal analysis. The time complexity of building a heap is $O(n)$ while HeapSort owns a time complexity of $O(nlog(n))$; thus, we mainly focus on the read and write operations in the sorting process. Assume that a Min-heap has the size of n, the distance from a node to the root is the height of the node, and l is the max node height, we can derive that $n \approx 2^{l+1}$. Suppose x is the height of current last node in the heap; when the node becomes the new root and is sifted down till heap property holds again, $O(x)$ reads and writes are done to the heap, $O(1)$ reads and writes for each level. In one level, each node has the same height, along with the same probability for being moved up (which incurs $O(1)$ times of reads and writes). Since level h has about 2^h nodes, the probability of a node being moved up is $\frac{1}{2^h}$, then the expectation of read/write operations on each node is $O\left(\frac{1}{2^h}\right)$. Therefore, after all the node is pop out, the total expectation of the read/write operations on each node in level h can be represented by (4).

$$E(h) = \sum_{i=h}^{l} O\left(\frac{1}{2^h}\right) 2^i = O\left(\frac{1}{2^h}\right) \sum_{i=h}^{l} 2^i = O\left(\frac{(2^{l+1} - 2^h)}{2^h}\right) = O\left(\frac{n}{2^h}\right) \quad (4)$$

That means the nodes closer to root have higher expectation of read and write. For example, the root node has an expectation of $O(n)$ while a leaf node has an expectation of $O(1)$. Using Formula (1), we can derive that the C^g node in level h, as represented by (5).

$$C_h^g = O\left(\frac{n}{2^h}\right)\left(r^p + w^p\right) - O\left(\frac{n}{2^h}\right)\left(r^d + w^d\right) = O\left(\frac{n}{2^h}\right)\left(r^p + w^p - r^d - w^d\right) \quad (5)$$

By leveraging the *optimal data placement scheme*, we can optimize HeapSort in the follow manner: place the nodes close to the heap root into DRAM and maintain other nodes in NVM.

Following the above idea, we devise a new sorting algorithm named NVMSort, as shown in Algorithm 1. Like HeapSort, NVMSort is also based on a heap data structure. However, differing from the traditional HeapSort, we divide the heap nodes in NVMSort into two parts, namely *NearRoots* and *NearLeafs*. *NearRoots* can fill up into

DRAM, which are placed in DRAM. Meanwhile, *NearLeafs* are organized in NVM before performing a classical HeapSort procedure.

In this part, we calculate the theoretical performance of NVMSort to prove that NVMSort outperforms the classical HeapSort in reducing both memory cost and writes to NVM. We assume that in the memory the proportion of DRAM is ϵ and we need to sort n elements. According to Formula (4), The total read-write operations of HeapSort is $\sum_{i=1}^{n} O\left(\frac{n}{2^{h(i)}}\right) = n \sum_{i=1}^{n} O\left(\frac{1}{2^{\lfloor \log(i) \rfloor}}\right) = O(n\log(n))$, where $h(i)$ is the height of node i, i.e., $h(i) = \lfloor \log(i) \rfloor$. Because HeapSort treats the hybrid memory as a uniform memory space, we can assume that the read/write operations are distributed evenly among the available memory space. Therefore, $O(n\log(n)) \cdot \epsilon$ read/write operations will occur on DRAM and $O(n\log(n)) \cdot (1 - \epsilon)$ operations will occur on NVM. According to Formula (2), the total memory cost of HeapSort can be calculated by (6):

$$
\begin{aligned}
C_{HeapSort} &= O(n\log(n)) \cdot \epsilon \cdot (r_d + w_d) + O(n\log(n)) \cdot (1 - \epsilon) \cdot (r_p + w_p) \\
&= O(n\log(n)) \cdot (\epsilon(r_d + w_d) + (1 - \epsilon)(r_p + w_p))
\end{aligned}
\tag{6}
$$

Algorithm 1. NVMSort

Input: an unsorted data array A
Output: a sorted array A in ascending order
1:      *NearRoot* = *DRAM* space;
2:      *NearLeaf* = (*A*.size * Node_Size − *NearRoot*) *NVM* space;
3   *heap_space* = {*NearRoot*, *NearLeaf*};
4:      // *make sure Nodes close to Root reside in NearRoot space*
5:      *heap* = Build_Max_Heap(*heap_space*, *A*);
7:      **for** i = *heap.length* downto 2
8:      | *A[i]* = *heap*[1];
9:      | *heap[1]* = *heap*[i];
10:     | *heap*.size = *heap*.size - 1;
11:     | Max_Heapify(*heap*);
12:     **end for**
13:     *A*[1] = *heap*[1];
14:     **return** A;

While for NVMSort, the read/write operations to DRAM are all absorbed by *Near-Roots*. The total number of operations is $\sum_{i=1}^{n \cdot \epsilon} O\left(\frac{n}{2^{h(i)}}\right) = n \sum_{i=1}^{n \cdot \epsilon} O\left(\frac{1}{2^{\lfloor \log(i) \rfloor}}\right) = O(n\log(n \cdot \epsilon))$. Similarly, the read/write operations to NVM are concentrated on *Near-Leaf*, which are $\sum_{i=n \cdot \epsilon+1}^{n} O\left(\frac{n}{2^{h(i)}}\right) = n \sum_{i=n \cdot \epsilon+1}^{n} O\left(\frac{1}{2^{\lfloor \log(i) \rfloor}}\right) = O\left(n\log\left(\frac{1}{\epsilon}\right)\right)$. Thus, we can calculate the total memory cost of NVMSort by (7).

$$
C_{NVMSort} = O(n\log(n \cdot \epsilon))(r_d + w_d) + O\left(n\log\left(\frac{1}{\epsilon}\right)\right)(r_p + w_p)
\tag{7}
$$

According to the analysis above, HeapSort incurs $O(n\log(n))$ writes to NVM while NVMSort only makes $O(n)$ NVM writes. To this end, we can see that NVMSort is more write-friendly to NVM. Due to the read and write asymmetry of NVM chips, especially the write latency w_p of NVM is typically one order higher than the read latency r_p, the reduction of NVM writes in NVMSort can result in significant performance improving, when compared with HeapSort. When combined reads and writes, NVMSort can reduce the total access time of NVM by $O\left(nlog\left(\epsilon \cdot n^{1-\epsilon}\right)\right)\left(r_p + w_p\right)$. In the following experiments, ei demonstrate the performance of NVMSort to support the above analysis.

4.2 Sort Join with NVMSort

In this section, we discuss the applicability of NVMSort in traditional sort join. Sort join is one of the commonly used join algorithms in modern relational DBMSs. As the relations to be joined are supposed to be in external storage, e.g., SSDs or magnetic disks, traditional sort join usually employs the merge sort algorithm to sort relations residing in disks. The basic process of the merge-sort join consists of the following steps:

(1) Read pages into the memory;
(2) Sort the tuples in the memory;
(3) Write the sorted tuples as a run into the disk;
(4) Read all the first pages in each run, merge in memory, and write out to the file.

NVMSort can be used to optimize step (2) in the merge-sort join algorithm. In the implementation of traditional merge-join, we usually utilize QuickSort as the main-memory sort algorithm, while in the NVM-based hybrid memory QuickSort is not the best choice, as we have discussed before. We will experimentally demonstrate that NVMSort is more efficient than QuickSort as well as HeapSort when applied into the merge-sort join algorithm.

5 Performance Evaluation

In this section, we evaluate the efficiency of NVMSort by comparing it with other sort algorithms. The competitors include two traditional sort algorithms, including HeapSort and QuickSort [8]. We measure the total run time of each algorithm along with its total reads and writes to DRAM and NVM. The results show that our NVMSort has the best performance with limited writes to NVM.

5.1 Settings

All the experiments are performed on an Intel Core i5-8500 3.0 GHZ CPU. This CPU has 6 physical cores, with a 9 MB L3 cache. Each core has a private 256 KB L2 cache and 32 KB L1d cache and 32 KB L1i cache. The operation system is Ubuntu 18.04 with the kernel version of 5.2.8. We use the open-source persistent memory development kit [16] to simulate NVM on Linux. This library maps a memory region to on-disk file and ensure the atomicity and persistency on read/write operation to that region. In order to simulate the read-write asymmetric of mainstream NVM, we follow the lead of the hardware community [17] and inject artificial 240 ns delays after each write operation to that persistent region.

Each algorithm is allocated with the same amount of memory. The ratio between DRAM and NVM space is set to same for each algorithm. However, the classical Heap-Sort and QuickSort are unaware of the read-write asymmetric property of NVM. They use the hybrid memory space as a uniform memory space. NVMSort treats DRAM and NVM differently. It splits the heap structure into two parts before sorting; the formal part is put into DRAM and the latter part is placed in NVM.

We use three datasets to test the sorting algorithms, scaling from 100k elements (denoted as 100K below), 1 million elements (denoted as 1M) to 10 million elements (denoted as 10M). All data elements are tuples with 8 bytes key and 8 bytes payload, which is generated randomly and stored in the hybrid memory space.

5.2 Sorting Performance

We mainly evaluate the run time of each sorting algorithm. In addition, we measure the write count on NVM during sorting, as reducing writes to NVM is one of the key objectives of NVMSort. In the following text, we will present the results on the small dataset as well as on the large dataset.

Scalability. Figure 2 shows the run time on the three datasets under different NVM write latency. In this experiment, the DRAM ratio is set to 0.2. We can see that NVMSort gets the best sorting performance when running on datasets scaling from 100K, 1M to 10M. Averagely, NVMSort is $1.5\times$ faster than HeapSort and QuickSort. As NVM write latency gets higher, NVMSort exhibits more advantage than its competitors. This experiment shows that NVMSort can well suit different sizes of datasets, showing good scalability for real applications. Therefore, in the following experiments, we use the 1M dataset by default.

NVM Writes. Table 2 summarizes the DRAM reads/writes and NVM reads/writes of NVMSort, HeapSort, and QuickSort, when running on the 1M dataset with the DRAM ratio on 0.2. We can see that NVMSort has the fewest NVM writes among all algorithms; thus it is more NVM friendly than the other two algorithms. As NVM has limited write endurance, reducing writes to NVM is a critical issue in the design of NVM-aware sorting algorithms. NVMSort also has fewer NVM reads than QuickSort and HeapSort. The DRAM accesses of NVMSort is comparable to that of HeapSort. Although the total DRAM reads/writes of NVMSort is a little more than QuickSort, QuickSort has much more NVM reads/writes than NVMSort, resulting in poor time performance of QuickSort as shown in Fig. 2. On the other hand, this study does not aim to optimize DRAM operations.

Sensitivity to the DRAM Ratio. In this experiment, we aim to measure the performance of NVMSort when varying the ratio of DRAM among the hybrid memory. The 1M dataset is used in this experiment. The NVM write latency is set to 300 ns. Figure 3 shows the run time of NVMSort and its two competitors, in which the DRAM ratio is varied from 0.1 to 0.4. Note that we do not increase the DRAM ratio to a much high value,

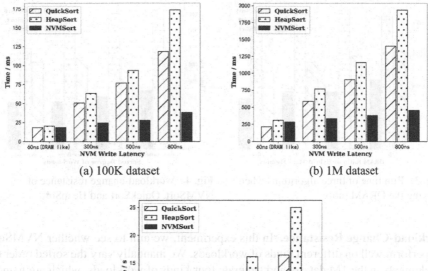

(a) 100K dataset　　　　　　　　(b) 1M dataset

(c) 10M dataset

Fig. 2. Run time on different scales of datasets

Table 2. Comparison of DRAM read/writes and NVM read/writes

	DRAM		NVM	
	Reads (10^6)	Writes (10^6)	Reads (10^6)	Writes (10^6)
QuickSort	11	56	41	22
HeapSort	24	83	87	30
NVMSort	102	35	9	3

because NVM has higher density than DRAM. Thus, it is not likely that the DRAM size is over the NVM size in the hybrid memory system. As shown in Fig. 3, NVMSort keeps relatively high and stable performance when varying the DRAM ratio. This is because NVMSort can choose optimal data placement according to the DRAM and NVM usage during the sorting process. On the contrary, both QuickSort and HeapSort do not make any optimizations for NVM, resulting poor time performance.

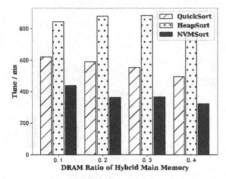
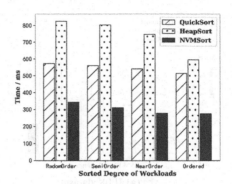

Fig. 3. Run time of three algorithms when varying the DRAM ratio

Fig. 4. Workload-change resistance of NVMSort, QuickSort and HeapSort

Workload-Change Resistance. In this experiment, we aim to see whether NVMSort can perform well on different kinds of workloads. We manually vary the sorted order of the elements in the 1M dataset and generate four kinds of workloads, which are named RandomOrder (totally unordered), SemiOrder (50% ordered), NearOrder (80% ordered), and Ordered (totally ordered). We then run NVMSort on these four kinds of workloads to verify its performance. Here, the DRAM ratio is also set to 0.2 and the NVM write latency is set to 300 ns. As Fig. 4 shows, NVMSort has stable time performance when the workload property changes from "totally unordered" to "totally ordered". In addition, NVMSort has the best time performance in all workloads, indicating that NVMSort is workload-aware due to its intrinsic optimal data placement scheme.

NVM Efficiency. NVMSort is designed for reducing both sorting time and NVM writes. The above experiments present detailed measurement of NVMSort with respect to different measures. However, as NVMSort is designed not only for high time performance but also for reducing NVM writes, it is not apparently to see the overall performance of NVMSort in terms of time performance and NVM friendliness.

Thus, we further propose a metric named *NVM Efficiency* that combines both run time and NVM write reduction. The *NVM Efficiency* as the overall performance is defined as follows:

$$NVM\ Efficiency = P/W_{NVM}, \ where \ P = 1/t \tag{8}$$

Here, P represents the time performance of a sorting algorithm (t is the run time of the algorithm), and W_{NVM} is the NVM writes caused by a sorting algorithm. The *NVM Efficiency* represents the time performance per NVM write, which implies that higher time performance or less NVM writes will result in high NVM efficiency. As higher time performance and less NVM writes is actually the design goal of this study, the *NVM Efficiency* is suitable for measuring the overall performance of a sorting algorithm.

Figures 5 shows the NVM efficiency on the 1M datasets. Here, the time performance P is calculated with the time granularity of second. In this figure, we can see that as the DRAM ratio gets higher, all the sort algorithms have better NVM Efficiency due to more DRAM space can reduces costly NVM writes. Particularly, when the DRAM ratio is 0.1, the NVM Efficiency of NVMSort is 6× higher than QuickSort and 13× than HeapSort. As the DRAM ratio goes up to 0.4, NVMSort is 12×/22× higher than QuickSort/HeapSort in term of NVM Efficiency. That clearly shows that the proposed NVMSort algorithms maintains the highest NVM efficiency in various DRAM ratio, meaning that NVMSort has the best overall performance.

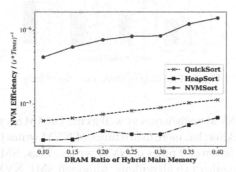

Fig. 5. NVM efficiency

To sum up, all the experiments show that our NVMSort is more efficient for the DRAM-NVM-based hybrid memory system compared with traditional sorting algorithms including HeapSort and QuickSort. This is mainly due to the new structural design in NVMSort, i.e., the heap-node organization and placement based on the optimal data placement model.

5.3 Join Performance

In this experiment, we compare the time performance as well as NVM writes of sort join algorithms. We implement the traditional merge-sort join algorithm and replace the in-memory sorting algorithm with NVMSort. This improved sort join is denoted as SMJ_NVMSort (meaning Sort-Merge-Join with NVMSort). Similarly, we get the other two sort join algorithms, which are denoted as SMJ_QuickSort and SMJ_HeapSort. In this experiment we also use the 1M dataset and the DRAM ratio is set to 0.2. The join algorithm runs on two relational tables R and S. Since we allocate total 12 MB memory (including 9.6 MB NVM and 2.4 MB DRAM, nearly 4:1) and assume that both relations cannot be hold in the memory, we set the size of R to 16 MB and the size of S to 160 MB. This setting is compatible with the general assumption of join algorithms in traditional relational database systems, in which the build relation R is smaller than the probe relation S.

The results are shown in Fig. 6. When the NVM write latency is close to that of DRAM, SMJ_NVMSort has similar performance with the other two algorithms. However, when the NVM write latency is set to be higher than that of DRAM, we can see that SMJ_NVMSort outperforms SMJ_QuickSort and SMJ_HeapSort. To this end, SMJ_NVMSort is more suitable for DRAM-NVM-based hybrid memory.

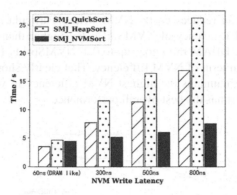

Fig. 6. Run time of three join algorithms

Table 3 shows the NVM reads/writes as well as the DRAM accesses of the three join algorithms. SMJ_QuickSort has more than 6 times of NVM writes than SMJ_NVMSort, while SMJ_HeapSort has over 9 times of NVM writes. Thus, SMJ_NVMSort is more NVM friendly than the other two algorithms. Although SMJ_NVMSort has a bit more DRAM accesses than its competitors, this does not influence the time performance of SMJ_NVMSort, as shown in Fig. 6. In addition, as DRAM has unlimited write endurance and is faster than NVM, increasing DRAM accesses is not acceptable in DRAM-NVM-based hybrid memory.

Table 3. Comparison of DRAM read/writes and NVM read/writes

	DRAM		NVM	
	Reads (10^6)	Writes (10^6)	Reads (10^6)	Writes (10^6)
SMJ_QuickSort	141	755	468	252
SMJ_HeapSort	202	724	1093	380
SMJ_NVMSort	1178	409	117	44

6 Conclusions

NVM has become an alternative of next-generation memories. It is a trend to construct hybrid memory systems composed of DRAM and NVM in the future. In this paper, we studied the fundamental sorting issue on DRAM-NVM-based hybrid memory and proposed an efficient sorting algorithm called NVMSort. NVMSort is based on the optimal data placement model that proposes to maintain highly-written data structures in DRAM and others in NVM. We theoretically proved the correctness and efficiency of the optimal data placement model, based on which NVMSort was presented. We further integrated NVMSort into the traditional merge-sort join algorithm. We conduct extensive experiments on various datasets with different settings. The results suggest the high performance and NVM friendliness of NVMSort and the new sort-join scheme.

There are some future research directions that are worth further investigating. First, it is a promising issue to consider efficient buffer management schemes to improve the join performance on NVM-based hybrid memory [18, 19]. Second, it is valuable to study other join algorithms such as hash join that runs on NVM-based hybrid memory [20]. Third, the architecture of hybrid memory systems has a big impact on fundamental algorithms. There are also other kinds of architecture for hybrid memory systems, e.g., the hybrid storage involving DRAM, NVM, and SSD [21]. Thus, we will investigate sorting and join algorithms on other kinds of hybrid-storage architecture in the future.

Acknowledgement. This paper is partially supported by the National Science Foundation of China under the grant number no. 61672479.

References

1. Garcia-Molina, H., Ullman, J.D., Widom, J.: Database System Implementation, 2nd edn. Prentice Hall, Upper Saddle River (2010)
2. Faerber, F., Kemper, A., Larson, P., Levandoski, J.J., Neumann, T., Pavlo, A.: Main memory database systems. Found. Trends Databases **8**(1–2), 1–130 (2017)
3. Renen, A., et al.: Managing non-volatile memory in database systems. In: SIGMOD, pp. 1541–1555 (2018)
4. Psaropoulos, G., Oukid, I., Legler, T., May, N., Ailamaki, A.: Bridging the latency gap between NVM and DRAM for latency-bound operations. In: DaMoN, 13:1–13:8 (2019)
5. Chen, K., Jin, P., Yue, L.: A novel page replacement algorithm for the hybrid memory architecture involving PCM and DRAM. In: Hsu, C.-H., Shi, X., Salapura, V. (eds.) NPC 2014. LNCS, vol. 8707, pp. 108–119. Springer, Heidelberg (2014). https://doi.org/10.1007/978-3-662-44917-2_10
6. Zhang, D., Jin, P., Wang, X., Yang, C., Yue, L.: DPHSim: a flexible simulator for DRAM/PCM-based hybrid memory. In: Chen, L., Jensen, C.S., Shahabi, C., Yang, X., Lian, X. (eds.) APWeb-WAIM 2017. LNCS, vol. 10367, pp. 319–323. Springer, Cham (2017). https://doi.org/10.1007/978-3-319-63564-4_27
7. Jin, P., Yang, C., Jensen, C.S., Yang, P., Yue, L.: Read/write-optimized tree indexing for solid-state drives. VLDB J. **25**(5), 695–717 (2016)
8. Cormen, T., Leiserson, C., Rivest, R., et al.: Introduction to Algorithms. MIT Press, Cambridge (2009)
9. Wang, Y., Li, K., Zhang, J., Li, K.: Energy optimization for data allocation with hybrid SRAM+NVM SPM. IEEE Trans. Circ. Syst. **65-I**(1), 307–318 (2018)
10. Salkhordeh, R., Mutlu, O., Asadi, H.: An analytical model for performance and lifetime estimation of hybrid DRAM-NVM main memories. IEEE Trans. Comput. **68**(8), 1114–1130 (2019)
11. Li, L., Jin, P., Yang, C., Wan, S., Yue, L.: XB+-tree: a novel index for PCM/DRAM-based hybrid memory. In: Cheema, M.A., Zhang, W., Chang, L. (eds.) ADC 2016. LNCS, vol. 9877, pp. 357–368. Springer, Cham (2016). https://doi.org/10.1007/978-3-319-46922-5_28
12. Chen, S., Gibbons, P.B., Nath, S.: Rethinking database algorithms for phase change memory. In: CIDR (2011)
13. Viglas, S.: Adapting the B+-tree for asymmetric I/O. In: ADBIS, pp. 399–412 (2012)

14. Viglas, S.: Write-limited sorts and joins for persistent memory. Proc. VLDB Endow. **7**(5), 413–424 (2014)
15. Blelloch, G.E., Fineman, J.T., Gibbons, P.B., Gu, Y., Shun, J.: Sorting with asymmetric read and write costs. In: SPAA, pp. 1–12 (2015)
16. OFTC: Persistent Memory Development Kit, 25 March 2020. https://pmem.io/pmdk/
17. Volos, H., Tack, A.J., Swift, M.M.: Mnemosyne: lightweight persistent memory. In: ASPLOS, pp. 91–104 (2011)
18. Wu, Z., Jin, P., Yang, C., Yue, L.: APP-LRU: a new page replacement method for PCM/DRAM-based hybrid memory systems. In: Hsu, C.-H., Shi, X., Salapura, V. (eds.) NPC 2014. LNCS, vol. 8707, pp. 84–95. Springer, Heidelberg (2014). https://doi.org/10.1007/978-3-662-44917-2_8
19. Ou, Y., Härder, T., Jin, P.: CFDC: a flash-aware buffer management algorithm for database systems. In: Catania, B., Ivanović, M., Thalheim, B. (eds.) ADBIS 2010. LNCS, vol. 6295, pp. 435–449. Springer, Heidelberg (2010). https://doi.org/10.1007/978-3-642-15576-5_33
20. Yang, L., Jin, P., Wan, S.: BF-join: an efficient hash join algorithm for DRAM-NVM-based hybrid memory systems. In: ISPA, pp. 875–882 (2019)
21. Jin, P., Yang, P., Yue, L.: Optimizing B+-tree for hybrid storage systems. Distrib. Parallel Databases **33**(3), 449–475 (2015)

Parallel SCC Detection Based on Reusing Warps and Coloring Partitions on GPUs

Junteng Hou[1,2][(✉)], Shupeng Wang[1], Guangjun Wu[1], Bingnan Ma[3][(✉)],
and Lei Zhang[1]

[1] Institute of Information Engineering, Chinese Academy of Sciences, Beijing, China
{houjunteng,wangshupeng,wuguangjun,zhanglei1}@iie.ac.cn
[2] School of Cyber Security, University of Chinese Academy of Sciences,
Beijing, China
[3] National Computer Network Emergency Response Technical Team/Coordination
Center of China, Beijing, China
mabingnan90@gmail.com

Abstract. Detecting strongly connected components (SCCs) in directed graphs is crucial for analyzing the structure of graphs. It can accelerate the process of large-scale graphs with applying GPUs to graph computation. However, GPU-based SCC detection still faces the challenge from graphs, especially real-world graphs, skewing and strict synchronization requirement. In this paper, we proposed a new paradigm of identifying SCCs with reusing warps and coloring partitions. The scheme of reusing warps assigns multiple vertices to each virtual warp at one time, which greatly reduces the number of warps that are applied but assigned few tasks. Furthermore, we proved that the partitions formed by coloring independently contains all SCCs in them, so we use coloring to identify more partitions and parallel detect SCCs in each partition to get more SCCs. Finally, the colored partitions are deeply combined with the following traversal to optimize the algorithm. We conduct extensive theoretical and experimental analysis to demonstrate the efficiency and accuracy of our approach. The experimental results expose that our approach can achieve 7.23×, 30.55×, 1.75× and 1.26× speedup for SCC detection using NVIDIA K80 compared with algorithms of Tarjan, Barnat, Hong and Slota respectively.

Keywords: Strongly connected components detection · GPUs · Reusing warps · Coloring partitions · Partitions combined with traversal

1 Introduction

Strongly connected component (SCC) is a significant graph structure in graph data management. In a directed graph, an SCC is a maximal subset of vertices

This work was supported by the National Natural Science Foundation of China (No. 61931019).

M. Qiu (Ed.): ICA3PP 2020, LNCS 12452, pp. 31–46, 2020.
https://doi.org/10.1007/978-3-030-60245-1_3

such that every vertex has at least one directed path to other vertices. Detecting SCCs is a fundamental graph algorithm for graph analytics [1], which is widely used in many applications including pattern matching [4], scientific computing [2] and model checking [3].

Most of state-of-the-art GPU-based SCC detection algorithms are improved from Forward-Backward (FB) algorithm [20] or color algorithm [8]. Mclendon III et al. found that there are lots of SCCs composed of one vertex (1-SCC) in most graphs and proposed FB-Trim algorithm [15] using Trim operation to quickly detect a large number of 1-SCCs. In real-world graphs, the scale of different SCCs varies greatly because of power-low distribution of vertices [10]. Li et al. [11] divided SCC detection into two stages. The great SCC containing a mass of vertices is identified at data-level parallel stage and the numerous small SCCs containing few vertices are identified at task-level parallel stage. Because there will be many non-connected partitions after detecting the great SCC, Hong et al. [12] added weakly connected component (WCC) detection to identify these partitions, then the SCC detection can be performed in parallel in each partition. In addition, they extended Trim operation from only detecting SCCs formed by one vertex to detecting SCCs formed by one or two vertices. The algorithm proposed by Li et al. [13] divided the graph into multiple partitions before detection, and detected SCCs in each partition in parallel, which enhanced the parallelism of the algorithm, but it introduced workload of processing boundary vertices. Considering the power-law properties of real-world graphs, Slota et al. [14] improved the color algorithm by using FB algorithm to detect the great SCC and then performing color algorithm. Devshatwar et al. [10] proposed selecting the vertex with the largest product of in-degree and out-degree as the pivot to ensure that the SCC detected at the first stage is the great SCC, and they adjusted thread assignment of each warp at different stages. This scheme performed well when applied to Slota's [14] and Hong's [12] algorithms.

We propose a new SCC detection scheme named FBw-Pc that can make full use of the assigned threads by reusing warps and deeply combining coloring partitions with FB algorithm. The contributions are shown as follows:

(1) We propose the scheme of reusing warps to detect the great SCC. In this method, we group several threads into a virtual warp, and assign multiple vertex tasks to each virtual warp. All threads in each virtual warp process adjacent vertices of each assigned vertex in parallel. This method can effectively reduce the number of assigned threads, and avoid uneven thread assignment caused by graph skewed to make full use of the assigned warps.

(2) We prove that the partitions formed by coloring independently contain all the SCCs in them, in other words, all vertices in a certain SCC are contained in the same partition formed by coloring. Accordingly, we identify more partitions by coloring and detect SCC in each partition in parallel. Compared with the method of WCC detection forming multiple partitions, the scheme of coloring partitions can not only form more partitions, but also detect one SCC in each partition at the same time.

(3) We adjust the calling frequency of coloring subprocess and FB subprocess to optimize assignment of partitioning and SCC detection. The method of coloring partitions can detect one SCC in each partition, but it can't guarantee that the maximum SCC in current partition is detected. We perform multiple FB operations after coloring operation to make full use of the formed partitions.

The rest of this paper is organized as follows. We illustrate the background in Sect. 2. Section 3 overviews our novel algorithm and presents the detail of our algorithm based on reusing warps and coloring partitions. Section 4 describes the experimental setup and results. Section 5 concludes.

2 Preliminaries

In this paper, we use $G = (V, E)$ to denote a directed graph, where V is vertex set and E is edge set. We consider the algorithm's performance on both synthetic graphs and real-world graphs. Real-world graphs are extracted from real-world application scenarios including: web links, social networks, citation networks, reinforcement learning, etc. [2]. Power-law distribution is the main feature of real-world graphs [16], which means that there is a great SCC with a mass of vertices and a large number of SCCs composed of few vertices, and the number of SCCs obeys power-law distribution. So real-world graphs are severely skewed by power-law distribution. Synthetic graphs are directly generated with custom size and structure to simulate real-world graphs. Georgia Tech. Graph generator (GTgraph) [17] is a commonly used graph generation tool, which can generate the following three types of graphs: Random, R-MAT [18] and SSCA#2 [19].

Most early algorithms of SCC detection are sequential algorithms includeing: Tarjan [5], Dijkstra [6], and Kosaraju [7]. Tarjan algorithm is one of the most efficient sequential algorithms of SCC detection for it only needs to traverse each edge once to detect all SCCs. Barnat et al. [8] summarized the parallel algorithms of SCC detection and divided them into three types: FB algorithm, color algorithm, Recursive OBF algorithm. Barnat et al. [9] performed comparative experiments on the above three algorithms and demonstrated that parallel algorithms, especially FB algorithm, are significantly faster than sequential algorithms on large-scale graphs. FB algorithm is the basis of most state-of-the-art parallel algorithms of SCC detection, which is based on divide and conquer scheme proposed by Fleischer et al. [20] according to the following Theorem 1.

Theorem 1. *Let $G = (V, E)$ be a directed graph with a vertex $u \in V$. Then $FW_G(u) \cap BW_G(u)$ is an SCC containing u. Moreover, every other SCC in G is contained in either $FW_G(u) \setminus BW_G(u)$, $BW_G(u) \setminus FW_G(u)$, or $V \setminus (FW_G(u) \cup BW_G(u))$.*

In Theorem 1, $FW_G(u)$ or $BW_G(u)$ is the vertex set formed by forward or backward traversal from vertex u. According to this theorem, it can get an SCC by intersecting the two sets formed by forward and backward traversal from any vertex. And the remaining SCCs are contained in three separate partitions, so the next SCC detection can be performed in parallel in each partition.

The color algorithm can also be used for SCC detection. First, the ID of each vertex is treated as color value; second, the color of current vertex's forward adjacent vertices is detected. If it is less than the color of current vertex, the current vertex's color will be modified to the smaller one; third, the above operations are iterated until the color value of all vertices no longer change. At this time, the graph is divided into many partitions according to different colors; forth, the vertex whose initial color is the same as its partition's color is selected as the pivot, and BFS is performed from this pivot in each partition, then the vertices traversed by each pivot form an SCC; finally, the colors of undetected vertices are reset to zero and the above operations are iterated until all vertices are detected.

3 The FBw-Pc Implementation

In this section, we first show the overall process of FBw-Pc algorithm, and introduce the great SCC detection algorithm optimized by reusing warps scheme. We prove that there is no SCC crossing the partitions formed by coloring, and then propose a new partition scheme based on coloring. Finally, our algorithm is optimized by adjusting the parameters combining coloring partitions and FB operations.

3.1 Overview

The rough process of FBw-Pc algorithm is shown in Algorithm 1, where each operation represents a subprocess of the algorithm. In the two phases of the algorithm, the great SCC is detected in the first phase and the remaining SCCs are detected in the second phase. In addition, we also optimize Trim operation, pivot selection program, and partitioning method. In our algorithm, we use compressed sparse row (CSR) format to save graphs. In addition to input parameters, we also applied for two other arrays of size equal to the number of vertices, where M is used to mark the vertex state, that is, whether the vertex is selected as pivot, detected by Trim, traversed forward or backward, etc. And P records partition ID of each vertex.

In Phase 1, we propose reusing warps scheme to detect the great SCC. First, **Trim1** is used to detect the SCC consisting of one vertex (1-SCC). Second, **Pivot-choose1** selects a pivot. Third, starting from the selected pivot, **FBw** traverses the graph using forward and backward BFS improved by reusing warps scheme. Finally, according to Theorem 1, it can obtain one SCC and three partitions from the traversal result of each pivot. **Update-state** analyzes the result, marks the vertices of obtained SCCs in V_{SCC}, marks the vertices of obtained partitions in P, and resets the state of undetected vertices to 0 in M. At the beginning of Phase 2, we first determine the necessity of Trim operation. If k_t is 0, it will not perform FB operation in Phase 2. The next algorithm is similar to color algorithm with partitions, which doesn't require Trim operations. Otherwise, we perform multiple trimming of 1-SCCs and one trimming of 2-SCCs (the SCCs consisting of two vertices) in **Trim**. The remaining is mainly composed of two loops. In the outer loop, **Pc** is based on coloring partitions scheme, which colors adjacent vertices of all undetected vertices to generate new partitions, and traverses the colored vertices backward to identify SCCs. In the inner loop, **Pivot-choose2** selects pivots and **FB** performs forward and backward BFS from these points. **Update-state** analyzes the result and marks the vertices of obtained SCCs and partitions. The detection restarts until it performs k_t times or all SCCs are detected.

The pivot selection schemes used in Phase 1 and Phase 2 are different. In Phase 1, it needs to detect the great SCC, so **Pivot-choose1** selects the vertex with largest product of in-degree and out-degree as the pivot. The rest SCCs are numerous but not large, so **Pivot-choose2** randomly selects a vertex as the pivot in each partition. The detection operations also take partition marks P in account, which means that only undetected vertices in the same partition can be detected into the same SCC. In **Trim1** or **Trim**, the vertices with zero in-degrees or out-degrees are detected as 1-SCCs and any two interconnected vertices with no other incoming or outgoing edges are detected as a 2-SCC.

Algorithm 1. FBw-Pc SCC Detection Algorithm

Input: G: the input graph stored in CSR format. V_{SCC}: An zero array with the size equals to the number of vertices. k_w: preset variable, the number of vertices assigned to each virtual warp in reusing warps scheme. k_t: preset variable, the number of FB operations performed after coloring partitions in Phase 2.

Output: V_{SCC}: an array, each element records the pivot of the corresponding vertex's SCC, which is also the ID of SCC.

```
 1: /* Phase 1 */
 2: Trim1 (G, V_SCC, M)
 3: Pivot-choose1 (G, M)
 4: do in parallel
 5:    FBw (G, M, k_w)
 6: end do
 7: Update-state (G, V_SCC, M, P)
 8: /* Phase 2 */
 9: if k_t ≠ 0 then
10:    Trim (G, V_SCC, M, P)
11: end if
12: repeat in parallel
13:    Pc (G, M, P)
14:    repeat in parallel
15:       Pivot-choose2 (G, M, P)
16:       FB (G, M, P)
17:       Trim1 (G, V_SCC, M, P)
18:       Update-state (G, V_SCC, M, P)
```

19:    **until** repeat k_t times or no SCC generated
20: **until** no SCC generated
21: **return** V_{SCC}

3.2   The Scheme of Reusing Warps

When detecting the great SCC, traditional algorithms assign one thread to each vertex. However, most vertices of the great SCC have large in-degree and out-degree, and it must iteratively process these vertices multiple times to complete the detection. In response to this problem, Devshatwar et al. proposed to assign a virtual warp including multiple threads to process each vertex. But the ratio of tasks assigned to each thread is the same as traditional algorithms, so the workload of each thread is still uneven caused by the skewed structure of graph. In addition, even though only one virtual warp is working, the other virtual warps in the same warp can't be released until the busy virtual warp completes its work. Although the vertices in great SCC account for a large proportion of all vertices, there are still not many vertices to be processed in each iteration. Therefore, traditional warp-based algorithms require increasing number of threads, but most threads are soon released after assignment because their corresponding vertices don't need to be processed.

We propose a new method called reusing warps, which assigns multiple vertices to each virtual warp. This method effectively reduces the number of assigned threads and makes the most of assigned threads. Specifically, we preset a threshold k_w at the beginning to represent the number of assigned vertices in each virtual warp. When detecting the great SCC, massive tasks of processing vertices are assigned to each virtual warp. In each virtual warp, we check whether the assigned vertices need to be processed and use all threads to process the vertices that need to be processed in parallel.

The reusing warps method can alleviate uneven workload when detecting the great SCC. For simplicity, suppose there are 16 threads (32 actually) in a warp, and each virtual warp has 4 threads. As shown in the following Fig. 1, 2 and Fig. 3, the curves in rectangles represent threads in warps, and the circles represent vertices that may be processed during forward or backward traversal, where the gray ones represent inactive vertices that do not need to be processed, the green ones represent active vertices, and the number in them indicate the number of their adjacent vertices.

Traditional scheme of thread assignment is shown in Fig. 1. All threads in a warp can't be released until no thread is occupied, so the running time of this warp is determined by vertex A_2, which requires 32 traversals. The virtual warp algorithm is shown in Fig. 3. In a certain warp, each virtual warp can only process one vertex, and the warp is released after all virtual warps are not occupied. Therefore, in order to process 16 vertices, it requires three additional thread rearrangements including warp application, virtual warp division, and thread assignment. Specifically, the first task of processing vertices requires 8 traversals determined by vertex A_2; the second task of processing vertices requires 4 traversals determined by vertex A_5; the third task of processing vertices requires 4 vertices determined by vertex A_{11}; the fourth task of processing vertices requires 2 vertices determined by vertex A_{16}, so the virtual warp algorithm requires 16 traversals and three additional thread rearrangements. Our method is shown in Fig. 2, and it has three advantages in compared with virtual warp scheme. First, it only needs 9 traversals determined by A_2 and A_4; second, it doesn't require additional thread rearrangement; third, the number of required threads is a quarter of virtual warp scheme. In actual SCC detection, there is lower proportion of vertices that need to be processed in each traversal. And our algorithm will be better with greater and more variable vertices' out-degree and in-degree.

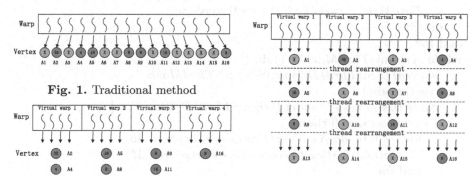

Fig. 1. Traditional method

Fig. 2. Our method

Fig. 3. Virtual warps method

We propose two schemes to accomplish reusing warps method, the straightforward scheme is using all threads in each virtual warp to find the vertices that need to be processed, then each thread traverses the adjacent vertices of one found vertex in parallel, and finally the above steps is iteratively performed. The second scheme is that all threads in each virtual warp parallel detect whether the vertices need to be processed, and parallel traverses the adjacent vertices of all vertices that need to be processed. The details of the second scheme are shown in Algorithm 2, which is the most important sub-process of **FBw** in Algorithm 1. We use current thread $curtThrdID$ to detect its corresponding vertices, and mark the vertex that needs to be processed with $ThrdMrk$. The internal function **syncAny** is used to synchronously detect the vertex marked in $ThrdMrk$ in the current block, and store it in the shared memory $sharedID$ of its corresponding virtual warp. **shft** broadcasts the vertex stored in $sharedID$ to all threads of each virtual warp. In **visit**, the threads access adjacent vertices of the broadcasted vertices in parallel and mark them in M. The above processes is iterated until all vertices are processed. This method can quickly skip vertices that don't need to be processed.

38   J. Hou et al.

Algorithm 2. Warp Reusing Algorithm

Input: $VertList$: an array, the ID of vertices assigned in each warp.
$VertNum$: the number of vertices assigned in each warp. $AdjList$: an array,
the adjacent vertices' ID of vertices assigned in each warp. $AdjNum$: the
adjacent vertices' number of vertices assigned in each warp. $BlockSize$: the
number of threads in each thread block. $WarpSize$: the number of threads in
each virtual warp. M: an array, the state of each vertex.

Output: M.

    __shared__ $sharedID[BlockSize/WarpSize]$
    for $i = 0$ to $VertNum/WarpSize$ **do**
      if $VertList[curtThrdID + i * WarpSize]$ needs process **then**
        $ThrdMrk := curtThrdID + i * WarpSize$
      end if
    end for
    while syncAny($ThrdMark > 1$) **do**
      $sharedID[curThrdID/WarpSize] := ThrdMark$
      Syncthreads()
      $curtProcess = shft(sharedID[curtThrdID/WarpSize])$
      if $curtProcess \neq 0$ **then**
        for $i = 0$ to $AdjNum[curtProcess]/WarpSize$ **do**
          visit($AdjList[curtProcess + i * WarpSize], M$)
        end for
      end if
      for $i = 0$ to $VertNum/WarpSize$ **do**
        if $VertList[curtThrdID + i * WarpSize]$ needs process **then**
          $ThrdMrk := curtThrdID + i * WarpSize$
        end if
      end for
    end while

3.3 The Scheme of Coloring Partitions

The traditional color algorithms can't make full use of the detected operations.
We propose the scheme of coloring partitions to accelerate SCC detection. In our
method, we take advantage of the partitions formed by coloring in each iteration
to increase SCCs generated in the next iteration. When coloring, there are many
partitions formed in each iteration, which will be released after the color values
of undetected vertices turn to zero in current iteration. It proves that the SCC
detection in each iteration can be performed in parallel in each partition formed
in previous iteration. As shown in Fig. 4, if partitions A and B are formed in a
iteration of coloring, we need to prove that the next iteration can be performed
separately in partitions A and B. It only needs to prove that there is no SCC
across partitions A and B. Proof is shown as follows, where $V_a \rightarrow V_b$ means V_a
can be connected to V_b by forward BFS.

Proof. Let P_a and P_b be the pivots of partitions A and B, and their ID value relationship is: $P_a < P_b$.

Assume that there is a vertex V_a in partition A and a vertex V_b in partition B, and V_a and V_b belong to the same SCC.

$\because V_a$ and V_b belong to the same SCC.

$\therefore V_a \to V_b, V_b \to V_a$.

$\because$ In partition A, $P_a \to V_a$.

$\therefore P_a \to V_b$.

$\therefore V_b \in A$, which does not match assumptions.

$\therefore$ There is no SCC across partitions A and B.

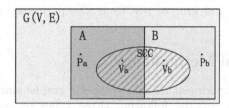

Fig. 4. Partitions formed in an iteration

According to the above proof, the SCC detection can be performed in parallel in the partitions formed by coloring, which greatly increases SCCs detected in each iteration. In traditional FB-Trim algorithm, after detecting the great SCC, it usually executes WCC to partition before performing other detections. Because WCC can aggregate all mutually connected vertices into the same partition, the remaining SCC detections can be performed in parallel in each partition. However, we have proved that the partitions formed by coloring can be used for parallel detection. Compared with partitions formed by WCC detection, coloring partitions scheme aggregates the vertices which the pivot can forward connect to into the same partition. Therefore, the partitions formed by our method is smaller. So it can form more partitions than WCC-based algorithms and detect more SCCs in the next iteration, and our algorithm requires fewer iterations. In addition, WCC can only be used for partitioning, but our method can detect the same number of SCCs as generated partitions. The specific operations are as follows: after the process of coloring partitions, the vertex whose initial color is the same as its partition's color can be forward connected to all vertices in the current partition. We can select this vertex as the pivot. According to Theorem 1, we only need to perform parallel backward BFS traversal from this pivot, then all traversed vertices form an SCC, and every other undetected SCCs are included in the rest of each partition.

3.4   Details of Parameter Optimization

In our algorithm, we have also fully considered detailed parameters including call frequency of coloring partition and Trim operations. In traditional algorithms, the partition operation is called only once between the first and second phases; the reason is that although detecting more partitions can increase SCCs detected in next iteration, it still takes time and doesn't directly generate SCCs. However, coloring partitions method can not only increases partitions, but also generates the same number of SCCs as partitions, so it is recommended to call coloring partitions multiple times. As shown in Algorithm 1, it presets an iteration threshold k_t. Coloring partitions operation is performed only on condition that detection operations are performed k_t times. Trim can detect SCCs consisting of one or two vertices within a short time, which is only performed at the beginning of most algorithms. Our algorithm not only performs Trim at the beginning, but also performs Trim of 1-SCCs multiple times to detect SCCs consisting of one vertex and Trim of 2-SCCs to detect SCCs consisting of two vertices between Phase 1 and Phase 2, and Trim of 1-SCCs is also performed in each iteration in Phase 2.

4 Experimental Methodology

Graphs used in our experiment include synthetic graphs and real-world graphs. The synthetic graphs are the following three types of graphs generated by **GTgraph** [17]: Random, R-MAT [18] and SSCA#2 [19]. Real-world graphs are selected from two commonly used benchmarks [10–12,14]: SNAP database [21] and Koblenz Network Collection database [22]. The details are shown in Table 1.

4.1 Experiment Setup

Table 1. Details of experimental graphs

Name	Vertices	Edges	SCC statistics			
			1-SCCs	2-SCCs	Remains	Total
soc-LiveJournal1	4,847,571	68,993,773	916,071	47,009	8,152	971,232
soc-pock	1,632,804	30,622,564	323,799	1,904	190	325,893
wiki-topcats	1,791,489	28,511,807	0	0	1	1
WikiTalk	2,394,385	5,021,410	2,281,311	529	39	2,281,879
youtube-links	1,138,499	4,942,297	602,051	8,696	2,675	613,422
baidu-internallink	3,966,925	13,820,833	1,475,852	23,433	3,719	1,503,004
synthetic-rmat	10,000,000	100,000,000	2,247,077	0	1	2,247,078
synthetic-random	10,000,000	100,000,000	876	0	1	877
synthetic-ssca	10,000,000	95,068,514	121,712	0	2,995	124,707
wikipedia-en	12,150,977	378,142,420	4,846,076	3,296	3,528	4,852,900
uk-2002	18,520,487	298,113,762	17,505,908	33,281	27,868	17,567,057

We compare six implementations including: (1) Tarjan: Tarjan's sequential SCC detection algorithm is a classic and representative sequential algorithm [5,9]; (2)

Barnat: Barnat's SCC detection method is a classical parallel algorithm, which is compared in many state-of-the-art GPU-based algorithms [9–14]; (3) Hong: Hong's parallel algorithm of SCC detection with WCC partitioning method [12]; (4) wHong: Devshatwar's improved parallel algorithm of SCC detection using virtual warp on the basis of Hong's algorithm [10]; (5) Slota: Slota's optimized color algorithm of parallel SCC detection [14]. (6) FBw-Pc: our parallel SCC detection algorithm with coloring partitions and reusing warps. We use gcc and nvcc with the -O3 optimization option for compilation along with -arch=sm_37 when compiling for the GPU. We execute all the benchmarks 10 times and collect the average execution time to avoid system noise.

4.2   Performance Analysis of Reusing Warps

In order to show that our reusing warps method can accelerate detecting the great SCC, we experimentally analyze the performance of different algorithms in detecting the great SCC. In various of compared parallel algorithms, Slota's algorithm [14] uses FB method to detect the great SCC in Phase 1 and iterative coloring to detect the remaining SCCs in Phase 2. Deshatwar et al.'s wSlota algorithm improved Slota's algorithm by adopting virtual warps method in Phase 1. For easy comparison, we adopt the reusing warps method mentioned in Sect. 3.2 to improve Phase 1 of Slota's algorithm and propose ReuseWarps1 and Reusing-Warps2 corresponding to two implementations of reusing warps methods. The running time in milliseconds of these four algorithms on different graphs is shown in Table 2.

Table 2. Running time (ms) of different algorithms

Name	Slota	wSlota	ReusingWarps1	ReusingWarps2
soc-LiveJournal1	215.075	221.727	200.538	**192.508**
soc-pock	104.735	89.440	**87.065**	87.938
wiki-topcats	415.820	380.122	263.125	**146.377**
WikiTalk	129.016	55.937	**55.108**	50.434
youtube-links	120.968	**52.340**	56.145	56.639
baidu-internallink	198.023	132.268	**126.782**	132.679
synthetic-rmat	346.220	379.750	385.638	**345.330**
synthetic-random	**344.003**	352.820	362.146	346.785
synthetic-ssca	**967.494**	4264.943	3721.599	1916.128
wikipedia-en	1977.462	2385.520	2041.199	**1405.830**
uk-2002	1660.969	1666.089	1709.536	**1600.710**

As shown in Table 2, our algorithms are better than other algorithms on most graphs, especially on wiki-topcats, WikiTalk, and youtube-links, where the acceleration ratio are 2.84, 2.56, and 2.14. Compared with Slota algorithm, our method takes longer time only on two synthetic graphs. This is because our scheme mainly improves threads utilization efficiency on skewed graphs, but the connections between vertices of synthetic graphs are randomly assigned under a certain rule, so the in-degree and out-degree of vertices are almost uniform, and the structure of synthetic graphs is not particularly skewed. Compared with wSlota algorithm improved by virtual warp method, our algorithm is only slightly slow on youtube-links with a seven percent speed difference. So our algorithms perform well on most graphs.

Table 3. Comparison of partitioning schemes

Name	Coloring partition			WCC partition		
	Partitions	SCCs	Iterations	Partitions	SCCs	Iterations
soc-LiveJournal1	8087	8087	6	7963	0	9
soc-pock	189	189	1	189	0	1
wiki-topcats	0	0	0	0	0	0
WikiTalk	38	38	1	38	0	1
youtube-links	2668	2668	2	2668	0	3
baidu-internallink	3685	3685	4	3658	0	5
synthetic-rmat	0	0	0	0	0	0
synthetic-random	0	0	0	0	0	0
synthetic-ssca	2923	2923	5	2917	0	6
wikipedia-en	3513	3513	2	3513	0	3
uk-2002	18421	18421	1773	18421	0	1671

4.3   Performance Analysis of Partitioning Method

In order to show that our scheme can generate more partitions and detect the same amount of SCCs as partitions, we performed comparative experiments on different partition schemes. The comparing algorithm is Hong algorithm [10] that adopts WCC partition method. To facilitate comparison, we use coloring partitions scheme to replace WCC partition scheme in Hong algorithm, and the rest of Hong algorithm are not modified. In Table 3, it shows the number of partitions and SCCs detected in partition operation of different algorithms, and the number of iterative detections in Phase 2 of thees algorithms. There are some real-world and synthetic graphs that don't require the operations in algorithm's Phase 2. For example, synthetic graphs of RMAT and Random only contain one large-scale SCC and a large number of 1-SCCs. This is the reason why most parallel algorithms accelerate faster on synthetic graphs than on real-world graphs. For the graphs that need to be processed in the algorithm's Phase

2, our algorithm can not only form more partitions, but also reduce iterations of the entire algorithm, and we can detect the same number of SCCs as partitions, so the coloring partitions scheme can dramatically accelerate the algorithm in Phase 2. In Fig. 5, the acceleration is also demonstrated by the overall running time comparison of different algorithms on graphs that require multiple iterative detection. On most graphs, our algorithm is faster that WCC-based algorithm, especially on un-2002, the speed-up ratio is 4.12×.

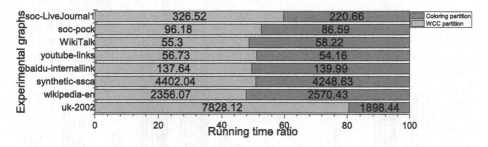

Fig. 5. Comparison of running time (ms)

4.4 Performance Analysis of the Entire Algorithm

In order to verify the acceleration of our algorithm, we compare the running time of our algorithm with other five algorithms on 11 graphs. The comparison results are shown in Fig. 6. On most graphs, our algorithm is significantly faster than other algorithms, and it achieves average accelerations of 7.23, 30.55, 1.75, 1.26, and 1.10 compared with Tarjan, Barnat, Hong, Slota, and wHong, respectively. The most obvious acceleration is on two synthetic graphs named synthetic-rmat and synthetic-random, where all parallel algorithms have more than 10 times acceleration compared with traditional sequential algorithm Tarjan. In these two graphs, Barnat achieves the highest acceleration about 20-fold, which surpasses other improved parallel algorithms. It can be seen from Table 1 that each of these two graphs only contains a great SCC and a large number of 1-SCCs, so it only needs to detect great SCC in Phase 1 and use Trim to detect 1-SCCs. The improved parallel algorithms based on Barnat optimize detection of the rest of SCCs besides the great SCC and 1-SCCs, which always introduce some workloads and result in a speed lower than Barnat. For another synthetic graph synthetic-ssca, the graph structure is not particularly skewed due to random assignment of its vertices connections. Therefore, wHong and FBw-Pc based on warps don't achieve enough acceleration.

Graph algorithms are mainly used to process real-world graphs. It can be seen from Fig. 6 that our algorithm performs well on all real-world graphs, especially on soc-LiveJournal1, soc-pock, and wikipedia-en with accelerations of more than

8 times. In real-world graphs, Barnat algorithm is almost slower than the traditional sequential algorithm Tarjan. Even on the two huge graphs of wikipedia-en and uk-2002, Barnat can't complete SCC detection. Therefore, our algorithm has a significant advantage on real-world graphs.

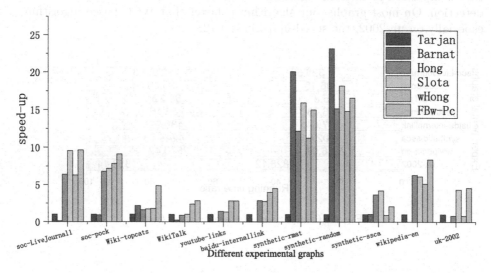

Fig. 6. Algorithms acceleration on Tarjan

5   Conclusion

The GPU-based parallel algorithms of SCC detection are significantly faster than traditional sequential algorithms when processing large-scale graphs, but current GPU-based algorithms still face the challenge of graph skewing and strict synchronization requirements of GPUs. This paper proposes an effective parallel algorithm to detect SCCs based on reusing warps and coloring partitions. In the algorithm based on virtual warps, it needs to apply for a large number of threads and the assigned threads are not fully utilized. The reusing warps method reduces number of applied threads and alleviates the problems of data skewing and thread underutilization. We prove that there is no SCC crossing partitions formed by coloring, and propose coloring partitions method to obtain more partitions after detecting the great SCC. The following SCC detection is directly parallel performed in partitions formed by coloring. Finally, the colored partitions are deeply integrated with forward and backward traversal to optimize the algorithm. The evaluations show that our algorithm outperforms state-of-the-art GPU-based methods on most graphs.

References

1. Zhang, Z., Yu, J.X., Qin, L., Chang, L., Lin, X.: I/O efficient: computing SCCs in massive graphs. VLDB J.—Int. J. Very Large Data Bases **24**(2), 245–270 (2015)
2. Xie, A., Beerel, P.A.: Implicit enumeration of strongly connected components and an application to formal verification. IEEE Trans. Comput. Aided Des. Integr. Circuits Syst. **19**(10), 1225–1230 (2000)
3. Orzan, S.: On distributed verification and verified distribution, Ph.D. dissertation, Center for Mathematics and Computer Science (2004)
4. Fan, W., Li, J., Ma, S., Wang, H., Wu, Y.: Graph homomorphism revisited for graph matching. Proc. VLDB Endow. **3**(1–2), 1161–1172 (2010)
5. Tarjan, R.: Depth-first search and linear graph algorithms. SIAM J. Comput. **1**(4), 146–160 (1972)
6. Dijkstra, E.W.: A Discipline of Programming, 1st edn. Prentice Hall, Englewood Cliffs (1976)
7. Cormen, T.H., Leiserson, C.E., Rivest, R.L., Stein, C.: Introduction to Algorithms, 3rd edn. The MIT Press, Cambridge (2009)
8. Barnat, J., Chaloupka, J., Van De Pol, J.: Distributed algorithms for SCC decomposition. J. Logic Comput. **21**(1), 23–44 (2011)
9. Barnat, J., Bauch, P., Brim, L., Ceska, M.: Computing strongly connected components in parallel on CUDA. In: Sussman, A., Mueller, F., Beaumont, O., Kandemir, M.T., Nikolopoulos, D. (eds.) IPDPS 2011, pp. 544–555. IEEE (2011)
10. Devshatwar, S., Amilkanthwar, M., Nasre, R.: GPU centric extensions for parallel strongly connected components computation. In: GPGPU@PPoPP 2016, pp. 2–11. ACM, Barcelona (2016). https://doi.org/10.1145%2F2884045.2884048
11. Li, P., Chen, X, Shen, J., Fang, J., Tang, T., Yang, C.: High performance detection of strongly connected components in sparse graphs on GPUs. In: PMAM@PPoPP 2017, pp. 48–57. ACM, Texas (2017)
12. Hong, S., Rodia, N.C., Olukotun, K.: On fast parallel detection of strongly connected components (SCC) in small-world graphs. In: SC 2013, pp. 1–11. ACM, Denver (2013)
13. Li, G., Zhu, Z., Cong, Z., Yang, F.: Efficient decomposition of strongly connected components on GPUs. J. Syst. Architect. **60**(1), 1–10 (2014)
14. Slota, G.M., Rajamanickam, S., Madduri, K.: BFS and coloring-based parallel algorithms for strongly connected components and related problems. In: 2014 International Parallel and Distributed Processing Symposium, pp. 550–559. IEEE (2014)
15. Mclendon Iii, W., Hendrickson, B., Plimpton, S.J., Rauchwerger, L.: Finding strongly connected components in distributed graphs. J. Parallel Distrib. Comput. **65**(8), 901–910 (2005)
16. Kumar, R., Novak, J., Tomkins, A.: Structure and evolution of online social networks. In: Proceedings of the 12th ACM SIGKDD International Conference on Knowledge Discovery and Data Mining, KDD 2006, pp. 611–617. ACM, New York (2006)
17. Bader, D.A., Madduri, K.: GTGraph: a synthetic graph generator suite, vol. 38 (2006)
18. Chakrabarti, D., Zhan, Y., Faloutsos, C.: R-MAT: a recursive model for graph mining. In: Proceedings of the 2004 SIAM International Conference on Data Mining, pp. 442–446. SIAM (2004)

19. Bader, D.A., Madduri, K.: Design and implementation of the HPCS graph analysis benchmark on symmetric multiprocessors. In: Bader, D.A., Parashar, M., Sridhar, V., Prasanna, V.K. (eds.) HiPC 2005. LNCS, vol. 3769, pp. 465–476. Springer, Heidelberg (2005). https://doi.org/10.1007/11602569_48
20. Fleischer, L.K., Hendrickson, B., Pınar, A.: On identifying strongly connected components in parallel. In: Rolim, J. (ed.) IPDPS 2000. LNCS, vol. 1800, pp. 505–511. Springer, Heidelberg (2000). https://doi.org/10.1007/3-540-45591-4_68
21. Leskovec, J., Krevl, A.: SNAP Datasets: Stanford Large Network Dataset Collection, vol. 6 (2014). http://snap.stanford.edu/data. Accessed 2 Feb 2020
22. Koblenz Network Collection. http://konect.uni-koblenz.de/. Accessed 2 Feb 2020

Procedure and Loop Level Speculative Parallelism Analysis in HPEC

Xinyi Wang[1] (ID), Yaobin Wang[1]([✉]), Ling Li[1], Yang Yang[2], Deqing Bu[1],
and Manasah Musariri[1]

[1] Southwest University of Science and Technology, Mianyang, China
wangxinyi_0102@163.com, wangyaobin@foxmail.com, liling82@mail.ustc.edu.cn,
dq_bu@foxmail.com, mmanasah@gmail.com
[2] Sichuan Institute of Computer Sciences, Chengdu, China
49799012@qq.com

Abstract. Although High Performance Embedded Computing(HPEC) has been effectively analyzed on different platforms, there is still room for an in-depth analysis of thread level speculation (TLS), especially at the procedure level. This paper explores the potential parallelism of HPEC from procedure and loop level TLS techniques, and designs the corresponding analysis mechanism and data structures. Our aim is to show the improved performance of various applications used in HPEC. Results from our experiments demonstrate that: 1) the performance of all applications was relatively good, the best tdfir application achieves 221.8x speedup in procedure level speculation whilst a ct application gets a 13x speedup in loop level speculation; 2) HPEC programs can be accelerated by effectively utilizing the computing resources of 16 to 32 cores; 3) Applications, that contain multiple non-severe data-dependency procedure calls, are more suitable for developing parallelism using procedure level TLS technology.

Keywords: Thread level speculation · Multi-core · HPEC · Data dependency · Dynamic profiling

1 Introduction

With the development of semiconductor technology, the number of cores on a single chip is continuing to increase with the trend of Moore's Law. Evidently, it has brought about more on-chip computing resources [1]. Our challenge however, emanates from how we can fully utilize these resources using our theoretical knowledge and practical skills.

In order to achieve this, serial programs need to be parallelized. But many applications are written on a single core, and multithreaded programs are difficult to develop and maintain [2]. To fix this problem, thread level speculation

Supported financially by the National Natural Science Foundation of China grants 61672438, Sichuan Science and Technology Plan Project 2019YJ0326, China Scholarship Council Project CSC201908510040.

was proposed. It is an automatic parallelization technology supporting multi-core processors [3]. TLS divides the serial code into multiple code segments, thereby executing on multiple cores. It automatically exploits the potential parallelism of the program, as a result improving the utilization of hardware computing resources on the chips [4]. For serial programs that are traditionally difficult to manually or automatically parallelize, TLS can also exploit its maximum potential parallelism. In addition, TLS allows code to be executed in parallel without relying on previous compile-time correlation analysis and does not violate the semantics of serial execution [5]. During speculative execution, each speculative thread executes a different part of the serial program. There is only one non-speculative thread which is allowed to commit the result to memory, all other threads are speculative [6].

Early evaluations of processor performance were usually based on comparing the number of instructions per unit of time. Its measurement standard was linear when systematically evaluating the influence of various factors of the processor. Therefore, the benchmark proposed in the 1980s was more recognized by the industry. HPEC benchmark suite, developed by MIT, aims to provide a well-defined benchmark for measuring the performance of its libraries and systems in the fields of signal processing, communication, and information and knowledge processing [7].Currently, it has been effectively analyzed in several studies [7–9], but has not been thoroughly analyzed using TLS technology.

In this paper, we propose a procedure and loop level speculative execution model for accelerating the HPEC programs. The aim of this experiment is to find a TLS method suitable for its performance improvement. Several programs were selected from it to analyze the impact of performance factors such as thread size and inter-thread data dependencies on potential parallelism of the programs. Finally, evaluating whether HPEC programs are suitable for TLS acceleration is based on the speculative execution results, providing advice on future chip designs.

The rest of this paper is organized as follows; related work is discussed in Sect. 2, speculative mechanism in Sect. 3, performance influencing factors in Sect. 4, experimental analysis in Sect. 5 and finally, we conclude in Sect. 6.

2   Related Work

In order to make better use of TLS technology, various thread speculation schemes have been introduced. For example, Multiscalar [10] proposed a comprehensive study on software and hardware support of TLS technology, using fine-grained threads. Hydra [11] adds hardware support for thread level speculation, has a primary cache through write operations, and allows all processor cores to monitor all writes executed. At present, by implementing TLS on the hardware-transactional-memory in [12], some loops can achieve up to 3.8x speedup. A TLS model for copy-on-write on cache is proposed in [13], which uses 8 and 16 speculative threads to test typical benchmark tests with an average acceleration of 5.69x and 10.04x respectively. In [14], a thread level speculation system using a virtual memory system and implemented entirely in kernel space is introduced, and the Cilk suite is tested with an average speedup of 7.8x. Our previous work [15] also analyzed the thread level speculative parallelism of block encryption algorithm. Therefore, thread level speculation technology can be effectively applied to the parallel analysis of a variety of serial programs.

Currently, the main research on HPEC is as follows. In [8], an efficient implementation of HPEC benchmark was developed on the GPU and DSP, comparing the performance and power results. DSP produced better results. Bergmann ported several HPEC programs to VSIPL ++ and tested them in parallel. In most cases, the acceleration results were linear or super-linear [7]. In [9], by using HPEC to evaluate the performance of OpenSPARC processors on Virtex-5 FPGAs, the conclusion is that replicating the floating-point units would not cost much and can have a significant impact on the performance of floating-point applications. However, HPEC has not been thoroughly studied by using TLS technology, so in this paper, we will discuss this area in detail.

3   Speculative Mechanism

3.1   Speculation Execution Model

In order to ensure the correctness of the serial semantics, all speculative threads cache the intermediate results in a separate state space, commit after the speculative execution is successful, and restart after the failure.

The loop level speculation execution model as shown in Fig. 1(a) shows a serial execution process of a traditional program, and Fig. 1(b) shows a speculative execution process. When the speculative execution begins, the main processor signals all the idle processors (speculative processors) to load each loop iteration by issuing a 'Loop_Start' signal. During the speculative execution, only the head processor can directly write to the memory. Then the head processor completes execution and commits, the next processor will become the new head processor and it will speculatively execute the new thread. After a processor completely executes the last loop iteration, it will send a 'Loop_End' signal to the other processors to end the speculative execution of the loop structure, and then the remaining code is executed by the main processor.

Figure 2 shows a procedure level speculative execution model. When executing to a procedure that needs to be speculatively executed, the speculative processor will create a new thread to execute the code. The current processor will continue to execute using the predicted value and creates an overhead area for comparing the predicted and return value. After the execution in the speculative processor is completed, the current processor will compare the predicted value with the return value in the speculative processor, if the comparison results are same, the program will continue to run, otherwise the program will use the return value of the speculative processor to correct and continue execution.

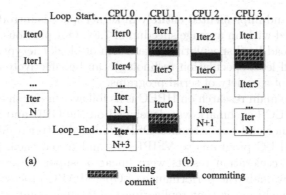

(a)   (b)

waiting
commit   commiting

Fig. 1. Speculative execution model of loops.

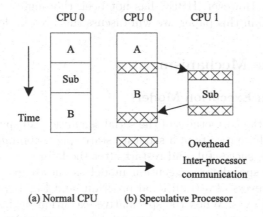

Overhead

Inter-processor
communication

(a) Normal CPU   (b) Speculative Processor

Fig. 2. Speculative execution model of procedures.

3.2 Profiling Mechanism

As shown in Fig. 3. First, the program is initially analyzed using the GNU Prof tool that comes with the Linux system. The tool can roughly analyze the function call relationship, the running time ratio of each procedure during running of the program. Combining the information obtained above, a 'hot' spot segment with a program running time of 5% or more is selected as a 'hot' spot profiling region and then, select the loops or procedures in the 'hot' spot region and compile it through a cross-compiler to get the binary executable file. Finally, the profiling tool analyzes the 'hot' spot area by identifying the 'jarl', 'jal' or 'jr' instructions and combining the candidate 'hot' spot area files, and the profiling results are obtained after quantitative analysis based on the key factors influencing performance.

This work disassembles the binary code file generated during profiling using the disassembly tool objdump4pisa of the PISA instruction set. Figure 4 shows a procedure profiling example. As can be seen from the figure, the procedure

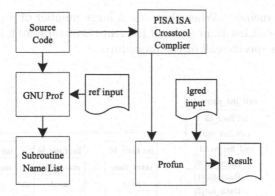

Fig. 3. Profiling mechanism.

return address is stored in register 31, and the global variable sum is stored in register 28. Before the operation on sum, it is loaded into register 2, and the result is saved to register 3 at the end of the operation.

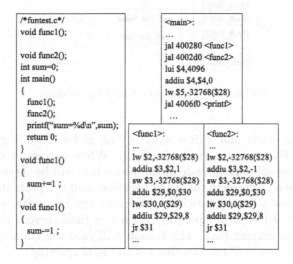

Fig. 4. Procedure profiling example.

3.3   Kernel Data Structure

Because the kernel data structure principle of procedure and loop is the same, the procedure is taken as an example, as shown in Fig. 5. The call_list structure records each call and return time of procedure to be profiled. Each procedure corresponds to a call_list_entry_t structure, and each call_list_entry structure contains n call_list structures. The call_list structure is used to save information about the function being called. The Hash_list structure is used to save write operations to the memory. When there are a large number of procedure calls in the program, the call_list_head and call_list_tail traversal search in the Hash_list can quickly find a specific call of the procedure.

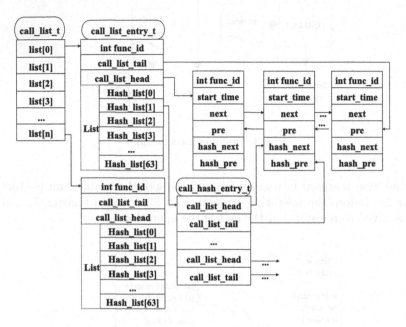

Fig. 5. Kernel data structure of procedure.

The program reads and writes data of the global data segments in the last_write_hash_list structure shown in Fig. 6. When executing the procedure to be profiled, the subsequent code of the procedure will be intercepted by the profiling tool when performing the read operation, and the profiling tool will get the time of the last write operation by retrieving the last_write_hash_list structure. If the time of last write operation is greater than current time, it will be replaced with the current time. The system will save the running time of the serial version. The ratio of the two times is the final speedup.

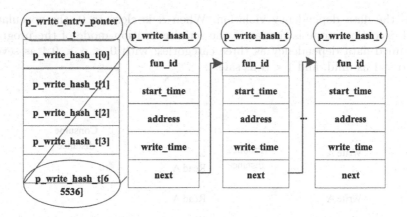

Fig. 6. Data structure of last_write_hash_list structure.

4   Impact Factors of TLS

This work summarizes the four impact factors of TLS performance.

Thread Size. We defined the size of a subroutine call or a loop execution instruction as thread size. If the thread size is too small, the overhead during speculative execution will increase and will affect the parallel effect. If it is too large, there may be an error violation, which may result in unnecessary speculative thread compression.

Speculative Parallelism coverage. According to Amdahl's law, the program speculative parallelism coverage is higher, the program parallelism is better.

Inter-thread Control Dependency Feature. If a control dependency violence occurs, the speculative thread needs to be restarted, which affects the performance of the speculative thread execution. Using value predict technology can effectively resolve inter-thread control dependencies.

Inter-thread Data Dependency Feature. This is the most critical performance factor of TLS technology. Thread level data dependency result in a series of overheads such as dependency violation detection, synchronization, fallback, and restart. We will explain it in detail as follows.

This work simplified the data dependency into the relationship between the data producer and consumer. The operation of producing data is treated as a write operation to the memory, and the operation of consuming data is regarded as a read operation to the memory. As shown in Fig. 7. We define the produce-distance as the number of instructions between the execution of the thread and the last write to the memory, defining the consume-distance as the number of instructions between the execution of the thread and the first read of the thread to the memory. And we use α to represent consume-distance/produce-distance, and use it to analyze the inter-thread data dependency. When $\alpha < 1$, data dependency violation occurs at this time, and the smaller the α value, the more serious the data dependency violation. When α is close to 0, the speculative thread execution mode is similar to the serial execution mode of the program. We defined data dependency as three categories, with $0 < \alpha < 0.4$ as severe, $0.4 < \alpha < 1$ as mild, and $1 < \alpha$ as safe.

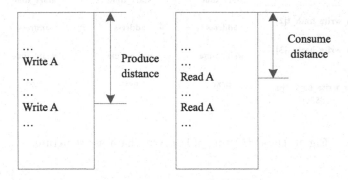

Fig. 7. Produce-distance & consume-distance.

In this paper, the average α we used is heavy data dependency as a proportion of all data dependencies. The smaller the average α, the smaller the data dependency violation. In addition, for procedure level TLS technology, the number of procedure calls is also one of the factors that affect its performance. The number of procedure calls in the execution process will affect the program performance improvement potential, which will affect the program acceleration effect.

5   Experiment Analysis

In this section, we will use the mechanism shown in Sect. 3 to analyze HPEC, experimental results of HPEC in combination with the influencing factors of TLS technology.

This work was done on x86-based Ubuntu 14.04. In order to make the parsing time better adapt to the compiler overhead, the parser uses a modified and augmented version of sim-fast based on the Simplescalar tool set function simulator, which executes one instruction per period. The cross compiler is the back-end modified gcc-2.7.2.3 provided by the Simplescalar tool set and the instruction set is PISA. We have selected 6 representative programs from HPEC benchmark suite, as shown in Table 1.

Table 1. Test programs.

Category	Benchmarks	Description
Signal Processing	fdfir	Basic operation of signal processing
	tdfir	Basic operation of signal processing
	cfar	Constant False Alarm Rate Detection
Communication	ct	Change the data storage order
Information and Knowledge Processing	pm	Pattern matching
	ga	Graph of the genetic algorithm

5.1   Results in Loop Level Speculation

The Fig. 8 shows the speedup of the programs running at different core numbers in the loop and procedure levels. Under the loop level speculation, we choose 2, 4, 8, 16, 32, 64 and infinite cores to accelerate.

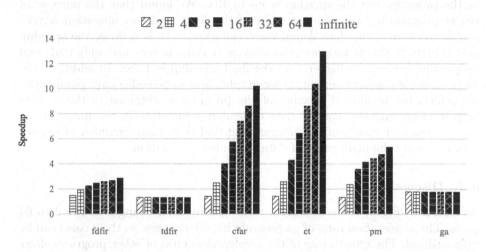

Fig. 8. Speedup of loop level.

The results of the loop level speculation as shown in Table 2. We can clearly see from it that the parallelism coverage is over 90%. Most of the applications in the HPEC benchmark suite have very high parallelism coverage. Therefore, parallelism coverage is not a major factor affecting the speedup at the loop level. The thread size of the test programs is not very large, both are about 100. And we can observe that there are data dependency violations in programs other than cfar. The average α values of fdfir, tdfir, and ct are large, which means that the proportion of serious data dependency violation is small. Therefore, the speedup of cfar is 10.2x, which is the result of the combination of high parallelism coverage and low data dependency violation. Similar to the situation is pm, which has a nice speedup of 5.3x.

Table 2. Impacting Factors In Loop And Procedure Level Speculation.

Benchmark	Parallelism coverage		Thread size		Average α	
	Loop	Procedure	Loop	Procedure	Loop	Procedure
cfar	100%	100%	3.4E+01	6.9E+08	0	0
ct	100%	100%	2.5E+01	1.2E+08	1	0
pm	100%	100%	1.7E+01	1.2E+12	0.01	0
tdfir	100%	100%	2.0E+02	3.2E+12	0.5	0
fdfir	100%	100%	4.1E+01	2.0E+05	0.99	0
ga	99%	99%	2.3E+01	1.8E+09	0.01	0.04

The acceleration potential of fdfir and tdfir is affected by data dependency violation, and their speedup is between 1.3x and 2.8x. The case of ct is quite special. The average α value of ct is 1, and there is only heavy data dependency in the program, but the speedup is up to 13x, We found that the purpose of the ct program is to change the order of the data, and this operation is very simple, so even if the data dependency violation in ct is serious, the speedup is still high. In the ga program, the average α value is very low, only 0.01, and its parallel coverage is higher, and the final speedup is 1.78x. In addition, the ga program is a genetic algorithm for graphs, and some codes have parallelism. Therefore, the maximum speedup of the program is obtained in the 2 cores loop level speculation. For the data dependency violation in the program, we analyze the source code of the program and find that a large number of pointer operations are the main cause of data dependency violation.

5.2 Discussion

As shown in Fig. 9, the acceleration ratio of loop level changes at the 2 to 64 cores. The acceleration ratio of ga program hardly changes, so the 2 cores can be fully utilized. The growth rate of the acceleration ratio of other programs slows down as the number of cores increases, and a large performance improvement can be achieved at 16 to 32 cores. After consideration, most of the applications here can effectively use 16 to 32 cores in TLS to achieve significant performance improvements.

As shown in the Fig. 10, we compared the speedup of the loop level and the procedure level in the infinite core. As can be seen from the figure, Cfar, ct, pm have no obvious speedup at the procedure level, while tdfir, fdfir, and ga archive higher speedup than the loop level. Tdfir achieves procedure level speedup of approximately 221.8x, making it particularly suitable for procedure level speculation, while other work [8] accelerates the tdfir program through the loop structure, and the resulting acceleration ratio is 28.88. Therefore, among the selected programs, cfar, ct and pm are more suitable to adopt loop level TLS technology to explore parallelism, while tdfir, fdfir and ga are more suitable to adopt procedure level.

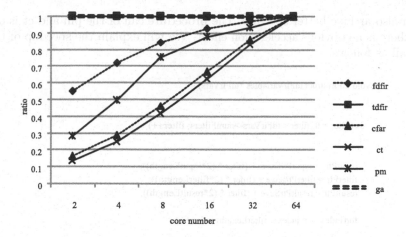

Fig. 9. Acceleration ratio change trend.

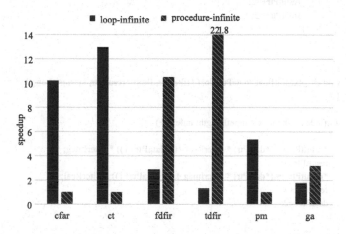

Fig. 10. Loop and procedure speedup comparison.

In order to understand the reason for the procedure level speedup, we will analyze the performance influencing factors as shown in Table 2. It shows that with the loop level, parallelism coverage is not a major performance influencing factor. The obvious difference is that the thread size is larger than the loop level and the average α value is almost zero. And we analyze the code and find that the 'hot' procedure can get more speedup in the loop. In addition, the number of procedure calls is also one of the performance factors of the procedure level. Most of the HPEC programs have better procedure level speedup. The main reason is that the number of procedure calls of the 'hot' procedure is high. The number of procedure calls is inversely proportional to the number of data dependencies between threads. There are a small number of data dependency violations in the ga can also archive better speedup. The procedure call of the pm and ct is only 1, so there is no obvious acceleration effect. We will explain the speedup of tdfir in detail as follows.

```
void tdFir(struct tdFirVariables *tdFirVars)
{
    ...
    for(filter = 0; filter < tdFirVars->numFilters; filter++)
    {

        inputPtr  = inputPtrSave  + (filter * (2*inputLength));
        filterPtr = filterPtrSave + (filter * (2*filterLength));
        resultPtr = resultPtrSave + (filter * (2*resultLength));

        for(index = 0; index < filterLength; index++)
            {
                elCplxMul(inputPtr, filterPtr, resultPtr, tdFirVars->inputLength);
                resultPtr+=2;
                filterPtr+=2;
            }
    }
    ...
}
void elCplxMul(float *dataPtr, float *filterPtr, float *resultPtr, int inputLength)
{
    ...
    for(index = 0; index < inputLength; index++)
    {
        *resultPtr += (*dataPtr) * filterReal - (*(dataPtr+1)) * filterImag;
        resultPtr++;
        *resultPtr += (*dataPtr) * filterImag + (*(dataPtr+1)) * filterReal;
        resultPtr++;
        dataPtr+=2;
    }
}
```

Fig. 11. Code segment of tdfir's 'hot' procedure.

The 'hot' procedure in tdfir is elCplxMul, as shown in Fig. 11. The procedure is in a double-layer loop, which makes the procedure call more times. And there is only one loop structure in the procedure. This loop operation is a simple complex multiplication operation. In addition, as shown in Table 2, the average α value of tdfir is 0, so there is no severe data dependency violation, and it has a more appropriate thread size. Based on the above factors, tdfir can achieve such a high speedup.

In summary, the combination of low average α values, high parallelism coverage, and high number of procedure calls results in better speedup for most programs.

In this paper, we compare the speedup between the loop level and the procedure level, and confirm that although the loop level is usually used as the main source of speculative parallelism, the procedure level can be used as an additional supplement to the loop level parallelism. Sometimes the procedure level gets more parallelism than the loop level, that is, some programs are more suitable for procedure level to achieve parallelism.

6   Conclusion

This paper presents how to explore speculative procedure and loop parallelism in HPEC applications, including the speculative execution model, profiling mechanism and performance influencing factors, etc. This paper makes the following main contributions:

1) We explored the procedure and loop level speculative potential parallelism for HPEC applications, and showed that most applications are suitable for TLS.
2) In HPEC applications, TLS technology can effectively utilize 16 to 32 cores to develop potential parallelism.
3) The procedure level can be used as an additional supplement to the loop level parallelism. Such applications, which contain multiple non-severe data-dependency procedure calls, are more suitable for developing parallelism using procedure level TLS technology.

In the future work, we will modify our profiling tools to make the experiment results more precise.

References

1. Ye, J.M., Yan, H., Hou, H., Chen, T.: Potential thread-level-parallelism exploration with superblock reordering. Computing **96**(6), 545–564 (2014). https://doi.org/10.1007/s00607-014-0387-8
2. Luo, Q., Rosu, G.: EnforceMOP: a runtime property enforcement system for multithreaded programs. In: Pezzè, M., Harman, M. (eds.) International Symposium on Software Testing and Analysis, ISSTA 2013, Lugano, Switzerland, 15–20 July 2013, pp. 156–166. ACM (2013)
3. Oplinger, J.T., Heine, D.L., Lam, M.S.: In search of speculative thread-level parallelism. In: Proceedings of the 1999 International Conference on Parallel Architectures and Compilation Techniques, Newport Beach, California, USA, 12–16 October 1999, pp. 303–313. IEEE Computer Society (1999)
4. Liu, B., Zhao, Y., Li, M., Liu, Y., Feng, B.: A virtual sample generation approach for speculative multithreading using feature sets and abstract syntax trees. In: Shen, H., Sang, Y., Li, Y., Qian, D., Zomaya, A.Y. (eds.) 13th International Conference on Parallel and Distributed Computing, Applications and Technologies, PDCAT 2012, Beijing, China, 14–16 December 2012, pp. 39–44. IEEE (2012)
5. Xekalakis, P., Ioannou, N., Cintra, M.: Combining thread level speculation helper threads and runahead execution. In: Gschwind, M., Nicolau, A., Salapura, V., Moreira, J.E. (eds.) Proceedings of the 23rd International Conference on Supercomputing, Yorktown Heights, NY, USA, 8–12 June 2009, pp. 410–420. ACM (2009)

6. Wang, Y., An, H., Liu, Z., Li, L., Huang, J.: A flexible chip multiprocessor simulator dedicated for thread level speculation. In: 2016 IEEE Trustcom/BigDataSE/ISPA, Tianjin, China, 23–26 August 2016, pp. 2127–2132. IEEE (2016)
7. Bergmann, J., Mccoy, D.: Sourcery VSIPL++ HPEC benchmark performance, pp. 308–314 (2006)
8. Mu, S., et al.: Evaluating the potential of graphics processors for high performance embedded computing. In: Design, Automation and Test in Europe, DATE 2011, Grenoble, France, 14–18 March 2011, pp. 709–714. IEEE (2011)
9. Mhaidat, K.M., Baset, A., Al-Khaleel, O.: OpenSPARC processor evaluation using Virtex-5 FPGA and high performance embedded computing (HPEC) benchmark suite. IJERTCS **5**(1), 61–74 (2014)
10. Sohi, G.S., Breach, S.E., Vijaykumar, T.N.: Multiscalar processors. In: Patterson, D.A. (ed.) Proceedings of the 22nd Annual International Symposium on Computer Architecture, ISCA 1995, Santa Margherita Ligure, Italy, 22–24 June 1995, pp. 414–425. ACM (1995)
11. Hammond, L., Hubbert, B.A., Siu, M., Prabhu, M.K., Chen, M.K., Olukotun, K.: The Stanford hydra CMP. IEEE Micro **20**(2), 71–84 (2000)
12. Salamanca, J., Amaral, J.N., Araujo, G.: Using hardware-transactional-memory support to implement thread-level speculation. IEEE Trans. Parallel Distrib. Syst. **29**(2), 466–480 (2018)
13. Wang, Q., Wang, J., Shen, L., Wang, Z.: A software-hardware co-designed methodology for efficient thread level speculation. In: 2017 IEEE International Conference on Computer and Information Technology, CIT 2017, Helsinki, Finland, 21–23 August 2017, pp. 184–191. IEEE Computer Society (2017)
14. Hammacher, C., Streit, K., Zeller, A., Hack, S.: Thread-level speculation with kernel support. In: Zaks, A., Hermenegildo, M.V. (eds.) Proceedings of the 25th International Conference on Compiler Construction, CC 2016, Barcelona, Spain, 12–18 March 2016, pp. 1–11. ACM (2016)
15. Wang, Y., An, H., Liu, Z., Zhang, L., Wang, Q.: Parallelizing block cryptography algorithms on speculative multicores. In: Wang, G., Zomaya, A., Perez, G.M., Li, K. (eds.) ICA3PP 2015. LNCS, vol. 9528, pp. 3–15. Springer, Cham (2015). https://doi.org/10.1007/978-3-319-27119-4_1

CTA: A Critical Task Aware Scheduling Mechanism for Dataflow Architecture

Yan Ou[1,2], Chongfei Shen[3], Yujing Feng[1], Xinxin Wu[1,2], Wenming Li[1], Xiaochun Ye[1,4], and Dongrui Fan[1,2(✉)]

[1] State Key Laboratory of Computer Architecture, Institute of Computing Technology, Chinese Academy of Sciences, Beijing 100190, China
{ouyan,fengyujing,wuxinxin,liwenming,yexiaochun,fandr}@ict.ac.cn
[2] School of Computer and Control Engineering, University of Chinese Academy of Sciences, Beijing 100049, China
[3] Beijing Smartchip Microelectronics Technology Company Limited, Haidian District, Beijing 100000, China
shenchongfei@sgitg.sgcc.com.cn
[4] State Key Laboratory of Mathematical Engineering and Advanced Computing, Wuxi 214125, Jiangsu, China

Abstract. Critical tasks directly affect the overall performance of a program, especially in dataflow architecture. The reason is that dependency between tasks in dataflow scenarios is much more complex than those in control-flow scenarios. However, previous works fail to sufficiently accelerate the critical task because most of them only applied optimization for static critical tasks, without considering the runtime status during execution. We propose a critical task aware (CTA) scheduling mechanism for dataflow architecture. By adopting co-optimization of hardware and software, higher execution priorities are assigned to the critical tasks for better scheduling. The experimental results show that our mechanism increases the computational performance by 14%–78%, and increases the power efficiency by 11%–41%.

Keywords: Critical tasks · Dataflow architecture · Scheduling mechanism · Execution priority · Power efficiency

1 Introduction

The critical task is one of the fundamental runtime characteristics and has a direct impact on the final output of a program [19,21,22]. It identifies the task with the longest execution time at a certain stage during execution while other tasks need to wait for its result for subsequent calculations. Previous works have been proposed to accelerate the critical tasks by adopting either hardware or software. For the hardware optimization, some architectures [7,27] support the simultaneous execution of multiple copies of the same program to improve the utilization of function units. For the software optimization, some instruction mapping algorithms [21,23,26] are used to map the tasks in critical path (static

© Springer Nature Switzerland AG 2020
M. Qiu (Ed.): ICA3PP 2020, LNCS 12452, pp. 61–77, 2020.
https://doi.org/10.1007/978-3-030-60245-1_5

critical tasks, SCTs) to adjacent processing elements (PEs) for reducing data transmission delay. However, these works focus only on the static characteristics of the critical tasks, without considering the runtime status during execution. The execution priorities of tasks in a program may not always rely on static-analyzed critical path, and a task in non-critical path may turn into a critical task (dynamic critical task, DCT) because of some dynamic characteristics, such as network congestion as well as resource competition. Therefore, the previous static optimization for the critical tasks cannot sufficiently accelerate the program due to the reason that dynamic features cannot statically predict.

Dataflow architecture [5] has been proved to be a promising alternative to accelerate high performance (HPC) and artificial intelligence (AI) applications [2,7,16,18,30], which can be reflected from the following 2 aspects. (1) Data are directly transferred between tasks rather than through shared-memory. (2) Any task can be executed as long as its operands are available. The state of art ASICs[6], such as TPU [10], CGRA [1,20,24] and Eyeriss [3], have borrowed basic ideas from dataflow execution model for exploiting both instruction level parallelism (ILP) and data level parallelism (DLP). However, the performance of dataflow architecture is more sensitive to the execution delay of the critical task than that of the control-flow architecture due to frequent interaction among tasks as well as complicated dynamic condition [9,12,15,17]. Therefore, the task scheduling mechanism has significance and reference value for other architectures if it can be applied to accelerating the critical tasks of dataflow architecture.

Both SCT and DCT are the important factors influencing the performance of dataflow program. SCT is determined during dataflow graph build-time, where the future execution environment is predicted by using the statistics collected during the past execution. All tasks in the static critical path are regarded as the SCTs. DCT is determined by the current state of a dataflow system. Such network congestion, resources competition or any other dynamic feature would turn a task in non-critical path into a critical task. As shown in Fig. 1 (a), we assume that Task1, Task2 and Task3 are the memory access tasks while other tasks are the computing tasks. Thus (Task0-Task1/Tasks2-Task4-Task5) is the critical path, in which these tasks are SCTs that are assigned higher execution priorities during build-time. However, Task3 may turn into a critical task if it is blocked by network congestion or cache missing. When this happens, the whole dataflow graph would be blocked if the low execution priority of Task3 is not boosted. Therefore, tracking DCTs and allocating resources for them can exploit potential performance of the program. In this work, we have counted the number of DCTs, which are not in static critical path but the dataflow graph is blocked by them during execution, of 7 HPC+AI workloads (further discussion in EVALUATION section) in dataflow architecture. As shown in Fig. 1 (b), statistics suggest that the DCT accounts for 16.70% of the total, which restricts the further improvement of performance. The proportion of DCT is influenced by dynamic runtime environment. The more complicated the runtime environment of program is, the greater the performance penalty implied by DCTs.

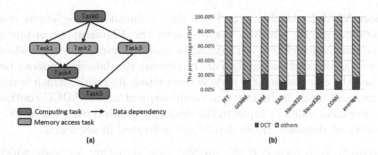

Fig. 1. (a) An example of a dataflow program. (b) The percentage of DCT.

In this paper, we propose a critical task aware (CTA) scheduling mechanism for dataflow architecture, which dynamically tracks the exact critical tasks, and efficiently assigns execution priorities for them. The experimental results show that our mechanism increases the computational performance by 14%–78%, and increases the power efficiency by 11%–41% compared with previous task scheduling mechanisms. Our contributions are summarized as follows.

- We put forward a new concept of *congested-degree*, which is used for measuring the congestion of data in a task.
- We propose an elastic handshake mechanism (EHM), which extends the traditional handshake mechanism for better exploiting ILP.
- We further propose a task scheduling mechanism (TSM) based on EHM, which is used for tracking the critical tasks and allocating computing resources.

The remainder of the paper is organized as follows. Section 2 introduces the background of dataflow architecture. Section 3 analyzes the key factors affecting the performance and illustrates a task scheduling algorithm. In Sect. 4, we describe CTA to support the algorithm. In Sect. 5, we present our evaluation platform and simulation results. In Sect. 6, we discuss related work, and in Sect. 7 we provide concluding remarks.

2 Dataflow Architecture

This section explains the architecture, execution model, and task format of a general purpose dataflow process unit(DPU), which resembles TRIPS [7], WaveScalar [24] and SPU [29].

Architecture. Figure 2 (a) illustrates the architecture of DPU, which includes PE array, scratchpad memory (SPM) and a micro controller (MICC). PE array consists of a grid of 16 PEs connected by 2D mesh network. SPM is responsible for storing instruction and data. MICC is responsible for managing execution process of DPU, including dispatching instruction, generating iteration index, and receiving the finish signal from PE array.

Execution Model. As shown in Fig.2 (b), a dataflow program is compiled into a dataflow graph, which is mapped into the PE array according to the specific mapping algorithm [4]. In a dataflow graph, a node represents a task that contains a set of instructions, and an arc represents the data dependency between tasks. When the calculation of a node is completed, its result, which is regarded as input for next tasks, is directly sent to subsequent nodes. MICC continuously generates iteration index as input to the dataflow graph during execution, while tasks of different dataflow graph depths are activated in sequence.

Task Format. Task format is the specific data structure for task, which contains the basic information of this task and the data dependencies to other tasks. Figure 2 (c) shows the basic format of dataflow task, where Opcode is the operation type. OpAddr$i$ is the address of (i)-th operand in local memory. DesAddr$j$ is the address of (j)-th subsequent tasks, which is jointly determined by PE location (PEPos), task sequence number (TaskNo) and operand sequence number (OperandNo).

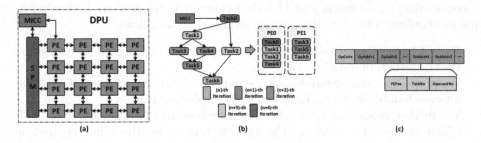

Fig. 2. (a) The architecture of DPU. (b) The execution model of DPU. (c) The basic format of dataflow task.

3   A Critical Task Aware Scheduling Algorithm

In this section, first of all, a quantitative analysis model is established, and the impact of some factors on the performance of dataflow execution model is analyzed. Secondly, a new concept called *congested-degree* is introduced, which is used for measuring the congestion of data in a task. Finally, we propose a critical task aware scheduling algorithm for accelerating the critical tasks.

3.1   Performance Model of Dataflow Execution

The execution model of dataflow task scheduling mechanisms is shown in Fig. 2 (b), dataflow graph is just like a big pipeline and the critical tasks determines the time interval of pipeline output. The performance of pipelined execution model is calculated as follows.

$$Performance = \frac{U}{(MAX(t_1, \cdots, t_N) + \Delta) \cdot (M - 1) + T_F} \tag{1}$$

t_n is the execution time of (n)-th critical task. N is the number of tasks, including SCTs and DCTs. Δ is the average wait time for the critical tasks to be scheduled. U is the total number of computations. M is the total number of iterations of the program. T_F is the filling time of the dataflow graph, which equals the time of the first iterations flows from the start node to the end node of dataflow graph.

For a given program, U, M and T_F are constant. t_n and N are variables and are affected by the runtime environment. For example, resource competition, data synchronization and network congestion will increase the t_n and may turn a task in non-critical path into a critical task. Δ is also a variable since numerous tasks are mapped into the small number of PEs, where tasks are processed in the intended order by the task scheduling mechanism. Therefore, if dataflow execution model wants to achieve optimal performance, there are 2 aspects to consider. (1) Dataflow graph can be executed in the pipeline model. (2) Task scheduling mechanism can track the critical tasks and preferentially allocate computing resources for them.

3.2   Task Scheduling Algorithm

In this section, we put forward a new concept of *Congested-degree* (*Cd*), which is used for measuring the congestion of data in a task. *Congested-degree* is calculated as follows.

$$Cd(v) = Sum_r(v) \cdot T(v) + \sum_{n=1}^{C(v)} \{Max(0, (T(v) - \mu \cdot n)) \cdot Sum_{ur}(v, n)\} \qquad (2)$$

$Cd(v)$ is the *congested-degree* of task v, with higher $Cd(v)$ value indicating the more congestion of data in task v, and the more likely this task is to be a critical task. Cd is comprised of 2 parts, current execution time and potential execution time, which represent the total processing time of iterations that is ready and is not ready, respectively.

$Sum_r(v) \cdot T(v)$ is the current execution time of task v, which represents the total required execution time of iterated data that is ready but has not been scheduled. $Sum_r(v)$ is the number of iterations that are ready. $T(v)$ is the expected execution time of task v (eg. 1 cycle for add operation, 6–15 cycle for division operation). It is noted that the iteration is considered to be ready only if all data belongs to this iteration has been received.

$\sum_{n=1}^{C(v)} \{Max(0, (T(v) - \mu \cdot n)) \cdot Sum_{ur}(v, n)\}$ is the potential execution time of task v. $Sum_{ur}(v, n)$ is the number of iterations that lacks n operands for ready. $C(v)$ is the total number of dependent operands for calculation. μ is the average transmission delay between two tasks. The potential execution time represents the potential required execution time of iterated data that is not ready while some of operands have been received. It is equal to the time of a single execution minus data transmission time of absent operands, and its value is equals to 0 when transmission time bigger than execution time. Algorithm 1 shows the pseudocode of the task scheduling algorithm.

Algorithm 1. A task scheduling algorithm.

Require: I-tasks of a PE, C-function units of a PE.
Ensure: M-tasks are actually selected to be executed.
 1: **for** $\forall i \in I$ **do**
 2:     $Cd_r(i)$ = The current execution time of task i
 3:     $Cd_{ur}(i)$ = The potential execution time of task i
 4:     **if** $Cd_r(i) > 0$ **then**
 5:         $S+ = \{i\}$; $Cd(i) = Cd_r(i) + Cd_{ur}(i)$
 6:     **end if**
 7: **end for**
 8: $Q = Top\_down\_Cd\_sort(S)$
 9: **for** $\forall c \in C$ **do**
10:     q = the task with the highest congested-degree value in sorted list Q
11:     $M+ = \{q\}$; $Update\_Cd(q)$
12: **end for**

List I contains all tasks that are mapped into this PE. List C contains all function units of the PE. List M contains tasks that are actually selected to be executed. List Q and S contain the candidate tasks that can be executed. $Top\_down\_Cd\_sort$ is responsible for sorting tasks according to the Cd. $Update\_Cd$ is responsible for updating the Cd value.

The task scheduling algorithm is made up of 3 steps. Firstly, task, whose data is ready $(Cd_r(i) > 0)$, is inserted into the candidate task list S, and its Cd is computed. Secondly, all candidate tasks are prioritized through the function $Top\_down\_Cd\_sort$. Thirdly, the task with the highest Cd is selected to be executed, then its available operands have changed, so its Cd value is updated by the function $Update\_Cd$. In this algorithm, current and potential execution time are not recalculated when a task scheduling event happens. Instead, they are, similar to Translation Lookaside Buffer (TLB), automatically updated by the hardware (further discussion in follow section). Then tasks with higher Cd are assigned higher execution priorities during execution by reference to the precomputed Cd.

4   CTA Architecture

In the previous section, we propose a task scheduling algorithm for tracking DCTs and allocating computing resources for them. In this section, we propose CTA to support this algorithm, which is composed of elastic handshake mechanism (EHM) and task scheduling mechanism (TSM). EHM allows multiple iterations of the same dataflow graph to be executed simultaneously, which makes DPU executing in pipeline mode. In addition, EHM also computes current potential execution time of each task, providing *congested-degree* value for task scheduling algorithm. TSM is responsible for assigning execution priorities for tasks. Candidate tasks is ranked by TSM based *congested-degree*, then DCTs are assigned higher execution priorities. Figure 3 illustrates the architecture of

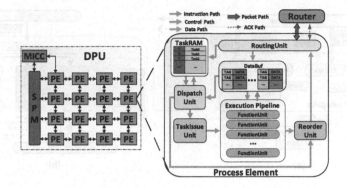

Fig. 3. The architecture of CTA in a PE.

CTA in a PE, where EHM is composed of DataBuf and RoutingUnit, TSM is composed of DispatchUnit and ReorderUnit.

4.1 Elastic Handshake Mechanism

In this section, we propose EHM to make up the shortage of HM. Firstly, each task in EHM has a memory for storing the different iterated data, which eliminates *bubbles* by HM and improves the utilization of function units. Secondly, to avoid the packet loss and network bandwidth wasted when the memory of subsequent task is full, **Counter** that implies how many results can be sent to subsequent tasks is added to task format.

EHM extends the format of the task by inserting flag (**Counter**) into each DestAddr, which represents how many results can be sent to the subsequent tasks. Each **Counter** in different DestAddr is independent. It decrements by 1 every time a result is sent, and increments by 1 when an acknowledgement (further discussion in follow section) is received. The task obtains permission to send results to DestAddrs only if all their **Counters** are bigger than 0. We suppose that the initial value of **Counter** equals the maximum data storage capacity of subsequent tasks.

EHM architecture consists of RoutingUnit and DataBuf. RoutingUnit is responsible for communicating with other PEs. DataBuf is responsible for storing operands of the task.

RoutingUnit. Figure 4 (a) shows the architecture of RoutingUnit, where ACK-Control is responsible for sending acknowledgement to the sender. Unpack is responsible for unpacking the network packet. Pack is responsible for packaging the result into the network packet. TxFifo is the first-in-first-out queue that is used to temporarily store the network packet that needs to be sent.

When a PE receives a network packet, Unpack parses the content of the packet and extracts the feature information. For a data packet, the feature

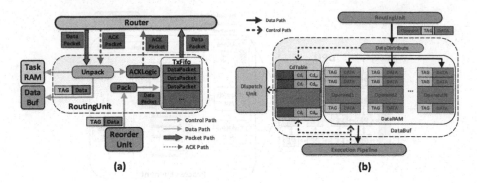

Fig. 4. (a) The architecture of RoutingUnit. (b) The architecture of Databuf.

information, including data and TAG (further discussion in follow section) is forwarded to DataBuf, then ACKControl sends the acknowledgement message (ACK) to the precedent task. For an ACK packet, the feature information, including TAG, is forwarded to TaskRAM to update the related **Counter**. The network supporting the transfer of backward ACK messages has a lower signaling overhead since ACK message does not need to transmit iterated data. Result is converted to network packet by Pack and sent to relevant address after it has been calculated.

DataBuf. Figure 4 (b) shows the architecture of DataBuf, where DataRAM is responsible for storing the iterated data and supporting both Execution Pipeline and RoutingUnit access. DataDistribute is responsible for distributing access requests to RAM slices. CdTable is responsible for maintaining the *congested-degree* of tasks.

EHM is a simple but reliable mechanism to ensure the execution correctness of iterations. It provide an asynchronous communication mechanism, where tasks communicates with each other indirectly through an intermediary (DataBuf). Task can send results to the subsequent tasks only if the related **Counter** is bigger than 0, without re-sending the redundant network packets (i.e. credit mechanism [8,24]).

4.2 Task Scheduling Mechanism Architecture

In this section, we propose a task scheduling mechanism (TSM) based on EHM, which is the hardware implementation for Algorithm 1. There are 2 factors should be taken into consideration. Firstly, Candidate tasks that meet the issue requirement are screened from the all tasks, then DCTs are assigned higher execution priorities. Secondly, since the execution time of some tasks is uncertain (i.e., memory access, division, non-linear function), the (n+1)-th iteration may be completed earlier than the (n)-th iteration. Therefore, TSM should provide a reorder mechanism where the execution results are submitted in the correct order. TSM is consisted of DispatchedUnit and ReorderUnit. DispatchUnit is

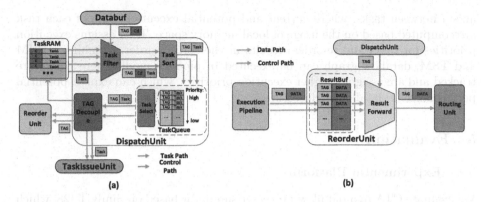

Fig. 5. (a) The architecture of DispatchUnit. (b) The architecture of ReorderUnit.

responsible for assigning execution priorities to tasks. ReorderUnit is responsible for re-ordering the execution result.

DispatchUnit. Figure 5 (a) shows the architecture of DispatchUnit, where TaskFilter is responsible for extracting the candidate tasks from the all. TaskSort is responsible for ranking candidate tasks according to the *congested-degree*, whose result is stored in TaskQueue. TaskSelect is responsible for selecting task to be executed. TAGDecouple is responsible for separating the TAG from iterated data.

DispatchUnit is responsible for assigning priorities to tasks. The whole procedure can be divided into 3 stages. (1) TaskFilter pairs tasks with their *congested-degree* and filter out the tasks that is not ready ($Cd_r = 0$), then the rest is the candidate task. (2) Candidate tasks are sorted by *congested-degree* in TaskSort. (3) Task with the highest *congested-degree* is selected by TaskSelect to be executed, then its TAG is separated by TAGDecouple and sent to ReorderUnit. As mentioned in the previous section, Task's *congested-degree* is automatically updated by CdTable and is stored in TaskQueue. Hence, InstSelect is able to directly use the pre-computed *congested-degree* value whenever a task scheduling event happens.

ReorderUnit. Figure 5 (b) shows the architecture of ReorderUnit, where ResultBuf is responsible for storing the execution result. ResultForward is responsible for forwarding result to RoutingUnit.

In pipeline execution model, the (n+1)-th iteration may be finished earlier than the (n)-th of the same task since the execution time of some tasks is uncertain (i.e. memory access, nonlinear function), which could result in faulty results (i.e. Read after Write). When an iterated data has been calculated by Execution pipeline, its results is forwarded to RoutingUnit based on the TAG received from DispatchUnit, which ensures results are submitted in the same order as they are dispatched by MICC.

CTA architecture is the hardware implementation of Algorithm 1, which is composed of EHM and TSM. EHM provides an effective communication mechanism between tasks, where current and potential execution time of each task are computed based on the usage of local memory space. TSM assigns execution priorities for tasks and re-orders results in the correct order. Combining EHM and TSM, dataflow graph can be executed in pipeline mode where DCTs are tracked and are assigned higher execution priorities, which can exploit potential performance of dataflow architecture.

5 Evaluation

5.1 Experimental Platform

We evaluate CTA in a dataflow processor simulator based on SimICT [28] which is a cycle accurate, fast and flexible simulation framework for large-scale architecture. The implementation of the simulator is strictly consistent with the design in Fig. 2. The configuration of DPU is shown in Table 1. DPU consists of 64 PEs constructed in the form of 8 × 8 array. Each PE contains 4 fused multiply-accumulate (FMACs) in double precision, 4 FMACs in single precision and 8 64-bit arithmetic logic unit (ALUs) in SIMD-4 model. In this paper, we also compare the power efficiency of DPU with NVIDIA Titan Xp GPGPU. We have compared the execution cycle on the simulator with RTL emulation, the results show that the average deviation is under 3%.

Table 1. Configuration of DPU

Module	Configuration
MICC	ARM, 1 GHz
PE	4 FMACs (single precision), 4 FMACs (double precision), 8 ALUs, SIMD4, 128 instructions, 1 GHz
2D mesh	1 cycle/hop, X-Y routing algorithm, 4 operand/memory-access networks, 4 ACK message networks
SPM	64 Bank, 16 MB
Peak performance	512GFLOPS for FP64, 512GFLOPS for FP32

5.2 Workloads

To test the efficiency of CTA, we select 7 typical HPC+AI applications, namely Fast Fourier Transform (FFT), Stencil2D, Stencil3D, Lattice-Boltzman Method (LBM), Dense Matrix Multiplication (GEMM), Sum of Absolute Differences (SAD), Convolution (CONV), where CONV is calculated in single precision while other workloads are calculated in double precision. We choose the CONV layers of AlexNet [11] in the Caffe model as the workload since it occupies about 85% computational time of the entire network processing [14]. These workloads are important and representative of different types of HPC+AI applications, and they also cover different computational density and memory access patterns. The specification of workloads is shown in Table 2, FLOP is the floating-point operations of workloads which is used for calculating performance of DPU, M is for the million. Each application is translated into dataflow graph through the compiler based on LLVM [13], due to the limitation of space, the detail of the compiler are not presented in this article.

Table 2. Specification of workloads

Workload	Scale	FLOP
FFT	16384 32 Points	13M
Stencil2D	256 × 256 100 time steps	41M
Stencil3D	64 × 64 × 64 50 time steps	92M
LBM	1024 Points 1000 time steps	16M
GEMM	1024 × 1024	2147M
SAD	1920 × 1072	4487M
CONV	The 1st convolution of AlexNet	211M

In this work, we implement and evaluate several task scheduling mechanisms (further discussion in RELATED WORK section), including static scheduling mechanism (Static), single graph multi flow (SGMF), credit scheduling mechanism (Credit), a pipelining loop optimization method (PLO), elastic handshake mechanism (EHM) and a critical task aware (CTA) scheduling mechanism.

5.3 Evaluation Metric

There are 2 types of calculation in DPU, namely, fix-point calculation and floating-point calculation. Fix-point calculation is responsible for computing the memory address and iteration index. Floating-point calculation is responsible for computing the kernel of workloads. Therefore, we use the Giga Floating-point Operations per Second (GFLOPS) as the performance metrics to evaluate different task scheduling mechanisms, which measures if the task scheduling mechanisms exploits sufficient parallelism and makes the best use of floating-point units.

We also have implemented RTL of DPU and synthesized the design with Synopsys Design Compiler (dc-2014.09-SP5) and TSMC 40 nm technology library, which meets timing 1 GHz with 30% of extra margin.

5.4 Result and Discussion

Performance. Figure 6 shows the performance of DPU of 7 HPC+AI workloads on different task scheduling mechanisms. The performance of CTA is 1.78x, 1.19x, 1.14x, 1.37x, 1.14x higher than that in Static, SGMF4, Credit, PLO and

EHM. Stencil-like applications (Stencil2D, Stencil3D and LBM) achieve better performance improvement than the others. The reason is that they are typical memory-intensive applications and the execution time of memory access instruction is hard to predict, making critical tasks even more difficult to be dynamically tracked. Thus for Stencil-like applications, accelerating the critical tasks efficiently reduces the congestion of iterated data, which decreases execution time of a single dataflow graph. However, GEMM has been optimized not obvious effect of other workloads to optimize effective, which can be attributed to the 2 reasons. (1) Most of instructions in GEMM are multiply-add instructions, which are regular and there are minimal execution time difference among them. (2) Tasks of GEMM are mapped to systolic array by using canonical mapping method, which effectively avoids network congestion as well as computing resources competition.

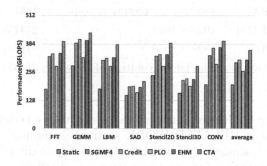

Fig. 6. Performance on different task scheduling mechanism.

As shown in Fig. 6, performance of Static is the lowest since only a single iteration can be executed at the same time, where most tasks would always be waiting, so the workload effectiveness is reduced. PLO fails to achieve high performance because 3 synchronous packet handshake is adopted in PLO before data transmission, which severely degrades the transmission efficiency between tasks. The other 3 task scheduling mechanisms (SGMF4, Credit and EHM) achieve relatively higher performance. SGMF4 provides enough independent dataflow graph to be scheduled. Once a dataflow graph is blocked by critical tasks, function units are enable to select other graphs to be scheduled, which makes up for the shortage of lacking iteration parallelism in Static. Credit offers an effective communication mechanism between tasks, where multiple iterations of the same dataflow graph can be executed simultaneously, just as in EHM. However, these mechanisms are all without considering the possible influence of DCTs, which restricts further improvement in performance.

Overhead of Area and Power. The comparison of area and power of DPU of the different task scheduling mechanisms are shown in Fig. 7 (a) and Fig. 7 (b), respectively. CTA requires extra on-chip logic in MICC, network of chip (NoC) and PE. The additional hardware overhead of MICC and NoC mainly comes from the ACK network as well as status registers (**Counter**) of EHM. The additional hardware overheads of PE comes from the data acknowledgement of EHM and task scheduling of TSM. In total, the area of CTA is 1.30x, 1.01x, 1.03x, 1.27x and 1.03x higher than that in Static, SGMF4, Credit, PLO, EHM. And the power of CTA is 1.26x, 1.01x, 1.01x, 1.23x and 1.04x higher than that in these mechanisms. Considering the performance gaining, the increment of the area and power is acceptable.

Figure 7 (c) and Fig. 7 (d) show the area and power breakdown of a single PE of CTA, respectively. We can see that, DataBuf consumes the majority (29.66% of PE area and 16.76% of PE power) of a single PE, since EHM provides enough memory (DataRAM) for storing iterated data to cover the transmission delay of handshake signal. The ACKControl, which is added hardware to support sending ACK message, consumes 14.68% of RoutingUnit area and 11.63% of RoutingUnit power. The area and power consumption of DispatchUnit, which is added for implementing the adjusted issue policy, and ReorderUnit, which is added for reordering execution results, are limited within 10% of the PE total.

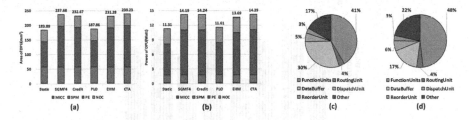

Fig. 7. (a) Area of DPU. (b) Power of DPU. (c) Area breakdown of a single PE of CTA. (d) Power breakdown of a single PE of CTA.

Power Efficiency. Figure 8 shows the power efficiency of the different task scheduling mechanisms, which is measured by performance per watt that exhibits the computational ability achieved with a unit power consumption. The power efficiency of CTA is 1.41x, 1.18x, 1.14x, 1.11x, 1.10x higher than that in Static, SGMF4, Credit, PLO and EHM. SGMF represents classical implementation that sacrifices space for improving performance, thus its power efficiency is quite low in spite of high performance. On the contrary, PLO, which merely adds handshake signal between tasks, can obtain performance improvement at low hardware cost. Both Credit and EHM can achieve high performance and power efficiency. The similarity is that both provide an efficient handshake mechanism for task communication, the difference is that for Credit, iterated data may be sent repeatedly when the memory of subsequent task is full, which adds further area and power overhead. Compared with other task scheduling mechanisms, CTA can achieve higher performance and power efficiency, which is mainly generalized in 2 points. (1) EHM enables multiple iterated data of the same dataflow graph to be executed simultaneously with minimal hardware overhead. (2) TSM tracks the critical tasks and preferentially allocates resource for them during execution, which improves the execution efficiency of critical tasks.

Comparison with GPU. We also have compared CTA with a modern GPU (NVIDIA Titan Xp, 12TFLOPS single-precision, 380GFLOPS double-precision, 16 nm FinFET, 250W TDP). As shown in Fig. 8, GPU can achieve high power efficiency in GEMM as well as CONV compared with other HPC workloads. The reason is that these GEMM and CONV are iteratively optimized by GPU as artificial neural nets becomes the hottest top research on AI, where the program such as resources competition and network congestion gets alleviating. DPU shares many common characteristics with GPU, both aims at accelerating iterated data. However, GPU runs in traditional control-flow fashion while DPU runs in dataflow fashion, which exploits more ILP, DLP and consumes less on-chip resources. As shown in Fig. 8, the average power efficiency of DPU is 3.23x higher than that in GPU. Note that even though the process of DPU 40 nm which is about three generations behind the GPU (16 nm).

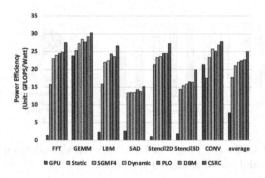

Fig. 8. Power efficiency of the different task scheduling mechanisms and GPU.

6 Related Work

The critical tasks directly affects the overall performance of the program, especially in dataflow architecture. The reason is that in dataflow architecture, the interaction among tasks is more frequent and the dynamic condition is more complicated. There are several mechanisms for task scheduling.

Static [7]. In order to prevent (n)-th iteration is overwritten by the (n+1)-th iteration. MICC in Static is prohibited to generate new iteration index back-to-back to dataflow graph. Thus, few tasks can be executed simultaneously while most tasks are waiting for the data to be ready. Although SGMF [27] extends a single graph to multiple independent graphs based on Static for improving the utilization of function units, the problem that the performance of a single graph is poor has not been solved.

Credit [8,24]. In this mechanism, each task has a RAM slice for storing the received data. Credit uses the credit (TAG-token) to differentiate among iterations. Once the RAM slice is full, task drops out the newly received iterated data and request its precedent task to resend data again until the RAM slice is available. Although Credit allows multiple iterations of the same graph to be executed in parallel, it may produce lots of redundant network packet, resulting in wasted bandwidth of network.

PLO [25]. Handshake mechanism is implemented in PLO based on Static. In this mechanism, an ACK message will be sent to the precedent task as soon as the data is consumed. A task is considered to be ready only if the iterated data from the precedent task and ACK message from the subsequent task are both ready. However, subsequent task is enable to send acknowledgement message to precedent task only if the storage space is available, *bubbles* are produced by synchronous handshake signal, which leads to the pipeline stall in dataflow graph.

Some instruction mapping algorithms have been proposed to accelerate the critical tasks of dataflow graph. In SPDI [15], instructions are sorted according to their maximum depths, and the instruction with higher depth is assigned higher execution priority. In SPS [4], routing distance among instructions is computed by anchor points, then instructions in the critical path should be optimized for communication while instructions in non-critical path should be optimized for execution.

Existing task scheduling mechanisms mainly focus on optimization of communication among tasks to exploit iteration level parallelism, but they fail to consider the impacts of the critical tasks. Existing instruction mapping algorithms are unable to dynamically track the critical tasks due to they have been almost no consideration for the possible influence of runtime environment during execution. Therefore, neither previous task scheduling mechanism nor instruction mapping algorithm can exploit the all potential performance of the critical tasks.

7   Conculusion

The critical tasks is the key factor affecting the performance of the program, especially in dataflow architecture. The reason is that the interaction among tasks is more frequent and the dynamic condition is more complicated in dataflow architecture. In this paper, we first put forward a new concept of *congested-degree* for measuring the congestion of data in a task. Then we propose a critical task aware (CTA) scheduling mechanism for dataflow architecture, which is consisted of 2 parts. (1) Elastic handshake mechanism (EHM) allows multiple iterated data of the same dataflow graph to be executed in parallel with minimal hardware overhead. (2) A task scheduling mechanism (TSM) tracks the critical tasks and assigns higher execution priorities for them. The effectiveness of CTA is tested by 7 HPC+AI workloads, the experimental results show that our mechanism increases the computational performance by 14%–78%, and increases the power efficiency by 11%–41% compared with previous task scheduling mechanisms. Finally, we believe that this study has use for reference meaning to other architectures which have borrowed dataflow or dynamic task scheduling ideas.

Acknowledgment. This work was supported by the project of the state grid corporation of China in 2020 "Integration technology research and prototype development for high end controller chip" under Grant No. 5700-202041264A-0-0-00.

References

1. Akbari, O., Kamal, M., Afzali-Kusha, A., Pedram, M., Shafique, M.: PX-CGRA: polymorphic approximate coarse-grained reconfigurable architecture. In: 2018 Design, Automation & Test in Europe Conference & Exhibition (DATE), pp. 413–418. IEEE (2018)
2. Chen, Y.H., Krishna, T., Emer, J.S., Sze, V.: Eyeriss: an energy-efficient reconfigurable accelerator for deep convolutional neural networks. IEEE J. Solid-State Circ. **52**(1), 127–138 (2016)
3. Chen, Y.H., Yang, T.J., Emer, J., Sze, V.: Eyeriss v2: a flexible accelerator for emerging deep neural networks on mobile devices. IEEE J. Emerg. Sel. Top. Circ. Syst. **9**(2), 292–308 (2019)
4. Coons, K.E., Chen, X., Burger, D., McKinley, K.S., Kushwaha, S.K.: A spatial path scheduling algorithm for edge architectures. ACM SIGOPS Oper. Syst. Rev. **40**(5), 129–140 (2006)
5. Dennis, J.B.: First version of a data flow procedure language. In: Robinet, B. (ed.) Programming Symposium. LNCS, vol. 19, pp. 362–376. Springer, Heidelberg (1974). https://doi.org/10.1007/3-540-06859-7_145
6. Fan, D., et al.: SmarCO: an efficient many-core processor for high-throughput applications in datacenters. In: 2018 IEEE International Symposium on High Performance Computer Architecture (HPCA), pp. 596–607. IEEE (2018)
7. Fu, H., et al.: Scaling reverse time migration performance through reconfigurable dataflow engines. IEEE Micro **34**(1), 30–40 (2013)
8. Hiraki, K., Sekiguchi, S., Shimada, T.: Efficient vector processing on a dataflow supercomputer sigma-1. In: Supercomputing 1988: Proceedings of the 1988 ACM/IEEE Conference on Supercomputing, vol. I, pp. 374–381. IEEE (1988)
9. Hoffmann, H.: Stream algorithms and architecture. Ph.D. thesis, Massachusetts Institute of Technology (2003)
10. Jouppi, N.P., et al.: In-datacenter performance analysis of a tensor processing unit. In: Proceedings of the 44th Annual International Symposium on Computer Architecture, pp. 1–12 (2017)
11. Krizhevsky, A., Sutskever, I., Hinton, G.E.: ImageNet classification with deep convolutional neural networks. In: Advances in Neural Information Processing Systems, pp. 1097–1105 (2012)
12. Kyriacou, C., Evripidou, P., Trancoso, P.: Data-driven multithreading using conventional microprocessors. IEEE Trans. Parallel Distrib. Syst. **17**(10), 1176–1188 (2006)
13. Lattner, C., Adve, V.: LLVM: a compilation framework for lifelong program analysis & transformation. In: International Symposium on Code Generation and Optimization, CGO 2004, pp. 75–86. IEEE (2004)

14. Long, J., Shelhamer, E., Darrell, T.: Fully convolutional networks for semantic segmentation. In: Proceedings of the IEEE Conference on Computer Vision and Pattern Recognition, pp. 3431–3440 (2015)
15. Nagarajan, R., Kushwaha, S.K., Burger, D., McKinley, K.S., Lin, C., Keckler, S.W.: Static placement, dynamic issue (SPDI) scheduling for edge architectures. In: Proceedings. 13th International Conference on Parallel Architecture and Compilation Techniques, PACT 2004, pp. 74–84. IEEE (2004)
16. Oriato, D., Tilbury, S., Marrocu, M., Pusceddu, G.: Acceleration of a meteorological limited area model with dataflow engines. In: 2012 Symposium on Application Accelerators in High Performance Computing, pp. 129–132. IEEE (2012)
17. Oskin, M.H., Swanson, S.J., Eggers, S.J.: Wavescalar architecture having a wave order memory, uS Patent 7,657,882, 2 February 2010
18. Pratas, F., Oriato, D., Pell, O., Mata, R.A., Sousa, L.: Accelerating the computation of induced dipoles for molecular mechanics with dataflow engines. In: 2013 IEEE 21st Annual International Symposium on Field-Programmable Custom Computing Machines, pp. 177–180. IEEE (2013)
19. Rahman, M., Venugopal, S., Buyya, R.: A dynamic critical path algorithm for scheduling scientific workflow applications on global grids. In: Third IEEE International Conference on e-Science and Grid Computing (e-Science 2007), pp. 35–42. IEEE (2007)
20. Sankaralingam, K., et al.: Exploiting ILP, TLP, and DLP with the polymorphous trips architecture. In: Proceedings of the 30th Annual International Symposium on Computer Architecture, pp. 422–433. IEEE (2003)
21. Schulz, M.: Extracting critical path graphs from MPI applications. In: 2005 IEEE International Conference on Cluster Computing, pp. 1–10. IEEE (2005)
22. Son, J.H., Kim, J.S., Kim, M.H.: Extracting the workflow critical path from the extended well-formed workflow schema. J. Comput. Syst. Sci. **70**(1), 86–106 (2005)
23. Son, J.H., Kim, M.H.: Analyzing the critical path for the well-formed workflow schema. In: Proceedings Seventh International Conference on Database Systems for Advanced Applications, DASFAA 2001, pp. 146–147. IEEE (2001)
24. Swanson, S., Michelson, K., Schwerin, A., Oskin, M.: WaveScalar. In: Proceedings. 36th Annual IEEE/ACM International Symposium on Microarchitecture, MICRO-36, pp. 291–302. IEEE (2003)
25. Tan, X., et al.: A pipelining loop optimization method for dataflow architecture. J. Comput. Sci. Technol. **33**(1), 116–130 (2018)
26. Tian, Y., Gu, Y., Ekici, E., Ozguner, F.: Dynamic critical-path task mapping and scheduling for collaborative in-network processing in multi-hop wireless sensor networks. In: 2006 International Conference on Parallel Processing Workshops (ICPPW 2006), pp. 8-pp. IEEE (2006)
27. Voitsechov, D., Etsion, Y.: Single-graph multiple flows: energy efficient design alternative for GPGPUs. ACM SIGARCH Comput. Archit. News **42**(3), 205–216 (2014)
28. Ye, X., Fan, D., Sun, N., Tang, S., Zhang, M., Zhang, H.: SimICT: a fast and flexible framework for performance and power evaluation of large-scale architecture. In: International Symposium on Low Power Electronics and Design (ISLPED), pp. 273–278. IEEE (2013)
29. Ye, X., et al.: An efficient dataflow accelerator for scientific applications. Future Gener. Comput. Syst. **112**, 580–588 (2020)
30. Ye, X., et al.: Applying CNN on a scientific application accelerator based on dataflow architecture. CCF Trans. High Perform. Comput. **1**(3–4), 177–195 (2019)

An Adaptive Thread Partitioning Approach in Speculative Multithreading

Yuxiang Li[1,2] , Zhiyong Zhang[1,2(✉)], and Bin Liu[3]

[1] Henan University of Science and Technology, Luoyang 471023, Henan, China
xidianzzy@126.com
[2] Henan International Joint Laboratory of Cyberspace Security Applications,
Luoyang 471023, Henan, China
[3] Northwest Agriculture and Forestry University, Yangling 471023, China
liubin0929@nwsuaf.edu.cn
http://www.sigdrm.org/~zzhang/

Abstract. Thread partition is a core part of Speculative Multithreading (SpMT) technique. The existing thread partition approaches mostly adopt one unique thread partitioning scheme for unknown programs, resulting in high misspeculation ratio, restricting the programs' speedup improvement due to inappropriate partitioning schemes. This paper which introduces an adaptive thread partition approach (AdapTPA), takes the relationship between program complexity and thread partitioning scheme as the research entry point, and uses the irregular programs as the research carrier, and utilizes formal analysis, probability statistics, mathematical modeling and simulation experiments to reveal the rule that program's characteristics affect speedup performance, and generates a compound thread partitioning scheme for one program, and selects and executes the most suitable thread partitioning scheme according to the runtime context and the program's complexity, so to achieve the expected maximum speedups. With the method of path statistics on one program's control flow graph, the program's complexity calculation model is set up; A candidate thread partitioning scheme set is constructed on the foundation of classical thread partitioning approaches; Using expert knowledge to guide production rules, a scheme selection mechanism that complies with program complexity is explored. Compared to the heuristic rules-based (HR-based) thread partitioning method, the experiment results show that AdapTPA delivers an average 18.24% performance improvement.

Keywords: Thread partition approach · Speculative multithreading · Expert knowledge

1 Introduction

Thread-Level Speculation (TLS), which is a speculative multi-threading technique [1,2], allows data execution between concurrent units to be aggressively

© Springer Nature Switzerland AG 2020
M. Qiu (Ed.): ICA3PP 2020, LNCS 12452, pp. 78–91, 2020.
https://doi.org/10.1007/978-3-030-60245-1_6

executed in parallel, overcoming the weakness that the traditional parallelization methods cannot effectively eliminate thread-level fuzzy dependence. Thread
partitioning is a key step in TLS, as it takes charge of the insertion of thread
partitioning statements. It is the core of programs' speculative parallelization,
which directly affects the speedup performance [3]. Therefore, the research on
thread partitioning approach is urgent. The existing thread partition methods
mainly include: thread partition method based on heuristic rules [4] and thread
partition methods based on machine learning [5], and et al. [6–10]. The former
determines the granularity of threads generated after the program is partitioned,
the data dependency between threads, and the spawning distance according to
heuristic rules, so as to determine the partition flags ((spawning point, sp) and
(control quasi-independent point, cqip)); The latter uses machine learning to
learn the knowledge of thread partition in the sample set, and predicts its partition scheme based on the characteristics of the new input program. However,
these two types of thread partition methods consider one program as a partition unit. A unified thread partition scheme is used for the procedure in the
program. The lack of a personalized thread partition scheme for the procedures
in the program results in the incomplete parallelism of some procedures in the
program, resulting that the performance of the program after parallelization can
not be maximized. Therefore, it is of great significance to carry out thread partition with the procedures in the program as the object, which can overcome the
shortcomings of traditional thread partition methods.

Based on the previous work, this paper intends to use an adaptive mechanism
to adaptively select the most suitable thread partitioning scheme based on the
program features and context, which can ensure the maximum performance of
the serialization program after parallelization, and can provide a new method
for multi-core processor design.

The remaining parts of this paper are organized as follows. In Sect. 2, we
first briefly present the motivation of AdapTPA; Sect. 3 presents the overall
framework of AdapTPA; Implementation of AdaTPA is shown in Sect. 4; Sect. 5
presents experiment and analysis; Sect. 6 shows conclusion and future work.

2   Motivation of AdapTPA

This paper brings an adaptive mechanism, proposing an adaptive thread
partition ing approach (AdapTPA) for irregular programs, aiming at achieving
the overall research goal of maximizing the speedup performance, and providing
the possibility for the wide application and healthy development of emerging
parallel technologies.

3   Overall Framework

The research framework to be adopted is shown in Fig. 1. The research framework regards the irregular serial program as the input, and establishes the program complexity calculation model, the generation of candidate thread partition

scheme, and the expert knowledge-based partition scheme selection as the main research points, and selects the most suitable thread partitioning scheme to perform the thread partition. The results are run on Prophet simulator to obtain speedups and programs' results.

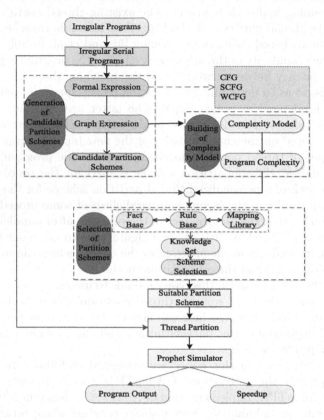

Fig. 1. Overall research framework. The framework includes three main parts: generation of candidate partition schemes, building of complexity model, and selection of partition schemes

3.1 Feature Extraction

Conventionally, compiler researchers have used fixed-length representations of the program's source code features or intermediate representations [11]. They are extracted from programs and collected during compilation time. Afterwards, we apply graph-based features to build WCFG for GbA. The feature graphs are generated from profiling pass, which extracts static and dynamic features. Figure 4 gives a simple description of feature extraction. The input programs are Olden benchmarks [12]. Thread granularity, load balance, data dependence, and control dependence are the main influence factors on program speedup. Hence, we take dynamic instruction number, DDD, DDC, loop branch probability, and critical path into account and regard them as program features. The specific features and descriptions are given in Table 1.

Table 1. Extracted features and descriptions

Features	Descriptions
Instruction number	Actual number of instructions in a basic block
DDC	Data dependence count between two basic blocks
DDD	Data dependence distance between two basic blocks
Loop branch probability	Probability for loop to jump to testing part of code
Branch probability	Probability for control flow to pass through a branch

3.2 Knowledge Expression

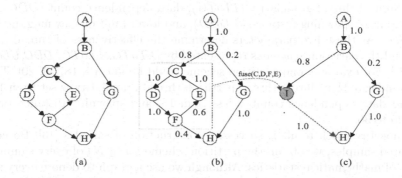

Fig. 2. Transformation from CFG (a) to WCFG (b), and finally to SCFG (c)

Deriving from the same input, train set and validation set are divided. Firstly, we use partition compiler (in the Prophet) to generate control flow graph (CFG) after an intermediate pass. Then, profiled feature information generated by profiling model are annotated to the CFG with a structural analysis method, so to generate WCFG. The weights of each edge are denoted with the relative branch probabilities. Then, a structural analysis traverses the CFGs of programs to WCFGs and also identifies loop regions. Then, the loop regions are induced into a super node with one entry and one exit node and WCFG traverses to super control flow graph (SCFG), where loop region is represented as an abstract node. Each node in the SCFG is either basic block or super basic block, which represents loop region. Figure 2(a) shows a CFG. After structural analysis as well as loop region induction, basic blocks: C, D, F and E in the dashed box of Fig. 2(b)

Table 2. Heuristic rules for thread partition

1.SP can appear anywhere in programs and behind a function call instruction as far as possible. In non-loop region, CQIP is located in the beginning of a basic block. CQIP is located in front of loop branch instructions in the last basic block in loop regions.
2.SP-CQIP pairs are located within the same loop or function. The number of dynamic instructions from SP to CQIP is between THREAD_LOWER_LIMIT and THREAD_UPPER_LIMIT.
3.Between two successive candidate threads, spawning distance is bigger than minimum DIS_LOWER_LIMIT.
4.Data dependence between two consecutive candidate threads is less than threshold DEP_THRESHOLD.
5.Between SP and CQIP, the number of function call instructions is less than the threshold CALL_LOWER.

are induced into a super basic block I (shown in Fig. 2(c)). AdapTPA represents the desired thread partition scheme with a vector $H=[H_1,\ H_2,\ H_3,\ H_4,\ H_5]$ and the sample partition scheme with $h_i = [h_{i1},\ h_{i2},\ h_{i3},\ h_{i4},\ h_{i5}](i \in N)$, which all include five thresholds: the upper limit of thread granularity ($ULoTG$), the lower limit of thread granularity ($LLoTG$), data dependence count (DDC), the upper limit of spawning distance ($ULoSD$), and lower limit of spawning distance ($LLoSD$). As these five parameters determine the effectiveness of thread partition, and the partition scheme is represented by $[ULoTG, LLoTG, DDC, ULoSD, LLoSD]$. For example, one partition scheme could be $[50, 10, 18, 30, 20]$. These values indicate that during thread partition thread granularity is set from 10 to 50, and data dependence count is less than 18, and spawning distance ranges from 20 to 30.

A novel research result is successful construction of samples [13]. Based on generated samples, we obtain the partition scheme $h_i(i \in N)$ of every sample by means of mathematical statistics. Although we use a graph to denote every sample, the node in the graph is represented by the first part of $T = \{X, H\}$, where X represents program features, and H denotes the optimal partition scheme, which are composed by five partition thresholds, namely $ULoTG$, $LLoTG$, DDC, $ULoSD$, $LLoSD$.

4   Implementation of AdapTPA

4.1   Building of Complexity Calculation Model

There are many program features that affect thread partitioning, such as data dependency, control dependency, number of branches, number of basic blocks, number of average dynamic instructions, nesting level of loop structure, number of procedure calls, and so on. The values of these features reflect the complexity of the program. Most of the existing thread partitioning methods can not fully consider the influence of program complexity on thread partitioning. Only the program features are selected as the input of the thread partitioning method. It is easy to cause the program features selected by different thread partitioning methods to be inconsistent, and the generated thread partitioning scheme is not accurate enough.

In the proposed program's complexity calculation model, the formal expression is firstly constructed, and the CFG diagram of the program is constructed with the basic block as the analysis unit. The feature values obtained by the program analysis are added to the CFG diagrams in the form of annotations to form the weighted control flow (WCFG); based on probability statistics and graph traversal, the complexity (sub-complexity) of possible paths on WCFG is calculated; finally, the overall complexity of the program is obtained by integrating sub-complexities. Figure 3 shows the flow chart for program complexity calculation, and Table 3 shows the pseudocode of complexity calculation.

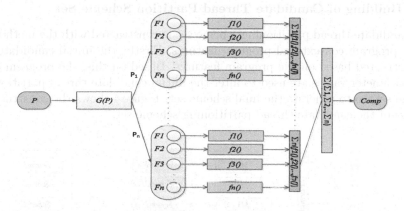

Fig. 3. Flow diagram of complexity calculation. The graph P is firstly represented by $G(P)$, the complexity of different part of P is separately calculated, then aggregated into $Comp$

In Fig. 3, P represents the input irregular serial program, $G(P)$ stands for WCFG, $F1 \sim Fn$ ($n \in$ N) stand for program features, and $f1() \sim fn()$ ($n \in$ N) stand for transfer function, $Comp1() \sim Compn()$ ($n \in$ N) represent the complexity of each path, and $Comp$ represents the total complexity of P. In the model, first, the unknown program P is formalized and converted into WCFG, i.e. $G(P)$; Secondly, feature extraction is performed on each possible path (from the head node to the tail node) in $G(P)$ respectively, using $F1 \sim Fn$ ($n \in$ N); Thirdly, using the conversion function $f1() \sim fn()(n \in$ N) to achieve the mapping of eigenvalues to complexity, for example, the complexity of basic blocks x is $0.01 \times x$, the complexity of loops y is $0.2 \times y$, etc; then, the complexity $Comp1() \sim Compn()$ of each path in $G(P)$ is calculated separately; Finally, for each path, the complexity is summarized to get the complexity of the program P.

Table 3. Computation of complexity

Input: irregular program P
Output: Complexity of program P
$G(P)$ = formalize(P);
Extract and express the characteristics of program's the i_{th} possible path (possibility P_i in Fig.4) with F1~Fn;
Set $f_1()$~$f_n()$ to be n transfer functions;
Set W_1~W_n to be n weight parameters for F_1~F_n;
Use function $Sub\_complexity(f_1,f_2,f_3,...,f_n)$ to compute the complexity of every possible path;
According to every Sub_complexity, compute the final complexity computation with function $Comp = Complexity(Sub\_complexity1,Sub\_complexity2,...,Sub\_complexityn)$;

4.2 Building of Candidate Thread Partition Scheme Set

The candidate thread partitioning scheme set is constructed with the method of fusing program context and program features. Firstly, the initial candidate set is constructed based on the program features. Based on this, the program context parameter values are used to filter the initial candidate thread partitioning scheme set, so to generate the final scheme set. Figure 4 shows the construction process of the candidate thread partitioning scheme set.

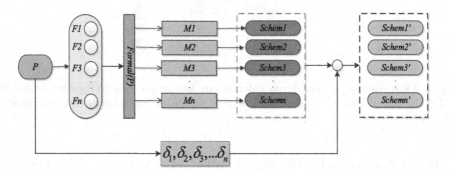

Fig. 4. Flow diagram of building candidate thread partition scheme set. The graphs of input programs are firstly formalized, then generated into different partition schemes

In Fig. 4, P stands for an irregular serial program, $F1$~Fn stand for program feature, $Formal(P)$ stands for formal expression of P, $M1$~$Mn(n \in N)$ stand for n classical thread partitioning approaches, and $Schem1$~$Schemn$ stand for n thread partitioning schemes. The number of thread partition method is set as follows: HR-based thread partition method ($M1$) is numbered 1, ML-based thread partition method ($M2$) is numbered 2, and the critical path-based thread partition method($M3$) is numbered 3, the full path-based thread partition method ($M4$) is numbered 4, the hybrid thread partition method ($M5$) is numbered

5, and so on. The path numbers are set as follows: the critical path number is numbered 1, and the other non-critical path numbers are $2\sim n$. The thread partition scheme consists of five main parameters: number of thread partition method, path number and thread partition algorithm (the five parameters are: Upper Limit of Spawning Distance (*ULoSD*), Lower Limit of Spawning Distance, (*LLoSD*), Data Dependence Count (*DDC*), Upper Limit of Thread Granularity (*ULoTG*), and Lower Limit of Thread Granularity (*LLoTG*)). By introducing the context parameter $\delta_1\sim\delta_n(n\in N)$, the thread partitioning method in this topic is context-aware, and the candidate thread partitioning scheme set can also capture the change of the program state.

4.3 Construction of Thread Partitioning Scheme Selection Mechanism in Line with Program Complexity

After calculating the program complexity and constructing the candidate thread partitioning scheme set respectively in the technical routes (1) and (2), based on the expert knowledge, the mapping rule set of scheme selection of "program complexity->thread partitioning scheme" is established; according to mapping rule and program complexity, execution context, the most suitable thread partitioning scheme in the candidate set is selected. Figure 5 shows the flow chart of thread partitioning scheme selection.

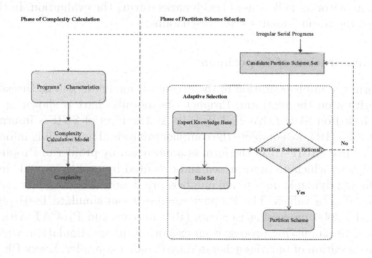

Fig. 5. Flow graph of thread partition scheme selection. The complexity of input program is firstly calculated, then partition schemes are generated by using expert knowledge

Rule sets are used to store expert knowledge for reasoning. In the rule set, the expert knowledge of the thread partitioning scheme selection mechanism is represented by a production rule (also called a mapping rule). The production rule divides the knowledge representation into two parts: premise and conclusion. The general form of expert knowledge production rule representation is IF <condition>, THEN <conclusion>, for example:

1. IF <$Comp \in [0.8, 1.0]$>, THEN <select $Schem1'$>;

2. IF <$Comp \in [0.6, 0.8)$>, THEN <select $Schem2'$>;

3. IF <$Comp \in [0.4, 0.6)$>, THEN <select $Schem3'$>;

4. IF <$Comp \in [0.2, 0.4)$>, THEN <select $Schem4'$>;

5. IF <$Comp \in [0.0, 0.2)$>, THEN <select $Schem5'$>;

where, $Schem1' \sim Schem5'$ is a partitioning scheme selected by the candidate thread partitioning scheme generated by the technical route (2), which is determined by the complexity and rules of the program. Some examples of generating mapping rules are given above.

5 Experiment and Analysis

In this section, the experimental setup is introduced, to provide details of the Prophet simulator as well as used benchmarks during the evaluation. In the last, we present the results' analysis and discussions.

5.1 Configuration of Experiment

We perform the implementation of the execution model as well as thread partition algorithm on the platform: Prophet (its module chart is shown in Fig. 6), which is based on SUIF/MACHSUIF [14]. At the level of SUIF's intermediate representation (IR), we complete the compiler analysis. The profiling information is produced from SUIF-IR in the form of annotation by profiler of Prophet. The SUIF programs which are interpreted and executed by profiler provide information, including dynamic instruction number, prediction of control flow path, and prediction of data values. The Prophet simulator can simulate 1~64 pipelined mips-based R3000 processing elements (PE) and we run ProCAT with 4 PEs or 8 PEs. This simulating process is an execution-driven simulation, which performs the execution of binaries generated by Prophet compiler. Every PE fetches and executes instructions from one thread, and orderly issues 4 instructions per cycle. Every PE owns a private multiversioned L1 cache, which has latency of 2 cycles. Speculative results of PEs are buffered and cache communication is performed via multiversioned L1 caches. With a snoopy bus, a write-back L2 cache is shared by the 8 PEs. The parameter configuration of simulator is shown in Table 4.

Olden benchmarks [15] and SPEC2000 [16] are used to evaluate ProCTA. As a popular benchmarks of studying irregular programs, Olden benchmarks process

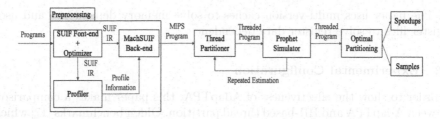

Fig. 6. Module chart of prophet. Programs are firstly transformed into MIPS codes, then partitioned into threaded programs, then run on Prophet simulator

complex control flows, pointer-intensive, as well as irregular data structures. The benchmarks own dynamic structures, e.g., trees, lists, and DAGs, et al, which are all difficult to get parallelized using conventional approaches.

Table 4. Configuration of prophet simulation (Per PE)

Parameters of Configuration	Value
Function Units	4 int ALU (1 cycle)
	4 int Mult/Div (3/12 Cycles)
	4 fp ALU (2 Cycles)
	4 fp Mult/Div (4/12 Cycles)
Spec. Buffer Size	Fully Associative 2KB (1 Cycle)
Bandwidth for Fetch,In-order Issue 4 Instructions	
and Commit Pipeline Stages	Fetch/Issue/Ex/WB/Commit
Architectural Registers	32 int and 32 fp
L1-Cache(Multiversioned)	4-Way Associative 64KB (32B/Block)
	Hit Latency 2
	LRU Replacement
L2-Cache	4-Way Associative 2MB (64B/block)
	5 hit latency, 80 cycles(miss)
	LRU replacement
Spawn Overhead	5 Cycles
Validation Overhead	15 Cycles
Local Register	1 Cycle
Commit Overhead	5 Cycles
k	5
Similarity Threshold	0.5

AdapTPA makes use of one leave-one-out cross-validation method to perform its results' evaluation. It means that the program which is to be partitioned is firstly moved from training set, and based on the left programs a prediction model is built. The method has an advantage that the prediction model never sees the programs to be partitioned before. The partition schemes for the left programs are built by applying the prediction model. Every program is performed with this process in turn.

The paper uses multi-version caches to solve memory dependence and uses register files to solve register data dependence.

5.2   Experimental Configuration

In order to show the effectiveness of AdapTPA, this paper makes a comparison between AdapTPA and HR-based thread partition. Olden benchmarks [17] which have complex data dependence and control dependence among basic blocks, are selected as the inputting programs. When we analyze the experimental results, we only compare the performance of the original HR-based thread partition approach and AdapTPA, and then we will analyze the experimental results, in which we only select several program analysis in the Olden benchmarks.

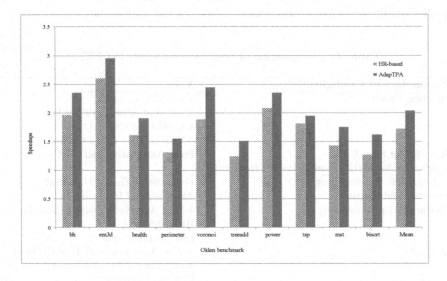

Fig. 7. Comparison diagram of speedups for olden benchmarks

The main data structure in program *bh* is a heterogeneous *octree*, which has very complex data dependence. Its parallelisms exist in and out of loop structures. For the heuristic rules, the same partition scheme is used to partition all the procedures in the *bh* program, and for the AdapTPA, the optimal partition scheme matching with the characteristic of every procedure in the program can be selected, and then the partition scheme is applied to the threads. However, due to the existence of more dependence, AdapTPA gains 19.54% performance improvement.

The main data structure of the program *em3d* is a single linked list, in which the loop structure occupies most of the total, and all the parallelism of program *em3d* comes mainly from the loop structure. Although AdapTPA can obtain the

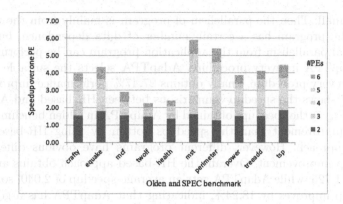

Fig. 8. Speedup comparison diagram of olden and SPEC2000 benchmarks over different PEs

partition scheme suitable for its own characteristics, the characteristic extraction of the loop is not enough. Finally, compared with the HR-based partition approach, 13.17% performance improvement is achieved.

The main data structure of the program *health* is a two-way linked list, which contains both loop and nonloop structure. In *health*, the loop structure is the main source of parallelism, and compared with the HR-based partition approach, you can obtain the partition scheme of *health* suitable for its characteristics. During the partition of loop partition, although the loops occupy most of the program, it has a large loop body and simple data dependence, so *health* gets 18.27% speedup improvement.

The main data structure of program *perimeter* is four fork tree, the program primarily contains loop structure, rather than nonloop structure. The parallelism of program mainly comes from the decomposition of function into multithreading. Because it is difficult to predict the return value of the function, the acceleration effect of these two approaches are not good. Compared with HR-based partition approach, AdapTPA selects the suitable partition scheme in line with its own characteristics, and the partition scheme is not affected by loops. The assessment models adopted by nonloops are used to find the better thread partition boundary for the current program, so the final execution performance improves 18.23%.

The main data structure of program *treeadd* is two fork tree, which is a simple program structure. In this structure, only four procedures are included, and the program does not contain any loop structure, so the parallelism comes from the nonloops. AdapTPA can select the appropriate partition scheme for every procedure, but there are many recursive function calls and data dependence in *treeadd*, and finally the program achieves 21.19% performance improvement.

The main data structure of the program *bisort* is two fork tree. Through the analysis of the source code, we can see that there are only three loops in the program, and only two loops are executed, and the granularity of the loop is relatively small. Then the parallelism of program is mainly from the nonloops, although the program has a certain number of data dependence, but mining the potential parallelism from the application program can be performed based on the AdapTPA in every procedure. AdapTPA selects the suitable partition scheme for every procedure, finally obtains 27.47% performance improvement.

Figure 7 shows the speedup comparisons between HR-based and AdapTPA. Seen from Fig. 7, the speedups obtained by AdapTPA in Olden benchmarks have a certain improvement than the speedups gotten by using HR-based thread partition approach. However, different programs have obvious differences in the speedup improvement. Overall, the HR-based approach obtains an average speedup of 1.725, while AdapTPA gets an average speedup of 2.040, so the average speedup improves by 18.24%, indicating that AdapTPA has a good effect on the program partition. Figure 8 shows the speedups of some SPEC2000 and Olden benchmarks on different number of cores.

6   Conclusion and Future Work

Based on the Prophet system, this paper proposes an Adaptive Thread Partition Approach (AdapTPA), and brings an adaption mechanism into thread partition. According to programs' characteristics, the complexity of unknown program is calculated, candidate thread partition scheme set is built, and the most suitable partition scheme is selected in accordance with programs' characteristics and running context. Finally, the program is executed on the Prophet simulator to verify its execution performance. Thread Level Speculation has been evolving many years, showing great advantages in making use of multicore resources. AdapTPA is proposed to handle the issue that conventional partitioning approaches can not generate the best partitioning scheme for unknown programs. The trend of adaptive thread partition falls on two parts: 1. more detailed candidate thread partitioning schemes are designed; 2. adaptive thread partition is implemented on hardwares.

Acknowledgement. We thank all members of Henan Joint International Research Laboratory of Cyberspace Security Applications for their great support, and give our best hope to them for their collaboration. We also thank reviewers for their careful comments and suggestions. The work was sponsored by National Natural Science Foundation of China Grant No. 61972133, Project of Leading Talents in Science and Technology Innovation for Thousands of People Plan in Henan Province Grant No. 204200510021, Henan Province Key Scientific and Technological Projects Grant No. 192102210130 and No. 202102210162, and Key Scientific Research Projects of Henan Province Universities Grant No. 19B520008.

References

1. Estebanez, A., Llanos, D.R., Gonzalez-Escribano, A.: A survey on thread-level speculation techniques. ACM Comput. Surv. (CSUR) **49**(2), 22 (2016)

2. Hammacher, C., Streit, K., Zeller, A., Hack, S.: Thread-level speculation with kernel support (2016)
3. Yu-Xiang, L.I., Zhao, Y.L., Liu, B., Shuo, J.I.: Optimization of thread partitioning parameters in speculative multithreading based on artificial immune algorithm. Front. Inf. Technol. Electron. Eng. **16**(3), 205–216 (2015)
4. Madriles, C., et al.: Mitosis: a speculative multithreaded processor based on pre-computation slices. IEEE Trans. Parallel Distrib. Syst. **19**(7), 914–925 (2008)
5. Li, Y., Zhao, Y., Wu, Q.: GbA: a graph-based thread partition approach in speculative multithreading. Concurrency Comput. Practice Experience **29**(21), e4294 (2017)
6. Qiu, M., Sha, E.H.M.: Cost minimization while satisfying hard/soft timing constraints for heterogeneous embedded systems. ACM Trans. Des. Autom. Electron. Syst. **14**(2), 25 (2009)
7. Qiu, M., Dai, W., Vasilakos, A.V.: Loop parallelism maximization for multimedia data processing in mobile vehicular clouds. IEEE Trans. Cloud Comput. **7**(1), 250–258 (2019)
8. Qiu, H., Noura, H., Qiu, M., Ming, Z., Memmi, G.: A user-centric data protection method for cloud storage based on invertible DWT. IEEE Trans. Cloud Comput. 1 (2019)
9. Li, J., Ming, Z., Qiu, M., Quan, G., Qin, X., Chen, T.: Resource allocation robustness in multi-core embedded systems with inaccurate information. J. Syst. Archit. **57**(9), 840–849 (2011)
10. Qiu, M., Chen, Z., Niu, J., Zong, Z., Quan, G., Qin, X., Yang, L.T.: Data allocation for hybrid memory with genetic algorithm. IEEE Trans. Emerg. Topics Comput. **3**(4), 544–555 (2015)
11. Monsifrot, A., Bodin, F., Quiniou, R.: A machine learning approach to automatic production of compiler heuristics. In: Scott, D. (ed.) AIMSA 2002. LNCS (LNAI), vol. 2443, pp. 41–50. Springer, Heidelberg (2002). https://doi.org/10.1007/3-540-46148-5_5
12. Olden, B.: benchmark suite v (2010)
13. Li, Y., Zhao, Y., Sun, L., Shen, M.: A hybrid sample generation approach in speculative multithreading. J. Supercomput. **75**(8), 4193–4225 (2017). https://doi.org/10.1007/s11227-017-2118-3
14. Wilson, R.P., et al.: SUIF: an infrastructure for research on parallelizing and optimizing compilers. ACM Sigplan Notices **29**(12), 31–37 (1994)
15. Rogers, A., Carlisle, M.C., Reppy, J.H., Hendren, L.J.: Supporting dynamic data structures on distributed-memory machines. ACM Trans. Program. Lang. Syst. (TOPLAS) **17**(2), 233–263 (1995)
16. Prabhu, M.K., Olukotun, K.: Exposing speculative thread parallelism in spec2000. In: Proceedings of the Tenth ACM SIGPLAN Symposium on Principles and Practice of Parallel Programming, pp. 142–152. ACM (2005)
17. Carlisle, M.C.: Olden: parallelizing programs with dynamic data structures on distributed-memory machines. Ph.D. dissertation, Princeton University (1996)

PMC-Based Dynamic Adaptive CPU and DRAM Power Modeling

Yunfang Zhang[✉] [iD], Yong Dong, Juan Chen, Zhixin Ou, and Yuan Yuan

National University of Defense Technology, Changsha, Hunan, China
zhangyunfang1995@163.com
https://www.nudt.edu.cn/

Abstract. The problem of high power consumption has become one of the main obstacles that affect the reliability, stability, and performance of high-performance computers. How to get the power of CPU and memory instantaneously and accurately is an important basis for evaluating their power's optimization methods. At present, much work has been done to model CPU and memory power using the performance monitoring counter (PMC). Most of these models are static, which fit and estimate the power of the corresponding CPU or memory by collecting and counting key performance monitoring events. However, when the performance behavior of the application changes dramatically with time, the accuracy of the real-time power measurement values will decline, because the performance monitoring values used in the power model can not fit the power values well in a long time. In order to solve this problem, we first analyze the changing features of application performance indicators when CPU or memory power changes, especially the correlation between PMC events and CPU and memory power, and then propose a dynamic adaptive power modeling method (DAPM) based on PMC events using dynamic adaptive technology, which is used for real-time power measurement of CPU and memory. The DAPM can realize the adaptive selection/matching of the model by introducing the power measurement data at the node level, and enhance the real-time power measurement accuracy by dynamically expanding the model library. Besides, the running cost of the DAPM is low. Compared with other PMC power models, DAPM can achieve lower CPU and DRAM power error rates. The error rates of three conventional PMC power models are Isci's model 7%(CPU), Singh's 7.2%(CPU), and Bircher's 6.7%(CPU) and 8.8%(DRAM), while the CPU error rate of DAPM is less than 2%, and the DRAM error rate is less than 5.5%.

Keywords: PMC · CPU power · DRAM power · Adaptive · Dynamic modeling · Power estimation

1 Introduction

Power consumption has become a critical bottleneck to improve the computer's performance and an essential factor to affect the reliability and stability of

© Springer Nature Switzerland AG 2020
M. Qiu (Ed.): ICA3PP 2020, LNCS 12452, pp. 92–111, 2020.
https://doi.org/10.1007/978-3-030-60245-1_7

the system. CPU and memory are important components in computing nodes. There are a lot of power optimization methods for CPU and memory [5, 8–10, 17, 18, 21, 24, 28, 39]. Power has become an important factor that affects the effect of CPU and memory power optimization to provide a high-precision and low overhead power measurement method. However, it is difficult to get the real-time power of CPU and memory accurately. The existing power measurement methods based on hardware and software models have problems. These problems lead to the high cost and low accuracy of the current power measurement technology, which affects the optimization method based on power measurement. Our work is to solve these problems by introducing a dynamic adaptive power modeling (DAPM) method.

At present, there are many hardware-based power measurement methods. Although these methods can obtain very accurate real-time power value, the deployment of power measurement instruments is complicated and sometimes limited by the actual system conditions. In general, the power monitor can be deployed on the mainboard to measure the current and voltage related to CPU and memory components, to obtain the real-time power value. When the monitor is deployed on the mainboard, a unique circuit design is often needed for the mainboard, which costs a lot. At the same time, it is difficult to use this method to deploy power monitoring for the running system.

Due to the complexity of the actual deployment of power measurement equipment in the system, some CPU manufacturers choose to provide users with some tools to help them obtain the powers of CPU and memory in real-time. It mainly uses some built-in power measurement interfaces to provide users with standardized interfaces to obtain power values, such as Intel's Running Average Power Limit (RAPL) [14], AMD's power management system [1], and so on. Although the real-time power measurement accuracy of such tools is often high that RAPL can get a power value every 1ms, but there is a big problem of running overhead. Moreover, the corresponding power management system is only applicable to the specified processor model. For example, RAPL can only be used for Intel's processor.

A more general approach is to model CPU and memory power. It is widely used to build the power model of CPU and memory based on PMC events. The real-time power value is obtained through the power model. At present, the common power models established by using performance monitoring counter (PMC) mainly include linear models [2, 3, 6, 7, 11, 16, 20, 22, 27], nonlinear models [4, 13, 30], artificial neural network algorithm models [31, 33, 34] et al. Once these models are established, the parameters are fixed without updating and adjusting, so the accuracy of the model depends on the data samples used to build the model to a large extent, which brings challenges to the accurate power measurement of applications with substantial changes in features.

The diversity of program behavior and the big fluctuation of real-time changes often significantly impact the accuracy of power measurement. Figure 1 gives a typical example. This example uses Isci's model [16] to measure the power of the twolf program in the SPEC test set. Figure 1(a) is the result of literature [16]. Figure 1(b) is on the platform Intel Xeon 2660 E5. The results of the above two results show that the program's behavior changes much, and the power model is difficult to capture the large fluctuation of the performance value accurately, so the accuracy of the power model is not high. In terms of the measurement accuracy of several more power models, the measurement errors of CPU power modeling obtained in [4,16,30] are all about 6% to 7%, and the measurement error of memory power obtained in the [4] is larger, at 8%. The measurement error of CPU power from 6% to 7% is sometimes catastrophic for evaluating optimization methods.

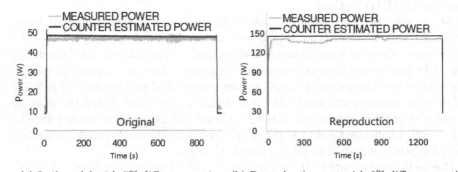

(a) Isci' model with 7% difference ratio   (b) Reproduction test with 6% difference ratio

Fig. 1. The power of twolf program in the reproduction of Isci's model

How to make our real-time power adapt to the diversity of program behavior and automatically match the large floating changes of program behavior in real-time are the problems to be solved to improve the power measurement accuracy. Dynamic adaptive technology is one of the effective methods to solve this problem. Dynamic adaptive technology has been widely used in many similar scenes. For example, Widyawan et al. [35] proposed a dynamic template adaptive matching algorithm to improve the accuracy of reference picture motion detection. Khavari et al. [19] similar proposed a real-time dynamic selection data compression algorithm to deal with the diversity of communication data modes between GPUs. Shiyou Qian et al. [26] similar proposed an adaptive dynamic script matching algorithm to improve the availability and efficiency of content-based publish/subscribe system. The above examples provide a way to solve the problem of power measurement accuracy by using dynamic adaptive technology. There is little research on the application of dynamic adaptive technology in power modeling and real-time monitoring.

We apply dynamic adaptive technology to improve the accuracy of power measurement. The basic idea is as follows: the diversity of program behavior is reflected in the program is divided into calculation-intensive and memory intensive. In the calculation intensive program, it is often necessary to consider the impact of different operation instruction types on the power of CPU and memory. However, in memory-intensive programs, the impact of cache failure on power is far more significant than that of CPU instruction type difference. By establishing the single models to distinguish different program types, it can better adapt to the diversity of program behavior changes and improve the accuracy of power measurement in the local range. Furthermore, the dynamic adaptive method is used to achieve the automatic matching of the single models with different program features, and improve the overall power measurement accuracy. Based on the above basic ideas, the main steps of our method are: First, to establish a large number of single models (mainly linear models with fewer parameters), which are applied to single programs' behavior features. Second, we judge the behavior features of the current program regularly in running the program and select the power model suitable for the features of the current program from many single models that have been established.

There are many difficulties in the application of the dynamic adaptive algorithm in power modeling. First, the program's features are complex and diverse, and there is no unified standard for its detailed classification. Any new program may introduce a new program feature. Therefore, we do not have a fixed set of models to choose from but need to build and expand the optional model library dynamically. Second, there are no clear indicators of program features for comparison and matching. It is difficult to judge whether a model is suitable for the current program features without comparing the actual power measurement results. Finally, compared with the static power model, the dynamic adaptive model will increase the system's overhead to a certain extent. We need to reduce the cost of the method to an acceptable level.

Because of the above difficulties, we propose an adaptive dynamic selection method to achieve high-precision power measurement. The main contributions of this paper are as follows:

- In this paper, we propose an adaptive dynamic power modeling method and build a dynamically expandable local model library which makes the dynamic model can contain a large number of local models suitable for different types of program features.
- We realized the accurate matching of program behavior features and local model by introducing node-level power data, and prove that the accuracy of dynamic selection model is high, the average error of CPU power in HPCC test set is within 2%, and the average error of memory power is within 5.5% through experiments.
- We analyzed the overhead of the dynamic selection power model by experiment, and it is proved that the online operation of this method has little effect on the computer's performance.

2   Related Work

As early as 2003, Canturk Isci and Margaret Martonosi [16] proposed a power model based on performance counters. They take Intel Pentium 4 CPU as the experimental object and subdivide it into 22 more fine-grained components, such as bus controller, all the caches, computing units, etc. They use more than ten different PMCs to build a power model for each component and get the total CPU power by adding up the output of all these models. In their paper, they claim that the average error of the model is 3.5W. Considering the typical power of the CPU in the experimental results is about 50W, the relative error is 7%. Isci's work only involves CPU power modeling, no memory, and other components. Rivoire et al. [29] analyzed the trade-off between the complexity and accuracy of PMCs based power models, as well as the limitations of various models. Their experimental results show that the average error of the general power model is about 5%. Karan Singh et al. [30] established a segmented CPU power model with four kinds of PMCs. When the PMC values are small, they replace the model's parameters with their logarithms and give a set of different coefficients. The error of this model is 7.2% under SPEC 2006 test set. Michael D. Powell et al. [25] established a Common Activity-based Model for Power (CAMP) model through eleven PMCs, with the final average error of 8%.

All of the above models are limited to the estimation of CPU power, while the research and modeling of memory and other components are relatively few. Bircher and John [4] built a power model for the whole computer system in their work. They divided the computer into six parts, including CPU and memory, and built a model for each part. They also use a dozen different PMCs. The average error of the model is 6.67% for CPU and 8.80% for memory. Except for this, we did not find any model to calculate the memory power directly. Some other models analyze the power of memory, but only use it as a part of a higher-level power monitoring system, and do not get the fine-grained power value. For example, in the model established by Robert Basmadjian et al. [2] for servers in the data center, power models have been established for CPU, memory, disk, and even fan in different types of servers. These models are quite rough and are used as part of the whole server power model, and they cannot accurately estimate component level power.

Also, many power models are not based on PMCs. These models are often used in large-scale computing centers, data centers, cloud servers, and others, which are similar to the application scenarios in this paper. Guo Ming Tang et al. [32] proposed a method to obtain the power of each server by decomposing the total power of the data center based on the state values of each server, such as CPU utilization, memory utilization, and other information. The claimed error is within 5%. However, their power decomposition granularity is only at the server level, not reach the component level. Besides, in their Blue Gene computer system, IBM acquires power through the deployment of hardware monitoring [40]. These methods have high costs and limited applications.

3   Methodology

3.1   Overview

In this paper, a PMC based, dynamic adaptive power modeling method (DAPM) is proposed to measure the power of CPU and memory. The basic idea of DAPM is to automatically select a local model that adapts to the current program features in the process of the program running.

As shown in Fig. 2(a), the DAPM method mainly includes the following four functional modules:

– *local model library*. It is used to store all alternative local models. The local model library will update dynamically with the running of the system.
– Model *Constructor*. It is used to initialize the local model library, build a new local model, and update the local model library.
– Automatic adaptive power model *Matcher*. It is used to determine whether the current program features match the current model and match a new model in the local model library if it has to be replaced.
– Hardware performance and power *Collector*. It is used to collect PMC and power and provide data for model matcher and model builder.

The flow diagram of DAPM is shown in Fig. 2(b). The specific steps are as follows:

– **Step 1**. When the system is running, *Collector* collects PMC and node power of the system every Δt. The current local model calculates CPU and memory power.
– **Step 2**. At intervals of ΔT, judge whether the current model still matches the current program features. If not, go to Step 3. Otherwise, go back to Step 1.
– **Step 3**. Select the model that matches the features of the current program in *local model library*. If a matching model is found, it is set as the current power model and go back to Step 1. Otherwise, go to step 4.
– **Step 4**. Send program samples that cannot match the model to the model *Constructor*, build a new model, and add it to the *local model library*. Set the new model as the current model and go back to Step 1.

Two issues need to be addressed:

– Q1: How to build local power models and what kind of models they are, and ensure that the local models to achieve sufficient accuracy.
– Q2: How to accurately select and match the local models when the system runs in real-time. This is the key issue.

Because of the above problems, the specific solutions and considerations in DAPM are as follows: We use the linear models with fewer parameters (no more than 6) as the local models. Among them, the parameters of the models are all from the PMCs provided by the current system. We will filter the PMCs with

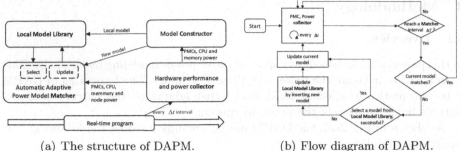

(a) The structure of DAPM.   (b) Flow diagram of DAPM.

Fig. 2. The structure and process of DAPM.

the strongest correlation with the program's power behavior to build the model. Because the same PMC has different effects on different program features, each local model is built with different PMCs. Specific modeling methods will be described in Sect. 3.2. The way to filter PMCs is described in Sect. 3.4.

3.2 Build Local Models

– **Step 1**. Select the appropriate PMCs. The standard of PMC selection is that they have strong linear correlation with the power of each components of the computer, and there should be linear independence between these PMCs. The specific selection method of PMC is described in Sect. 3.3. For CPU and memory, we select several different PMCs, which are recorded as $c_{i,1}, c_{i,2}, ..., c_{i,n_i}$, where n_i is the number of PMCs selected for the ith component. When i is 0, it represents the whole node; when it is 1, it represents CPU; when it is 2, it represents memory.
– **Step 2**. Run the sample program. During the operation of the program, the CPU, memory and node power values and all PMCs selected in the previous step are collected at the interval of time Δt. The number of times of data acquisition is recorded as t, and the vector composed of t values of the sth PMC of the ith component is recorded as Cs, i, and the power is recorded as P_i. The node power vector is recorded as P_0.
– **Step 3**. We use the classical least square method [36] to calculate the coefficient vector that used for calculate the ith component power by PMCs, which is recorded as W_i:

$$W_i = (X_i^T X_i)^{-1} X_i^T P_i \tag{1}$$

where $X_i = [1, C_{0,i}, C_{1,i}, C_{2,i}]$, and 1 is the constant term vector, corresponding to constant coefficient in the model. When $i = 0$, it is for the node power model, and X is the matrix of all PMCs. W_i is the $n_i + 1$-dimensional coefficient vector.
– **Step 4**. From the coefficient vector obtained in the previous step, we can establish a linear model for CPU, memory and node power respectively:

$$P_i^{model} = X_i^T W_i \tag{2}$$

where X_i is the vector of the current PMC values.

3.3 Select PMC

The selection of PMCs is crucial for the establishment of the model. For each local linear model, the higher linear correlation between PMCs and power means a single model's higher accuracy. At the same time, the PMCs should be relatively independent. Otherwise, the two PMCs with strong correlation contain similar information, which is not conducive to improving the accuracy of the model but will increase the cost of calculation. In view of these considerations, we choose PMCs as follows:

- **Step 1**. Run the sample program on the modeling node to collect all available PMC, CPU, memory, and node power provided by the platform.
- **Step 2**. Calculate the absolute value of the Pearson correlation coefficient [37] of all PMCs with CPU and memory power, respectively. Select all PMCs with a correlation coefficient greater than 0.5 as the reserved PMCs for CPU and memory.
- **Step 3**. Calculate the correlation coefficient between every two reserved PMCs. For any two PMCs, whose absolute correlation coefficient is greater than 0.5, delete the PMC whose correlation coefficient with CPU or memory is smaller than another.

Finally, all the remaining PMCs are used as power modeling parameters. Among them, the PMCs used to model node power is the union of the PMCs used to model CPU and memory. We will verify and analyze the effectiveness and availability of PMCs selected dynamically by this method in Sect. 4.3.

3.4 Dynamic Model Matching and Updating

We introduce node-level power data to match models for different programs. In model matching, the node power model results in each local model are compared with the measured node power to determine whether the model is suitable for the current program. The specific steps are as follows:

- **Step 1**. PMC data is collected by *Collector* once per interval Δt. The CPU, memory and node power are calculated by the current local power model.
- **Step 2**. Note that the actual measured node power is P_0. The error $MRE = |P_0^{model} - P_0|/P_0$ between the measured P_0 and the calculated P_0^{model}. Set the threshold value MRE_{std}. If $MRE \leq MRE_{std}$, and the program has not run to the given interval ΔT since the last adaptive selection of the model, then the CPU and memory power obtained in Step 1 is the measurement result and returns to Step 1. Otherwise, go to Step 3.
- **Step 3**. The average error of each model is calculated by the following formula:

$$MRE_k = \frac{1}{t} \sum_{j=1}^{t} \frac{|P_{k,0,j}^{model} - P_{k,0,j}|}{P_{0,j}} \tag{3}$$

where k is the serial number of each model in the *local model library*. $t = \Delta T/\Delta t$ is the number of collected samples since the last adaptive model selection. $P_{k,0,j}^{model}$ is the node power of the jth time point calculated by the kth model. $P_{0,j}$ is the measured node power at the jth time point.

- **Step 4**. Compare the sizes of all MRE_k, and select the kth model with minimized MRE_k. That is to say, the current computer program is the most similar to the kth program. If $MRE_k \leq MRE_{std}$, select the kth model as the currently applicable model, and calculate the CPU and memory power from it, and return to Step 1. Otherwise, it will be considered that there is no suitable model in the *local model library*, and go to Step 5.
- **Step 5**. The *Constructor* runs the currently running program, builds a new local model, and adds it to the local model library. The method of building a new model is described in Sect. 3.2. Set the current applicable model as the new model. Go back to Step 1.

3.5   Utilization of DAPM

In this section, we introduce how to use the DAPM to predict CPU and DRAM's power. Firstly, we add an additional node in the cluster that needs to measure CPU and memory power, and its hardware is the same as other nodes in the cluster. Secondly, we deploy tools to measure CPU and memory power consumption on this node, such as Baseboard Management Controller (BMC) [41]. Thirdly, we select PMCs by following the steps introduced in Sect. 3.3. Fourthly, we build an *initial local model library* by following the steps introduced in Sect. 3.2. The above operations only need to be done once. When the system runs, we collect the selected PMCs and estimate the power of the CPU and DRAM by formula 2. At regular intervals, we check the accuracy of the current local model, change it with another, or build a new local model if it is not suitable for the current system state. The details have been introduced in Sect. 3.1.

DAPM has three main advantages: First, it avoids deploying power monitoring tools (like BMC) for each node in the cluster. Second, we do not need to analyze their architectures for different CPUs to build complex power models. Instead, we build many simple linear models automatically and dynamically. Third, the accuracy of DAPM will be better than traditional static power models. We will verify this in the next section.

4   Experiment

In this section, we use DAPM to measure the actual power. Section 4.1 describes the experimental platform and the test workloads used. Section 4.2 shows the measurement results and accuracy of the DAPM method. In Sect. 4.3, we carried out several groups of comparative experiments to verify each specific step of DAPM. Section 4.5 evaluates the cost of DAPM.

4.1 Platform and Configuration

We experimented with DAPM on two servers with dual Intel Xeon E5 2660 CPUs on each. Each CPU has 10 cores with 2.6 GHz main frequency and 3.3 GHz super frequency. Its summit power is 105 W [15]. Each server uses a 128G DDR4 memory. The servers run CentOS 7.4 operating system, and the Linux kernel version is 3.10.0. We use DAPM to measure the CPU and memory powers of one of the servers. The *local model library* and adaptive power model *Matcher* in DAPM also run on this node. Another server is used to deploy the model *Constructor* in the DAPM framework. Both servers use `perf` [38] to collect PMC data. We use the running average power limit (RAPL) [14] tool provided by Intel CPU to measure the real-time powers of CPU and DRAM noted as P_{real}. We take the average relative error (MRE) between the real-time power of the server to be tested and the measured power of DAPM P_{DAPM} as the index to evaluate the effect of DAPM, where:

$$MRE = \sum_{j=1}^{t} \frac{|P_{real} - P_{DAPM}|}{P_{real}} \qquad (4)$$

The RAPL data on another server is used for the modeling input of the model builder (P_i) in the formula (1).

In the experiment, we used 5 representative programs in the HPCC test set and 29 basic programs in the SPEC 2006 test set. Among them, HPCC is a test set specially designed to test the performance of high-performance computer systems [23]. We use the following 5 programs in HPCC: `RandomAccess`, `PTRANS`, `DGEMM`, `STREAM`, and `FFT`.

SPEC [12] is a commonly used computer performance test set. Its basic set includes 12 integer and 19 floating-point operation test programs. We use the SPEC programs to build the initial *local model library*, and use HPCC to test the effect of DAPM.

We use GCC v4.8.2 to compile all the DAPM code and optimize it with the "-O3" option. We set the relevant parameters in DAPM described in Sect. 3.1 as follows:

- Time interval for dynamic matching of model ΔT. The frequency of regular automatic adjustment does not need to be too high. We set this parameter to 60 s.
- Sampling interval Δt. On the premise that the result is accurate enough, and on the other hand, to make sure that reading PMC itself will not generate much overhead, we set this parameter to 1 s.
- The node power error threshold MRE_{std}. It is used for determining whether the model needs to be rematched. We set the standard at 50%.

4.2 Overall Results

The SPEC programs are used to build the initial *local model library*, and the HPCC programs are used as the test program. The specific process of building the *local model library* will be detailed in Sect. 4.3. We run five HPCC programs for two loops. During the first loop, the power measurement effect of DAPM on programs that have never run on the system is checked. During the second loop, the effect of DAPM on the programs that have been run on the system is checked.

The results are shown in Fig. 3. When the five programs run for the first time (0 s, 280 s, 470 s, 1480 s, 1710 s), there will be a significant deviation between the power measured by DAPM and the real-time power (the blue circles in the figure). That is due to the change of program features, where the original model is no longer suitable for the new program feature, and the process of reselecting the model has a short lag. After that, in about 60s-90s, the deviation decreased, but still existed (the red circles). That is because there is no model in the current *local model library* suitable for the features of the current program. The model *Constructor* is now building a new local model. After the new model is built and added to the *local model library*, the measurement results of DAPM are consistent with the real-time power.

When the five programs run for the second time (1900 s, 2200 s, 2370 s, 3380 s, 3620 s), the waveform of DAPM still has a significant deviation instantaneous value due to program feature conversion. However, the measured results of DAPM are consistent with real-time power. There is no longer a deviation of about 60s because the models that meet the features of the programs have been added to the *local model library*. Hence, there is no need to model again. In the whole process, the average error of CPU power is 1.74%, and the average error of memory power is 5.16%.

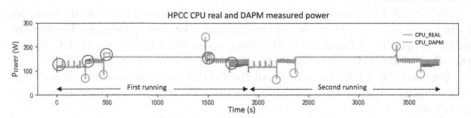

(a) The modeling result of CPU power.

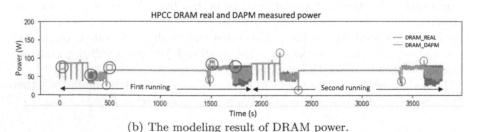

(b) The modeling result of DRAM power.

Fig. 3. The result of DAPM of HPCC programs. CPU_REAL refers to the power measured by RAPL.

In order to further test the effectiveness of each step in DAPM, we test the effect of the method of building local model in Sect. 4.3. In Sect. 4.4, by setting different matching model intervals ΔT, the most optimal ΔT that makes the method achieve the best effect is verified. Then, the initial *local model library* is expanded to avoid the situation where DAPM needs to update the *local model library* in the running process, to minimize the error in the running process of the whole program. The accuracy in this ideal situation will be compared with that in the general situation where the *local model library* needs to be updated sometimes in this section.

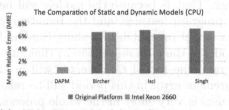
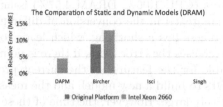

(a) CPU Power Modeling Comparison between DAPM and other models.

(b) DRAM Power Modeling Comparison between DAPM and other models.

Fig. 4. The comparison of traditional static models and DAPM. Note that Isci [16] and Singh [30] didn't give DRAM model.

4.3 Local Model Effect Examination (Q1)

In this section, we accurately describe the construction process of the local model and test the effect of the local model. After running all SPEC programs continuously on the model *Constructor*, we segment all PMCs and power data collected in ΔT(60 s) intervals. Each segment of data is used as input to build a local model. Each local model includes a node, CPU, and memory power models. All SPEC programs run for about 600 min. Therefore, we get 600 local models as our initial *local model library*.

First, we test the reliability of the PMCs selected by our method. In building the models, we found that the following PMCs are the most commonly selected that they account for 95% of all local model selection parameters:

PMCs that are most commonly used as CPU power model parameters: uncore_imc_1/cas_count_read, uncore_imc_1/clockticks, L1-dcache-load -misses, cpu-cycles. PMCs that are most commonly used as DRAM power model parameters: branch-loads, cache-misses.

We analyzed the linear correlation coefficients between the six PMCs mentioned above and CPU, memory powers, and PMCs themselves. The results are shown in Tables 1 and 2. The tables show that the absolute value of the correlation coefficients between PMCs, CPU, and memory powers are all greater than

0.5, which means strong correlation, and the correlation coefficients between PMCs are all below 0.5.

To verify our local models' validity, we compare the error between the output of each local model and the measured power. These local models include the models built by SPEC programs in the initial *local model library*, and the models updated dynamically during the real-time running of HPCC. As shown in Fig. 5, the minimum error of local models can reach 0.32%, and the maximum error is not exceeding 0.8%, which shows that our local models can achieve high accuracy.

Compared with the accuracy of the local models (average error of CPU power 0.53%, DRAM 0.8%), the average error in the dynamic power monitoring process (CPU 1.06%, DRAM 4.57%) is larger. That is because the models will be rematched in the conversion of different program features, and the rematching process has a short lag, which leads to a model mismatch in a short time and increases the error. Also, if there is no model in the *local model library* that suitable for the features of the current program, the model *Constructor* will need 60 s to build a new model, and the measurement error in this period will be relatively large. However, the time of these cases is very short in the whole process, and the models matched at other times are correct. Therefore, the method for model building and matching is effective.

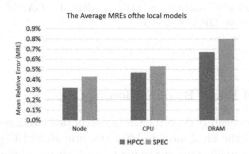

Fig. 5. From this figure, we can see that the MREs of local modes are all smaller than 1%.

Table 1. Pearson coefficients between selected PMCs and CPU, memory powers.

Component	PMC	Linear coefficient
CPU	cas_count_read1	0.77
	clockticks1	−0.65
	L1-dcache-load-misses	0.80
	cpu-cycles	−0.55
MEM	branch-loads	−0.91
	cache-misses	0.58

Table 2. Pearson coefficients between every 2 PMCs.

PMC	cr	ctk	Ldlm	ccs	bl	cm
cas_count_read1	1.00	−0.25	0.31	−0.42	−0.44	−0.42
clockticks1	−0.25	1.00	−0.46	0.34	−0.44	−0.44
L1-dcache-load-misses	0.31	−0.46	1.00	−0.43	0.34	−0.44
cpu-cycles	−0.42	0.34	−0.43	1.00	−0.54	−0.44
branch-loads	−0.44	−0.44	0.34	−0.54	1.00	−0.45
cache-misses	−0.42	−0.44	−0.44	−0.44	−0.45	1.00

4.4   Dynamic Model Matching Effect Examination (Q2)

By adjusting ΔT to different sizes, we carried out several experiments to verify
the best value of ΔT that makes the dynamic matching effect of DAPM opti-
mal. We set $\Delta T = 10\,\text{s}$, $\Delta T = 30\,\text{s}$, and $\Delta T = 120\,\text{s}$, respectively. The other
experimental conditions are the same as Sect. 4.2, and the running results are
shown in Fig. 6.

As can be seen from Fig. 6, when ΔT is small (10 s and 30 s, (a) (b) (c) (d)
in Fig. 6), there is a lot of jitter in DAPM measurement results, mainly concentrated in RandomAccess (0–280 s, 1900–2180 s), PTRANS (280–470 s, 2180–
2450 s) and STREAM (1480–1720 s, 3380–3620 s) programs operation interval.
That is because ΔT is too short, making the information used for local model
matching and modeling too little to cover the current program features, so
DAPM mistakenly believes that the program features are changing all the time,
so it constantly produces fluctuations.

When the ΔT is too large (120 s, (e) (f) in Fig. 6), compared with the result
that when $\Delta T = 60\,\text{s}$ in Sect. 4.2 (CPU 1.74%, memory 5.16%), the accuracy
decreases (CPU 2.24%, memory 6.50%). Because when the model needs to be
updated, the model *Constructor* needs to take a longer time to construct a new
model. During this time, the measured values are biased, so the overall accuracy
decreases. Therefore, it is reasonable to use 60 s as the interval of model adaptive
matching and updating.

The effect of DAPM is closely related to the construction of the initial *local
model library*. It can add new models to the *local model library* in time through
dynamic adaptive strategy when the initial *local model library* is not very sufficient, and achieve high accuracy of power measurement, which also shows the
adaptability and effectiveness of DAPM method. The example in Fig. 7 illustrates this point. At this time, the initial *local model library* contains enough
benchmark, and the model trained by the HPCC benchmark is already in the
initial *local model library*, which is called *Model**. When the real-time running
program is in HPCC, since the initial *local model library* already contains the
target requirements, the power measurement accuracy will be very high. At this
time, the CPU power measurement error rate is 1.36%, and the error rate of
DRAM power measurement is 4.34%. This value should be considered as the

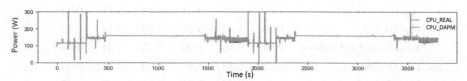

(a) DAPM CPU Power monitoring with $\Delta T = 10s$, Mean Relative Error=3.46%

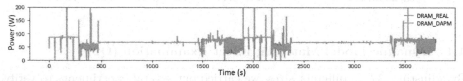

(b) DAPM DRAM Power monitoring with $\Delta T = 10s$, Mean Relative Error=10.23%

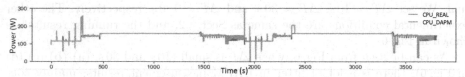

(c) DAPM CPU Power monitoring with $\Delta T = 30s$, Mean Relative Error=2.71%

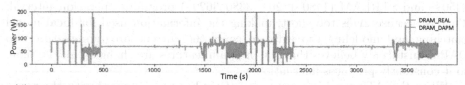

(d) DAPM DRAM Power monitoring with $\Delta T = 30s$, Mean Relative Error=7.08%

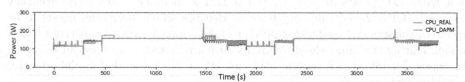

(e) DAPM CPU Power monitoring with $\Delta T = 120s$, Mean Relative Error=2.24%

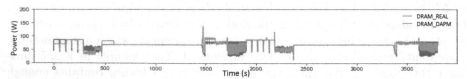

(f) DAPM DRAM Power monitoring with $\Delta T = 120s$, Mean Relative Error=6.50%

Fig. 6. Three groups of DAPM power monitoring results with different ΔTs.

ideal value that can be found. Based on this benchmark, we will compare the CPU and DRAM power measurements with this standard. Assuming that the initial *local model library* established in the previous Sect. 4.2 is called $Model_0$, it can be seen that although $Model_0$ lacks some models to be matched compared with *model**, the dynamic update of the *local model library* by DAPM method can also achieve an ideal average power measurement error rate. At this time, the CPU power measurement error rate is 1.74%, and the DRAM power error rate is 5.16% (Sect. 4.2). It can be seen that the error rate is quite near to that of the ideal initial *local model library* (CPU 1.36%, memory 4.34%).

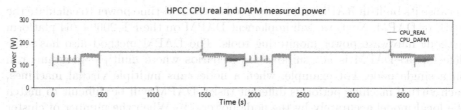

(a) The modeling result of CPU power in the best situation. Its Mean Relative Error is 1.36%

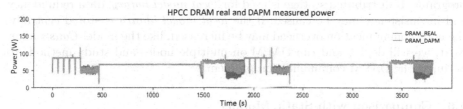

(b) The modeling result of DRAM power in the best situation. Its Mean Relative Error is 4.34%

Fig. 7. The result of DAPM of HPCC programs in the best situation.

4.5 DAPM Cost and Defects Analysis

In order to evaluate the effect of DAPM on node performance, we designed a control experiment. When DAPM is used, and DAPM is not used, five programs of HPCC test set are run on four different nodes (two Intel Xeon 2640 CPU servers and two Intel Xeon 2660 servers) with the same data scale, and their running time is counted. The result is shown in Fig. 8. After joining DAPM, the running time of all programs increased by 0.2% in the worst case, while that of some programs did not increase. The average running time increased by 0.1%. That shows that DAPM has little effect on program performance.

The main application scenario of DAPM is a platform without hardware power monitoring tools. We use the Intel Xeon CPU platform in our experiment

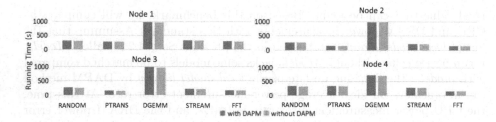

Fig. 8. The running times of HPCC programs on several nodes.

because its built-in RAPL tool can be used as the real-time power to calculate the MRE of DAPM. Next, we will implement DAPM on the FT-2000+/64 platform without hardware power monitoring tools. The DAPM method also has some defects: (a) DAPM is not suitable for scenarios where multiple programs run on a single node. For example, when a node runs multiple virtual machines, each virtual machine performs different tasks, DAPM will be difficult to match the local model accurately by the node power. (b) When the number of cluster nodes increases, the communication between model *Constructor* and each node will increase, which may reduce the performance of communication-intensive programs. If distributed storage is used for *local model library*, data redundancy and inconsistency may be caused. If the *local model library* is stored centrally, the node communication overhead may be increased, like the model *Constructor*. Next, we will deploy and run DAPM on multiple nodes and study methods to optimize the DAPM communication problem.

4.6 Comparison with Static Models

As this paper is the first work of modeling component power by adaptive dynamic method, we compare DAPM with the existed static power models. To get a better comparison, we reimplement three classic power models [4,30] and [16] on our platform. All of these are well-known PMC-based power-modeling works. We have introduced the details of these three models in Sect. 2. Isci et al. [16] used 9 PMCs to build a CPU power model, Singh [30] used 4, and Bircher [4] used 4 to construct two models for CPU and DRAM respectively. We select the corresponding PMCs from the set provided by our system and construct the models the same with these three works. The error rates of three conventional PMC power models are Isci's model 7%(CPU), Singh's 7.2%(CPU), and Bircher's 6.7%(CPU) and 8.8%(DRAM), while the CPU error rate of DAPM is less than 2%, and the DRAM error rate is less than 5.5%. It can be seen that the DAPM improves the accuracy of the static power models significantly.

5   Conclusion

By analyzing the existing CPU and memory power models, we point out the accuracy problems of the static power models in real-time power measurement.

Based on this, a dynamic adaptive power modeling method (DAPM) based on PMC events is proposed to measure the real-time power of CPU and memory by analyzing the features of program performance indicators changing with the power of CPU and memory, as well as the correlation analysis. The whole DAPM framework mainly includes establishing the local model library, dynamic model matching, and selection process. Compared with the traditional static power models, the DAPM proposed reduces the CPU power's measurement error by 7%, memory power by 3%. The error of the CPU power of DAPM range from 1.36% to 1.74% and the error of memory power is less than 5.5%. DAPM only increases the system running cost by 0.1%, which has little impact on the program's running performance.

Acknowledgements. This work is supported in part by the Advanced Research Project of China under grant number 31511010203 and the Research Program of NUDT grant number ZK18-03-10.

References

1. AMD: BIOS and Kernel Developer's Guide (BKDG) for AMD Family 14h Models 00h–0Fh Processors, January 2011
2. Basmadjian, R., Ali, N., Niedermeier, F., de Meer, H., Giuliani, G.: A methodology to predict the power consumption of servers in data centres. In: Proceedings of the 2nd International Conference. e-Energy 2011. ACM, New York, May 2011
3. Bellosa, F.: The benefits of event—driven energy accounting in power-sensitive systems. In: Proceedings of the 9th Workshop on ACM SIGOPS European Workshop (2000)
4. Bircher, W.L., John, L.: Complete system power estimation using processor performance events. IEEE Trans. Comput. **61**(4), 563–577 (2010)
5. Chou, C., Bhuyan, L.N., Wong, D.: μdpm: dynamic power management for the microsecond era. In: 2019 IEEE International Symposium on High Performance Computer Architecture (HPCA), pp. 120–132 (2019)
6. Economou, D., Rivoire, S., Kozyrakis, C., Ranganathan, P.: Full-system power analysis and modeling for server environments. In: Proceedings of the Workshop on Modeling, Benchmarking, and Simulation, pp. 70–77 (2006)
7. Fan, X., Weber, W.D., Barroso, L.A.: Power provisioning for a warehouse-sized computer (2007)
8. Feihao, W., et al.: A holistic energy-efficient approach for a processor-memory system. Tsinghua Sci. Technol. **24**(4), 468–483 (2019)
9. Gholkar, N., Mueller, F., Rountree, B., Marathe, A.: PShifter: Feedback-based dynamic power shifting within HPC jobs for performance. In: Proceedings of the 27th International Symposium on High-Performance Parallel and Distributed Computing. New York, NY, USA (2018)
10. Hanson, H., et al.: Processor-memory power shifting for multi-core systems (2012)
11. Heath, T., Diniz, B., Horizonte, B., Carrera, E.V., Bianchini, R.: Energy conservation in heterogeneous server clusters. In: Proceedings of the 10th ACM SIGPLAN Symposium on Principles and Practice of Parallel Programming (PPoPP 2005), pp. 186–195. ACM (2005)

12. Henning, J.L.: SPEC CPU2006 benchmark descriptions. ACM SIGARCH Comput. Archit. News **34**(4), 1–17 (2006)
13. Wang, H., et al.: Distributed systems meet economics: pricing in the cloud. In: Proceedings of the 2nd USENIX Conference on Hot Topics in Cloud Computing (2010)
14. Intel: Intel 82599 10 GbE Controller Datasheet, November 2019
15. Intel: Intel xeon processor e5–2660 v3 (2020). https://ark.intel.com/content/www/us/en/ark/products/81706/intel-xeon-processor-e5-2660-v3-25m-cache-2-60-ghz.html
16. Isci, C.: Runtime power monitoring in high-end processors: methodology and empirical data. In: Proceedings of International Symposium on Microarchitecture (2003)
17. Juan, C., et al.: Analyzing time-dimension communication characterizations for representative scientific applications on supercomputer systems. Front. Comput. Sci. **13**(6), 1228–1242 (2019)
18. Khatamifard, S.K., Wang, L., Das, A., Kose, S., Karpuzcu, U.R.: Powert channels: a novel class of covert communication exploiting power management vulnerabilities. In: 2019 IEEE International Symposium on High Performance Computer Architecture (HPCA), pp. 291–303 (2019)
19. Khavari Tavana, M., Sun, Y., Bohm Agostini, N., Kaeli, D.: Exploiting adaptive data compression to improve performance and energy-efficiency of compute workloads in multi-GPU systems. In: 2019 IEEE International Parallel and Distributed Processing Symposium (IPDPS), pp. 664–674 (2019)
20. Lee, B.C., Brooks, D.M.: Accurate and efficient regression modeling for microarchitectural performance and power prediction. ACM SIGOPS Oper. Syst. Rev. **40**(5), 185
21. Lefurgy, C., Wang, X., Ware, M.: Power capping: a prelude to power shifting. Cluster Comput. **11**, 183–195 (2008)
22. Lewis, A., Ghosh, S., Tzeng, N.F.: Run-time energy consumption estimation based on workload in server systems. In: Workshop on Power Aware Computing and Systems, HotPower 2008, 7 December 2008, San Diego, CA, USA, Proceedings (2008)
23. Luszczek, P.R., Bailey, D.H., Dongarra, J.J., Kepner, J., Takahashi, D.: S12-the HPC challenge (HPCC) benchmark suite. In: Proceedings of the ACM/IEEE SC2006 Conference on High Performance Networking and Computing, 11–17 November 2006, Tampa, FL, USA (2006)
24. Patel, T., Tiwari, D.: Perq: fair and efficient power management of power-constrained large-scale computing systems. In: Proceedings of the 28th International Symposium on High-Performance Parallel and Distributed Computing. New York, NY, USA (2019)
25. Powell, M.D., Biswas, A., Emer, J.S., Mukherjee, S.S., Yardi, S.M.: CAMP: a technique to estimate per-structure power at run-time using a few simple parameters. In: 15th International Conference on High-Performance Computer Architecture (HPCA), 14–18 February 2009, Raleigh, North Carolina, USA (2009)
26. Qian, S., et al.: Adjusting matching algorithm to adapt to dynamic subscriptions in content-based publish/subscribe systems. In: 2018 IEEE International Conference on Parallel Distributed Processing with Applications (2018)
27. Basmadjian, R., de Meer, H.: Evaluating and modeling power consumption of multi-core processors. In: Proceedings of the 2012 3rd International Conference on Future Energy Systems: Where Energy, Computing and Communication Meet (2012)

28. Raghavendra, R., Ranganathan, P., Talwar, V., Wang, Z., Zhu, X.: No "power" struggles: coordinated multi-level power management for the data center. In: Proceedings of the 13th International Conference on Architectural Support for Programming Languages and Operating Systems. New York, NY, USA (2008)
29. Rivoire, S., Ranganathan, P., Kozyrakis, C.: A comparison of high-level full-system power models. In: Workshop on Power Aware Computing and Systems, HotPower 2008, 7 December 2008, San Diego, CA, USA, Proceedings (2008)
30. Singh, K., Bhadauria, M., McKee, S.A.: Real time power estimation and thread scheduling via performance counters. ACM SIGARCH Comput. Archit. News **37**(2), 46 (2009)
31. Song, S., Su, Chunyi, C., Kirk, W.: A simplified and accurate model of power-performance efficiency on;emergent GPU architectures, pp. 673–686 (2013)
32. Tang, G., Jiang, W., Xu, Z., Liu, F., Wu, K.: Zero-cost, fine-grained power monitoring of datacenters using non-intrusive power disaggregation (2015)
33. Tiwari, A., Laurenzano, M.A., Carrington, L., Snavely, A.: Modeling power and energy usage of HPC kernels. In: IEEE International Parallel and Distributed Processing Symposium Workshops and Phd Forum (2012)
34. Wang, H., Cao, Y.: Predicting power consumption of GPUs with fuzzy wavelet neural networks. Parallel Comput. **44**, 18–36
35. Widyawan, Z.M.I., Nugroho, L.E.: Adaptive motion detection algorithm using frame differences and dynamic template matching method. In: Ubiquitous Robots and Ambient Intelligence, International Conference (2012)
36. Wikipedia: Least squares (2020). en.wikipedia.org/wiki/Leastsquares
37. Wikipedia: Pearson correlation coefficient (2020). http://en.wikipedia.org/wiki/Pearson_correlation_coefficient
38. Wikipedia: Perf wiki (2020). perf.wiki.kernel.org/index.php/MainPage
39. Yong, D., Juan, C., Yuhua, T., Junjie, W., Huiquan, W., Enqiang, Z.: Lazy scheduling based disk energy optimization method. Tsinghua Sci. Technol. **25**(2), 203–216 (2020)
40. Yoshii, K., Iskra, K., Gupta, R., Beckman, P., Coghlan, S.: Evaluating power-monitoring capabilities on IBM Blue Gene/P And Blue Gene/Q. In: 2012 IEEE International Conference on Cluster Computing (CLUSTER) (2012)
41. Zhang, J., Huo, H., Fang, Q., Zhang, D.: System and method for managing baseboard management controller (2008)

ParaCA: A Speculative Parallel Crawling Approach on Apache Spark

Yuxiang Li[1,2], Zhiyong Zhang[1,2(✉)], DanMei Niu[1,2,3], and Junchang Jing[1,2]

[1] Henan University of Science and Technology, Luoyang 471023, Henan, China
xidianzzy@126.com
[2] Henan International Joint Laboratory of Cyberspace Security Applications,
Luoyang 471023, Henan, China
[3] State key Laboratory of Networking and Switching Technology, Beijing University
of Posts and Telecommunications, Beijing, China
http://www.sigdrm.org/~zzhang/

Abstract. The World Wide Web today is growing at a phenomenal rate. The crawling approach is of vital importance to improve the efficiency of crawling the web. The existing sequent crawling algorithms are mostly time consuming and do not support large data well. In order to improve parallelism and efficiency of crawler on distributed network environments, based on the software thread-level speculation technique, this paper raises a speculative parallel crawler approach (ParaCA) on Apache Spark. By analyzing the process of web crawler, the ParaCA firstly hires a function to divide a crawling process into several subprocesses which can be implemented independently and then spawns a number of threads to speculatively crawl in parallel. At last, the speculative results are merged to form the final outcome. Comparing with the conventional parallel approach on multicore platform, ParaCA is very efficiency and obtains a high parallelism degree by making the best of the resources of the cluster. Experiments show that the proposed approach could achieve a significant speedup improvement with compare to the traditional approach in average. In addition, with the growing number of working nodes, the execution time decreases gradually, and the speedup scales linearly. The results indicate that the crawling efficiency can be significantly enhanced by adopting this speculative parallel algorithm.

Keywords: Crawling approach · Speculative multithreading · Apache Spark

1 Introduction

Crawlers [1] are generally used in search engines to find information that is of interest to users quickly and efficiently from vast internet information. Crawlers are also used to collect the research data that researchers need. For data acquisition needs, Xu et al. [2] put forward the strategy of fusing different acquisition

© Springer Nature Switzerland AG 2020
M. Qiu (Ed.): ICA3PP 2020, LNCS 12452, pp. 112–124, 2020.
https://doi.org/10.1007/978-3-030-60245-1_8

programs, the fusion strategy can quickly and efficiently collect large amounts of data. However, with the advent of big data and Web 2.0, all kinds of information on multimedia social networks have exploded. The efficiency and update speed of single crawlers have been unable to meet the needs of users. Using parallel technology for crawlers can effectively improve the efficiency of crawlers [3], in a big data environment, parallel crawlers are implemented in a distributed architecture, because distributed crawlers are more suitable for large data environments than stand-alone multicore parallel crawlers. Xia et al. [4], used distributed web crawler framework and techniques to collect data from social networking site Sina Weibo to monitor public opinion and other valuable findings, overwhelming traditional web crawlers in terms of efficiency, scalability, and cost, greatly improving the efficiency and accuracy of data collection. Su et al. [5], put forward a web crawler model of fetching data speedily based on Hadoop distributed system in view of a large of data, a lack of filtering and sorting. The crawler model will transplant single-threaded or multi-threaded web crawler into a distributed system by way of diversifying and personalizing operations of fetching data and data storage, so that it can improve the scalability and reliability of the crawler.

A good approach to process large-scale data is to make use of enormous computing power offered by modern distributed computing platforms (or called big data platforms) like Apache Hadoop [6] or Apache Spark [7,8]. These popular platforms adopt MapReduce, which is a specialization of the split-apply-combine strategy for data analysis, as their programming model. A standard MapReduce program is composed by pairs of Map operation (whose job is to sorting or filtering data) and Reduce operation (whose job is to do summary of the result by Map operations), and a high parallelism of MapReduce model is obtained by marshalling the operation pairs and performing them in distributed servers in parallel. The platforms that adopt MapReduce model often split the input, turn the large scale problems into sets of problems with small-scale, and then solve the problem sets in a parallel way. At present, many resource-intensive algorithms are successfully implemented to these big data platforms and achieve a better performance [9–11], and it is a good way to enhance conventional algorithms' efficiency. However, there exists a problem that the inherent dependence in the conventional crawling approach affects the effect of parallelism. So, designing and implementing a parallel crawling approach with high efficiency on Big Data platforms becomes very essential. By this method, the crawling web data can be split into some data blocks and then captured in parallel. After crawling the webs, the results are validated at a certain point, the proper results would be submitted and the improper ones would be recalculated. Even though false parallelization may occur, it is still a good method to leverage the crawling performance.

The remaining parts of this paper are organized as follows: we first briefly describe the execution model and related work in Sect. 2; we present Spark-based distributed parallel crawler framework in Sect. 3; based on the framework, experimental results and analysis are shown in Sect. 3.2.

2 Spark-Based Speculative Parallel Crawler Framework

The distributed crawler system used in this paper adopts a master-slave structure. That is, one master node controls all slave nodes to perform crawl tasks. The master node is responsible for allocating tasks and ensures load balancing of all slave nodes in the cluster. The used allocation algorithm is to calculate the hash value of the host corresponding to each URL, and then divide the URL of the same host into a partition. The purpose of this is to have the URL of the same host crawled on one machine. Distributed crawlers can be viewed as a combination of multiple centralized crawler systems. Each slave node is equivalent to a centralized crawler system. These centralized crawler systems are controlled and managed by a master node in a distributed crawler system.

From the diagram of distributed parallel crawler framework in Fig. 1, we can see that the main part of the framework includes master node for task generation, task allocation and scheduling, and the master node's control and management of the entire system (such as the depth of crawler, configuration of update time, system startup and stop etc.); The crawler cluster is responsible for parallel downloading pages; the Map/Reduce function module is responsible for parsing pages, optimizing links, and web page updates; Message middleware is responsible for communication and collaboration between master node, crawler nodes, and clusters (e.g. log management, data exchange and maintenance between clusters, etc.); and Distributed File System (HDFS) for data storage.

2.1 Speculative Parallel Crawling

To process large-scale data stored in HDFS in parallel, Hadoop offers a parallel computing framework called Map/Reduce. Spark is founded on Hadoop. The framework effectively manages and schedules nodes in the entire cluster to complete the programs' parallel execution and data processing and allows every slave node to localize calculation data on the local node as much as possible.

As can be seen from Fig. 1, the core of the entire crawler system can be divided into three modules, including download, parsing and optimization. Every module is an independent function module, and every module corresponds to a Map/Reduce process.

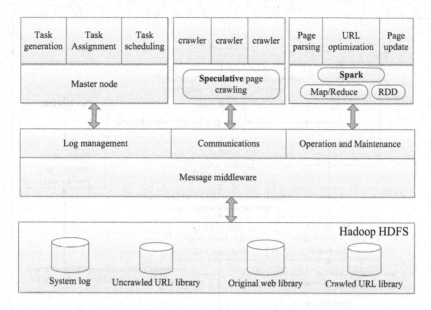

Fig. 1. Speculative crawler framework diagram

- The download module can download web pages in parallel. Specific download is completed in the Reduce phase, and multi-threaded download is used.
- Parsing module can analyze downloaded pages in parallel, extract the link out. The module not only needs a Map stage to complete the goal, but also limits the type of links to prevent the extracted links to other sites through the rules.
- The optimization module can optimize the collection of links in parallel and filter out duplicate links.

It can be seen that the parallelism of the Web crawler system is achieved through these three parallelizable modules, which are essentially implemented through the parallel computing framework of Map/Reduce.

At the beginning of the Map/Reduce task, the input data is split into several slices, with a default of 64 MB for every slice. Every piece is processed by a Map process, a crawler can open multiple Map processes at the same time. After the output of all Map is combined, according to the partitioning algorithm, the URL of the same site is assigned to a partition, so that the URL of the same site can be crawled on the same machine, the tasks of every partition are processed by a Reduce process, several partitions have several Reduces which perform parallel processing, while a crawler can also open multiple Reduce processes. Finally, the results of the parallel execution are saved to HDFS.

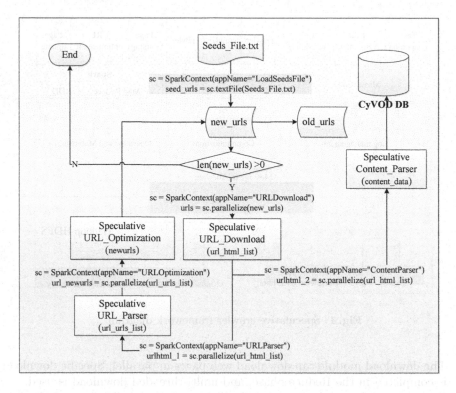

Fig. 2. Speculative parallel crawler framework diagram with spark

Figure 2 shows a parallel crawler framework diagram on Spark, in which three critical modules are *URL_Download()*, *URL_Parser()*, *URL_Optimization()*. "*Seeds_File.txt*" is the source of URLs, which are used to the process of initial crawler. From the file of "*Seeds_File.txt*", one new URL is extracted and then used to judge whether its length is larger than 0 or not. If the length of extracted URL is greater than 0, the next step is to download its web page. The downloaded pages need to be parsed, including *URL_Parser* and *Content_Parser*. The parsed contents need to be optimized, so to filter the improper URLs. "sc = *SparkContext(appName="URLDownload")*" and "*urls = sc.parallelize(new_urls)*" are used to realize the parallelization of download process. After this process, "*url_html_list*" is generated and used for parsing URLs and contents.

Figure 3 shows a parallel download diagram with Spark, in which three critical processes are included, *i.e.* judgement of urls, using SparkContext to get *sc*, using *parallelize()* to realize parallelization.

Figure 4 presents a parallel parser diagram with Spark, in which three critical processes are also included, *i.e.* URL_download, using SparkContext to get *sc*, using *parallelize()* to realize parallelization of *url_html_list*.

Similar to Fig. 3 and Fig. 4, Fig. 5 shows the process of parallelizing URL_op timization, including URL_parser, obtaining *sc*, using *parallelize(url_urls_list)* to realize the parallelization. The function *collect()* is used to obtain new *url_list*, and the function *distinct_urls* is used to remove the repetitive urls.

The specific processes of parallelizing crawling can be reduced to be:

(1) Collecting a set of seeds. First, for each crawler target to collect a URL seed as the entrance link to download data, and then the files of seeds from the local file system upload to input folder of hadoop cluster distributed file system, input folder always holds the URL to be crawled by the current layer. At the same time, the setting layer which has been crawled is 0;

(2) Judge whether the list to be fetched in the input folder is empty. If yes, skipping to (7); otherwise, executing (3);

(3) Speculatively download pages in parallel. And save the original page to the html folder in HDFS, html folder holds raw web pages of every layer;

(4) Speculatively parse pages in parallel. Extract the eligible links from the crawled pages in the html folder and save the results to the output folder in HDFS. The output folder always stores the outgoing links that are parsed at the current level;

(5) Speculatively optimize outgoing links in parallel. Filter out the crawled URLs from all the parsed URLs in the output folder, and save the optimized results to input folder in HDFS for the next crawl;

(6) Judge whether the number of crawled layers is less than the parameter depth. If yes, "crawled layers" increase by 1, return (2); otherwise enter (7);

(7) Combine the pages crawled by every layer and remove the duplicate crawled pages. The results are still stored in the html folder;

(8) According to the webpage crawling plan, these pages with the crawling task at the moment are added to the crawling list;

(9) Further parse webpages content. Analyze the content of the webpages in parallel, and then parse out the required attribute information from the merged and duplicated webpages. The attribute information required by the system includes title, publishing time, copyright owner, text, and video source;

(10) According to the attribute information which is parsed out, further screening is done. If the attribute information satisfies the user rules, such as the publication date in the last 7 days and the content of the text related to the scientific and technical information. It will upload the attribute information that meets the conditions, including the URLs, to the server database; Otherwise, give up.

2.2 Speculative Parallel Algorithm Based Map/reduce

Definition 1: Crawler $= \{c_1, c_2, ..., c_n\}$: represents a collection of crawler nodes in a cluster. c_i represents the i_{th} crawler node. The maximum number of Map processes and the maximum number of Reduce processes which a reptile node can open are determined by the number of processors on the node.

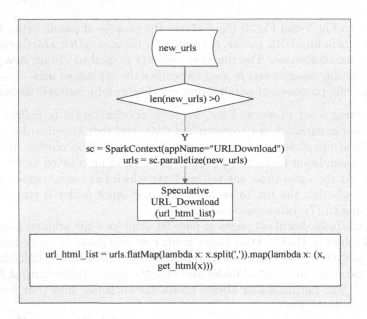

Fig. 3. Speculative parallel download with spark

Definition 2: $\{split_0,\ split_1\ ,...,\ split_{m-1}\}$: represents a collection of file slices. A slice is handled by a Map process.

Definition 3: $\{part_0,\ part_1\ ,...,\ part_{k-1}\}$: represents a collection of file partitions. A partition is handled by a Reduce process.

Assuming $m = 2n$, $k = n$, then parallel algorithm based Map/Reduce is shown in Algorithm 1.

In Table 1, a parallel algorithm (Algorithm 1) based Map/Reduce is specifically introduced. Firstly, an input-file is splitted into m-1 parts; Then, send an adjacent slices ($split\_k$, $split\_k+1$, where $k\%2==0$) to c_k, and $split\_k \sim split\_k+1$ are all defined above; Next, map processes are performed to process these adjacent slices, and combine all map output to *Inter-results*; Then, use the partitioner to partition *Inter-results* to $part\_0 \sim part\_k - 1$ ($k \in$ N); Next, send $part\_i$ to $c\_i + 1(i \in$ N), and open a process $part\_j \Rightarrow partition\_j$ ($j \in$ N) (Fig. 6 and Tables 2, 3, 4 and 5).

3   Experiment and Analysis

3.1   Experiment Configuration

This section will present a performance evaluation of ParaCA. Table 6 shows the specific configuration of experiment environment.

Table 1. Algorithm 1: speculative parallel algorithm based map/reduce

---
Algorithm 1:Speculative Parallel algorithm based Map/Reduce

Input: Input-File
Output: Output-File
1: Begin
2: Split Input-File into m slices =>{*split0, split1, ..., splitm-1*};
3: send *split0, split1* to *c1* , open two Map processes to process this two slices;
4: and send *split2, split3* to *c2* , open two Map processes to process this two slices;and ....;
5: and send *splitm-2, splitm-1* to *cn* , open two Map processes to process this two slices;
6: Combine all Map output => Inter-results;
7: partitioner(Inter-results) => {*part0, part1, ..., partk-1*};
8: send *part0* to *c1* , open a Reduce process to process part0 => *partition_0*;
9: and send *part1* to *c2*, open a Reduce process to process part1 => *partition_1*;
10: and ....;
11: and send *partk-1* to *cn* , open a Reduce process to process *partk-1*
12: => *partition_n-1* ; *Output-File* = {*partition_0,partition_1,...,partition_n-1*};
13: End

Table 2. Algorithm 2: speculative download of URL based spark

---
Algorithm 2:Speculative Download of URL based Spark

---
def *urls_download*(urls):
sc = *SparkContext*(*appName=*"URLDownload");
new_urls = *sc.parallelize*(urls);
url_html_list = *new_urls.flatMap*(lambda x: x.*split*(',')).map(lambda x:
(x, *get_html*(x)));
output = *url_html_list.collect*();
sc.stop();
print('urls_download: %s ' % *len*(output));
return *output*;

3.2 Analysis of Experimental Results

Figure 5 shows a time comparison between sequential crawling and parallel crawling. The left blue line represents the time of sequential crawling while the right red line represents the time of parallel crawling. Figure 7 shows a comparison of crawling websites between sequential crawling and parallel crawling. The left orange line represents the number of sequential crawling websites while the right orange line represents the number of parallel crawling websites. Figure 8 shows the changing of core number.

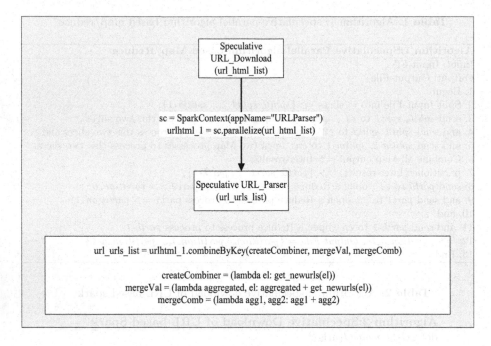

Fig. 4. Speculative parallel parser with spark

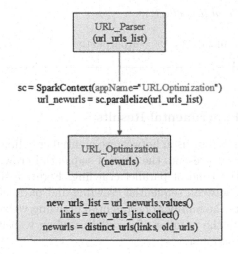

Fig. 5. Speculative parallel optimization with spark

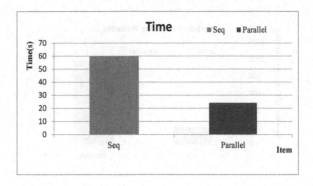

Fig. 6. Comparison of crawling time

Table 3. Algorithm 3: speculative parallel parser of URL based spark

Algorithm 3:Speculative Parallel Parser of URL based Spark

```
def url_parser(url_html_list):
sc = SparkContext(appName="URLParser");
url_htmls = sc.parallelize(url_html_list);
createCombiner = (lambda el: get_newurls(el));
mergeVal = (lambda aggregated, el: aggregated + get_newurls(el));
mergeComb = (lambda agg1, agg2: agg1 + agg2);
new_urlss = url_htmls.combineByKey(createCombiner, mergeVal, mergeComb);
output = new_urlss.collect(); sc.stop();
print('url_parser: %s % len(output), output);
return output;
if_name_ == "_main_":
sc = SparkContext(appName="LoadSeedsFile");
seeds_file = "file:///home/hadoop/CyCrawler/News_CyCrawler/Spider/seeds-file.txt";
lines = sc.textFile(seeds_file);
root_urls = lines.flatMap(lambda x: x.split(' ')).distinct();
urls = root_urls.collect();
sc.stop();
url_html_list = urls_download(urls);
url_parser(url_html_list);
```

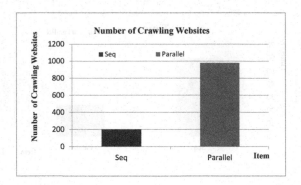

Fig. 7. Comparison of crawling websites (Color figure online)

Table 4. Algorithm 4: speculative parallel content parser based spark

Algorithm 4:Speculative Parallel Content Parser based Spark

```
def content_parser(url_html_list):
sc = SparkContext(appName="ContentParser");
url_htmls = sc.parallelize(url_html_list);
createCombiner = (lambda el: get_content(el));
mergeVal = (lambda aggregated, el: aggregated + get_content(el));
mergeComb = (lambda agg1, agg2: agg1 + agg2);
new_data = url_htmls.combineByKey(createCombiner, mergeVal, mergeComb);
output = new_data.collect();
sc.stop();
return output;
if _name_ == "_main_":
old_urls = [];
sc = SparkContext(appName="LoadSeedsFile");
seeds_file = "file:///home/hadoop/CyCrawler/News_CyCrawler/Spider/video-file.txt";
lines = sc.textFile(seeds_file);
root_urls = lines.flatMap(lambda x: x.split(' ')).distinct();
urls = root_urls.collect();
print(urls);
sc.stop();
for url in urls:
old_urls.append(url);
  print("======");
print(urls);
url_html_list = urls_download(urls);
content_parser(url_html_list);
```

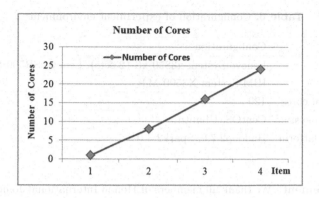

Fig. 8. Number of cores

Table 5. Algorithm 5: speculative parallel optimization of URL based spark

---

Algorithm 5:Speculative Parallel Optimization of URL based Spark

```
def distinct_urls(links, old_urls):
new_urls = [];
num = 0;
for j in range(len(links));
if links[j] = None;
num = num + len(links[j]);
for url in links[j];
if url not in old_urls;
if len(url)>0;
new_urls.append(url);
print('before url_optimi: %s', %num);
return list(set(new_urls));
def url_optimi(url_urls_list, old_urls);
sc = SparkContext(appName="URLOptimization");
url_urls_list2 = sc.parallelize(url_urls_list);
new_urls_list = url_urls_list2.values();
links = new_urls_list.collect();
sc.stop();
urls = distinct_urls(links, old_urls);
print('after url_optimi: %s ' % len(urls), urls);
return urls;
if _name_== "_main_";
old_urls = [];
sc = SparkContext(appName="LoadSeedsFile");
seeds_file = "file:///home/hadoop/CyCrawler/News_CyCrawler/Spider/seeds-file.txt";
lines = sc.textFile(seeds_file);
root_urls = lines.flatMap(lambda x: x.split(' ')).distinct();
urls = root_urls.collect();
print(urls);
sc.stop();
for url in urls:
old_urls.append(url);
url_html_list = urls_download(urls);
url_newurls_list = url_parser(url_html_list);
url_optimi(url_newurls_list, old_urls);
```

Table 6. Configuration of experiment environment

Item	Configuration
Servers	Lenovo System x3850 x6 (2 sets), Lenovo SR590 (2 sets), IBM System X3500 M4;
Number of cores	120
Operation system	CentOS, Ubuntu
Parallel platform	Hadoop2.7.0, Spark2.3.0

Acknowledgement. We thank all members of Henan International Joint Laboratory of Cyberspace Security Applications for their great support, and give our best hope to them for their collaboration. We also thank reviewers for their careful comments and suggestions. The work was sponsored by National Natural Science Foundation of China Grant No. 61972133, Project of Leading Talents in Science and Technology Innovation for Thousands of People Plan in Henan Province Grant No. 204200510021, Henan Province Key Scientific and Technological Projects Grant No. 192102210130 and No. 202102210162, and Key Scientific Research Projects of Henan Province Universities Grant No. 19B520008.

References

1. Zhou, D.M.: Survey of high-performance web crawler. Comput. Sci. (2009)
2. Yan-Fei, X.U., Liu, Y., Wen-Peng, W.U.: Research and application of social network data acquisition technology. Comput. Sci. (2017)
3. Guo, R., Wang, H., Chen, M., Li, J., Gao, H.: Parallelizing the extraction of fresh information from online social networks. Future Gen. Comput. Syst. **59**(C), 33–46 (2016)
4. Xia, J., Wan, W., Liu, R., Chen, G., Feng, Q.: Distributed web crawling: a framework for crawling of micro-blog data. In: International Conference on Smart and Sustainable City and Big Data, pp. 62–68 (2016)
5. Su, L., Wang, F.: Web crawler model of fetching data speedily based on hadoop distributed system. In: IEEE International Conference on Software Engineering and Service Science, pp. 927–931 (2017)
6. Honnutagi, P.S.: The hadoop distributed file system. Int. J. Comput. Sci. Inf. Technol. (2014)
7. Shoro, A.G., Soomro, T.R.: Big data analysis: Apache spark perspective, vol. 15 (2015)
8. Zaharia, M., et al.: Apache spark: a unified engine for big data processing. Commun. ACM **59**(11), 56–65 (2016)
9. Qi, R.Z., Wang, Z.J., Li, S.Y.: A parallel genetic algorithm based on spark for pairwise test suite generation. J. Comput. Sci. Technol. **31**(2), 417–427 (2016)
10. Qiu, H., Gu, R., Yuan, C., Huang, Y.: YAFIM: a parallel frequent itemset mining algorithm with spark. In: Parallel & Distributed Processing Symposium Workshops, pp. 1664–1671 (2014)
11. Shi, K., Denny, J., Amato, N.M.: Spark PRM: Using RRTs within PRMs to efficiently explore narrow passages. In: IEEE International Conference on Robotics and Automation, pp. 4659–4666 (2014)

A Multi-threaded Algorithm for Capacity Constrained Assignment over Road Networks

Abhishek Mishra[1], Venkata M. V. Gunturi[1(✉)], and Sarnath Ramnath[2]

[1] IIT Ropar, Rupnagar, Punjab, India
{2018csm1002,gunturi}@iitrpr.ac.in
[2] St. Cloud State University, St. Cloud, USA
sarnath@stcloudstate.edu

Abstract. Input to the capacity constrained assignment (CCA) problem over road networks consists of the following: (a) a road network represented as a directed graph; (b) a set of public service units (e.g., flu-clinics, schools) as vertices in the graph and; (c) a set of demand locations (e.g., people or school children) also as vertices in the graph. In addition, each service center is also associated with a notion of capacity and a penalty which is incurred if it gets overloaded. Given the input, the goal of CCA problem is to determine a mapping between the set of demand vertices and the set of service centers. The objective here is to generate a mapping which minimizes the sum of the total distance between demand vertices and their associated service centers, and the total penalty incurred. CCA problem has value addition potential in the domain of urban planning. CCA problem can be reduced to min-cost bipartite matching. However, optimal algorithms for min-cost bipartite matching do not scale beyond graphs of size few thousand nodes. Moreover, its non-trivial to parallelize optimal algorithms for min-cost bipartite matching due to their inherent iterative nature. The current relevant work in the area of parallel algorithms is limited to problems like finding max-flow and maximum cardinality matching, which are fundamentally different than min-cost bipartite matching. In this paper, we propose a novel assignment subspace re-organization based approach (ASRAC) for the CCA problem. ASRAC can load-balance and take full advantage of multi-core systems to speed-up execution. Our experimental results indicate that our proposed algorithm (ASRAC) can scale up to large graphs while maintaining better solution quality over alternative approaches.

1 Introduction

Input to our problem consists of the following three things. First, a road network represented as a directed graph $G(V, E)$ with V being the set of vertices and E being the set of edges. Second, a set S ($S \subset V$) of service centers (e.g. schools, hospitals, etc.). Each service center $s_i \in S$ is associated with an positive integer capacity and a notion of "penalty" which denotes the *"cost"* that must be paid

© Springer Nature Switzerland AG 2020
M. Qiu (Ed.): ICA3PP 2020, LNCS 12452, pp. 125–142, 2020.
https://doi.org/10.1007/978-3-030-60245-1_9

to overload the particular service center. And, the third input to our problem is a set D $(D \subset V - S)$ of demand vertices. Given the input, the problem of *Capacity Constrained Assignment* (CCA) determines an mapping between the set of demand vertices and the set of service centers. Here, each demand node is allotted to only one service center. The objective is to determine a mapping which minimizes the sum of the following two terms: (1) total distance between the demand vertices and their allotted service center and, (2) total penalty incurred (if any) while overloading the service centers.

Problem Importance: The Capacity Constrained Assignment (CCA) problem finds its application in the area of urban planning. More specifically, we can use CCA problem while defining the zones of operation of public service units such as schools (refer catchment area [15]), walk-in pathogen testing centers (in case of continuous monitoring of infectious diseases such as COVID-19) and, infectious disease clinics. The key aspect over here being that each of the previously mentioned type of service centers is associated with a general notion of capacity. This capacity dictates the number of people (demand) that can be accommodated (comfortably) each day, week or during any specific duration of time (e.g., typical duration of sickness of patients). In addition, the quality of service at any of these service centers is expected to degrade if significantly more number of people (beyond its capacity) are assigned to it. This aspect is modeled as penalty. Determining the region of operation under such constraints can be mapped to our CCA problem.

Computational Challenges of CCA Problem. An instance of the Capacity Constrained Assignment (CCA) problem can be theoretically reduced to an instance of the min-Cost bipartite matching problem [2]. However, optimal algorithms for min-cost bipartite matching fail to scale up for large problem instances. For instance, our experiments revealed that, for a CCA problem instance with just 13 service centers and around 4000 demand vertices, optimal algorithm for min-cost bipartite matching takes around 7 h for execution and occupies about 40 GB in RAM. From this, one can easily expect that optimal algorithm for min-cost bipartite matching algorithm cannot scale-up to large problem instances.

Limitations of Work Done in Parallel Algorithms: Parallel techniques for flow related problems has been an active area of research in the high performance computing community. Over the years, researchers have developed parallel techniques for maximum flow (e.g. [8,13]), maximum cardinality matching (e.g., [1,4,9]) and perfect matching [5]. However, these works cannot be generalized for our CCA problem. Note that our CCA problem reduces to min-cost bi-partite matching problem which has a fundamentally different computational structure due to presence of weights on edges. There has been some work [11] on parallel algorithms for min-cost flows. However, it assumes presence of PRAM, whereas, several modern computers only have SDRAMs.

Limitations of Work Done in Spatial Data Analytics Researchers in the area of spatial data analytics have been studying variations of the assignment problem on spatial networks since long (e.g., [14,18,19]). While different works have used different terminology, they nevertheless they share a commonality in their problem definition. Earlier works such as [14] assume that the service centers have infinite capacity, an assumption which not suitable in many real-world scenarios. Later works in this area [18,19] did consider the notion of capacities, but they did not consider the concept of penalties. They perform allotments (of demand nodes) in an iterative fashion as long as there exists a service center with available capacity. However, this assumption is also not always desirable in developing nation scenarios where the total demand is usually more than the total capacity of service centers.

There have been other works (e.g., [7,10,16,17])) which considered the assignment problem in different contexts. For e.g., [7,17] considered the problem in the context of allocating data in a main memory hierarchy. [10] considered it in the context of allocating resources in heterogeneous multi-core systems. And [16] explored the allocation problem in V2X networks.

In our work, we address a more general problem of assignment by considering both capacity constraints and overload penalties. In our CCA problem, we allow the allotments to go beyond the capacities of the service centers. And after a service center is full, it uses the *overload penalties* for guiding the further allotments and load sharing.

Our Contributions: This paper makes the following contributions:

(a) Propose the novel concept of assignment subspace re-organization for the CCA problem. The key idea to is to determine paths for re-organizing the partially constructed solution in what we refer to as the *assignment subspace tree*.
(b) Propose a novel multi-threaded algorithm called *ASRAC* for the CCA problem which is based on the concept of assignment subspace re-organization.
(c) ASRAC is adaptive in the sense that it can automatically distribute its available workload among all the available threads.
(d) Evaluate ASRAC analytically via time complexity analysis.
(e) Evaluate our ASRAC algorithm experimentally using road network datasets obtained from OpenStreetMaps and compare it with alternative approaches.

Outline. The rest of the paper is organized as follows. Section 2 discusses the basic concepts and presents the problem formally. We present our proposed approach in Sect. 3. Section 4 presents a detailed time complexity analysis of our proposed ASRAC algorithm. In Sect. 5, we evaluate our proposed approach experimentally and compare it with alternative approaches.

2   Basic Concepts and Problem Definition

Definition 1. Road Network: *is modeled as weighted directed graph $G(V, E)$ with a set of vertices V and a set of edges E. Road intersections are modeled as vertices and road segments (between two intersections) are modeled as directed edges. Each edge $e = (u, v) \in E$ is associated with a cost w_e which represents the cost to reach vertex v from vertex u.*

Definition 2. Service Center (s_i) *is a vertex in the input road network* G. *It represents a public service unit of a particular kind such as schools, clinics or hospitals in a city.*

Definition 3. Demand Vertex (d_i) *is also a vertex in the input road network* G. *Demand vertex represents the location of a unit population which is interested in accessing the previously defined service center. For simplicity, we assume that all the demand exists on the vertices of the graph. We can trivially generalize to case where demand exist on edges by creating dummy vertices at that location.*

Definition 4. Capacity of a service center (c_{s_i}) *is the prescribed amount demand that a service center* s_i *can accommodate. Real world examples of capacity include aspects such as infrastructure limitations (e.g., number of faculty and lab facilities), number of beds in an isolation ward of the hospital, etc.*

Definition 5. Penalty function of a service center: *Each service center* s_i *is associated with penalty function* $p_{s_i}()$. *This function returns the "extra cost" that must be paid for every new assignment to* s_i *after it has exhausted its capacity* c_{s_i}. *If the allotment is done within the capacity of* s_i, *then no penalty needs to be paid. Examples of penalty cost in real world could be things like cost to add additional infrastructure and/or faculty in a school.*

This function takes into account the current status of s_i *(i.e., number of demand nodes which have been already added to* s_i*) and then returns a penalty for the* j^{th} *assignment to* s_i *beyond its capacity.* $p_{s_i}()$ *returns only positive values and is monotonically increasing over the* j *(*$1 \leq j \leq (|D| - c_{s_i})$, *$|D|$ is the number of demand nodes in the problem). The intuition behind penalty functions being that one may have to add increasingly more resources to a school (or a hospital) as the overloading keeps increasing.*

2.1 Problem Statement

We now formally define the problem of capacity constrained assignment (CCA) by detailing the input, output and the objective function:
Given:

- A road network represented as a directed $G(V, E)$, where each edge $e \in E$ has a positive cost.
- A set of service centers $S = \{s_1, ..., s_{n_s}\}$ where $S \subset V$. Each service center s_i has a capacity c_{s_i} and a penalty function p_{s_i}.
- A set of demand vertices $D = \{d_1, ..., d_{n_d}\}$ where $D \subset V - S$.

Output: An allotment consisting of pairs $< d_k, s_j >$. Here, the demand vertex d_k has been assigned to the service center s_j. A demand vertex can be assigned to only one service center. Note that we use the terms allotment and assignment interchangeably in this paper (Table 1).

Objective Function:

$$Min\left\{ \sum_{\substack{s_i \in Service \\ Centers}} \left\{ \sum_{\substack{d_j \in Demand\ vertices \\ allotted\ to\ s_i}} SCost(d_j, s_i) \right\} + Total\ Penalty\ across\ all\ s_i \right\}$$

(1)

Table 1. Notations used in the paper

Symbol used	Meaning
D	Input set of demand vertices
S	Input set of service centers
$\mathcal{A}$	An allotment containing pairs $< d_k, s_j >$. Here, demand vertex d_k is allotted to service center s_j. Note that we use the terms allotment and assignment interchangeably
$SCost(dn, s_j)$	Shortest distance from demand vertex dn to service center s_j
$\Psi(\mathcal{A}, s*)$	Assignment subspace tree of the allotment $\mathcal{A}$. Nodes of this tree are service centers and edges represent transfer of demand vertices across service centers. Service center $s*$ is root
$v^*(s_i, s_j)$	Best transfer demand node $\mathbf{v^*(s_i, s_j)}$ for transfer from s_i to s_j

3 Proposed Approach

This section details our proposed approach and is organized as follows: Sect. 3.1 presents a brief overview of our solution. In Sect. 3.2 we formally define our key idea of *assignment subspace tree*. Section 3.3 details the notion of *cascades* in a *assignment subspace tree*. Cascades are basically potential paths for re-organization in a partial solution. We present details of our proposed algorithm ASRAC in Sect. 3.4.

3.1 Overview

Our proposed approach constructs the solution incrementally. And, after each new assignment it re-organizes the partially constructed solution so as to improve the objective function value. This way we are able to balance the trade-off between computational efficiency and solution quality. Our approach starts by computing the shortest distance between all pairs of demand vertices and service centers. Following this, we determine the closest service center for each demand vertex and put the pairs <demand vertex d_i – closest s_i> pairs into a min-heap. This heap is ordered on the distance to the closest s_i. In each iteration, the demand vertex to be processed next is determined via the extract-min operation on the previously mentioned heap. Let <d_i, $closest\ s_j$> be the result of the extract-min at any intermediate stage of the algorithm. d_i is assigned to the service center s_j. This assignment could have happened under following two cases: **Case (a)** s_i had free space (in terms of capacity) or; **Case (b)** s_i was already full. Case (a) is quite straightforward and no further operations are needed. Note that in this case, we also need not pay any penalty for this assignment. However, in case(b), after assigning d_i to s_j, the algorithm undertakes the process of *re-organizing* the current solution with an intention of lowering the current objective function value. Note that in case (b), the objective function would increase by the quantity $\delta = SCost(d_i, s_j) + Penalty(s_i)$ after the assigning d_i to s_j. We now briefly present our idea of re-organization at high level.

Re-organization of an Assignment: Consider Fig. 1 which illustrates a partially constructed solution for a sample problem instance. Here, vertices, $S1$, $S2$, $S3$ and $S4$ are service centers. The figure also details the total capacity and penalty values for each of the service centers. For sake of ease of understanding, we assume that penalty function is just an integer (instead of a monotonically increasing function as described in Definition 5).

In the partial assignment shown in Fig. 1, the first few demand vertices (A, B, D, E and J) have already been processed. All of these were processed via the previously described case(a) where the service center of choice had free capacity. Demand vertices which are allotted a service center are filled using the same color as that of their allotted service center. For e.g., demand vertices A, B are assigned to service center $S1$. Nodes which are not yet allotted are shown without any filling.

Service Center	Capacity	Penalty
S1	2	1
S2	1	15
S3	1	5
S4	1	15

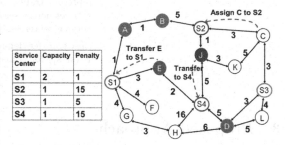

Fig. 1. A partially constructed assignment on a sample input. Processed demand nodes are filled with the same color as their respective service center. (Best in color) (Color figure online)

Now consider the pair $<C, S2>$ to be processed next. Here, $S2$ is the closest service center for the demand vertex C, however it is already full. So, when C is allotted to $S2$, the objective function value would increase by $SCost(C, S2)$ + penalty at $S2$ ($3 + 15 = 18$). As mentioned earlier, at this stage our algorithm tries to re-organize the current allotment with the goal of lowering the objective function value as much as possible. Re-organization is done by undertaking a series of adjustments to the current solution. In each adjustment, we transfer a demand vertex which is already assigned to a service center to another one.

For instance, consider the dotted arrows in Fig. 1. After assigning C to $S2$, we first transfer J to $S4$ (from $S2$) and then, transfer E to service center $S1$ (from $S4$). Total increase in objective function after the assignment and re-organization process is just 9 (as compared to increase of 18 units in case of no follow-up re-organization). Details of the change in objective function value after each step are given below:

(**Step-0 Assignment**) Assign $<C$ to S2$>$: $SCost(C, S2)$ + Penalty(S2) $\implies$ $3 + 15 = 18$
(**Step-1 Transfer**) Transfer $<J$ from S2 to S4$>$: $-$Penalty(S2) $-$ $SCost(J, S2)$ + $SCost(J, S4)$ + Penalty(S4) $\implies$ $-15 - 1 + 5 + 15 = 4$
(**Step-3 Transfer**) Transfer $<E$ from S4 to S1$>$: $-$Penalty(S4) $-$ $SCost(E, S4)$ + $SCost(E, S1)$ + Penalty(S1) $\implies$ $-15 - 2 + 3 + 1 = -13$
Total increase in objective function: assignment cost + re-organization cost $\implies$ $18 - 9 = 9$

Recall that the penalty costs associated with any service depends on its current total load (beyond its capacity). Thus, the penalty costs mentioned in step 2 and step 3 previously are not independent of each other. They were added (and subtracted) by keeping a note of the current total load on each service center as the demand vertices were being transferred.

The sequence of transfers illustrated above was an instance of what we refer to as a *cascade* in this paper. In our actual implementation of the concept, in each iteration, we determine the best cascade for re-organization. And this cascade is determined by exploring, what we refer to as, the *assignment subspace tree* (details in Sect. 3.2). This exploration is done by using multiple threads. Note that the notion of a cascade was originally proposed in [12]. In this paper, we are employing the concept in a much more sophisticated manner.

3.2   Assignment Subspace Tree

A key step of the algorithm involves determining a suitable *cascade* for the re-organization process after each of the assignment steps which lead to penalty. This cascade is determined by exploring a larger space of potential adjustments represented in an *assignment subspace tree*. Following is a formal definition of the concept.

Definition 6. *Assignment subspace tree* $\mathbf{\Psi}()$: *Consider an instance of problem with service centers* $S = \{s_1, ..., s_{n_s}\}$. *Assume any intermediate stage of the algorithm, where a demand vertex* d_{new} *was added to service center* $s*$ $(s* \in S)$. *Let* $\mathcal{A}$ *be the current allotment comprising of tuples of the form* $< d_i, s_j >$ *(demand node* d_i *is assigned to* s_j*). The allotment subspace tree* $\Psi(\mathcal{A}, s*)$ *is defined as follows:*

- *Nodes of the tree are the service centers in* S.
- $s*$ *forms the root (i.e., level 0) of the tree.*
- *At level 1, we have a node for each of the service centers* $S - \{s*\}$. *In other words, each of the service centers in* $S - \{s*\}$ *is a child node of* $s*$.
- *For any node* s_j *at level 1, all service centers in* $S - \{s*, s_j\}$ *would be its children at level 2.*
- *Likewise, for any arbitrary node* s_t *in level* k, *all service centers in* $S - \{nodes$ *in root to* s_t *path$\}$ would be its children at level* $k + 1$.
- *The edge between a service center* s_i *(at level* t*) and its child* s_j *(at level* $t + 1$*) corresponds to the transfer of the best transfer-demand vertex* $v*(s_i, s_j)$ *(Definition 7). Cost of this edge is defined as* $SCost(v*(s_i, s_j), s_j) - SCost(v*(s_i, s_j), s_i)$. *Here,* $v*(s_i, s_j)$ *is a demand vertex which is currently assigned to* s_i *in* $(\mathcal{A})$, *but is most suitable for transfer to* s_j *according to Definition 7.*

Definition 7. *Best transfer-demand vertex* $\mathbf{v}*(\mathbf{s_i}, \mathbf{s_j})$ *for transfer from* s_i *to* s_j: *Let* $\mathcal{A}$ *be the current allotment comprising of tuples of the form* $< d_i, s_j >$ *(demand vertex* d_i *is assigned to* s_j*). For any ordered pair of service centers* (s_i, s_j), $v*(s_i, s_j)$ *is the demand vertex* d_i *(currently assigned to* s_i*) that has the minimum value for the term* $SCost(d_i, s_j) - SCost(d_i, s_i)$ *amongst all* d_i *currently assigned to* s_i. *The expression* "$SCost(d_i, s_j) - SCost(d_i, s_i)$" *denotes the cost of transferring* d_i *from* s_i *to* s_j.

Note that according to Definition 6, service centers are never repeated on any root to leaf path in the assignment subspace tree. We make this particular design decision in our proposed approach with an intention of reducing the search space.

One of the key tasks while building $\Psi(\mathcal{A}, s*)$ involves determining the best transfer-demand vertex for each of the parent-child pairs of service centers in the tree. And computing best transfer-demand vertex for a pair (s_i, s_j) involves finding a minimum (refer Definition 7) over all possible transfers between service centers s_i and s_j. This would be an computationally intensive task if we have a large number of service centers and/or the size of allotment is large. To this end, we use the concept of *boundary vertices* to minimize the number of potential transfers considered while determining the best transfer-demand vertex between any pair of service centers in $\Psi()$. The notion of boundary vertices was originally proposed in [12]. We are replicating it here for sake of convenience. Definition 8 presents the notion of boundary vertices formally.

Definition 8. *Boundary Vertices of a service center* s_i *(B_{s_i}) is a set of demand vertices assigned to the service center* s_i *such that each vertex in* B_{s_i} *has at-least one of the following three properties: (a) an outgoing edge to a vertex allotted to a different service center* s_j, *(b) an outgoing edge to a different service center* s_j *or, (c) an outgoing edge to an unprocessed demand vertex.*

Using Boundary vertices while constructing $\Psi()$: Consider the construction of the assignment subspace tree ($\Psi(\mathcal{A}, s*)$) after a demand vertex was just added to the service center $s*$. Let $\mathcal{A}$ be the current allotment and, s_i be an arbitrary non-leaf service center node in ($\Psi(\mathcal{A}, s*)$). Further assume that ω number of demand vertices are currently assigned to s_i in $\mathcal{A}$.

Now for determining the best transfer-demand vertex to any child s_j of s_i (in $\Psi(\mathcal{A}, s*)$), we need to evaluate all of the ω demand nodes (currently assigned to s_i) as potential candidates to be transferred to s_j. Basically, we would compute the expression "$SCost(d_i, s_j) - SCost(d_i, s_i)$" (referred to as the transfer cost) for each of the d_i's currently assigned to s_i and then, choose the d_i which has the minimum value. However, as mentioned earlier, this could be computationally intensive.

To this end, we propose to consider only the boundary vertices of s_i while determining the best transfer-demand vertex to any of its child node s_j. The intuition behind this is that any demand vertex which is inside the "assignment region" of s_i would have to travel through a boundary vertex (of s_i) to reach another service. As a result, such demand vertices would have a higher value of transfer cost than the boundary vertices. Correctness of this statement can trivially deduced from Lemma 1 in [12]. Detailed proof is not included here due to lack of space.

3.3 Cascades in Assignment Subspace Tree

Structure of a Cascade: Any path which starts at the root and terminates on any node (either internal or a leaf node) in an assignment subspace tree is a cascade. For instance, consider a cascade $c_i = \; < p, q >$ in an arbitrary assignment subspace tree ($\Psi(\mathcal{A}, S_a)$) which starts at service center S_a (root of $\Psi(\mathcal{A}, S_a)$) and passes through the service centers S_b and S_c. Here, p is a boundary vertex of S_a which was chosen as the best transfer (Definition 7) to S_b. Similarly, q was the best transfer demand vertex for transferring from S_b to S_c. Now, the total cost of c_i ($Cost(c_i)$) is given by the following expression.

$$Cost(c_i) = -p_{s_a} - SCost(p, S_a) + SCost(p, S_b) + \cancel{p_{s_b}} - \cancel{p_{s_b}} - SCost(q, S_b)$$
$$+ SCost(q, S_c) + p_{s_c}$$

Here, p_{s_a} refers to penalty paid at S_a (similar is the case with terms p_{s_b} and p_{s_c}). Note that the above expression of $Cost(c_i)$ models the case where each of service centers S_a, S_b and S_c are overloaded by at least one. Thus, care must be taken in this regard to get valid costs.

Beneficial Cascade: A cascade whose total cost is less than zero is termed as a beneficial cascade. For our previous example, cascade c_i is said to be a beneficial cascade, iff $Cost(c_i) + p_{s_a} + SCost(dn, S_a) < p_{s_a} + SCost(dn, S_a)$. Here, dn was the new demand node which was assigned to S_a. The most beneficial cascade in $\Psi(\mathcal{A}, S_a)$ would be the one whose cost is most negative.

3.4   ASRAC Algorithm

This section details our proposed Assignment Subspace tree Re-organization based Approach for CCA (ASRAC) problem. As one may imagine, a trivial approach for determining the most beneficial cascade (in each iteration) would be to *create* the entire assignment subspace tree and then use a modified version of the depth first search to determine the most beneficial cascade. In this modified version of depth first search, we would maintain the cost of the current best beneficial cascade seen so far. This approach, however, is not scalable to any real world road networks. To this end, our approach makes following four design decisions: (a) we construct the assignment subspace tree on-the-fly as we search for most beneficial cascade. (b) We use multiple threads to help in the process. (c) At any internal node s_i (including root service center) in $\Psi()$, we only enumerate best α number of children of s_i (more details later). In other words, we limit the breadth of search in $\Psi()$ using parameter α. (d) we also limit the depth of search with parameter β.

Algorithm 1 details the pseudo-code for our proposed ASRAC algorithm. The algorithm initializes in lines 1–4. We first determine the shortest distance (on basis of edge costs) between all pairs of demand vertices and service centers. Following this, we determine the closest service center for each demand vertex and build a min heap ($ClosestSC$) with pairs <demand-vertex, closest service center>. $ClosestSC$ is ordered on the distance to the closest service center of each demand vertex.

Lines 5–33 represent the main *while* loop of the ASRAC algorithm. This while loop is executed till $ClosestSC$ heap is not empty. The demand vertex to be processed next is obtained by extract-min operation (Line 6).

Algorithm 1. Assignment Subspace Re-organization based Approach for CCA

Input: (a) α: allowed breadth of the search in assignment subspace tree ($\alpha \leq |S| - 1$);
(b) β: allowed length of cascade ($\beta \leq |S| - 1$); (c) τ: Number of threads available
Output: Assignment $\mathcal{A}$ with objective function value Δ

```
 1: Determine shortest distance between all pairs of demand vertices and service centers
 2: For each demand vertex d_i ∈ D determine the closest service center
 3: Initialize ClosestSC heap with all <demand vertex-closest service center> pairs
 4: Initialize assignment A ← NULL
 5: while ClosestSC heap is not empty do
 6:     < d_i, s* > ← extract-min operation on ClosestSC heap
 7:     Add < d_i, s* > to the current assignment A and increment Δ by SCost(d_i, s*)
 8:     if s* has vacancy then
 9:         Decrement capacity of s*
10:     else
11:         Increment Δ by penalty corresponding to s*
12:         Initialize the list of seeds Γ
13:         Γ ← Determine α seeds
14:         if α ≤ τ then /*allowed breadth of the search is less than #threads*/
15:             Λ ← Create α threads
16:         else/*allowed breadth of the search is more than threads*/
17:             Λ ← Create τ threads
18:         end if
19:         Create a job queue JQ with seeds present in Γ. /*Each seed is a job*/
20:         while JQ has unfinished jobs do
21:             Assign the next unfinished job to an available thread T_i /*T_i ∈ [1,..,|Λ|] */
22:             <s*,ssc,bcost,bpath> ← unexplored seed in Γ
23:             Initialize global variables TiCost − 0 and TiPath=NULL for thread T_i
24:             φ ← Penalty seen at service center s*
25:             Call Algorithm 2 with parameters <s*,ssc,bcost,bpath>, α, β, φ
26:         end while
27:         Barrier to ensure that all threads in Λ terminate
28:         < CCost*, BestCas >← Cascade with lowest cost amongst all cascades
            determined by different threads
29:         if CCost* < 0 then /*The best cascade was indeed beneficial */
30:             Re-organize A according to BestCas and update Δ value using CCost*
31:         end if
32:     end if
33: end while
```

Let the result of extract-min operation be a demand vertex d_i and its closest service center be $s*$. d_i is assigned to $s*$ and current allotment data-structure ($\mathcal{A}$) is updated accordingly. This new assignment could fall in one of the following two cases: (a) $s*$ had the required capacity or (b) $s*$ was already full. Case (a) is trivial, we just decrement the available capacity of $s*$ and increment the objective function value by $SCost(d_i, s*)$. On the other hand, case (b) involves re-organization of the current allotment $\mathcal{A}$ using the most beneficial cascade in $\Psi(\mathcal{A}, s*)$. We now detail this aspect next.

ASRAC algorithm determines the most beneficial cascade (in $\Psi(\mathcal{A}, s*)$) in lines 12–28. As mentioned earlier, ASRAC does not enumerate the entire $\Psi(\mathcal{A}, s*)$ beforehand. Rather, the enumeration and exploration happens simultaneously in order to achieve good performance. The algorithm first determines the seeds (which would be expanded in parallel using threads). After the seeds the generated, each of them become a job in the job queue. This job queue is processed using threads. Each seed is given to a thread. Any thread which finishes its job is assigned an unprocessed seed from the job queue.

A seed in the job queue represents a path in $\Psi(\mathcal{A}, s*)$ from root ($s*$) to an internal node u. The sub-tree under u is explored (using Algorithm 2) to determine the most beneficial cascade in that particular sub-tree. Note that the cost of this cascade would include the cost of the seed as well. Beneficial cascades determined from each seed are compared to determine the most beneficial cascade $BestCas$ in the entire $\Psi(\mathcal{A}, s*)$ (Line 28 in Algorithm 1). Following this, the current assignment $\mathcal{A}$ is re-organized according to the $BestCas$ and the objective function value is updated accordingly. We now explain our basic seed generation process and the procedure of searching the sub-tree (under a seed) for the beneficial cascade (Algorithm 2).

Seed Generation Process: The algorithm basically picks the α best children of root service center $s*$ in $\Psi(\mathcal{A}, s*)$ as seeds. At the root level (i.e., Level 0) in $\Psi(\mathcal{A}, s*)$, the service center $s*$ can theoretically have edges to all service centers $s_i \in S - \{s*\}$. This would be $|S| - 1$ number of edges. Now, we need to select α number of edges from these. This is done as follows. We determine the best transfer-demand vertex between $s*$ and every other service center $s_i \in S - \{s*\}$. Then, a min heap (called as $BestSeeds$) is created which is ordered on the cost of transferring the best transfer-demand vertex (refer Definition 7) from $s*$ to the respective service center. The top α items from this heap become the seeds.

Note that this approach can be trivially adapted to generate seeds according to the number of available threads. In this adapted version, we would first generate α best children (at level 1 of $\Psi()$) of root, and then proceed to generate $|S| - 2$ children (at level 2 of $\Psi()$) of these α children. We then combine the edges enumerated to create paths of length 2 edges from $s*$ in $\Psi()$. From these, $\alpha(|S| - 2)$ candidate paths, we can select top k according to the number of threads available. Note that for any realistic problem setting, $\alpha(|S| - 2)$ would come out to be in hundreds (considering $\alpha \geq |S|/2$). We did not present a pseudo-code for seed generation algorithm to maintain simplicity and clarity of presentation.

Searching a Sub-tree for Beneficial Cascade (Algorithm 2): ASRAC uses Algorithm 2 to search the sub-tree under a seed. Algorithm 2 essentially follows a recursive depth first strategy. It enumerates (portions of $\Psi(\mathcal{A}, s*)$ under a seed) and explores (for most beneficial cascade) simultaneously in order to achieve good performance. And during the search, it maintains the lowest cost beneficial cascade seen so far. During the search, it also keeps checking for any free threads in the system (refer line 25 in Algorithm 2). **If free threads are found, then the algorithm internally divides the search amongst the available threads.**

Algorithm 2. Determine Most Beneficial Cascade in Sub-tree of $\Psi()$

Input: (a) $<X$ service centers to ignore, next service center nsc, $bcost$, $cpath>$, (b) α: allowed breadth of the search in assignment subspace tree ($\alpha \leq |S| - 1$), (c) β: allowed length of cascade ($\beta \leq |S| - 1$), (d) ϕ base penalty seen at root of $\Psi()$

Output: Returns the most beneficial cascade in the sub-tree rooted at nsc.

```
 1: X ← X ∪ {nsc}
 2: current cost of cascade ccost ← bcost + penalty at nsc (if any) - φ
 3: if ccost < TiCost then
 4:    TiCost = ccost and TiPath = cpath
 5: end if
 6: if ccost > current best cascade cost TiCost then
 7:    Return
 8: end if
 9: if nsc had available capacity then
10:    Return
11: end if
12: if β == 0 then
13:    Return
14: end if
15: Initialize BestSucessor min heap
16: for all service center sᵢ ∈ S − X do
17:    ω(nsc, sᵢ) ← Cost of transferring best transfer-demand node (v*(nsc, sᵢ)) from
       nsc to sᵢ.
18:    Push the tuple < sᵢ, ω(nsc, sᵢ), v*(nsc, sᵢ) > into BestSucessor heap
19: end for
20: i = 1
21: while i ≤ α And BestSucessor not empty do
22:    < sᵢ, ω(nsc, sᵢ), v*(nsc, sᵢ) > ← extract min from BestSucessor
23:    nbcost ← bcost + ω(nsc, sᵢ)
24:    cpath ← cpath ∪ < v*(nsc, sᵢ), sᵢ >
25:    Recursively call Algo 2 with parameters < X, sᵢ, nbcost, cpath >, α, φ, β − 1
       /*If free threads are available then each recursive call in the loop can be taken
       forward by different thread. */
26:    Increment i
27: end while
28: Return
```

Enumeration Strategy: At any given internal node u in $\Psi()$ (initially this would be the service center node from the seed), the algorithm determines the *best children* of u (Lines 15–24) to roll forward the recursion (Line 25). For determining the best children, the algorithm uses a strategy similar to that of the seed determination algorithm discussed previously. However note that, we can enumerate a maximum of only α children at any stage. Moreover, according to the definition of $\Psi()$, no service center is allowed to repeat on any root to node path in $\Psi()$. Thus, as the algorithm proceeds deep into the recursion, it would progressively enumerate fewer children.

Termination Condition: With the intention of making the depth first search faster, we have implemented the following three termination conditions for backtracking the search: (a) We reach a service center which has free space (i.e., it not overloaded). (b) Total number of transfers already planned in this cascade become larger than a certain threshold (parameter β in Algorithm 2). (c) The cost of the current cascade (accumulated till now) becomes greater than cost of the current best beneficial cascade.

4   Analytical Evaluation

We first derive the time complexity of the seed generation process and Algorithm 2. Following this, we detail the time complexity of ASRAC. Let the input road network $G(V, E)$ contain n vertices and m edges. Furthermore, assume that S is set of service centers and D is set of demand vertices given in the input. Typically, in any realistic problem scenario, $|D|$ is $O(n)$ and $|S|$ is much less than $|D|$. Thus, $|S|$ can be considered as constant $O(1)$.

Finding Seeds from $\Psi()$ for Parallel Exploration: This algorithm determines α number of seeds from $\Psi()$ for parallel exploration. As mentioned earlier, these seeds are basically the best α children of root service center $s*$ in $\Psi()$. The algorithm computes the seeds in three major steps. First, it determines the best transfer-demand vertex for all pairs $<s*,s_i>$ ($\forall s_i \in S - \{s*\}$). This would take a maximum of $O(n|S|)$ time. Note that this is an absolute worst-case upper bound and it happens when $s*$ has $O(n)$ boundary vertices. Following the first step, the algorithm prepares potential seeds using the best transfer-demand vertices determined previously and puts them in a min-heap (*BestSeeds* heap). This would take a total of $O(|S|)$ time (if heap is constructed in a single pass). And lastly, the algorithm performs α number of extract-min operations on *BestSeeds* heap to determine the seeds. This would take $O(\alpha \log |S|)$ time. Therefore, the total time complexity of this process is upper bounded by $O(n|S| + \alpha \log |S|)$ time. Given that $|S|$ is considered constant and $\alpha < |S|$, the overall complexity becomes $O(n)$.

Algorithm 2: Finding Most Beneficial Cascade in a Sub-tree of $\Psi()$: This algorithm simultaneously enumerates and explores a sub-tree under $\Psi()$. During the execution at any intermediate stage, the algorithm can enumerate a maximum of α children and no service center is allowed to repeat on any root to node path. In worst case, $\alpha = |S|$. The total number of nodes in any particular subtree under the root in $\Psi()$ can be computed by series: $1 + (|S| - 2) + (|S| - 2)(|S| - 3) + \ldots + |S|!/(|S| * (|S| - 1))$, which is of the order $O(|S|^{|S|})$. At each node, the algorithm selects a maximum of α best children to take forward the recursion. This process is very similar to selecting seeds at root. Thus it can take maximum of $O(n)$ time. Therefore, the total running time of Algorithm 2 is upper bounded by $O(n|S|^{|S|})$. Note that, though $|S|$ is constant, the term $|S|^{|S|}$ is still quite significant.

Algorithm 1: ASRAC Algorithm: The ASRAC algorithm begins by computing shortest path distances between all pairs of demand vertices and service centers. This step takes $O(n^3)$ time and was done using using floyd-warshall algorithm [3]. Note that we used floyd-warshall algorithm as a black-box, one can replace it with any other shortest path algorithm (e.g., [6]). Following this, the algorithm determines the closest service center for each demand node which takes $O(n|S|)$ time. The algorithm then prepares the *ClosestSC* heap (Line# 3 in Algorithm 1) which takes $O(|D|)$ $(O(n))$ time.

From step 5 on wards, the algorithm enters the main *while* loop which has $|D|$ $(O(n))$ iterations. Following key tasks (from the perspective of time complexity) are done in each iteration of the loop: (1) extract-min from *ClosestSC* heap; (2) create α number of seeds and put them in a job queue; (3) threads use Algorithm 2 to determine the most beneficial cascade in the sub-trees defined by seeds in the job queue. Task (1) would take $O(\log n)$ time. Task (2) would take $O(n)$ time. Task (3) would take $O(\alpha n|S|^{|S|})$ time as we have α jobs in the job queue. To summarize, the total complexity of the ASRAC algorithm would be $O(n^3) + O(n) + O(n|S|) + O(n\log n + n^2 + n^2|S|^{|S|})$. This would be $O(n^3 + n^2|S|^{|S|})$.

5   Experimental Evaluation

We used the road network of New Delhi, India in our experiments. This dataset was obtained from OpenStreetMaps [www.openstreetmap.org]. Algorithms were implemented in Java 1.8 and Java Threads were used for creating threads in ASRAC algorithm. Experiments were conducted on a Ubuntu machine (with 90 GB RAM) capable of 100 threads (multiple Intel Xeon E-8870 v3 CPUs).

Candidate Algorithms: Following 3 algorithms were compared in experiments.

1. **Our proposed ASRAC algorithm:** We ran ASRAC algorithm by varying its α. Max length of cascade was #service centers-1.
2. **LoRaL algorithm [12]:** We ran LoRaL (a serial algorithm) by setting its k to the number of service centers ($|S|$). This means that LoRaL would explore $|S|$ number of potential cascades to choose the best for re-organization. Note that this is the maximum amount of exploration that LoRaL algorithm can perform on any CCA problem instance. In other words, **the solution quality of LoRaL reported in this section is the best that can be attained by LoRaL**. Max length of cascade is set to #service centers-1.
3. **Min-Cost Bipartite Matching:** Procedure given in [2]

In our experiments, we randomly chose some vertices from the graph and made them service centers. The remaining vertices were designated as demand vertices (each with unit demand). The number of service centers was varied in the experiments according to ratio $|D| : |S|$. A ratio of 600 : 1 implies that we have one service center for every 600 demand vertices. For a given number of service centers and demand, the total capacity across all service centers is given by $\theta * |D|$. This total capacity is distributed amongst $|S|$ service centers randomly. For sake of easy interpretation of results, instead of generating monotonic functions, we assume that penalty value of a service center is a random value between 1 and μ. In our experiments, $\mu = 200$ or $\mu = 500$. We pre-computed the shortest path distances between all pairs of demand vertices and service centers.

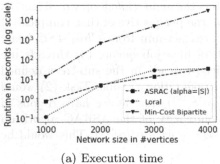

#vertices in graph	ASRAC (alpha=|S|)	LoRaL (k=|S|)	Min-cost Bipartite matching
1000	106796	106796	106643
2000	177758	177814	177814
3000	350925	351170	349861
4000	356461	357565	355524

(a) Execution time   (b) Final Objective function value

Fig. 2. ASRAC vs Min-Cost Bipartite Matching algorithm. $|D| : |S|$ was 300 : 1

Comparing ASRAC with Min-Cost Bipartite Matching: Fig. 2 illustrates these results. Here, $\theta = 0.50$ and $\mu = 200$. Following conclusions were drawn from this experiment: (a) Optimal algorithm for min-cost bipartite matching does not scale beyond graphs containing 4000 vertices. Even its main memory requirements grew very fast. It requires greater than 90 GB RAM for a CCA problem instance with 4000 demand vertices and 40 service centers. (b) ASRAC is much closer to optimal solution than LoRaL and continues to be closer to optimal solution (than LoRaL) as the number of vertices in graph increase.

Comparing ASRAC with LoRaL Algorithm: Fig. 3 illustrates these results. Here, the input graph had 23,000 vertices, total capacity parameter $\theta = 0.70$ and $\mu = 500$. Following conclusions were drawn from this experiment: (1) ASRAC obtains significantly lower values of objective function without spending too much time in execution (refer third column in Fig. 3(b)). As Fig. 3(b) shows, ASRAC's final objective function value was lower by at least 20,000 units. (2) Increasing α increases run-time but also improves solution quality. On graph with 65,000 vertices ($|D| : |S| = 800 : 1$, $\theta = 0.70$ and $\mu = 500$) **ASRAC obtained a objective function value which was lower (than LoRaL) by almost 125000 units.** Those results are not shown due to lack of space. Runtime of LoRaL was around 40 s in this experiment, which is less than ASRAC. However,

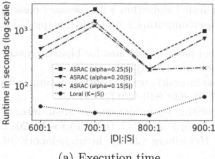

(a) Execution time

|D|:|S|	LoRaL (k=|S|)	ASRAC (alpha=0.25|S|)	Difference in Objective value LoRaL - ASRAC
600:1	3619817	3599597	20220
700:1	3384470	3336413	48057
800:1	4380739	4354375	26364
900:1	5084700	5055659	29041

(b) Final Objective function value

Fig. 3. Comparing ASRAC with LoRaL algorithm. #vertices in graph 23,000.

the key thing to note is the drastic improvement in solution quality of ASRAC over LoRaL. And the dominance increases with increase in graph size.

Conclusion: Capacity constrained assignment on road networks is a societally important problem. CCA problem can be reduced to min-cost bipartite matching problem. However, optimal algorithms for min-cost bipartite matching cannot scale to large problem instances. Our proposed ASRAC algorithm is able to scale-up while maintaining better solution quality over alternative approaches.

Acknowledgement. Microsoft Azure credits (AI for Health scheme), Mr Kapish Malik, DST (ECR/2016/001053) and IIT Ropar.

References

1. Azad, A., Buluç, A.: Distributed-memory algorithms for maximum cardinality matching in bipartite graphs. In: Proceedings of the IPDPS, pp. 32–42 (2016)
2. Bortnikov, E., et al.: The load-distance balancing problem. Networks **59**(1), 22–29 (2012)
3. Cormen, T.H., Stein, C., Rivest, R.L., Leiserson, C.E.: Introduction to Algorithms, 2nd edn. McGraw-Hill Higher Education, New York (2001)
4. Deveci, M., Kaya, K., Uçar, B., Çatalyürek, Ü.V.: GPU accelerated maximum cardinality matching algorithms for bipartite graphs. In: Wolf, F., Mohr, B., an Mey, D. (eds.) Euro-Par 2013. LNCS, vol. 8097, pp. 850–861. Springer, Heidelberg (2013). https://doi.org/10.1007/978-3-642-40047-6_84
5. Fenner, S., Gurjar, R., Thierauf, T.: A deterministic parallel algorithm for bipartite perfect matching. Commun. ACM **62**(3), 109–115 (2019)
6. Geisberger, R., Sanders, P., Schultes, D., Delling, D.: Contraction hierarchies: faster and simpler hierarchical routing in road networks. In: McGeoch, C.C. (ed.) WEA 2008. LNCS, vol. 5038, pp. 319–333. Springer, Heidelberg (2008). https://doi.org/10.1007/978-3-540-68552-4_24
7. Hu, J., et al.: Towards energy efficient hybrid on-chip Scratch Pad Memory with non-volatile memory. In: 2011 Design, Automation Test in Europe, pp. 1–6 (2011)

8. Jiang, J., Wu, L.: Two-stage distributed parallel algorithm with message passing interface for maximum flow problem. J. Supercomputing 1–19 (2014). https://doi.org/10.1007/s11227-014-1314-7

9. Langguth, J., et al.: On parallel push-relabel based algorithms for bipartite maximum matching. Parallel Comput. **40**(7), 289–308 (2014)

10. Li, J., et al.: Resource allocation robustness in multi-core embedded systems with inaccurate information. J. Syst. Archit. **57**(9), 840–849 (2011)

11. Lingas, A., Persson, M.: A fast parallel algorithm for minimum-cost small integral flows. In: Kaklamanis, C., Papatheodorou, T., Spirakis, P.G. (eds.) Euro-Par 2012. LNCS, vol. 7484, pp. 688–699. Springer, Heidelberg (2012). https://doi.org/10.1007/978-3-642-32820-6_68

12. Mehta, A., Malik, K., Gunturi, V.M.V., Goel, A., Sethia, P., Aggarwal, A.: Load balancing in network Voronoi diagrams under overload penalties. In: Hartmann, S., Ma, H., Hameurlain, A., Pernul, G., Wagner, R.R. (eds.) DEXA 2018. LNCS, vol. 11029, pp. 457–475. Springer, Cham (2018). https://doi.org/10.1007/978-3-319-98809-2_28

13. Nagy, N., Akl, S.G.: The maximum flow problem: a real-time approach. Parallel Comput. **29**(6), 767–794 (2003)

14. Okabe, A., et al.: Generalized network Voronoi diagrams: concepts, computational methods, and applications. Intl. J. GIS **22**(9), 965–994 (2008)

15. Parsons, E., et al.: School catchments and pupil movements: a case study in parental choice. Educ. Stud. **26**(1), 33–48 (2000)

16. Qiu, H., et al.: An efficient key distribution system for data fusion in V2X heterogeneous networks. Inf. Fusion **50**, 212–220 (2019)

17. Qiu, M., et al.: Data allocation for hybrid memory with genetic algorithm. IEEE Trans. Emerg. Topics Comput. **3**(4), 544–555 (2015)

18. Leong, H.U., et al.: Optimal matching between spatial datasets under capacity constraints. ACM Trans. Database Syst. **35**(2), 9:1–9:44 (2010)

19. Yang, K., et al.: Capacity-constrained network-Voronoi diagram. IEEE Trans. Knowl. Data Eng. **27**(11), 2919–2932 (2015)

A Dynamic Scheduling Strategy of ADMM Sub-problem Optimization Algorithm Based on Hierarchical Structure

Jiawei Ji[ID], Yongmei Lei[✉], and Shenghong Jiang

School of Computer Engineering and Science, Shanghai University, Shanghai, China
lei@shu.edu.cn

Abstract. The Alternating Direction Method of Multiplier (ADMM) is a simple algorithm to resolve decomposable convex optimization problems, especially effective in solving large-scale problems. However, this algorithm suffers from the straggler problem its updates have to be synchronized. Therefore, the asynchronous ADMM algorithm is proposed. However, the convergence speed of the ADMM algorithm is not very satisfactory. In this paper, we propose a dynamic scheduling strategy for sub-problems-automatically calling different algorithms at different iteration periods of each iteration, and combining this strategy with a hierarchical communication structure. The experiments based on ZiQiang 4000 cluster experimental environment show that the dynamic scheduling strategy based on hierarchical communication structure can solve the ADMM sub-problem and effectively improve the convergence speed and communication efficiency of the algorithm.

Keywords: ADMM · Asynchronous · Dynamic scheduling strategy · Hierarchical communication structure · Distributed optimization

1 Introduction

Machine learning has achieved unprecedented success in many fields, which has completely changed the development direction of artificial intelligence and triggered the arrival of the era of big data. One of the most challenging problems is solved by distributed machine learning.

In general, many distributed machine learning algorithms can be attributed to the global variable consistency optimization problems, such as Linear regression [1], Logistic regression [2] and Support Vector Machine (SVM) [3], and [4] proved that these problems can be effectively solved by ADMM. ADMM algorithm is a simple way to solve the problem of decomposable convex optimization, especially to deal with the problem of large-scale productive. The ADMM algorithm equably decomposes the objective function into several sub-problems for solving, and then solves each sub-problem in parallel.

Supported by the Natural Science Foundation of China under grant No. U1811461.

The running time of ADMM algorithm is mainly composed of communication time and calculation time. In the implementation of consistency problem, due to the existence of consistency conditions, the nodes must be synchronized. In a distributed cluster, synchronous operations between nodes mean a lot of communication. Obviously, a lot of time will be spent on communication. Therefore, performing communication operations at the right time and the right communication structure are key to reduce runtime.

Currently, there are generally three consistency models, namely BSP [5], ASP and SSP [6]. In terms of the distributed implementation of ADMM algorithm, [4] solved the global consistency problem in two distributed programming environments – MPI and MapReduce through ADMM algorithm. [2] actually solved the regular Logistic regression problem of L2 on Hadoop by using ADMM algorithm. However, the basic principle of these methods are based on the algorithm implementation of BSP model. This will undoubtedly cost a lot of time for synchronization. To alleviate this problem, [7] proposed asynchronous ADMM algorithm(AD-ADMM) for global consistency problem, and used asynchronous SSP model to implement ADMM algorithm to solve the problem of large-scale distributed global consistency optimization. Unlike the synchronous BSP model, the asynchronous model allows the nodes to work more independently, is more flexible in their communication strategies. The asynchronous approach is more efficient than the synchronous one. However, most asynchronous ADMM algorithms in [8] et al. use Master-slave mode, which is not efficient in multi-core distributed clusters. In order to address this problem, [9] proposed a group-based ADMM (GADMM). [10] designed a communication structure for ADMM algorithm.

In addition, the convergence speed of the algorithm has a great impact on the calculation speed. From the perspective of the ADMM algorithm itself, the penalty term ρ has a significant effect on the number of convergence times of the algorithm. We will not discuss it in this paper. On the other hand, considering the optimization algorithm for solving the x-subproblem of ADMM, without considering the communication, during a single iteration of the ADMM algorithm, the computational overhead required for the x update is the largest. If the most suitable optimization algorithm can be automatically selected at the right time, the number of convergence times and the convergence time of the ADMM algorithm may be improved. And there is no in-depth research on this aspect, which is vital for algorithm speed.

In this paper, we propose a dynamic scheduling optimization algorithm based on hierarchical communication structure to solve the x-subproblem of ADMM. This method improves the communication efficiency and has a faster convergence speed. The hierarchical communication structure improves the communication efficiency of the algorithm, and the dynamic scheduling strategy accelerates the convergence speed of ADMM. In this paper, the optimization algorithm is mainly selected by residuals. And this dynamic scheduling strategy can call a suitable optimization algorithm according to different convergence stages in each iteration. This paper tests the dynamic scheduling strategy on the Shanghai University ZiQiang 4000 cluster system. The test results show the acceleration effect of the dynamic scheduling strategy based on the hierarchical communication structure.

2 Problem Description and Related Work

2.1 Distributed Alternating Direction Method of Multipliers

In general, many distributed machine learning problems can be expressed as the following global consistency optimization problems:

$$\min f(\mathbf{x}) = \sum_{i=1}^{N} f_i(x_i) \tag{1a}$$

$$\text{s.t. } x_i - z = 0, i = 1, \ldots, N \tag{1b}$$

where $\mathbf{x} \in R_n, f_i : R_n \rightarrow R \cup \{+\infty\}$, z is a consistency variable. Constraints indicate that all local variables x_i need to be equal to each other. In (1a) the objective function $f(x)$ is divided into N parts, so the solution of the problem can be realized in a distributed environment with N processes.

The iterative formula for solving the problem (1) by ADMM algorithm is as follows:

$$x_i^{k+1} = \underset{x_i}{\operatorname{argmin}} \left(f_i(x_i) + y_i^{kT} \left(x_i - z^k \right) + \frac{\rho}{2} \left\| x_i - z^k \right\|_2^2 \right) \tag{2}$$

$$z^{k+1} = \underset{z}{\operatorname{argmin}} \left(f_i \left(x_i^{k+1} \right) + y_i^{kT} \left(x_i^{k+1} - z^k \right) + \frac{\rho}{2} \left\| x_i - z^k \right\|_2^2 \right)$$
$$= \frac{1}{N} \sum_{i=1}^{N} \left(x_i + \frac{1}{\rho} y_i \right) \tag{3}$$

$$y_i^{k+1} = y_i^k + \rho \left(x_i^{k+1} - z^{k+1} \right) \tag{4}$$

Where y_i is a local Lagrangian dual variable and $\rho > 0$ is a penalty term parameter.

It can be found that the update of x_i^{k+1} and y_i^{k+1} can be performed independently and in parallel on separate processes, and the consistency variable z needs to aggregate the local variables to update in each iteration. Every update of the global parameter z requires all parameters x_i^{k+1} and y_i^{k+1}. Therefore, the computing performance of each Slave process in the cluster has a great impact on the overall iteration speed.

In practical applications, the global parameter update strategy of BSP is very limited. First, the performance limitation of a node in the physical cluster may affect the overall speed of the cluster; Secondly, the synchronized parameter update strategy causes a lot of time. These problems in the BSP model can be alleviated by the SSP model.

2.2 Asynchronous Distributed Alternating Direction Method of Multipliers

AD-ADMM [9] use SSP's parameter synchronization strategy for z update. That is, the AD-ADMM algorithm is implemented in a distributed environment by means of limited asynchronous and local updates, which not only reduces time overhead, but also ensures convergence, and also improves fault tolerance to a certain extent.

AD-ADMM algorithm divides the processes into a Master and N Workers. The Master does not have to wait for all Workers, but receives parameters sent by A(0 < A < N) Workers to update the global variable z. To ensure convergence, a limited asynchronous strategy is used in AD-ADMM. The algorithm sets a clock value for each process. After each iteration of the process, the clock value is increased by one. For a Worker whose clock value is greater than the time window $\tau > 0$, the Master needs to wait for the Worker to finish sending.

The update process of Master and Workers in the AD-ADMM is given in Table 1.

Table 1. Asynchronous Distributed ADMM (AD-ADMM).

AD-ADMM
Processing by the Master:
1: **initialize:** z, k=0, $d_1 = d_2 = \cdots = d_N = 0$.
2: **broadcast** z to all Workers.
3: **repeat**
4:   **wait** until receiving $\{\widehat{x}_i, \widehat{y}_i\}$ from Workers i, $i \in A_k$ such that $|A_k| \geq A$ and $\forall i \in A_k^c, d_i < \tau$.
5:   **update**
$$x_i^{k+1} = \begin{cases} \widehat{x}_i, i \in A_k \\ x_i^k, i \in A_k^c, \end{cases} \quad y_i^{k+1} = \begin{cases} \widehat{y}_i, i \in A_k \\ y_i^k, i \in A_k^c, \end{cases} \quad d_i = \begin{cases} 1, i \in A_k \\ d_i + 1, i \in A_k^c, \end{cases}$$ $$z^{k+1} = \underset{z}{argmin}(f_i(x_i^{k+1}) + y_i^{(k+1)T}(x_i^{k+1} - z) + \frac{\rho}{2}\|x_i^{k+1} - z\|_2^2).$$
6:   **broadcast** z^{k+1} to the Workers i, $i \in A_k$.
7:   **set** $k \leftarrow k + 1$.
8: **until** the stopping criterion is satisfied.
9: **output** z^k.
Processing by Worker i:
1: **initialize:** $x_i, y_i, k_i = 0$.
2: **repeat**
3:   **wait** until receiving $\widehat{z}$ from Master.
4:   **update**
$$x_i^{k_i+1} = argmin_{x_i}(f_i(x_i) + y_i^{k_i T}(x_i - \widehat{z}) + \frac{\rho}{2}\|x_i - \widehat{z}\|_2^2),$$ $$y_i^{k_i+1} = y_i^{k_i} + \rho(x_i^{k_i+1} - \widehat{z}).$$
5:   **sent** $\{x_i^{k_i+1}, y_i^{k_i+1}\}$ to Master.
6:   **set** $k_i \leftarrow k_i + 1$.
7: **until** the stopping criterion is satisfied.

where k is the clock value of the Master and k_i is the local clock value of each Worker. The $\{d_1, d_2 \ldots d_N\}$ in the Master records the clock values of the N Workers that reached the last time. A_k is the set of Workers that reach the Master when the Master clock value is k, and A_k^c is the complement of A_k, that is, the set of Workers that did not reach the Master at the k-th iteration.

2.3 Hierarchical Communication Structure

As shown in Fig. 1, the Hierarchical Communication Structure(HCS) divides the process into two categories, Master and Worker. Since each Master needs to communicate with a part of Workers, in order to reduce the number of communication between nodes during the Master-Worker communication process, HCS associates processes with each node according to the Round Robin algorithm, and sets the first process of each node as a Master, Called SubMaster. In addition, a Master process is set up in the cluster to communicate with each SubMaster.

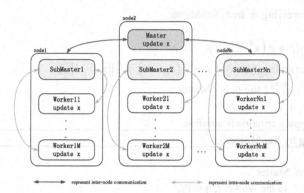

Fig. 1. Hierarchical communication structure

Combining HCS structure and AD-ADMM algorithm, HAD-ADMM (AD-ADMM based on HCS) is proposed. HAD-ADMM divides the algorithm into three parts: Worker, SubMaster and Master. Worker updates y_{ij} and x_{ij}, and then passes x_{ij} and y_{ij} to SubMaster of this node. The SubMaster performs decentralized updates through the parameters passed by Worker in this node, then sends the updated parameter $z_i^{k_i+1}$ to the Master. After the Master receives $z_i^{k_i+1}$ from each node, it aggregates them, calculates the final global variable z^{k+1}, and sends it to each SubMaster.

Similar to AD-ADMM, HAD-ADMM also sets a clock value for all processes. The update process of Master, SubMaster and Workers in the AD-ADMM is given in Table 2.

Table 2. AD-ADMM based on HCS (HAD-ADMM)

HAD-ADMM:

Processing by Worker j on SubMaster i:

1: **initialize:** $x_{ij}, y_{ij}, k_i = 0$.

2: **repeat**

3: **wait** until receiving $\hat{z}$ from SubMaster.

4: **update**

$$y_{ij}^{k_i+1} = y_{ij}^{k_i} + \rho\left(x_{ij}^{k_i+1} - \hat{z}\right).$$

$$x_{ij}^{k_i+1} = \operatorname{argmin}_{x_{ij}}(f_{ij}(x_{ij}) + y_{ij}^{k_i T}(x_{ij} - \hat{z}) + \frac{\rho}{2}\|x_{ij} - \hat{z}\|_2^2).$$

5: **sent** $\{x_{ij}^{k_i+1}, y_{ij}^{k_i+1}\}$ to i-th SubMaster.

6: **set** $k_i \leftarrow k_i + 1$.

7: **until** the stopping criterion is satisfied.

Processing by SubMaster i:

2: **repeat**

3: **wait** $\hat{z}$ from Master.

4: **broadcast** $\hat{z}$ to the Worker j, $j \in P_i$.

5: **wait** $\{\{\widehat{x_{i1}}, \widehat{y_{i1}}\}, \{\widehat{x_{i2}}, \widehat{y_{i2}}\}, \cdots, \{\widehat{x_{iM_i}}, \widehat{y_{iM_i}}\}\}$ from all Workers in P_i.

6: **compute:**

$$z_i^{k_i+1} = \frac{1}{N}\sum_{j=1}^{M_i}(x_{ij}^{k_i+1} + \frac{1}{\rho}y_{ij}^{k_i+1}).$$

7: **send** $z_i^{k_i+1}$ to Master.

8: **set** $k_i \leftarrow k_i + 1$.

9: **until** the stopping criterion is satisfied.

Processing by Master:

1: **initialize:** $z, k=0, d_{1'} = d_{2'} = \cdots d_{N'} = 0$.

2: **broadcast** z to all SubMasters.

3: **repeat**

4: **wait** until receiving $\{\widehat{z_1}, \cdots, \widehat{z_{N_n}}\}$ from SubMaster i, $i \in A_k'$ such that $|A_k'| \geq A_n$ and $\forall i \in A_k^c, d_{i'} < \tau$.

5: **update** $z_i^{k+1} = \begin{cases} \widehat{z_i}, i \in A_k' \\ z_i^k, i \in A_k'^c \end{cases}$ $d_i = \begin{cases} 1, i \in A_k' \\ d_i + 1, i \in A_k'^c \end{cases}$ $z^{k+1} = \sum_{i=1}^{N_n} z_i^{k+1}$.

6: **broadcast** z^{k+1} to the SubMaster i, $i \in A_k'$.

7: **set** $k \leftarrow k + 1$.

8: **until** the stopping criterion is satisfied.

9: **output** z^k.

Where P_i represents the set of Worker processes in the i-th node. A_k' ($A_n \leq A_k' \leq N_n$) is the set of SubMasters that reach the Master when the clock value is k. $\{d_{1'} = d_{2'} = \cdots d_{N'}\}$ is the number of clock values from the last time that each SubMaster reached the Master.

3   Dynamic Scheduling Strategy Based on HAD-ADMM

This section discusses the influence of the optimization algorithm for solving the x-subproblem on the convergence of the algorithm. Some simple simulation experiments are used to illustrate the effect of different sub-problem optimization algorithms on the number of convergences of the algorithm. Then according to the characteristics of different optimization algorithms, a dynamic scheduling strategy based on HAD-ADMM is proposed to improve the convergence speed of ADMM algorithm. And the scheduling basis is introduced.

3.1   Optimization Algorithm for Solving Sub-problem

From the foregoing formulas (2), (3), and (4), it can be found that the update of x is a problem of solving the optimal value, the update of y is a simple gradient ascent process, and the update of z is an average process. For the optimal value problem of the x-subproblem, it is mostly solved by other optimization algorithms. Such as gradient descent, conjugate gradient method, Newton method or quasi-Newton method. The common point of these algorithms is that the algorithm process is an iterative solution process. Therefore, without considering communication, during the ADMM algorithm iteration, the update cost of x is the largest. So the solution of the x-subproblem also greatly affects the running time of the ADMM algorithm.

The optimization algorithms for solving sub-problems have their own characteristics. This section uses a small two-node cluster with two processes running on each cluster. For the same ADMM algorithm problem, the L-BFGS and conjugate gradient method are used to solve.

This section uses a small two-node cluster with two processes running on each cluster. For the same ADMM algorithm problem, the L-BFGS and conjugate gradient method are used to solve.

As can be seen from Fig. 2, the effects of different optimization algorithms are different in different convergence stages. The FRCG algorithm converges faster at the early stage of the iteration, while the L-BFGS algorithm converges faster at the later stage of the iteration.

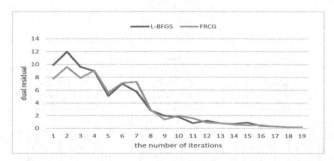

Fig. 2. Algorithm iteration process of solving x-subproblem with different optimization algorithms

It was also found in the experiment that in the case of the same optimization algorithm, the time required for one iteration of different processes is different. The time required for iteration using the L-BFGS algorithm is shown in Fig. 3.

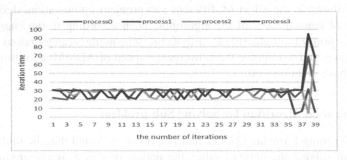

Fig. 3. The running time of solving x-subproblem for different processes in the iteration process of L-BFGS

These running time differences may be due to different performance of different nodes, or due to different training data on different processes. Therefore, the optimization algorithm for solving the x-subproblem also has a great influence on the convergence time of the ADMM algorithm. In practical applications, if the most suitable optimization algorithm can be automatically selected at the right time, the number of convergence times and the convergence time of the ADMM algorithm may be improved.

3.2 Dynamic Scheduling Strategy Based on HCS

The main idea of the dynamic scheduling strategy is to use different optimization algorithms at different periods of each iteration.

The basic idea is shown in Fig. 4. Specifically, the strategy of dynamic scheduling optimization algorithms can be divided into the following five steps.

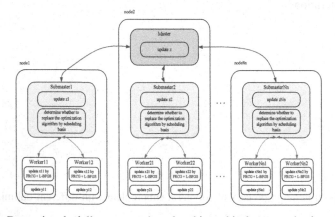

Fig. 4. Dynamic scheduling strategy based on hierarchical communication structure

First, divide all Worker nodes into groups. We set a node as a group and every group contains multiple processes with each process allocated by round robin Round Robin algorithm. And sets the first process of each group as a Master, Called SubMaster. In addition, a Master process is set up in the cluster to communicate with each SubMaster.

Secondly, the local variable x_i of each Worker process is updated with different optimization algorithms at different times of the iteration. In this paper, we only use two scheduling algorithms, using the FRCG algorithm early in the iteration, and using the L-BFGS algorithm when it is close to convergence.

Subsequently, how to determine whether it is time to replace the algorithm is an important step in this strategy. Combined with the HCS and AD-ADMM synchronization strategy, the selected process can be incorporated into the update of the SubMaster. The SubMaster process judges whether it is time to replace the algorithm according to the proposed decision strategy, and updates the local global variable z_i.

And then the Master process receives z_i from each SubMaster process to update the global variable z, and then sends the latest global variable z to the corresponding SubMaster process for the next iteration. Finally, each SubMaster process sends all parameters and replacement sub-problem optimization algorithm flags to the Worker processes in its own group.

3.3 The Dynamic Scheduling Basises

The ADMM algorithm mainly uses the primal residual r^k and dual residual s^k to make convergence judgments:

$$\left\| r^k \right\|_2^2 = \sum_{i=1}^{N} \left\| x_i^k - z^k \right\|_2^2 \tag{5}$$

$$\left\| s^k \right\|_2^2 = N\rho^2 \left\| z^k - z^{k-1} \right\|_2^2 \tag{6}$$

Therefore, the primal residual and dual residual can measure the degree of convergence of the current variable. This paper also uses the residuals as the basis for dynamic scheduling, and uses the residuals to determine the degree of convergence. The FRCG algorithm is used before reaching a certain degree of convergence, and the L-BFGS algorithm is used after reaching a certain degree of convergence. There are different choices based on different residuals:

(1) primal residual r^k: The primal residual is in the form of the sum of the squares of the L2 norms of the differences between each sub-variable of x^k and z^k. Because the dynamic scheduling process is combined with the algorithm update of the Sub-Master part, at this time, the SubMaster has not yet performed the sending operation to the Master, so z^k is unknown. Therefore, the intermediate result z_i of the local update of z in SubMaster can be used instead of z^k in the calculation formula of the primal residual. At this time, the local primal residual calculation formula is:

$$\left\| r_{ij}^k \right\|_2^2 = \left\| x_{ij}^k - z_i^k \right\|_2^2 \tag{7}$$

Where, the subscript of r_{ij} indicates the j-th Worker on the i-th node. The dynamic scheduling strategy of the SubMaster part based on the primal residual is shown in Table 3:

Table 3. ADMM with dynamic scheduling strategy based on primal residual (DSP-ADMM).

DSP-ADMM: Processing by SubMaster i:
1: **initialize:** $x_i, y_i, P_i, k_i = 0, F_j = 1, \varepsilon_{DSP}$.
2: **repeat**
3:     **wait** $\{\{x_{i1}, y_{i1}\}, \{x_{i2}, y_{i2}\}, ..., \{x_{iM_i}, y_{iM_i}\}\}$ from all Workers in P_i .
4:     **compute:**
$\quad\quad z_i^{k_i+1} = \frac{1}{N_n} \sum_{j=1}^{M_i} (x_{ij}^{k_i+1} + \frac{1}{\rho} y_{ij}^{k_i+1}).$
$\quad\quad r_j^k = \left\| x_{ij} - z_i^{k_i+1} \right\|_2^2.$
5:     **if** $r_j^k < \varepsilon_{DSP}$ then F_j=0   else F_j=1
6:     **send** $z_i^{k_i+1}$ to Master.
7:     **wait** $\hat{z}$ from Master.
8:     **broadcast** $\hat{z}, F_j$ to the Worker j ,j $\in P_i$.
9:     **set** $k_i \leftarrow k_i + 1$.
10: **until** the stopping criterion is satisfied.

Where F_j is the flag bit of the replacement sub-problem optimization algorithm, and ε_{DSP} is the set threshold. When the primal residual is less than this value, it is determined that it reaches the middle and late iterations, and L-BFGS algorithm can be used instead.

(2) dual residual s^k: The main form of the dual residual is the sum of squares of the L2 norm of the difference between the result of this iteration z^k and the result of the previous iteration z^{k-1}. As with the previous primal residual as the basis for decision-making, z^k is unknown. At this time, z^k can be replaced by the calculation result x_{ij} of each Worker process on the same node as SubMaster. The formula for calculating the local dual residuals is:

$$\left\| s_{ij}^k \right\|_2^2 = N\rho^2 \left\| x_{ij}^k - z^{k-1} \right\|_2^2 \tag{8}$$

The dynamic scheduling strategy of the SubMaster part based on the dual residual is shown in Table 4:

Table 4. ADMM with dynamic scheduling strategy based on dual residual (DSD-ADMM).

DSD-ADMM: Processing by SubMaster i:

1: **initialize:** $x_i, y_i, P_i, k_i = 0, F_j = 1, \varepsilon_{DSD}$.

2: **repeat**

3:    **wait** $\{\{x_{i1}, y_{i1}\}, \{x_{i2}, y_{i2}\}, \ldots, \{x_{iM_i}, y_{iM_i}\}\}$ from all Workers in P_i .

4:    **compute:**
$$s_j^k = \left\| x_{ij} - z_i^{k_i} \right\|_2^2.$$
$$z_i^{k_i+1} = \frac{1}{N_n} \Sigma_{j=1}^{M_i} (x_{ij}^{k_i+1} + \frac{1}{\rho} y_{ij}^{k_i+1}).$$

5:    **if** $s_j^k < \varepsilon_{DSP}$ then F_j=0 else F_j=1

6:    **send** $z_i^{k_i+1}$ to Master.

7:    **wait** $\hat{z}$ from Master.

8:    **compute:**
$$z_i^{k_i} = \hat{z}.$$

9:    **broadcast** $\hat{z}, F_j$ to the Worker j,j $\in P_i$.

10:   **set** $k_i \leftarrow k_i + 1$.

11: **until** the stopping criterion is satisfied.

Where ε_{DSD} is the set threshold. When the dual residual is less than this value, it is determined that it reaches the middle and late iterations, and L-BFGS algorithm can be used instead.

The primal residual is the derivative of y in the augmented Lagrangian function, so it can represent the convergence of y. The dual residual is derived from the partial derivative of x, so it can represent the convergence of x. When the algorithm reaches convergence, x and y converge at the same time. The above two strategies use a single residual to make optimization algorithm decisions. Therefore, which strategy is better needs to be verified through experiments.

In fact, a better strategy means that its corresponding variables have a greater impact on the convergence of the ADMM algorithm. From the point of view of the ADMM algorithm iteration stop condition, y does not participate in the calculation of the two residuals, so from the perspective of the stop condition, x has a greater influence on the speed of algorithm convergence than y. From the calculation formula of primal residual and dual residual, the calculation method in the decision strategy of DSP-ADMM only includes the parameter value in one node, while the calculation method of DSD-ADMM takes into account the result of the previous iteration.

In addition, the algorithm of the Worker part also needs to be modified accordingly. Workers on the same node need to determine whether they need to replace the sub-problem optimization algorithm, so in addition to receiving the global variable z^{k+1}, the Worker also needs to receive the flag bit F_j of the sub-problem optimization algorithm determined by this iteration. The Worker part of the dynamic selection optimization strategy is shown in Table 5:

Table 5. ADMM with dynamic scheduling strategy (DSP-ADMM, DSD-ADMM).

DSP-ADMM, DSD-ADMM:
Processing by Worker j on SubMaster i:
1: **initialize:** $x_{ij}, y_{ij}, k_i = 0$.
2: **repeat**
3:    **wait** until receiving $\hat{z}, F_j$ from SubMaster.
4:    **update**
if $F_j = 1$ then choose L-BFGS else choose FRCG
$y_{ij}^{k_i+1} = y_{ij}^{k_i} + \rho\left(x_{ij}^{k_i+1} - \hat{z}\right),$
$x_{ij}^{k_i+1} = \underset{x_{ij}}{\text{argmin}}(f_{ij}(x_{ij}) + y_{ij}^{k_i T}(x_{ij} - \hat{z}) + \frac{\rho}{2}\|x_{ij} - \hat{z}\|_2^2).$
5:    **send** $\{x_{ij}^{k_i+1}, y_{ij}^{k_i+1}\}$ to i-th SubMaster.
6:    **set** $k_i \leftarrow k_i + 1$.
7: **until** the stopping criterion is satisfied.

During different iterations, the Worker process uses different optimization algorithms. In this paper, each Worker process uses FRCG to calculate the allocated data fragments in the early stage of the iteration, and uses L-BFGS to calculate the data fragments in the later stage of the iteration. Different Workers on the same node use different training data, but the update in the ADMM algorithm uses only one Worker's parameters at a time on each node. Therefore, in fact, each iteration of the ADMM algorithm is equivalent to using ideas similar to the batch algorithm.

4   Experiments and Result Analysis

This section mainly conducts convergence and performance tests on different dynamic scheduling algorithms designed based on two residuals, mainly for large-scale sparse logistic regression problems. The data set consists of 43264 samples and a dimensional space of 10000000. The nodes of the cluster are connected by Infiniband. Each node has 16 cores and 64 GB of memory. A total of 16 computing nodes were used in this experiment. The algorithm is implemented using C++ and MPICH.

The logistic regression problem can be expressed as:

$$\min_{w_i} \sum_{i=1}^{N} \sum_{j=1}^{m} \log\left(1 + \exp\left(-y_j x_j^T w_i\right)\right) \tag{9a}$$

$$\text{s.t. } w_i - z = 0, i = 1, \ldots, N \tag{9b}$$

Where, w_i is the model parameter to be solved, x_i represents the characteristics of a sample, y_i represents the sample label, m is the sample size, and the data set is evenly divided into N nodes. On this basis, the algorithm's convergence stop criterion can be expressed as:

$$\|r^k\|_2 \leq \varepsilon^{abs}\sqrt{m} + \varepsilon^{rel}\max\{\|w^k\|_2, \|z^k\|_2\} \tag{10}$$

$$\left\| s^k \right\|_2 \leq \varepsilon^{abs} \sqrt{n} + \varepsilon^{rel} \left\| y^k \right\|_2 \tag{11}$$

Where, n is the number of features in the data set, ε^{abs} means absolute error, and ε^{rel} means relative error. In this experiment, ε^{abs} is set to 10^{-2} and ε^{rel} is set to 10^{-4}. The algorithm uses the asynchronous communication mode based on the SSP model and the penalty parameter $\rho = 1$. In addition, in this experiment, ε_{DSD} is set to 1, ε_{DSP} is set to 2.

In addition to DSP-ADMM and DSD-ADMM, this section also runs the AD-ADMM algorithm for solving sub-problems with the L-BFGS algorithm and the AD-ADMM algorithm for solving sub-problems with the FRCG algorithm as comparative experiments. In the comparative experiment using a single algorithm, the number of Workers on each node is set to 2, which is consistent with the communication structure of DSP-ADMM and DSD-ADMM.

4.1 Convergence Testing and Analysis

Figure 5 shows the variation of the dual residuals and primal residuals of the four algorithms with the number of iterations when using eight computing nodes.

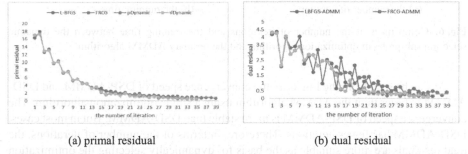

(a) primal residual   (b) dual residual

Fig. 5. Comparison of primal residual and dual residual using dynamic selector sub-problem optimization algorithm and ordinary ADMM algorithm

It can be seen from the figure that the residual rate of DSD-ADMM and DSP-ADMM is much faster than the ADMM algorithm of a single optimization algorithm in most cases. The residual is not just the smaller value of the iteration results of the two single algorithms, but it is smaller than the minimum value of the iteration results of the two single algorithms during operation. To some extent, it accelerates the convergence speed of the ADMM algorithm.

According to the optimization algorithm that tracks the dynamically selected sub-problems during the operation of the algorithm, the time for changing to the L-BFGS algorithm for different processes is different. As analyzed in the previous section, this should be related to different data allocated by different processes. From the perspective of the overall convergence of the algorithm, L-BFGS converges faster than FRCG, but in different iteration stages, the convergence speed of L-BFGS is not always faster. Therefore, the strategy of dynamically selecting the sub-problem optimization algorithm can decide which algorithm to use for iterative calculation according to different convergence stages.

4.2 Performance Testing and Analysis

This experiment is to test the performance of the algorithm, using a total of 16 computing nodes. It mainly includes two experiments: (1) compare the iteration number and iteration time of the four algorithms under different node numbers. (2) compare the running time of the four algorithms under different asynchronous conditions (An, τ);

The number of iterations to achieve convergence under different numbers of nodes for 4 different algorithms is shown in Fig. 6(a). In addition to the number of iterations, the running time of the algorithm is also an important comparison indicator. The running time comparison of the 4 algorithms is shown in Fig. 6(b).

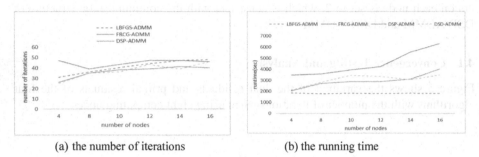

(a) the number of iterations   (b) the running time

Fig. 6. Comparison of the number of iterations and the running time between the dynamic selection sub-problem optimization algorithm and the ordinary ADMM algorithm

As can be seen from the Fig. 6(a), the convergence speed of DSP-ADMM and DSD-ADMM is faster than the ADMM algorithm using a single optimization algorithm. The convergence speed of DSP-ADMM is more stable than DSD-ADMM. But in most cases, DSD-ADMM has fewer iterations. Therefore, in terms of the number of iterations, the dual residuals are more suitable as the basis for dynamically selecting the optimization algorithm.

As can be seen from the Fig. 6(b), since the data set used in this section is more suitable for the L-BFGS algorithm, the FRCG algorithm has the most iterations, so it also has the longest running time. Although DSD-ADMM and DSP-ADMM have more operations than the ordinary ADMM algorithm, they do not increase the running time compared to the ordinary ADMM algorithm due to the use of the hierarchical communication structure(HCS). And the running time of DSP-ADMM is not much different from that of DSD-ADMM.

The setting of the finite asynchronous condition (An, τ) of the AD-ADMM has a certain effect on the convergence. This paper uses 8 nodes for different asynchronous conditions for experiments, and found that DSP-ADMM is very sensitive to these parameters in the experiments. It is difficult to achieve convergence in a limited number of iterations. Therefore, this paper only analyzes the experimental results of DSD-ADMM.

The Fig. 7 shows the change of primal residual and dual residual under different asynchronous conditions (An, τ). When An = 8 or τ = 1, the asynchronous ADMM algorithm at this time is equivalent to the synchronous ADMM algorithm. It can be seen from the Fig. 7 that when An = 4, there is similar convergence, this is, the value of

τ has little effect on convergence. This may be due to the higher performance of the experimental environment. In addition, when $\tau = 8$ and An has a small value, both the dual residual and the primal residual require more iterations to reach convergence, which may be caused by a high degree of asynchrony.

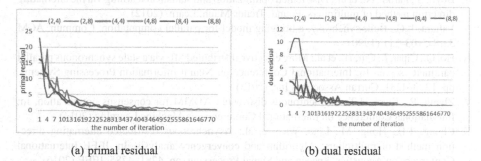

(a) primal residual   (b) dual residual

Fig. 7. Under different asynchronous conditions, the primal residual and dual residual of DSD-ADMM change with the number of iteration

5 Conclusion

This paper proposes a dynamic scheduling strategy based on a hierarchical communication structure to solve the x-subproblem of the ADMM algorithm to speed up the convergence speed. And two dynamic scheduling strategies based on primal residual and dual residual are designed. In addition, this paper also analyzes the convergence and performance of the algorithm, and concludes that the dual residual is more suitable as the dynamic scheduling basis of the optimization algorithm. Experiments conducted in the Ziqiang 4000 cluster experiment environment show that the dynamic scheduling strategy proposed in this paper can reduce the number of iterations and accelerate the convergence speed of the ADMM algorithm.

And this strategy provides support for the engineering implementation of algorithm components, can avoid the problem that the ADMM algorithm cannot converge due to inappropriate preset optimization algorithms. Of course, this proposed method still has many areas for improvement. For example, the timing of switching to an optimization algorithm may also be set using the value of the objective function.

Acknowledgment. This research was supported in part by the National Natural Science Foundation of China under grant No. U1811461 and ZiQiang 4000 cluster experimental environment of Shanghai University.

References

1. Mateos, G., Bazerque, J.A., Giannakis, G.B.: Distributed sparse linear regression. IEEE Trans. Sig. Process. **58**(10), 5262–5276 (2010)

2. Lubell-Doughtie, P., Sondag, J.: Practical distributed classification using the Alternating Direction Method of Multipliers algorithm. In: IEEE International Conference on Big Data, pp. 773–776. IEEE (2013)
3. Forero, P.A., Cano, A., Giannakis, G.B.: Consensus-based distributed support vector machines. J. Mach. Learn. Res. **11**(3), 1663–1707 (2010)
4. Boyd, S., Parikh, N., et al.: Distributed optimization and statistical learning via the alternating direction method of multipliers. Found. Trends Mach. Learn. (2012)
5. Valiant, L.G.: Bulk-synchrony: a bridging model for parallel computation. Commun. ACM **33**(8), 103–111 (1990)
6. Ho, Q., Cipar, J., Cui, H., et al.: More effective distributed ML via a stale synchronous parallel parameter server. In: International Conference on Neural Information Processing Systems, pp. 1223–1231. Curran Associates Inc. (2013)
7. Zhang, R., Kwok, J.T.: Asynchronous distributed ADMM for consensus optimization. In: International Conference on International Conference on Machine Learning (2014)
8. Chang, T.H., Hong, M., Liao, W.C., et al.: Asynchronous distributed alternating direction method of multipliers: Algorithm and convergence analysis. In: IEEE International Conference on Acoustics, Speech and Signal Processing, pp. 4781–4785. IEEE (2016)
9. Wang, H., Gao, Y., Shi, Y., Wang, R.: Group-based alternating direction method of multipliers for distributed linear classification. IEEE Trans. Cybern. **47**(11), 3568–3582 (2017)
10. Fang, L., Lei, Y.M.: An asynchronous distributed ADMM algorithm and efficient communication model. In: Dependable, Autonomic and Secure Computing, 14th International Conference on Pervasive Intelligence and Computing, 2nd International Conference on Big Data Intelligence and Computing and Cyber Science and Technology Congress (DASC/PiCom/DataCom/CyberSciTech), 2016 IEEE 14th International Conference, pp. 136–140. IEEE, August 2016

An Improved Heterogeneous Dynamic List Schedule Algorithm

Wei Hu[1], Yu Gan[1(✉)], Yuan Wen[2(✉)], Xiangyu Lv[1], Yonghao Wang[3], Xiao Zeng[1], and Meikang Qiu[4]

[1] Wuhan University of Science and Technology, Wuhan, China
rorchach2k@gmail.com
[2] Trinity College Dublin, Dublin, Ireland
weny@tcd.ie
[3] Birmingham City University, Birmingham, UK
[4] Harrisburg University Science and Technology, Harrisburg, USA

Abstract. Scheduling algorithm impacts system substantially in terms of throughput and load balance. Traditional methods rely on static criteria, such as earliest finishing time, critical path, and the importance of the nodes, to prioritise workloads towards various hardware settings. In practice, however, a global static scheduling method often works suboptimally given the dependence complexity among tasks and the performance diversity on separate hardware configurations. To cope with such issue, in this paper, we propose an improved heterogeneous dynamic list scheduling algorithm (IHDSA) to balance workload across heterogeneous cores and optimize communication overhead adaptively. The proposed algorithm performs three steps. First, it transforms the DAG task graph into a list and marks job status. Then, it calculates the shortest completion time of three distinctive scheduling schemes and selects the best solution among the three. Finally, it sets up thresholds for computing units and monitors the status to balance the usage of those cores. In our experiment, the IHDSA adaptive scheduling improves the performance significantly over the static counterpart.

Keywords: Heterogeneous computing systems · Dynamic scheduling · Load balance

1 Introduction

The heterogeneous computing system (HECS) becomes an increasingly popular platform due to its overwhelming advantage in performance and power consumption. A typical system generally consists of computing units with separate architectures integrated into the same package or connected via bus/internet [10,16,23]. The architectural heterogeneity presents flexibility for compute-intensive workloads of separate type to perform while increases the scheduling complexity given a multi-task environment [2,15,21]. The capability of using

© Springer Nature Switzerland AG 2020
M. Qiu (Ed.): ICA3PP 2020, LNCS 12452, pp. 159–173, 2020.
https://doi.org/10.1007/978-3-030-60245-1_11

resource properly and maintaining the throughput is an enduring goal for task scheduling.

The conventional scheduling algorithms apply static heuristics to allocate jobs to separate computing units. For simple implementations, the scheduler migrating a task or part of it from one processor to another if the target shorts of workloads. A more sophisticated mechanism would consider task dependency by analyzing DAG (Directed Acyclic Graph) of the application to guarantee a subsequent task performing on the same device it parent job does [4,9,17]. The performance is thereby maintained because of the enhanced data locality and elimination of unnecessary communication. Some schedulers also prioritize jobs depending on their importance and attribute to change the order of the execution [1,22]. Such a design is common in real-time systems in which some deadline must meet to promise the correctness of the functionality.

Though the global static method works well in many cases, the system often suffers suboptimal in performance because of the complexity in practice. In this paper, we propose a new method IHDSA (Improved Heterogeneous Dynamic List Schedule Algorithm) to improve workload balance and resource utilization adaptively. The algorithm contains three parts. It first describes the task dependency in a DAG, in which nodes represent jobs while edges depict the relationship between nodes. Then, the scheduler calculates the performance of three different schemes by traversing the DAG. Such schemes include replication-based mapping, interval insertion, and predecessor/successor-based mapping. Finally, a threshold is set each processing unit. The runtime, therefore, balances the workload by monitoring the status of each processor and the preset threshold. In our experiment, we examined the performance of our method against other static approaches. The IHDSA outperforms those static methods continually in a wide range of task numbers.

The paper makes the following contributions:

- It proposes a dynamic scheduling framework.
- Three adaptive scheduling methods have been examined by traversing a task DAG.
- The workload is balanced by the combination of adaptive scheduling schemes and hardware monitoring under preset thresholds.

2 System Model

We present the definition of models that are used in our scheduling algorithm in the following section. The models describe the processor, application, and task mappings.

2.1 Processor Model

We use a tuple $PM = (P, Rate)$ to model heterogeneous processors with separate architectures. It is a tuple with two vector elements that abstract the hardware units and communication overhead.

The vector P is defined as $P = (p_1, ..., p_i, ..., p_m)$, where i indicates the ith processor and m represents the number of processors.

The vector $Rate = (..., rate_{i,j}, ...)$ describes the communication cost between cores. It contains m^2 elements. For any given $rate_{i,j}$, it indicates the communication, such as data movement, taking place between processor i and j, where the moving direction is from i to j.

The processor topology used in this paper is fully connected where any two cores are directly linked together. We select this type of topology for the complexity reason. Other topologies can be viewed as special cases of the fully connected by adding extra constraints over communication cost to make some path unreachable [18]. In this topology mode, we also define the communication cost to be 0 if the predecessor job and its successor has been allocated to the same core, as there are no needs for data migration.

2.2 Application Model

Applications are represented in directed acyclic graphs (DAG), in which vertices stand for the tasks and edges between nodes describe the dependency. Data movements follow the direction of the graph, from the source vertex to the sink. We use a quadruple to model the application on heterogeneous platforms by extending the DAG with hardware information. We define the quadruple as $G = (V, E, W, C)$, where V, E, W and C represents vertix, edge, performance, and communication respectively.

The vertex is defined as an n dimension vector $V = (v_1, ..., v_i, ..., v_n)$. It represents an application with n tasks that can be scheduled independently. The DAG can be updated dynamically by adding new vertices to the graph.

The vector $E = (e_{1,2}, ..., e_{i,j}, ...)$ denotes the directed edges. An edge $e_{i,j}$ identifies the dependency between v_i and v_j. To be specific, it indicates that the output of v_i is the input of v_j.

W is a matrix with dimension of $n \times m$, where n is the amount of tasks while m is the number of hardware units. Each element $w_{i,j}$ in the matrix presents the execution of task v_i on processor p_j.

$C = \{c_{i,j} | i, j \in [1, n]\}$ is also a two-dimensional matrix, where each $c_{i,j}$ represents the communication overhead between v_i and v_j. By its definition, we force $c_{i,j}$ to be 0 if task v_i and v_j have been allocated to the same processor, because of no communication across cores needed.

We also defined $Pre(V_i)$ and $Suc(V_i)$ as the predecessor and successor of v_i, respectively. According to dependency, a successor task cannot be performed ahead of the completion of its predecessors. The scheduler will force all tasks following this rule by imposing the order of the jobs and monitoring their execution status.

Finally, we define the node with no predecessor in the DAT to be the entry node, and we mark such a node as v_{entry}.

2.3   Dynamic Task Mapping Model

The DTMM (Dynamic Task Mapping Model) is a mapping scheme to issue tasks to processors. It uses heuristics guided by task status to insert them to a list from which jobs are dequeued to devices. We mark the task status by a label that uses 1 to represent a completion task, 0 to indicate a ready state in the waiting list, and −1 to describe a holding state that can proceed unless some conditions satisfied.

Two queues are used in mapping the workload. The global queue is formed from flattening the DAG. Nodes from the DAG are inserted to the queue in order. Before scheduling, the scheduler first marks the status of the entry node as 0 to make it ready to map while labels all the rest nodes as −1 because they have to wait until the entry task finished. Once a task completed, its status will be changed from 0 to 1 and all its successors status switched from −1 to 0. Then the scheduling keeps going until all tasks are finished. A centralized global queue can introduce unbalance among processors because all of them fetch jobs at various speed from the queue, and the task dependency can prohibit job from going to the idle cores. For this reason, we introduce the local queue to balance workloads across cores. The local queue is tied up to processors. We use a threshold to limit the number of tasks that can be allocated to a specific core. Once the number goes beyond the threshold, the processor will stop accommodating new jobs.

3   The Proposed Algorithm

The algorithm of IHDSA consists of three parts, including establishing the task list, adapting multi-strategy mapping, and setting up threshold boundaries for workload balance. All there three parts are detailed in the following subsections.

3.1   Task List Establishment

The algorithm traverses the DAG to form a task list. Each vertex in the DAG maps to a node in the task list. Every node has a label to mark its status. The status 0 means a ready state, 1 and −1 stands for the completion and holding state, respectively. At the beginning stage, we change the status of the entry node from −1 to 0, and the states of all its successor nodes will be updated after the entry task finished. Then the update of the task status keeps going until the last one completed.

3.2   Multi-strategy Task Mapping

In this stage, each task will be assigned to the most suitable processor. The scheduler finds out all tasks with status 0 in the task list, and calculate the earliest completion time, then select the one with the earliest completion time. The earliest start and finish time on core P are expressed as $EST_p(V_i)$ and $EFT_p(V_i)$ separately. The definition is shown in formula 1 and 2.

$$EST_p(V_{entry}) = 0 \tag{1}$$

$$EFT_p(V_i) = max(EFT_p(V_i) + E(V_i, V_j), EFT_p(V_k), Avail(P)); k \neq j \quad (2)$$

In above equations, V_j and V_k are the predecessor nodes of V_i, and $Avail(P)$ represents the time available on processor P.

Hybrid Task Mapping Approach. In a heterogeneous multi-core environment, a task can experience better performance on some processing units than the others. The table scheduling algorithm will sort the task priority according to the order of their presence in the task list. In the process of scheduling, we need to consider the relationship between the predecessor and successor of tasks because of data dependence. Task replication technology adopts the key precursor technology of current tasks so that the predecessor and successor tasks are executed on the same processing unit for reducing communication overhead. The interval insertion technology improves the utilization by detecting the idle time of each processing unit, then migrating the tasks from other cores to the idle one to reduce the standby time, hence avoiding the wastage of resources. We propose our hybrid task scheduling strategy based on task replication and interval insertion.

Interval insertion technology finds the processing unit with idle time slots and inserts the task into the free interval of the processing unit once satisfying the priority requirements of the task list. We used $AST_{insert}(V_i, P_j)$ to denote the starting time of a task V_i and EFT_{insert} to denote the earliest finish time based on interval insertion technique. The Eq. 3 shows the definition.

$$EFT_{insert}(V_i) = \min_{0 \leq j \leq m} \{AST_{insert}(V_i, P_i) + W_{i,j}\} \quad (3)$$

Task replication technology copies the predecessor tasks of the current job to the same processing unit. It saves the communication time between the predecessor and the successor tasks hence improves the utilization of the processing unit. The Eq. 4 presents the definition.

$$EFT_{copy}(V_i) = \min_{0 \leq j \leq m} \{AST_{copy}(V_i, P_i) + W_{i,j}\} \quad (4)$$

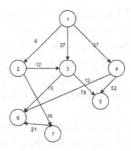

Fig. 1. DAG diagram based on task replication and interval insertion

Table 1. Task execution time on different processors

Processing unit	1	2	3	4	5	6	7
P1	17	59	22	55	13	25	24
P2	63	40	26	28	28	17	21

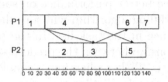

Fig. 2. Scheduling graph based on task replication and interval insertion

Figure 1 shows a practical example of applying the above task mapping method. The execution time of tasks on processor *P1* and *P2* is provided in Table 1. In this example, *task 3* is the precursor of *task 6*. It means that *task 6* has to wait until *task 3* finished its data transmission, then it can start to execute.

Similarly, *task 5* is the successor of *task 4*. It requires the communication completion of *task 4* has to come earlier than the execution of *task 5*.

Therefore processor *P1* becomes idle after finishing *task 4*. At the same time, *task 5* is waiting for the data migrating from *P1* to *P2*. In this case, we can duplicate *task 3* and migrate it to the *P1* to take over the idle time slot. Also, since *task 3* and its successor *task 6* are allocated to the same processor, the subsequent communication overhead has been optimized afterwards, which allows *task 6* to complete on time. Figure 2 shows the task mapping and data dependencies.

Another option is interval insertion, it inserts *task 6* into the free time slot on processing unit *P2*, and the completion time can also be optimized. We compare the results of task replication and interval insertion and choose the method more suitable for the actual environment.

$$EFT_{EX} = min\{EFT_{insert}(V_i), EFT_{copy}, EFT_p(V_i)\} \tag{5}$$

Predecessor-Based Task Mapping Technique. Most of the traditional table scheduling algorithms are based on the earliest completion time, and seldom consider the interaction between tasks with data dependency. As a result, there is a significant amount of idle time in each processing unit. So our algorithm aims to reduce the overall scheduling time by increasing the utilization rate of these idle periods as much as possible in the resource-intensive multi-task scheduling system.

In this section, we propose a task mapping strategy based on the predecessor node. The current task and the predecessor task are assigned to the same processor to avoid transferring data between the predecessor and successor, and the communication cost is therefore eliminated. Such a model faces the risk of allocating all tasks to the same processor while leaving the rest cores unused. For this reason, we introduce a constraint condition. We calculate the difference between EFT_{PA} and EFT_{SA} to ensure the result smaller than the average weight $\overline{W}$. This mechanism prevents allocating all tasks to the same core. The Eq. 6 and 7 show the definition. EFT_{PA} represents the earliest completion time and V_{trans} indicates the last task that has been communicated with V_i.

$$EFT_{PA}(V_i) = \max_{0 \le j \le m} \{Avail(PA), AFT(V_{par}, P_k) + C(V_{trans}, V_i)\} + W(V_i, PA) \tag{6}$$

$$\overline{W} = \sum_{j=1}^{m} W_{i,j}/m \tag{7}$$

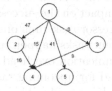

Fig. 3. DAG diagram based on the earliest completion time of predecessor nodes.

Table 2. Task execution time on different processors

Processing unit	1	2	3	4	5
P1	13	50	25	30	25
P2	26	65	30	11	43

A practical example is exhibited in Fig. 3, 4 and Table 2, based on the precursor task mapping method. As shown in the figure, the execution time of *task 4* on processing unit *P1* is shorter than on *P2*. In this case, processing unit *P2* would be selected according to traditional scheduling strategies, because the earliest completion time is sooner. But in practice, doing so results in a lot of idle time on *P2*. Again, comparing the execution time of *task 4* on the two processing units, it is found that in fact, the completion time of *task 4* on processing unit *P1* is only slightly shorter than that of *P2*, and there will not be so much idle time for processing on *P1*. From the result, *task 5* can be completed in advance, but the scheduling length is shorter.

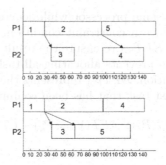

Fig. 4. Scheduling chart based on the earliest completion time of predecessor nodes

Successor-Based Task Mapping Technique. Although the mapping strategy based on predecessor reduces the idle time between predecessor and successor tasks, it does not consider the impact on the successor tasks of current tasks. The mapping method based on successor task runs on the standard processing unit. If the current task and subsequent tasks happen to run on the most appropriate processor core, then the communication overhead of both sides can be ignored, and more resources can be saved for upcoming tasks. $DR(V_i)$ denotes the time when all the data required by V_i is ready. The standard successor node of V_i is represented by $LB(V_i)$. The earliest finish time obtained by successor-based approach can be calculated by Eq. 8.

$$EFT_{SA} = max\{Avail(SA), DR(V_i)\} + W(V_i, SA) + W(LB(V_i), SA) \qquad (8)$$

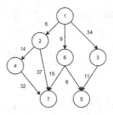

Fig. 5. DAG diagram based on task allocation scheme of successor nodes.

Table 3. Task execution time on different processors

Processing unit	1	2	3	4	5	6	7
P1		15	89	17	19	30	21	7
P2		60	5	24	96	17	15	29

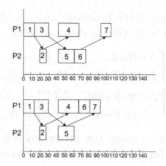

Fig. 6. Scheduling graph based on task allocation scheme of successor nodes.

Figure 5, 6 and Table 3 give an example based on this mapping method. When scheduling according to the earliest completion time, it is obvious that *task 6* will finish earlier on *P2*, so *task 6* should be executed on *P2*, but doing so will cause a large amount of idle time on the processor core. If *task 6* is assigned to *P1*, the task completion time is only slightly longer than that on processing unit *P2*, but the idle time is greatly reduced. The final result is that *task 7* can be completed in advance and the overall scheduling length of task is shortened.

Final Mapping Strategy. As mentioned above, IHDSA algorithm has three strategies: task replication-based mapping, interval insertion mapping, predecessor-based mapping and successor-based mapping. Before task scheduling, the earliest completion time of tasks is calculated by Eq. 2. Then tasks scheduling methods based on task replication and interval insertion is calculated by Eq. 5. Tasks using a mapping strategy based on the predecessor task is calculated by Eq. 6, and tasks mapping strategy based on the successor task is calculated by Eq. 8. The scheme with the earliest and the shortest completion time is then selected as the final mapping strategy.

3.3 Workload Balancing Among Processors

A workload threshold is set for each processor core. The status of processors is collected on-the-fly, and the tasks assigned to the corresponding processor is controlled dynamically to balance the number of tasks among distinctive processors. The equation is show as follows. $L_{max}(P)$ represents the load limit of the processor and $N(P)$ indicates the current load of the processor.

$$N(P) = \frac{L(P)}{L_{max}(P)} \tag{9}$$

The lower and upper limit of the processor threshold are represented by N_{min} and N_{max} respectively. The equation is show as follows.

$$Loadstate = \begin{cases} No\ load, & N(P) = 0 \\ Light\ load, & N(P) \in (0, N_{min}] \\ fitness\ load, & N(P) \in (N_{min}, N_{max}) \\ heavy\ load, & N(P) \geq N_{max} \end{cases} \tag{10}$$

According to the equation, when the processor core is in the state of heavy load, the processor core is forbidden to accept new tasks until it recovers to the state of light load.

4   Experiment

4.1   Performance Metrics

In the evaluation of heterogeneous multi-core dynamic task scheduling algorithm, the completion time of tasks in each processing unit, the time complexity of the algorithm and the load capacity between processing units should be taken into account. Scheduling algorithm not only needs to ensure the scheduling efficiency of the system but also needs to consider shortening the scheduling length to reduce the communication overhead between tasks. In this paper, we choose the schedule length ratio (SLR) and the speedup as performance evaluation parameters.

The time from the beginning to the completion of the task is defined as the scheduling length makespan. The calculation formula is shown in Eq. 11. Where $load(P_j)$ represents the processor load.

$$makespan(T) = max(load(P_j)) \tag{11}$$

The processor load is calculated by Eq. 12. The sum of the tasks allocated to P_j is the load of processing unit P_j. The x stands for the tasks ot P_i while the y represents the jobs mapped to P_j.

$$Load(P_j) = \sum_{i=1}^{x} C_{i,j} + \sum_{k=1}^{y} C_{k,j} \tag{12}$$

The SLR is defined as the minimum possible value of scheduling length divided by the scheduling length.

$$SLR = \frac{makespan}{\sum\limits_{V_i \in CP_{min}} \min\limits_{1 \leq j \leq m} W_{i,j}} \tag{13}$$

Speedup is used to measure the parallel degree of multi-core processor. It is the ratio of task completion time on a single processing unit over task completion time on a multi-core. The calculation is shown in Eq. 14.

$$Speedup = \frac{\min\limits_{1 \leq j \leq m} \sum\limits_{1 \leq j \leq n} W_{i,j}}{makespan} \tag{14}$$

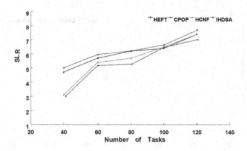

Fig. 7. SLR comparison of different task size.

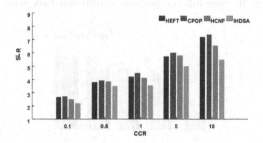

Fig. 8. SLR for varing CCR.

4.2   Result and Analysis

We use TGFF to generate DAG graphs randomly in our experiment to simulate the dependence and communication between tasks. The reason we used random graphs is to avoid the bias from the specific applications. The results are, therefore, more robust to various type of workloads.

The size of the workload is defined as N which has a range between 40 to 120, $N = \{40, 60, 80, 100, 120\}$.

The heterogeneity factor H represents the degree of difference in execution time $H = \{0.1, 0.25, 0.5, 0.75, 1.0\}$. A higher ratio means processing time of given task sets is more diverged across separate microprocessors.

The parallel factor α reflects the height and width of a DAG. $\alpha = \{0.5, 1.0, 2.0\}$.

CCR is the cost ratio of communication and computation. $CCR = \{0.1, 0.5, 1.0, 5.0, 10.0\}$.

Figure 7 shows the impact of the number of tasks on the performance of the scheduling algorithm. It can be seen from the figure that the SLR and hcnf algorithms of IHDSA algorithm are similar when the number of tasks is small, but still smaller than the other two algorithms. With the number of tasks increased, the SLR of the four algorithms gradually increases accordingly. At this time, the SLR of IHDSA algorithm is smaller than the other three algorithms. It can be seen that the IHDSA algorithm performs outstanding scheduling in the system with a large number of tasks and a large density of resources.

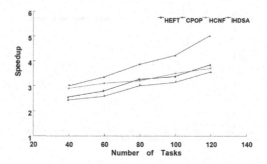

Fig. 9. Speedup comparison of different task size.

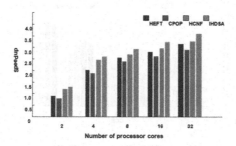

Fig. 10. Speedup of processors with different number of cores.

Figure 8 presents the impacts of CCR on SLR across various algorithms. CCR is the ratio of task communication cost and calculation cost. When the communication cost is greater than the execution time, CCR becomes larger, the idle time segment on each processing unit increases, and the scheduling efficiency of the algorithm decreases, so the SLR of the four algorithms in the figure decreases accordingly. With the increasing of CCR, there is an obvious trend that the SLR of IHDSA algorithm is increasingly smaller than the other three algorithms.

Figure 9 presents the impact of the task sizes across separate algorithms. In our experiment, regardless of task sizes, the performance of IHDSA algorithm always outperforms the other three algorithms. Such a trend become more obvious with the number of tasks increases. Due to the communication overhead raises with the number of tasks, subsequent tasks need to wait for more for their predecessor tasks before they can start to execute, which leads to an increasing idle time caused by waiting on each processing unit. Since IHDSA algorithm has the highest utilization for idle time, it has the highest efficiency among all four algorithms.

Figure 10 shows the speedup scalability with the number of processors. In general, more cores present higher performance; however, such an increment is not linear with the number of cores. In our experiment, the speedup of IHDSA algorithm is continuously higher than the other three algorithms. This demonstrates that IHDSA algorithm has higher efficiency and better performance in the heterogeneous multi-core environment.

5   Related Work

The existing static heuristic table scheduling algorithm has some shortcomings in the priority calculation and task allocation stage.

HEFT, a static heuristic scheduling algorithm [19], takes into account the execution length of backward critical path of task nodes, and has higher scheduling efficiency. HEFT algorithm takes the earliest completion time of nodes as the priority weight to schedule [12, 14]. In the task allocation stage, interval insertion technology [8] is used to improve the scheduling, which makes the nodes obtain less task completion time and lower time complexity. However, the algorithm may lead to the low priority of some key tasks, which may reduce the utilization of the processor [19].

CPOP algorithm is also based on a priority list to assign tasks [20]. In this algorithm, the sum of the critical path length from the entry node to the current node and the length of the critical path between the exit node and the current node are taken as weights, and tasks are assigned to the processor core according to the weight. CPOP ensures that the key nodes will get priority scheduling by giving the highest priority to critical path nodes in real-time scheduling, but the priority of non critical path nodes will inevitably be reduced. When the communication cost of tasks between different processors is large, the delay effect of non critical path nodes will become more obvious, and the performance of the algorithm will also decline accordingly [6, 7].

The HCNF algorithm can also improve the scheduling efficiency by scheduling the critical path nodes first [3]. Moreover, when the HCNF algorithm encounters the task with high communication overhead, it can effectively improve the utilization of processor idle time and further shorten the scheduling length. However, the priority allocation standard used by HCNF algorithm in the stage of constructing the list is the average cost of task nodes on each processor core, which fails to give full play to the efficiency advantage of heterogeneous multi-core processors. If there is a huge difference in the cost of task nodes on different processor cores, the performance of the algorithm will not be very good.

This paper systematically studies the scheduling model and algorithm in heterogeneous multi-core environment. The research shows that the status of multi-core processor in the field of processor research is increasing, and the excellent performance of heterogeneous multi-core processor will make it become more and more important to perform scheduling tasks. In the second part of this paper, based on the above research, a dynamic task mapping model DTMM [5, 11, 13] is proposed, which uses global queue and bureau force column to schedule tasks, and sets processing threshold for each processor core to monitor the load of all units in real time. In the third part of this paper, a new dynamic task algorithm based on DTMM model is designed, which is the improved heterogeneous dynamic list scheduling algorithm (IHDSA). In the fourth part of this paper, experiments are designed and compared with the new IHDSA algorithm and the three classical scheduling algorithms. It is proved that IHDSA algorithm has higher utilization of processor idle time and better performance when executing scheduling tasks with relatively high computing density.

6   Conclusion

Based on the in-depth research of heterogeneous multi-core scheduling environment, this paper analyzes the advantages and disadvantages of the existing task scheduling algorithm and proposes the heterogeneous multi-core dynamic scheduling algorithm IHDSA. The algorithm is divided into three parts: task list building, task scheduling and scheduling optimization. In the list building part, DAG graph is used to transform the list with status bits, which determines the priority of tasks and represents the scheduling status of each task in the list with status bits. In the task scheduling part, three new task mapping strategies are adopted. The core idea is to steal the idle time of the processing unit, which reduces the communication overhead between tasks. Finally, the load threshold is set for the processing unit in the scheduling optimization part, and the load status of each processing unit is monitored through the load threshold to achieve load balancing. By comparing the IHDSA algorithm with three traditional scheduling algorithms, it is proved that the speedup and SLR performance of IHDSA algorithm should be better.

Acknowledgement. This work was supported by Science Foundation Ireland grant 13/RC/2094 to Lero - The Irish Software Research Centre.

References

1. Ashouraie, M., Navimipour, N.J.: Priority-based task scheduling on heterogeneous resources in the expert cloud. Kybernetes **44**(10), 1455–1471 (2015)
2. Augonnet, C., Thibault, S., Namyst, R., Wacrenier, P.: StarPU: a unified platform for task scheduling on heterogeneous multicore architectures. Concurr. Comput. Pract. Exp. **23**(2), 187–198 (2011)
3. Baskiyar, S., SaiRanga, P.C.: Scheduling directed a-cyclic task graphs on heterogeneous network of workstations to minimize schedule length. In: Proceedings of the 2003 International Conference on Parallel Processing Workshops 2003, pp. 97–103 (2003)
4. Bittencourt, L.F., Sakellariou, R., Madeira, E.R.M.: Dag scheduling using a lookahead variant of the heterogeneous earliest finish time algorithm. In: 2010 18th Euromicro Conference on Parallel, Distributed and Network-based Processing, pp. 27–34 (2010)
5. Carvalho, E., Calazans, N.L.V., Moraes, F.G.: Dynamic task mapping for MPSoCs. IEEE Des. Test Comput. **27**, 26–35 (2010)
6. Castrillon, J., Tretter, A., Leupers, R., Ascheid, G.: Communication-aware mapping of KPN applications onto heterogeneous MPSoCs. DAC Design Automation Conference **2012**, 1262–1267 (2012)
7. Dathathri, R., et al.: Gluon: a communication-optimizing substrate for distributed heterogeneous graph analytics. In: PLDI, pp. 752–768. ACM (2018)
8. Gao, K., Suganthan, P., Chua, T., Chong, C., Cai, T., Pan, Q.K.: A two-stage artificial bee colony algorithm scheduling flexible job-shop scheduling problem with new job insertion. Expert Syst. Appl. **42**(21), 7653–7663 (2015). https://doi.org/10.1016/j.eswa.2015.06.004

9. Zhao, H., Sakellariou, R.: Scheduling multiple DAGs onto heterogeneous systems. In: Proceedings 20th IEEE International Parallel Distributed Processing Symposium, p. 14 (2006)
10. Khokhar, A.A., Prasanna, V.K., Shaaban, M.E., Wang, C.: Heterogeneous computing: challenges and opportunities. IEEE Comput. **26**(6), 18–27 (1993)
11. Middendorf, L., Zebelein, C., Haubelt, C.: Dynamic task mapping onto multicore architectures through stream rewriting. In: 2013 International Conference on Embedded Computer Systems: Architectures, Modeling, and Simulation (SAMOS), pp. 196–204 (2013)
12. Munir, E.U., Mohsin, S., Hussain, A., Nisar, M.W., Ali, S.: SDBATs: a novel algorithm for task scheduling in heterogeneous computing systems. In: 2013 IEEE International Symposium on Parallel Distributed Processing, Workshops and PhD Forum, pp. 43–53 (2013)
13. Möller, L., Indrusiak, L.S., Ost, L., Moraes, F., Glesner, M.: Comparative analysis of dynamic task mapping heuristics in heterogeneous NoC-based MPSoCs. In: 2012 International Symposium on System on Chip (SoC), pp. 1–4 (2012)
14. Nasonov, D.A., Visheratin, A.A., Butakov, N., Shindyapina, N., Melnik, M., Boukhanovsky, A.: Hybrid evolutionary workflow scheduling algorithm for dynamic heterogeneous distributed computational environment. J. Appl. Log. **24**, 50–61 (2017)
15. Bajaj, R., Agrawal, D.P.: Improving scheduling of tasks in a heterogeneous environment. IEEE Trans. Parallel Distrib. Syst. **15**(2), 107–118 (2004)
16. Rogers, P.: Heterogeneous system architecture overview. In: Hot Chips Symposium, pp. 1–41. IEEE (2013)
17. Sakellariou, R., Zhao, H.: A hybrid heuristic for DAG scheduling on heterogeneous systems. In: Proceedings of the 18th International Parallel and Distributed Processing Symposium, 2004, p. 111 (2004)
18. Saldaña, M., Shannon, L., Chow, P.: The routability of multiprocessor network topologies in FPGAs. In: SLIP, pp. 49–56. ACM (2006)
19. Samadi, Y., Zbakh, M., Tadonki, C.: E-HEFT: enhancement heterogeneous earliest finish time algorithm for task scheduling based on load balancing in cloud computing. In: HPCS, pp. 601–609. IEEE (2018)
20. Shekar, V., Izadi, B.: Energy aware scheduling for DAG structured applications on heterogeneous and DVS enabled processors. In: International Conference on Green Computing, pp. 495–502 (2010)
21. Wen, Y., O'Boyle, M.F.P.: Merge or separate?: multi-job scheduling for OpenCL Kernels on CPU/GPU platforms. In: GPGPU@PPoPP, pp. 22–31. ACM (2017)
22. Xu, Y., Li, K., Khac, T.T., Qiu, M.: A multiple priority queueing genetic algorithm for task scheduling on heterogeneous computing systems. In: 2012 IEEE 14th International Conference on High Performance Computing and Communication, 2012 IEEE 9th International Conference on Embedded Software and Systems, pp. 639–646 (2012)
23. Zarkesh-Ha, P., Davis, J.A., Meindl, J.D.: Prediction of net-length distribution for global interconnects in a heterogeneous system-on-a-chip. IEEE Trans. Very Large Scale Integr. (VLSI) Syst. **8**(6), 649–659 (2000)

FastThetaJoin: An Optimization on Multi-way Data Stream θ-join with Range Constraints

Ziyue Hu[1,2], Xiaopeng Fan[1], Yang Wang[1(✉)], and Chengzhong Xu[3]

[1] Shenzhen Institutes of Advanced Technology, Chinese Academy of Sciences,
Shenzhen, China
yang.wang1@siat.ac.cn
[2] University of Chinese Academy of Sciences, Beijing, China
[3] University of Macau, Macau, China

Abstract. In this paper, we propose FastThetaJoin, an optimization technique for θ-join operation on multi-way data streams, which is an essential query often used in many data analytical tasks. The θ-join operation on multi-way data streams is notoriously difficult as it always involves tremendous shuffle cost due to data movements between multiple operation components, rendering it hard to be efficiently implemented in a distributed environment. As with previous methods, FastThetaJoin also tries to minimize the number of θ-joins, but it is distinct from others in terms of making partitions, deleting unnecessary data items, and performing the Cartesian product. FastThetaJoin not only effectively minimizes the number of θ-joins, but also substantially improves the efficiency of its operations in a distributed environment. We implemented FastThetaJoin in the framework of Spark Streaming, characterized by its efficient bucket implementation of parameterized windows. The experimental results show that, compared with the existing solutions, our proposed method can speed up the θ-join processing while reducing its overhead; the specific effects of the optimization is correlated to the nature of data streams–the greater the data difference is, the more apparent the optimization effect is.

Keywords: θ-join · Theta join · Multi-way data streams · Data streams · Spark streaming

1 Introduction

Data stream processing as one of the important big data technologies are often used to perform a series of operations on a continuous stream of data for online query service in real-time, and join is one of the key stream operations used to detect the scenarios that satisfy certain conditions. However, due to its high cost, the efficiency of the join operation among multiple data streams has always been a pragmatic concern, which has drawn great attentions from both academia and industry [16].

© Springer Nature Switzerland AG 2020
M. Qiu (Ed.): ICA3PP 2020, LNCS 12452, pp. 174–189, 2020.
https://doi.org/10.1007/978-3-030-60245-1_12

θ-join, denoted as $S \bowtie R$ $(A\theta B)$, is a special kind of join operations between data set R and data set S, wherein a new set of tuples from the generalized Cartesian product of the two data sets is generated to satisfy certain conditions in data analysis and processing, here A and B are attributed in data set R and S, respectively, and θ is a join condition represented by '$<$', '$\leq$', '$\geq$', and '$>$'. In particular, if θ is '$=$', it is called equivalent join.

Many previous studies have been conducted on join in general and θ-join in particular, but they often suffer some limitations. There are some works using the MapReduce framework to process the joins, which might cause huge I/O overheads [4] as many iterative calculations are involved. This paper [11] can join multiple relational tables in a single MapReduce task. However, because intermediate results are stored on HDFS and then accessed from HDFS immediately, there will be lots of I/O overheads. Many works only focus on equivalent join [7,17]. Some research can support θ-join, but they need to change the framework [12,19]. Most studies are based on static datasets, and few are used for streaming calculations. This work [15] carefully studies the implementation of each type of join on MapReduce. The main idea is to map the tuples of two datasets into a matrix, then divide it according to the size of the matrix and the number of reducers. And the components of the same region are on the same reducer to implement the Cartesian products. However, this method does not support θ-join of multiple tables.

Some work can solve the trouble of θ-join in multiple tables. This research [20] proposed an algorithm based on a symmetric replication connection. The main idea is to slice multiple relational tables, scan the tuples in each table at map phase, and correspond to the fragment in which they are located. At the shuffle phase, each fragment is copied to the corresponding node in a one-to-many manner. The tuples that match θ condition are joined at the reduce side. However, as the number of relational tables increases, this method will result in a large number of unnecessary data movements.

In order to solve the tricky trouble of θ-join in multi-way data streams and minimize data transmission overheads during the shuffle phase, we propose Fast-ThetaJoin in this paper, an optimization method which partitions based on the range of data value, then adopts a special filter operation before shuffle and do Cartesian product between part of Partitions. The optimization of our method is mainly in two aspects. On the one hand, we can filter quite many unnecessary data without sort before data transmission. One the other hand in the shuffle phrase, we need not use one-to-all ways to send partitions to all the nodes. Because we know the size relationships between the partitions in our method, we just send some partition to part of the nodes. With the optimization from the above two aspects, we can achieve the least Cartesian product operation. So the number of θ connections is greatly reduced, and the calculation speed is improved as well. The method proposed in this paper is experimentally evaluated on a Spark Streaming computing platform.

In summary, we make the following contributions in this paper:

1. **We propose FastThetaJoin—an optimized method of multi-way data stream's θ-join.** It can calculate not only two-way data streams but also multi-way data streams on a distributed realtime computation system in an efficient way.
2. **We propose a specific filter strategy without sorting.** Better than the most advanced method CFS, the strategy proposed doesn't have global sorting time overhead, which is $O(n \log n)$. In this way, we just need $O(1)$ to delete the most apparently unsatisfactory data and then delete others by small range traversal. This is because when we partition, we make the partitions ordered and the partitions unordered. This strategy can try its best to reduce the overhead of data transmission.
3. **We propose an efficient cross product strategy for θ-join.** In the shuffle phrase, we needn't use one-to-all ways to send partitions to all nodes. Because we know the size relationship between partitions in our method, we just send some partition to part of nodes. This strategy can go to a further step to reduce the overhead of data transmission.

In the remainder of this paper, we introduce some related work in Sect. 2 and describe the detailed design of FastThetaJoin in Sect. 3. We present the experiments in Sect. 4, and conclude the paper in the last section.

2 Related Work

In the current mainstream research, MapReduce is mainly used to handle θ-join operation with load balancing in network as a primary consideration. When the data set becomes large, a large number of intermediate results result in high communication overhead. Especially in the iterative calculation, MapReduce stores each intermediate result on HDFS, and the next calculation needs to read from HDFS firstly and then recalculated, resulting in many unnecessary I/O overheads. In order to solve it, some techniques are optimized based on MapReduce, using one MapReduce task as much as possible.

1-bucket-theta [15] and its derivative technologies. 1-bucket-theta is the most common method of evaluating a single θ-join before. 1-bucket-theta algorithm is based on the MapReduce framework. This method sorts the dataset according to the specified attributes first. Then divide data according to fixed number k, that is from the $(0 * k + 1)th$ to $(1 * k)th$ are divided into the same partition, $(1 * k + 1)th$ to $(2 * k)th$ are divided into the same partition, and so on. After that distribute partitions to the corresponding node in the cluster. Next is the Cartesian product. Because the size relationship between partitions is unknown, each partition in different datasets has to do Cartesian product. Finally, select the result according to the θ condition.

Figure 1(a) takes the θ-join of two static datasets as an example. θ is '≥' means to select items pair in which the first dataset is bigger than the second. The row indicates that the specified attribute value after sorted in the first dataset is {1, 2, 3, 4, 5, 6, 7, 8, 9}, and the column indicates that the specified attribute value after sorted in the second dataset is {4, 5, 6, 7, 8, 9, 10, 11, 12}.

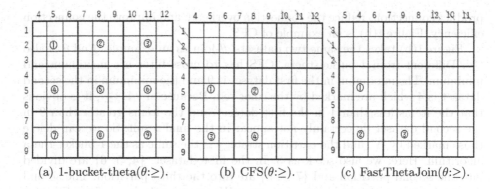

(a) 1-bucket-theta($\theta:\geq$).   (b) CFS($\theta:\geq$).   (c) FastThetaJoin($\theta:\geq$).

Fig. 1. Algorithm comparison

According to the above rules, here we set k to 3. In the first dataset the first to the third, i.e. $\{1, 2, 3\}$, are divided into the first partition, the forth to the sixth, i.e. $\{4, 5, 6\}$, are divided into the second partition, and the seventh to the ninth, i.e. $\{7, 8, 9\}$, are divided into the third partition. The second dataset is also partitioned in the same way. Because the size relationship during partitions is unknown, each partition has to do the Cartesian product with other partitions in another dataset. So we can get nine parts of Cartesian product. Finally, we have to select the final result one by one according to theta conditions in the nine parts.

A multi-way θ-join [21], based on 1-bucket-theta, uses Hilbert curves to achieve chained connections. Because Hilbert curves require the same size of the dataset, they cannot be used when joining tables of different sizes. MRJs (MapReduce Jobs) [18] decomposes the θ-join into non-equal joins and non-equal joins. In each θ-join process, the 1-bucket-theta method is also used. Z. Khayya proposed a θ-join method based on sorting, permuting arrays, and bit arrays [12]. It puts the columns to be joined into the sorted array and rearranges the tuples using the permutation array to make the order. This method requires fundamental modifications to the parallel framework and is not convenient for practical applications. Due to the disadvantages of MapReduce in iterative calculations, some methods deal with θ-join in other frameworks, such as Storm [13] or Spark [17], they also need to change the framework to add new features.

The latest technology is the Cross Filter Strategy (CFS) [14]. This method is also based on 1-bucket-theta, but these two algorithms have significant differences in two ways. On the one hand, CFS is based on the spark framework instead of MapReduce. On the other hand, CFS deletes the items in the two datasets that are obviously not satisfied θ-join condition after sorting. Then Similarly, every k data are divided into the same partition, and then the data are distributed to reducer in the cluster. The advantage of deleting some data that is obviously not satisfied is to reduce the amount of data in the Cartesian product connection, reduce the connection overhead, and improve the connection efficiency. Figure 1(b) is an example of CFS.

Figure 1(b) takes the same two datasets' θ-join as an example. And θ is also '$\geq$'. Different from 1-bucket-theta, CFS has to cross filter use maximum and minimum. The maximum in the first dataset is 9 and the minimum in the second dataset is 4 which are marked by the color red. Because of θ is '$\geq$', it is obvious that the items less than 4 in the first dataset and the items greater than 9 in the second dataset do certainly not appear in the final result. So we can delete those marked by '\', and the partition strategy is the same as the 1-bucket-theta algorithm. Here, we also set k as 3. In the first dataset, {4, 5, 6} are divided into the second partition and {7, 8, 9} are into the third partition. The second dataset is also partitioned in the same way. We can intuitively see from Fig. 1(a) and Fig. 1(b) that the Cartesian product of CFS is less than 1-bucket-theta. Using the same datasets, CFS just needs four parts.

For streams, there have been some studies in recent years. Some studies have focused on converting batch processing to stream processing. They try to treat static data sets as micro-batch processing, which can be approximated as stream processing to speed up processing efficiency [1,2]. This method can also first use the windows to collect a small batch of data items for processing, and then iterate the next batch. As long as the window size is selected properly, it can be treated as a stream. In fact, in some scenarios that do not require one-by-one processing, the window mechanism is also a good choice. Continuous Query Language, or CQL for short [5,9], gives a semantics of a sliding-window stream join by regarding it as a relational join view over the sliding windows, each of which contains the bag of tuples in the current window of the respective stream.

FastThetaJoin proposed in this paper has been implemented in the Spark Streaming framework. The experimental results show that compared with the existing solutions, the proposed method can go a further step to reduce Cartesian product, reduce network transmission cost, reduce shuffle cost, speed up θ-join processing, improve the performance of θ-join.

3   FastThetaJoin Design

In this section, we will describe in detail the design of FastThetaJoin. In the front 3 parts, we will describe the case of two data streams. In the last part, we will illustrate how to implement FastThetaJoin in multi-way data streams.

> Step 1: Get the maximum and minimum of two streams respectively (line 2 in Algorithm 1).
> Step 2: Get the range of each partition. One of the input parameters is partition number. We can combine the maximums and minimums to calculate the partition range. And we can estimate the average load.
> Step 3: Each stream is partitioned by range (lines 3 to 6 in Algorithm 1).
> Step 4: Rough Filter. Use maximums and minimums do cross filter according to θ. This step can be divided into the following four cases (lines 7 to 25 in Algorithm 1).

1) If θ is '>'. The minimum of stream 2 is used as the initial filter condition to delete the data in stream 1 that is not larger than it. The maximum of the data stream 1 is used as the initial filter condition to delete the data in stream 2 that is not less than it.

2) If θ is '<'. The maximum of stream 2 is used as the initial filter condition to delete the data in stream 1 that is not less than it. The minimum of the stream 1 is used as the initial filter condition to delete the data in stream 2 that is not larger than it.

3) If θ is '≥'. The minimum of the data stream 2 is used as the initial filter condition to delete the data in data stream 1 that is smaller than it. The maximum of the data stream 1 is used as the initial filter condition to delete the data in data stream 2 that is larger than it.

4) If θ is '≤'. The minimum of the data stream 2 is used as the initial filter condition to delete the data in data stream 1 that is larger than it. The maximum of the data stream 1 is used as the initial filter condition to delete the data in data stream 2 that is smaller than it.

Step 5: The Cross Product phase. Because we know the range of each partition. So we just need part of partitions to do the Cartesian product with each other according to θ.

Step 6: Use θ to filter the result of step 5. Delete data that does not satisfy the condition. The final result after θ-join is obtained.

Each stream has its own window size, the size can be set by the input parameters. The sliding interval of each data stream window is determined by the input parameters. Split each stream by window size. We define the data in the current window of the first data stream as dataset R and the data in the current window of the second data stream as dataset S. B is one of the common attributes in two data streams. Of course, the two streams also can contain other attributes except for B. Since many of the existing methods are very suitable for equivalence join, and the method proposed in this paper, FastThetaJoin, has a more obvious advantage in non-equivalent connections. Therefore, this article only discusses the situation that Join condition θ belongs to $\{$'<', '≤', '≥', '>'$\}$.

3.1 Filter Strategy

In Pre-process, we make the map of join attribute as the key, and make all attributes merged as values, i.e. every element is stored by ⟨key, valve⟩. First, we have to get the maximum and minimum numbers in each set. Then using

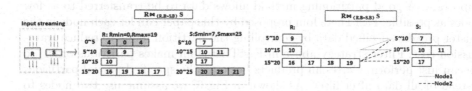

Fig. 2. Rough filter phrase.   Fig. 3. Cross product.

the maximum or minimum number in each stream to filter roughly before cross product. In the map phase, data sets R and S are mapped based on the range of the fingerprint of B respectively. Take Fig. 2 as an example to elaborate on this process in detail.

We assume that stream 1 contains two attributes A, B, and stream 2 contains two attributes B, C. Dataset R is the data to be processed in stream 1, denoted as $R(A, B)$; dataset S is the data to be processed in the stream 2, denoted as $S(B, C)$, and the join condition is: $R \bowtie_{(R.B>S.B)} S$, i.e. B in R is bigger than the B in S. The value of B in dataset R contains {4, 0, 4, 6, 9, 10, 16, 19, 18 17} (repeated number is because of different timestamp or other attribution.). S contains {7, 10, 11, 17, 20, 23, 21}. $Rmin$ is the minimum value in R. $Rmax$ is the maximum value in R. $Smin$ is the minimum value in S. In this example, $Rmin = 0$, $Rmax = 19$, $Smin = 7$ and $Smax = 23$. Assuming partition number is 4 here. We can partition as follows. For dataset R, the range in $[0, 5)$ contains {0, 1, 4}; the range of $[5, 10)$ contains {6, 9}; the range of $[10, 15)$ contains {10}; the range of $[15, 20)$ contains {16, 19, 18, 17}. Use the same strategy to map data set S. Because θ is '≥', for dataset R, only keep the data larger than $Smin$. And for dataset S, only keep the data smaller than $Rmax$. Since $Smin = 7$ belongs to the second partition $[5, 10)$, we just need $O(1)$ to find the corresponding partition. Then all partitions which are less than the second partition in R are directly deleted. For data belongs to partition $[5, 10)$, delete the value which is less than $Smin$. Since $Rmax = 19$ belongs to the third partition $[15, 20)$, directly delete all partitions which are bigger than the third partition in S. The time is also just $O(1)$. For data whose partition belongs to $[15, 20)$, delete data smaller than $Rmax$ and this is just traversing one by one within a very small area. In summary, delete the data represented by the shaded portion in Fig. 2 before the Cartesian product phrase.

The filter strategy between two data streams proposed by this paper is shown in Algorithm 1. Better than the most advanced method CFS, the strategy proposed does not have global sorting time overhead which is $O(n(\log n))$. We just only use $O(1)$ to delete great majority unnecessary data. Time of $O(m)$ to traverse one by one within a very small area and m are far less than the total number of data sets.

3.2   Cross Product Strategy

In distributed computing, communication is costly. Therefore, overall performance can be greatly improved by controlling the data distribution to achieve minimal network transmission. During θ-join, reasonable partitioning is very important. A good partitioning method allows data to be transferred to as few nodes as possible. FastThetaJoin proposed by this paper can let each node in the cluster process specified data by controlling the partition mode to reduce transmission cost. Filter strategy above in FastThetaJoin makes partitions order, so we can just perform Cartesian products between the specified partial partitions instead of all data after filter. As shown in Fig. 3, we assume use two nodes to calculate. To reduce data transmission, data in data set R with the range $[5, 10)$,

---

Algorithm 1. FilterStrategy

Require: *firstStream* including attr. B, *secondStream* including attr. B, function θ, *windowSize*, *windowSliding*.

Ensure: filtered pairs collection of *firstStream* P_R, filtered pairs collection of *secondStream* P_S.

1: get data sets R and S according to *windowSize*;
2: get *Rmin,Rmax,Smin,Smax*;
3: $P_R \leftarrow firstStream$.map{hash(_.B),_.B};
4: $P_R \leftarrow P_R$.groupbyKey;
5: $P_S \leftarrow secondStream$.map{hash(_.B),_.B};
6: $P_S \leftarrow P_S$.groupbyKey;
7: $P_R \leftarrow P_R$.filter{ record $\Rightarrow$
8: **if** θ is '>' or '$\geq$' **then**
9:    record.key $\geq$ hash($Smin$)
10: **else if** θ is '<' or '$\leq$' **then**
11:    record.key $\leq$ hash($Smax$)
12: **end if**}
13: $P_R \leftarrow P_R$.filter{record $\Rightarrow$
14: **if** record.key $=$ hash ($Smin$) **then**
15:    **if** θ is '>' **then**
16:       record.value $> Smin$
17:    **else if** θ is '$\geq$' **then**
18:       record.value $\geq Smin$
19:    **else if** θ is '<' **then**
20:       record.value $< Smax$
21:    **else if** θ is '$\leq$' **then**
22:       record.value $\leq Smax$
23:    **end if**
24: **end if**}
25: P_S do similar filter operations with P_R;
26: return P_R,P_S;

[10, 15) is sent to Node1 and data with a range of [15, 20) is sent to Node2. All data in data set S is sent to Node2 and data in the range [5, 10), [10, 15) is sent to Node1. In this way, it only needs 6 times in transmission of data. CFS and 1-bucket-theta needs $3 \times 3 = 9$ times because both of them are not sure which partition is bigger.

Compared with the current state-of-the-art method CFS, FastThetaJoin can reduce the transmission of data in distributed computing. And further reduces the number of Cartesian products after filtering. The advantage is more obvious in the massive data θ connection. FastThetaJoin can go to a further step to reduce final filtering overhead. We will use Fig. 1(c) to elaborate.

In order to intuitively reflect the difference between FastThetaJoin and CFS, Fig. 1(c) uses the same datasets with Fig. 1(b) and θ is also '$\geq$'. FastTheta-Join looks a little bit similar to CFS that both of them need to filter before cross product, but they are completely different. There are two main differences: partition method and the phrase of the cross product. FastThetaJoin gets maximum and minimum of two data streams respectively, then calculates the range of every partition, divides by range later, performs cross-filtering using maximums and minimums, after that calculates Cartesian product between partial partitions and gets result according to θ finally. After cross-filtering, CFS has to calculate four parts of Cartesian product because the size relationship between partitions is unknown. In FastThetaJoin, after cross-filtering, we can know the first partition is the range [4, 6] which is smaller than the second partition of the second stream. So we need not calculate the Cartesian product between them. So FastThetaJoin can reduce network transmission, reduce shuffle cost, go a step further to reduce Cartesian product count, speed up θ-join processing, improve the performance of θ-join.

3.3 Data Skew Processing Strategy

FastThetaJoin also considers data skew. According to the rate of data flow, window size, and partition numbers, it is easy to estimate the expected numbers of data in every partition. Once the number of data in some partition is exceeded estimates, we judge it is skewed. Then the skewed data will be repartitioned to balance the load. Until the skew process is completed, execute the above filter strategy, cross product strategy, and other operations. We use Fig. 4 as an example to describe how to process skewed data in detail.

Fig. 4. Data skew processing (flag indicates if skew happened)

MaxSize indicates the maximum amount of data that each node can process under best performance. $MaxSize = \lceil windowSize/partitionNumber \rceil$. ActualSize indicates the current actual data numbers of the node. $flag$ is used to judge whether skew occurs. $flag = \lceil ActualSize * 1.0/MaxSize \rceil$. If $flag$ is greater than 0, it means skew occurs and we need to repartition data into $flag$ partitions. In Fig. 4, we assume that ActualSize = 700, MaxSize = 300. So $flag = \lceil 700 * 1.0/300 \rceil = 3$. $flag > 0$ indicates data skew. According to data skew processing strategy proposed, we will repartition these data into 3 partitions whose size is 233, 233, 234 respectively.

3.4 θ-join of Multi-way Data Streams

The method proposed in this paper is not only applicable to two-way data streams, but also to multi-way data streams. For θ-join between multi-way data streams, the multi-way data streams may be paired into two pairs. Then we execute the above Step1 to Step6, and multiple pairs are concurrently executed. The results are as the input of the next round until there is only one output. For data that does not pair to pairs, leave it to the next round to be the input of the next round directly. For example, when we need to calculate θ-join during n streams, we can Calculate every two adjacent first. E.g (R1.a1 θ R2.a2) $\bowtie$ (R2.a1 θ R3.a2) $\bowtie ... \bowtie$ (Rn-1.ai θ Rn.aj), where R1, R2 ... Rn respectively represents different streams and a1, a2 ... ai, aj respectively represents the attribute that corresponding stream contains.

FastThetaJoin based on Four-way Data Streams is showed in Algorithm 2.

Algorithm 2. FastThetaJoin based on Four-way Data Streams

Require: *firstDataStream*,   *secondDataStream*,   *thirdDataStream*   and *forthDataStream* all includes attribution B, join condition θ_1, θ_2, θ_3, *windowSize*, *windowSliding*.
Ensure: θ-join results collection $(P_R.B \ \theta_1 \ P_S.B).B \ \theta_2 \ (P_L.B \ \theta_3 \ P_T.B).B$;
  1: get data sets R, S, T, L according to *windowSize*;
  2: $P_{R,S}$ = filtered R and S by FastThetaJoin;
  3: $P_{L,T}$ = filtered L and T by FastThetaJoin;
  4: $P_{(R,S),(L,T)}$ = filtered $P_{R,S}$ and $P_{L,T}$ by FastThetaJoin;
  5: return $P_{(R,S),(L,T)}$;

---

4 Experiment Evaluation

In the experiment, we use three other methods for comparison. For θ-join between two data streams, the compared algorithms include CrossFilterStrategy(CFS), SparkSQL, DirectCrossProduct and FastThetaJoin proposed. For multi-way θ-join, we only compare CFS with FastThetaJoin proposed and the reason will be elaborate later. CFS is the newest algorithm in the existing literature. SparkSQL refers to use Spark functional programming API to write SQL for θ-join [6]. DirectCrossProduct refers to do Cartesian product directly without any optimization, and finally filter to achieve θ-join. In the experiment, we use the window mechanism. After fetching data from the data source, the data is cut into limited blocks for processing along time boundaries. These time-based sliding-window has used widely [8, 10].

4.1 Experiment Setup

All the experiments are conducted on a spark streaming cluster that consists of 11 servers. The required components versions are Spark 2.3.0, Scala 2.11, Kafka 2.11, JDK 1.8.0, and Hadoop 2.7.4. Each node is AMD Opteron (TM) Processor 6238 with 32GB of memory with installed CentOS Linux release 7.4.17.

We modify an open-source stream generator project on GitHub as the data source [3]. The input parameters of the stream generator includes the number of streams, window time span, window sliding interval, data distribution type of each streaming. In fact, window time span and window sliding interval are used to calculate the fetching rate in each streaming. Each data produced by stream generator contains at least following attributes: topic, timestamp, a set of attributes.

4.2   Two-Way Data Stream

Each stream is set a window to indicate items that need to be calculated currently. The major factors which affect θ-join include: 1) window size which shows how long we need to intercept the data at one time; 2)window sliding intervals which represent the interval of time window sliding; 3)partition number. In fact window size and window sliding decide how much data to process in a batch. So in this section, we take the above factors as input parameters to test the performance of each algorithm. One of input parameters is the value of θ which belongs to $\{`<`, `\leq`, `\geq`, `>`\}$. Here, we assume θ is `>`. So we need to find out all tuples where attribute B in the first Stream is bigger than that in the second Stream. We have two evaluation metrics in the experiment: running time and the cross product number.

First, we set each window to contain 1000 items and the partition number as 10. The experimental results are shown in Fig. 5. We show 8 groups of experiments. The abscissa indicates the group of experiments. The ordinate indicates the running time in ms. As shown in Fig. 5(a), DirectCrossProduct has the longest execution time which is almost 3× than SparkSQL and far greater than the other two methods. SparkSQL has the second longest running time and is also far bigger than CFS and FastThetaJoin proposed in this paper. Because the difference between CFS and FastThetaJoin is not obvious in Fig. 5(a). So we extract their results into Fig. 5(b). As shown in Fig. 5(b), FastThetaJoin is faster than the state-of-the-art method CFS.

It is easy to understand why the running time of DirectCrossProduct and SparkSQL is so high. Literally, DirectCrossProduct does not have any optimization. Each data item in one stream has to make a Cartesian product with a data item belonging to other stream, which has $O(nm)$ Cartesian products. Although the SparkSQL framework has some optimizations, SparkSQL performs the Cartesian product when θ condition is satisfied, each number is actually compared with others in other streams. So SparkSQL is always faster than DirectCrossProduct. Both the other two methods filter before the Cartesian product phrase. They discard some elements that obviously do not satisfy the condition, reducing the transmission overhead and the number of Cartesian product calculations. Consequently, the running time is naturally short.

Compare with CFS, the reason why FastThetaJoin proposed in this paper is faster is that it has no global sorting time of $O(n \log n)$. Rather, we only use $O(1)$ to delete the most useless data items. And the Cartesian products are calculated between selected partitions in FastThetaJoin, instead of involving all partitions

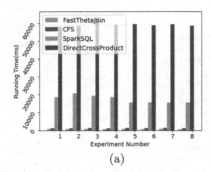 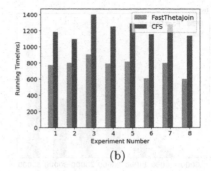

(a)   (b)

Fig. 5. Performance test of two streams. (The abscissa represents the experiment group, and the ordinate represents the running time)

as in CFS. From the above two aspects, FastThetaJoin can go to a further step to reduce the filtering overhead and the execution time as well.

To explore the relationships between different window sizes and running time, we slowly increased the window size. In the beginning, we set window size as 1000 and measured the running time with respect to this window, and made ten experiments as a group, taking the average as the result of the group. Then we increased the window size gradually, and counted the results of CFS and FastThetaJoin. Since CFS and FastThetaJoin are significantly faster when the window size is just 1000 as shown in Fig. 5(a), we only compare FastThetaJoin with CFS when the window size is more than 1000. Figure 6 shows the relationships between the window size and the running time. The x-axis represents different window sizes while y-axis indicates the running time in ms. The experimental results show that FastThetaJoin has a shorter running time and a faster calculation speed than CFS, especially when the window is larger, the advantage of FastThetaJoin is more obvious.

In fact, the window size should be set reasonably. Not the bigger the better. There are two main reasons. On the one hand, if the window size is very large, the intermediate results will be too many to result in OOM errors. On the other hand, in real-time stream computing, the larger the window, the more data need to be processed and the processing time is longer. This can result in data processing time much longer than the data generation time, the two rates do not match, and ultimately affect performance.

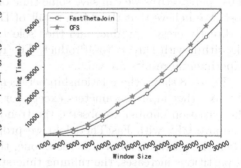

Fig. 6. Comparison of FastThetaJoin and CFS under different window sizes.

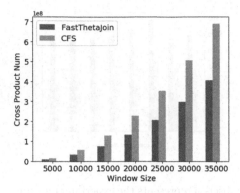

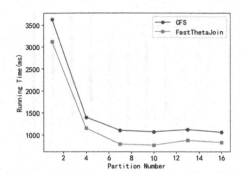

Fig. 7. Comparation of cross product number under different window size.

Fig. 8. Relationship between partition number and running time.

Figure 7 shows cross product numbers under different window sizes between CFS and FastThetaJoin. Vertical coordinate representation cross product number. We first set window size as 1000, get the average result of a group and alter window size. As shown in Fig. 7, when the window size is fixed, the cross product number of FastThetaJoin is less than that of CFS. This shows that FastThetaJoin has the least join cost. And for both CFS and FastThetaJoin, running time increases as the window increases. This also shows that the window size should not be too large from another aspect.

The experimental results show that, as expected, FastThetaJoin proposed in this paper is faster than the latest method CFS. There are two main reasons. On the one hand, there is no global sorting time so we can save $O(n \log n)$ time. Although we need $O(n)$ to traverse to get maximum and minimum and time of $O(m)$ to traverse one by one within a very small area, m is far less than n. On the other hand in the shuffle phase, because the partitions are ordered, not all partitions need to calculate Cartesian products with each other. From these perspectives we can save some time of transmission and Cartesian product cost. We believe that the θ-join result of DirectCrossProduct is correct because it directly crosses product and then select. We compare θ-join results of every algorithm with DirectCrossProduct. The results are the same. So all the above experimental results are valid.

Figure 8 shows the relationship between partition number and running time. We fix other input parameters except for partition numbers. Then we change the partition number to observe the trend of running time. Similarly, we only compare CFS with FastThetaJoin we proposed in this article. As we can see, when the partition number is set as one, the speed is very slow. As the number of partitions increases, the running time stabilizes after a steep drop. This phenomenon is reasonable. When the number of partitions is small, the node load is large, so the processing is slow. As the number of partitions increases, several nodes execute concurrently, so the processing speed becomes faster, running time gets small. We have an estimated minimum overhead. Therefore, below a certain set threshold, as the number of nodes increases, other redundant nodes do not participate in the execution, so the total execution time remains stable.

4.3 Multi-way Data Stream

In this section, we generate 4 streams and set every window size is 1000 and the partition number is 10. For multi-way θ-join, we compare CFS with FastThetaJoin proposed. The general idea is to divide them into pairs and execute multi-

Table 1. θ-join result of multi-way data stream

θ	FastThetaJoin	CFS	Improve ratio (%)
>, ≥, <	2480	3746	33.8
≥, >, <	2224	3592	38.1
<, >, <	2277	3536	35.6

ple pairs concurrently. The results are as the input of the next round until there is only one output. As shown in Table 1, the elements in the first column in turn represent the θ condition between the four streams. The second column indicates the running time of the FastThetaJoin algorithm in ms. The third column indicates that the CFS algorithm running time in ms. The last column shows how much the efficiency of FastThetaJoin has improved than CFS. θ_1 is '>', θ_2 is '≥', θ_3 is '<'. In order to improve the efficiency of concurrency, speed up the processing, we can divide the application into the following 3 blocks (P_R.B > P_S.B).B ≥ (P_L.B < P_T.B).B. As we can see in the test result, in multi-way θ-join queries, FastThetaJoin is also faster than CFS.

5 Conclusion

This paper proposes FastThetaJoin, an optimization technique for θ-join operation on multi-way data streams in massive streaming data analysis and processing. The main idea is to filter unnecessary items before cross-product and make fewer data to participate in the Cartesian product. We have two main innovations to implement our ideas. One is the partition strategy and the other is filter strategy. This method partitions the data according to the range of the specified attribute to make an order between partitions, and disorder within partitions. The filter strategy makes partial partitions participate cross product instead of all. With this design, we can not only speed up the deletion in filtering, but also quickly partition the stream in parallel distributed computing to further reduce the number of Cartesian product connections. Our experimental results show that the larger the gap between different data streams is, the better the performance of the method proposed in this paper is. In particular, when the data flow increases, the performance improvement is also obvious.

Acknowledgment. This work is supported in part by Key-Area Research and Development Program of Guangdong Province (2020B010164002), Shenzhen strategic Emerging Industry Development Funds (JCYJ20170818163026031), and also in part by National Natural Science Foundation of China (61672513).

References

1. https://www.oreilly.com/ideas/the-world-beyond-batch-streaming-101
2. https://www.oreilly.com/ideas/the-world-beyond-batch-streaming-102
3. https://github.com/YongyiZhou/Multiway-Stream-Generator/blob/master/src/main/java/DSMain.java
4. Afrati, F.N., Ullman, J.D.: Optimizing joins in a map-reduce environment. In: Proceedings of the 13th International Conference on Extending Database Technology, pp. 99–110. ACM (2010)
5. Arasu, A., Babu, S., Widom, J.: The CQL continuous query language: semantic foundations and query execution. VLDB J. **15**(2), 121–142 (2006)
6. Armbrust, M., et al.: Spark SQL: relational data processing in spark. In: Proceedings of the 2015 ACM SIGMOD International Conference on Management of Data, pp. 1383–1394. ACM (2015)
7. Blanas, S., Patel, J.M., Ercegovac, V., Rao, J., Shekita, E.J., Tian, Y.: A comparison of join algorithms for log processing in MapReduce. In: Proceedings of the 2010 ACM SIGMOD International Conference on Management of Data, pp. 975–986. ACM (2010)
8. Carney, D., et al.: Monitoring streams: a new class of data management applications. In: Proceedings of the 28th International Conference on Very Large Data Bases, pp. 215–226. VLDB Endowment (2002)
9. Golab, L., Özsu, M.T.: Update-pattern-aware modeling and processing of continuous queries. In: Proceedings of the 2005 ACM SIGMOD International Conference on Management of Data, pp. 658–669. ACM (2005)
10. Hammad, M.A., Aref, W.G., Elmagarmid, A.K.: Stream window join: tracking moving objects in sensor-network databases. In: 15th International Conference on Scientific and Statistical Database Management 2003, pp. 75–84. IEEE (2003)
11. Jiang, D., Tung, A.K., Chen, G.: MAP-JOIN-REDUCE: toward scalable and efficient data analysis on large clusters. IEEE Trans. Knowl. Data Eng. **23**(9), 1299–1311 (2010)
12. Khayyat, Z., et al.: Lightning fast and space efficient inequality joins. Proc. VLDB Endow. **8**(13), 2074–2085 (2015)
13. Lin, Q., Ooi, B.C., Wang, Z., Yu, C.: Scalable distributed stream join processing. In: Proceedings of the 2015 ACM SIGMOD International Conference on Management of Data, pp. 811–825. ACM (2015)
14. Liu, W., Li, Z., Zhou, Y.: An efficient filter strategy for theta-join query in distributed environment. In: 2017 46th International Conference on Parallel Processing Workshops (ICPPW), pp. 77–84. IEEE (2017)
15. Okcan, A., Riedewald, M.: Processing theta-joins using MapReduce. In: Proceedings of the 2011 ACM SIGMOD International Conference on Management of Data, pp. 949–960. ACM (2011)
16. Olston, C., Reed, B., Srivastava, U., Kumar, R., Tomkins, A.: Pig latin: a not-so-foreign language for data processing. In: Proceedings of the 2008 ACM SIGMOD International Conference on Management of Data, pp. 1099–1110. ACM (2008)
17. Xie, D., Li, F., Yao, B., Li, G., Zhou, L., Guo, M.: Simba: efficient in-memory spatial analytics. In: Proceedings of the 2016 International Conference on Management of Data, pp. 1071–1085. ACM (2016)
18. Yan, K., Zhu, H.: Two MRJs for multi-way theta-join in MapReduce. In: Pathan, M., Wei, G., Fortino, G. (eds.) IDCS 2013. LNCS, vol. 8223, pp. 321–332. Springer, Heidelberg (2013). https://doi.org/10.1007/978-3-642-41428-2_26

19. Yang, H.C., Dasdan, A., Hsiao, R.L., Parker, D.S.: Map-reduce-merge: simplified relational data processing on large clusters. In: Proceedings of the 2007 ACM SIGMOD International Conference on Management of Data, pp. 1029–1040 (2007)
20. Zhang, C., Li, J., Wu, L., Lin, M., Liu, W.: SEJ: an even approach to multiway theta-joins using MapReduce. In: 2012 Second International Conference on Cloud and Green Computing, pp. 73–80. IEEE (2012)
21. Zhang, X., Chen, L., Wang, M.: Efficient multi-way theta-join processing using MapReduce. Proc. VLDB Endow. 5(11), 1184–1195 (2012)

A Distributed Framework for Online Stream Data Clustering

Jiafeng Ding[1], Junhua Fang[1(✉)], Pingfu Chao[2], Jiajie Xu[1], PengPeng Zhao[1], and Lei Zhao[1]

[1] Department of Computer Science and Technology,
Soochow University, Suzhou, China
20185227068@stu.suda.edu.cn, jhfang@suda.edu.cn,
{xujj,ppzhao,zhaol}@suda.edu.cn
[2] The University of Queensland, Brisbane, Australia
p.chao@uq.edu.au

Abstract. The recent prevalence of positioning sensors and mobile devices generates a massive amount of spatial-temporal data from moving objects in real-time. As one of the fundamental processes in data analysis, the clustering on spatial-temporal data creates various applications, like event detection and travel pattern extraction. However, most of the existing works only focus on the offline scenario, which is not applicable to online time-sensitive applications due to their low efficiency and ignorance of temporal features. In this paper, we propose a distributed streaming framework for spatial-temporal data clustering, which accepts various clustering algorithms while ensuring low resource consumption and result correctness. The framework includes a dynamic partitioning strategy for continuous load-balancing and a cluster-merging algorithm based on convex hulls [10], which guarantees the result correctness. Extensive experiments on real dataset prove the effectiveness of our proposed framework and its advantage over existing solutions.

Keywords: Real-time cluster analysis · Distributed stream processing · Spatial-temporal data mining · Top-k query · Parallel computing

1 Introduction

Recently, the ubiquity of GPS-equipped devices enables the track of users' travel behaviors from various sources, like mobile phones, vehicles, and other wearable devices. With the large-scale spatial-temporal data available, many applications

This work is partially supported by NSFC (No.61802273), the Postdoctoral Science Foundation of China under Grant (No. 2017M621813), the Postdoctoral Science Foundation of Jiangsu Province of China under Grant (No. 2018K029C), and the Natural Science Foundation for Colleges and Universities in Jiangsu Province of China under Grant (No. 18KJB520044).

M. Qiu (Ed.): ICA3PP 2020, LNCS 12452, pp. 190–204, 2020.
https://doi.org/10.1007/978-3-030-60245-1_13

focus on analyzing the user's travel patterns or predicting the traffic flows, which are based on the data clustering process. Data clustering has been playing a crucial role in numerous applications, such as pattern recognition [20], information retrieval [13], medical sciences [14], social network analysis [17] and image processing [15].

Generally speaking, the timestamp attribute carried by spatial-temporal data determines it is time-efficient, that is, the value of spatial-temporal data will decrease with the passage of time. Furthermore, applications based on these spatial-temporal data are naturally time sensitive, like identifying the topic changes in social networks [5] and detecting real-time road congestion in the transportation system [18]. Real-time data clustering is the basis operation of many time-dependent processes in such online scenario. However, despite a significant number of clustering algorithms proposed in the last few decades, most of them cannot be applied to an online scenario as they fail to achieve interactive performance. Clustream algorithm [3] is one of the most classical stream clustering algorithm. It use the two phase patterns, including online and offline, to cluster the data. However, the influence of out-of-date data points cannot be eliminated in the online clustering process, which greatly reduces the clustering effect of the algorithm. Several other algorithms [6,12] introduced recently target at the online streaming environment with decent efficiency. However, they either do not support sliding-window queries or cannot cluster large-scale data sets into arbitrary-shape clusters. A fast large-scale trajectory clustering [21] has been proposed to efficiently identify k "representative" paths in a road network, however, it only certain k values can be found, not arbitrary values.

Traditional stream clustering solutions are either use micro-batch centralized computing mode and optimized by attenuation factor, which is difficult to support the increasing data volume, or use classical two-stage clustering, including online and offline, but cannot eliminate the impact of expired points. According to our observation, since the data stream is usually continuous, disordered, and infinite, the main challenges in online distributed stream processing system are three-fold:

- **Dynamic:** The dynamic features of data stream come from two aspects: (1) the dynamic data volume means the size of incoming data is dynamic and unpredictable, (2) the dynamic value distribution implies the distribution of the incoming data values may change over time. Both features require the distributed framework to be load-adaptive and load-balancing.
- **Timeliness:** The in-built temporal feature of the stream data defines the importance of real-time processing. In other words, the value of the stream data decays quickly as time goes. Therefore, the stream processing framework should ensure low processing latency.
- **Efficiency:** Efficiency requires the system to minimize resource consumption while ensuring high throughput and low latency when dealing with dynamic workloads. Considering the spatial feature of our clustering problem, merely partitioning the data by their semantic features may lead to absurd physical data localization, which causes extra resource consumption. Therefore, a smart partitioning strategy is necessary.

To address the aforementioned challenges, in this paper, we propose a new distributed framework for online clustering analysis, termed as RT-Clustering. Different from existing works, our framework deals with real-time streaming data as the data source to quickly calculate the maximum k clustering regions. We eliminate the impact of expired points by filtering them in time. In addition, we ensure the balance of data distribution by a feedback mechanism. Besides, our framework supports real-time data clustering based on the sliding window and we optimize our solution by a cluster-merging process to further improve the efficiency. Overall, our contributions are listed as follows:

- We propose a distributed framework for online data clustering. The framework accepts most of the existing clustering algorithms while ensuring the load-balancing and efficiency in a large-scale distributed environment.
- We propose a rule-based adaptive clustering strategy that utilizes the historical cluster results for incoming data distribution. Our proposed strategy ensures the load-balancing and efficiency under dynamic data input.
- We propose a cluster-merging algorithm to refined the distributed clustering results. Based on the idea of the convex hull, our cluster-merging algorithm guarantees the result correctness while minimizes the system resource consumption.
- Extensive experiments are conducted on a real data set to prove the effectiveness of our proposed framework and optimization algorithms in terms of efficiency and correctness.

The rest of the paper is organized as follows: We first formally define the problem and introduce our proposed framework in Sect. 2, then we elaborate our rule-based adaptive clustering strategy and the cluster-merging optimization in detail in Sect. 4. The experiments are demonstrated in Sect. 5. We review the related work in Sect. 6 and draw conclusion in Sect. 7.

2   Preliminaries

The main notations used throughout this paper and their respective definitions are listed in Table 1.

Table 1. The notations.

Notations	Definitions
p	A point of trajectory
θ	The threshold of the density cell
γ	The threshold of the sparse cell
P	Points of a trajectory
$\mathbb{S}$	Input data stream
k	Number of top requested results
W	Sliding window
$T=< P_1, P_2..., P_m, ... >$	The trajectory consisting of indefinite points
$\tau =< T_1, T_2, ..., T_n >$	Data set consisting of N trajectories

2.1 Clustering Algorithm

To help understand and facilitate subsequent description, we briefly introduce DBSCAN with the example of Fig. 1. The common definitions of DBSCAN are as follows:

- ϵ**−Neighborhood:** Consider a set P of points, the ϵ-neighborhood of $p \in P$, denoted as $N_\epsilon(p)$, is a set of the points whose distances from p are within ϵ, i.e., $N_\epsilon(p) = \{q|d(q,p) \leq \epsilon, q \in P\}$. For example, in Fig. 1., the grey circle of each point is its ϵ−Neighborhood, e.g, $N_\epsilon(O_3)$ =$\{O_2, O_4, O_6\}$.
- **Core Point:** A point p is a core point if at least $minPts$ point u satisfy $d(p, u) \leq \epsilon$, where d(p, u) denotes the distance between p and u. If p is not a core point, it may be an edge point. Consider the example in Fig. 1., we set $minPts = 3$, we can find that O_3, O_4 have three points within their ϵ−Neighborhood. So they are considered as core objects.
- **Density Reachable Point:** A point p is density reachable from location v if there exist a sequence of locations $x_1, x_2, ..., x_t(t \geq 2)$ conform that (i) $x_1 = p$ and $x_t = u$; (ii) x_i $(1 \leq i < t)$ are core points; and (iii) $d(x_i, x_i + 1) \leq \epsilon(1 \leq i < t)$. DBSCAN utilizes the density-reachability to assemble a large dense region by a iterative process. Specially, while $t = 2$, we call it **Directly Density-Reachable** since they are not connected by any core point. Considering the example data set in Fig. 1. again, O_1 is directly density-reachable from O_2, but O_1 is not directly density-reachable from O_6 since the distance between O_1 and O_6 is greater than ϵ, but O_1 is density-reachable from $\{O_2, O_3, O_4, O_5, O_6\}$.

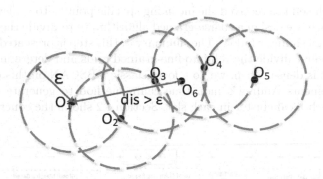

Fig. 1. Example Of DBSCAN

2.2 Problem Description

Based on the model of stream data and its workload, we define the real-time cluster mining problem, with the objectives on (i) dynamic workload among the processing slots, (ii) perform the clustering algorithm rapidly, and (iii) ensure semantic correct in the share-nothing environment. Specifically, to achieve the above goal, we formulate the description as an optimization problem: Given a time-series data stream $\mathbb{S}$, which contains trajectory set $\tau =< T_1, T_2, ..., T_n >$ and each trajectory, $T =< P_1, P_2..., P_m, ... >$, contains uncertain numbers points; a distance measure function D; a sliding window W; a user-defined parameter k, the aim of function RT-Clustering is searching the top-k clusters which cover the k-largest areas. The formula of real-time clustering is shown as below:

$$Top - k_{clusters} = RT - Clustering(\mathbb{S}, D, k, W)$$

Note that k is defined by users. Besides, we use Euclidean distance as the distance function because of the properties of spatiotemporal data and use DBSCAN as the clustering algorithm in this paper.

3   Solution Overview

The processing framework of RT-Clustering is shown in Fig. 2, and can be decomposed into the following three steps:

- Step1: We first divide the initial region into fixed cells in advance and each cell is a processing unit. Next, as shown in step1 of Fig. 2, we get the latest cluster results to generate a partition strategy and allocate the fixed cells according to the strategy in each cell. Then, we mark each point in the data stream $\mathbb{S}$ according to its geographic position. A key-based partition algorithm would be implemented to assign the points to the corresponding cell. We will present the detailed dynamic data partition in Sect. 4.1.
- Step2: each cell has received the incoming specific points. To achieve a speedy performance, we design a coarse-grained algorithm to re-divide the cell, which is based on a sliding window. The summary of this step is presented in step2 of Fig. 2. We first divide the cell into fine-grained grids and determine each grid whether it is dense. Then, we do a local classic DBSCAN on this dense grids and implements Andrew's monotone chain method to generate the convex hull for each local cluster in each slot. Section 4.2 shows the concrete details.

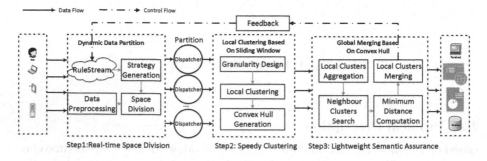

Fig. 2. Overview Of RT-clustering

– Step3: all the local clusters need to be merged to generate the final clusters. Step3 of Fig. 2 shows the main procedure. We first collect all local clusters and their convex hulls. Next, we merge the local clusters by calculating the minimum distance min_{dis} between convex hulls and compare with the ϵ. If $min_{dis} < \epsilon$ and the anti-pair is the form of point-point, they belong to the same cluster and we merge them. We will present the details in Sect. 4.3.

4   Real-Time Distribute Clustering

4.1   Dynamic Data Partition

To maximize throughput and minimize computation latency, we need to balance the workload of each cell. A dense area may contain several grids in RT-Clusting, so we compute the dense area and set parameters to determine whether the range needs to be optimized. The dense area in this paper is defined as shown below:

Definition 1 *(Dense Area).* At time t, for a cell c, given the clusters at time t, we definite the dense area that

$$DenseArea(c,t) = Area(clusters,t) \cap Area(c,t)$$

where Area(clusters,t) is the area of clusters at time t, and Area(c,t) is the area of each cell. Based on the above definition, the density coefficient can be defined as follows:

Definition 2 *(D-cell).* At time t, for a cell c, the density coefficient

$$D(c,t) = \frac{DenseArea(c,t)}{Area(c,t)} \tag{1}$$

Algorithm 1 shows the procedure of the strategy generation algorithm. It contains two parts: strategy generation and points partition. Our basic idea is that using the latest cluster results to predict the current dense area and assign more nodes to process. Then we divide them into more ranges and each range can be seen as a cell. However, to reduce the computation, we only re-divide the original ranges rather than re-divide the whole area. In the first stage, line 1–11 of the Algorithm 1 present that we use the Definition 2 to determine these grids. If the area is detected as a dense area, it will be divided into several ranges. Those sparse and transitional areas are maintained with the latest state. In the second stage, as shown line 12–16 of the Algorithm 1, all points in the stream do a map operation according to their location information and are distributed into the corresponding cells by their cell *id*.

Algorithm 1. Strategy Generation Algorithm

Input:
    The data stream $\mathbb{S}$;
Output:
    each keyedPoint P_{keyed};
1: clusters = Feedback()
2: **for** each cell c in region **do**
3:     $f(c,t) = \frac{Area(c,t) \cap Area(clusters,t)}{Area(c,t)}$
4:     **if** $f(c,t) > \theta$ **then**
5:         cell_list $\leftarrow$ divide(c)
6:     **else if** $\gamma < f(c,t) < \theta$ **then**
7:         l_{cell} = transitional; cell_list $\leftarrow$ c
8:     **else**
9:         l_{cell} = sparse; cell_list $\leftarrow$ c
10:    **end if**
11: **end for**
12: **for** each point in Stream $\mathbb{S}$ **do**
13:    cid = Location(point)
14:    map $\leftarrow$ < cid, point >
15:    **return** P_{keyed} = keyedDistribution(map)
16: **end for**

4.2 Local Clustering Algorithm

In order to enable local tasks to detect the clusters quickly, we design a coarse-grained algorithm to re-divide the cells here. Figure 3 shows an example of dynamic data partition. The blue range is the Intersect area. Suppose that $\theta = 0.8$ and $\gamma = 0.2$, we can easily find that the C_5 is dense, C_2, C_4, C_8 may be transitional grids and C_1, C_3, C_6, C_7, C_9 are sparse grids through computation according to the Definition 2. Then we divide C_5 into smaller cells and use more slots to process it to improve the speed of calculation. In this example, we divide it into C_{5_1}, C_{5_2}, C_{5_3} and C_{5_4}. These cells are placed in slots by a hash-based method. Finally, the points in the data stream $\mathbb{S}$ will be marked with the *id* of their cell based on their location information. In Fig. 3.(a), P_1, P_2 and P_3 are the incoming points which carry coordinate information. Fig. 3.(a) is the cluster result at last time. Figure 3.(c) is the partition strategy of Fig. 3.(a) and Fig. 3.(d) is the partition strategy of Fig. 3.(b). After dividing, the range of sparse and transitional grids don't change, but the dense grids have been divided into smaller range. In this stage, we can easily determine that $P_1(1,0)$ belongs to C_2. $P_2(1.2, 1.3)$ is sent to C_{5_2} rather than C_7 and $P_3(2,2)$ is sent to C_9.

After points are distributed to correlative cells, we perform the local clustering algorithm in this step. To achieve a speedy performance and minimize network traffic, as shown in Algorithm 2, we design a micro-coarse-grained partitioning on these cells and maintain a map to record the number of points in micro-grids. We first divide the grid internally, and each micro-grid represents a macro point. It should be noted that the size of the grids can be adjusted,

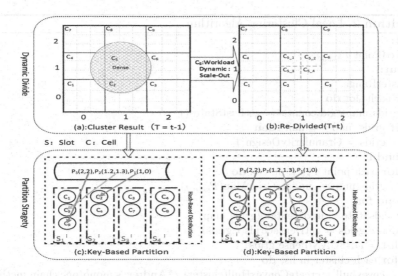

Fig. 3. Example of dynamic data partition

and the smaller the mesh, the higher the accuracy. Then we need to determine whether each grid is dense or sparse. There are two situations that must be considered. The first stage is that the cell is sparse, and it maintains the latest grids state. We only need to update the corresponding number. The other scene is that the cell is dense, and it has been re-divided. We need to recompute the micro-coarse-grained grids. Next, we perform a classic DBSCAN on these dense grids. Finally, we implement Andrew's monotone chain method to generate the convex hull of the finite set of points of each cluster in the two-dimensional Euclidean space.

4.3 Global Clustering Algorithm

As mentioned above, the processing slots are share-nothing. In this case, we need to recover the state between the points for cluster merging. Algorithm 3 presents the lightweight cluster merging algorithm. We first construct a list of all local clusters and their corresponding convex hulls from Algorithm 2. Next we traverse the list and use the rotate jam algorithm to compute the minimum distance, which is shown as Line 8~19 in Algorithm 3. For two convex hulls P and Q, firstly, we calculate the minimum value of convex hull P in the y-axis direction as yminP, and the maximum value of convex hull Q in the y-axis direction as ymaxQ. Secondly, we establish two horizontal straight lines L_P, L_Q next to yminP, ymaxQ. Then, we take them to face different directions, e.g., one is upwards, and another is down. At this time, they form an antipodal pair. Thirdly, we calculate the distance of (yminP, ymaxQ) and record it as a minimum value. Fourthly, we turn the two straight lines clockwise until one of the edges that meet the convex hull. As long as a straight line meets an edge, we

---

Algorithm 2. Local Clustering Algorithm

Input:
   keyedPoint P_{keyed};
Output:
   clusterRange
1: **for** each slot **do**
2:    grids_map<gid,count> = getLastStates()
3:    **if** grids_map == null **then**
4:       grids = GranularityDesign()
5:    **end if**
6:    **for** each point P_{keyed} in cell **do**
7:       grids_map ← updateGridsMap(P_{keyed})
8:    **end for**
9:    GridsState = updateState(grids_map)
10:   dense_grids = getDenseGrids(ℓ,grids)
11:   list = DBSCAN(dense_grids)
12:   **for** each cluster in list **do**
13:      covexhull = createConvexHull(cluster) /*Andrew's monotone chain method*/

14:      **return** clusterRange ← <cluster,covexhull>
15:   **end for**
16: **end for**

---

need to calculate the distance between the new vertex-vertex antipodal, compare it with the minimum, and update it. We repeat steps 3 and 4 until the two lines return to the starting position and get the minimum distance. If the minimum distance is less than ϵ, we treat them as one cluster and merge them. After this computing procedure, we will get the list of final clusters that distance between any two of them is greater than ϵ. Finally, we get the top-k largest cluster of the list as the final result.

5 Experiment

5.1 Experimental Setup

Environment: We implement and conduct the experiment on Apache Flink1.9 [1]. The algorithm is implemented in Java, and the Flink system is deployed on a cluster of machines, each of which runs CentOS 7 operating system and is equipped with 32 Intel core processors (E7-8860 at 2.20 GHz) with 128 cores and 256 GB RAM.

Data Set: We use a real-world GPS trajectory data set of Beijing Taxi, which contains more over 13000 trajectories, to test our proposed algorithm. Each trajectory record includes the taxi ID, the timestamp of the event and the position(latitude and longitude).

Algorithm 3. Global Merging Algorithm

Input:
    clusterRange
Output:
    Top-$k_{clusters}$
1: initialize clusterRangeList
2: **for** each clusterRange **do**
3:     clusterRangeList: add(clusterRange)
4: **end for**
5: **for** i =1 to length(clusterRangeList) **do**
6:     $c_i \leftarrow clusterRange_i$; $points_i$ = getPoints(c_i)
7:     **for** j = i+1; j<length(clusterRangeList);j++ **do**
8:         shortestDis = 0;
9:         $c_j \leftarrow clusterRange_i$; $points_j$ = getPoints(c_j)
10:        $ymin_i$ = getMinY($points_i$); $ymax_j$ = getMinY($points_j$)
11:        $line_i$ = buildVerticalXLine($ymin_i$); $line_j$ = buildVerticalXLine($ymax_j$)
12:        **while** $line_i$ or $line_j$ does not back to starting point **do**
13:           clockwiseRotate($line_i$); clockwiseRotate($line_j$)
14:           **if** $line_i$ contains points in $points_i$ or $line_j$ contains points in $points_j$ **then**
15:              dis = getDistance($line_i,line_j$)
16:              **if** dis < shortestDis **then**
17:                 shortestDis = dis
18:              **end if**
19:           **end if**
20:           **if** shortestDis < ϵ **then**
21:              merge(c_i,c_j); clusterRangeList : remove(c_j)
22:           **end if**
23:        **end while**
24:     **end for**
25: **end for**
26: **return** Top-k(clusterRangeList,k)

Performance Metrics: We evaluate resource utilization and system performance through the following metrics: i)Throughput(T) is the average result number of tuples processed by the system per second. ii)Execution time (E) is the average computation time in each window during the entire process. iii)Score(S) is the sum of intra-cluster distance variances according to the formula:

$$score = \sum_{i=1}^{n} \sigma_i^2$$

where n is the number of clusters and σ_i^2 is the variance of intra-cluster distances of cluster c_i.

Performance Parameters: We use the following parameters for comparison experiments. i)ϵ is the parameter of the classic DBSCAN algorithm. ii)P is the number of Parallelisms. iii)k is the value of Top-k that defined by users. IV)$size$ is the scale of the data set, which is shown in Table 2.

Table 2. The DataSet

Dataset	Points	Size	Dataset	Points	Size
sort200	330557	14.2 MB	sort500	827239	35.4 MB
sort800	1325742	56.6 MB	sort1000	1659489	70.8 MB
sort1200	1984567	84.2 MB	sort1500	2329284	97.8 MB
sort2000	2868541	122 MB			

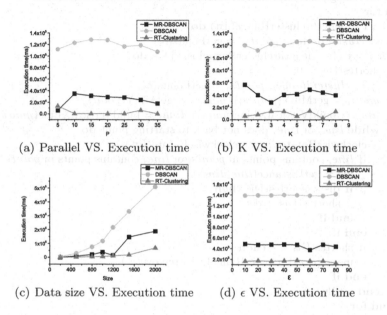

(a) Parallel VS. Execution time   (b) K VS. Execution time

(c) Data size VS. Execution time   (d) ϵ VS. Execution time

Fig. 4. Performance of execution time

Performance Experiment: We observe the performance of the approach under different parameter values. Since there are several parameters in our frame, we change one or several of them to measure it. At the same time, we compare our approach with the MR-DBSCAN [22] and the traditional DBSCAN algorithm. We deploy them on Flink [1] for comparison.

5.2   Performance on Real Data

Firstly, we deploy the experiment of execution time under different parameters. Figure 4(a). shows the execution time while the parallelism P varying from 5 to 35. We can clearly see that the execution time of the classic DBSCAN algorithm is the longest since the P does not affect it. However, the execution time of MR-DBSCAN and RT-Clustering is decreasing with P increasing, and our algorithm is better than MR-DBSCAN. Figure 4(b) shows the execution time under the parallelism K from 1 to 8. Obviously, our algorithm is almost three times better than MR-DBSCAN and 13 times better than DBSCAN. Figure 4(c) shows the

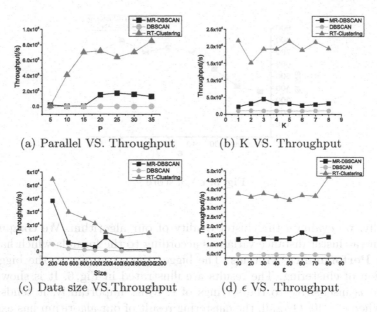

(a) Parallel VS. Throughput   (b) K VS. Throughput

(c) Data size VS.Throughput   (d) ϵ VS. Throughput

Fig. 5. Performance of throughput

effect of data set size on execution time. From the figure, we can see that as the data set grows, so does the execution time. This is due to the larger data set, which requires more data to be processed. The execution time of traditional DBSCAN is exponential growth, and that of MR-DBSCAN is growing linearly. However, our algorithm is growing smoothly, and the speed of growth is far less than others. Figure 4(d) shows the effect of the parameter of DBSCAN-ϵ. The setting of ϵ vars 10 m to 80 m. We can find that the time cost fluctuates as ϵ grows. Overall, our algorithm, which is based on convex hull, has less execution time and greater advantages.

Then, we evaluated the impact of each factor on throughput. We see the impact of parallelism on throughput in Fig. 5(a). We find that the parallelism does not affect the throughput of traditional DBSCAN since it is centralized. The throughput of MR-DBSCAN is increasing with the growth of P. However, it's speed is slow, and when $P > 20$, the speed is down. On the contrary, the throughput of our algorithm is always increasing. Figure 5(b) shows the impact of K on throughput. With K growing, the throughput slightly decreases, and it is not obvious since with K growing, it needs more computation, but it just is a small part of the whole framework. Figure 5(c) shows the impact of the size of data on throughput. From the figure, we can see that the throughput of all three algorithms is decreasing while the size is increasing. This is due to that it needs more time to compute with the bigger data size, and the throughput is reduced. Figure 5(d) shows the impact of ϵ on throughput. With ϵ growing, the throughput of the three algorithms is fluctuating and no significant change. The throughput is insensitive to the settings of ϵ. Overall, Our algorithm has greater throughput than others and has better cost performance.

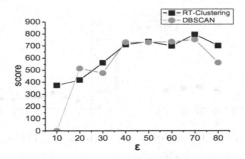

Fig. 6. ϵ VS. score

Finally, we evaluate the cluster quality of our algorithm. We compute the sum of intra-cluster distance variances according to the formula which has been given in **Performance Metrics**. The bigger the score means, the bigger the fluctuation of clustering. The results are illustrated in Fig. 6. It is shown that the score is insensitive to the settings of ϵ. Most importantly, it tends to be steady when $\epsilon > 40$. Overall, the clustering result of our algorithm has excellent quality.

6   Related Work

Distributed Stream Processing. With the increasing number of data sources and the increasing requirements of real-time data analysis, the processing of stream data becomes more and more important. Decades ago, Mitch Cherniack [8] describe scalable distributed stream processing to fit a large class of new applications for which conventional DBMSs fall short. Tyler Akidau proposes that a fundamental shift of approach is necessary to deal with these evolved requirements in modern data processing [4]. Especially, several open-source distributed batch-stream processing platforms have been proposed, which provide two different types of processing. In a tuple process, each incoming record is processed as soon as it arrives without waiting for other records. Storm [2] and Flink [1] support this type of processing. But not limited to this, our methods and techniques are generic and, therefore, easily applicable to other distributed flow processing platforms.

Distributed Cluster Analysis. Cluster analysis has received much attention due to numerous attractive properties. In the past few decades, with the large-scale increase in the amount of data, the traditional centralized clustering algorithm gradually failed to meet the computing requirements, and a large number of distributed clustering algorithms were proposed. SDBC [11] gives a quality criterion to select objects as local representatives and generate global approximate DBSCAN clusters. RP-DBSCAN [19] uses a cell-based random partitioning scheme together with a two-level cell dictionary to find out approximate

DBSCAN clusters. However, all the algorithms can not present the clustering results dynamically.

Online Cluster Analysis. Cluster stream data methods have been proposed in recent years. Gong [9] proposes a stream clustering algorithm EDMStream by exploring the evolution of density mountain that is used to abstract the data distribution. Nasir et al. [16] designed a series of randomized routing algorithms to balance the workload of stream processing operators. Chen proposes D-stream [7], which is a framework for clustering stream data using a density-based approach. It adopts a density decaying technique to capture the dynamic changes of a data stream, but the algorithm can not maintain the complete status information.

7   Conclusion

In this paper, we identify and solve the real-time cluster problem. To support the rapidly coming and massive data, we propose a real-time parallel distributed clustering algorithm that utilizes a rule stream to assist space-based partition to ensure workload balance. Besides, we build a convex hull for each cluster to compute the minimum distance of convex hulls for merging. Moreover, we conduct extensive experiments on real data sets to demonstrate that RT-Clustering is much more efficient and stable. In this paper, we study the clustering of points, but not limited to this, we would like to explore trajectory clustering in a real-time environment in further work.

References

1. Apache Flink Project. http://flink.apache.org/
2. Apache Storm Project. http://storm.apache.org/
3. Aggarwal, C.C., Philip, S.Y., Han, J., Wang, J.: A framework for clustering evolving data streams. In: Proceedings 2003 VLDB Conference, pp. 81–92. Elsevier (2003)
4. Akidau, T., et al.: The dataflow model: a practical approach to balancing correctness, latency, and cost in massive-scale, unbounded, out-of-order data processing. Proc. VLDB Endow. **8**(12), 1792–1803 (2015)
5. Chen, H., Yin, H., Li, X., Wang, M., Chen, W., Chen, T.: People opinion topic model: opinion based user clustering in social networks. In: Proceedings of the 26th International Conference on World Wide Web Companion, pp. 1353–1359 (2017)
6. Chen, L., Gao, Y., Fang, Z., Miao, X., Jensen, C.S., Guo, C.: Real-time distributed co-movement pattern detection on streaming trajectories. Proc. VLDB Endow. **12**(10), 1208–1220 (2019)
7. Chen, Y., Tu, L.: Density-based clustering for real-time stream data. In: Proceedings of the 13th ACM SIGKDD International Conference on Knowledge Discovery and Data Mining, pp. 133–142. ACM (2007)
8. Cherniack, M., et al.: Scalable distributed stream processing. In: CIDR, vol. 3, pp. 257–268 (2003)

9. Gong, S., Zhang, Y., Ge, Yu.: Clustering stream data by exploring the evolution of density mountain. Proc. VLDB Endow. **11**(4), 393–405 (2017)
10. Graham, R.L.: An efficient algorithm for determining the convex hull of a finite planar set. **1**(4), 132–133 (1972)
11. Januzaj, E., Kriegel, H.-P., Pfeifle, M.: Scalable density-based distributed clustering. In: Boulicaut, J.-F., Esposito, F., Giannotti, F., Pedreschi, D. (eds.) PKDD 2004. LNCS (LNAI), vol. 3202, pp. 231–244. Springer, Heidelberg (2004). https://doi.org/10.1007/978-3-540-30116-5_23
12. Lee, P., Lakshmanan, L.V., Milios, E.E.: Incremental cluster evolution tracking from highly dynamic network data. In: 2014 IEEE 30th International Conference on Data Engineering, pp. 3–14. IEEE (2014)
13. Liang, S., Yilmaz, E., Kanoulas, E.: Dynamic clustering of streaming short documents. In: Proceedings of the 22nd ACM SIGKDD International Conference on Knowledge Discovery and Data Mining, pp. 995–1004. ACM (2016)
14. Madeira, S.C., Oliveira, A.L.: Biclustering algorithms for biological data analysis: a survey. IEEE/ACM Trans. Comput. Biol. Bioinf. (TCBB) **1**(1), 24–45 (2004)
15. Naik, D., Shah, P.: A review on image segmentation clustering algorithms. Int. J. Comput. Sci. Inform. Technol. **5**(3), 3289–93 (2014)
16. Nasir, M.A.U., Morales, G.D.F., Kourtellis, N., Serafini, M.: When two choices are not enough: balancing at scale in distributed stream processing. In: 2016 IEEE 32nd International Conference on Data Engineering (ICDE), pp. 589–600. IEEE (2016)
17. Pham, M.C., Cao, Y., Klamma, R., Jarke, M.: A clustering approach for collaborative filtering recommendation using social network analysis. J. UCS **17**(4), 583–604 (2011)
18. Rempe, F., Huber, G., Bogenberger, K.: Spatio-temporal congestion patterns in urban traffic networks. Transp. Res. Procedia **15**, 513–524 (2016)
19. Song, H., Lee, J.-G. RP-DBSCAN: a superfast parallel DBSCAN algorithm based on random partitioning. In: Proceedings of the 2018 International Conference on Management of Data, pp. 1173–1187. ACM (2018)
20. Sturn, A., Quackenbush, J., Trajanoski, Z.: Genesis: cluster analysis of microarray data. Bioinformatics **18**(1), 207–208 (2002)
21. Wang, S., Bao, Z., Culpepper, J.S., Sellis, T., Qin, X.: Fast large-scale trajectory clustering. Proc. VLDB Endow. **13**(1), 29–42 (2019)
22. He, Y., Tan, H., Luo, W., Feng, S., Fan, J.: MR-DBSCAN: a scalable MapReduce-based DBSCAN algorithm for heavily skewed data. Front. Comput. Sci. **8**(1), 83–99 (2014)

End-System Aware Large File Transfer Solution for Rich Media Applications over 5G Mobile Networks

Xukang Lyu[1](✉) and Chase Q. Wu[2]

[1] School of Computer Software, College of Intelligence and Computing, Tianjin University,
Tianjin 300354, People's Republic of China
lvxukang@tju.edu.cn

[2] Department of Computer Science, New Jersey Institute of Technology,
Newark, NJ 07103, USA
chase.wu@njit.edu

Abstract. In general, rich media applications demand high bandwidth to transfer large files over wide-area connections. The arrival of 5G will overcome the limitations of existing networks in support of these network-intensive applications because they provide high bandwidth reaching up to multiple Gbps for large file transfer. One of the main challenges to maximize and stabilize goodput is explicitly managing the randomness inherent in high-speed mobile networks. The AIMD-based TCP cannot make full utilization of bandwidth over links with high Bandwidth Delay Product. The inefficiency of transport control protocol can not satisfy the media applications transport requirements even over a high bandwidth end-to-end connection. We conducted an extensive analytical study of the design and implementation issues of high-speed data transfer methods, especially the impact of application-level receive buffer and the background workloads on the data transfer performance. We developed an optimized large file transfer solution based on this analysis to achieve the maximal bottleneck goodput. We showed that the proposed transport protocol achieves better performance compared to state-of-the-art transport methods.

Keywords: Transport control · Big data transfer · Mobile networks · 5G

1 Introduction

Future mobile technologies enable rich media applications such as mobile virtual reality/augmented reality (AR/VR), as well as 4 K/8 K and 360° video streaming to deliver high quality services. Rich media applications require huge amounts of data to be transferred and shared around the world. The success of these applications relies on fast accesses to the media content, i.e., videos 8 K or VR. The arrival of 5G will overcome the limitations of existing networks in support of these network-intensive applications because they provide high bandwidth reaching up to multiple Gbps for large file transfer. In fact, the importance of 5G has been well recognized, and many projects are currently underway to develop such capabilities around the world. One of such initiatives

© Springer Nature Switzerland AG 2020
M. Qiu (Ed.): ICA3PP 2020, LNCS 12452, pp. 205–218, 2020.
https://doi.org/10.1007/978-3-030-60245-1_14

is Turkey's first 5G cloud application [1]. This application tested the remote access of AR and VR based educational content over a 5G cloud connection with Huawei. With this work that focuses on education, this application provided low latency high quality image transfer over 5G network on subjects such as biology, chemistry and electronics which are especially visual and experience based. Educational content is just one of the many use areas of 5G technology [17]. Given high-speed mobile networks, transport protocols are the key to delivering the provisioned bandwidths to the applications.

A main challenge in high-speed mobile network is to maximize goodput[1] by adapting transport parameters automatically. The AIMD-based TCP cannot make full utilization of bandwidth over links with high Bandwidth Delay Product. The inefficiency of transport control protocol can not satisfy the media applications transport requirements even over a high bandwidth end-to-end connection. A lot of works have been done to solve the challenges associated with transferring data over long-haul networks. In [7], new congestion control algorithms were proposed to improve the performance of TCP. In order to overcome the performance limitation of TCP, researchers tried to design application protocols based on UDP over high BDP networks. These protocols include QUIC [10], UDT (UDP-based Data Transfer) [9], PLUT [14,20], and many others [11,15,18,21]. The high bandwidth available in 5G mobile networks will shift the bottleneck of end-to-end data transfer from the network to the end systems. However, the existing transport protocols based on TCP or UDP are not optimized for handling system dynamics like shared CPUs, network and disk I/O, memory management, etc. Therefore, end users have not been able to see the corresponding performance improvement in their applications.

Although some efforts attempt to address the issue of flow control in a bottlenecked end system, their proposed solutions underestimate the capacity of the end system, which generates a suboptimal feedback rate for flow control in high-speed networks. These protocols usually employ single thread to receive packets arriving from the high-speed network. This design cannot take advantage of modern multi-core architecture, and leads to a waste of end system resources and a lower estimated bottleneck rate. Therefore, it is critical to explore the adoption of mulitple UDP connections to improve both goodput and resources utilization. On one hand, these UDP-based transport protocols can obtain more aggregate CPU cycles on a shared end system with single processor to improve the application goodput by implementing a parallel receiving strategy than a single receiving process in the presence of concurrent workloads. On the other hand, the advance of multi-core processors make it possible to run different packet receiving processes on different cores, which can improve the aggregate application goodput. Thus, new transport approaches in the context of multiple parallel connections need to be developed to estimate the best rate at which the end system can consume packets coming from the network. This rate which we call maximal bottleneck rate will be sent back to the sender for source rate control.

Previous protocols with multiple connections didn't account for the impact of the background workloads, which vary during the data transfer period. The parallelism tuning mechanism needs to be adjusted to achieve the maximal bottleneck rate by taking

---

[1] Goodput only counts the user payload and is equivalent in value to throughput if packet duplicates and protocol headers are negligible.

the varying background workloads into consideration. As shown in [16], the control theory has been applied successfully in modeling systems with unpredictable workloads.

We extented our previous analysis to cover the impact of application-level receive buffer and the background workloads on the data transfer performance. We further refined the dynamic buffer management scheme and parallelism mechanism based on this analysis to achieve the maximal bottleneck goodput. We evaluated the performance of PLUT over a local 10 Gbps network connection, and showed the performance superiority of PLUT over existing methods in different use cases.

2   Maximizing Transport Performance over High-Speed Connections

The main challenge of transport control is to maximize goodput over high-speed connections by automatically adapting different transport parameters. The estimation of those parameters are impacted by both network and system dynamics. Many UDP-based transport protocols use the manual parameter tuning in high-performance networks. In [12,20], we proposed *Peak Link Utilization Transport* (PLUT) with a performance-adaptive flow control mechanism, which regulates the activities of both the sender and receiver in response to system dynamics and automates the rate stabilization for throughput maximization using stochastic approximation methods.

2.1   Performance Model for Data Receiver

We conduct our following analysis based on Linux kernel. In [13], we illustrated the effects of concurrent background workloads on the performance of UDP-based transport protocols. We modeled the packet receiving process from socket receive buffer to the user application as two special cases of death/birth process based on the assumptions on packet arrival and consumption. According to the analysis in [13] and [12], we observed that if the data receiving rate did not match with the sending rate, it would cause either packet loss or wasted system resources. Even though the multiple connections could further improve the goodput compared with single receiving process, the application does not achieve optimal performance in either cases.

Application-Level Receive Buffer.   Another source of packet loss occurs when the application-level receive buffer fills up. RAID offers much better write speed with multiple written in parallel, which significantly reduces the packet loss in the application-level buffer, than that of a single disk. To instantiate our analysis, we consider one application-level receive buffer with one data receiving process writing data into this buffer. But there are multiple disk write processes running in parallel to consume data arriving from the kernel as shown in Fig. 1. In this subsection, we provide an initial analysis of the impact of this model on the data transfer performance.

A disk write process can only handle one packet at a time and hence, it is either in a "busy" or an "idle" state. If all disk write processes are busy upon the arrival of a packet, the newly arriving packet is buffered, assuming that application-level receive

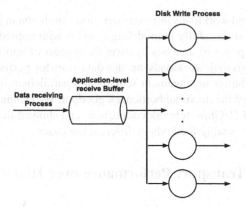

Fig. 1. Single buffer with parallel disk write process.

buffer space is available. When the packet currently in process is finished, one of the waiting packets is selected for service according to a queueing discipline.

Assume that the packets are saved to application-level receive buffer by the data receiving process at a mean Poisson rate of $r(recv)$. And the packets size are fixed and less than the Maximum Transfer Unit (MTU) with the header length HDR. We further assume that the effective service time is exponentially distributed and its mean is $\frac{1}{r(disk)}$ for packet processing carried out by each disk write process. Let m be the number of disk write process which is less than the number of Disk Controllers in RAID, and M be the application-level receive buffer size in bytes and B be the maximum number of packets in this buffer. Thus,

$$B = \lceil \frac{M}{MTU - HDR} \rceil, \tag{1}$$

and $B > m$. We denote the quantity $\frac{r(recv)}{m*r(disk)}$ by β, i.e. $\beta = \frac{r(recv)}{m*r(disk)}$ and the quantity $\frac{r(recv)}{r(disk)}$ by ρ, i.e. $\rho = \frac{r(recv)}{r(disk)}$. Based on the above assumptions, this data processing flow that employs multiple disk write processes to consume arriving data is an M/M/m/B queuing system which can be modeled as a death birth process [4–6,8], as shown in Fig. 2.

As to the application-level receive buffer, we represent the process states of this model by the number of packets as n. The number of packets n in the buffer is used to represent the corresponding states n. The state-space is finitely limited by the buffer size.

The steady-state probability P_n of this process being in state n is given by:

$$P_n = \begin{cases} \frac{r(recv)^n}{n!r(disk)^n} \cdot P_0 & 0 \leq n \leq m-1, \\ \frac{r(recv)^n}{m!m^{n-m}r(disk)^n} \cdot P_0 & m \leq n \leq B, \\ 0 & n > B. \end{cases} \tag{2}$$

And, because $\beta = \frac{r(recv)}{m*r(disk)}$,

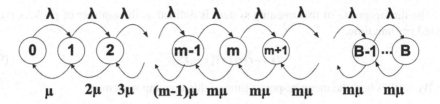

Fig. 2. M/M/m/B state transition diagram for modeling packet processing.

$$P_n = \begin{cases} \frac{(m\beta)^n}{n!} \cdot P_0 & 0 \le n \le m-1, \\ \frac{\beta^n m^m}{m!} \cdot P_0 & m \le n \le B, \\ 0 & n > B. \end{cases} \tag{3}$$

Using the boundary condition:

$$P_0 + \Sigma_{n=1}^{B} P_n = 1, \tag{4}$$

we obtain

$$P_0 = \begin{cases} \frac{1}{1 + \frac{(1-\beta^{B-m+1})(m\beta)^m}{m!(1-\beta)} + \Sigma_{n=1}^{m-1} \frac{(m\beta)^n}{n!}} & \beta \ne 1, \\ \frac{1}{1 + \frac{m^m}{m!}(B-m+1) + \Sigma_{n=1}^{m-1} \frac{(m)^n}{n!}} & \beta = 1. \end{cases} \tag{5}$$

The number of packets n in the queuing system is defined as the number of packets in queue n_q plus that of in service n_s. The mean number of packets in the queuing system is given by:

$$E[n] = \sum_{n=1}^{B} nP_n. \tag{6}$$

The mean number of packets in the application-level receive buffer is given by:

$$E[n_q] = \sum_{n=m+1}^{B} (n-m)P_n. \tag{7}$$

Because the arrivals of packets are constrained by the size of application-level receive buffer, packets can be received only when buffer is available. The effective packet arrival rate $\tilde{r}(recv)$ is less than $r(revc)$:

$$\tilde{r}(recv) = \sum_{n=0}^{B-1} r(recv)P_n$$
$$= r(recv)(1 - P_B). \tag{8}$$

And the difference $r(recv) - \tilde{r}(recv)$ is the packet loss rate.

The throughput G of this queuing system is defined as the number of packets processed per unit time.

$$G = r(recv)(1 - P_B). \tag{9}$$

By Little's law, the mean response time of this queuing system is:

$$R = \frac{E[n]}{\tilde{r}(recv)}.$$

The mean waiting time in the queue is:

$$W_q = \frac{E[n_q]}{\tilde{r}(recv)}.$$

The utilization of each disk write process is:

$$E[U] = \frac{\tilde{r}(recv)}{m * r(disk)}$$
$$= \beta(1 - P_B). \tag{10}$$

The probability of having n or more packets in the application-level buffer is given by:

$$P(\geq n \; packets \; in \; buffer) = \Sigma_{j=n}^{B} P_j$$
$$= \Sigma_{j=n}^{B} \frac{(1 - \beta) \cdot \beta^j}{1 - \beta^{B+1}}$$
$$= \frac{\beta^n - \beta^{B+1}}{1 - \beta^{B+1}}, \; \beta \neq 1 \tag{11}$$

Applying the similar comparative analysis in [12] on single and parallel disk write processes, it is easy to show that parallel disk write processes have a better performance in throughput and mean response time than single disk write process. Empirical speaking, it is usually best to use two disk write processes on a machine without RAID.

Let t be the time in seconds. The time to deplete M when the disk write process runs out of its time slice is given by:

$$t = \frac{M}{r(recv)}. \tag{12}$$

On the other hand, the time to deplete M when the CPU time is available to write the arriving packets into disk is given by:

$$t = \frac{M}{r(recv) - r(disk)}. \tag{13}$$

At time t, the application-level receive buffer is not able to accept any new packets and thus will have to drop them. The depleted buffer results in the drop of the packets coming from the kernel. It is often the case that $r(disk)$ is much less than $r(recv)$. An overly high sending rate will cause this application-level buffer to grow at a rate relative to the difference between $r(recv)$ and $r(disk)$. If this mismatch continues, packet loss will inexorably occur due to finite buffer sizes. Therefore, any protocol attempting to prevent this must communicate with the sender to make sure that the sender adjusts its sending rate ahead of overflowing this receive buffer. However, most of these existing UDP-based transport protocols adopt a static buffer scheme, which fixes the application-level buffer size initially negotiated by the users. Such a static buffer scheme is problematic due to the latency involved with this type of synchronous communication. The receiver would potentially have to drop more than 80,000 packets of size 1500 bytes on a 10 Gbps link with 100 ms round-trip time. In order to accommodate this type of synchronous communication, we adopt a dynamic buffer adaptation approach to increase the buffer size based on the $r(recv)$, RTT and the available free memory, when the buffer occupancy goes beyond a predefined threshold.

2.2 Peak Link Utilization Transport

Rate Control for Data Sender. At the sender side, fixing the sending rate at a certain value does not guarantee the optimal goodput at the receiver side considering both network and host dynamics. We apply a dynamic version of Robbins-Monro method [19] to adjust the source rate to achieve the target goodput $g^*(k)$ at the receiver:

$$\hat{r}_S(k+1) = \hat{r}_S(k) - \rho_k[\hat{g}_R(k) - \hat{g}^*(k)], \tag{14}$$

where the time step adjustment coefficient is given by $\rho_k = b/k^\gamma$ for $0.5 < \gamma < 1.0$ and $b > 0$, a suitably chosen constant. This rate control mechanism automatically adjusts the sending rate to match up with the maximum attainable goodput.

Automatic Parallelism Tuning Mechanism. We employ parallel UDP connections in one single data transfer to achieve better resource utilization and throughput. However, the aggregated throughput does not always increase along with the number m of concurrent UDP connections caused by the increased overhead on end systems. In [12], we employ an automatic parallelism tuning mechanism to dynamically change the number of parallel UDP connections. Let $G(m)$ be the goodput measured at a certain interval, and m_{-1} be the number of parallel UDP connections used for the previous data partition transfer, R be the sending rate for the newly established UDP connection, R_{br} be the bottleneck rate of the data receiver, and C be the link bandwidth. We take the following steps to adjust the number of parallel UDP connections:

(a) Initialize parameters:

$$m \leftarrow m_0,$$
$$R \leftarrow R_{br},$$
$$G(m_{-1}) \leftarrow 0,$$

where m_0 is set to 1 as the initial number of parallel UDP connections.

(b) Transfer partitions of data through each UDP channel and measure the aggregated goodput $G(m)$.

(c) If the following inequality is satisfied, then terminate the algorithm:

$$G(m) < G(m_{-1});$$

otherwise, set

$$R \leftarrow \min(C - G(m), \ R_{br}).$$

and proceed to Step (d).

(d) Increase the number of parallel UDP connections by 1, set the sending rate of the newly added UDP channel to be R, and return to Step (b).

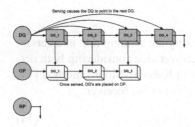

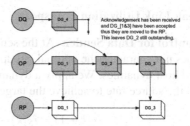

Fig. 3. Initial buffer states.   **Fig. 4.** Buffer states after receiving acknowledgements.

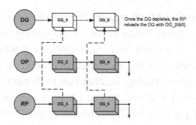

Fig. 5. Buffer states after reloading.

ACK Interval Control for Goodput Maximization. At the data receiver, we maximize the application goodput, i.e. $GP_T(T^*) = \max_T GP_T(T)$, by adjusting the ACK interval $T(k) = T^*$ according to Kiefer-Wolfowitz Stochastic Approximation (KWSA) and Simultaneous Perturbation Stochastic Approximation (SPSA) methods.

In high-speed transport control, it is difficult to satisfy the requirements of SPSA on collecting the number of measurements before adjusting the control parameters. We maximize the goodput at the receiver by employing a special form of SPSA to $T(k)$ as follows:

$$\hat{T}(k+1) = \hat{T}(k) - s_k \hat{G}\mathcal{P}(T(k)), \tag{15}$$

where $\hat{T}(k)$ represents the acknowledge interval at time step k, $\hat{G}\mathcal{P}(T(k))$ represents the SP approximation to the gradient of goodput acknowledge regression $GP_T(.)$ and s_k is a constant coefficient. And s_k satisfies the following property: $s_k \to 0$ as $k \to \infty$, $\sum_{k=1}^{\infty} s_k = \infty$ and $\sum_{k=1}^{\infty} s_k^2 < \infty$. Based on the analysis, it is asymptotically and probabilistically guaranteed for T^* to converge to optimal, which is independent of the underlying probability distributions.

Dynamic Buffer Adaptation. At the data receiver, we use the level of buffer occupancy as key performance indicator to assess its processing capacity. Initially at the receiver side, PLUT allocates a buffer of datagrams statically. The merits of this scheme are: 1) simple implementation; 2) index with random access; 3) no allocation/deallocation of heap-dynamic memory. When the disk wirte rate does not match up with the packet arriving rate, the static buffer could be "overflowing". When the circular buffer has no free slots for saving datagrams from the network, the receiver will stop, hence drastically reduces the aggregated application throughput. We maximize the PLUT throughput at the receiver by adopting a dynamic buffer management method.

To solve the above Datagram Buffer Management (DBM) problem, we proposed a Three Tier Dynamic Queuing Buffer (3TDQB) and a dynamic buffer management method. These three tiers of 3TDQB are one Linked Queue, and two Linked Lists, which implement Datagram Queuing(DQ) Buffer, Outstanding Pointer (OP) Buffer, and Reload Pointer (RP) Buffer respectively. At the start of the protocol, PLUT dynamically allocate DQ buffer, while PLUT only save the pointers to the space in DQ buffer in OP buffer and RP buffer. As shown in Fig. 3, the size of DQ buffer will not change once it was created. Each datagram will travel through three tiers consecutively before being flushed, reloaded and returned to the DQ. Datagrams waiting to be written into the disk reside in the DQ. As shown in Fig. 3, once the datagram is served, it will be placed in the OP buffer by being linked through a pointer. The datagram then waits for acknowledgment from the disk writing threads. As shown in Fig. 4, the datagram will be placed in the RP buffer for flushing of the written data and reloading of the new data after being acknowledged. As shown in Fig. 5, after the DQ depletes of all datagrams, the RP then reloads the DQ with an address change thus creating a new queue without new memory allocation. The worst case scenario of the 3TDQB is that all remaining datagrams in the DQ has been served, for some reason the OP has not yet received an acknowledgment, and the RP is empty. This is resembling to a buffer overflow, which is similar to the static buffer method. To deal with the overflow, the 3TDQB will continue to allocate datagrams from the heap while waiting for acknowledgemen from the disk writing threads. Thus the receiver never reducs the datagram receiving rate.

3   Implementation and Experimental Results

3.1   Protocol Implementation

We implement four types of acknowledgment packets: Next (NXT), Retransmission (RXM), Timeout Next (TNT), and Timeout Retransmission (TMO). For each normal ACK control period, if PLUT receives all datagrams so far in continuity, it generates an "NXT" ACK and sends it to the sender; otherwise if some datagrams are lost (i.e., "holes" in the datagram checklist), the receiver generates a "RXM" ACK, and sends it to the sender with a list of lost datagram sequence numbers. If the receiver receives no datagram within a certain period of time, it triggers a timeout event. In this case, if the receiver receives all datagrams so far in continuity, it sends a "TNT" ACK. Otherwise, if there are missing parts in the datagram checklist, the receiver sends out a "TMO" ACK with the lost datagram sequence number list. The receiver calculates the current instant goodput for all types of ACK, and sends it to the sender as part of the acknowledgment. On the sender side, PLUT applies rate control using the goodput measurements enclosed in the acknowledgment for each incoming acknowledgment. At the send, we implement a proportional datagram allocation mechanism for parallel transfer. Once the number of parallel UDP connections is maximized, the rest of datagrams is divided by the sender into several partitions, one for each connection. The size of each partition is proportional to the goodput measurement of the related UDP channel.

PLUT Monitor provides a layer between the operating system and the transfer protocol. PLUT needs the state information from operating system for flow control. PLUT uses LibGTop library [2] to obtain the required state information. LibGTop library reads the virtual file system /proc which contains the current state of the kernel, and stores the information of currently running processes. PLUT first reads the state (running or sleeping) and PID for each process from /proc. Then PLUT checks the state of these processes to estimate changes. If a process is always ready to run when its state is checked, it is considered as CPU-bound. If the state of this process is different from "running" [3], it is considered as IO-bound.

3.2   Performance Evaluation

We compare the performance of PLUT, and UDT (version 4.4) over a local 10Gbps connection. This connection was setup between two Linux boxes with kernel 4.11.10, each of which is equipped with one 1 GigE NIC and one 10 GigE NIC, dual quad-core 2.4 GHz Xeon(R) CPU, 24 GBytes of RAM. We use netem to emulate the link delay, packet loss, duplication and re-ordering of high-speed networks. The packet loss rate in optical fiber cables is very small according physical measurements. We always set the packet loss rate to be 10^{-5} in our experiments.

Case A: 10 Gbps, Memory-to-Memory Transfer with Different Link Delay. We conduct data transfer experiments by varying link RTT from 0.04 to 250 ms to explore the impact of link RTT on the performance of PLUT. Since the network delay is not constant, each selected RTT in the experiment varies based on a normal distribution. The average throughput performance measurements for PLUT and UDT are plotted in Fig. 6, where both protocol's throughput decrease along with the increase of RTT.

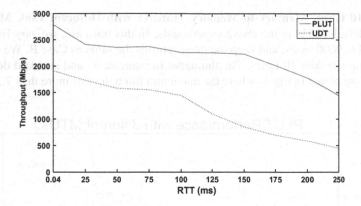

Fig. 6. Performance comparison over a 10 Gpbs link with different RTT.

Case B: 10 Gbps, Memory-to-Memory Transfer with Concurrent Background Workloads. We design a CPU-bound program named Burncpu to emulate concurrent host background workloads. In this experiment, we set the RTT to be 125 ms. And 10 s after the data transfer begins, we execute the first concurrent Burncpu process at the data receiver, which will run 25 s. And 15 s after the data transfer begins, we execute the second concurrent Burncpu process, which will run 30 s. Figure 7 plots the corresponding PLUT and UDT throughput measurements. These measurements clearly shows that the performance of each transport method is significantly impacted by the amount of concurrent background workloads. UDT also adapts to the workload changes by adopting a more conservative rate control scheme than PLUT. UDT has a relative unstable throughput measurement, because the packet loss rate increases when several CPU-bound processes compete for CPU resources.

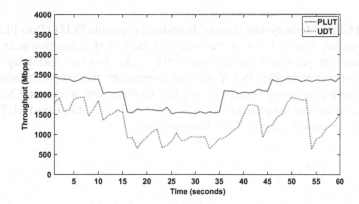

Fig. 7. Memory to memory performance comparison over 10 Gpbs link with background workloads.

Case C: 10 Gbps, Memory-to-Memory Transfer with Different Link MTU. We used the default MTU in the above experiments. In this test case, we vary link MTU from 3000 to 9000 bytes, and keep the other settings the same as Case B. We use each MTU to transfer data 10 times. The througput measurements and standard deviations for PLUT are ploted in Fig. 8, where the maximum throughput is more than 7.3 Gbps.

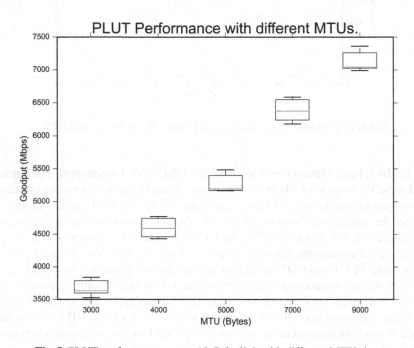

Fig. 8. PLUT performance over a 10 Gpbs link with different MTU sizes.

Case D: 10 Gbps, Memory-to-Memory Transfer. For paralle PLUT (Para-PLUT) performance, we run Para-PLUT over the same local dedicated connection as in Case B. In our experiment, we use netem to emulate the packet loss rate and delay. Figure 9 plots the corresponding Para-PLUT meomry-to-memory throughput measurements. The maximum throughput of Para-PLUT reaches 3.1 Gbps, which is around 20% higher than the maximum throughput reached by PLUT. In all cases, the Para-PLUT method consistently outperforms than PLUT and other protocols.

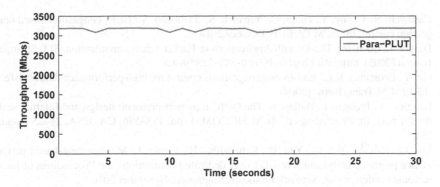

Fig. 9. Memory to memory performance comparison over 10 Gpbs link.

4 Conclusion

We conducted an extensive analytical study of the design and implementation issues of high-speed data transfer methods, and develop an optimized large file transfer solution to overcome the performance limitations of current protocols. PLUT adopts an innovative flow control mechanism to automatically adjust the activities of both the data sender and receiver and stabilizes the sending rate to maximize throughput using stochastic approximation methods. We extented our previous analysis to cover the impact of application-level receive buffer and the background workloads on the data transfer performance. We further refined the dynamic buffer management scheme and parallelism mechanism based on this analysis to achieve the maximal bottleneck goodput. The experimental results collected over a local 10Gbps connection shows the performance superiority of PLUT. There is still a need for considerable further research in this area. For example, the proposed protocol needs to be further evaluated in real-life 5G mobile networks. In the future, we will do more in-depth research on these aspects.

Ackowledgement. This work is sponsored by National Natural Science Foundation of China under Grant No. 61402328.

References

1. https://www.huawei.com/en/press-events/news/2019/2/turk-telekom-huawei-5g-cloud-vertical-sector-application
2. Libgtop. ftp://ftp.gnome.org/pub/GNOME/sources/libgtop/2.0
3. Beltran, M., Guzman, A.: A new CPU availability prediction model for time-shared systems. IEEE Trans. **57**(7), 865–875 (2009)
4. Bhat, U.: An Introduction to Queueing Theory. SIT. Birkhäuser Publications, Boston (2008). https://doi.org/10.1007/978-0-8176-4725-4_9
5. Bocharov, P., D'Apice, C., Pechinkin, A., Salerno, S.: Queueing Theory. SIT. Springer, New York (2009). https://doi.org/10.1007/978-0-387-49706-8_9
6. Bolch, G., Greiner, S., Meer, H., Trivedi, K.: Queueing Networks and Markov Chains. Wiley (2006)

7. Cardwell, N., Cheng, Y., Gunn, C., Yeganeh, S., Jacobson, V.: BBR: congestion-based congestion control. In: ACM QUEUE, October 2016
8. Daigle, J.: Queueing Theory with Applications to Packet Telecommunication. SIT. Springer, Boston (2005). https://doi.org/10.1007/0-387-22859-4_8
9. Gu, Y., Grossman, R.L.: End-to-end congestions control for high performance data transfers. IEEE/ACM Trans. Netw. (2004)
10. Langley, A., Iyengar, J., Bailey, J.: The QUIC transport protocol: design and Internet-scale deployment. In: Proceedings of ACM SIGCOMM, pp. 183–196, CA, USA, 21–25 August 2017
11. Liu, Q., Rao, N., Wu, Q., Yun, D., Kcttimuthu, R., Foster, I.: Measurement-based performance profiles and dynamics of UDT over dedicated connections. In: Proceedings of International Conference on Network Protocols, Singapore, November 2016
12. Lu, X., Wu, Q., Rao, N.S.V., Wang, Z.: On parallel UDP-based transport control over dedicated connections. In: Proceedings of the 2010 IEEE Global Communications Conference, Miami, FL, 6–10 December 2010
13. Lu, X., Wu, Q., Rao, N., Wang, Z.: On performance-adaptive flow control for large data transfer in high speed networks. In: Proceedings of the 28th IEEE International Performance Computing and Communications Conference, AZ, 14–16 December 2009
14. Lyu, X., Wu, C.Q., Rao, N.: An integrated high-performance transport solution for big data transfer over wide-area networks, pp. 1661–1668, June 2018
15. Nine, M., Guner, K., Huang, Z., Wang, X., Xu, J., Kosar, T.: Big data transfer optimization based on offline knowledge discovery and adaptive sampling. In: IEEE International Conference on Big Data, MA, USA, 11–14 December 2017
16. Patrasand, P., Banchs, A., Serrano, P.: A control theoretic approach for throughput optimization in IEEE 802.11e EDCA WLANs. Mob. Netw. Appl. **14**(6), 697–708 (2009)
17. Qiu, T., Liu, X., Li, K., Hu, Q.: Community-aware data propagation with small world feature for Internet of vehicles. IEEE Commun. Mag. **56**(1), 86–91 (2018)
18. Rao, N., et al.: Experiments and analyses of data transfers over wide-area dedicated connections. In: Proceedings of the 26th International Conference on Computer Communications and Networks, Canada, August 2017
19. Spall, J.: Introduction to Stochastic Search and Optimization: Estimation, Simulation, and Control. Wiley Pub. (2003)
20. Wu, Q., Rao, N., Lu, X.: On transport methods for peak utilization of dedicated connections. In: Proceedings of the 6th International Conference on Broadband Communications, Networks, and Systems, Marid, Spain, 14–17 September 2009
21. Yun, D., Wu, C., Rao, N., Settlemyer, B., Lothian, J., Vishwanath, V.: Profiling transport performance for big data transfer over dedicated channels. In: Proceedings of the 2015 International Conference on Computing, Networking and Communications, Anaheim, USA, February 2015

Broad Learning System
with Proportional-Integral-Differential
Gradient Descent

Weidong Zou[1], Yuanqing Xia[1], Weipeng Cao[2(✉)], and Zhong Ming[2]

[1] School of Automation, Beijing Institute of Technology, Beijing, China
[2] College of Computer Science and Software Engineering, Shenzhen University,
Shenzhen, China
caoweipeng@szu.edu.cn

Abstract. Broad learning system (BLS) has attracted much attention
in recent years due to its fast training speed and good generalization
ability. Most of the existing BLS-based algorithms use the least square
method to calculate its output weights. As the size of the training data set
increases, this approach will cause the training efficiency of the model to
be seriously reduced, and the solution of the model will also be unstable.
To solve this problem, we have designed a new gradient descent method
(GD) based on the proportional-integral-differential technique (PID) to
replace the least square operation in the existing BLS algorithms, which
is called PID-GD-BLS. Extensive experimental results on four benchmark data sets show that PID-GD can achieve faster convergence rate
than traditional optimization algorithms such as Adam and AdaMod,
and the generalization performance and stability of the PID-GD-BLS
are much better than that of BLS and its variants. This study provides a
new direction for BLS optimization and a better solution for BLS-based
data mining.

Keywords: Broad learning system · Neural networks with random
weights · Proportional-integral-differential · Randomized algorithms ·
Optimization algorithms

1 Introduction

Broad learning system (BLS), derived from the random vector functional-link
neural network (RVFL) [3,19,25], was proposed by Chen CLP et al. in 2017
[6] to alleviate the low training efficiency of traditional deep learning models. In
essence, BLS is a typical neural network with random weights (NNRW) [4], which

This work was supported by National Key Research and Development Program
of China (2018YFB1700400), Opening Project of Shanghai Trusted Industrial Control Platform (TICPSH202003008-ZC), National Natural Science Foundation of
China (61836005, 61672358), and Guangdong Science and Technology Department
(2018B010107004).

© Springer Nature Switzerland AG 2020
M. Qiu (Ed.): ICA3PP 2020, LNCS 12452, pp. 219–231, 2020.
https://doi.org/10.1007/978-3-030-60245-1_15

uses a non-iterative training mechanism to make the model have both extremely fast training speed and good generalization ability [6]. Compared with other NNRW algorithms such as RVFL, BLS adopts a feature extraction mechanism with twice random mapping, which makes it better to transform data features internally. In recent years, BLS has attracted much attention and various BLS-based variants and applications have been proposed [7,10,12,23].

In BLS, there are two hidden layers in its network structure. The neurons in the first hidden layer and the second hidden layer are called feature nodes and enhancement nodes, respectively. All feature nodes and enhancement nodes are fed into the output layer in a fully connected way. During model training, input data is first randomly mapped to a high-dimensional space by the feature nodes and then abstracted again by the enhanced nodes. The input weights of the feature nodes and the enhancement nodes are randomly generated according to certain rules and remain unchanged throughout the training process of the model, while the output weights of the model are obtained analytically.

Specifically, given a BLS with p feature nodes and d enhancement nodes, the output weights (denoted as β) of the model are generally calculated by $\beta = \mathbf{G}^{\dagger}\mathbf{Y} = [\mathbf{Z}^p \mid \mathbf{H}^d]^{\dagger}\mathbf{Y}$, where $\mathbf{G}^{\dagger} = [\mathbf{Z}^p \mid \mathbf{H}^d]^{\dagger}$ is the Moore-Penrose generalized inverse of the hidden layers output matrix, $\mathbf{Y}$ is the true labels of training samples, $\mathbf{Z}^p$ and $\mathbf{H}^d$ are the outputs of the p feature nodes and the d enhancement nodes, respectively [6]. Chen CLP et al. have proved theoretically that BLS can achieve universal approximation capability [7]. In other words, BLS can approach any continuous target function with any precision.

From the above training mechanism of BLS, it can be known that BLS avoids the practice of adjusting all parameters through iterative iterations (e.g., the gradient descent [18]) adopted by traditional deep learning algorithms, thereby making the model training efficient. Moreover, compared with other least square methods, using the Moore-Penrose generalized inverse to solve the system of linear matrix equations can ensure that BLS can always obtain the optimal solution [21]. Many literatures have shown that BLS can achieve better generalization ability and faster training speed than traditional randomized networks such as the back-propagation algorithm (BP [20]) in application scenarios with relatively small data sets.

However, when the training data set is large (i.e., big data scenarios), using BLS will face two troublesome problems: (1) it takes too much memory and training time to calculate the Moore-Penrose generalized inverse of the hidden layer output matrix; (2) the generalization performance of the model is extremely sensitive to the number of hidden layer neurons. These two problems will cause the training efficiency of BLS to be greatly reduced and the stability of the model cannot be guaranteed.

To solve the above problems, we propose an enhanced BLS algorithm that uses an improved gradient descent method to solve the output weights of the model. Specifically, we combine the proportional-integral-differential technique (PID) [16] and traditional gradient descent (GD) and propose a new optimization algorithm called PID-GD, and then apply it to solve the output weights of BLS. We call the new algorithm PID-GD-BLS. The new training mechanism can effectively alleviate the above-mentioned problems and make PID-GD-BLS more suitable for big data scenarios than the existing BLS algorithms.

The contribution of this study can be summarized as follows.

(1) A novel optimization algorithm (i.e., PID-GD) was proposed. It can be used not only for BLS, but also for other neural networks such as convolutional neural networks (CNN) [13] and recurrent neural network (RNN) [14]. We have experimentally proved that PID-GD has a faster convergence rate than other popular optimization algorithms such as Adam [16]and AdaMod [9].

(2) An improved BLS algorithm called PID-GD-BLS was proposed. PID-GD-BLS uses the above optimization algorithm to solve the output weights of the model, which can not only make the model converge faster, especially in big data scenarios, but also make the performance of the model more stable. We have experimentally proved that the prediction ability and stability of PID-GD-BLS exceed BLS and its variants.

(3) We have shown the effectiveness of PID-GD and PID-GD-BLS with extensive experimental results. These two results can provide researchers with new directions to design advanced optimization algorithms and BLS algorithms.

The rest of this paper is organized as follows. In Sect. 2, we briefly introduce the training mechanisms of BLS and PID. In Sect. 3, we present the details of the proposed optimization algorithm (i.e., PID-GD) and the improved BLS algorithm (i.e., PID-GD-BLS). We show the experimental results and the corresponding analysis in Sect. 4. In Sect. 5, we conclude this paper.

2 Preliminaries

2.1 Broad Learning System (BLS)

The network structure of BLS is shown in Fig. 1, where $\mathbf{X}$ and $\mathbf{Y}$ refer to the input and output of the model, $\mathbf{a}^e$ denotes the input weights between the input layer and the feature layer, $\mathbf{a}^m$ denotes the input weights between the feature layer and the enhancement layer, β^e refers to the input weights between the feature layer and the output layer, and β^m refers to the input weights between the enhancement layer and the output layer. z_i and h_j refer to the output of the i-th feature node and the j-th enhancement node, respectively.

In traditional neural networks such as convolutional neural networks (CNN) [5,13] and and recurrent neural networks (RNN) [14], the parameters in the network are solved iteratively by gradient descent and its variant algorithms; while in BLS, $\mathbf{a}^e$ and $\mathbf{a}^m$ are randomly assigned and remain unchanged throughout model training, while β^e and β^m are obtained analytically.

Specifically, given a training data set $\{\mathbf{X}, \mathbf{Y}\} \in \Re^{(l+t) \times N}$, where l is the feature dimension of input data, t is the class number of classification problems, and N is the number of training samples. The mathematic model of a BLS [6]

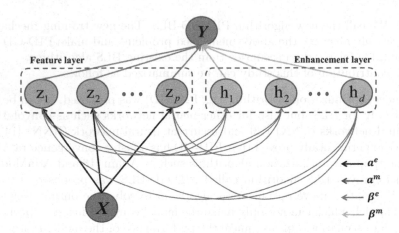

Fig. 1. The network structure of BLS.

with p feature nodes and d enhancement nodes can be expressed as follows:

$$\mathbf{F}(\mathbf{X}) = \sum_{i=1}^{p} \beta_i^e \mathbf{z}_i(\mathbf{a}_i^e \cdot \mathbf{X} + b_i^e) + \sum_{j=1}^{d} \beta_j^m \mathbf{h}_j(\mathbf{a}_j^m \cdot \mathbf{Z} + b_j^m), \qquad (1)$$

where b_i^e and b_j^m are the thresholds of the i-th feature node and the j-th enhancement node, respectively; $\mathbf{z}(\cdot)$ and $\mathbf{h}(\cdot)$ are the activation functions of the feature nodes and the enhancement nodes, respectively.

The output weights of the BLS model can be obtained by using the following equations:

$$\mathbf{Z} = [\mathbf{z}_1, ..., \mathbf{z}_p], \mathbf{H} = [\mathbf{h}_1, ..., \mathbf{h}_d], \qquad (2)$$

$$\boldsymbol{\beta} = [\beta_1^e, ..., \beta_p^e, \beta_1^m, ..., \beta_d^m] = \mathbf{G}^\dagger \mathbf{Y} = [\mathbf{Z}^p \mid \mathbf{H}^d]^\dagger \mathbf{Y}, \qquad (3)$$

where $\mathbf{G}^\dagger = [\mathbf{Z}^p \mid \mathbf{H}^d]^\dagger$ is the Moore-Penrose generalized inverse of the hidden layer output matrix.

2.2 Proportional-Integral-Differential (PID)

PID is a feedback control technique commonly used in the automatic control system [1,15], which can make the system run smoothly by dynamically adjusting the contribution of different components of the system and its implementation is simple and widely applicable. PID controller requires that the control action is proportional to the current error, the integral of past errors with time, and the error derivative representing future trends. The schematic diagram of PID controller is shown in Fig. 2.

Given the state of controlled object $\vartheta(t)$ and reference state ϑ_r at time t, then the system error can be expressed as $e(t) = \vartheta_r - \vartheta(t)$. Suppose that the control

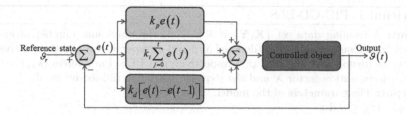

Fig. 2. The schematic diagram of PID controller.

input of the controlled object is $u(t)$, the mathematic model of PID controller can be expressed as follows [15]:

$$u(t) = k_p e(t) + k_i \sum_{j=0}^{t} e(j) + k_d [e(t) - e(t-1)], \tag{4}$$

where k_p, k_i, and k_d are gain coefficients on the P (Proportional), I (Integral), and D (Differential) terms, respectively.

It can be inferred from (4) that k_p, k_i, and k_d control the contributions of present, past, and future errors to the current correction, PID controller can simultaneously adjust system errors in different periods (both long-term and short-term), providing a direction for us to optimize the system from a global perspective. In the next section, we introduce how to combine PID with traditional gradient descent algorithm to improve the convergence rate of the algorithm.

3   BLS with PID-based Gradient Descent (PID-GD-BLS)

Given a training data set $\{\mathbf{X}, \mathbf{Y}\} \in \Re^{(l+t) \times N}$ and a BLS [6] with p feature nodes and d enhancement nodes, the cost function of BLS can be expressed as follows:

$$\ell(\mathbf{X}, \mathbf{Y}) = \frac{1}{2} \|\mathbf{G} \cdot \boldsymbol{\beta} - \mathbf{Y}\|^2 + \frac{\lambda}{2} \|\boldsymbol{\beta}\|^2, \tag{5}$$

where $\| \cdot \|$ represents 2-Norm, λ is the regularization factor and satisfies $\lambda > 0$, $\mathbf{G}$ is a $N \times (p+d)$ matrix with respect to the output of the hidden layers, $\boldsymbol{\beta}$ is a $(p+d) \times 1$ vector with respect to the output weights of BLS, and $\mathbf{Y}$ is a $N \times t$ matrix with respect to the labels of training samples.

The gradient of the cost function of BLS at time-step t can be obtained using the following equation:

$$\mathbf{g}_t = \nabla_\beta \mathcal{L}_t(\boldsymbol{\beta}_{t-1}) = \mathbf{G}^T \cdot \mathbf{G} \cdot \boldsymbol{\beta}_{t-1} - \mathbf{G}^T \cdot \mathbf{Y} + \lambda \boldsymbol{\beta}_{t-1}. \tag{6}$$

Combining this gradient with the PID mechanism mentioned in Sect. 2.2, one can get a new gradient representation, which we call PID-based gradient descent method (PID-GD). The PID-GD can be expressed as follows:

$$\mathbf{u}_t = k_p \mathbf{g}_t + k_i \sum_{j=1}^{t} \mathbf{g}_j + k_d (\mathbf{g}_t - \mathbf{g}_{t-1}). \tag{7}$$

Algorithm 1. PID-GD-BLS

Input: A training data set $\{\mathbf{X}, \mathbf{Y}\} \in \Re^{(l+t) \times N}$, the maximum number of feature nodes and enhancement nodes (denoted as p and d respectively) and their activation functions (denoted as $\phi(\cdot)$ and $\varphi(\cdot)$ respectively) in BLS. Parameters (k_p, k_i, k_d) in PID, regularization factor λ, and the step size η are determined by hand.
Output: The parameters of the model.
for $q = 1$; $q \leq p$ **do**
    Randomly assign the input weights and hidden biases of feature nodes $(\mathbf{a}_q^e, b_q^e)$;
    Calculate the output of the q-th feature node:
    $\mathbf{z}_q = \phi(\mathbf{a}_q^e \cdot \mathbf{X} + b_q^e)$;
end for
Calculate the output matrix of the feature nodes:
$\mathbf{Z} = [\mathbf{z}_1, ..., \mathbf{z}_p]$;
for $k = 1$; $k \leq d$ **do**
    Randomly assign the parameters connecting feature nodes to enhancement nodes $(\mathbf{a}_k^m, b_k^m)$;
    Calculate the output of the k-th enhancement node:
    $\mathbf{h}_k = \varphi(\mathbf{a}_k^m \cdot \mathbf{Z} + b_k^m)$;
end for
Calculate the output matrix of the enhancement nodes:
$\mathbf{H} = [\mathbf{h}_1, ..., \mathbf{h}_d]$;
Use the PID-GD to calculate the output weights of BLS: Given the maximum number of iterations M and set $\beta_0 = \mathbf{0}$.
for $t = 1$; $t \leq M$ **do**
    $\mathbf{g}_t = \nabla_\beta \mathbf{f}_t(\beta_{t-1}) = \mathbf{G}^T \cdot \mathbf{G} \cdot \beta_{t-1} - \mathbf{G}^T \cdot \mathbf{Y} + \lambda \beta_{t-1}$;
    $\mathbf{u}_t = k_p \mathbf{g}_t + k_i \sum_{j=1}^{t} \mathbf{g}_j + k_d(\mathbf{g}_t - \mathbf{g}_{t-1})$;
    $\beta_t = \beta_{t-1} - \eta \mathbf{u}_t$;
end for
Set the output weights of BLS to β_t
return All the parameters of the model.

Then, one can update the output weights of BLS using the following equation:

$$\beta_t = \beta_{t-1} - \eta \mathbf{u}_t, \tag{8}$$

where η is the step size.

According to the above introduction, here we summarize the PID-GD-BLS as Algorithm 1.

From Algorithm 1, one can observe that PID-GD-BLS uses a completely different approach from traditional BLS to solve the output weights of the model, that is, a PID-based gradient descent method. This method can avoid the difficulty of solving the Moore-Penrose generalized inverse in the case of big data and provide a stable solution for the model.

Now, we analyze the computational complexity analysis of PID-GD-BLS according to the method of [17]. For a PID-GD-BLS with p feature nodes and d enhancement nodes, according to Algorithm 1, $\mathbf{G}^T \cdot \mathbf{G}$ requires $(p + d)^2 N$

multiplications, $\mathbf{G}^T \cdot \mathbf{Y}$ requires $(p+d)N$ multiplications, formulae (6), (7), and (8) require $3(p+d)$ additions. Then, if the number of iterations is M, the total number of multiplications and additions for training the PID-GD-BLS model is $(p+d)[(p+d)N + N + 3M]$.

4   Simulation Experiments and Discussions

In this section, we evaluate the convergence ability of the proposed PID-based gradient descent method (i.e., PID-GD) and the effectiveness of using PID-GD to solve the output weights of BLS (i.e., PID-GD-BLS).

All the experiments were conducted in PC with Intel(R) Core(TM) i7-3520M CPU with NVIDIA NVS 5400M and 8 GB RAM. The randomized parameters of BLS were assigned from $[-1,1]^l \times [-1,1]$ under a uniform distribution, the hyperbolic tangent function was chosen as the activation function of feature nodes and enhancement nodes, and the regularization factor λ was set to 1.

4.1   Convergence Rate Comparison of PID-GD, Adam, and AdaMod

Adam [16] and AdaMod [9] are the most commonly used optimization algorithms in deep learning algorithms. Their convergence ability has been verified in many algorithms [2,11,22]. Here we use Adam, AdaMod, and the proposed PID-GD to solve the output weights of BLS respectively, and analyze the convergence rate of these three optimization algorithms.

Our experiments were conducted on four benchmark data sets from UCI machine learning repository[1], which are Airfoil Self-Noise, Concrete Compressive Strength, Geographical Original of Music, and Yacht Hydrodynamic. The number of feature nodes and enhancement nodes in BLS was set to 150 and 100, respectively. The exponential decay rates of first moment estimate and second raw moment estimate for Adam and AdaMod were set to $\beta_1 = 0.9$ and $\beta_2 = 0.999$, and the discount factor of AdaMod was set to $\beta_3 = 0.91$. The perturbation value of Adam and AdaMod was set to $\varepsilon = 0.001$. The step size for Adam, AdaMod, and PID-GD was set to $\eta = 0.0001$. The gain coefficients of PID-GD were set to $k_p = 0.052$, $k_i = 0.00001$, and $k_d = 0.002$. Each experiment was performed 100 times and the average 2-Norm of the first-order gradient of the cost function for the BLS model is shown in Fig. 3.

From Fig. 3, one can observe that in most tasks (3/4), our proposed PID-GD can achieve a faster convergence rate than Adam and AdaMod, which implies that the method of controlling gradient descent using PID controller is feasible and effective.

The experimental result on the data set Yacht Hydrodynamic shows that the convergence rate of PID-GD is shower than that of other optimization algorithms. For this phenomenon, a speculative explanation is that the number of training

---

[1] UCI machine learning repository: http://archive.ics.uci.edu/ml/index.php.

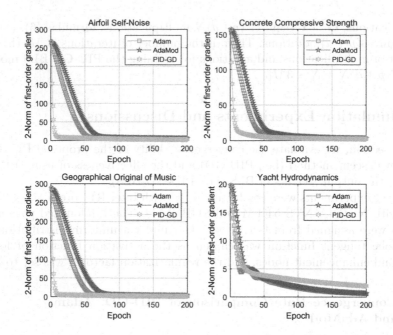

Fig. 3. Convergence comparison of PID-GD, Adam, and AdaMod on four benchmark data sets.

samples in this data set is too small (only 308 instances), which makes it difficult for PID-GD to establish the error feedback system over time, so it is difficult for PID-GD to control the gradient efficiently. This provides an idea for us to improve PID-GD in the future, that is, to optimize PID-GD according to the scale of training sets.

4.2 Performance Comparison of PID-GD-BLS, BLS, and FBLS

In the section, we compare the performance of PID-GD-BLS, BLS [6], and fuzzy BLS (FBLS) [10] on the above mentioned four benchmark data sets. Note that FBLS is one of the most advanced BLS variants.

For FBLS, the number of fuzzy subsystems and rules in each fuzzy subsystem were set to 50 and 2, respectively. For BLS, FBLS, and PID-GD-BLS, the number of feature nodes was set to 50, while the number of enhancement nodes was determined from $[50, 60, 70, 80, 90, 100, 110, 120, 130, 140, 150]$ one by one. The regularization factor of BLS and FBLS was set to $\lambda = 420$. The source code of all the algorithms used in this experiment has been published to GitHub.[2]

In our experiments, the Root Mean Square Error (RMSE), Mean Absolute Percentage Error (MAPE) [8], and the Ratio of Standard Deviation (RSD) [24]

---

[2] The source code of all the algorithms used in our experiment: https://github.com/Wepond/PID-GD-BLS.

were chosen as the performance indicators. They can be obtained by using the following equations:

$$RMSE = \sqrt{\frac{\sum_{i=1}^{m}(u_i - \hat{u}_i)^2}{m}}, \tag{9}$$

$$MAPE = \frac{1}{m}\sum_{i=1}^{m}|\frac{\hat{u}_i - u_i}{\hat{u}_i}| \times 100\%, \tag{10}$$

$$RSD = \sqrt{\frac{\sum_{i=1}^{m}(u_i - \tilde{u})^2}{\sum_{i=1}^{m}(\hat{u}_i - \bar{u})^2}}, \tag{11}$$

where
  u_i refers to the predicted value,
  $\hat{u}_i$ refers to the true value,
  $\tilde{u}$ refers to the average of the predicted values,
  $\bar{u}$ refers to the average of the true values,
  and m is the number of samples.

Note that RMSE and MAPE can effectively reflect the prediction ability of the model. The smaller the values, the better the generalization ability of the model. RSD can reflect the stability of the model. The larger the value, the better the stability of the model. In our experiments, the given experimental results are the average results of ten-fold cross-validation, which are shown in Fig. 4, 5, 6 and Fig. 7.

Taking Fig. 4 as an example, one can observe that the RMSE and MAPE of PID-GD-BLS are much smaller than those of BLS and FBLS, which means that the PID-GD-BLS model has better generalization ability than other models on this data set. In terms of the RSD index, the corresponding value of the PID-GD-BLS model is higher than that of BLS and FBLS, which implies that the stability of the PID-GD-BLS model exceeds that of the BLS and FBLS models. Similar experimental phenomena can also be observed from Fig. 5, 6 and Fig. 7.

The above experimental phenomena show that using the PID-GD to calculate the output weights of BLS is feasible and can effectively improve the generalization ability and stability of the model.

Remark: A speculative explanation for the effectiveness of PID-GD-BLS. The original BLS uses the least square method to solve its output weights, which can get a good analytical solution when the number of training samples and the number of hidden nodes in the network structure are relatively small. However, if these two conditions cannot be satisfied in practice, this approach may cause the instability and large errors in the solution of the linear system, which then will seriously reduce the generalization ability of the model. The proposed PID-GD avoids these problems. Regardless of the size of the data set and the number of hidden nodes, it can get a relatively stable solution by iterative calculation. Therefore, PID-GD-BLS can always achieve higher prediction accuracy and better stability than BLS in our experiments.

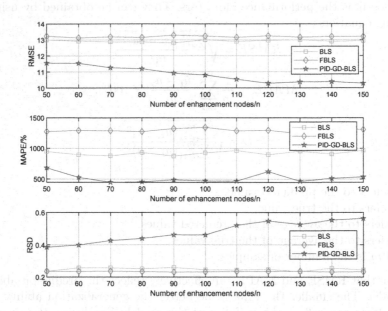

Fig. 4. Performance of PID-GD-BLS, BLS, and FBLS on Yacht hydrodynamic.

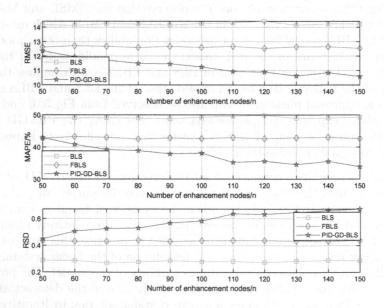

Fig. 5. Performance of PID-GD-BLS, BLS, and FBLS on concrete compressive strength.

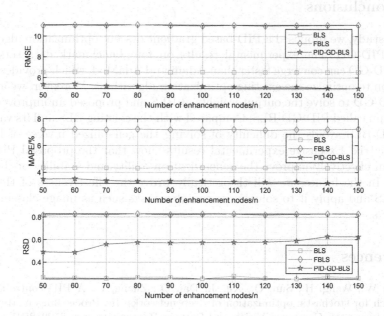

Fig. 6. Performance of PID-GD-BLS and FBLS on airfoil self-noise.

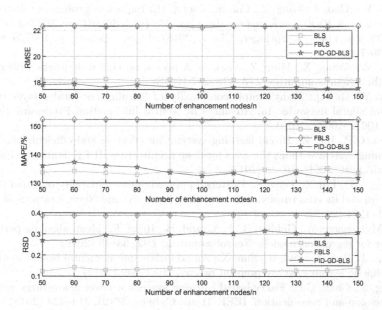

Fig. 7. Performance of PID-GD-BLS, BLS, and FBLS on geographical original of music.

5   Conclusions

In this study, we proposed a PID-based gradient descent optimization algorithm called PID-GD. The experimental results on four benchmark data sets show that PID-GD can converge faster than Adam and AdaMod, which provides a new direction to optimize the existing optimization algorithms. Moreover, we applied the PID-GD to solve the output weights of BLS and proposed an improved BLS algorithm called PID-GD-BLS. Compared with the existing BLS and its variants, PID-GD-BLS avoids the difficulty of solving the generalized inverse of hidden layer matrix. Extensive experimental results show that the proposed PID-GD-BLS can effectively improve the generalization ability and stability of the BLS model. In the future, we will theoretically prove the effectiveness of the PID-GD-BLS and apply it to solve more complex tasks such as image classification problems.

References

1. An, W., Wang, H., Sun, Q., Xu, J., Dai, Q., Zhang, L.: A PID controller approach for stochastic optimization of deep networks. In: Proceedings of the IEEE Conference on Computer Vision and Pattern Recognition, pp. 8522–8531 (2018)
2. Bian, J., et al.: Unsupervised scale-consistent depth and ego-motion learning from monocular video. In: Advances in Neural Information Processing Systems, pp. 35–45 (2019)
3. Cao, W., Gao, J., Ming, Z., Cai, S., Zheng, H.: Impact of probability distribution selection on RVFL performance. In: Qiu, M. (ed.) SmartCom 2017. LNCS, vol. 10699, pp. 114–124. Springer, Cham (2018). https://doi.org/10.1007/978-3-319-73830-7_12
4. Cao, W., Wang, X., Ming, Z., Gao, J.: A review on neural networks with random weights. Neurocomputing **275**, 278–287 (2018)
5. Cao, Y., Gu, Q.: Tight sample complexity of learning one-hidden-layer convolutional neural networks. In: Advances in Neural Information Processing Systems, pp. 10612–10622 (2019)
6. Chen, C.P., Liu, Z.: Broad learning system: an effective and efficient incremental learning system without the need for deep architecture. IEEE Trans. Neural Netw. Learn. Syst. **29**(1), 10–24 (2017)
7. Chen, C.P., Liu, Z., Feng, S.: Universal approximation capability of broad learning system and its structural variations. IEEE Trans. Neural Netw. Learn. Syst. **30**(4), 1191–1204 (2018)
8. De Myttenaere, A., Golden, B., Le Grand, B., Rossi, F.: Mean absolute percentage error for regression models. Neurocomputing **192**, 38–48 (2016)
9. Ding, J., Ren, X., Luo, R., Sun, X.: An adaptive and momental bound method for stochastic learning. arXiv preprint arXiv:1910.12249 (2019)
10. Feng, S., Chen, C.P.: Fuzzy broad learning system: a novel neuro-fuzzy model for regression and classification. IEEE Trans. Cybern. **50**(2), 414–424 (2018)
11. Gulrajani, I., Ahmed, F., Arjovsky, M., Dumoulin, V., Courville, A.C.: Improved training of Wasserstein GANs. In: Advances in Neural Information Processing Systems, pp. 5767–5777 (2017)

12. Guo, H., Sheng, B., Li, P., Chen, C.L.P.: Multiview high dynamic range image synthesis using fuzzy broad learning system. IEEE Trans. Cybern. **PP**(99), 1–13 (2019)
13. He, Y., Liu, P., Wang, Z., Hu, Z., Yang, Y.: Filter pruning via geometric median for deep convolutional neural networks acceleration. In: Proceedings of the IEEE Conference on Computer Vision and Pattern Recognition, pp. 4340–4349 (2019)
14. Kerg, G., et al.: Non-normal recurrent neural network (NNRNN): learning long time dependencies while improving expressivity with transient dynamics. In: Advances in Neural Information Processing Systems, pp. 13591–13601 (2019)
15. Khan, A.H., Shao, Z., Li, S., Wang, Q., Guan, N.: Which is the best PID variant for pneumatic soft robots? an experimental study. IEEE/CAA J. Autom. Sin. **7**(2), 451–460 (2020)
16. Kingma, D.P., Ba, J.: Adam: a method for stochastic optimization. In: International Conference on Learning Representations, pp. 1–15 (2015)
17. Lai, X., Cao, J., Huang, X., Wang, T., Lin, Z.: A maximally split and relaxed ADMM for regularized extreme learning machines. IEEE Trans. Neural Netw. Learn. Syst. **PP**(99), 1–15 (2019)
18. Mandt, S., Hoffman, M.D., Blei, D.M.: Stochastic gradient descent as approximate Bayesian inference. J. Mach. Learn. Res. **18**(1), 4873–4907 (2017)
19. Pao, Y.H., Takefuji, Y.: Functional-link net computing: theory, system architecture, and functionalities. Computer **25**(5), 76–79 (1992)
20. Rumelhart, D.E., Durbin, R., Golden, R., Chauvin, Y.: Backpropagation: the basic theory. In: Backpropagation: Theory, Architectures and Applications, pp. 1–34 (1995)
21. Tanabe, K.: Conjugate-gradient method for computing the Moore-Penrose inverse and rank of a matrix. J. Optim. Theory Appl. **22**(1), 1–23 (1977)
22. Vaswani, A., et al.: Attention is all you need. In: Advances in Neural Information Processing Systems, pp. 5998–6008 (2017)
23. Xu, M., Han, M., Chen, C.L.P., Qiu, T.: Recurrent broad learning systems for time series prediction. IEEE Trans. Cybern. **50**(4), 1405–1417 (2020)
24. Ye, G.Y., Xu, K.J., Wu, W.K.: Standard deviation based acoustic emission signal analysis for detecting valve internal leakage. Sens. Actuators A: Phys. **283**, 340–347 (2018)
25. Zhang, L., Suganthan, P.N.: A comprehensive evaluation of random vector functional link networks. Inf. Sci. **367**, 1094–1105 (2016)

Accelerating *De Novo* Assembler WTDBG2 on Commodity Servers

Ming Dun[1], Yunchun Li[1,2], Xin You[2], Qingxiao Sun[2], Zerong Luan[4],
and Hailong Yang[2,3(✉)]

[1] School of Cyber Science and Technology, Beihang University, Beijing 100191, China
[2] School of Computer Science and Engineering, Beihang University,
Beijing 100191, China
hailong.yang@buaa.edu.cn
[3] State Key Laboratory of Mathematical Engineering and Advanced Computing,
Beijing University of Technology, Beijing 100083, China
[4] College of Life Sciences and Bioengineering, Beijing University of Technology,
Beijing 100083, China

Abstract. *De novo* genome assembly reconstructs the chromosomes from massive relatively short fragmented reads and serves as fundamental for studying new species where there is no reference genome. Wtdbg2 is a *de novo* assembler for long reads that is up to hundreds of kilobases. It is based on fuzzy-Bruijn graph (FBG) and is ten times faster than the cutting-edge assemblers such as Canu. However, the performance of wtdbg2 still requires further improvement: *1)* it requires up to terabytes of memory to compute the assembly, which is infeasible to run on commodity server; *2)* it requires tens of hours for assembling on large datasets such as genomes of *homo sapiens*. To address the above drawbacks, we propose several optimization techniques for accelerating wtdbg2 on commodity server, including a memory auto-tuning scheme, sequence alignment optimization and intermediate result elimination in the output procedure. We compare the optimized wtdbg2 with the original implementation and two cutting-edge assemblers on real-world datasets. The experiment results demonstrate that optimized wtdbg2 achieves maximum and average speedup of 2.31× and 1.54× respectively. In addition, our proposed optimization reduces the memory usage of wtdbg2 by 39.5% without affecting the correctness.

Keywords: Genome assembly · wtdbg2 · Performance optimization · Computational biology · Auto-tuning · Load balance

1 Introduction

De Novo genome assembly aims to generate a new genome from DNA fragments (named as reads) without the reference genome. It is of great significance in bioinformatics for identifying previously uncharacterized genomes [23] and analyzing the structural genomic changes [22]. Moreover, with the prosperities and

© Springer Nature Switzerland AG 2020
M. Qiu (Ed.): ICA3PP 2020, LNCS 12452, pp. 232–246, 2020.
https://doi.org/10.1007/978-3-030-60245-1_16

advances of DNA sequencing technologies from Oxford Nanopore Technologies (ONT) and Pacific Bioscience (PacBio), the length of reads has been increased up to several hundreds of base pairs (bps). The most popular assembly methods such as *de Bruijn* Graph are developed for short read assembly in second-generation DNA sequencing. However, recent research works have taken efforts to modify the *de Bruijn* Graph for long and error-prone reads in next-generation sequencing such as A-Bruijn graph in Flye [13] and fuzzy-Bruijn graph (FBG) in wtdbg2 [22].

Wtdbg2 [22] is one of the fastest status-quo assemblers for long noisy reads and has been adopted in Novel Sequence Insertion (NSI) tools such as rCANID [12] in addition to analysis for genome datasets [24]. Wtdbg2 is based on FBG that wraps each 256bp of reads into a unit named as bin and utilizes hash lists to encode the reads. In wtdbg2, a k-bin of FBG contains K consecutive bins. To improve the error toleration of FBG, a vertex can represent various k-bins if aligned together in sequence alignment routine. For the algorithm details of FBG, the readers can refer to [22].

However, to further improve the performance of wtdbg2, there are two major challenges to be addressed. The first challenge is the prohibitive memory consumption generated during the execution of wtdbg2. For instance, the wtdbg2 takes up to 1,788 GB memory on *Axolotl* [5], thus it is infeasible to assemble large datasets on commodity servers that usually contains less than 512 GB memory. The second challenge is the tremendous computation power required by wtdbg2. For instance, the wtdbg2 takes up to tens of hours for mammalian genome assembly on datasets such as *homo sapiens*.

Therefore, to alleviate the memory consumption and to further enhance the performance of wtdbg2, we propose a memory auto-tuning scheme based on regression model to reduce memory usage. In addition, we improve the parallelization of sequence alignment routine, which is one of the major bottlenecks, by enhancing the thread efficiency and load balance in wtdbg2. Moreover, we optimize the output procedure by eliminating the redundant intermediate results. We compare our optimization strategies with the original wtdbg2 and other two status-quo assemblers Canu and Flye under five real-world genome datasets and the results show that the maximum acceleration rate is of $2.31\times$, with the average acceleration rate is of $1.54\times$ and the memory cost can be reduced up to 39.5%, without affecting the correctness. With our proposed optimizations, massive parallel genome assembling that is previously infeasible with wtdbg2 on commodity server now can be run successfully.

- We perform comprehensive analysis of the performance bottlenecks in wtdbg2 and identify that the all-vs-all sequence alignment and output procedure are the major hotspots in large genome assembly.
- We propose a memory auto-tuning scheme based on regression model to alleviate the memory usage of wtdbg2. In addition, we accelerate the sequence alignment and optimize the output procedure to further improve the performance of wtdbg2.

– We evaluate our optimized wtdbg2 with five real-world genome datasets and compare to the original wtdbg2 as well as two cutting-edge genome assemblers. The experiment results demonstrate our optimization strategies are effective to reduce the memory usage in addition to improve the performance of wtdbg2 with correctness validation.

This paper is organized as follows. In Sect. 2, we present the background of genome assembly and wtdbg2. We present the bottleneck analysis and optimization strategies of wtdbg2 in Sect. 3. The detailed implementations are described in Sect. 4. Section 5 presents the evaluation results of the optimized wtdbg2. Section 6 describes the related work. Section 7 concludes this paper.

2 Background

2.1 Genome Assembly

Genome assembly is a computational representation of a genome sequence. Since we are not able to sequence along the whole length of DNA, genome assembly provides a computational method to reconstruct a DNA sequence from a large set of short reads of sequenced DNA fragments, which may be overlapped by each other. Moreover, the *de novo* genome assembler is a type of assembler aiming at assembling short reads to construct full-length sequences of DNA without any reference template. Among the methodologies for implementing a *de novo* genome assembler, *De Bruijn* graph is one of the most popular method, which aligns *k-mers* based on $k - 1$ sequence conservation to create contigs. The short k-mers allow fast hashing to decrease the computationally intensity and enhance the overall performance.

Nowadays, a number of software such as Canu [14] and MECAT [25] is developed for *de novo* assembly. Most of them reported that *sequence alignment* process, which includes *k-mer* counting and contig generation, is the major bottleneck of these existing assemblers [10] and consumes large memory footprint when parallelized on shared-memory system.

2.2 Wtdbg2

Wtdbg2 [22] is an efficient long-read genome assembler. It implements fuzzy-Bruijn graph (FBG) which extends the basic ideas of De Bruijn graph (DBG) to support long noisy reads on shared memory system with *pthread* parallelization. Similar to DBG, a "base" in FBG is a 256bp bin and a k-mer in FBG consists of k consecutive bins on reads. FBG is made up of vertices of k-bins and edges between two vertices which indicates their adjacency on a read. The difference is that a single vertex may represent different k-bins if they are aligned together based on all-vs-all read alignment, which tolerates errors in noisy long reads.

As shown in Fig. 1, the workflow to obtain the constructed contigs of input reads contains four steps: (a) binning and pairwise alignment, (b) graph construction, (c) graph clearing and (d) consensus. In the first stage, input reads are all loaded from files to memory and each base is encoded with 2 bits. For all-vs-all read alignment, a Smith-Waterman-like dynamic programming is applied. After alignment, FBG is then constructed by adding vertices according to the obtained all-vs-all alignments and edges of two vertices if they are in the same read. In the phase of graph clearing, wtdbg2 retains only one edge if there are multiple edges between two k-bins. Besides, edges covered by less than 3 reads are omitted. The resulting FBG is then consensused to obtain the final long contigs.

In assembling, the execution of wtdbg2 is split into two phases. The first phase is binning and pairwise alignment, whose results are written to disk as intermediate results. The second phase includes graph constructing, graph cleaning and consensus, which consumes large memory footprint.

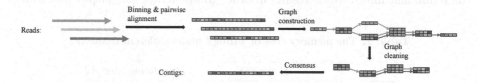

Fig. 1. The overview of the long-read assembling process in wtdbg2.

3 Methodology

In this section, we first provide the analysis for bottleneck as well as for the memory consumption of wtdbg2 to guide our optimization. Then we present the strategies for reducing the memory usage and optimizing execution hotspots, including the sequence alignment and output procedure.

3.1 Bottleneck Analysis

The reported [5] memory consumption of wtdbg2 under different datasets are shown in Table 1. Since most of the commodity servers contain less than 256 GB memory, the large mammalian datasets especially the genome of *homo sapiens* cannot be directly assembled by wtdbg2 on the commodity servers. Though the original wtdbg2 provides a variable *kbmparts* to control the memory usage, it still requires manual tuning, and thus facing the risk of assembly failure due to memory overflow.

We profile the execution of wtdbg2 using HPCToolkit [6]. The result of bottleneck analysis is shown in Fig. 2, where the execution time for output procedure is excluded. The description of the datasets is shown in Table 2. It is clear that the KBM indexing is the major bottleneck of wtdbg2 under small datasets such as *E.coli* and *C.elegans*. However, when the dataset's genome size increases, especially in the genomes of *homo sapiens*, the sequence alignment process dominates the execution time and becomes the bottleneck of wtdbg2. Since the large datasets usually take hours to be processed and thus worth the efforts for acceleration, we focus on optimizing the sequence alignment routine to improve the performance of wtdbg2 on large datasets.

In addition, the size of output contigs of wtdbg2 can take up to 30 GB under dataset *Human HG00733*. On the other hand, it can take more than 41.3% of entire execution time to write the contigs under dataset *E.coli* (e.g., 2.03 s of the overall 4.91 s). What's more, the performance of the output procedure also needs to be improved to further accelerate wtdbg2. Moreover, the wtdbg2 also writes the intermediate alignment results to the hard disks, whose size is at the same order of magnitude as the contigs. These intermediate alignment results are redundant during the execution. Therefore, we propose to eliminate the redundant intermediate results in order to accelerate the output procedure.

Table 1. The memory cost of wtdbg2 under different datasets.

	C.elegans	D.melanogaster A4
Memory cost (GB)	1.0	19.4
	Human NA24385	Human HG00733
Memory cost (GB)	112.9	338.1

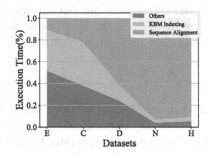

Fig. 2. The result of bottleneck analysis of wtdbg2. E, C, D, N and H represents the datasets: *E.coli*, *C.elegans*, *D.melanogaster A4*, *Human NA24385* and *Human HG00733*. The time for output is excluded.

3.2 Memory Auto-Tuning

We develop an auto-tuning scheme based on linear regression model to satisfy the memory demand of the large genome datasets. We identify that the highest memory usage is at the phase of sequence alignment where wtdbg2 easily crashes due to memory shortage at large datasets. Another execution phase that is unlikely to crash due to memory shortage is the reads collecting phase whose memory usage approximately equals to the size of the input dataset. We also notice that there is a parameter *kbm_parts* in wtdbg2 to control the number of iterations of alignment, which exhibits a negative linear relationship with memory usage in sequence alignment routine. Based on the above observations, we develop a regression model to estimate the memory usage of sequence alignment based on the memory measurement of reads collecting. At the reads collecting phase, the model estimates how much sequence alignment would exceed the system memory capacity and then adjusts the *kbm_parts* automatically to satisfy the memory demand. The regression model is built on the statistics collected in previous execution of wtdbg2. Since running the regression model only requires monitoring the system memory usage and executing a simple linear function, the overhead of memory auto-tuning is negligible.

3.3 Sequence Alignment Optimization

Thread Efficiency. We improve the parallelization of sequence alignment routine in wtdbg2 to accelerate the major bottleneck under large datasets. In the original wtdbg2, as shown in Fig. 3a, after processing the alignment with threads at the granularity of a single read, the alignment results of each read are merged into the graph in the master thread. In addition, the read assignment among threads is also performed in the master thread. After analyzing the performance of sequence alignment, the result merging function *map2rdhit_graph* dominates the execution time of the master thread, which increases the time delay between two reads alignment for a thread and thus decreases the efficiency of multi-threading. Since wtdbg2 adopts the shared-memory parallelization using pthread, we distribute the work of result merging among threads by leveraging the mutual exclusive locks [16] as shown in Fig. 3a, which improves the multi-threading efficiency of the sequence alignment routine.

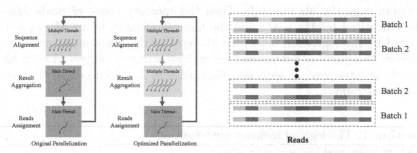

(a) Original and optimized parallelization of sequence alignment

(b) Batched read assignment scheme in sequence alignment

Fig. 3. The illustration of optimization strategies in sequence alignment routine of wtdbg2.

Load Balance. When the sequence alignment routine analyzes a specific read R_i in wtdbg2, it only compares R_i with reads that have larger identifiers. Hence, as the identifier of the R_i increases, the number of reads that R_i needs to be aligned with decreases. Since a thread only performs sequence alignment for reads in ascending order in wtdbg2, load imbalance becomes severe as the identifier of the read increases. Moreover, reducing the number of assignments can alleviate the load imbalance between master and slave threads. However, fewer assignments means more reads in each assignment, which in turn increases the load of slave threads.

To address load imbalance among threads and to mitigate the load in the master thread, we develop a batched read assignment scheme for the sequence alignment. The proposed allocation scheme is shown in Fig. 3b, where *batchsize* is set to 4. We partition all the reads into M *batches*, each with N reads. For the *i*th *batch*, the index of its reads ranges from Ni to $Ni + \frac{N}{2}$ in ascending or descending order. Thus, the number of reads each thread needs to compare is the same theoretically, which also decreases the number of read assignments in the master thread. Since the solution space of M is huge and the best M is different for various datasets, we develop a linear regression model to automatically search for the optimal setting of M.

4   Implementation

In this section, we provide the implementation details of the optimization methodologies proposed in Sect. 3.

4.1   Memory Auto-Tuning

The processing logic of memory auto-tuning is shown in Algorithm 1. After the reads collecting process finishes, our implementation uses function *check_meminfo* to obtain the total memory M_T and available memory M_A of the system from the file *meminfo*. Then the memory usage of reads M_{READ} can be calculated by $M_T - M_A$ and the highest memory usage M_{MAX} can be estimated by the regression model. At the end of the auto-tuning process, the parameter *kbm_parts* is determined using *kbm_parts* $= \frac{M_{MAX}}{M_T}$. The value of *kbm_parts* is later used in sequence alignment to avoid exceeding the system memory constraint.

Algorithm 1. The logic of memory auto-tuning

1: Input: regression model $f : M_{READ} \rightarrow M_{MAX}$
2: Output: *kbm_parts*
3: $(M_T, M_A) \leftarrow$ check_meminfo()
4: $M_{READ} \leftarrow M_T - M_A$
5: $M_{MAX} \leftarrow f(M_{READ})$
6: *kbm_parts* $\leftarrow \frac{M_{MAX}}{M_T}$

4.2 Sequence Alignment Optimization

Thread Efficiency. As shown in Algorithm 2, for each read R in the sets of read $\mathbf{R}$, the master thread finds an idle slave thread k and then assigns the sequence to it. The thread k first encodes its read R with a specific hash list to get the binned sequence BS_k and then performs the sequence alignment between BS_k and other binned sequences whose indices are bigger than the index of R. After the sequence alignment, it generates the alignment results including the cigars $cigars_k$ and the overlapped bins with other sequences $hits_k$. With the results generated, the thread k requests for the mutex lock and merges the local results to the FBG graph. The reason for applying mutex lock is to avoid writing conflict among multiple threads. Once the result merging finishes, the thread k releases the lock and becomes an idle thread. Then the master thread continues to assign remaining reads to thread k unless all reads have already been assigned. The parallel result merging method reduces the assignment delay for the slave threads as well as alleviating the load for the master thread.

Algorithm 2. The logic of thread efficiency optimization

1: Input: Reads $\mathbf{R}$
2: Output: cigars g_cigars, hits g_hits
3: **for** each $R \in \mathbf{R}$ **do**
4:     $k = $ get-idle-thread-id()
5:     assign R to thread k
6:     $BS_k \leftarrow$ query_index(R, k)
7:     $(cigars_k, hits_k) = $ kbm_alignment(BS_k, k) /*the new intermediate result in thread k*/
8:     get-mutex_lock(k)
9:     $(g_cigars, g_hits) = $ result-gather($cigars$, $hits$, $cigars_k$, $hits_k$)
10:    free-mutex_lock(k)
11:    turn-idle-thread(k)
12: **end for**

Load Balance. The implementation of batched read assignment is shown in Algorithm 3. The variable *batchtime* is calculated by the regression model based on the number of reads $\mathbf{R}.size$, which is then used to determine the optimal setting of *batchsize*. After the number of batched reads (*batchsize*) is determined, the master thread computes the indices of reads that belong to a specific *batch* i. Once the *batch* is assigned, thread k will finish the sequence alignment process for the reads in batch i, generating the cigars as well as hits and merging them to the graph.

4.3 Output Optimization

We use a condition parameter *WITHALIGNMENT* to control the output of intermediate alignment results written to hard disks. Once the *WITHALIGNMENT = true*, the output of intermediate results is skipped. In addition, we use another condition parameter *PRINTGRAPH* to control the output of runtime information such as that for FBG graph.

Algorithm 3. The logic of load balance optimization

```
 1: Input: all the reads R
 2: Output: cigars g_cigars, hits g_hits
 3: batchtime ← f₁(R.size)
 4: batchsize ← R.size/(batchtime × ncpu − 1)
 5: for i ← 0 → batchtime × ncpu do
 6:     batchᵢ ← get_newbatch()
 7:     k = get-idle-thread-id()
 8:     BSₖ ← query_index_batch(batchᵢ, k)
 9:     (cigarsₖ, hitsₖ) = kbm_alignment_batch(BSₖ, k) /*the new intermediate result in thread
        k*/
10:     turn-idle-thread(k)
11: end for
```

5 Evaluation

We implement the proposed optimizations for wtdbg2 and compare its performance with the original wtdbg2 and two state-of-art sequence assemblers under five real-world datasets.

5.1 Experiment Setup

Assemblers For Comparison. We choose two cutting-edge genome assemblers Canu [14] and Flye [13] for comparison. These two assemblers use both PacBio and Oxford Nanopore datasets as input. However, the assemble and consensus processes are integrated in Canu and Flye, whereas they are separate in wtdbg2. Canu is based on MinHash Alignment Process (MHAP) [8] and Celera Assembler [9], whereas Flye is based on A-Bruijn Graph.

Genome Datasets. We choose five datasets to compare the performance of different genome assemblers, including E.coli [5], C.elegans [1], D.melanogaster ISO1 [2], Human NA24385 [4] and Human HG00733 [3]. The details of the datasets are shown in Table 2. For convenience, we abbreviate the datasets to **E, C, DISO, NA** and **HG** respectively.

Experiment Platform. We conduct all the experiments on a commodity server that is memory constrained. The server has 2 Intel Xeon E5-2680v4 CPUs, each with 14 hyper-threaded cores. The memory capacity of the server is 256 GB, which is much smaller than the fat node with 2 TB memory used in the experiment of original wtdbg2 [22].

Evaluation Criteria. We first compare the memory usage of wtdbg2 using memory auto-tuning optimization with the original wtdbg2 to measure the effectiveness of the memory auto-tuning. To evaluate the performance improvement, we compare the optimized versions of wtdbg2 with Canu and Flye. We only measure the execution time of assemble process in Canu and Flye. We choose wtdbg2 applied with memory auto-tuning optimization as the baseline since the original wtdbg2 fails to run on large datasets such as *HG* due to out of memory error. For validation of correctness, we use QUAST [11] with assembly indicators including N50, NGA50, genome fraction and total genome length. All assembly indicators derived for both original and optimized wtdbg2 are based on pre-polished assembly results.

Table 2. The genome datasets.

Dataset	Dataset type	Coverage	Genome size
E.coli	PacBio RSII	20×	4.6 MB
C.elegans	PacBio RSII	80×	100 MB
D.melanogaster ISO1	Oxford Nanopore	32×	144 MB
Human NA24385	PacBio CCS	28×	3 GB
Human HG00733	PacBio Sequel	93×	3 GB

5.2 Performance Analysis

Memory Auto-Tuning. The memory usage of the optimized wtdbg2 normalized to the original wtdbg2 is shown in Fig. 4. We adopt the memory usage of the original wtdbg2 reported from [5], which indicates the dataset *HG00733* consumes 338.1 GB memory that exceeds the memory capacity of our experiment server. From Fig. 4, we can see that the regression model used in the optimized wtdbg2 can successfully predict the memory usage and adjust the parameter *kbm_parts* accordingly so that it satisfies the memory demand from large datasets. Specifically, for dataset *HG00733*, the memory auto-tuning scheme reduces the memory demand by 39.5% compared to the original wtdbg2, and thus enables a successful execution on this dataset.

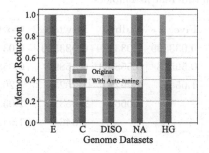

Fig. 4. The memory usage after applying the memory-auto tuning optimization on wtdbg2.

Performance Improvement. The performance comparison results between the optimized wtdbg2 and two state-of-art assemblers Canu and Flye are shown in Fig. 5. The metrics for assembly quality including N50, NGA50, genome fraction and total genome length compared to the reference genome are shown in Table 3, Table 4, Table 5 and Table 6, respectively. The missing bar in Fig. 5 and the '–' symbol in Table 3, 4, 5 and 6 indicate the execution failure due to out of memory error or extreme long execution time (e.g., more than 7 days). From Fig. 5, we can see that the optimized wtdbg2 with both sequence alignment and output optimizations achieves the best performance under all datasets. Based on the results from Table 3, 4, 5 and 6, we can see that the optimized wtdbg2 achieves comparable or even better assembly quality when compared to the original wtdbg2. The highest speedup achieved by A-wtdbg2 is 2.31× under dataset **NA** and the average speedup of A-wtdbg2 is 1.54× compared to the baseline. The reason for A-wtdbg2 to achieve the lower speedup with datasets **E** and **C** is that the sequence alignment routine takes up a small portion of the entire execution time on these datasets, which is also described in the bottleneck analysis in Sect. 3.1.

Table 3. The N50 (million base pairs (Mbps)) of the assembly results from different assemblers.

Datasets	Canu	Flye	wtdbg2	M-wtdbg2	O-wtdbg2	A-wtdbg2
E	4.6680	4.6371	4.6719	4.6358	4.6358	4.6341
C	2.0681	2.8587	1.9726	1.9726	1.9726	1.9718
DISO	4.2986	16.431	6.2255	6.2258	6.2259	6.2268
NA	–	–	15.512	15.512	15.512	15.511
HG	34.637	–	–	21.293	26.354	24.699

Table 4. The NGA50 (million base pairs (Mbps)) of the assembly results compared to reference genome from different assemblers.

Datasets	Canu	Flye	wtdbg2	M-wtdbg2	O-wtdbg2	A-wtdbg2
E	0.033895	0.033539	0.033539	0.033540	0.033539	0.033935
C	0.56166	0.56513	0.558,99	0.55899	0.55899	0.55752
DISO	1.0913	1.6581	1.2923	1.2923	1.2038	1.1852
NA	–	–	2.0664	2.0531	2.0531	2.0714
HG	2.4485	–	–	1.6464	1.9423	1.9534

Table 5. The genome fraction(%) of the assembly results compared to reference genome from different assemblers.

Datasets	Canu	Flye	wtdbg2	M-wtdbg2	O-wtdbg2	A-wtdbg2
E	73.451	73.478	72.97	72.992	73.056	72.994
C	99.685	99.647	98.662	98.66	98.662	98.686
DISO	93.516	93.273	88.361	88.35	88.322	88.411
NA	–	–	87.607	87.705	87.705	88.417
HG	91.726	–	–	87.284	88.39	88.39

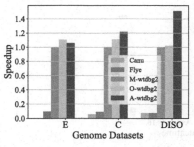

(a) Small Non-mammalian Datasets

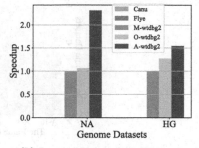

(b) Large Mammalian Datasets

Fig. 5. Performance comparison among optimized wtdbg2, Canu and Flye, where M-wtdbg2, O-wtdbg2 and A-wtdbg2 is the original wtdbg2 applied with memory auto-tuning optimization (baseline), both memory and output optimizations and all proposed optimizations, respectively. Performance comparison among optimized wtdbg2, Canu and Flye, where M-wtdbg2, O-wtdbg2 and A-wtdbg2 is the original wtdbg2 applied with memory auto-tuning optimization (baseline), both memory and output optimizations and all proposed optimizations, respectively.

Table 6. The total genome length (million base pairs(Mbps)) of the assembly results from different assemblers.

Datasets	Canu	Flye	wtdbg2	M-wtdbg2	O-wtdbg2	A-wtdbg2
E	4.6476	4.6371	4.6358	4.6719	4.6719	4.6702
C	108.19	102.46	106.10	106.09	106.09	106.31
DISO	140.57	144.43	136.90	136.86	136.78	136.82
NA	–	–	2,749.3	2,752.6	2,752.4	2,774.5
HG	3,038.5	–	–	2,775.5	2,804.1	2,802.3

5.3 Parameter Sensitivity Analysis

To better understand the impact of parameter *batchtime* described in Sect. 4.2 on the performance of wtdbg2, we adjust the setting of *batchtime* within the range of A ±16 under four different datasets, where A is the setting after applying memory auto-tuning optimization described in Sect. 4.2. From Figure 6 we can see that when the *batchtime* increases, which in turn decreases the *batchsize*, the performance of wtdbg2 usually gets improved. The reason for that is a finer-grained partitioning improves the load balance, while at the same time hardly increases the load for threads in a single execution of alignment. Moreover, we notice that the performance impact of *batchtime* varies across different datasets and thus an auto-tuning scheme is required to search for the optimal *batchtime*.

6   Related Work

As genome assembly is widely used to obtain genome information, various genome assemblers are developed to construct assembly graphs from a large set of

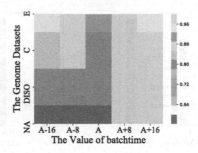

Fig. 6. The performance impact of parameter *batchtime* under different datasets, where A equals the performance when applying memory auto-tuning optimization. The results under each dataset have been normalized by the best one.

reads, including A-Bruijn assembly graph [13] and de Bruijn graph. Among them, de Bruijn graph (DBG) is the most commonly used method for genome assembling. Among recent implementations on a single machine, IDBA-UD [19] algorithm can reconstruct long contigs with higher accuracy through multi-thread parallelism, which applies progressive relative depth, local assembly and error correction. In addition, Canu [14] is developed for scalable and accurate long-read assembly. It implements adaptive k-mer weighting and repetitive separation methods, and parallelizes the overlap computation into multiple jobs and merges these results with parallel bucket sort algorithm. On the other hand, an open-source *de novo* genome assembly, MECAT [25], combines fast mapping, error correction and de novo assembly. MECAT indices reads with hash tables and accelerates the computation by sampling with sliding window and thus reduces the number of searched k-mers from the degree of sampling number.

Recently, research works are proposed to scale genome assembly to multiple nodes. Pan *et al.* [17] develop distributed memory parallel hash tables for DNA k-mer counting and evaluate their methods on 4,096 cores of the NERSC Cori supercomputer. In addition, Pakman [10] is one of the most recent parallel implementations, which enables large-scale genome assembly with distributed memory parallelism up 8K cores. In addition to parallel I/O and load-balanced counting of k-mers, Pakman proposed a new type of graph named PakGraph for better parallelism. Other optimizations such as dynamic memory allocation and runtime power control [20,21] can also be applied to further accelerate genome assembly.

Moreover, there are research works focusing on accelerating the genome sequencing or assembly on modern GPU. Nvidia provided its own CUDA library NVBIO [18] for sequence analysis with high throughput. To outperform NVBIO, Ahmed *et al.* proposed specially designed APIs (GASAL [7]) to provide GPU accelerated kernels for local, global and semi-global alignment routines, which achieved notable speedup. In addition, CUDASW++ 3.0 [15] accelerates Smith-Waterman algorithm by the use of CPU and GPU SIMD operations as well as the collaborated processing on CPU and GPU. However, the above GPU acceleration methods are not applicable for wtdbg2 due to its implementation with hash list and varying read length.

7 Conclusion

In this paper, we analyze the hotspots of wtdbg2, and identify that sequence alignment and output procedure are the two major performance bottlenecks. In addition, we also reduce the prohibitive memory usage of the original wtdbg2. Specifically, we propose a memory auto-tuning scheme to satisfy the memory demand when running wtdbg2 with large datasets on commodity servers. We also propose sequence alignment optimization to improve the multi-threading efficiency and load balance. Moreover, we apply output optimization to eliminate the redundant intermediate alignment results to further improve the performance of wtdbg2. The experiment results demonstrate that the optimized wtdbg2 achieves a maximum speedup of 2.31× and an average speedup of 1.54×. In addition, our optimization reduces the memory usage by 39.5% without affecting the correctness.

Acknowledgment. This work is supported by National Key Research and Development Program of China (Grant No. 2016YFB1000304), National Natural Science Foundation of China (Grant No. 61502019), and the Open Project Program of the State Key Laboratory of Mathematical Engineering and Advanced Computing (Grant No. 2019A12).

References

1. C.elegans genome dataset (2019). http://datasets.pacb.com.s3.amazonaws.com/2014/c_elegans
2. D.melanogaster iso1 genome dataset (2019). https://www.ebi.ac.uk/ena/data/view/SRR6702603
3. Human hg00733 genome dataset (2019). https://www.ebi.ac.uk/ena/data/view/SRR7615963
4. Human na24385 genome dataset (2019). https://ftp-trace.ncbi.nlm.nih.gov/giab/ftp/data/AshkenazimTrio/HG002_NA24385_son/PacBio_CCS_15kb
5. wtdbg2 (2019). https://github.com/ruanjue/wtdbg2
6. Adhianto, L., et al.: HPCTOOLKIT: tools for performance analysis of optimized parallel programs. Concurr. Comput. Pract. Exp. **22**(6), 685–701 (2010)
7. Ahmed, N., Mushtaq, H., Bertels, K., Al-Ars, Z.: GPU accelerated API for alignment of genomics sequencing data. In: 2017 IEEE International Conference on Bioinformatics and Biomedicine (BIBM), pp. 510–515. IEEE (2017)
8. Berlin, K., Koren, S., Chin, C.S., Drake, J.P., Landolin, J.M., Phillippy, A.M.: Assembling large genomes with single-molecule sequencing and locality-sensitive hashing. Nat. Biotechnol. **33**(6), 623 (2015)
9. Denisov, G., et al.: Consensus generation and variant detection by Celera assembler. Bioinformatics **24**(8), 1035–1040 (2008)

10. Ghosh, P., Krishnamoorthy, S.: PaKman: scalable assembly of large genomes on distributed memory machines. In: 2019 IEEE International Parallel and Distributed Processing Symposium (IPDPS) (2019)
11. Gurevich, A., Saveliev, V., Vyahhi, N., Tesler, G.: QUAST: quality assessment tool for genome assemblies. Bioinformatics **29**(8), 1072–1075 (2013)
12. Jiang, T., Fu, Y., Liu, B., Wang, Y.: Long-read based novel sequence insertion detection with rCANID. IEEE Trans. Nanobiosci. **18**(3), 343–352 (2019)
13. Kolmogorov, M., Yuan, J., Lin, Y., Pevzner, P.A.: Assembly of long, error-prone reads using repeat graphs. Nat. Biotechnol. **37**(5), 540 (2019)
14. Koren, S., Walenz, B.P., Berlin, K., Miller, J.R., Bergman, N.H., Phillippy, A.M.: Canu: scalable and accurate long-read assembly via adaptive k-mer weighting and repeat separation. Genome Res. **27**(5), 722–736 (2017)
15. Liu, Y., Wirawan, A., Schmidt, B.: CUDASW++ 3.0: accelerating smith-waterman protein database search by coupling CPU and GPU SIMD instructions. BMC Bioinform. **14**(1) (2013). Article number: 117. https://doi.org/10.1186/1471-2105-14-117
16. Nichols, B., Buttlar, D., Farrell, J., Farrell, J.: Pthreads Programming: A POSIX Standard for Better Multiprocessing. O'Reilly Media Inc., Sebastopol (1996)
17. Pan, T.C., Misra, S., Aluru, S.: Optimizing high performance distributed memory parallel hash tables for DNA k-mer counting. In: SC 2018: International Conference for High Performance Computing, Networking, Storage and Analysis, pp. 135–147. IEEE (2018)
18. Pantaleoni, J., Subtil, N.: NVBIO: a library of reusable components designed by NVIDIA corporation to accelerate bioinformatics applications using CUDA (2014). http://nvlabs.github.io/nvbio
19. Peng, Y., Leung, H.C., Yiu, S.M., Chin, F.Y.: IDBA-UD: a de novo assembler for single-cell and metagenomic sequencing data with highly uneven depth. Bioinformatics **28**(11), 1420–1428 (2012)
20. Qiu, M., et al.: Data allocation for hybrid memory with genetic algorithm. IEEE Trans. Emerg. Top. Comput. **3**(4), 544–555 (2015)
21. Qiu, M., Ming, Z., Li, J., Liu, S., Wang, B., Lu, Z.: Three-phase time-aware energy minimization with DVFS and unrolling for chip multiprocessors. J. Syst. Archit. **58**(10), 439–445 (2012)
22. Ruan, J., Li, H.: Fast and accurate long-read assembly with wtdbg2. Nat. Methods **17**(2), 155–158 (2020)
23. Simpson, J.T., Durbin, R.: Efficient de novo assembly of large genomes using compressed data structures. Genome Res. **22**(3), 549–556 (2012)
24. Wenger, A.M., et al.: Highly-accurate long-read sequencing improves variant detection and assembly of a human genome, p. 519025. bioRxiv (2019)
25. Xiao, C.L., et al.: MECAT: fast mapping, error correction, and de novo assembly for single-molecule sequencing reads. Nat. Methods **14**(11), 1072 (2017)

Typing Everywhere with an EMG Keyboard: A Novel Myo Armband-Based HCI Tool

Zongkai Fu[1,2], Huiyong Li[1], Zhenchao Ouyang[1,2,3,5(✉)], Xuefeng Liu[1], and Jianwei Niu[1,2,3,4]

[1] State Key Laboratory of Virtual Reality Technology and Systems, Beihang University, Xueyuan Road #37, Haidian, Beijing 100191, China

[2] Hangzhou Innovation Institution, Beihang University, Chuanghui Street #18, Binjiang, Hangzhou 310000, Zhejiang, China
ouyangkid@buaa.edu.cn

[3] Beijing Advanced Innovation Center for Big Data and Brain Computing (BDBC), Beihang University, Xueyuan Road #37, Haidian, Beijing 100191, China

[4] Zhengzhou University Research Institute of Industrial Technology, Zhengzhou University, Zhengzhou 450001, China

[5] Nanhu Laboratory, Jiaxing 314000, Zhejiang, China

Abstract. To enhance users' experience of inputting characters on mobile devices with small screens, this paper designed a novel virtual keyboard used on mobile devices. In particular, we introduce a novel virtual keyboard based on the MYO armband which is able to capture the electromyogram (EMG) signals of users when typing on any surfaces (such as human body or normal desktop). The actions of the three fingers are mapped to the nine keys of the T9 keyboard. After that, the signals of finger motions are translated into key sequences of the T9 keyboard. However, the identification of continuous finger motions is a critical challenge. To address the challenge, we convert the EMG signals in time domain into a 3D time-frequency map (each channel corresponds to the EMG unit of a frequency-domain feature), and extract the convolutional features with a 4-layer CNN (Convolutional Neural Network) module, an im2col module of Optical Character Recognition (OCR) and a Long Short-Term Memory (LSTM) module, and the final result is achieved as a probability graph of finger gestures. The Connection Temporal Classification (CTC) algorithm is adopted to find the best gesture sequence from the probability map. Experimental results show that our method can effectively identify different key sequences at three different input speeds with an average accuracy of 85.9%, and the integration testing with different volunteers shows that our method can achieve an average typing speed of 15.7 Word-Per-Minute (WPM).

Keywords: EMG · Keyboard · Deep learning · CTC · Mobile device

Supported by Hangzhou Innovation Institution, Beihang University.

M. Qiu (Ed.): ICA3PP 2020, LNCS 12452, pp. 247–261, 2020.
https://doi.org/10.1007/978-3-030-60245-1_17

1   Introduction

The physical keyboard is a commonly-used device for Human-Computer Interaction (HCI), but it is always placed in a fixed place and too large to bring out. In the context of Mobile HCI, the devices should have the characteristics of small-size, easy-to-carry and the ability of computation and communication. As a result, virtual keyboard is introduced which allows users type through different equipment and methods. The most common virtual keyboard of the smartphone and the pad is set in the screen of the devices, and the user enter text through touching the corresponding screen area of the key. However, the size of these virtual keyboards is usually small which makes the user easily touch the wrong areas. Furthermore, the virtual keyboard needs to occupy an extra display area which is very expensive resources in mobile devices.

Electromyogram (EMG) is a technique for evaluating and recording the electrical activity from muscles. Conventionally, the EMG is used in two fields, the medical field which diagnoses muscular disease and assists disabled people in controlling prostheses [1,2], and the muscle computer interaction (MCI) field [3]. The previous researches demonstrated the feasibility of using EMG for MCI [4] and many researchers have been devoted to design and implement different gesture recognition to communicate with the computer [3]. In the application of text-entry, the automatic recognition of sign language is often used to transform the EMG signal into different words [5,6]. However, these methods require the users to be familiar with the sign language gesture in advance, and can't give the users a real experience like using the keyboard. Considering that many people are already familiar with the input habits under keyboard layout, it is also necessary for designing keyboard layout-based input method with EMG devices.

In this paper, we present a virtual keyboard that is portable and brings the users an experience like using the physical keyboard with common EMG device. We map 9 finger gestures to 9 keys of the T9 keyboard, therefore the user can enter texts/orders through performing the gesture of typing on any surfaces. Furthermore, the proposed system is also convenient because the whole system is based on a MYO armband and the smart device itself. The MYO armband can measure eight channels of EMG signals and communicate with mobile devices through Bluetooth. The most challenging part is how to recognize accurately the finger gestures through raw flexible EMG signals. To address this problem, we design a lightweight hybrid neural network model that combines 4-layer convolutional neural network (CNN) module, an im2col module of optical character recognition (OCR), and a long short-term memory (LSTM) module. By organizing a figure gesture data set with key labels, we can train the model in 'end-to-end' pattern. The output of the network is defined as probability map of continuous gesture sequence instead of a single key to improve the input efficiency. Evaluation resulting from several volunteers shows that the presented system can achieve high word-per-minute (WPM) speed at 15.7.

The remainder of this paper is organized as follows. Section 2 introduces related work in gesture recognition based on myoelectric signal and text entry system on wearable device. Section 3 presents the overview of the EMG keyboard system. Section 4 describes the main principle of the gesture sequence model. Section 5 presents the experiments' result and analysis. Finally, Sect. 6 concludes the paper.

2 Related Work

We review the related works from the following two aspects: the EMG signals-based gesture classification and wearable device-based text entry system.

2.1 Gesture Recognition by EMG

Huding [7] developed a myoelectric patterns classification scheme which consisted of segment feature extraction module, network training module and pattern classification module. He used several simple time domain features (i.e. Mean Absolute Value, Zero Crossings, Slope Sign Changes and Waveform Length) as the feature set and sent them to an artificial neural network for training and classification. His system could recognize the diverse set of the myoelectric patterns produced by intact and amputated musculature with an average accuracy of 84%. Latter, many researches [8, 9] sought to improve the accuracy of gesture recognition through changing the classifier based on Hudgins's scheme. Yet, it is hard to make further progress only using the time domain features as the feature set. Author of [10] used the wavelet based features as the feature set and send it to the Latent Dirichlet Allocation (LDA) classifier to recognize six hand gestures, and got an average accuracy of 98%. Besides, researchers [11–13] used the combination of time domain and frequency domain features as the feature set and send them to the different classifiers. Among them, the author [13] selected the sixteen features (i.e. Integral of Absolute Value, Difference of Absolute Mean Value, k-th order Zero-Crossings, Mean Frequency, WaveletMom, etc.) as feature set and got an accuracy of 99.4% to recognize nine American Sign Language (ASL) gestures. As for the Deep Learning-based methods, CNN is the most commonly used model. The author of [14] built a CNN-based model to solve the inter-user variability and proved that the features learned by CNN had a better performance than hand-crafted features in the task of detecting target EMG patterns across different users. After that, the authors of [14–16] took the problem of gesture recognition as instantaneous EMG image classification and used the CNN architecture to transform the signal picture into feature maps to identify. Also, YuNan [17] used LSTM-based network to capture temporal dependencies of EMG data, and could produce better accuracy of 84% in Ninapro database. In general, DL-methods relieve people from signal engineering and experience, and extract more information from the raw EMG data.

2.2 Wearable Device-Based Text Entry System

BlindType [18] is an eyes-free thumb-based text entry system which consists of two statistical decoding algorithms (i.e., the absolute algorithm and the relative algorithm) based on the position of touch endpoints of the thumb. The two decoding algorithms use regression to model the button touching pattern on three kinds of statistical features, i.e., the size and location as well as the standard deviation of the touch endpoints. The whole system can offer an average 17–23 WPM depending on the related algorithms. In TipText [19], the author presents a new text entry technique using micro thumb-tip gestures on a miniature QWERTY keyboard on one of the first segment of user's finger. The austhor tries different keyboard layouts for optimizing the efficiency and balancing among three aspects: layout learnability, key size and word disambiguation. The final TipText is a 2×3 grid of PET film. Real-world evaluation with different participants shows the system can achieve an average text entry speed of 11.9 WPM. Compass [20] is a non-touch bezel-based text entry technique, it provides a solution for text entry on rotational interface (e.g., smartwatches) without using the touchscreen. To enter text, the users rotate the bezel to select keys on the circular keyboard and the position of each key is dynamically optimized after each key selection according to the probability of the next key. The participants achieve an average 12.5 WPM using this text entry technique.

As for the existing research, EOG and EMG based virtual Keyboard [21] is the most relevant work to ours. It is a two-step keyboard that the users first gaze the position of the desired letter on the screen, and then click the navigated letter. Eye-gaze direction is obtained from the electrooculogram (EOG) signal and the click action is classified by the EMG signal. Their virtual keyboard system links the human's bio-signal and the computer, and can be used for disabled and aged persons. However, the users' faces need to be connected with a lot of sensors during operation, and the users' WPM is not very high due to the extra effort of the two-step selection. As a result, EOG and EMG based virtual keyboard is not convenient to use in our daily life.

3   System Overview

In this section, we introduce the system from system architecture and keyboard design.

3.1   System Architecture

This section presents the design of our EMG signal based virtual keyboard. The novel system captures the continuous EMG signal from user's arm, recognizes the EMG signal with a neural network and then outputs the character sequence. As shown in Fig. 1, our system consists of four main components: Data Preprocessing, Feature Extraction, Gesture Recognition and KeyCode to Character. Generally, the signal is collected by the Myo armband and sent to the data-preprocessing module through Bluetooth. In the data-preprocessing module, the high pass filter is first used to get the EMG signal between 10 Hz and 200 Hz, and further extract the active segments of gestures based on the short-term energy of the denoised signal. Then the feature extraction module splits the signal into frames with a 500 ms-sliding window, and performs short-time Fourier transform

(STFT) on each frame. To prevent spectrum leakage [22], we add the Handing window on the signal and pad the length of the frame to 256 through filling the signal with zero. After that, the gesture sequence recognition module recognizes finger gesture and converts each gesture to the key code. Finally, the keycode to character module finds all the possibilities of the key code sequence from the probability graph, and print them out for the users.

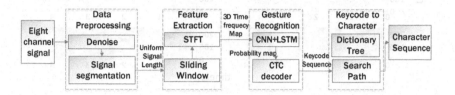

Fig. 1. Architecture of the system.

3.2 Keyboard Design

The study is to design a keyboard based on the EMG signal. We need to define the different finger gestures which can be decoded into characters. As we have only one Myo armband, we can't simulate standard 'qwerty' keyboard. In our first opinion, we would like to refer to one-handed keyboard layout and fingering to design the gesture. We give up because we get a bad test result with low accuracy. Finally, we decide to use T9 as our keyboard layout. T9 is a predictive text technology for mobile phones, standing for text on 9 keys. Each key represents three or four symbols and characters. We have defined nine sets of finger gestures and each set of actions corresponds to an area in one of the T9 keyboard area. The user needs to use three fingers when performing our defined actions, namely the index finger, the middle finger and the ring finger. Our keyboard fingering is described as the index finger controls left column of the keyboard, middle finger controls the middle column of the keyboard and index finger controls the right column of the keyboard. Each column is divided into three positions, so our designed gestures can be interpreted as tapping in the surface with index, middle and ring finger in three positions: extended, standardized and curled.

4   Gesture Sequence Model

4.1   Model Architecture

The input of the model is an 8-channel time-frequency map which contains an uncertain number of gestures. As shown in Fig. 2, it is an example of time-frequency map corresponding to the gesture sequence (**T4, T8, T2, T5**). There are four darker areas in each channel of the time-frequency map corresponding

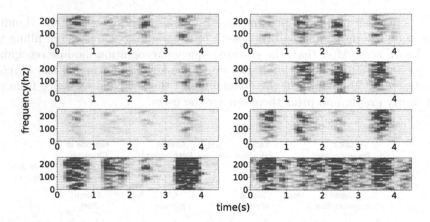

Fig. 2. An example of the eight-channel 3D time-frequency map transformed from the continuous EMG signal. The X axis represents the time and the Y axis represents the frequency.

to four gestures, and the image characteristics corresponding to each gesture are different. While, the lighter area in the time-frequency map represents the inactive segment of the signal which can be considered as the blank label. The purpose of our model is to detect all the gesture blocks in the image and then identify them, which is similar to the work of OCR (Optical Character Recognition) on the image. The model architecture is shown in Fig. 3. A hybrid network is used to combine the CNN layers and the LSTM layers. The CNN layers are used to extract the spatial features of the time-frequency map, and the LSTM layers are used to learn the temporally local information relationships of the feature sequence.

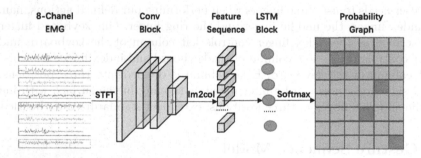

Fig. 3. Architecture of the hybrid network combining CNN network and LSTM network.

First, we design a lightweight CNN network with reference to Lenet5 [23] to extract spatial features on the original image. The network consists of two convolution layers and one pooling layer. We use the 5*5 kernel and the 3*3 kernel for convolving and use the 2*2 kernel for pooling. After the calculation of the above three layers, the original input is transformed into the feature-map1 of size 10*16*32. As the X axis of the input represents time, the gesture area in the time-frequency map should be identified horizontally and each column of the feature map can be treated as a feature vector. To enhance the ability of the feature vector expressing the feature map, we employ an im2col module using a sliding window of size 3*3 to store all the values into the feature vector. As a result, the size of the feature vector is increased by nine times. Finally, we flatten the feature map of eight channels to get the feature sequence of 32*1440 size, and we input this feature sequence into the LSTM units. The motivation of using LSTM is to take temporally local information over several frames of sequence into account, as the character sequence of the model output is related to its context. In addition, the LSTM network overcomes the problems of gradient disappearance and gradient explosion generated by the RNN network through adding input gates, output gates, and forgetting gates on each unit. Finally, we use the 'softmax' function to activate the sequence of the LSTM network output, and get a probability graph of size 32*10, where 32 is the width of the feature-map3, and 10 is the sum of the number of pre-defined gesture labels and blank label.

4.2   Model Training

Our hybrid neural network is trained as frame-level classifiers in gesture sequence recognition which requires a separate training target for every frame. However we can't align each frame of the sequence to each frame of the target sequence. The CTC (Connectionist Temporal Classification) algorithm [24] solves that problem through calculating the sequence which has the largest probability to transform to the target sequence. To calculate the network loss, we need to calculate $p(l \mid x)$ which is the probability of output as label l for the input x. $p(l \mid x)$ can be computed by the formula as follows:

$$p(l \mid x) = \sum_{B(\pi_{1:t}=l_{1:s})} \prod_{t=1}^{T} y_{\pi_t}^t \qquad (1)$$

This formula calculates the sum of probabilities of all the path that can be converted to the target sequence by the CTC algorithm mentioned above, where π_t represents the state of the path π at time t. The aim of the model training is to maximize the probabilities of all the correct sequence which can transform to the target sequence simultaneously, so we should maximize the value of $p(l \mid x)$ in our system.

4.3   Model Predict

The predict result is the path that has the highest score along the timeline from left of the probability graph to right of the probability graph. To improve the accuracy of finding the right path, we add the language model to the decoding algorithm:

$$C = argmax(\alpha * (P_{em}(C \mid X)) + \beta * (P_{lm}(C \mid X)))) \tag{2}$$

C is the final result sequence and X is the input sequence. We use the beam search algorithm which can only select several nodes with the highest score in each time-step.

Considering the calculation speed and the model size, we choose N-GRAM model as the language model which is based on statistical language model. The probability that the next letter appears is affected by the first few letters when the first few letters are given because people have greater possibility typing a word instead of a disordered character sequence. According to the 'chain rule', we calculate the language model score as follows:

$$p(w_1, w_2...w_n) = \prod_{i=2}^{n} p(w_i \mid w_{i-1}w_{i-2}) \tag{3}$$

The EMG model score is calculated as follows:

$$pr_{emg}(t) = \prod_{i=1}^{t} A(i, k_i) \tag{4}$$

A is the probability graph mentioned above. The process to find the best path can be concluded as calculating the score at each time step and adding the scores to the existing path. In the search process, we select the top n paths with the highest scores as potential paths after calculating the score for each time step and use them for the next time step calculation. In the last step we select the path with the highest score as the predict result.

5   Experimental Evaluation

5.1   Goals

In this section, we describe our experimental steps and results. Our goal is to train a model which converts the continuous EMG signal into the character sequence. Since the application in the paper is a virtual keyboard, the gestures we preset are finger-tapping. Besides, the keyboard ensures that people can input text in high speed which makes the interval between the gestures very short. As a result, the model we propose should either recognize the finger gesture or identify the gesture sequence when the user performs in high frequency.

5.2   Procedure

The participant performs preset gestures through watching the sequence content on the screen when collecting data. As they neither get the display of real-time input feedback, nor are familiar with the keyboard layout and fingering, the wrong button is often pressed resulting in the EMG signal not corresponding to the label. To collect data accurately, we draw a virtual keyboard on the plane, including 9 areas and convert the English characters sequence into tap area code in advance, so that the participant can tap the correct area accurately when performing the finger gestures. The start time and stop time of performing gestures is controlled by the operator. The whole data collection session is divided into collecting training data and collecting test data. The participant need to take a break to relax the muscles after they input 50 words.

We choose '1-billion-word-language-modeling-benchmark-r13output' as dataset and select 2000 words from the dataset for experiment. 1000 of these words are used to train the model and the rest are used to validate the model. The experiment is divided into two levels as the character-level and the word-level. The test data consists of 1000 words which is divided into 20 sessions equally. The participant needs to perform gesture at different speed levels: fast, mid and slow. The participant's tapping speed (WPM) is between 20–25 in the fast level, 15–20 in the mid level, and 5–15 in the slow level.

5.3   Experiment Result

We implement four models to evaluate the experimental results.

1. SVM: It combines the frequency and time domain features containing Zero-Crossing rate, Integral of Absolute Value, AR model coefficients, Mean Frequency into feature vectors and sends them to the SVM classifier. The purpose of adding this model is to compare the differences between features extracted by CNN and hand-crafted features.
2. CNN: The structure of this model is same as the CNN network architecture in the hybrid network mentioned in the section IV. Instead of converting the feature map into a feature sequence for transmission to the LSTM unit, the CNN model flattens the feature map and sends it to the fully connected layer for classification directly.
3. CNN+LSTM without CTC: The architecture of this model is same as the CNN+LSTM with CTC. We just flatten the LSTM output sequence. As a result, this model can only recognize the single gestures.
4. CNN+LSTM with CTC: It is an end-to-end model which uses the hybrid network architecture mentioned in the section IV and the CTC algorithm to identify the gesture sequence.

Accuracy Formula. We use the LER (label-error-rate) [24] to evaluate the experiment result which can be calculated as follows:

$$LER(p, q]) = \frac{ED(p, q)}{|q|} \tag{5}$$

$ED(p, q)$ is the minimum number of insertions, substitutions and deletions required to change the input sequence p into the target sequence q. And the final accuracy can be calculated as

$$Accuracy = 1 - LER \qquad (6)$$

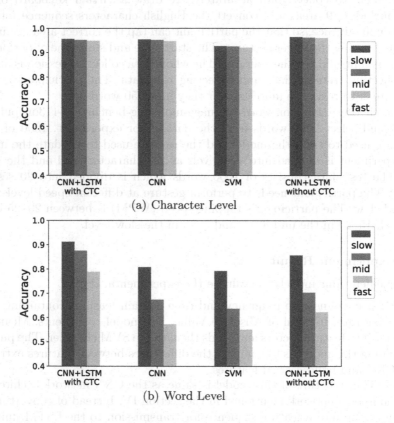

Fig. 4. The experiment result of four designed models in three typing speed levels.

Character-Level Recognition. As shown in Fig. 4(a), the four models have high accuracy over 90% to classify single character. While, the SVM model has a slightly lower accuracy than the other three models, which can be interpreted as the CNN networks learn better features to represent the original signal than the hand-crafted features. Besides, the accuracy of typing at three speed levels is stable and maintained at a high value, which shows that the four models we implement can accurately recognize nine finger-tapping gestures at different speed levels.

Word-Level Recognition. As shown in Fig. 4(b), the result of the 'CNN+LSTM with CTC' model is better than the other three model at three different speed levels. At the same time, the accuracy of the 'CNN+LSTM with CTC' model shows stable performance, while the accuracy of the other three models are greatly reduced as the tapping speed levels change from slow to fast.

5.4 Analysis

The chief difference between 'CNN+LSTM with CTC' model and the other three is that it is an end-to-end model and does not need to align the label of each frame, instead of extracting the active segmentations from the continuous EMG signal to recognize the single gesture. We think that extracting signal inaccurately is the main factor to cause this experiment result. We try to adjust the size of the sliding window because the energy spectrogram can represent the amount of energy at the current moment when the window becomes smaller. Figure 5 is the result of word-level experiment using different sliding window size in 'CNN+LSTM' model. It can be observed from the Fig. 5(a) that the relationship between the accuracy of model recognition and the size of the sliding window is not consistent at the three typing speed levels. As shown in Fig. 5(b), the number of gesture detected by the model decreases when the size of the sliding window enlarges. Therefore, it is difficult to select the suitable window size, the offset threshold value and the onset threshold value to extract the signals in different tapping speed levels accurately.

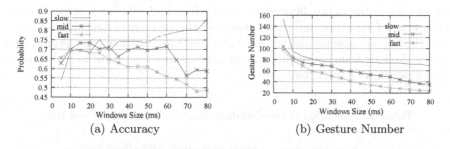

(a) Accuracy   (b) Gesture Number

Fig. 5. Effect of window size on gesture sequence recognition.

The 'CNN+LSTM with CTC' model makes convolution on the input and encodes it into the sequence that computes the state of each frame in the continuous EMG signal which ensures that it does not lose any information of the EMG signal. At the same time, we don't need to pre-align each frame of data when training the model which makes the dataset correct. The 'CNN+LSTM with CTC' model predicts the gesture sequence as a series of spikes, seperated by the inactive segmentation.

5.5   Add the Language Model to the Decoding Process

It is effective to add the language model to the decode process which can improve the accuracy of experimental results. The experiment result of adding the language model in the decoding process is shown in the Table 1. The 'CNN+LSTM with CTC' model have a slight improvement in the recognition accuracy comparing to decoding process without the language model.

Table 1. The accuracy of 'CNN+LSTM with CTC' using language model

	Slow	Mid	Fast
Use language model	0.934	0.889	0.813
Dont't use language model	0.912	0.876	0.789

5.6   Keycode to Character

We compared the speed of using the dictionary tree to find paths and traversing the English dictionary violently to find all the possible English strings that the key code sequence can transform. As shown in Table 2, the run time of searching decoding result in the dictionary tree is much lower than the violent search in the dictionary, while its load time and memory usage is much higher than the violent search method. It is mainly due to the cost of generating the dictionary tree for each string in the dictionary which brings high efficiency in the search at the expense of load time and space. However, the sacrifice of load time and space is worthwhile because the excessive delay in decoding the key code sequence will prevent the user from converting gesture actions into text input in real time and bring a poor user experience.

Table 2. Evaluation of two decoding algorithms running in real time

	Load time	Run time	Ram usage
Violent search	2.80 ms	43.98 ms	856 kb
Dictionary tree search	62.91 ms	0.61 ms	91264 kb

5.7   Real Time Performance

In the end, we integrated codes of all the modules and realize them into real-time application. The running time of each module is shown in Table 3. The sum of the running time of all the modules is compressed with 600 ms which will bring some delay when user performs the typing gesture to make text entry. While it

Table 3. Evaluation of time cost

	Time
Data preprocess	46 ms
CNN+LSTM	541 ms
CTC decoding	144 ms
T9 keycode transformation	6 ms

can still meet the real-time demand as the human and computer interfaces. The participant performs the typing experience assessment in the real-time virtual keyboard system. The evaluation process is divided into two sessions as adapting to the keyboard and the typing speed test. In the first session, participants need to be proficient in nine typing gestures and control the speed of performing gestures to reduce the error of the gesture sequence recognition. The second session is divided into eight blocks. The participant types a paragraph of text consisting of about 300 words in each block. The participant needs to retype the text if the gesture sequence model fails to predict the true key code sequence. The participant's typing speed tends to stabilize and can reach 15.7 WPM.

6   Conclusion

This paper presents the EMG keyboard, a new technology of HCI tools to entry text into computer based on the EMG signal. We choose the T9 as the keyboard layout and design nine finger-tapping gestures which are similar to tapping the T9 keyboard on the mobile phone. We evaluate four models through the character-level experiment and the word-level experiment and finally choose the 'CNN+LSTM with CTC' model to recognize the gesture sequence as it has the stable high accuracy over 80% in three typing speed levels. Considering of the user's experience and the compute cost, we use the dictionary tree search algorithm to transform T9 keyboard code into English character. We implement the all modules into the real-time application and the participant's typing speed can reach 15.7WPM (word-per-minute). We also plan to change the current keyboard layout from T9 to 'qwerty' to further improve the typing speed.

In general, the EMG keyboard has the same or even higher typing speed and accuracy compared to other text-entry technologies in the previous work. However, the EMG keyboard can't replace the physical keyboard with 'qwerty' layout as the WPM of the physical keyboard can reach as high as 60. While, the EMG keyboard can be regarded as a portable and friendly keyboard due to the benefit of mobile, wearable and eyes-free text entry. In the future, we will change our virtual keyboard layout from T9 to 'qwerty' which will improve the typing speed ant bring a more realistic experience like typing on the physical keyboard to users.

Acknowledgment. This work has been supported by National Natural Science Foundation of China (61772060, 61976012, 61602024), Qianjiang Postdoctoral Foundation (2020-Y4-A-001), and CERNET Innovation Project (NGII20170315).

References

1. Barry, D.T., Gordon, K.E., Hinton, G.G.: Acoustic and surface EMG diagnosis of pediatric muscle disease. Muscle Nerve Off. J. Am. Assoc. Electrodiagn. Med. **13**(4), 286–290 (1990)
2. Hernandez Arieta, A., Katoh, R., Yokoi, H., Wenwei, Y.: Development of a multi-DOF electromyography prosthetic system using the adaptive joint mechanism. Appl. Bionics Biomech. **3**(2), 101–111 (2006)
3. Saponas, T.S., Tan, D.S., Morris, D., Balakrishnan, R., Turner, J., Landay, J.A.: Enabling always-available input with muscle-computer interfaces. In: Proceedings of the 22nd Annual ACM Symposium on User Interface Software and Technology, pp. 167–176. ACM (2009)
4. Saponas, T.S., Tan, D.S., Morris, D., Balakrishnan, R.: Demonstrating the feasibility of using forearm electromyography for muscle-computer interfaces. In: Proceedings of the SIGCHI Conference on Human Factors in Computing Systems, pp. 515–524. ACM (2008)
5. Li, Y., Chen, X., Tian, J., Zhang, X., Wang, K., Yang, J.: Automatic recognition of sign language subwords based on portable accelerometer and EMG sensors. In: International Conference on Multimodal Interfaces and the Workshop on Machine Learning for Multimodal Interaction, p. 17. ACM (2010)
6. Wu, J., Tian, Z., Sun, L., Estevez, L., Jafari, R.: Real-time American sign language recognition using wrist-worn motion and surface EMG sensors. In: 2015 IEEE 12th International Conference on Wearable and Implantable Body Sensor Networks (BSN), pp. 1–6. IEEE (2015)
7. Hudgins, B., Parker, P., Scott, R.N.: A new strategy for multifunction myoelectric control. IEEE Trans. Biomed. Eng. **40**(1), 82–94 (1993)
8. Farry, K.A., Walker, I.D., Baraniuk, R.G.: Myoelectric teleoperation of a complex robotic hand. IEEE Trans. Robot. Autom. **12**(5), 775–788 (1996)
9. Englehart, K., Hudgins, B., Stevenson, M., Parker, P.A.: Classification of transient myoelectric signals using a dynamic feedforward neural network. In: Proceedings World Congress Neural Networks (1995)
10. Englehart, K., Hudgins, B., Parker, P.A., et al.: A wavelet-based continuous classification scheme for multifunction myoelectric control. IEEE Trans. Biomed. Eng. **48**(3), 302–311 (2001)
11. Phinyomark, A., Phukpattaranont, P., Limsakul, C.: Feature reduction and selection for EMG signal classification. Expert Syst. Appl. **39**(8), 7420–7431 (2012)
12. Kunapipat, M., Phukpattaranont, P., Neranon, P., Thongpull, K.: Sensor-assisted emg data recording system. In: 2018 15th International Conference on Electrical Engineering/Electronics, Computer, Telecommunications and Information Technology (ECTI-CON), pp. 772–775. IEEE (2018)
13. Kosmidou, V.E., Hadjileontiadis, L.J., Panas, S.M.: Evaluation of surface EMG features for the recognition of American sign language gestures. In: 2006 International Conference of the IEEE Engineering in Medicine and Biology Society, pp. 6197–6200. IEEE (2006)

14. Park, K.-H., Lee, S.-W.: Movement intention decoding based on deep learning for multiuser myoelectric interfaces. In: 2016 4th International Winter Conference on Brain-Computer Interface (BCI), pp. 1–2. IEEE (2016)
15. Becker, V., Oldrati, P., Barrios, L., Sörös, G.: Touchsense: classifying finger touches and measuring their force with an electromyography armband. In: Proceedings of the 2018 ACM International Symposium on Wearable Computers, pp. 1–8. ACM (2018)
16. Atzori, M., Cognolato, M., Müller, H.: Deep learning with convolutional neural networks applied to electromyography data: a resource for the classification of movements for prosthetic hands. Front. Neurorobotics 10, 9 (2016)
17. He, Y., Fukuda, O., Bu, N., Okumura, H., Yamaguchi, N.: Surface EMG pattern recognition using long short-term memory combined with multilayer perceptron. In: 2018 40th Annual International Conference of the IEEE Engineering in Medicine and Biology Society (EMBC), pp. 5636–5639. IEEE (2018)
18. Yiqin, L., Chun, Yu., Yi, X., Shi, Y., Zhao, S.: BlindType: eyes-free text entry on handheld touchpad by leveraging thumb's muscle memory. Proc. ACM Interact. Mob. Wearable Ubiquit. Technol. 1(2), 18 (2017)
19. Xu, Z. et al.: TipText: eyes-free text entry on a fingertip keyboard (2019)
20. Yi, X., Yu, C., Xu, W., Bi, X., Shi, Y.: COMPASS: rotational keyboard on non-touch smartwatches. In: Proceedings of the 2017 CHI Conference on Human Factors in Computing Systems, pp. 705–715. ACM (2017)
21. Dhillon, H.S., Singla, R., Rekhi, N.S., Jha, R.: EOG and EMG based virtual keyboard: a brain-computer interface. In: 2009 2nd IEEE International Conference on Computer Science and Information Technology, pp. 259–262. IEEE (2009)
22. Testa, A., Gallo, D., Langella, R.: On the processing of harmonics and interharmonics: using Hanning window in standard framework. IEEE Trans. Power Deliv. 19(1), 28–34 (2004)
23. LeCun, Y., Bottou, L., Bengio, Y., Haffner, P., et al.: Gradient-based learning applied to document recognition. Proc. IEEE 86(11), 2278–2324 (1998)
24. Graves, A., Fernández, S., Gomez, F., Schmidhuber, J.: Connectionist temporal classification: labelling unsegmented sequence data with recurrent neural networks. In: Proceedings of the 23rd International Conference on Machine Learning, pp. 369–376. ACM (2006)

Accelerating Pattern Matching on Intel Xeon Phi Processors

Victoria Sanz[1,2]([✉]), Adrián Pousa[1], Marcelo Naiouf[1],
and Armando De Giusti[1,3]

[1] III-LIDI, School of Computer Sciences, National University of La Plata,
La Plata, Argentina
{vsanz,apousa,mnaiouf,degiusti}@lidi.info.unlp.edu.ar
[2] CIC, Buenos Aires, Argentina
[3] CONICET, Buenos Aires, Argentina

Abstract. Pattern matching algorithms are used in several areas such as network security, bioinformatics and text mining. In order to provide real-time response for large inputs, high-performance systems should be used. However, this requires adapting the algorithm to the underlying architecture. Intel Xeon Phi processors have attracted attention in recent years because they offer massive parallelism, good programmability and portability. In this paper, we present a pattern matching algorithm that exploits the full computational power of Intel Xeon Phi processors by using both SIMD and thread parallelism. We evaluate our algorithm on a Xeon Phi 7230 Knights Landing processor and measure its performance as the data size and the number of threads increase. The results reveal that both parallelism methods provide performance gains. Also they indicate that our algorithm is up to 63x faster than its serial counterpart and behaves well as the workload is increased.

Keywords: Pattern matching · Intel Xeon Phi Knights Landing processor · SIMD instructions · Thread-level parallelism · Aho-Corasick

1 Introduction

Pattern matching algorithms locate some or all occurrences of a finite number of patterns (pattern set or dictionary) in a text (data set). These algorithms are key components of DNA analysis applications [1], antivirus [2], intrusion detection systems [3,4], among others. In this context, the Aho-Corasick (AC) algorithm [5] is widely used because it efficiently processes the text in a single pass.

Nowadays, the amount of data that need to be processed grows very rapidly. This has led several authors to investigate the acceleration of AC on emerging parallel architectures. In particular, researchers have proposed different approaches to parallelize AC on shared-memory architectures, distributed-memory architectures (clusters), GPUs, multiple GPUs and CPU-GPU heterogeneous systems [6–14].

© Springer Nature Switzerland AG 2020
M. Qiu (Ed.): ICA3PP 2020, LNCS 12452, pp. 262–274, 2020.
https://doi.org/10.1007/978-3-030-60245-1_18

In recent years, Intel Xeon Phi systems have attracted attention of researchers because of their massive parallelism. These systems support multiple programming languages and tools for parallel programming, thus they provide better programmability and portability than a system using GPUs. There are two generations of Xeon Phi: Knights Corner (KNC) and Knights Landing (KNL) [15,16]. The former has limitations similar to those of GPUs: it is a PCIe-connected coprocessor with limited memory. The latter is also available as a standalone processor, therefore it does not have the limitations of the previous model. Furthermore, it is binary compatible with prior Intel processors.

In general, a Xeon Phi is composed of multiple cores. Each core has 1 or 2 vector processing units (VPUs), depending on the model. To take full advantage of its computational power, applications must exploit thread-level and SIMD parallelism. This is a challenge for developers since the code must be changed in order to launch multiple parallel tasks and expose vectorization opportunities.

So far, little work has been done to accelerate Aho-Corasick on Xeon Phi. In [17] the authors present a parallel AC algorithm and test it on a Xeon Phi KNC coprocessor. Briefly, the algorithm consists of dividing the text among threads, then each thread processes its segment in a vectorized way. Similarly, in [18] the authors propose an AC algorithm for Xeon Phi coprocessors, which uses a strategy that increases cache locality during the pattern matching process. The strategy is based on partitioning the set of patterns into smaller subsets and executing several independent instances of the matching procedure on the entire text (i.e., one instance for each subset of patterns). However, this proposal does not take advantage of the VPUs of the coprocessor.

In this paper, we present an AC algorithm that exploits the full computational power of Intel Xeon Phi processors by using both SIMD and thread parallelism. We evaluate our algorithm on a Xeon Phi 7230 KNL processor and show the performance gain when using SIMD instructions and threads respectively. Furthermore, we measure the performance of our algorithm as the data size and the number of threads increase, and study its behaviour. The results reveal that both parallelism methods improve performance. Also they indicate that our algorithm is up to 63x faster than its serial counterpart and behaves well as the workload is increased.

Our work differs from previous studies in two ways. First, our proposed algorithm is based on the Parallel Failureless Aho-Corasick (PFAC) algorithm [7], which efficiently exploits the parallelism of AC. PFAC is suitable for many-core architectures and it was first implemented for GPUs and multicore systems; in the last case vectorization techniques were not applied. Second, in contrast to previous works [17,18], we evaluate the performance and scalability of our proposed algorithm on a Xeon Phi KNL processor.

The rest of the paper is organized as follows. Section 2 provides some background information on pattern matching algorithms and Xeon Phi processors. Section 3 describes our parallel algorithm for pattern matching on Xeon Phi processors. Section 4 shows our experimental results. Finally, Sect. 5 presents the main conclusions and future research.

2   Background

This section describes the AC and PFAC algorithms, the Intel Xeon Phi processor and how to program it.

2.1   The Aho-Corasick Algorithm

The Aho-Corasick (AC) algorithm [5] has been widely used since it is able to locate all occurrences of user-specified patterns in a single pass of the text. The algorithm consists of two steps: the first is to construct a finite state pattern matching machine; the second is to process the text using the state machine constructed in the previous step. The pattern matching machine has valid and failure transitions. The former are used to detect all user-specified patterns. The latter are used to backtrack the state machine, specifically to the state that represents the longest proper suffix, in order to recognize patterns starting at any location of the text. Certain states are designated as "output states" which indicate that a set of patterns has been found. The AC machine works as follows: given a current state and an input character, it tries to follow a valid transition; if such a transition does not exist, it jumps to the state pointed by the failure transition and processes the same character until it causes a valid transition. The machine emits the corresponding patterns whenever an output state is found. Figure 1 shows the AC state machine for the pattern set {he, she, his, hers}. Solid lines represent valid transitions and dotted lines represent failure transitions.

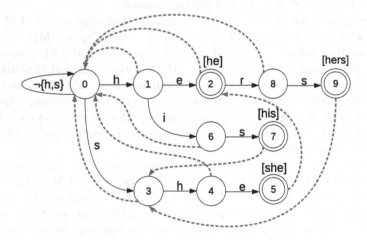

Fig. 1. AC state machine for the pattern set {he, she, his, hers}

2.2   The Parallel Failureless Aho-Corasick Algorithm

The Parallel Failureless Aho-Corasick (PFAC) algorithm [7] efficiently exploits the parallelism of AC and therefore is suitable for many-core architectures. PFAC assigns each position of the text to a particular thread. For each assigned position *start* (task), the thread is responsible for identifying the pattern beginning at that position. For that, the thread reads the text and traverses the state machine, starting from the *initial state*, and terminates immediately when it cannot follow a valid transition; at that point, the thread registers the longest match found. Note that all threads use the same state machine. Since PFAC does not use failure transitions, they can be removed from the state machine. Figures 2 and 3 give an example of the Failureless-AC state machine and the parallelization strategy of PFAC, respectively. Algorithm 1 shows the PFAC code executed by each thread for each assigned task.

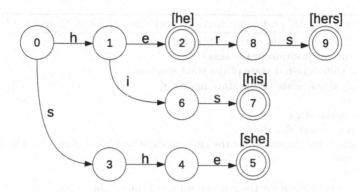

Fig. 2. Failureless-AC state machine for the pattern set {he, she, his, hers}

2.3 Intel Xeon Phi Knights Landing Processor

The Intel Xeon Phi KNL processor [16,19] is composed of many tiles that are interconnected by a cache-coherent, 2D mesh interconnect. Each tile consists of two cores, two vector-processing units (VPUs) per core, and a 1MB L2 cache shared between the two cores. Each core supports 4 hardware threads. Figure 4 illustrates this architecture.

Regarding the memory hierarchy, the KNL processor has two types of memory: (1) MCDRAM, a high-bandwidth memory integrated on-package, and (2) DDR, a high-capacity memory that is external to the package. The MCDRAM can be used as a last-level cache for the DDR (cache mode), as addressable memory (flat mode), or as a combination of the last two modes (hybrid mode) - i.e. using a portion of the MCDRAM as cache and the rest as standard memory. When using the flat or hybrid mode, the access to MCDRAM as memory requires programmer intervention.

Furthermore, the KNL processor can be configured in different modes (cluster modes) in order to optimize on-chip memory traffic.

The KNL processor supports all legacy x86 instructions, therefore it is binary-compatible with prior Intel processors, and incorporates AVX-512 instructions.

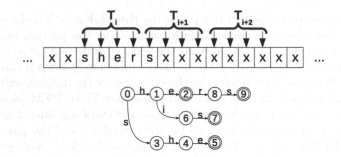

Fig. 3. Parallelization strategy of PFAC

Algorithm 1. PFAC code, executed by each thread for each assigned task

---
{*start* :: initial position of the task in the text}
{*pos* :: current position in the text}
{*initial state* :: initial state of the state machine}
{*state* :: current state of the state machine}
1: pos = start
2: state = initial state
3: **while** pos < text size **do**
4:    **if** there is no transition for the current state and input character **then**
5:       break
6:    **end if**
7:    state = next state for the current state and input character
8:    **if** state is an output state **then**
9:       store the pattern found at the position start in output[start]
10:    **end if**
11:    pos = pos + 1
12: **end while**

The efficient use of cores and VPUs is critical to obtain high performance. To that end, the programmer must divide the work into parallel tasks or threads, and organize their code to expose vectorization opportunities. Finally, vectorization can be achieved manually, by writing assembly code or using vector intrinsics, or automatically, relying on compiler optimizations.

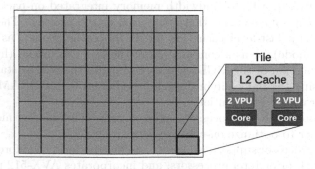

Fig. 4. Intel Xeon Phi KNL processor

3   Pattern Matching on Xeon Phi Processors

This section describes our strategy for parallelizing pattern matching on Xeon Phi processors and some implementation details.

3.1   Parallelization Strategy

Figure 5 depicts the parallelization strategy of PFAC_VEC, our pattern matching algorithm for Xeon Phi based on PFAC. Recall that the strategy followed by PFAC divides the input text into as many tasks as the text size; each task involves identifying the pattern starting at a certain text position; all tasks are equally distributed among threads and each thread follows a sequential control flow to process its tasks. In contrast, the strategy of PFAC_VEC splits the input text into blocks of tasks; each block is composed of a fixed number of consecutive text positions or tasks (this number is referred to as *block size*); all blocks are equally distributed among threads; then each thread processes each assigned block in a vectorized way, i.e. all its tasks are simultaneously solved.

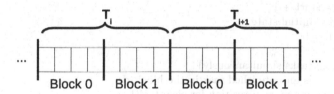

Fig. 5. Parallelization strategy of PFAC_VEC

The distribution of blocks among threads is done through the OpenMP *for* work-sharing construct. On the other hand, vectorization will be performed automatically by the compiler. However, the code executed by each thread to process a block of tasks must meet some requirements to be automatically vectorized [19]. Thus, it requires programmer intervention. The next section presents the implementation details of that code.

3.2   Implementation Details

Algorithm 2 shows the code executed by each thread to process a block of tasks in a vectorized way, taking advantage of the VPUs of the core. Note that all tasks of a block (*block size* in total) are solved simultaneously by applying the same operation to multiple data items. Remember that each task involves identifying the pattern that starts at a certain block position.

The thread first creates vectors of length *block size* to store the current position in the text, the current state of the state machine and the pattern found so far, for each task. The initialization of these vectors is vectorizable (Algorithm 2, Lines 1–5).

Then, the thread processes the tasks until all are finished (Algorithm 2, Lines 6–15). In each iteration of the outer loop, a state transition is carried out for each task. Specifically, the operations of reading (Algorithm 2, Line 8), determining the next state (Algorithm 2, Line 9), verifying if the reached state is an output state (Algorithm 2, Lines 10–12), and incrementing the current position to point to the next character (Algorithm 2, Line 13) are performed in a vectorized way.

Finally, the patterns found in this block are recorded in the output vector in a vectorized manner (Algorithm 2, Lines 16–18).

Algorithm 2. PFAC_VEC code, executed by each thread for each assigned block

---

{*start* :: initial position of the block in the text}
{*initial state* :: initial state of the state machine}
{*pos* :: vector containing the current position in the text for each task}
{*state* :: vector containing the current state of the state machine for each task}
{*result* :: vector containing the longest match found for each task}
{For each task, set pos, state and result to their initial values}
1: **for** i = 0 **to** block size - 1 **do** {Vectorizable loop}
2:    pos[i] = start + i
3:    state[i] = initial state
4:    result[i] = 0
5: **end for**
   {Process all tasks simultaneously}
6: **while** there is an unfinished task in the block **do**
       {Perform one more transition for each task}
7:    **for** i = 0 **to** block size - 1 **do** {Vectorizable loop}
8:       inputchar[i] = text[pos[i]]
9:       state[i] = next state for state[i] and inputchar[i]
10:       **if** state[i] is an output state **then**
11:          store the pattern found at the position "start+i" in result[i]
12:       **end if**
13:       pos[i] = pos[i] + 1
14:    **end for**
15: **end while**
16: **for** i = 0 **to** block size - 1 **do** {Vectorizable loop}
17:    output[start + i] = result[i]
18: **end for**

In order to guarantee the correct execution of the algorithm, lines 8 and 9 must not fail or have any effect when the current position in the text of task i exceeds the size of the text (pos[i] $\geq$ text size) and when there is not a valid transition for the current state and character of task i, respectively.

To solve the first problem (Line 8), an additional number of special characters was appended to the input text. This number is equal to the length of the longest pattern in the dictionary and represents the maximum amount of characters that can be processed when solving a task.

To solve the second problem (Line 9), a *dead state* was added to the state machine, i.e. a nonaccepting state that goes to itself on every possible input symbol. Also, additional transitions were added to this state from each other state q, on all input symbols for which q has no other transition. Figure 6 shows the Failureless-AC state machine for the pattern set {he, she, his, hers}, which includes the *dead state*.

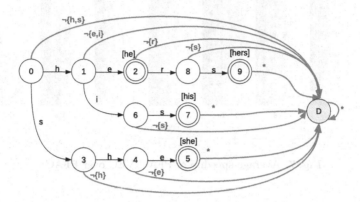

Fig. 6. Modified Failureless-AC state machine for the pattern set {he, she, his, hers}

Now, it can be said that the exit condition of the while loop in Line 6 involves determining if there is a task that has not reached the *dead state* yet. This operation is done by iterating over the vector of current states (the loop is vectorizable).

The characteristics of the proposed algorithm enable the Intel compiler to vectorize the code automatically. In particular, it reports an estimated potential speedup of 10.5x, 10x and 21.3x for the loops in Lines 1, 7, and 16 respectively. Furthermore, it shows a potential speedup of 15.3x for the exit condition of the while loop in Line 6.

4   Experimental Results

Our experimental platform is a machine with an Intel Xeon Phi 7230 (KNL) processor and 128 GB DDR4 RAM. This processor has 64 1.30 GHz cores and 16 GB MCDRAM. Each core supports 4 hardware threads, thus the processor supports 256 threads in total. The processor is configured in flat mode. We use the numactl command to place all data in MCDRAM.

Test scenarios were generated by combining four English texts of different sizes with four English dictionaries with different number of patterns. All the texts were extracted from the British National Corpus [20]: text 1 is a 4-million-word sample (21 MB); text 2 is a 50-million-word sample (268 MB); text 3 is a 100-million-word sample (544 MB); text 4 is a 200-million-word sample

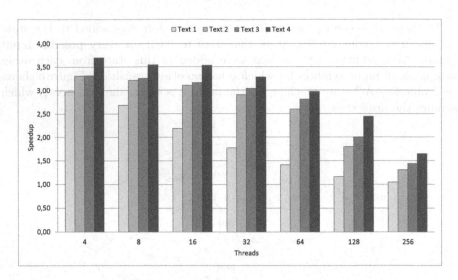

Fig. 7. Average speedup of PFAC_VEC over PFAC

(1090 MB). The dictionaries include frequently used words: dictionary 1 with 3000 words; dictionary 2 with 100000 words; dictionary 3 with 178690 words; dictionary 4 with 263533 words.

Our experiments focus on the matching step since it is the most significant part of pattern matching algorithms. For each test scenario, we ran each implementation 100 times and averaged the execution time.

First, we ran the single-threaded non-vectorized Failureless Aho-Corasick algorithm (SFAC), provided by Lin et al. [7], and the single-threaded vectorized version (SFAC_VEC), presented in this paper. In the last case, different values of *block size* (64, 128, 256) were used. For all test scenarios, the value of *block size* that produces the best result is 128. Using that value, SFAC_VEC is up to 3.6x faster than SFAC, and provides an average acceleration of 3.5 for each text. These results demonstrate that our vectorization technique provides performance gains.

Next, we ran the multi-threaded non-vectorized code (PFAC), provided by Lin et al. [7], and the multi-threaded vectorized code (PFAC_VEC), presented here, for different number of threads (4, 8, 16, 32, 64, 128, 256) and affinity settings [21] (none or default, scatter, compact, balanced). The value of *block size* used by PFAC_VEC is 128.

Regarding the affinity settings, the compact affinity gives the worst results for all tests. The other affinities perform similarly for different combinations of test scenarios and number of threads, except for the largest text considered and few threads, for which the none affinity provides the best execution times. Therefore, from now on, the none affinity will be used.

Afterwards, we evaluated the performance (Speedup[1]) of PFAC_VEC over PFAC and SFAC_VEC, respectively. For each text and number of threads, the average speedup is shown. This is because the speedup does not vary significantly with the dictionary.

Figure 7 illustrates the average speedup of PFAC_VEC over PFAC, for different texts and number of threads. It can be seen that the speedup ranges between 1.05 and 3.70. Higher speedup values are obtained when using fewer threads, and as this parameter increases the speedup decreases. However, for a fixed number of threads, the speedup tends to increase with the size of the text. These results reveal that our vectorization technique is also effective to reduce parallel execution times.

Figure 8 shows the average speedup of PFAC_VEC over SFAC_VEC, for different texts and number of threads. Note that for a fixed text, the speedup first increases with the number of threads and then decreases. However, for a fixed number of threads the speedup always increases with the size of the text. From these results we conclude that PFAC_VEC behaves well as the workload is increased.

Figure 9 shows the best performance (average speedup) of PFAC_VEC over SFAC_VEC for each text, and the number of threads that provides this value. As it can be observed, the number of threads that provides the best performance depends on the text. PFAC_VEC achieves an average speedup of 8.53 for text 1, 38.49 for text 2, 49.03 for text 3 and 62.88 for text 4, using 32, 64, 128 and 128 threads respectively.

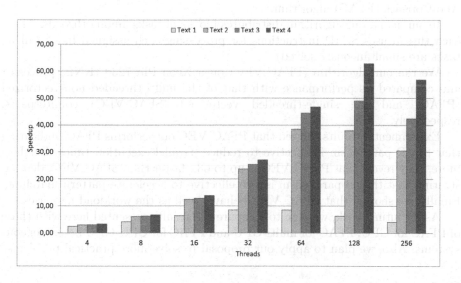

Fig. 8. Average speedup of PFAC_VEC over SFAC_VEC

---

[1] The Speedup of B over A is defined as $\frac{T_A}{T_B}$, where T_A is the execution time of Algorithm A and T_B is the execution time of Algorithm B.

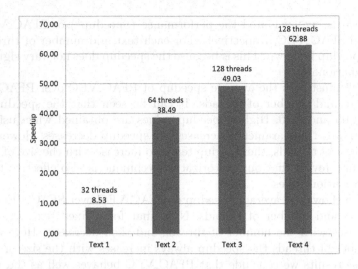

Fig. 9. Best performance (speedup) of PFAC_VEC for each text

5   Conclusions and Future Work

In this paper we presented a novel pattern matching algorithm that efficiently exploits the full computational power of Intel Xeon Phi processors by using both SIMD and thread parallelism. Our proposal is based on the Parallel Failureless Aho-Corasick (PFAC) algorithm.

In summary, our algorithm distributes blocks of tasks among threads. Then, each thread uses SIMD instructions for processing each assigned block (all its tasks are simultaneously solved).

We ran our algorithm (PFAC_VEC) on a Xeon Phi 7230 (KNL) processor and compared its performance with that of the multi-threaded non-vectorized (PFAC) and the single-threaded vectorized (SFAC_VEC) counterparts, respectively.

Experimental results showed that PFAC_VEC outperforms PFAC, indicating that SIMD parallelism is effective to reduce parallel execution times. Furthermore, they reveal that PFAC_VEC is up to 63x faster than SFAC_VEC, demonstrating that thread parallelism is also effective to accelerate pattern matching. Finally, we showed that PFAC_VEC behaves well as the workload increases.

As for future work, we plan to compare the results presented here with those of PFAC for GPU, PFAC for multi-GPU and PFAC for CPU-GPU heterogeneous systems. Also, we plan to apply our proposal to solve more practical problems.

References

1. Tumeo, A., Villa, O.: Accelerating DNA analysis applications on GPU clusters. In: IEEE 8th Symposium on Application Specific Processors (SASP), pp. 71–76. IEEE Computer Society, Washington D.C. (2010)

2. Clamav. http://www.clamav.net
3. Norton, M.: Optimizing pattern matching for intrusion detection. White Paper. Sourcefire Inc. https://www.snort.org/documents/optimization-of-pattern-matches-for-ids
4. Tumeo, A., et al.: Efficient pattern matching on GPUs for intrusion detection systems. In: Proceedings of the 7th ACM International Conference on Computing Frontiers, pp. 87–88. ACM, New York (2010)
5. Aho, A.V., Corasick, M.J.: Efficient string matching: an aid to bibliographic search. Commun. ACM **18**(6), 333–340 (1975)
6. Tumeo, A., et al.: Aho-Corasick string matching on shared and distributed-memory parallel architectures. IEEE Trans. Parallel Distrib. Syst. **23**(3), 436–443 (2012)
7. Lin, C.H., et al.: Accelerating pattern matching using a novel parallel algorithm on GPUs. IEEE Trans. Comput. **62**(10), 1906–1916 (2013)
8. Arudchutha, S., et al.: String matching with multicore CPUs: performing better with the Aho-Corasick algorithm. In: Proceedings of the IEEE 8th International Conference on Industrial and Information Systems, pp. 231–236. IEEE Computer Society, Washington D.C. (2013)
9. Herath, D., et al.: Accelerating string matching for bio-computing applications on multi-core CPUs. In: Proceedings of the IEEE 7th International Conference on Industrial and Information Systems (ICIIS), pp. 1–6. IEEE Computer Society, Washington D.C. (2012)
10. Lin, C.H., et al.: A novel hierarchical parallelism for accelerating NIDS using GPUs. In: Proceedings of the 2018 IEEE International Conference on Applied System Invention (ICASI), pp. 578–581. IEEE (2018)
11. Soroushnia, S., et al.: Heterogeneous parallelization of Aho-Corasick algorithm. In: Proceedings of the IEEE 7th International Conference on Industrial and Information Systems (ICIIS), pp. 1–6. IEEE Computer Society, Washington D.C. (2012)
12. Lee, C.L., et al.: A hybrid CPU/GPU pattern-matching algorithm for deep packet inspection. PLoS ONE **10**(10), 1–22 (2015)
13. Sanz, V., Pousa, A., Naiouf, M., De Giusti, A.: Accelerating pattern matching with CPU-GPU collaborative computing. In: Vaidya, J., Li, J. (eds.) ICA3PP 2018. LNCS, vol. 11334, pp. 310–322. Springer, Cham (2018). https://doi.org/10.1007/978-3-030-05051-1_22
14. Sanz, V., Pousa, A., Naiouf, M., De Giusti, A.: Efficient pattern matching on CPU-GPU heterogeneous systems. In: Wen, S., Zomaya, A., Yang, L.T. (eds.) ICA3PP 2019. LNCS, vol. 11944, pp. 391–403. Springer, Cham (2020). https://doi.org/10.1007/978-3-030-38991-8_26
15. Chrysos, G.: Intel Xeon Phi X100 family coprocessor - the architecture. The first Intel Many Integrated Core (Intel MIC) architecture product. White Paper. Intel Corporation (2012). https://software.intel.com/en-us/articles/intel-xeon-phi-coprocessor-codename-knights-corner
16. Sodani, A., et al.: Knights landing: second-generation Intel Xeon Phi product. IEEE Micro **36**(2), 34–46 (2016)
17. Memeti, S., Pllana, S.: Accelerating DNA sequence analysis using Intel(R) Xeon Phi(TM). In: Proceedings of the 2015 IEEE Trustcom/BigDataSE/ISPA, pp. 222–227. IEEE (2015)
18. Tran, N., et al.: Cache locality-centric parallel string matching on many-core accelerator chips. Sci. Program. **2015**(1), 937694:1–937694:20 (2015)
19. Jeffers, J., et al.: Intel Xeon Phi Processor High Performance Programming: Knights Landing Edition, 2nd edn. Morgan Kaufmann Publishers Inc., San Francisco (2016)

20. The British National Corpus, version 3 (BNC XML edition). Distributed by Bodleian Libraries, University of Oxford, on behalf of the BNC Consortium (2007). http://www.natcorp.ox.ac.uk/

21. Vladimirov, A., et al.: Parallel Programming and Optimization with Intel Xeon Phi Coprocessors. Handbook on the Development and Optimization of Parallel Applications for Intel Xeon Processors and Intel Xeon Phi Coprocessors. Colfax International, Sunnyvale (2015)

Redistributing and Optimizing High-Resolution Ocean Model POP2 to Million Sunway Cores

Yunhui Zeng[1,2,3], Li Wang[1,2,3], Jie Zhang[1,2,3]([✉]), Guanghui Zhu[1,2,3],
Yuan Zhuang[1,2,3], and Qiang Guo[1,2,3]

[1] Qilu University of Technology (Shandong Academy of Sciences), Jinan 250014, China
{zengyh,wangl,zhangjie,zhugh,zhuangy,guoqiang}@sdas.org
[2] Shandong Computer Science Center (National Supercomputing
Center in Jinan), Jinan 250014, China
[3] Shandong Provincial Key Laboratory of Computer Networks, Jinan 250014, China

Abstract. The high-resolution CESM is widely applied in climate simulations, while a simulation speed of 5.0 simulated years per day has traditionally been considered the minimum necessary for long-term simulations. When Sunway TaihuLight supercomputer was open, the atmosphere model CAM5, one of CESM's major component models, was already ported. But the ocean model POP2, another major component model, has not been fully done yet as known. In this paper, the high-resolution POP2 coupled in CESM is fully ported to Shenwei many-core infrastructure. Although many methods accumulated, there are still some new challenges when it comes to POP2. If just simply translated, its performance may not be well to support long-term simulations. In order to achieve high performance, three stages are adopted. Firstly, the original POP2 is ported with athread interface and fine-grained optimized to Shenwei many-core. Secondly, the grid decomposition is redesigned, and a new slave-core partition method is proposed to solve the problem that some two-dimension array related kernels after athreaded may be insignificant or even false speedup under large scale processes. Then many two-dimension array related kernels in POP2 are effectively redistributed to slave-cores. Lastly, some case-oriented skills are intensively utilized as necessary supplements. Some experiments show that the simulation speed of the finally optimized POP2 in high-resolution CESM G-compset is over 5.5 simulated years per day under 18,300 processes with 1,189,500 cores, compared with 1.43 simulated years per day of the original version, and its speed-up ratio is still over 3.8.

Keywords: Ocean model · Many-core · Scalability · Slave-core partition

1 Introduction and Motivation

High-resolution global climate models have become increasingly important in recent years as a means for understanding climate variability and projecting future climate change. While a simulation rate of 5.0 simulated years per day (SYPD) has traditionally been considered the minimum necessary to carry out long-term climate simulations [1].

© Springer Nature Switzerland AG 2020
M. Qiu (Ed.): ICA3PP 2020, LNCS 12452, pp. 275–289, 2020.
https://doi.org/10.1007/978-3-030-60245-1_19

With the development of supercomputer, simulating global climate variability in the past serial centuries with the high-resolution model can be implemented. The Sunway TaihuLight system is the China's first top one system completely based on homegrown infrastructure with a peak performance greater than 100PFlops many-core processors. It is based on the SW26010 processor shown in Fig. 1. The processor chip is composed of 4 core groups, each of which includes a Management Processing Element and 64 Computing Processing Elements.

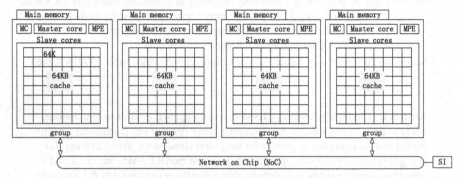

Fig. 1. A schematic illustration of the architecture of the Sunway SW26010 CPU.

The CESM model, whose development is centered at the National Center for Atmospheric Research (NCAR), is one of the most widely used global climate models [2]. It is a fully-coupled climate system model with components such as atmosphere, ocean, sea-ice and land. In order to achieve the simulation rate of 5.0 SYPD, so it is needed to effectively port CESM to national Shenwei many-core infrastructure. A high-resolution CESM test is simulated 20 model-days under 44,004 Sunway MPI processes, with atmosphere component CAM5 (Community Atmosphere Model) 29,000 processes and ocean component POP2 (Parallel Ocean Program) 15,004 processes. The run time is listed in Table 1, we can see the CAM5 and the POP2 are the two most time-consuming component models.

Table 1. Run time of a high-resolution CESM B-compset test on Sunway master-core.

Component model	Time cost (seconds)	Simulation speed (SYPD)
Atmosphere	3151.971	1.49
Ocean	4676.214	1.01
Sea-ice	759.367	6.20
Land	241.365	19.50
River runoff	0.989	4759.27

The simulation speed of POP2 in the default high-resolution G-compset (POP2 and CICE4 are active) of CESM under 18,300 processes is just 1.43 SYPD without output, when the CESM is simply ported to Shenwei master-core. If more processes are used, the improvement of simulation speed is fairly limited. So there is a large performance gap to 5.0 SYPD for long-term climate simulations with CESM. In this paper, the ocean component model POP2 of CESM is mainly focused on.

The POP is an ocean general circulation model popular for ocean and climate research [3]. It can be run in either serial or parallel mode, and can use OpenMP thread-based parallelism, message-passing (MPI) or a hybrid of them. Along with the widespread use of POP model, it has become an important part of the large-scale parallel Benchmark test program based on different machines and realized varying degrees of acceleration [4]. On graphics processing unit (GPU), the maximum speedup ratio was more than 2.2 based on POP [5, 6]. Zhu and Zhao [7] researched the application of SIMD optimization in POP model, and they shew that in high resolution the speedup ratio decreases to 1.17 from 1.34 while the threads increase from 1 to 8. Zhao and Lei [8] ported POP model to the Sunway BlueLight multi-core supercomputer, their results proved that POP in high-resolution could reach linear speedup ratio within 5,000 cores.

In order to improve the simulation speed, Dennis [11] described a new space-filling curve, the Cinco curve. Werkhoven and Maassen [9] proposed a hierarchical load-balancing block-partitioning scheme to take both processor load and the communication hierarchy of the target heterogenous machine into account for the POP model code on GPUs. Hu and Huang [10] implemented a preconditioned Chebyshev-type iteration method in POP2 which developed an effective block preconditioner based on the Error Vector Propagation method (EVP) to attain a competitive convergence rate. They demonstrated that the improved barotropic solver could result in a 5.2x speedup of the barotropic mode in high resolution POP2 on 16,875 cores, and a corresponding 1.7x speedup of the overall POP2 simulation.

As the China's largest computing resource, the Sunway TaihuLight system has been applied to many scientific applications. The CAM model in CESM was mapped to the millions cores on the Sunway TaihuLight system and achieved a simulation speed of 3.4 SYPD [13]. The first general purpose graph processing framework ShenTu was implemented with on-chip sorting, supernode routing and degree-aware messaging, which were tightly combined to the processors and memory hierarchy and system infrastructure of Sunway TaihuLight supercomputer [14]. Duan and Gao [15] redesigned the Molecular Dynamics (MD) tool LAMMPS on Sunway TaihuLight. Furthermore, they improved the transcendental functions, such as pow and exp, and saved lots of space on LDM by eliminating lookup table with a little performance cost. Hu [16] proposed the DGDFT method with a two-level parallelization strategy that can scale up to 8,519,680 cores (131,072 core groups) on the Sunway TaihuLight supercomputer for studying electronic structures of two-dimensional (2D) metallic graphene systems that contain tens of thousands of carbon atoms.

All the above works are very helpful for us, but they are not enough when POP2 in high-resolution mode is focused. The difficulties of this work majorly include, a) the diversity of POP2's hotshot, which results that many source files need to be analyzed and translated; b) the implementation of all of the three-dimensional tendency terms in the baroclinic portion, which is computed on two-dimensional horizontal slices to reduce memory use [17], while this leads to little data size at large scale for each process and it is just opposite to the performance of Shenwei many-core; c) the system architecture challenges of Shenwei many-core that has many slave-cores while little local storage LDM and limited shared memory bandwidth.

Accordingly, it is impossible to achieve 5.0 SYPD for POP2 if just with only one simple method, and a comprehensive porting with deep optimizations from high level algorithm design to low level coding is required. In this paper, three stages are adopted. The first stage is to port POP2 to many-core with the athread programming interface of Sunway TaihuLight, and optimize with some popular skills. Disappointedly, the resulted version only got a simulation speed of about 2.6 SYPD without output. Furthermore, the achieved performances of many two-dimension (2D) array related kernels after athreaded are insignificant, or even negative speedup under large scale processes. So some new methods and deep optimization are demanded. The second stage is to redesign the grid decomposition and select the block distribution method, then effectively redistribute many 2D-array computing kernels to slave-cores by proposing a new slave-core partition method. After the above two stages, the performance of high-resolution POP2 is highly improved, but there is still a gap to the aimed simulation speed. The last stage is to apply some case-oriented techniques, which are quite effective for improving the speedup ratio.

Some contributions of this paper include:

1) The simulation speed of the finally optimized POP2 model is over 5.5 SYPD without output, under the high-resolution G-compset of CESM. It could strongly support the CESM to simulate long-term climate changes.
2) A new slave-core partition method is proposed, which can be regarded as a Multiple Program Multiple Data (MPMD) programming pattern and can ensure the large scalability of many 2D-array computing kernels in POP2. Furthermore, the flow chart of tavg output, advection and diffusion kernels is redesigned to compute in parallel mode, while not the original serial mode.
3) The speedup ratio of the optimized POP2 model to the original version is still over 3.8x when it is scaled up to 18,300 MPI processes, i.e., 1,189,500 cores. At 1,024 MPI processes scale, the speedup ratio is over 5.4x.

The rest of this paper is organized as follows. Section 2 lists out the test cases of POP2. Section 3 introduces some major methods for porting and optimizing. Section 4 proposes an effective slave-core partition method which can effectively scales up the 2D computing in many-core. Section 5 reports some simulation speeds and acceleration results.

2   High-Resolution Case of POP2 in CESM

The CESM is a fully-coupled global model that provides state-of-the-art computer simulations of the Earth's past, present, and future climate states. As the ocean model of the CESM, POP2 is a level-coordinate ocean general circulation model that solves the three-dimensional primitive equations for ocean dynamics. The POP2 model can predict and simulate the ocean state at any given moment driven by atmospheric force, which can obtain a series of ocean phenomena at various space-time scales.

The flow chart of the POP2's main program is shown in the Fig. 2, which is divided into two major stages: initialization and calculation. In the initialization part, the codes mainly initialize the ocean parameters and provide the initial values needed during the model running period. The calculation part is a series of iterative executions. Each iteration completes the calculation of prediction variables at a prediction time, which mainly includes four portions: integrating baroclinic equations, solving barotropic equations, updating the block ghost cells and updating the prediction variables. In our tests, the second-order centered advection scheme for computing the advective term and the K-profile parameterization (KPP) [18] for computing the vertical diffusivity κ (VDC in the codes) and viscosity μ (VVC in the codes) are selected, to solve the baroclinic primitive equations. For solving the elliptic system of barotropic equations, the Preconditioned Classical Stiefel Iteration solver (PCSI) [10, 19] is chosen to gain efficiency improvements.

The test cases are a Gedanken experiment with the initial velocity is zero, the initial temperature and salt are interpolated from Levitus data, the forced field is calculated by NCEP related data. The test cases use a high-resolution tripole grid at a nominal 0.1° horizontal resolution and 62 vertical levels. The number of horizontal grids is 3600*2400. In this G-compset, it integrates 500 steps per day and continuous integration for five model days. The output of variables includes sea surface temperature (SST), sea surface height (SSH), speed in all directions (UVEL, VVEL, WVEL) and some other variables.

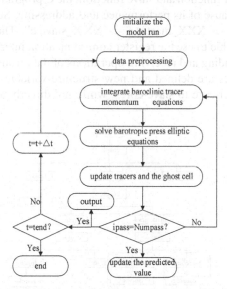

Fig. 2. Time-level iteration and major computations of the POP2 model.

In this paper, the G-compset cases with three different MPI process scales are set up, Test1 with 1,024 processes, Test2 with 16,800 processes, Test3 with 18,300 processes, and the same options for the remainder of the settings. Test1 and Test2 are applicable for small and large scale simulations with model variables validation and computing performance test, and Test3 is for the simulation speed test after the POP2 model is fully redistributed and optimized.

3   Porting and Optimizing to Many-Core

The porting of the CESM started from the Intel platform to the Sunway system. The major work is to resolve compatibility issues on the Sunway platform and standard Fortran compilers. With computing resources support from NSCC-Wuxi, stable runs on Sunway TaihuLight of the POP2 started from October 1st, 2018.

The code structure of POP2 model is shown in Fig. 3. Under the timing result of CESM, the main hotshots of the test cases are distributed in baroclinic, vertical_mix with kpp, horizontal_mix with del4 and barotropic portions. So we much focus on the following Fortran files to be translated into athreaded codes, such as step_mod, POP_SolversMod, baroclinic, vertical_mix, vmix_kpp, advection, hmix_del4.

In this paper, the Athread programming interface is applied for good performance and flexible coding, instead of using OpenACC as that in [12], which is directive-based with a simple approach to accelerators without significant programming effort but its performance is fairly limited. At the beginning, some hotshots of POP2, such as vmix_kpp, were tested with OpenACC directives while the simulation speed of POP2 was increased by less than 30%. The Athread interface is one of the thread-like programming interfaces, and it also can deliver performance portability across the popular platforms just replaced with some thread-special functions.

For coding convenience and clarity, a three-layer code style is used. For the newly added athreaded master function and slave function, the C programming language rather Fortran is adopted because of its performance and addressing. So the athreaded POP2 looks like "XXX.F90 -> XXX_master.c -> XXX_slave.c". The C language is used here because it is possible to use the register communication interface of Taihu Light to realize the efficient sending and receiving operations of data from each core.

New structure types are defined and new structure variables are declared to pass parameters from master-core to slave-cores, while not directly access memory which

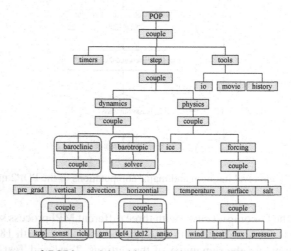

Fig. 3. Code structure of POP2 model, which contains three different parameterizations for computing the vertical diffusivity and viscosity, and four for computing the horizontal.

may be fairly low performance, especially in 2D-array computing kernels. Also, the improved transcendental functions that Duan and Gao [15] proposed are called.

Furthermore, some optimization methods on MPI communication such as that Zhao and Lei [8] and Zhang and Zhao [4] proposed are applied. In the current version of CESM, the routines for updating halo regions (ghost cells) using MPI calls have been replaced by the new modules from NCAR. In addition, PIO has been applied, so the I/O optimization is not addressed here, although it is another important issue.

Beyond the popular tips known, some other skills are also used. Such as loop fusion and function inline, loop collapse and data tiled and register communication to stencil-like computing.

4 Refactoring and Redistributing

4.1 Refactoring 2D-Array Computing Flow

In POP2, the baroclinic portion of the codes is the most computationally intensive, which computes all of the three dimensional tendency terms. In order to reduce memory usage and improve simulation performance, most tendencies are computed on two dimensional horizontal slices [17], and they have been implemented in the current version of POP2. Of course it may cause a decrease in communication efficiency and an increase the probability of communication collision.

While for Shenwei many-core infrastructure, the above strategy just brings some extra false effect to POP2, even in high-resolution mode. The slave-cores in one SW 26010 many-core processor have high computing capability, but all the 64 slave-cores in one core-group share 4 common data buses to memory tunnels [23], so competitions and collisions may happen when many slave-cores access some variables in master-core simultaneously, and the whole performance may decrease suddenly. The popular idea is to increase the ratio of computing load to memory access in two ways, one is to read or write larger data block than 256 bytes between LDM and DRAM [14], another is to parallel with only part of slave-cores, while the latter results in slave-cores waste.

But for POP2 in high resolution under large scale processes, especially, for 2D-array computing kernels, the size of 2D arrays are fairly small, they are not suitable to be distributed to many slave-cores. Otherwise, the performance of the athreaded 2D-array computing kernels may be worse than the original codes (Fig. 4). In fact, the 2D data distributed to each process is less than $24 \times 24 \times 8$ Bytes, the whole data block distributed to each slave-core is less than 72 bytes. So it is a problem to distribute 2D-array computing tasks to slave-cores effectively.

After the 2D-array computing related kernels in POP2 codes carefully analyzed, a slave-core partition-based parallel method is proposed, which can distribute different independent 2D-array computing kernels to run at different slave-core partitions simultaneously. That is to say, a Multiple Program Multiple Data (MPMD) programming pattern is implemented through spawning multiple tasks to multiple Shenwei slave-core partitions. The original 2D-array computing kernels are independent and parallel in essence, but they have to run serially just because of the traditional Single Program Multiple Data (SPMD) programming restriction in Fig. 5 (left). Through this slave-core partition method, each 2D-array kernel can be loaded to a different slave-core partition

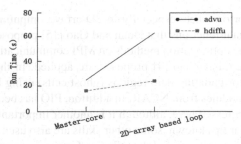

Fig. 4. The athreaded 2D-array computing kernels get negative speedup to the original master-core codes under Test2, and the performance of advu is much worse than that of hdiffu.

at the same time in Fig. 5 (right), thus the slave-cores are fully utilized, and the data size distributed to each slave-core is not too small. In this way, it is naturally hoped to achieve a good performance.

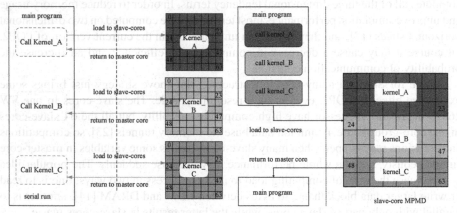

Fig. 5. The original serial flow of a computing kernel on SW 26010 many-core (the left), and a parallel flow of it with slave-core partition (the right).

To implement the above slave-core partition method, there are two kind of interface functions available on Sunway TaihuLight supercomputer. One is athread_create and athread_wait, another one is athread_spawn and athread_join, and each one has its special advantages and disadvantages. Both of them need to redesign the different kernel codes into one large kernel. If the kernels are located in different source files, this would bring much extra work. For athread_create and athread_wait, tasks are distributed by calling the athread_create function one by one in master.c (see Fig. 6(a)), which is much flexible in coding, while its performance is limited. For athread_spawn and athread_join, tasks are distributed by calling athread_spawn function only once in master.c, and the slave-core needs to judge and match its own task (see Fig. 6(b)), while its performance is much effective than that of the athread_create and athread_wait. So finally, the athread_spawn and athread_join mode is applied.

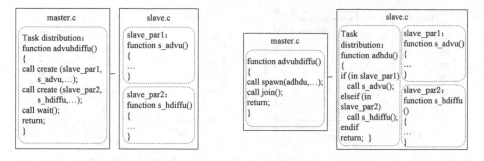

(a) athread_create + athread_wait.   (b) athread_spawn + athread_join.

Fig. 6. Skeleton of the two different methods with athread programming interface to implement slave-core partition on SW 26010 many-core.

Tavg Variables Output. These codes are centrally distributed in the subroutine baroclinic_driver of baroclinic.F90 and the subroutine advt of advection.F90. Most of the tavg variables are independent and each output calls the same subroutine accumulate_tavg_field, so they can be written out in parallel mode with any order.

In this case, the main flow of the codes is kept. The dominant work to do is to select a good partition with balanced number of slave-cores, especially, when there are a lot of tavg variables to output. So different slave-core partitions can get a fairly well load balance. The performance of the tavg variables output optimized with slave-core partitions in subroutine baroclinic_driver is given in Fig. 7(a).

It is shown that the speedup ratios of Test1 with the optimized kernel to the original kernel is more than 7.0x, the speedup ratios of Test2 with the optimized kernel to the original kernel is more than 5.0x. So the speedup ratio is lessened when the processes grows larger, while it still has positive speedup.

Right Terms Computing in the Primitive Equations. After time discretization in POP2 model, the resulting baroclinic momentum equations also include some right terms to be computed, such as momentum advection, horizontal friction, vertical friction and metric terms [3]. Almost each one of the right terms is implemented in an individual subroutine, and when it comes to computing, the corresponding subroutines are called in subroutine clinic of baroclinic.F90.

For this case, in order to apply the slave-core partition method, it almost needs to redesign the flow chart thoroughly to spawn many subroutines at once time, which brings a lot of extra work. Nevertheless, the total run time of these 2D-array computing kernels gets a positive speedup in Test2 with large processes, and some kernels' costs are hidden by the dominant kernels.

The whole performance of advu and hdiffu optimized by the slave-core partition is given in Fig. 7(b). It is shown that the speedup ratios of both Test1 and Test2 with the optimized kernels to the original kernels are still more than 3.0x.

The slave-core partition is also applied for tracer computing, which includes advt, hdifft, vdifft, etc.

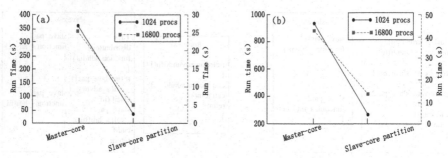

Fig. 7. Performances of slave-core partitions on SW26010 under Test1 with 1,024 processes and Test2 with 16,800 processes. (a) Costs of tavg variables' output in baroclinc.F90. (b) Total costs of advu and hdiffu with significant improvement compared with that of 2D-array based loop in Fig. 4.

Under the above slave-core partition method, the flow chart of advection and diffusion computing kernels in baroclinic portion is redesigned, and the corresponding F90 source files are thoroughly refactored. So these major kernels are newly computed in parallel mode, while not the original serial mode. The original and redesigned flow charts are schemed in Fig. 8.

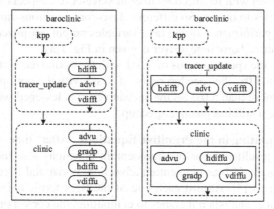

Fig. 8. Original and refactored flows of some major portions in baroclinic, applied with slave-core partitions so many 2D-array kernels are computed in parallel mode.

After the above, the scalability and simulation speed of high-resolution POP2 model is greatly improved, but there is still a gap to the demanded simulation speed. When the main flow chart of POP2 is focused again, it is found that there are still some optimization ways for the high-resolution Tests under consideration. Although case-oriented, they are effective in improving simulation speed.

4.2 Grid Decomposition and Block Distribution

In POP2, the full horizontal domain with size (nx_global, ny_global) is broken into many subdomains or blocks. The mapping of blocks distributed to processes or nodes can be performed using either a space-filling curve (SFC) distribution to try to give all processes an equal amount of work and just compute blocks without land, or a Cartesian distribution to ensure that the block's north, south, east and west neighbors remain the nearest neighbors. The SFC distribution draws a curve through the set of blocks in a way, which maintains locality and can more effectively balance both the computation load and the communication cost. While the SFC algorithm is supported for all resolutions, it is particularly effective at reducing the cost of high-resolution simulations.

In our three tests, the SFC is set as the distribution method of blocks to processes for load balancing, and different block sizes can result in different performances. If condition 'POP_AUTO_DECOMP = true' is satisfied, the original version of POP2 divides the grid automatically (nxblock = 24, nyblock = 28) and the simulation speed is just 1.43 SYPD with 18,300 processes under Test3. The active ocean blocks that involved the solution are about 70% of the total 18,000 blocks that divided from global grids under the nxblock = 24 and nyblock = 28, so the processor did not achieve the maximum utilization and the simulation speed can not reach the expected performance. After carefully analyzed, the numbers of active ocean blocks and processes are about equal, then the processes can drive to the highest utilization possible. So the block size is manually assigned as nxblock = 20 and nyblock = 24, the active ocean blocks is 18,285 that equal to number of processes, and the simulation speed with 18,300 processes rises to 1.89 SYPD. This improvement may be similar to that in Dennis [11].

Moreover, there are three distribution methods of blocks to processes that can be selected in POP2, and different distributions can be specified for the baroclinic portion and the barotropic portion. Different distribution methods have different data distributions on processes. So this would bring an additional step to match the blocks of the barotropic portion to the blocks of the baroclinic portion before solving the barotropic press elliptic equation, and verse visa after the solver finished. Two parts are related with the block distribution match, one is passing variables from the baroclinic portion to the barotropic portion, and another is updating variables back from the barotropic portion to the baroclinic portion after the barotropic solver has finished. At large process scale, this is very exhaustive. Under the test case, the baroclinic distribution is the same as the barotropic distribution, the above block distribution exchange between the barotropic portion and the barotropic portion could be omitted.

According to the above idea, the POP_SolversMod.F90 is modified. After tested under Test3, the time cost of barotropic portion is decreased from 116 s to only 56 s with a 2.0x speedup ratio.

5 Speedup Ratio and Scalability

The high-resolution POP2 coupled in CESM model is ported to the Sunway TaihuLight supercomputer. Some new changes of CESM can be found in Meehl and Yang [20]. Based on the ocean circulation model POP2 successfully ported, some speedup ratio tests for the master-core version of POP2 are carried out.

The test results are shown in Table 2. For different process numbers, the global domain with 3600*2400 grids is decomposed into many blocks with different block sizes. The simulation speeds of POP2 model in the high-resolution CESM G-compset almost approach a linear trend within 18,300 MPI processes, in which the speedup ratio increases as the number of processes increases.

Table 2. Test results under different cases of the original POP2 on Shenwei master-cores.

Grid size: 3600 * 2400 * 62, AUTO_DECOMP = true
time step: dt_count = 500, simulated days = 5

No.	Number of processes	Block size	Computing time (s)	Simulation speed (SYPD)
1	1024	76*52	9624.670	0.104
2	6124	28*54	1896.319	0.495
3	9196	24*54	1850.520	0.627
4	12248	34*34	1441.764	0.803
5	15004	29*29	1519.944	0.90
6	16800	28*28	898.075	1.26
7	18300	24*28	785.698	1.43

With the finally athreaded and optimized POP2 model in CESM high-resolution G-compset running for 5 simulated days, a simulation speed of 0.57 SYPD without output based on 1,024 Shenwei MPI processes is achieved, and the speedup ratio to the original Shenwei master-core version of POP2 is over 5.4x.

With the finally athreaded and optimized POP2 in CESM high-resolution G-compset for 5 simulated days, a simulation speed of 5.5 SYPD without output based on 18,300 Shenwei processes is achieved, and the speedup ratio to the original Shenwei master-core version of POP2 is still over 3.8. The costs of different major kernels are shown in Fig. 9. BAROCLINIC, BAROTROPIC and STEP are the remained three largest timing-cost parts in the finally optimized POP2. Also, the timing costs of ADV_MOMENTUM, VMIX_E_MOMEMTUM, VMIX_I_ MOMENTUM, HMIX_T_DEL4, VMIX_E_TRACER and VMIX_I_TRACER are hidden by slave-core partition method.

While improving the simulation speed of POP2 model, we must guarantee the model's accuracy. Firstly, the values of the computed or updated variables in each computing kernel are checked after it is ported or optimized, with them of the original master-core version, to make consistent in the precision range. In fact, the variables' values in some athreaded kernels have no difference to the original version, and most of the relative differences of the variables' values in the athreaded kernels are less than 10^{-15}, and several relative differences of them are between 10^{-11} and 10^{-15}.

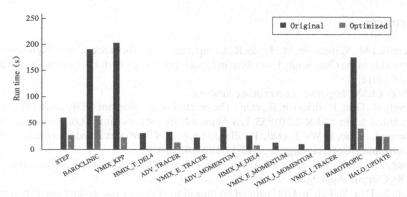

Fig. 9. Timing costs of major kernels in the original and optimized POP2 based on 18,300 MPI processes with SW26010 many-core, for each item the left bar is the cost of the original version of POP2 and the right bar is that of the finally ported and optimized version.

6   Conclusion and Discussions

The ocean model POP2 coupled in CESM is successfully ported to Sunway TaihuLight supercomputer. Firstly, the POP2 full horizontal domain is broken into many blocks of defined size and get a good load balancing performance. Moreover, by applying fine-grained athreaded programming techniques, proposing a slave-core partition method and redesigning the flows of some main 2D-array computing kernels, combining some additional strategies for the underlying G-compset case, the simulation speed of the finally optimized POP2 model can achieve 5.5 SYPD without output in high-resolution CESM G-compset, and its speedup ratio to the original Shenwei master-core version is more than 3.8x with 18,300 processes. It could strongly support the CESM to simulate long-term climate changes.

There are still some chances to improve the MPI communications in the baroclinic portion and the barotropic portion, and to readjust the PCSI solver for Shenwei many-core. So a more larger simulation speed may be expected. Furthermore, the output scheme of time-averaged history files can reduce the simulation speed. So optimizing the output scheme may be another effective means.

Through the athreaded POP2 model, it shows that the athread interface is more efficient in performance than the OpenACC interface under Shenwei many-core infrastructure. But the workload of coding is large and tiresome, and it is much more for mixed computation with many different-dimension arrays and for slave-core partition. A code auto-generation tool is important for further research, for example, similar to Muranushi [21] and Zhu [22].

Acknowledgments. We thank the anonymous referees for their valuable comments and suggestions to improve this paper. This research is supported by the Key R & D program of Ministry of Science and Technology of China (2016YFB0201100), Shandong Province Innovative Public Service Platform Project (2018JGX109), Major projects of Aoshan Science, Technology and Innovation Program (2018ASKJ01) and the "Colleges and Universities 20 Terms" Foundation of Jinan City, China (2018GXRC015).

References

1. Dennis, J.M., Vertenstein, M., Jacob, R.: Computational performance of ultra-high-resolution capability in the Community Earth System Model. Int. J. High Perform. Comput. Appl. **26**(1), 5–16 (2012)
2. About CESM. http://www.cesm.ucar.edu/about
3. Smith, R., Gent, P., Briegleb, B., et al.: The parallel ocean program (POP) reference manual. Technical report LAUR-10-01853. Los Alamos National Laboratory, Los Alamos (2010)
4. Zhang, L., Zhao, J., Wu, J., et al.: Parallel computing of POP ocean model on quad-core Intel Xeon cluster. Comput. Eng. Appl. **45**(5), 189–192 (2009)
5. Song, Z., Liu, H., Lei, X., et al.: The application of GPU in ocean general circulation mode POP. Comput. Appl. Softw. **27**(10), 27–29 (2010)
6. Guo, S., Dou, Y., Lei, Y.: GPU parallel optimization of the oceanic general circulation model POP. Comput. Eng. Sci. **34**(8), 147–153 (2012)
7. Zhu, R., Zhao, W., Chen, D.: The application of the SIMD optimization in ocean general circulation model POP. In: International Conference on Computer Science and Service System, Nanjing, China, pp. 1749–1753 (2012)
8. Zhao, W., Lei, X., Chen, D., et al.: Porting and application of global eddy-resolving parallel ocean mode POP to SW supercomputer. Comput. Appl. Softw. **31**(5), 42–45 (2014)
9. Werkhoven, B., Maassen, J., Kliphuis, M., et al.: A distributed computing approach to improve the performance of the Parallel Ocean Program (v2.1). Geosci. Model Dev. **7**, 267–281 (2014)
10. Hu, Y., Huang, X., Baker, A., et al.: Improving the scalability of the ocean barotropic solver in the community earth system model. In: Proceedings of SC 2015, pp. 15–20. ACM, Austin (2015)
11. Dennis, J.: Inverse space-filling curve partitioning of a global ocean model. In: IEEE International Parallel & Distributed Processing Symposium, pp. 1–10. IEEE, Long Beach (2007)
12. Fu, H., Liao, J., Xue, W., et al.: Refactoring and optimizing the community atmosphere model (CAM) on the Sunway TaihuLight supercomputer. In: Proceedings of SC 2016. IEEE, Salt Lake City (2016)
13. Fu, H., Liao, J., Ding, N., et al.: Redesigning CAM-SE for peta-scale climate modeling performance and ultra-high resolution on Sunway TaihuLight. In: Proceedings of SC 2017. ACM, Denver (2017). https://doi.org/10.1145/3126908.3126909
14. Lin, H., Zhu, X., Yu, B., et al.: ShenTu: processing multi-trillion edge graphs on millions of cores in seconds. In: Proceedings of SC 2018. IEEE, Dallas (2018)
15. Duan, X., Gao, P., Zhang, T., et al.: Redesigning LAMMPS for peta-scale and hundred-billion-atom simulation on Sunway TaihuLight. In: Proceedings of SC 2018. IEEE, Dallas (2018)
16. Hu, W., et al.: High performance computing of DGDFT for tens of thousands of atoms using millions of cores on Sunway TaihuLight. Sci. Bull. (2020). https://doi.org/10.1016/j.scib.2020.06.025
17. Jones, P.W., Worley, P.H., Yoshida, Y., et al.: Practical performance portability in the Parallel Ocean Program (POP). Concurr. Comput. Pract. Exp. **17**, 1317–1327 (2005)
18. Large, W., McWilliams, J., Doney, S.: Oceanic vertical mixing: a review and a model with a nonlocal boundary layer parameterization. Rev. Geophys. **32**(4), 363–403 (1994)
19. Huang, X., Tang, Q., Tseng, Y., et al.: P-CSI v1.0, an accelerated barotropic solver for the high-resolution ocean model component in the Community Earth System Model v2.0. Geosci. Model Dev. **9**(11), 4209–4225 (2016). https://doi.org/10.5194/gmd-9-4209-2016
20. Meehl, G., Yang, D., Arblaster, J., et al.: Effects of model resolution, physics, and coupling on southern hemisphere storm tracks in CESM1.3. Geophys. Res. Lett. https://doi.org/10.1029/2019GL084057

21. Muranushi, T., Hotta, H., Makino, J., et al.: Simulations of below-ground dynamics of fungi: 1.184 pflops attained by automated generation and autotuning of temporal blocking codes. In: Proceedings of SC 2016, pp. 23–33, Salt Lake City, USA (2016)
22. Zhu, X., Zeng, Y., Wei, Y., et al.: An auto code generator for stencil on SW26010. In: IEEE 21st International Conference on High Performance Computing and Communications, pp. 182–190. IEEE, Zhangjiajie (2019)
23. Chen, J.: Research on algorithm design and optimization methods of molecular biology applications for the domestic Sunway manycore system. Doctorial dissertation, University of Science and Technology of China, Hefei, China (2019)

Performance Optimization for Feature Extraction Section of DeepChem

Ke Zhan[1], ZhongHua Lu[1(⊠)], and YunQuan Zhang[2(⊠)]

[1] Computer Network Information Center, Chinese Academy of Sciences, Beijing, China
zhankecas@gmail.com, zhlu@sccas.cn
[2] State Key Laboratory of Computer Architecture,
Institute of Computing Technology, Chinese Academy of Sciences, Beijing, China
zyq@ict.ac.cn

Abstract. Based on the popular deep learning technique, the authors at Stanford implement DeepChem as an open source methods for the research in the fields of drug discovery, biology and so on. For the performance problem of training process of DeepChem neural network, this paper rebuilds the original serial feature extraction algorithm of DeepChem and optimizes the rebuilt serial algorithm based on the multiple processes algorithm. The experiment results show that the parallel algorithm achieves 15.38× speedup at the best compared with the serial algorithm. For the future work, first, in addition to the *multiprocessing* package, the other packages such as *concurrent, subprocess* and so on could be considered to optimize the feature extraction algorithm; Second, the serial and parallel algorithms run slower when the data block size is 150 compared with the other block sizes, optimization of training process performance for the smaller data block size is the second direction in future work.

Keywords: DeepChem · Feature Extraction · Multiple Processes Algorithm

1 Introduction

Computers and humans are appropriate to different kinds of work constitutionally. For instance, computing the product of two large numbers is very easy for computers, but this task is very difficult for humans; in other words, image segmentation is a ordinary task for humans, but this segmentation task is very difficult for computers [1]. As recently as about 10 years ago, the deep learning algorithms make great breakthrough. Some few cases lie in our understanding of human neural networks. The typical example is convolutional neural networks (CNN) for image recognition [2–4]. The structure of CNN was stimulated by the neurons organization experiment [5]. However, humans don't understand the biological networks completely. The fact that humans only have limited knowledge of how the brain really works doesn't affect the great progress in artificial

M. Qiu (Ed.): ICA3PP 2020, LNCS 12452, pp. 290–304, 2020.
https://doi.org/10.1007/978-3-030-60245-1_20

neural networks. As the increasing application of neural network in the fields such as natural language processing, computer vision and so on, humans expect that neural network will have higher accuracy and faster speed.

A neural network model could be regarded as a imitation of the learning process in living organism, there is also one more direct comprehension of neural network as computational graphs. Researchers have proposed an efficient entropy-aware I/O framework DeepIO [6] for large scale deep learning on high performance computing system. The overall goal of this framework is to coordinate the use of memory, communication and I/O resources for efficient training of datasets. The evaluation results show that DeepIO bring better performance than the memory-based storage systems obviously. The same team proposes Multi-Client DeepIO [7] to support multiple clients on a single node. To support multiple clients on a single node, the authors modify the DeepIO so that the server on a node dispatches the elements to all read buffers and handle requests from all clients in a round robin manner.

Deep learning is still in its preliminary stage among a whole variety of technologies used in the fields of drug discovery and design [8–10]. Because of unique properties of deep learning, it has been applied not only in the prediction of compound activity and toxicity, bur also in the generation of virtual compound library. The predictive ability of deep learning has no obvious advantage compared with the traditional machine learning methods. But deep learning can extract features automatically, doesn't require manual summarization of data features. Deep learning is developing rapidly, and methods such as unsupervised learning that don't rely on a large number of labeled data are undergoing more perfect gradually. It is expected to better assist in the development of new drugs.

Deep learning is one popular method in the fields of mining existing knowledge and finding the relevance of existing knowledge in massive data [11,12]. High-throughput imaging is usually used for screening chemical compounds based on morphologic characteristics transformation. The set of features which include morphological shape, intensity, spatial metrics, can be used as an image-based compound fingerprint. The compound fingerprints can group compounds and genes based on pharmacological mechanism and functional similarity respectively.

The researchers use Macac method to maps all tasks to the same latent space. For the Oncology project, in the glucocorticoid receptor high-throughput imaging assay, 60000 compounds are ranked, 124 compounds are submicromolar hits, the results show that 50-fold enrichment over the initial high-throughput screen; for the central nervous system project, 500000 image-annotated compounds are used to predict activity. 289-fold enrichment over the hit rate of the initial high-throughput screen because of 36 compounds are submicromolar hits finally [13].

DeepChem [14] is an open source software package based on TensorFlow [15], which aims to make the use of deep learning more popular in the fields of drug discovery and biology. The typical applications include creating graph convolutional models on the Tox21 dataset [16], building regression models for predicting the Ki of ligands to a protein, using machine learning technology to model protein-ligand affinity and so on. The molecules are usually represented as Simplified Molecular-Input Line-Entry System (SMILES) [17–20] format in the field of chemical informatics. The traditional deep learning algorithms accept fixed length datasets as the input. To build deep learning model from SMILES format datasets, researchers must extract available information into a fixed dimensional representation. These extraction processes are called featurization. Feature extraction operations will be invoked in the early stage during the run time of deep learning models. For the large size files, the feature extraction operations will take a long time. This paper optimizes the feature extraction operations of DeepChem using multiple processes algorithm. We test the performance of the serial algorithm and the parallel algorithm using four groups experiment. The experiment results show that when the data block size is set to 8192, the parallel algorithm achieves 15.38× speedup at the best.

The main contributions of this paper are:

- **First**: we rebuild the serial algorithm of feature extraction section of DeepChem. The nested structured class DataLoader and Python *generator* statement are reconstructed to facilitate parallelization.
- **Second**: we optimize the rebuilt serial algorithm using multiple processes algorithm. The experiment results show that the performance improves 15.38× at the best.
- **Third**: two different packages of *threading* and *multiprocessing* in Python are used to implement the parallel feature extraction algorithm in this paper, the performance has no improvement when using *threading* package, but for the *multiprocessing* package, the performance of parallel algorithm improves significantly.

The following sections of this paper are organized as: The Sect. 2 introduces the related optimized work for DeepChem and the motivation of this paper. The Sect. 3 summarizes the rebuilt serial algorithm and the optimized parallel algorithm. The Sect. 4 describes the experiment environment and the experiment results, gives explanations to the experiment results. The Sect. 5 draws a conclusion and discusses the future work. The section acknowledgment expresses gratitude.

2   Related Work and Motivation

Researchers can train bigger models and use larger datasets by using faster training methods [21]. Google brain proposes data echoing algorithm [22] to speed up training. The data echoing paper gives two directions to make neural network training faster, first, make the non-accelerator work faster; second, reduce the amount of non-accelerator work needed to achieve the desired predictive performance. This paper optimizes the feature extraction section of DeepChem to speed up the training which belongs to the first direction, make the non-accelerator work faster. We use multiple processes algorithm to divide file, every process handle each section of the file independently to speed up the feature extraction operations.

For the implementation of the parallel feature extraction algorithm, we compare the performance results using the *threading* package [23–25] and the *multiprocessing* package respectively. The results show that the parallel algorithm performance improves 15.38× when using *multiprocessing* package but the performance has no improvement when using *threading* package.

3 Algorithms and Implementations

The primal serial program is from github website of DeepChem [26]. The serial featurization algorithm is summarized as the following Algorithm 1. The main function calls the unlabelled_test function firstly in the line 28. From the line 23 to the line 25 in the function of *unlabelled_test*, reads the input data from CSV format file. Initializes the class object *featurizer* and *loader* from the class *ConvMolFeaturizer* and *CSVLoader* respectively. The line 26 calls the class function *featurize* with the input file as the parameter.

Algorithm 1. Serial Featurization Algorithm

Input: Large Size CSV format Data
Output: DiskDataset
 1: Class DiskDataset(Dataset):
 2:     Initialize
 3:     def CreateDataset(shard_generator):
 4:         metadataRows ← []
 5:         FOR shard_num, (X, y, w, ids) in enumerate(shard_generator)
 6:             metadataRowsAppend(DiskDatasetWriteDataToDisk)
 7:         ENDFOR
 8:         metadata_df ← construct_metadata(metadata_rows)
 9:         save_metadata(metadata_df)
10:         return DiskDataset(data_dir)
11: Class DataLoader():
12:     def featurize(shard):
13:         Initialize
14:         def shard_generator():
15:             FOR shard_num, shard in enumerate()
16:                 X, valid_inds ← self.featurize_shard(shard)
17:                 ids ← ids[valid_inds]
18:                 y, w ← convert_df_to_numpy
19:                 yield X, y, w, ids
20:             ENDFOR
21:             return DiskDataSetCreateDataset(shard_generator)
22: def unlabelled_test():
23:     inputfile ← Large Size CSV format Data
24:     featurizer ← ConvMolFeaturizer()

```
25:    loader ← data_loader.CSVLoader(featurizer = featurizer)
26:    loader.featurize(inputfile)
27: main()
28:    unlabelled_test()
```

From the line 12 to 21, lists the definition of *featurize* function. This function performs a feature extraction operation on each data block, the data block is represented by the shard in DeepChem. From the line 14, the *shard_generator* function is nested in the function *featurize*. And there is one yield statement in the function *shard_generator*. The yield statement represents that this *shard_generator* function is a generator. For the *generator* in Python, it is a different function from the normal function which includes yield statement and returns a *generator* iterator. The generator function is originally a function provided by Python to facilitate the use of iterator. *iterable* is one object which implement the *_iter_()* method. This method will return one *iterator* object. For the *iterable*, it is an object which returns one member at one time. Some *iterable* store all the values in the memory, for example the *list*, some *iterable* don't store all the values in the memory, for example, the *iterator*. The *iterator* is an object which implement the *_iter_* and the *_next_* methods. The *_iter_* method returns the *iterator* object itself. And the *_next_* method includes two operations: first, update the status of the *iterator*, make it point to the next item for the next invoke; second, return the current values. These two peculiarities that nested function and generator make it difficult to execute parallel operations for the original serial algorithm. For the convenience of parallelization, we rebuild this *featurize* function. The feature extraction results are saved in the function of *CreateDataset*.

From the line 3 to 10, defines the function of *CreateDataset*. For every data block, the *CreateDataset* function appends the generated information to the list, constructs the metadata and saves the results data to the appointed directory. This *CreateDataset* function provides the method to verify the correctness of parallel algorithm. We check the saved results between serial algorithm and parallel algorithm, the saved results are the same. The same results represent the correctness of the parallel algorithm.

For the parallel feature extraction Algorithm 2, from the line 44 to 46, the algorithm initializes the class objects and variables. The variable shard_size in line 46 sets the size of data block, which represents the file line number handled by one feature extraction operation. shard_size could be set to different values. Different size of one shard can affect the program performance. The affected results will be discussed in the experiment Sect. 4.

The line 47 calls the function *featurize* to extract the feature information from the data block. The rebuilt *ParaCreateDataset* function is invoked by the function *featurize* in the line 42.

Algorithm 2. Parallel Featurization Algorithm

Input: Large Size CSV format Data
Output: DiskDataset

```
 1: Class DiskDataSet():
 2:    Initialize
 3:    def ShardGeneratePROCESS(inputfile, shard_size):
 4:       StartBlockId ← set the data block start id handled by current process.
 5:       EndBlockId ← set the end id.
 6:       shard_num ← 0
 7:       FOR dataframe in pandas.read_csv(intputfile, shard_size)
 8:          IF(shard_num ≥ StartBlockId and shard_num ≤ EndBlockId):
 9:             df ← df.replace(numpy.nan, regex=True)
10:             X, valid_inds ← featurize_smiles_df(shard)
11:             ids ← ids[valid_inds]
12:             basename ← shard-str(shard_num)
13:             write_data_to_disk(base_name, data_dir)
14:          ELSE:
15:             shard_num += 1
16:          ENDIF
17:       ENDFOR
18:       return
19:    def ParaCreateDataset(inputfile, ProcessNum, shard_size):
20:       FileNum ← length(inputfile.ReadLines)
21:       IF(FileNum - 1) %the shard_size == 0:
22:          BlockNum ← (FileNum - 1)  / shard_size
23:       ELSE:
24:          BlockNum ← int( (FileNum - 1)  / shard_size ) + 1
25:       ENDIF
26:       ProcessBlockNum ← blocks number handled by every process except
          the last process
27:       LastProcessBlockNum ← blocks number handled by the last process
28:       ShardGProcesses = []
29:       FOR i in range(ProcessNum):
30:          ShardGWriteP = Process(target = ShardGeneratePROCESS)
31:          ShardGWriteP.start()
32:          ShardGProcesses.append(ShardGWriteP)
33:       ENDFOR
34:       FOR process in ShardGProcesses:
35:          process.join()
36:       ENDFOR
37:       construct_metadata(metadata_rows)
38:       save_metadata(data_dir)
39:       return DiskDataset(data_dir)
40:    def featurize():
```

```
41:     Initialize
42:        return ParaCreateDataset(inputfile, ProcessNum, shard_size)
43: main()
44:     feature ← featureloader.ConvMolFeaturizer()
45:     featureloader.CSVLoader(featurizer = feature)
46:     shard_size ← set one number
47:     loader.featurize(shard_size)
48:     return
```

From the line 19 to 39, *ParaCreateDataset* starts and synchronizes the multiple processes. From the line 20 to 27, compute the total file number and the blocks number to prepare for the task partition. From the line 29 to 33, start multiple processes, one process handle one section of the original file, several processes handle several sections simultaneously, every process runs independently. When the maintenance overhead of processes is less than the overhead required for performance improvement, the performance of algorithm will improve. From the line 34 to 36, synchronize the multiple processes to ensure that all the processes complete their own tasks. The target function *ShardGeneratePROCESS* is defined in the line 3 to 18. The line 4 to 6 set the range of data block identifier. Every process reads the different section of the original file and extracts feature information from each data block independently in the line 7 to 17.

For example, for the input dataset file, the parallel algorithm divides the file according to the file size and the number of CPU cores, assigns the task in a balanced manner. Since the parallel algorithm is constructed based on the rebuilt serial algorithm, the parallel algorithm runs correctly.

4   Experiment

DeepChem is implemented by the language of Python. There are *threading*, *multiprocessing, concurrent, subprocess, sched* and *queue* packages to support the concurrent operations in Python language [27–29].

For the *threading* package, there are two methods to run a separate thread, first, passing a callable object to the class constructor; second, overriding the *run* method in a subclass. Once the main program creates a thread object, invokes the *start* method to start the activity of the thread. Global Interpreter Lock (GIL) is used in the CPython interpreter as one synchronization mechanism. The GIL will make the interpreter run multiple threaded program more convenient, but at the expense of parallel performance degradation at the same time. Despite the impact of the GIL, for the I/O intensive algorithm, the *threading* package could improve the performance of the Python program [23,24].

The *multiprocessing* package could run on cross platforms such as UNIX and Windows. This package spans and manages processes using the API functions similar to the *threading* package. There are also objects such as *Pool* in the *multiprocessing* package but the *threading* package have no such objects. The *Pool* object provides an advantageous method to parallelize the execution of a function by the way of multiple input values. At the mean time, the *multiprocessing* package uses subprocesses to substitute for threads which avoid the affect of the Global Interpreter Lock.

We compare the experiment results of using the *threading* package and the *multiprocessing* package, the results demonstrate that the performance of parallel algorithm has no improvement when using the *threading* package, even the performance reduces. So we choose the *multiprocessing* package as the ultimate package to implement the Algorithm 2: Parallel Featurization Algorithm.

We compare the intermediate results of the original serial algorithm and the parallel algorithm, the comparison results show that the intermediate results of the two are the same. This comparison results verify the correctness of the parallel algorithm.

Based on the analysis of the experiment results, we find that the parallel algorithm performance improves $15.38\times$ at the best when the algorithm runs on the computing node2 and the data block size is 8192 for the 885 MB *File4*.

4.1 Experiment Environment

We use two computing nodes as the experiment platforms. The detailed configured information of the two nodes are listed in the Table 1. The main difference between the two nodes is the number of CPU cores. The first node has 24 logical CPU cores. The second node has 40 logical CPU cores.

Four files are used to test the performance of serial feature extraction algorithm and parallel feature extraction algorithm.

The dataset file is downloaded using this program [30]. The data size of the original file is only 708 KB, we copy the file contents to generate three larger files to test the performance of the serial and parallel algorithms. With the addition of the original file, we name these files as File1, File2, File3, File4 respectively. The size of the other three generated files are 14 MB, 222 MB, 885 MB respectively. For every file, we run the serial algorithm and parallel algorithm on two computing nodes which we name these computing nodes as node1, node2 respectively.

4.2 Experiment Results and Analysis

We sum up the performance experiment results as the following four groups.

First File Results
The size of the first original file is 708 KB. Serial algorithm and parallel algorithm are run on the two computing nodes. The experiment results are listed in the Fig. 1.

For the example of the Fig. 1, the *Serial* in the X axis represents the serial algorithm. The other numbers such as 2, 4 and so on represent the processes number when the parallel feature extraction algorithm runs. The first computing node has 24 logical CPU cores, and the second computing node has 40 logical CPU cores, so we set 24 processes and 40 processes at most on the first node and second node respectively. The Y axis represents the run time of the serial

Table 1. Computing nodes configuration

Config[a]	ComputeNodes	
	node1	node2
Arch[b]	x86_64	x86_64
CPUModelName[c]	E5-2620	E5-2640
CacheSize[d]	15360 KB	25600 KB
PhysicalCPUs[e]	2	2
Cores[f]	6	10
Processors[g]	24	40
Memory[h]	264037876 KB	264039828 KB
OS[i]	16.04	16.04
Kernel[j]	4.4.0-169	4.4.0-96

[a]Config: Configure information of the two computing nodes.
[b]Arch: Architecture of the CPU.
[c]CPUModelName: Intel(R) Xeon(R) CPU.
[d]CacheSize: Cache size of CPU.
[e]PhysicalCPUs: Number of physical CPUs.
[f]Cores: Number of CPU cores.
[g]Processors: Number of logical CPUs.
[h]Memory: Memory capacity.
[i]OS: Ubuntu version.
[j]Kernel: OS Kernel version.

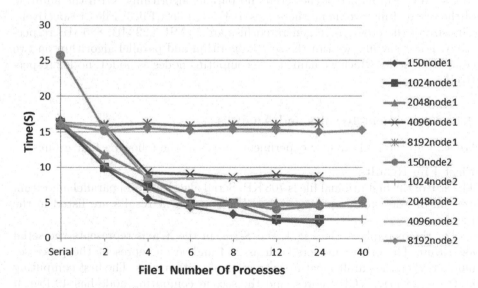

Fig. 1. File1 Performance Results Running On node1 and node2

and parallel algorithms indicated by the seconds. The sideward legend *150node1* represents the current curve stem from the experiment results when the serial and parallel algorithms run on the node1 and the size of the data block is 150. The other legends have the similar meaning. For the Fig. 2, 3, 4, they have the similar signifier.

For the curve *150node2* in the Fig. 1, when the block size is 150 and the serial algorithm runs on the computing node2, the expended time is 25.795 s which is the maximal expended time. For the curve *8192node1* and the curve *8192node2*, whether comparing serial algorithm and parallel algorithm, or comparing the number of different processes in parallel algorithm, there is no significant change in performance. Because for these two curves, the size of the data block is 8192, the size of the *File1* is 10000, the algorithms use up to two processes to complete all the tasks. Given more processes, it does not help to improve the performance of the algorithms. For the curves of *4096node1* and *4096node2*, when the number of processes is 4, the performance is improved compared to 2 processes. When the number of processes is greater than 4, there is no significant change in performance. The reason is similar as the curves *8192node1* and *8192node2*. For the curve *150node1*, the performance continues to improve as the number of processes increases. But for the curve *150node2*, when there is 40 processes, the performance decreases compared with the performance when the number of processes is 24. The reason is that when the size of data block is 150, there is only 67 data blocks. But there is 40 processes, the system need to provide resources for the processes running. The number of data blocks is small, the advantages of multiple processes can not be exerted. So compared with the 24 processes, the performance is lower when there are 40 processes.

Second File Results
For the Fig. 2, except the curve *150node2*, the results of the other curves show that the performance improves as the number of the processes increases. When the number of processes is 40 and the data block size is 150, the spent time is 101.098 s, but when there is 24 processes, the spent time is 89.788 s. The performance reduces as the number of processes increases. At the same time, for the computing node2, we find that the serial algorithm in which the data block size is 150 spends more time than other serial algorithm in which the data block size is greater than 150, and for every different processes number, the parallel algorithm runs lower when the data block size is 150 than the other cases when the data block size is larger than 150. These two findings are verified in the Fig. 3 and 4.

Third File Results
For the Fig. 3, there are similar results with the Fig. 2. As the number of processes increases, the run time of the algorithm continues to decrease except for the situation when the number of the processes is 40 on the computing node2 (1323.529 s, curve *150node2*). Especially, for the curve *150node2*, compare the situation when the processes number is 40 (1323.529 s) with the situation when the processes number is 24 (522.203 s), the difference between the two time values is more distinctive than the difference in the same curve of the Fig. 2 (101.098 s and 89.788 s).

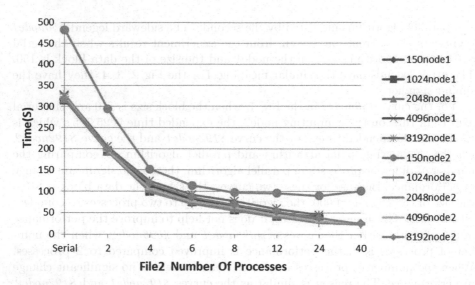

Fig. 2. File2 Performance Results Running On node1 and node2

Fourth File Results

For the Fig. 4, whether for the computing node1 or the computing node2, as the processes number increase, the algorithm run time continues to decrease except for the curve *150node2*. When the processes number is 24, the run time increases (6135.319 s) compared with the processes number 12 situation (6019.363 s).

Experiment Results Analysis

Based on the above four groups of experiment results, we summarize the results to the Table 2. For different files and different data block size, we calculate the performance improvement for computing node1 and node2 respectively. For example of the first value *7.62* in the Table 2, we compute the value based on the Fig. 1, the serial algorithm spent 16.656 s when run on the computing node1, for different processes number, the minimum running time of the parallel algorithm is 2.187 s, based on the division between these two values, the performance improves 7.62×. For the other files, the other data block sizes, the other computing node, we compute the performance improvement using the same method. Combining all the data in the Table 2, we find that the maximum performance improvement is **15.38×** when the block size is 8192 and implement the feature extraction for the 885 MB *File4* on the computing node2.

We implement the parallel algorithm using Python *multiprocessing* package. We compare the performance of parallel algorithm using similar package such as *threading*, the results show that the performance improves 15.38× at the best when using the *multiprocessing* package, but the performance has no improvment when using *threading* package. For the other similar packages, for example, *concurrent*, *subprocess*, we will verify the performance for the future work.

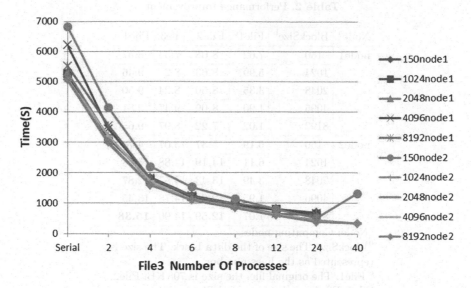

Fig. 3. File3 Performance Results Running On node1 and node2

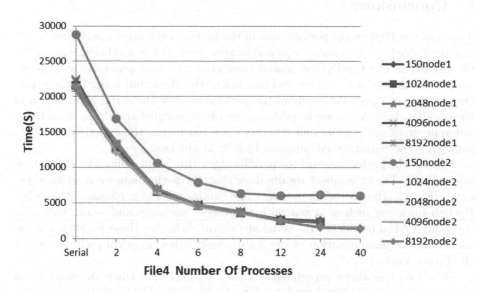

Fig. 4. File4 Performance Results Running On node1 and node2

Table 2. Performance improvement

Node[a]	BlockSize[b]	File1[c]	File2	File3	File4
node1	150	7.62	8.08	7.54	8.57
	1024	5.99	8.63	8.2	9.46
	2048	3.35	8.50	8.24	9.50
	4096	1.90	8.09	9.47	9.81
	8192	1.02	7.22	8.97	9.65
node2	150	6.19	5.37	13.07	4.78
	1024	6.11	14.19	14.88	15.08
	2048	3.49	13.43	14.57	12.87
	4096	1.95	12.26	14.78	13.45
	8192	1.07	12.59	14.90	**15.38**

[a]Node: Computing nodes.
[b]BlockSize: The size of the data block. The size is represented as the lines number.
[c] File1: The original file, the size is 708 KB. File2, File3, File4 are the generated files.

5 Conclusions

This paper optimizes the performance of the feature extraction operations which can be viewed as the making non-accelerator work faster method proposed by Google brain. For the Python nested class structure and *generator* statement, we rebuild the serial algorithm and parallelize the algorithm using Python *multiprocessing* package. We compare the performance of the serial algorithm and the parallel algorithm, also the performance of the parallel algorithm when using different processes number and different data block size, the experiment results show that the performance improves 15.38× at the best.

We test the performance of the parallel algorithm implemented by the *threading* package. The experiment results show that the performance has no improvement. Except the *threading* package and *multiprocessing* package, the other Python packages such as *concurrent, subprocess* packages and so on, they also could be used to parallelize the serial algorithm. Whether these packages can be useful for the parallelization of the feature extraction algorithm will be one of the future work.

Based on the above experiment results, we find that when the data block size is set to 150, for the three large files *Fiel2, Fiel3, Fiel4*, whether the serial algorithm or the parallel algorithm under different data block sizes, the algorithm run slower. Optimization of training process performance for the smaller data block size should also be considered for the future work.

Acknowledgment. We thank the platform provided by Computer Network Information Center, Chinese Academy of Sciences. Professor ZhongHua Lu and Professor YunQuan Zhang give their great support to this paper. We thank their guidance and suggestions. This paper is supported by Informatization Projects of the Chinese Academy of Sciences, Development of Operation Management and Application Software Environment for XiongAn (HengShui) Advanced Supercomputing Center (Grant No. XXH13515).

References

1. Taigman, Y., Yang, M., Ranzato, M.A., Wolf, L.: DeepFace: closing the gap to human-level performance in face verification. In: Proceedings of the IEEE Conference on Computer Vision and Pattern Recognition, pp. 1701–1708 (2014)
2. Krizhevsky, A., Hinton, G.: Learning multiple layers of features from tiny images (2009)
3. Krizhevsky, A., Sutskever, I., Hinton, G.: ImageNet classification with deep convolutional neural networks. In: Advances in Neural Information Processing Systems, pp. 1097–1105 (2012)
4. Krizhevsky, A.: One weird trick for parallelizing convolutional neural networks. arXiv preprint arXiv:1404.5997 (2014)
5. Hubel, D., Wiesel, T.: Receptive fields of single neurones in the cat's striate cortex. J. Physiol. **148**(3), 574–591 (1959)
6. Zhu, Y., et al.: Entropy-aware I/O pipelining for large-scale deep learning on HPC systems. In: IEEE 26th International Symposium on Modeling, Analysis, and Simulation of Computer and Telecommunication Systems (MASCOTS) (2018)
7. Zhu, Y., et al.: Multi-client DeepIO for large-scale deep learning on HPC systems. In: Proceedings of the International Conference on High Performance Computing, Networking, Storage and Analysis (SC 2018) (2018). Regular Poster
8. Kitchen, D.B., Decornez, H., Furr, J.R., Bajorath, J.: Docking and scoring in virtual screening for drug discovery: methods and applications. Nat. Rev. Drug Discov. **3**, 935–949 (2004)
9. Schneider, G., Fechner, U.: Computer-based de novo design of drug-like molecules. Nat. Rev. Drug Discov. **4**, 649–663 (2005)
10. Cumming, J.G., Davis, A.M., Muresan, S., Haeberlein, M., Chen, H.: Chemical predictive modelling to improve compound quality. Nat. Rev. Drug Discov. **12**, 948–962 (2013)
11. He, K., Zhang, X., Ren, S., Sun, J.: Deep residual learning for image recognition. In: Proceedings of the IEEE Conference on Computer Vision and Pattern Recognition, pp. 770–778 (2016)
12. Goyal, P., et al.: Accurate, large minibatch SGD: training imagenet in 1 hour. arXiv preprint arXiv:1706.02677 (2017)
13. Simm, J., et al.: Repurposing high-throughput image assays enables biological activity prediction for drug discovery. Cell Chem. Biol. **25**, 611–618 (2018)
14. https://github.com/deepchem/deepchem/tree/master/deepchem
15. https://www.tensorflow.org/
16. https://tripod.nih.gov/tox21/challenge/
17. https://www.daylight.com/dayhtml/doc/theory/theory.smiles.html
18. Neglur, G., Grossman, R.L., Liu, B.: Assigning unique keys to chemical compounds for data integration: some interesting counter examples. In: Ludäscher, B., Raschid, L. (eds.) DILS 2005. LNCS, vol. 3615, pp. 145–157. Springer, Heidelberg (2005). https://doi.org/10.1007/11530084_13
19. https://en.wikipedia.org/wiki/Simplified_molecular_input_line-entry_system

304   K. Zhan et al.

20. Jastrzebski, S., Lesniak, D., Czarnecki, W.M.: Learning to SMILE(S). In: International Conference on Learning Representations (2016)
21. Li, M., Zhang, T., Chen, Y., Smola, A.J.: Efficient mini-batch training for stochastic optimization. In: Proceedings of the 20th ACM SIGKDD International Conference on Knowledge Discovery and Data Mining, pp. 661–670. ACM (2014)
22. Choi, D., Passos, A., Shallue, C.J., Dahl, G.E.: Faster neural network training with data echoing. Google Brain. arXiv Preprint. arXiv:1907.05550 (2019)
23. Zhan, K., Piao, A.H.: Optimization of Ceph reads/writes based on multi-threaded algorithms. In: 18th IEEE International Conference on High Performance Computing and Communications (2016)
24. Zhan, K., Xu, L., Yuan, Z., Zhang, W.: Performance optimization of large files writes to Ceph based on multiple PipeLines algorithm. In: 16th IEEE International Symposium on Parallel and Distributed Processing with Applications (2018)
25. Islam, N.S., Wasi-ur Rahman, Md., Lu, X., Panda, D.K.: High performance design for HDFS with byte-addressability of NVM and RDMA. In: Proceedings of the 2016 International Conference on Supercomputing, p. 8. ACM (2016)
26. https://github.com/deepchem/deepchem/blob/master/deepchem/data/tests/test_data_loader.py
27. https://docs.python.org/3.7/library/threading.html#modulethreading
28. https://docs.python.org/3.7/library/multiprocessing.html
29. https://docs.python.org/3.8/whatsnew/changelog.html#python380final
30. https://deepchem.io/docs/notebooks/Multitask_Networks_on_MUV.html

Principal Component Analysis for Fingerprint Positioning

Yang Zhang[1], Qianqian Ren[1(✉)], Jinbao Li[2(✉)], and Yu Pan[1]

[1] Department of Computer Science and Technology, Heilongjiang University, Harbin 150080, China
renqianqian@hlju.edu.cn
[2] Shandong Artificial Intelligence Institute, Qilu University of Technology (Shandong Academy of Science), Jinan 250014, China
lijinb@sdas.org

Abstract. Due to the universal deployment of wireless LANs and the demands of indoor location based services, Wi-Fi fingerprinting has been investigated recently for localization. Regarding the influence of environment factors on RSSI measurements, this paper first presents fingerprint quantization method to form a quantized fingerprint database via threshold comparison. Then, we implement principal component analysis on the quantized fingerprint and generate dimensionality reduction fingerprint database. Based on these fingerprint, target localization is implemented. Finally, a comprehensive set of simulations are presented. We study and compare the localization accuracy under different fingerprint databases.

Keywords: Indoor localization · Fingerprinting · PCA

1 Introduction

In recent years, indoor localization has attracted widespread attention and many localization methods have been proposed [1]. However, the environment factors such as noise, signal fluctuation and obstacles presence propose challenges for the study of indoor localization. Among the existing localization techniques, Wi-Fi fingerprinting has been investigated and attracted continuous attention [2–5].

Wi-Fi fingerprinting is generally consisted of two stages, that's offline stage and online stage. During the offline stage, reference points(RPs) collect RSSI measurements of all the detected Wi-Fi signals emitted by access points (APs) and organized them in the of form RSSI vectors, which are stored as fingerprint database. During online stage, an interested target collects the vector of RSSI measurements and compares it with fingerprints in the database with a certain similarity metric, i.e. the Euclidean distance [1, 6]. A subset of RPs whose fingerprints are most similar with the RSSI vector of the target are chosen to participate in localization.

In this paper, we investigate the problem of Wi-Fi fingerprint-based positioning and proposed a principal component analysis (PCA) based positioning algorithm. In the offline stage, we first construct a RSSI fingerprint database. Regarding the influence of

M. Qiu (Ed.): ICA3PP 2020, LNCS 12452, pp. 305–313, 2020.
https://doi.org/10.1007/978-3-030-60245-1_21

environment factors on RSSI measurements, we quantize RSSI fingerprint using threshold comparison. Based on which, we further implement principal component analysis on quantized fingerprint to form PCA fingerprint database. The contributions of the paper is as followings:

(1) We construct a network model and collect RSSI measurements to form a RSSI fingerprint database.
(2) Considering the influence of environment factors on RSSI measurements and localization results, we propose a fingerprint quantitation scheme to further form a quantized fingerprint database.
(3) We implement PCA on quantized fingerprint and choose partial elements to form a reduced-dimension fingerprint database. Based on which, the target localization algorithm is implemented.
(4) We construct a simulation platform and implement the proposed algorithm, the experimental results show the excellent performance of our algorithm. Moreover, we compare the localization accuracy under different fingerprint databases.

This paper is organized as follows. In Sect. 2, we review the related work of localization algorithms. We present in Sect. 3 the deployment of Wi-Fi fingerprint-based systems. In Sect. 4, we describe the fingerprint database formation methods for positioning in detail. We conclude by briefly our work in Sect. 5.

2 Related Work

Since RSSI is easy to measure, no additional hardware configuration is required, and the implementation is simple, RSSI-based moving target tracking and positioning algorithms have been widely applied in recent years. We will make a brief review to the work related to this paper.

Xue et al. propose a method based on RSSI mean value to locate the target [7], which solves the problem of signal instability existing in the traditional RSSI-based positioning technology. However, in some cases, the selected RSSI measurement value is easily affected by the environment, resulting in greater RSSI mean value error. A partitioning and positioning solution based on support vector machine (SVM) is presented in [8]. This solution uses RSSI measurements collected in the real environment to perform multi-classification SVM algorithm. A moving target tracking and positioning algorithm is given in [9], which measures the RSSI values between the moving target and the fixed sensor nodes. The algorithm selects and weights RSSI measurements according to the strength of RSSI measurements. Moreover, a propagation model is used to convert the RSSI measurements into distance to estimate the position of the moving target timely.

Indoor radio transmission is relatively complex, as RSSI measurements are often noisy and fluctuant. A large number of RSSI measurements are needed to ensure localization accuracy, which inevitably leads to the increasing of computation and transmission amount. In order to balance the accuracy and computation cost, RSSI quantization based localization is proposed. Studies in [10] have found that RSSI value can be reduced from 8 bits to 2 bits without affecting positioning accuracy. Fuzzy logic is used in [11] to divide RSSI values into four regions, which reduces the uncertainty generated by changes in indoor environment. The method in [12] applies quantization to fingerprint positioning and used more beacon nodes to compensate the loss of quantization accuracy. The localization accuracy under different RSSI quantization schemes including global 1-bit quantization, local 1-bit quantization and global 2-bit quantization is analyzed in [13], and it is concluded that using local 1-bit quantization lead to substantial performance over a global 1-bit quantization [13].

3   System Model

In this section, we describe the deployment of a Wi-Fi indoor positioning system. The RSSI measurements are collected in the hall of a laboratory building without too many obstacles, thus the radio transmission environment is relatively stable. The deployment of APs and RPs are shown in Fig. 1. To reduce the configuration cost, the Wi-Fi APs are not densely installed. In our experiments, eight APs are installed, six APs are on the boundary of the monitoring area and two APs are in the middle. The signals of these APs can cover the whole monitoring area. In the deployment, the monitoring area is further divided into multiple cells, and there exists one RP at the center of each grid cell.

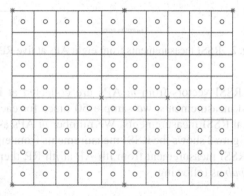

Fig. 1. A site map of a Wi-Fi indoor positioning system

4   Generating Fingerprint Database for Localization

In this section, we aim to generate three kinds of fingerprint database for localization, the first is fingerprint database on raw RSSI vectors, the second is fingerprint database on quantized RSSI vectors and the third is PCA based fingerprint database. Wi-Fi fingerprint database formation is conducted in three phases.

(1)   RSSI vectors of all the detected Wi-Fi signals from APs at multiple RPs are collected to form the original fingerprint database.
(2)   A quantized fingerprint database is formed via threshold comparison.

(3) Due to the signal contributed by each AP to RPs is different, we conduct PCA on quantized fingerprint database and select the components with higher contribution to generate PCA based fingerprint database.

4.1 RSSI Fingerprint Database

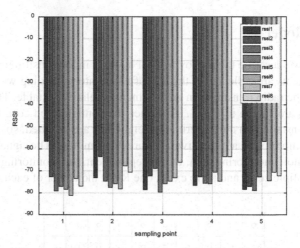

Fig. 2. RSSI measurements from eight APs at five RPs

In offline phase, RSSI vectors from m APs are collected at RPs. To reduce the influence of abnormal measurements, we sample 50 times for each AP and calculate the average as the final RSSI measurements. We number m APs as 1, 2, 3, ..., $m - 1$ and m, respectively. Figure 2 shows the RSSI measurements from eight APs at five RPs. Let $rssi_{ij}$ denote the RSSI measurement from ith AP at jth RP, then we can denote the fingerprint database $fpdb$ as following.

$$fpdb_{n \times m} = \left\{ \begin{array}{l} rssi_{11}, \ rssi_{12} \ \ldots \ rssi_{1m} \\ rssi_{21} \ \ rssi_{22} \ \ldots \ rssi_{21} \\ \qquad \qquad \ldots \\ rssi_{n1} \ \ rssi_{n1} \ \ldots \ rssi_{nn} \end{array} \right\} \tag{1}$$

In online phase, we use the weighted k-nearest neighbors scheme to locate the target. Given the RSSI vector $T = \{t_1, t_2, ..., t_m\}$ the target sampled from APs, we compare T with the stored fingerprint using the similarity metric of Euclidean distance, which is defined as following:

$$d_j = \sqrt{\sum_{i=1}^{m} (rssi_{ji} - t_i)^2} \tag{2}$$

Let $D_k = \{d_1, d_2, ..., d_k\}$ denote the distance set of k-nearest neighbor fingerprints, we calculate the weights of k-nearest neighbors as:

$$\omega_j = \frac{1 \big/ d_j^2}{\sum_{i=1}^{k} d_j^2} \tag{3}$$

Where w_j is the weight value of jth nearest neighbor. Thus, the localization result is:

$$(x, y) = \sum_{i=1}^{k} \omega_i \times (x_i, y_i) \tag{4}$$

4.2 Quantized Fingerprint Database

Traditional fingerprinting positioning is mostly based on the RSS signal vectors [1, 2]. As some environment factors such as multi-path effects may bring measurement noise, the RSSI vectors may not illustrate the distance between the target and APs. To reduce the influence and achieve higher accuracy, we try to quantize RSSI vectors and form quantized fingerprinting. It can provide relatively reliable localization results.

Given a threshold R_{thr}, RSSI values can be quantized as 0 or 1. Let R_{min} represent the smallest RSSI value in the fingerprint database and R_{max} represent the greatest RSSI value. The quantized RSSI can be denoted as:

$$rssi'_{ji} = \begin{cases} 0, & R_{\min} \le rssi_{ji} < R_{thr} \\ 1, & R_{thr} \le rssi_{ji} < R_{\max} \end{cases} \tag{5}$$

Based on quantized RSSI results, we further select two representative RSSI values for the two intervals 0 and 1, denoted as R_0 and R_1. We use R_0 and R_1 to represent 0 and 1 in the quantitation result.

4.3 PCA Fingerprint Database

In the procedure of fingerprint quantitation, we quantize each RSSI value in the fingerprint database in the same way, which means that the contribution of each AP to target localization is the same. Considering the the distance from each AP to the target is different, we can weight the contribution of each AP. In this section, we first conduct principal component analysis on the quantized fingerprint database, and select a group of principal components that have greater impact on localization accuracy to generate a new fingerprint database. Then, we locate the target on the new fingerprint database.

Principal Component Analysis (PCA) is a common data set dimensionality reduction algorithm. It aims to map the n-dimensional features onto the k-dimensional features, and these k new features are irrelevant orthogonal features. It looks for k orthogonal coordinate axes in the original data feature space. The first coordinate axis is the direction that maximizes the variance when original data is projected onto the coordinate axis. The second coordinate axis is the direction orthogonal to the first coordinate axis that maximizes the variance when original data is projected onto the coordinate axis, and so on. Finally, n orthogonal coordinate axis can be obtained. We find that the first k coordinate axes already contain most variance, so we only keep the first k coordinate axes and ignore the others, that's we reduce the original data from n-dimensional features to k-dimensional features.

We observe that signal patterns in the fingerprint database are correlated, that's temporal or spatial correlated. Moreover partial signals overlap. Based on above observations, we try to use principal component analysis to identify several irrelevant comprehensive indicators to represent most values in the fingerprint database. We assume that the quantized fingerprint database is $fpdb'$, then the comprehensive indexes obtained via PCA can be denoted as:

$$[I_1 I_2 \ldots I_n] = fpdb'_{n \times m} \times \begin{bmatrix} a_{11} & a_{12} & \ldots & a_{1n} \\ a_{21} & a_{22} & \ldots & a_{2n} \\ & & \ldots & \\ a_{m1} & a_{m2} & \ldots & a_{mn} \end{bmatrix} \tag{9}$$

Where I_n denotes the coordinate value of the quantized fingerprint projected onto the nth coordinate axis $(a_{1n}, a_{2n}, \ldots, a_{8n})$, that's the nth eigenvalue. In order to represent as much information as possible, comprehensive indexes need to satisfy the following the conditions:

(1) n axis vectors are orthonormal basis;
(2)   The variance of $I_1 \, I_2 \ldots I_n$ goes down, and it is independent of each other.

In this section, we choose the top k composite indexes with the highest contribution in the total variance as the principal components of localization.

5   Simulations and Analysis

We construct a simulation platform in the hall of a laboratory building, where eight APs are installed, six APs are on the boundary of the monitoring area and two APs are in the middle. Eighty RPs are deployed in the monitoring area. We use a personal laptop computer as a server to implement localization algorithm, running windows 8 operating system. The simulation platform and the whole localization procedure are implemented using Python language.

In the experiments, we first analyze the influence of parameters including k value choosing (k-nearest neighbor) and threshold on localization accuracy. Then, we generate the fingerprint database after dimensionality reduction. Finally, we compare the influence of three kinds fingerprint on localization accuracy.

5.1   Impact of k on Localization Accuracy

In this section, we analyze the influence of k value on localization accuracy. We first choose a good k value on the original fingerprint database via experience, and then we implement quantization and PCA using this k value. Figure 3 shows the localization error varying k from 1 to 12. It is obvious that the localization error is the greatest when k is 1 and 12, respectively. That's because if k is too small, there is not enough information to produce a good location estimation. As the increasing of k, localization error increases because the algorithm will take into account much irrelevant sample information, which makes the localization estimation deviate from the real position. According to the figure, we can obtain the optimal localization result when $k = 4$.

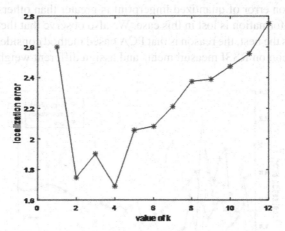

Fig. 3. The influence of k on localization accuracy

5.2 Impact of Threshold on Localization Accuracy

In this section, we study the influence of threshold values on localization accuracy. Figure 4 shows the localization error when the threshold changes from -85 to -55. It

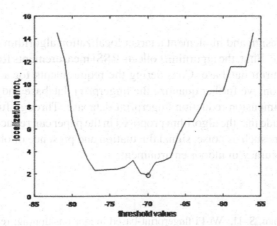

Fig. 4. The influence of threshold on localization accuracy

can be seen that the localization error is smaller when the threshold value ranges from −77 to −68. When the threshold is −70, the localization error is the smallest.

5.3   Localization Accuracy Comparison Under Different Fingerprint Database

In this section, we implement localization algorithm under original fingerprint database, quantized fingerprint database and PCA fingerprint database, respectively. Moreover, we compare the localization accuracy under these fingerprint databases. It is shown in Fig. 5 that the localization error of quantized fingerprint is greater than others, that's because partial location information is lost in this case. We also observe that the performance of PCA fingerprint is the best, the reason is that PCA based method considers the difference of APs' contribution on RSSI measurements and assign different weights for them.

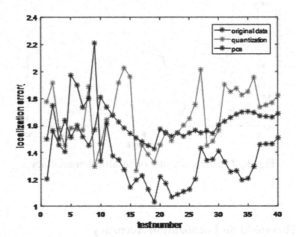

Fig. 5. Comparison of localization errors of three fingerprint algorithms

6   Conclusion

In this paper, we design and implement a target localization algorithm on three types of fingerprint database. First, the algorithm collects RSSI measurements from RPs and form the original fingerprint database. Considering the requirements for accurate and cost-effective localization, we further quantize the fingerprint database and implement PCA on it to generate dimension reduction fingerprint database. Through fully experimental analysis, we conclude that the algorithm proposed in the paper can reduce the influence of environment factors such as noise, signal fluctuation and presence of obstacles, improve the localization accuracy in indoor environment.

References

1. He, S., Gary Chan, S.-H.: Wi-Fi fingerprint-based indoor positioning: recent advances and comparisons. IEEE Commun. Surv. Tutorials **18**(1), 466–490 (2016)

2. Wu, C., Yang, Z., Liu, Y., Xi, W.: WILL: Wireless Indoor Localization without site survey. IEEE Trans. Parallel Distrib. Syst. **24**(4), 839–848 (2013)
3. Liu, H., et al.: Push the limit of WiFi based localization for smartphones. In: Proceedings of ACM MobiCom, pp. 305–316, September 2012
4. Sun, W., et al.: MoLoc: on distinguishing fingerprint twins. In: Proceedings of IEEE ICDCS, pp. 226–235, July 2013
5. Xiao, Z., et al.: Non-line-of-sight identification and mitigation using received signal strength. IEEE Trans. Wirel. Commun. **14**(3), 1689–1702 (2015)
6. Bahl, P., Padmanabhan, V.N.: RADAR: an in-building RF-based user location and tracking system. In: Proceedings of IEEE INFOCOM, pp. 775–784 (2000)
7. Xue, W., Qiu, W., Hua, X., et al.: Improved Wi-Fi RSSI measurement for indoor localization. IEEE Sens. J. **17**(7), 2224–2230 (2017)
8. Chriki, A., Touati, H., Snoussi, H.: SVM-based indoor localization in wireless sensor networks. In: 2017 13th International Wireless Communications and Mobile Computing Conference (IWCMC), pp. 1144–1149. IEEE (2017)
9. Barsocchi, P., Lenzi, S., Chessa, S., et al.: A novel approach to indoor RSSI localization by automatic calibration of the wireless propagation model. In: VTC Spring 2009-IEEE 69th Vehicular Technology Conference, pp. 1–5. IEEE (2009)
10. Gao, W., Nikolaidis, I., Harms, J.J.: RSSI quantization for indoor localization services. In: 2017 IEEE 28th Annual International Symposium on Personal, Indoor, and Mobile Radio Communications (PIMRC), pp. 1–7. IEEE (2017)
11. Hernández, N., Herranz, F., Ocaña, M., et al.: Wifi localization system based on fuzzy logic to deal with signal variations. In: 2009 IEEE Conference on Emerging Technologies & Factory Automation, pp. 1–6. IEEE (2009)
12. Mizmizi, M,. Reggiani, L.: Design of RSSI based fingerprinting with reduced quantization measures2016 International Conference on Indoor Positioning and Indoor Navigation (IPIN), pp. 1–6. IEEE (2016)
13. Yong, A., Nikolaidis, I., Harms, J.J.: Localization sensitivity under RSSI quantization. In: ICC 2019-2019 IEEE International Conference on Communications (ICC), pp. 1–6. IEEE (2019)

Priority Based Service Placement Strategy in Heterogeneous Mobile Edge Computing

Meiyan Teng[1], Xin Li[1,2,3]($\boxtimes$), Xiaolin Qin[1], and Jie Wu[4]

[1] CCST, Nanjing University of Aeronautics and Astronautics, Nanjing, China
{myteng,lics,qincs}@nuaa.edu.cn
[2] State Key Laboratory for Novel Software Technology,
Nanjing University, Nanjing, China
[3] Collaborative Innovation Center of Novel Software Technology
and Industrialization, Nanjing, China
[4] Center for Networked Computing, Temple University, Philadelphia, USA
jiewu@temple.edu

Abstract. Mobile Edge Computing (MEC) is a promising method to reduce service delay by computational ability at the edge nodes. However, the limited resources at the edge nodes make it hard to response various services simultaneously. Hence, it is challenging to utilize the limited edge resources to host various service and reduce service response time. In this paper, we investigate the service placement problem such that the average service response time is minimized, which affects the user experience significantly. We define the priorities for nodes and services according to their contribution values, which indicates the influence for reducing service response time. Then, we propose a priority placement (2P) algorithm by taking both priority properties and local optimization into account. We conduct extensive simulations and the experimental results show that the 2P algorithm can reduce the average service response time by 23%–46%, which indicates the 2P algorithm has better performance in reducing response time compared to the classical methods.

Keywords: Heterogeneity · MEC · Service placement · Low-latency · Data-intensive

1 Introduction

Nowadays, the scale and functions of the Internet of Things (IoT) have increased dramatically, heralding the arrival of the Internet of Everything. In the near future, the global IoT will expand to tens of billions of application devices [1]. Latency-sensitive application devices are on the rise, with devices such as augmented reality (AR) and driverless cars requiring real-time data processing.

© Springer Nature Switzerland AG 2020
M. Qiu (Ed.): ICA3PP 2020, LNCS 12452, pp. 314–329, 2020.
https://doi.org/10.1007/978-3-030-60245-1_22

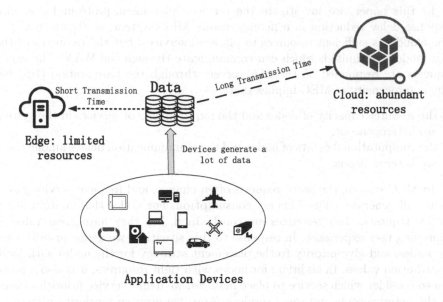

Fig. 1. Overview of cloud-edge in IoT environment.

With the rapid development of IoT technology, traditional cloud is no longer suitable for IoT applications [2].

Mobile edge computing (MEC), which places servers at the edge of the network to offload cloud resources and reduce response delay, has drawn more and more attention [3]. It allows real-time analysis, testing, optimization on edge servers and sends data that needs to be processed centrally to the cloud server [4]. This technology not only reduces the burden on the cloud server, but also improves user experience. Edge computing is not a substitute for the cloud computing paradigm [5], but adds another layer of computing where it is close to the user.

However, a research institution predicts that by 2035, there will be 54 million driverless cars in the world [6]. Cameras on driverless vehicles that capture road conditions in real time generate about 1 GB of data per second [7], even boeing 787 will produce 5 GB data per second [8]. The service requests for such data-intensive applications not only put higher demands on latency, but also take up a lot of resources. Although the edge computing model has reduced the bandwidth pressure in network transmission and achieved low-latency response of services, the computing capacity and storage capacity of edge nodes are limited compared with those of cloud. This defect causes the services of numberous applications to be unresponsive and reduces user experience, as shown in Fig. 1. In this case, the problem of service placement needs to take both the cloud and edge nodes into account and makes a trade-off between them to achieve the goal of minimum response time.

In this paper, we investigate the services placement problem for service response delay reduction in a heterogeneous MEC system, as shown in Fig. 2. The cloud has sufficient resources to place all services, but the resources of the edge nodes are limited, which can communicate through the WAN. The user's requests are responded on the edge servers through the base station (BS) [9]. The heterogeneity of MEC implies that:

- the resource capacity of nodes and the requirements of services in the system are heterogeneous;
- the computational delays of nodes and the communication delays among them are heterogeneous.

In MEC system, the users' requests often change, and frequent service placement will generate a lot of energy consumption. For nodes that contain many service requests, their resources are constrained, but they have great value in improving user experience. In response to this situation, we define priorities for the nodes, and give priority to the placement strategy for the nodes with large contribution values. In addition, for nodes with tight resources, it is also a problem to consider which service to place first. So, we define service priorities based on the heterogeneous nature of services. And, we propose a priority placement (2P) algorithm to get a placement strategy with the lower system latency. The main contributions in this paper are as follows:

1) We make a detailed description of the research problem, construct a heterogeneous MEC system model, set system parameters, formulate the above characteristics and research problem.
2) We propose a priority placement (2P) algorithm to achieve services placement at MEC. The main idea is to set priorities based on service load distribution and set an upper limit on the number of identical service replicas to attain the goal of minimizing system response time.
3) We conduct extensive simulations, taking into account the priorities affected by load factors, as well as conducting comparative experiments. The results show that our algorithm has a significant improvement in reducing response delay under different parameters.

The rest of the paper is organized as follows. We summarize the related work in Sect. 2. Then, we describe the system model and formulize the service placement problem in Sect. 3, and propose our priority placement algorithm in Sect. 4. Next, we evaluate the performance of our algorithm in Sect. 5. Finally, we give conclusion remarks in Sect. 6.

2   Related Work

The research on service placement has attracted a lot of attention in recent years and there have been many research results. The most direct approach is to place the service on the local edge server, and [10] shows that this approach is feasible. In the case of adequate node resources, this strategy minimizes response time in the system, but if the resources are limited, most services will only be placed in the centralized cloud. Therefore, this strategy is obviously not the most efficient.

The authors in [11] study the scenario of overlapping node coverage and use random rounding technology to solve the service placement method under resource constraints. The service placement strategy studied in [12] and [13] fails to take the limitations of node capacity, computing power and transmission bandwidth into account, so the strict resource constraints of edge nodes cannot be captured. In our paper, the scenario we studied is that node coverage does not overlap and resources are limited.

Xu et al. [14] weigh the delay and cost in the homogeneity MEC environment to obtain the placement strategy. He et al. [15] adopt greedy strategy, and the algorithm performance is better under the condition of homogeneous service performance. But the resource requirements for services are different and the performance is heterogeneous in general.

The authors in [16] adopt the way of node division to select the service with the greatest reward in a heterogeneous MEC system. This algorithm results in multiple placements of the same service replica on the uniform node. And frequent service placements increase system power consumption. And [17] adopts the linear programming method to jointly consider the service placement and request scheduling policies, but it is no work to consider the uneven load distribution of services in the case of heterogeneous MEC systems, which will have a great impact on the placement results.

In our paper, we define priorities for nodes and services based on the heterogeneous characteristics of nodes and services, then we propose a priority placement (2P) algorithm.

3   Problem Statement

3.1   Scenario and Notation

We focus on service placement issues within the heterogeneous MEC system, as shown in Fig. 2. The system is composed of some edge nodes and a remote centralized cloud. For the cloud, there are sufficient computing and storage resources to place services, which will generate faster computing speed. However, due to the distance from the data source and the limitation of communication bandwidth, the communication delay is high. So, it is suitable to place large services with insensitive latency. For edge nodes, they have limited resources to place services, which results in a longer calculation delay, but a lower communication delay. They are suitable to place small delay-sensitive services. The node characteristics and service requirements in MEC system are different. In this part, we use special symbols to represent the heterogeneous characteristics.

The set of nodes in the system is represented by $\mathbf{N}$, and the cloud is signed by N_0. For each node $n \in \mathbf{N}$ has a special performance $\langle R_n, \Gamma_n \rangle$.

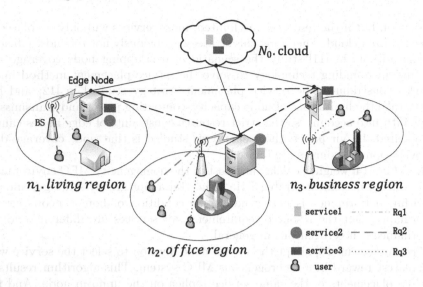

Fig. 2. System model.

- R_n represents the resource capacity of node n;
- Γ_n represents the communication delay time between node n and the cloud N_0.

The set of services in the system is represented by **S**, and the performance attributes $\langle r_l, P_l \rangle$ of each service $l \in$ **S** are different.

- r_l represents the resources that service l needs to consume when the service responds;
- P_l represents the number of replicas of service l in the system.

In addition to the above characteristics, load is also an important factor. Node load refers to the number of users in the node. The service load refers to the number of service requests. The load of the nodes varies in different time periods, and the service load is unevenly distributed. For example, during working hours, there are more users in the office region than in living and business regions, and the number of requests for various types of services in the office region is also higher than in other regions. In other time periods, different phenomena will occur. In this article, we use Φ to represent service load distribution.

- $\Phi_{l,n} \in \Phi$ represents the load of service l within the coverage of node n.

Requests of services can be scheduled among edge nodes. For example, l_2 is not placed on n_3, so the users in this region send a request Rq_2 for l_2, which cannot be satisfied on the local server. The request is dispatched to node n_1 or n_2 in other regions through WAN. Based on the above characteristics, we set the total response delay as **T**.

- $T^l_{m,n} \in \boldsymbol{T}$ represents the total response delay time when the request Rq_l within n region is scheduled to be served on node m.

The transmission time of the same request from n_1 to n_2 is the same as that from n_2 to n_1, but the computing power of n_1 and n_2 are different. To describe this characteristic of response time $\boldsymbol{T}$, we set

$$T^l_{m,n} = \partial^l_{m,n} + \beta^l_m \qquad (1)$$

- $\partial^l_{m,n}$ represents the communication delay time when the request Rq_l within n region is scheduled to be served on node m, $\partial^l_{m,n} = \partial^l_{n,m}$, and $\partial^l_{n,n} = 0$;
- β^l_m represents the computing time of service l on node n.

According to the above description of the total response delay $\boldsymbol{T}$, we know that the response time of the service is related to which node the service responds to. We measure the efficiency of various algorithms by the response time of service requests. However, it is closely related to the service placement strategy.

3.2   Problem Formulation

We aim to minimize the service request response time according to a service placement strategy in a heterogeneous MEC system. So, we use vector $\boldsymbol{X}$ to represent the service placement scheme in the system. For $\forall l \in \boldsymbol{S}, m \in \boldsymbol{N}$,

$$x_{l,n} = \begin{cases} 1, & \text{service } l \text{ placed on the node n} \\ 0, & \text{otherwise} \end{cases} \qquad (2)$$

The response time of system request is represented by vector $\boldsymbol{Y}$, which is determined by placement scheme $\boldsymbol{X}$. We stipulate that if there are nodes that have placed service l, the requests for l will be scheduled to a node m with the shortest response delay. Otherwise, it will be scheduled to the cloud. It can be formalized as follows:

$$y_{l,n} = \Theta\left(\sum_{m=0}^{|N|} x_{l,m} = 0 \right) \times \varGamma_n + \Theta\left(\sum_{m=0}^{|N|} x_{l,m} \neq 0 \right) \\ \times \min\left\{ T^l_{m,n} | x_{l,m} = 1 \right\}, \quad m, n \in \boldsymbol{N}, l \in \boldsymbol{S} \qquad (3)$$

Where, if $\boldsymbol{E}$ is true, $\Theta(\boldsymbol{E}) := 1$; otherwise, $\Theta(\boldsymbol{E}) := 0$. $y_{l,n}$ represents the response time of the request for service l within the scope of node n.

The purpose of our research is to improve the quality of service. In this paper, our research goal is to minimize the response time of the system, which is expressed as follows:

$$\min \sum_{l=0}^{|S|} \sum_{n=0}^{|N|} y_{l,n} \qquad (4)$$

$$\text{s.t.} \sum_{l=0}^{|s|} x_{l,n} \cdot r_l \leq R_n \quad \forall n \in \mathbf{N} \tag{4.1}$$

$$x_{l,n} \in \{0,1\} \quad \forall l \in \mathbf{S}, \forall n \in \mathbf{N} \tag{4.2}$$

Where the constraint (4.1) means that the services placed on each edge node should not exceed its capacity.

From the above problem description, we learned that service placement is really about allocating limited node resources to services. We predict that service placement problem is an NP-hard problem [18]. Next, we will show the hardness of this problem.

Theorem 1. *The service placement in a heterogeneous MEC system is NP-hard.*

Proof. We will prove the theorem by a special case, which is constructed with the following assumptions: 1) We set the case with one cloud and one edge node, and 2) Only one replica is placed for each service. In this case, the best strategy is to place services on this edge node to lessen response time. However, the available resource in edge node is limited, so we need to choose some services to occupy the node resources such that the response delay is minimized. This problem can be inferred from the typical knapsack problem.

The typical knapsack problem can be formalized as follows. Given a set of items $A = \{a_i, 0 \leq i \leq n\}$, where the weight and the value of a_i is w_i and v_i, respectively. The problem is to select a subset A_s such that the total weight does not exceed the capacity W of knapsack and the total value is maximized. Then, in the scenario we constructed, the set of services is $S = \{s_l, 0 \leq l \leq n\}$, then for each service s_l, let the requirement of s_l be r_l, and the response time of s_l be y_l. We aim to select a subset S_s of services such that the total requirement does not exceed the resource capacity R of edge node and the response time is minimized.

If there is a strategy to select a subset A_s such that $\sum_{a_i \in A_s} w_i \leq W$, and $\sum_{a_i \in A_s} v_i$ is maximized. For each item $a_i \in A_s$, we can select a service $s_l \in S$ with $r_l = w_i$ and $\frac{1}{y_l} = v_i$. The services that are selected have the lowest response delay. In addition, if we select a subset S_s of S to minimize the total response time, we can get the subset A_s to maximize the value. Because the knapsack problem is NP-hard, we conclude that the services placement problem in MEC system is NP-hard. □

4   Service Placement Strategy

To approximately minimize the response time of the service, we set priorities for all nodes and services according to the scene parameters, then determine the placement scheme according to the priority factors, so it is called priority placement (2P) algorithm. Firstly, the node with the highest priority is selected

Algorithm 1. Total Delay Algorithm

Input: Node set $\mathbf{N}(n \in \mathbf{N})$, attributes: $\langle R_n, \Gamma_n \rangle$; Service set $\mathbf{S}(l \in \mathbf{S})$. attributes: $\langle r_l, P_l \rangle$; Schedule delay $\mathbf{T}$; Service load distribution $\boldsymbol{\Phi}$; Upper limit of replicas θ.

1: $\mathbf{Q} = averageDelay(\mathbf{T}, \boldsymbol{\Phi})$;
2: $\mathbf{G} = nodeCandidateSet(\mathbf{Q})$;
3: $\mathbf{L} = initServiceCandidateSet(\mathbf{G}, 1)$;
4: $\mathbf{X} = servicePlacement(\mathbf{L}, \mathbf{G}, \mathbf{Q}, r, R, \boldsymbol{\Phi}, \theta)$;
5: **for** each $l \in \mathbf{S}$ **do**
6:     $P_l = getSum(\mathbf{X})$;
7:     **if** $P_l == 0$ **then**
8:         **for** each $n \in \mathbf{N}$ **do**
9:             $y_{l,n} = \Gamma_n$;
10:        **end for**
11:    **else**
12:        **for** each $n \in \mathbf{N}$ **do**
13:            $y_{l,n} = minResponseTime(\mathbf{X}, \mathbf{T})$;
14:        **end for**
15:    **end if**
16: **end for**
17: **for** each $l \in \mathbf{S}$ **do**
18:    **for** each $n \in \mathbf{N}$ **do**
19:        $t = t + y_{l,n} \cdot \Phi_{l,n}$;
20:    **end for**
21: **end for**
Output: Total delay time is t.

---

to study the placement strategy. Then we choose the service from the service candidate set of the node in order of the priority to place. In the process of service placement, we adjust the priority of nodes and services according to the deployed situation, make a dynamic priority definition. How to define node priority and service priority is a key point in the 2P algorithm. Next, we will introduce them one by one.

4.1 Total Delay Algorithm

The process of finding the total delay time in the system is described in Algorithm 1. Firstly, the input variables are defined. We set the attributes of nodes and services as well as the upper limit of replicas θ. Through the network prediction, we can get the distribution of total response time and service load in a period of time, which is represented by the $\mathbf{T}$ and $\boldsymbol{\Phi}$, separately. According to the input variables, we obtain the node priority through the three-step method of lines 1–3 in Algorithm 1, which is introduced in detail in Sect. 4.2. Next, taking the variables obtained from lines 1–3 as input, line 4 adopts the $servicePlacement()$ function to obtain the placement strategy $\mathbf{X}$; see Algorithm 2 for details.

Afterwards, based on placement scheme $\mathbf{X}$, lines 5–16 calculate the response time of the service request within the scope of each node. The number of replicas of each service in the system is calculated by the $getSum()$ function. If $P_l = 0$, the system does not have a replica of service l, so the request of service l in each region must be responded to in the cloud, then $y_{l,n} = \Gamma_n$. If $\mathrm{P}_l \neq 0$, it indicates that there is at least one replica of this service in the system. Therefore, for requests in each region, we can define the $minResponseTime()$ function on the basis of Eq. 3 to get the minimum response time $y_{l,n}$. Finally, lines 17–21 calculate the total response time $\mathbf{Y}$ for all requests.

4.2 Node Priority

Uneven load distribution will cause a phenomenon that the more load of the node would lead to a higher value of the contribution to the research target, but its resources are relatively tight. Our research goal is to minimize the total response time of all services in the system (e.g. Eq. 4). Therefore, in order to better achieve the research goal, we properly defined the node priority after three steps of calculation.

Firstly, we assume that all requests of service l in the system are responded to node m. According to the input variables $\boldsymbol{\Phi}$ and $\mathbf{T}$, we can get the average response time of service l placed at node m, which is defined as follows:

$$Q_m^l = \frac{\sum_{n=0}^{|N|} T_{m,n}^l \cdot \Phi_{l,n}}{\sum_{n=0}^{|N|} \Phi_{l,n}}, l \in \mathbf{S}, m, n \in \mathbf{N} \tag{5}$$

In the same way, we calculate the value of other services on each node. In line 1 of Algorithm 1, we utilize the $averageDelay(\mathbf{T}, \boldsymbol{\Phi})$ function to implement Eq. 5 and obtain matrix $\mathbf{Q}(Q_m^l \in \mathbf{Q}, l \in \mathbf{S}, m, n \in \mathbf{N})$, which is an important factor in the 2P algorithm.

Secondly, because the average delay time of each request Rq_l that was scheduled to diverse node is different $\left(Q_m^l \neq Q_n^l, l \in \mathbf{S}, m, n \in \mathbf{N}\right)$, we find the ideal node set of service according to the sequence of $\mathbf{Q}$, as follows:

$$G_l = \left\{ n_{p_1}, n_{p_2} \cdots n_{p_i} \cdots | Q_{n_{p_1}}^l \leq Q_{n_{p_2}}^l \leq \cdots \leq Q_{n_{p_i}}^l \leq \cdots, n_{p_i} \in \mathbf{N} \right\} \tag{6}$$

Where $G_{l,1} = n_{p_1}$ indicates that the first ideal node of service l is n_{p_1}, $G_{l,2} = n_{p_2}$ indicates that the second ideal node of service l is n_{p_2}, ..., and $G_{l,i} = n_{p_i}$ indicates that the ith ideal node of service l is n_{p_i}.

In line 2 of Algorithm 1, we adopt the $nodeCandidateSet(\mathbf{Q})$ function to obtain the two-dimensional matrix $G_{\mathbf{S} \times |N|}$, which represents the ideal node sequence for each service. The matrix $\mathbf{G}$ is composed of variables $G_{l,i} \in \mathbf{N}$, where $l \in \mathbf{S}, i \in |\mathbf{N}|$. Furthermore, we get a verdict from the two-dimensional matrix $\mathbf{G}$:

- row vector $G_{l,|\mathbf{N}|}$ shows the ideal node sequence of service l;
- column vector $G_{\mathbf{S},i}$ shows the ith ideal node of all services.

Thirdly, we use function $initServiceCandidateSet()$ based on $\mathbf{G}$ to obtain the initial service candidates set $\mathbf{L}$ of each node, as shown in Algorithm 1 in line 3. It is defined as follows:

$$L_n = \{l_{p_1}, l_{p_2}, \cdots l_{p_i} \cdots | G_{l_{p_i},1} = n, l_{p_i} \in \mathbf{S}\} \qquad (7)$$

Algorithm 2. Service Priority Placement Algorithm

Input: Average delay $\mathbf{Q}$; node candidate set $\mathbf{G}$; service candidate set $\mathbf{L}$; attributes of nodes $\langle R_n, \Gamma_n \rangle$; attributes of services $\langle r_l, P_l \rangle$; service load distribution $\mathbf{\Phi}$; Upper limit of replicas θ .

1: **while** (! isEmpty (**L**)) **do**
2:     $e \leftarrow \text{argmax}_{n \in N} |L_n|$;
3:     Order $L_e = \{l_{p_1}, l_{p_2}, \cdots, l_{p_k}\}$,so that $\Omega_{l_{p_i}} \geq \Omega_{l_{p_{i+1}}}, \forall i < k$;
4:     **for** each $l_{p_i} \in L_e$ **do**
5:         **if** $r_{l_{p_i}} \leq R_e$ **then**
6:             $x_{l_{p_i},e} = 1$;
7:             $R_e \leftarrow R_e - r_{l_{p_i}}$;
8:             $P_{l_{p_i}} + +$;
9:         **else**
10:             $x_{l_{p_i},e} = 0$;
11:         **end if**
12:         **if** $P_{l_{p_i}} < \theta$ **then**
13:             $e' \leftarrow findNextNode(\mathbf{G}, l_{p_i}, e)$;
14:             $update(\Omega_{l_{p_i}})$;
15:             $L_{e'} \leftarrow L_{e'} \cup l_{p_i}$;
16:         **end if**
17:     **end for**
18:     $Clear(L_e)$;
19: **end while**

Output: Service placement strategy is $\mathbf{X}$.

Initially, we put all services whose ideal node is n in the set L_n, and define them as the candidate service set of node n. The set L_n will be dynamically adjusted according to the placement situation. The more services in the set L_n, the higher the value of node n. So, the modulo $|L_n|$ is defined as the value of node n, also known as the priority.

4.3 Priority Placement Algorithm

Algorithm 2 describes the service priority placement function *service Placement()* in detail. The main idea of the algorithm is to select services in order from the service candidate set of the node e with the highest value (priority) for placement strategy, and then add the service to the set L'_e of its sub-ideal node e'.

Firstly, for the initial service candidate set **L** of each node, loop lines 1–18 until the service candidate set L_n of all nodes is null. In circulation, we select the node with the highest priority, which is marked as e, as shown in line 2 of Algorithm 2. Then we order the services in L_e based on service priority Ω. We consider the service loads, the number of replicas, the average response time and the response time gap with the sub-ideal node to define the variable Ω, as follows:

$$\Omega_{l,e} = \frac{\Delta Q_l + k_1 \cdot \sum_{n=0}^{|N|} \Phi_{l,n}}{Q_e^l + k_2 \cdot P_l}, l \in \mathbf{S}, e, n \in \mathbf{N} \tag{8}$$

Suppose that node e is the ith ideal node of service l, $G_{l,i} = e$. So, Q_e^l is the average response time of all requests that are scheduled to node e. If $G_{l,i+1} = e'$, then $\Delta Q_l = Q_{e'}^l - Q_e^l$. In addition, $\sum_{n=0}^{|N|} \Phi_{l,n}$ is the total load of the service l. P_l represents the number of replicas of service l. k_1 and k_2 is a parameter. In Eq. 8, it is known that the time gap and service loads are greater, the service priority is higher. And the greater the average delay time and the number of replicas are, the lower the priority is.

Next, lines 5–11 implement placement in order of priority. If the remaining resources of the node meet the requirements of the service, place it, and make $x_{l,e} = 1$. Otherwise, $x_{l,e} = 0$. Then, we decide whether to continue the placement according to the upper limit of service replicas θ on lines 12–16. If the number of service replicas in the system does not reach the upper limit θ, it will be added to the service candidate set $L_{e'}$ of the next ideal node e' that we use $findNextNode()$ function to find out. In this case, the priority of service and node e' will be adjusted dynamically. We update service priority $\Omega_{l,e'}$ and node priority $|L_{e'}|$ according to Eq. 8. Finally, through the $clear()$ function shown in line 18, we clean up the services in the L_e set. Continue the placement process in loops 1–18 until all nodes have zero priority.

5   Evaluation

In this section, we evaluate our algorithm. We take the greedy algorithm as the baseline and conduct comparative experiments for two significant factors. The basic strategy of the greedy algorithms is to place the service with the shortest latency on each node.

A comparison algorithm studies load factors, which we call the non-loaded algorithm. The non-loaded algorithm does not take into account the number of service requests in each node, which changes the average delay variable and service priority variable as follows:

$$Q_e'^l = \frac{\sum_{n=0}^{|N|} T_{e,n}^l}{|N|} \tag{9}$$

$$\Omega_{l,e}' = \frac{\Delta Q_l'}{Q_e'^l + k \cdot P_l} \tag{10}$$

Another comparison algorithm studies the service priority factor, which we call unitary priority placement (u2P) algorithm. The u2P algorithm redefines the service priority in Eq. 8 as opposed to the 2P algorithm, set

$$\Omega''_{l,e} = Q^l_e \qquad (11)$$

We have carried out a detailed simulation experiment and compared the experimental results.

5.1 Simulation Settings

There is only one cloud and N edge nodes in the simulation system, and the resource capacity of the nodes is heterogeneous. We use the unit time (*time slots*) to express the response delay time, and the communication delay between the nodes and the cloud is set to $\Gamma_n \in [50, 60]$. The scheduling delay matrix $\partial^l_{m,n}(m \neq n)$ is set to be within the range of $[5, 15]$, which is a symmetric matrix. The response delay jitter is set to $\beta^l_m \in [1, 5]$.

There are 100 services in the simulation system, and the amount of resource block occupied by each service is set to $r_l \in [2, 14]$. The loads of the services within each node are distributed within $[0, 100]$, which obeys the Gaussian distribution. We changed the three variables that are the number of nodes N, the resource capacity of nodes R_n and the upper limit of replicas θ. And we carried out three groups of comparative experiments.

In order to analyze our experimental results more clearly, we counted the average delay of each request, which is a measure to evaluate the superiority of the algorithms. At the same time, we calculate the reduction rate of the 2P algorithm in average response time relative to other algorithms. Supposing that the average response time of each request in the 2P algorithm is t_0, while the other three comparison algorithms are t_i, the reduction ratio is

$$\Upsilon = 1 - \frac{t_0}{t_i} \qquad (12)$$

5.2 Results Comparing

In Fig. 3, we set the number of nodes as 9 and the upper limit of replicas as 2 in the system, but the amount of node resources increases gradually. According to Fig. 3(a), we find that the delay time of all algorithms decreases with the increase of node resources. When the system node resources are sufficient, the response time remains stable. In addition, we also found that the greedy algorithm has the longest response time, the non-loaded algorithm has a smaller response time than the u2P algorithm, and the 2P algorithm has the shortest response time.

According to Fig. 3(b), we get the response time reduction rate of the 2P algorithm compared with other algorithms. Within the ranges of $[60, 90]$ and $[90, 120]$, the response time of the 2P algorithm is about 46% lower than that of the greedy algorithm. Compared with the non-loaded algorithm, the response time was reduced by about 23%. Compared with the u2P algorithm, the response time is reduced by 37%. However, within the range of [1, 30] and [150, 180], the time reduction rate of the 2P algorithm is relatively low. This is mainly because resources are at two extremes (too scarce or too abundant).

In Fig. 4, we set the node resource to be within [50, 100] and the upper limit of replicas as 2 in the system. We infer that the 2P algorithm performs significantly better than the other three algorithms from Fig. 4(a). The response time of the non-loaded algorithm is lower than that of the u2P algorithm, indicating that the service load distribution variable has a greater impact on the response time. When the number of nodes increases gradually, the response time of all algorithms will decline significantly. The response time of the 2P algorithm is always the lowest.

In Fig. 4(b), we know that when there are two nodes in the system, the response time of the 2P algorithm decreases less than that of the comparison algorithm. This is mainly because the number of nodes is too small, resulting in an excessive shortage of system resources, which is consistent with the conclusion in Fig. 3(b). However, with the increase of system resources, the reduction rate of the 2P algorithm increases gradually.

In Fig. 5, we set the system to 12 nodes with each resource capacity in the [50, 100] range and change the upper limit of replicas. From Fig. 5(a), the 2P algorithm has the shortest response time and the best performance based on stability. The greedy algorithm has the highest response time and the lowest performance. The delay time of the non-loaded algorithm and the u2P algorithm is larger than that of the 2P algorithm, but less than that of the greedy algorithm.

From Fig. 5(b), we see that when the upper limit of the replica is 1, the reduction rate of the 2P algorithm is less than 10%. This is because the resource is very sufficient and is a limit state, which is consistent with the conclusion of Fig. 3(b). When the upper limit of replicas is in the range of [2, 6], the response time reduction rate of the 2P algorithm is about 50% relative to the greedy algo-

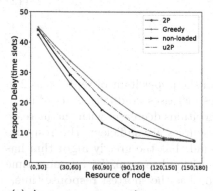 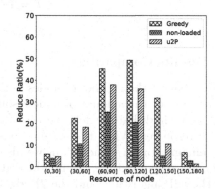

(a) Average response time per request   (b) The reduction ratio of the 2P algorithm

Fig. 3. Change the node capacity R_n, where N = 9 and $\theta = 2$

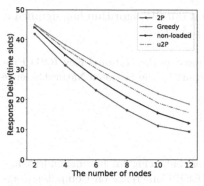
(a) Average response time per request

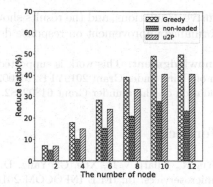
(b) The reduction ratio of the 2P algorithm

Fig. 4. Change the number of system nodes N, where $\theta = 2$ and $R_n \in [50, 100]$

rithm. Compared with the non-loaded algorithm, the response time is reduced by about 23%. Compared with the u2P algorithm, the response time is reduced by about 24%.

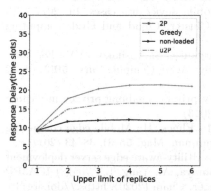
(a) Average response time per request

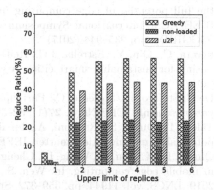
(b) The reduction ratio of the 2P algorithm

Fig. 5. Change the upper limit of service replicas θ, where $N = 12$ and $R_n \in [50, 100]$

6 Conclusion

In this paper, we investigate the service placement for response delay reduction in a heterogeneous MEC system. We set priorities for the nodes and services, which can be adjusted dynamically according to the placed state. To model the priority, we analyzed several factors such as service load distribution or delay time, obtained the service placement strategy in order of priority. We conduct extensive simulations, and the results show that our 2P algorithm has significant performance improvement on response delay reduction.

Acknowledgment. This work is supported in part by the National Key R&D Program of China under Grant 2019YFB2102002, in part by the National Natural Science Foundation of China under Grant 61802182.

References

1. Yu, R., Kilari, V.T., Xue, G., Yang, D.: Load balancing for interdependent IoT microservices. In: IEEE INFOCOM 2019 - IEEE Conference on Computer Communications, pp. 298–306 (2019)
2. Shi, W., Cao, J., Zhang, Q., Li, Y., Xu, L.: Edge computing: vision and challenges. IEEE IoT J. **3**(5), 637–646 (2016)
3. Li, Y., Wang, S.: An energy-aware edge server placement algorithm in mobile edge computing. In: 2018 IEEE International Conference on Edge Computing (EDGE), pp. 66–73 (2018)
4. Ali, S.S.D., Ping Zhao, H., Kim, H.: Mobile edge computing: a promising paradigm for future communication systems. In: TENCON 2018–2018 IEEE Region 10 Conference, pp. 1183–1187 (2018)
5. Reiter, A., Prünster, B., Zefferer, T.: Hybrid mobile edge computing: unleashing the full potential of edge computing in mobile device use cases. In: 2017 17th IEEE/ACM International Symposium on Cluster, Cloud and Grid Computing (CCGRID), pp. 935–944 (2017)
6. Perera, C., Qin, Y., Estrella, J.C., Reiff-Marganiec, S., Vasilakos, A.V.: Fog computing for sustainable smart cities: a survey. ACM Comput. Surv. **50**(3), 1–43 (2017)
7. De Cristofaro, E., Soriente, C.: Participatory privacy: enabling privacy in participatory sensing. IEEE Netw. **27**(1), 32–36 (2013)
8. Taleb, T., Dutta, S., Ksentini, A., Iqbal, M., Flinck, H.: Mobile edge computing potential in making cities smarter. IEEE Commun. Mag. **55**(3), 38–43 (2017)
9. Qiu, J., Li, X., Qin, X., Wang, H., Cheng, Y.: Utility-aware edge server deployment in mobile edge computing. In: Wen, S., Zomaya, A., Yang, L.T. (eds.) ICA3PP 2019. LNCS, vol. 11944, pp. 359–372. Springer, Cham (2020). https://doi.org/10.1007/978-3-030-38991-8_24
10. Ha, K., et al.: Adaptive VM handoff across cloudlets. Technical report CMU-CS-15-113 (2015)
11. Poularakis, K., Llorca, J., Tulino, A.M., Taylor, I., Tassiulas, L.: Joint service placement and request routing in multi-cell mobile edge computing networks. In: IEEE INFOCOM 2019 - IEEE Conference on Computer Communications, pp. 10–18 (2019)
12. Wang, S., Zafer, M., Leung, K.K.: Online placement of multi-component applications in edge computing environments. IEEE Access **5**, 2514–2533 (2017)
13. Wang, L., Jiao, L., He, T., Li, J., Mühlhäuser, M.: Service entity placement for social virtual reality applications in edge computing. In: IEEE INFOCOM 2018 - IEEE Conference on Computer Communications, pp. 468–476 (2018)
14. Xu, J., Chen, L., Zhou, P.: Joint service caching and task offloading for mobile edge computing in dense networks. In: IEEE INFOCOM 2018 - IEEE Conference on Computer Communications, pp. 207–215 (2018)

15. He, T., Khamfroush, H., Wang, S., La Porta, T., Stein, S.: It's hard to share: joint service placement and request scheduling in edge clouds with sharable and non-sharable resources. In: 2018 IEEE 38th International Conference on Distributed Computing Systems (ICDCS), pp. 365–375 (2018)
16. Pasteris, S., Wang, S., Herbster, M., He, T.: Service placement with provable guarantees in heterogeneous edge computing systems. In: IEEE INFOCOM 2019 - IEEE Conference on Computer Communications, pp. 514–522 (2019)
17. Farhadi, V., et al.: Service placement and request scheduling for data-intensive applications in edge clouds. In: IEEE INFOCOM 2019 - IEEE Conference on Computer Communications, pp. 1279–1287 (2019)
18. Li, X., Lian, Z., Qin, X., Abawajyz, J.: Delay-aware resource allocation for data analysis in cloud-edge system, pp. 816–823 (2018)

VTC: A Scheduling Framework Between Soft Real-Time and Hard Real-Time on Multimedia OS

Wei Hu[1,2], Hongqiang Zheng[1,2(✉)], Yonghao Wang[3], Yi Guo[1,2], and Jing Wu[1,2]

[1] The College of Computer Science and Technology, Wuhan
University of Science and Technology, Wuhan, China
zhengzhq@wust.edu.cn
[2] Hubei Province Key Laboratory of Intelligent Information Processing and Real-Time
Industrial System, Wuhan, China
[3] The Digital Media Technology Lab, Birmingham City University, Birmingham, UK

Abstract. With the rapid development of real-time multimedia interactive applications, the traditional universal scheduling framework has limitation when using in the scheduling of multimedia tasks. Multimedia tasks are soft real-time tasks and requiring to be completed as many as possible. With the increasingly strict requirements of multimedia scheduling, the QoS of original scheduling model dissatisfy such requirements. This paper proposes a new multimedia system scheduling framework between hard real-time and soft real-time, which called the "virtual task supplementary scheduling framework" (VTC), a multimedia task processing for real-time systems. The new scheduling framework extends the traditional periodic task scheduling, and on this basis, it has timing determinism and scheduling predictability. The simulation results show that it can achieve more stable, no packet loss, non-disruptive real-time performance, and can meet the extremely strict requirements of professional multimedia.

Keywords: QoS · Scheduling framework · Multimedia tasks · VTC

1 Introduction

With the widespread use of digital music processing technology in music processing and live, the delay of audio signals has become a common problem in digital systems, such as online games, live broadcasts, video conferences and so on. Generally, the industry think that multimedia tasks are periodic tasks on soft real-time system and it usually characterized by high load in streams. The user experience is related to the task delay, so the Real-time audio/video transmission delays have exceptionally high requirements [1]. Some scheduling frameworks adopt "best-effort" strategies for media tasks currently [2]. Although the delay has been reduced, it is not enough [3]. The existing scheduling frameworks not make good use of the periodic of the tasks to reduce the delay further.

This work was sponsored by Key Project of Hubei Provincial Department of Education under Granted No. D20181103.

M. Qiu (Ed.): ICA3PP 2020, LNCS 12452, pp. 330–343, 2020.
https://doi.org/10.1007/978-3-030-60245-1_23

For the past decades, audio signal processing has used cutting-edge research techniques for multi-channel feature extraction in machine learning, artificial intelligence, sound psychology and physiological models [4]. The use of multi-tasking-based general computers and networks as an audio processing platform has become inevitable. Research shows that modern multi-core computers and general-purpose multitasking operating systems deal with low latency in high-load, multi-tasking situations [6]. However, the existing computer hardware and software architecture has uncertainty in dealing with low-latency tasks such as multimedia tasks [5]. The problem of "how to make the computer architecture better support the stability and certainty of low latency" becomes the focus of research and breakthrough.

Currently, the scheduling framework that customize for multimedia task scheduling is dissatisfied. According to the characteristics of multimedia tasks, this paper proposes a new multimedia task scheduling framework titled as the virtual task supplementary scheduling framework (VTC). Periodic real-time tasks and non-periodic tasks reduced the period of the task set by supplementing the virtual tasks, thereby reducing the delay and solving the packet loss phenomenon, and satisfying the time certainty. The rest of this paper is organized as follows. The second part introduce the periodical tasks and the scheduling algorithm. The third part proposes the scheduling theory and task model. The fourth part describes VTC. Finally is the results of experiment and performances.

2 Related Work

Real-time scheduling refers to determining when and on which processor for a series of tasks under the limited system resources (such as CPU), and allocating the resources required for the task to run to ensure its time constraint (it also is named deadline) [7]. According to the different focuses of real-time systems, real-time scheduling has multiple classifications. The importance of real-time requirements for tasks can be divided into hard real-time scheduling and soft real-time scheduling. In soft real-time scheduling, when the system is overloaded, the task is allowed to miss the deadline [8]. For hard real-time scheduling, the task must be executed within its deadline, otherwise the task failed.

Compared with the general-purpose operating system, the real-time operating system is optimized in the worst case, while pursuing the logical correctness and time correctness of the calculation [9]. In real-time operating systems, the correctness of the calculation depends not only on the logical result of the calculation, but also on the time at which the calculation results are generated [10]. The task scheduling framework is an important means to ensure real-time system time constraints.

Real-time scheduling research and application has been a long time, and it has achieved a lot in real-time system [11]. For example, hard real-time applications require time requirements to be fully met, otherwise it will cause major safety accidents and even cause major loss of life and property and ecological damage, such as in the system of aerospace, military, nuclear industry and other key industries [12]. While soft real-time tasks such as system monitoring and information acquiring present time requirements, occasional violations of such requirements by real-time tasks do not have a serious impact on system operation and the run-time environment.

Traditionally, multimedia systems have been attributed to soft real-time scheduling. We believe that in the real-time multimedia field, it is neither hard real-time scheduling nor soft real-time scheduling, but both. We do not want any tasks with negative effects to be lost in real-time audio and video signal transmission, while reducing the delay as much as possible within the limits acceptable to the human ear.

Real-time scheduling has important characteristics such as time constraints, predictability, and reliability. Real-time tasks have time constraints (task execution deadlines), which have been characterized in hard real-time and soft real-time. Predictability means that the system can judge the execution time of real-time tasks and determine whether the time limit of the tasks can be met. Because of the rigor of time constraint for real-time scheduling, predictability is an important performance requirement for real-time systems. Most real-time systems require high reliability, requiring the system to work properly or avoid losses in the worst case (such as high CPU usage).

Most of the real-time task models are based on the periodic task model proposed by Layland and the sporadic task model proposed by Mok. However, both scheduling models are difficult to accurately describe the soft real-time characteristics of multimedia tasks, and it is difficult to meet the high requirements of current users for multimedia scheduling.

Rate-Monotonic Scheduling (RMS) was proposed by Liu and Layland in 1973 to apply a static priority scheduling framework for preemptive hard real-time periodic tasks. The RMS scheduling strategy is that the priority of the task is assigned by task period. It determines the scheduling priority according to the length of the task's execution cycle. The tasks with small period and higher priority; otherwise, the tasks with longer period and the priority is lower. However, the time to obtain data from the outside is uncertain in multimedia tasks, such as compression and decompression of video streams. If each frame is decompressed and displayed as one job, then the operation is not strictly compressed. The RMS scheduling model is difficult to obtain the characteristics of task. But the sporadic task model is also difficult to describe the characteristics of task, because task has no statistical regularity in the long time execute.

Periodic tasks (including periodic real-time tasks and aperiodic tasks) have strict deadlines in real-time multimedia tasks, while non-periodic tasks have lower response time requirements. Therefore, how to balance these two tasks is the key. This paper proposes a VTC scheduling model that distinguishes between periodic real-time tasks and non-periodic tasks. Within an acceptable delay range, it not only ensures that the periodic real-time tasks are not lost, but also performs the execution of the aperiodic tasks.

3   Literature Review

3.1   Scheduling Model

After the RMS scheduling framework was proposed, researchers studied many excellent scheduling frameworks based on RMS scheduling ideas and strategies. The following is a monotonic task model for audio scheduling.

In real-time system, a task consists with a sequence of jobs. For rate monotonic tasks, we have $\tau = \{\tau_1, \tau_2, \tau_3, \cdots \tau_n\}$, where τ is a task set that contains n different tasks. Each task can be defined as $\tau_i = \{C_i, T_i, D_i, P_i\}$, the C_i is worst-case execution time, the T_i is the period of τ_i, the D_i is the deadline of τ_i, the P_i is the priority of τ_i. For each task the CPU utilization is $U_i = C_i/T_i$, so the total utilization is

$$U = \sum\nolimits_{i=1}^{\infty} U_i \tag{1}$$

Normally, the $D_i = T_i$. The rate monotonic scheduling (RMS) framework assigns the task priority according to the task period. Task with shorter period has higher priority. Liu and Layland proved that the sufficient condition of scheduling of RMS is

$$U \le n\left(2^{1/n} - 1\right) \rightarrow 0.6931 \tag{2}$$

where n is the number of tasks. This condition is sufficient but not necessary based on the tasks are preemptible. Lehoczky 1989 shows the sufficient and necessary condition of schedulability of RMS with less upper limit CPU utilization but more complex schedulability test formula.

Lehoczky et al. conducted a more in-depth study based on the research of Liu and Layland. In 1989, the necessary and sufficient conditions for the schedulability determination of RMS scheduling framework were proposed.

For the task set $\tau = \{\tau_1, \tau_2, \tau_3, \cdots \tau_n\}$, we have:

$$W_i(t) = \sum\nolimits_{j=1}^{i} C_j t / T_i \tag{3}$$

Represents the cumulative demand for the CPU for the first i tasks of the task between time periods [0, t], where time 0 is the critical time (assuming i tasks are ready at the same time before the time). Then:

$$L_i(t) = W_i(t)/t \tag{4}$$

$$L_i = min_{\{0 < t \le T_i\}} L_i(t) \tag{5}$$

$$L = max_{\{1 \le i \le n\}} L_i \tag{6}$$

The above definitions give the necessary and sufficient conditions for the RMS scheduling framework.

3.2 TDCS Scheduling Theory

Time deterministic cyclic scheduling is a new design of multimedia real-time scheduling mechanism which is called TDCS, and it mainly for live multimedia tasks processing. The multimedia task includes periodic tasks and aperiodic tasks. For periodic tasks, it adopts a time division method, which is divided by the least common multiple of all task periods, and all tasks are rearranged in each segment after segmentation. The idle time in each segment after the split is performed could execute the non-periodic tasks. The advantage of TDCS is that the delay and schedulability in the scheduling process can be estimated in advance. However, the time division method tends to cause the segmentation length to be very large and impractical, resulting in a very unsatisfactory delay in the scheduling process. The scheduling framework proposed in this paper solves the high delay phenomenon in TDCS and has the advantages of TDCS time determinism.

4 Proposed Scheduling Framework

Virtual task supplementary scheduling framework (VTC) is sourced from traditional periodic scheduling theory. It uses the monotonic task model mentioned in Sect. 3.1. VTC tiers multimedia tasks: periodic soft real-time tasks and acyclic tasks. It follows the RMS priority policy in the scheduling priority of periodic tasks, and uses idle time to complete scheduling for non-periodic tasks, so that system resources are used as much as possible.

4.1 The Principle of VTC

The basic principle of the timing correctness of VTC is to divide the time according to the ARINC 653 standard. In the ARINC 653 standard, the operating system is classified as a separate hypercycle segment or a time domain master cycle driven by a precise underlying system timer. In VTC, we chose to define this period as TC. In theory, time segmentation can be the least common multiple (LCM) of different task cycles, but this can be very large and impractical. In our design, time segmentation is the cycle of the task of the largest cycle in the task set, the periodic task will go through two buffer systems, one input "mapping buffer" and one output "de-jitter buffer", the previous buffer is used to convert to any length time period you want, The second buffer is to render the task as its own source code rate.

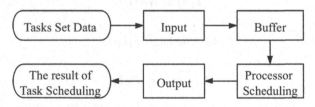

Fig. 1. The process of VTC

The time division mentioned above for VTC is divided by the period of the task with the longest task concentration period. In the entire data processing process, it is scheduled in units of time-divided time segments. That is, during the task set input process, a buffer area is passed. When the length of the cached task reaches the length of the splitting period, the cached time period task is released and passed to the processor for processing. After the processor scheduling is completed, the output is processed. The data is shown in Fig. 1. The entire input task, cache task, release task, scheduled task, and output task are all layered simultaneously, except that the time segment of the first period needs to wait for data buffering, and the subsequent system scheduling is synchronized.

4.2   The Framework of VTC

We define the least common multiple for all task periods as the single scheduling period (also named TL) and divided the TL with the short period (TC), which is the maximus period in the tasks set. We define the following formula:

$$T_L = LCM\{T_i\} \tag{7}$$

$$T_c = \max\{T_i\} \tag{8}$$

$$f_i = \frac{T_L}{T_i} \tag{9}$$

$$f_{Ci} = \left\lceil \frac{T_C}{T_i} \right\rceil \tag{10}$$

$$N = \frac{T_L}{T_C} \tag{11}$$

f_{Ci} is the frequency of τ_i in the ideal case of TC. In actual task scheduling, the frequency of tasks that may exist in each TC does not reach the ideal situation. In this case, virtual tasks need to be used to make up. We define f_{Vi} as the number of τ_i needs to be filled in the entire task set period, specifically defining the following formula:

$$f_{Vi} = f_{Ci} \times N - f_i \tag{12}$$

According to the formula (12), we can know that when f_{Vi} is 0, the task set can convert the period of the task set to Tc without adding virtual tasks. When f_{Vi} is not 0, the task set needs to convert its period to Tc by adding the method area of the virtual task. Therefore, it can be divided into the following two cases:

- **Need to add virtual tasks**
  According to the formula (9), (10) and (11), f_{Vi} cannot be zero only when T_c is not an integer multiple of T_i, that is, the period of T_i cannot be divided by T_c. Here is a practical example to test: there are 3 tasks, the period is $\{10, 15, 25\}$, so TL = 150, TC = 25, fi = $\{15, 10, 6\}$, f_{Ci} = $\{3, 2, 1\}$, N = 6, we then can find f_{Vi} = $\{3, 2, 0\}$.In this case, we can add 3 virtual task1, 2 virtual task2. The method of specifically calculating the virtual task position within a TL period is as follows:
  Let n be the index of the TC, n_i^j is the number of virtual tasks that task i needs to add in the jth TC, then:

a. First, calculate the number of τ_i in the first nth T_c: N_i^n, P_i is to determine whether the nth TCs are multiples of T_i. If P_i is greater than 0, according to (10), I need to add an extra one when calculating N_i^n.

$$P_i = (T_c \times n) \bmod T_i \tag{13}$$

$$N_i^n = \begin{cases} (T_c \times n)/T_i + 1 & P_i > 0 \\ (T_c \times n)/T & P_i = 0 \end{cases} \tag{14}$$

b. Second, calculate the number of τ_i in the first $(n-1)$th TC: N_i^{n-1}. Q_i is to determine whether the $(n-1)$ TCs in front are multiples of T_i. If Q_i is greater than 0, according to (10), I need to add an extra one when calculating N_i^{n-1}.

$$Q_i = (T_c \times (n-1)) \bmod T_i \tag{15}$$

$$N_i^{n-1} = \begin{cases} (T_c \times (n-1))/T_i + 1 & Q_i > 0 \\ (T_c \times (n-1))/T & Q_i = 0 \end{cases} \tag{16}$$

c. Third, get the number of τ_i in the nth TC: $\left(N_i^n - N_i^{n-1}\right)$.

d. fourth, get the number of virtual tasks that τ_i needs to add in the jth TC: n_i^j.

$$n_i^j = f_{Ci} - \left(N_i^n - N_i^{n-1}\right) \tag{17}$$

- **No need to add virtual tasks**

  Only when T_c is an integer multiple of T_i, f_{Vi} is zero, that is, T_c is divisible by the period of T_i. For example, there are 3 tasks, the period is $\{10, 15, 30\}$, so TL $= 30$, TC $= 30$, $f_{Vi} = \{0, 0, 0\}$. So, we can come up with a supplementary location for each virtual task in this example, as shown in the following Table 1:

Table 1. The position of the complement task

Task/position	1	2	3	4	5	6
Task1	0	1	0	1	0	1
Task2	0	0	1	0	0	1
Task3	0	0	0	0	0	0

After the addition of the virtual task, the long-period TL was divided into the same six components, so the period of the task set TL could be shortened to the TC.

4.3 The Process of Scheduling

In the VTC scheduling process, the task set is scheduled in cycles. On the timeline, after a TC buffer is completed, the data of the entire TC is released and then scheduled. Figure 2 is a scheduling diagram. Task1, task2 and task3 named T1, T2 and T3. The width of task represents the execution time of task. The task appears only once with period. The period of task3 is the buffer period, and the virtual tasks are complemented in the scheduling process. Each TC is allocated with the RMS small-cycle priority policy, so that the scheduling is orderly and regular.

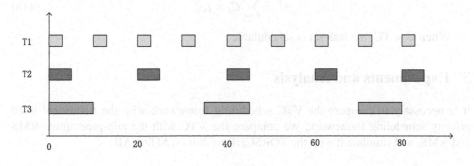

Before scheduling

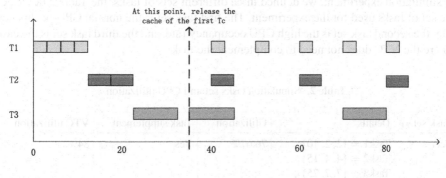

Scheduling

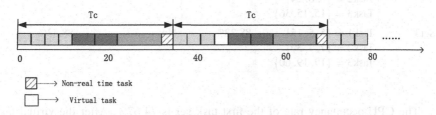

After scheduling

Fig. 2. The process of scheduling

4.4 Feasibility

Just like the traditional RMS, EDF and other scheduling framework, VTC also has its scheduling limits. The difference is that the scheduling limits of VTC fluctuate according to the number of virtual tasks. When the number of virtual tasks is 0, the CPU occupancy rate can reach 100%. At this time, the VTC can still be scheduled; when the number of virtual tasks is not 0, the CPU occupancy rate is less than 100%, and the task scheduling has a limit. The bounding formula is as follows:

Let C be the total execution time of the real-time task set, we have:

$$C = \sum C_i \times f_{Ci} \tag{18}$$

When C < TC, the task set is schedulable.

5   Experiments and Analysis

It is necessary to compare the VTC scheduling framework with the traditional fixed priority scheduling framework. We compare the VTC with the non-preemptive RMS and RMS, and simulate it with the TORSCHE toolkit on MTALAB.

5.1   Schedulability Simulation

In simulation experiment, we defined three different sets of tasks. the Table 2 describes the set of tasks used for the experiment. The first task set is the normal CPU occupancy rate, the second task set is the high CPU occupancy rate, and the third task set is the case where the VTC does not need to complement the task.

Table 2. Simulation tasks set and CPU utilization

Task set	Details	Utilization	Task supplement	VTC utilization
Set1	Task1 = {2, 2, 10} Task2 = {4, 4, 15} Task3 = {7, 7, 25}	76.67%	Yes	84%
Set2	Task1 = {3, 3, 12} Task2 = {6, 6, 24} Task3 = {15, 15, 36}	91.67%	Yes	100%
Set3	Task1 = {5, 5, 15} Task2 = {7, 7, 30} Task3 = {19, 19, 60}	88.33%	No	88.33%

The CPU occupancy rate of the first task set is 74.67%. After the virtual task is supplemented by VTC scheduling, the CPU occupancy rate is 84%. The simulation results are shown in Fig. 3. In the first subgraph, the three original real-time tasks of Set1 and the background task AF are displayed, and the second one shows the task scheduling of the RMS, which appears during the scheduling process. Interrupts and delays, but can be completed within the deadline. The third sub-picture shows the NP-RMS task schedule. Although there is no task interruption, even the highest priority task1 has some delay. The last sub-picture shows the task scheduling of VTC. It uses the scheduling scheme of "low priority period is the cache period", and the red part is the virtual supplementary task. It takes the CPU as a real-time task and sacrifices part of the CPU. The occupancy rate makes the task set cycle become the cycle of Task3, which reduces the delay of the task, and there is no interruption in the process of task scheduling.

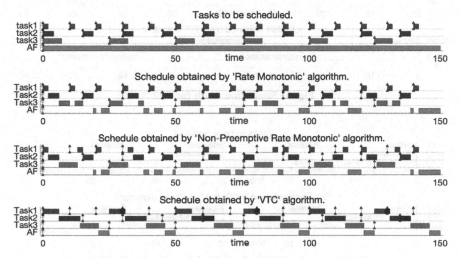

Fig. 3. Simulation result for Set1 (Color figure online)

Figure 4 shows that the CPU share of the task set is 91.67%. After the virtual task is supplemented by VTC scheduling, the CPU occupancy rate is 100%. The simulation results show that the RMS can complete the scheduling, but there is an interruption, NP-RMS can also complete the scheduling, but will exceed the deadline, there is no interruption in the VTC scheduling process, the task is scheduled in advance, and does not exceed the deadline.

The CPU occupation rate of task set 3 is 88.33%. The virtual task is not needed in the VTC scheduling process. Therefore, the CPU occupancy rate after VTC scheduling is still 88.33%, which is a special case of VTC scheduling. The simulation results are shown in Fig. 5. The RMS can be scheduled, but as always there are a lot of interrupts. NP-RMS can't complete scheduling, and there is a packet loss phenomenon, which will cause some important tasks to be completed. VTC can still complete the scheduling stably. At this time, the scheduling advantage of VTC is initially displayed.

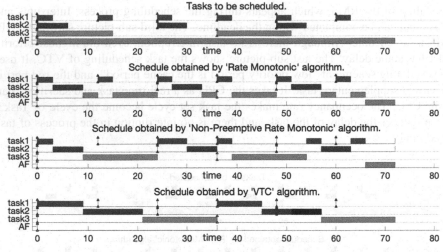

Fig. 4. Simulation result for Set2

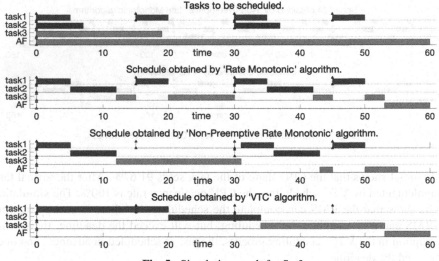

Fig. 5. Simulation result for Set3

5.2 The Comparison of Delay

We compared the delays of Set1 to RMS, NP-RMS, and VTC scheduling framework. The comparison is the delay between the initial task time and the actual execution time of the task.

Figure 6 shows the delay comparison of Task1 of Set1. It can be seen that the highest priority task has no delay after RMS scheduling, and both NP-RMS and VTC have small delays. Figure 7 shows the delay comparison of task2. The three scheduling frameworks have very small delays. Figure 8 shows the delay comparison of task3. The RMS and

NP-RMS delays are within 5 ms, and the VTC delay is within 10 ms. Acceptable delay. Based on the comparison of the above three, it can be found that VTC is comparable to RMS and NP-RMS in terms of delay.

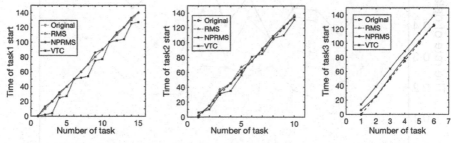

Fig. 6. Delay of Task1   **Fig. 7.** Delay of Task2   **Fig. 8.** Delay of Task3

5.3 Delay Rate

The VTC adopts a pre-scheduled method, and the scheduling is not started until the entire task set period is cached. We believe that the task is in the scheduling process, if it is executed before the initial execution time of the task, it has a positive meaning to the scheduling system, and the delay to start execution has a negative impact. We propose the concept of delay rate, which is the percentage of the number of tasks that are delayed to execute in a task set cycle as a percentage of the total number of tasks. The formula for calculating the delay rate is as follows:

$$P = \frac{\sum_1^n f_i^{delay}}{\sum_1^n f_i} \times 100\% \tag{19}$$

In formula (19), f_i^{delay} is the number of delays of the task i, and P is the delay rate. In order to more intuitively reflect the low latency of the VTC model, we randomly generated 100 sets of data sets, and their CPU occupancy increased gradually from 30% to 90%. The test results are shown in Fig. 9.

We can see that with the increase of CPU occupancy, the delay rate of NP-RMS transmission mode begins to increase more and more, and the delay rate of RMS and VTC models is relatively stable, but the overall delay rate of RMS. Higher than the VTC model, this shows that the VTC scheduling model has significant advantages in scheduling delay.

5.4 Scheduling Complexity

Interrupts are a feature of the RMS scheduling framework that cannot be eliminated. However, due to the non-preemptive nature of the scheduling process, the NP-RMS scheduling framework often leads to high latency or even loss of tasks. Therefore, comparing VTC with RMS and NP-RMS for interrupts and losses is not fair enough for RMS and NP-RMS. After all, RMS and NP-RMS are not framework for multimedia scheduling. The above comparison is only to show that VTC will not be interrupted and

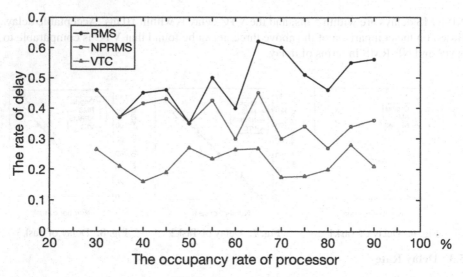

Fig. 9. The rate of delay

lost during the scheduling process. In terms of delay, VTC is comparable to RMS and NP-RMS. After all, in the field of multimedia scheduling, interruption, loss, and delay are very influential in multimedia effects.

VTC is a scheduling framework for audio/video multimedia. We need a way to judge multimedia scheduling. We propose the concept of scheduling complexity. In the system scheduling process, after the multimedia data is input, the data needs to be cached and then processed. When the CPU processes this data, we think that the difficulty of CPU scheduling is different for different scheduling framework. For example, when the same set of data is cached, the general scheduling framework RMS and NP-RMS cache holds the period TL, while the VTC caches a small period TC, which causes the amount of data to be different when the CPU processes. Interrupt, loss, and unstable delays can make system hardware scheduling more difficult when the CPU processes tasks for each cycle. In the scheduling process of VTC, the delay is stable, no interruption, no loss, which will greatly reduce the scheduling complexity of the system.

5.5 Advantages and Disadvantages of VTC

VTC has advantages in scheduling multimedia tasks such as audio/video, especially in real-time processing extensions or hybrid critical systems in general-purpose operating systems. (1). Real-time tasks can complement virtual tasks, and non-real-time tasks can be scheduled in idle time, resulting in extremely high CPU occupancy. (2). Timing determinism of multimedia cycle tasks, delay can be calculated in advance. (3). Simple schedulability judgment, so that schedulability can be judged before scheduling. (4). The flexible virtual task complement scheme reduces the cycle of the task set from TL to TC, which greatly reduces the delay. (5). Multimedia tasks can be driven by low-level system timers instead of software interrupts, reducing small delays.

Similar to RMS and EDF, VTC relies on the correct worst-case execution time of the task. If the worst-case execution time is inaccurate, it will affect the schedulability of the overall task, which is a disadvantage of VTC.

6 Conclusions

In this paper, we propose a scheduling framework for the multimedia field for low-latency audio and video processing. In the traditional concept of round-robin scheduling, VTC flexibly redistributes real-time tasks by adding virtual tasks, which makes the task set have a lower cycle, and the system can be scheduled for high CPU task sets. Finally, the experimental simulation shows that VTC has the advantages of non-interruption, no packet loss, and delay similar to the traditional scheduling framework even for high CPU occupancy task scheduling, and the scheduling complexity is lower than the traditional RMS and NP-RMS scheduling framework.

References

1. Tullimas, S., Nguyen, T., Sen-Ching, C.: Multimedia streaming using multiple TCP connections. In: Proceedings of 24th IEEE International Conference on Performance, Computing, and Communications, pp. 215–223 (2005)
2. Koodli, R., Puuskari, M.: Supporting packet-data QoS in next-generation cellular networks. IEEE Commun. Mag. **39**, 180–188 (2001)
3. Tu, W., Jia, W.: Adaptive playout buffer for wireless streaming media. In: Proceedings on 12th IEEE International Conference Networks, pp. 19 1–195 (2004)
4. Steinbach, E.G.: Adaptive playout for low latency video streaming. In: Proceedings of the International Conference on Image Processing, Thessaloniki, Greece (200 1)
5. Wang, Y., Grant, J., Foss, J.: Flexilink: a unified low latency network architecture for multichannel live audio. In: Audio Engineering Society Convention 133. Audio Engineering Society (2012)
6. Song, Y., Wang, Y., Bull, P., Reiss, J.D.: Performance evaluation of a new flexible time division multiplexing protocol on mixed traffic types. In: 2017 IEEE 31st International Conference on Advanced Information Networking and Applications (AINA), pp. 23–30. IEEE (2017)
7. Laoutaris, N., Stavrakakis, I.: Adaptive playout strategies for packet video receivers with finite buffer capacity. In: Proceedings of IEEE ICC 2001, Helsinki, Finland, June 200 1
8. Sha, L., Abdelzaher, T., Aerzean, K.-E., et al.: Real time scheduling theory: a historical perspective. Real-Time Syst. **28**(3), 101–155 (2004)
9. Liu, C.L., Layland, J.W.: Scheduling algorithms for multiprogramming in a hard-real time environment. In: Readings in Hardware/Software Co-Design. Kluwer Academic Publishers (2001)
10. Park, M.: Non-preemptive fixed priority scheduling of hard real-time periodic tasks. In: Shi, Y., van Albada, G.D., Dongarra, J., Sloot, P.M.A. (eds.) ICCS 2007. LNCS, vol. 4490, pp. 881–888. Springer, Heidelberg (2007). https://doi.org/10.1007/978-3-540-72590-9_134
11. Nasri, M., Brandenburg, B.B.: Offline equivalence: a non-preemptive scheduling technique for resource-constrained embedded real-time systems. In: Real-Time and Embedded Technology and Applications Symposium. IEEE (2017)
12. Ma, T., Wang, Y., Hu, W.: Evaluation of flexilink as unified real-time protocol for industrial networks. In: IEEE International Conference on Big Data Science and Engineering. IEEE (2018)

A BSP Based Approach for NFAs Intersection

Cheikh Ba$^{(\boxtimes)}$ ⓘ and Abdoulaye Gueye ⓘ

LANI - Université Gaston Berger, Saint-Louis, Senegal
cheikh2.ba@ugb.edu.sn, ablayesat@gmail.com

Abstract. Large NFAs are automata that cannot fit in a single computer, or the computations would not fit within the computer RAM, or may take a long time. We describe and implement a BSP solution of such large NFAs intersection. Our method avoids producing unreachable states of the product automaton, contrary to previous solutions. These solutions, based on MapReduce for instance, process all the Cartesian product of inputs NFAs. A running example is provided with execution details. Finally, complexity analysis is given. This work will help in bringing out relevant programming artifacts for our long term goal that consists of a high level distributed graph language.

Keywords: BSP · MapReduce · Large NFAs intersection · Complexity

1 Introduction

Nondeterministic Finite-state Automata (NFAs) are simple, yet powerful devices that model a plethora of computationally oriented phenomena. They are ubiquitous in computer science, and have been studied since the 1950s. Applications include pattern matching, natural language processing, speech recognition, token passing networks, compilers, web services composition, among others domains.

Some of the problems related to NFA that have been studied in the literature are intersection, determinization and minimization. For instance, one of the main issues for the NFA intersection problem is that the size of the output NFA is the product of the size of all inputs NFAs, mainly when the NFAs cannot fit in a single computer, or the large computations would not normally fit within the computer RAM. So, in the case of very large automata or graphs, distributed approach such as disk-based parallel algorithm, MapReduce parallel programming model and memory based distributed system are used.

In this paper we investigate the novel use of BSP [17] (bulk synchronous parallel) abstract computer to implement large NFAs intersection. Contrary to previous and classic approaches for NFAs intersection, we don't need to process the product of the size of all inputs NFAs. Indeed, testing for emptiness of the intersection of a set languages represented by NFAs is known to be PSPACE-complete [11]. The reason is that the most commonly used algorithm is the Cartesian construct. If there are m input NFAs each having n states, the product

© Springer Nature Switzerland AG 2020
M. Qiu (Ed.): ICA3PP 2020, LNCS 12452, pp. 344–354, 2020.
https://doi.org/10.1007/978-3-030-60245-1_24

NFA will have n^m states. It therefore would be important to come up with good distributed algorithms, that avoid at best processing useless product states.

Note that this work is one of the first steps of a long term goal, that aims a high level language for graph-like structure processing, a kind of distributed graph language that hides, the most, the distributed aspect.

The rest of this paper is organized as follows: Sect. 2 gives some related works and Sect. 3 presents technical definitions, namely Finite State Machines, MapReduce paradigm and BSP model. In Sect. 4 we describe our proposition and it analysis is given in Sect. 5. Conclusions are drawn in Sect. 6.

2 Related Works

Big data are data sets that are too large or complex to be dealt with by traditional data-processing application software. It's about data storage, analysis, sharing, visualization, querying and so on. A common solution requires the use of a large number of computers for a distributed storage and parallel processing.

This section will point out the existence of two families of solutions. The first one concerns from-scratch solutions in which, for each need, one have to become distributed systems expert to build the infrastructure to handle it. The second one is related to solutions with a higher level of abstraction that hides machines coordination complexity to create building blocks for programmers who just happen to have lots of data to store, or lots of data to analyze, or lots of machines to coordinate, and who don't have the time, the skill, or the inclination to program low level tasks. We place our present work in the second family.

One of the most important algorithms for automata processing is the minimization of a deterministic finite automaton (DFA). Some parallel algorithms consider shared memory machines [14,16]. These algorithms are applicable for tightly coupled parallel machines with shared RAM with heavy use of random access. In addition, work in [14] used a 512-processor CM-5 supercomputer to minimize a DFA with 525,000 states. In the case when the DFA considered is very large, a disk storage may be needed. In this context, [15] exhibits a parallel disk-based algorithm that uses a cluster of 29 commodity computers to produce an intermediate DFA with almost two billion states and then continues by producing the corresponding unique minimal DFA with less than 800,000 states.

The aforementioned propositions belong to the first family of solutions.

Recently, Google introduced Hadoop distributed platform [2], which has become the de facto standard for processing big data, and its MapReduce paradigm as a parallel programming model [4]. Its goal is to simplify parallel processing by offering two simple interfaces: *map* and *reduce*. It achieves data parallel computation by partitioning the data randomly to machines and executing the map and reduce functions on these partitions in parallel. Hadoop works by connecting many different computers, hiding the complexity from the user, as if it works with one giant computer. Since then, several MapReduce solutions for general graph problems has emerged [3,8,9,12]. In the case of large automata, works in [8,9] offer algorithms for intersection and minimization with MapReduce.

Although MapReduce is able to express many common graph algorithms, it has been recognized that it is not suitable for graph algorithms that are iterative, due to excessive I/O with HDFS (Hadoop distributed file system) and data shuffling at every iteration. Since the MapReduce model does not provide natural support for iteration, this can only be reproduced by scheduling several consecutive jobs, resulting in significant overhead. That is why several in-memory graph processing frameworks such as Google's Pregel [13], Spark/GraphX [7], and PowerGraph [6] are proposed to speed up the execution of iterative graph algorithms. Most of these frameworks follow a vertex-centric programming model. For instance, in frameworks such as Google's Pregel, based on BSP model [17], each vertex receives messages from its incoming neighbors, updates its state, and then sends messages to outgoing neighbors in each iteration.

However, these frameworks are not yet exploited for the case of automata. We then propose the novel use of BSP model for large NFAs intersection. Even though [9] proposes a one-iteration MapReduce solution (which is not the case for most of graph algorithms, such as DFA minimization [8]), they process the product of the size of all inputs NFAs, including potential useless product states.

3   Terminology and Background

3.1   Finite State Automata

In this section we introduce two kinds of *Finite State Automata* (FSA), namely *Deterministic* (DFA) and *Non-deterministic* (NFA) *Finite state Automaton*.

A DFA consists of a finite set of states with labelled and directed edges between pairs of states. For each state, there is at most one outgoing edge labeled by a given letter from the alphabet (*determinism*). The DFA accepts a word if the letters of the word determine transitions from the *initial state* to a *final* or *accepting state*. Formally, a DFA is a 5-tuple $A = (\Sigma, Q, i, \delta, F)$ where Σ is the input alphabet, Q is the set of states, $i \in Q$ is the initial state, and $F \subseteq Q$ is the final states. $\delta : Q \times \Sigma \to Q$ is the *transition function*, which decides which state the control will move to from the current state upon consuming a symbol.

An NFA is similar except that, for a given state, there may be more than one choice for outgoing edges with the same label (*non-determinism*). The NFA accepts a word if there exists a choice of transitions from the initial state to some final state. Formally, the only difference from a DFA is that $\delta : Q \times \Sigma \to 2^Q$.

The *language* accepted by a FSA A, denoted $L(A)$, is the set of words accepted by A. NFAs and DFAs are equivalent and a language is *regular* if and only if it is accepted by some FSA. Figure 1 shows two FSAs on $\Sigma = \{a, b\}$: a DFA which accepts words containing an even number of symbols (Fig. 1-a), and an NFA which accepts words starting with "a" and ending with "b" (Fig. 1-b).

It is well known that regular languages are closed under several operations. In particular, given NFAs $A_1 = (\Sigma, Q_1, i_1, \delta_1, F_1)$ and $A_2 = (\Sigma, Q_2, i_2, \delta_2, F_2)$,

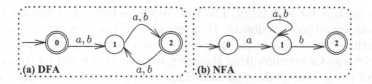

Fig. 1. An example of two FSA.

an *intersection* NFA A, such that the language $L(A)$ is $L(A_1) \cap L(A_2)$, can be computed by the *Cartesian construct* $A = A_1 \otimes A_2 = (\Sigma, Q_1 \times Q_2, \delta, (i_1, i_2), F_1 \times F_2)$ where $\delta : (Q_1 \times Q_2) \times \Sigma \to 2^{Q_1 \times Q_2}$, with $(q_1, q_2) \in \delta((p_1, p_2), \sigma)$ if and only if, for a given $\sigma \in \Sigma$, $q_1 \in \delta_1(p_1, \sigma)$ and $q_2 \in \delta_2(p_2, \sigma)$.

3.2 MapReduce and NFAs Intersection

We assume familiarity with the MapReduce model. Nevertheless we recall some important concepts. Google's MapReduce programming model [4] is based on two functions *map* and *reduce* that the programmer is required to implement. Their signatures are `map`: $\langle K_1, V_1 \rangle \to \{\langle K_2, V_2 \rangle\}$ and `reduce`: $\langle K_2, \{V_2\} \rangle \to \{\langle K_3, V_3 \rangle\}$. The initial data is stored on HDFS, and each mapper is responsible for a chunk of the input. In a round of MapReduce, the mappers emit a list of key-value pairs $\langle K, V \rangle$. This list is partitioned by the MapReduce framework depending on the values of K. All pairs having the same value of K belong to the same group $\langle K, [V_1, \cdots, V_l] \rangle$, and this group will be sent to the same reducer.

Works in [8,9] are the only propositions we know that use high level distributed platform for problems related to automata. In the case of NFAs intersection $A_1 \otimes \cdots \otimes A_m$, works in [9] present three variants of a solution using the Cartesian construct. For instance, considering the mapping based on symbols, we have one reducer for each of the alphabet symbols. From transition (p_i, σ, q_i) of NFA A_i, the mapper will generate the key-value pair $\langle \sigma, (p_i, \sigma, q_i) \rangle$. In this way, after having received inputs $\{(p_i, \sigma, q_i)\}$ for $i = 1, \cdots, m$, each corresponding reducer will output transitions $\{((p_1, \cdots, p_m), \sigma, (q_1, \cdots, q_m))\}$.

However, this way of doing is costly since if we have m input NFAs, each having n states, their method will produce an NFA with n^m states, especially as many of these states may be useless. A useless state is a state that cannot be reached from the initial state, or a state from which a final state cannot be reached (dead state). By the way, authors planned in future work to *"investigate reducing the number of states in the product automaton, either by eliminating all or part of the useless states or by determinizing and minimizing the automaton"* [9]. This goal is likely to be costly or difficult to implement with MapReduce programming model.

3.3 BSP Model

Bulk Synchronous Parallel [17] is a parallel programming model with a message passing interface, developed to address the problem of parallelizing jobs across multiple workers for scalability. The BSP model is defined as the combination of the following three attributes: (i) a number of components for performing computations, (ii) a router for delivering messages between components, and (iii) *supersteps* for synchronizing the computations performed by components.

Pregel [13] is one of the first BSP implementations that provides a native API specifically for programming graph algorithms, while abstracting away the details of underlying communication. The computing paradigm can be characterized as "think like a vertex". Graph computations are specified in terms of what each vertex has to compute; edges are communication channels from one vertex to another. In each superstep, a vertex can execute a user-defined function called `compute()`, send or receive messages to any other vertex with a known ID, and change its state from active to inactive. Synchronization barrier ensures that a message issued in superstep S will be available at its destination vertex in superstep $S + 1$. A vertex may vote to halt in any superstep (if it calls `voteToHalt()`) and is woken up when it receives a message. Pregel terminates once all vertices are inactive and no further messages are to be passed.

Note that Google's implementation of Pregel is not publicly available, but there exist several open source alternatives, such as Apache Giraph [1].

4   Solution for NFAs Intersection

Given m NFAs, Algorithm 1 gives the `compute()` function of our proposition to process product automaton $A = A_1 \otimes \cdots \otimes A_m$. We recall that this function will be executed by every active vertex during the system run. Since we are dealing with automata, states and transitions are respectively represented by Pregel's vertices and edges. A vertex can represent a *simple* state q_i (a state that belongs to input automaton A_i), or a *product* state $\otimes p = (p_1, \cdots, p_m)$ of the product or output automaton A. Algorithm 1 consists of an alternating of two kinds of supersteps, namely PRODUCTION_SUPERSTEP (even) and FEED_SUPERSTEP (odd). It is in even supersteps that new product states will be created, as well as their incoming transitions. In order to create the following product states, two supersteps later, a new product state $\otimes p = (p_1, \cdots, p_m)$ needs to ask for data (transitions) from each of its component states p_i, to be provided in the next (thus odd) superstep. In fact, FEED_SUPERSTEPs are devoted to sending data from simple states p_i to requesting product states $\otimes p_j$.

In the first superstep (line 3), all (simple) states are active, but only initial ones will execute code before voting to halt (lines from 4 to 9). The first product state is then created (line 7), that is, the initial product state $\otimes p_0 = (i_1, \cdots, i_m)$, with no outgoing transition yet. In fact, in order to create outgoing transitions and corresponding target states, $\otimes p_0$ needs information from each of its component (i_k). For this purpose, and according to the BSP model (Sect. 3.3), $\otimes p_0$ has to ask for information from each i_k first (line 8). Then, this request will be received and processed by each i_k in the following superstep (FEED_SUPERSTEP). Finally, $\otimes p_0$ will receive requested data in the next superstep (PRODUCTION_SUPERSTEP), that is, two supersteps later from the present one.

Apart from the first superstep, only "fed" product states $^\otimes p_i$ will execute code (from line 12 to line 27) in a PRODUCTION_SUPERSTEP. A "fed" product state is a state that requested data, two supersteps earlier, from each of its component simple states, and that just receives them (line 15). Then, state $^\otimes p$ checks if there is a common outgoing transition (symbol) from each of its component states p_i (line 16); in other words, if $^\otimes \Sigma \neq \emptyset$. In such a case, there is a creation of a new transition, with the common symbol, from $^\otimes p$ to each new product state $^\otimes q$ that matches (line 22). Thus, state $^\otimes q$ is a new destination vertex (line 21), if it is not previously created. Finally, as for the initial product state (line 8), state $^\otimes q$ will ask for information from its component states (line 23), in order create, two supersteps later, its potential output transitions and target states.

At last, as said earlier, FEED_SUPERSTEPs are devoted to sending data from simple states p_i to requesting product states $^\otimes p_j$ (line 32). In fact, only simple states that receive requests (from product states) are active in odd supersteps.

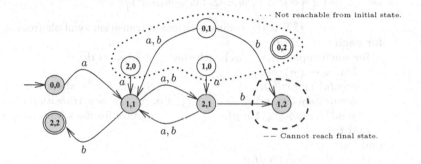

Fig. 2. Product automaton of FSA in Fig. 1.

Example 1. Let us consider the two FSA in Fig. 1, namely DFA A_1 and NFA A_2. The intersection or product automaton $A = A_1 \otimes A_2$ is depicted in Fig. 2. The solution in [9], described in Sect. 3.2, would produce all the nine product states: useful states as well as useless ones. Useful states are darkened, except the last one. Useless states are those that are not reachable from the initial state (dotted line), and one dead state (dashed line), namely state $(1, 2)$.

In contrast, our solution produces only darkened states, including a single useless one. Table 1 gives some details of the execution. Only even supersteps (PRODUCTION_SUPERSTEP) are represented. However, we give some information related to the two first supersteps. In superstep 0, each of the two actives initial states 0_{A_1} and 0_{A_2} asks for the creation of the initial product state $^\otimes p_0 = (0, 0)$ (line 7). Then, $^\otimes p$ requests information from 0_{A_1} and 0_{A_2} (line 8). In this way, in superstep 1 (FEED_SUPERSTEP), only states 0_{A_1} and 0_{A_2} are active. Both received request from $^\otimes p$ (line 31) and they, in turn, send information to $^\otimes p$ (line 32): 0_{A_1} sends "$(0, \{a, b\}, 1)$" and 0_{A_2} sends "$(0, \{a\}, 1)$". ∎

Algorithm 1. $compute(vertex, messages)$

1: **if** (PRODUCTION_SUPERSTEP) **then**
2:   {**Even** superstep}
3:     **if** ($superstep = 0$) **then**
4:         **if** ($vertex.isInitialSimpleState()$) **then**
5:                                 {$vertex$ represents simple state i_1 or $i_2 \cdots$ or i_m}
6:             Let ${}^{\otimes}p_0 \leftarrow (i_1, \cdots, i_m)$;   {Initial product state.}
7:             $createVertex({}^{\otimes}p_0)$;   {Idempotent operation.}
8:             $sendMessage({}^{\otimes}p_0.\text{ID}, i_k)$ {Asks for data from each i_k.}
9:         **end if**
10:     **else**
11:   {$vertex$ represents a product state ${}^{\otimes}p$}
12:         **if** ($message \neq \emptyset \wedge !vertex.isVisited()$) **then**
13:                                 {${}^{\otimes}p$ receives data it asked for, 2 supersteps earlier.}
14:             Let ${}^{\otimes}p \leftarrow (p_1, \cdots, p_m)$;
15:             Let $\delta_i \leftarrow getEdges(p_i.\text{ID}, messages)$ {Gets out-transitions for each p_i}
16:             Let ${}^{\otimes}\Sigma \leftarrow \bigcap_i \left\{ \sigma \in \Sigma \mid \exists q_i \in Q_i : q_i \in \delta_i(p_i, \sigma) \right\}$;
17:                                 {${}^{\otimes}\Sigma$ is the common outgoing-transition symbols from p_i. }
18:             **for each** $\sigma \in {}^{\otimes}\Sigma$ **do**
19:                 **for each** tuple $(q_1, \cdots, q_n)$ such that $q_i \in \delta_i(p_i, \sigma)$ **do**
20:                     Let ${}^{\otimes}q \leftarrow (q_1, \cdots, q_m)$;
21:                     $createVertex({}^{\otimes}q)$;   {No effect if ${}^{\otimes}q$ already exists. }
22:                     $createEdge({}^{\otimes}p, \sigma, {}^{\otimes}q)$;   {(${}^{\otimes}p, \sigma, {}^{\otimes}q$) is a new transition of A. }
23:                     $sendMessage({}^{\otimes}q.\text{ID}, q_i)$;   {Asks for data from each q_i.}
24:                 **end for**
25:             **end for**
26:             $vertex.setVisited(TRUE)$;
27:         **end if**
28:     **end if**
29: **else**
30:                     {FEED_SUPERSTEP. **Odd** superstep. $vertex$ represents a simple state}
31:     **for each** ${}^{\otimes}p.\text{ID}$ **in** $messages$ **do**
32:         $sendMessage(vertex.getEdges(), {}^{\otimes}p.\text{ID})$ {Sends message to requesting ${}^{\otimes}p$}
33:     **end for**
34: **end if**
35: $vertex.voteToHalt()$;

---

5   Analysis

In this section, we attempt to give state complexity as well as computational one. State complexity concerns the number of states that will be generated by our solution, no matter they are useful or not. Computational complexity concerns the numbers of iterations and message exchanges between processes. We have implemented our method with Giraph [1] and tested it with several inputs.

Table 1. Execution of Algorithm 1 for the intesection of the two FSA in Fig. 1.

	Superstep 0	Superstep 2	Superstep 4	Superstep 6	Superstep 8	END
Active vertices	All simple states of A_1 and A_2	$(0,0)$	$(1,1)$	$(2,1),(2,2)$	$(1,1),(1,2)$	$\emptyset$
$\otimes_\Sigma$	–	$\{a\}$	$\{a,b\}$	$\{a,b\}$	$\emptyset$	–
New states	$(0,0)$	$(1,1)$	$(2,1),(2,2)$	$(1,2)$	$\emptyset$	–
New transitions	–	1 new: $(0,0) \xrightarrow{a} (1,1)$	3 new: $(1,1) \xrightarrow{a,b} (2,1)$ $(1,1) \xrightarrow{b} (2,2)$	3 new: $(2,1) \xrightarrow{a,b} (1,1)$ $(2,1) \xrightarrow{b} (1,2)$	$\emptyset$	–

5.1 State Complexity

Our main contribution in relation to the MapReduce solution (Sect. 3.2) is the state complexity. Intrinsically, our solution cannot produce a state that is not reachable from the initial one, since it starts by the initial state $(i_1, \cdots, i_m)$ and moves forward to the future new state only if a common symbol can be read. We can clearly characterize these unreachable states in Example 1: if we regard the intersection A as a 3×3 matrix, then all states in the first row and in the first column are unreachable from $(0,0)$.

However, our solution may produce some states from which we cannot reach the final state. It is presently the case in Example 1 with the single dead state $(1,2)$. We could easily remove a dead state that is at the end of a dead path. The state just has to remove itself if it is not final and has no outgoing transition, as for state $(1,2)$. But we may have dead states that have outgoing transitions to some others dead states. For this case, the solution for deleting these states is likely to make Algorithm 1 confused and may lead to many more supersteps.

More generally, our method can be especially efficient in the case of multiple intersection. We recall that if there are m input NFAs each having n states, the product NFA will have n^m states, even if this product is empty. For instance, let us consider automaton A_3, having n states and whose accepted words don't start with symbol "a". Obviously, the language $L(A_1 \otimes A_2 \otimes A_3)$ is empty. Our solution will stop at third superstep instead of producing $3 \times 3 \times n$ useless states.

Formally, several works give study of state complexity of regular language intersections. For instance, [10] investigate a tight bound for intersection of finite languages and stipulate that, given two minimal DFAs A and B, $mn - 3(m + n) + 12$ states are necessary and sufficient in the worst case for the intersection of $L(A)$ and $L(B)$, where A and B have respectively m and n states. Later, [5] improves this upper bound by first computing a lower bound on the number of unreachable states. Nevertheless, these works concern minimal DFAs while our work is more general, and a direct adaptation may not be evident. That's why a similar formal study should be done for NFAs in our distributed context.

5.2 Computational Complexity

The cost of our solution is the number S of required supersteps multiplied by the cost of a superstep.

Number of Supersteps. A PRODUCTION_SUPERSTEP of our proposition corresponds exactly to an iteration of the well-known *Breadth-First Search* (BFS) algorithm for traversing or searching tree or graph data structures. The time complexity of BSF is $O(|V| + |E|)$ since all vertices ($|V|$ in number) and all edges ($|E|$ in number) will be explored in the worst case. Then, the number S of required supersteps is bounded by the double of the number of states of the largest input automaton plus one. In other words, $S \leq 2 * (\max_i |Q_i| + 1)$. This is due to the nature of automata intersection and the way Algorithm reffunct-compute moves forward to search next product states with alternating supersteps.

Cost of a Superstep. The cost of a superstep is determined as the sum of three terms; the cost of the longest running local computation, the cost of maximum communication between processors, and the cost of the barrier synchronization. The costs are computed in terms of abstract parameters which model the number of processors p, the cost of barrier synchronization l, the cost of the running local computation w_i, the number h_i of messages send or received by process p_i, and g the ability of a communication network to deliver data, defined such that it takes time $h_i g$ for a process p_i to deliver h_i messages. Thus, the cost of one superstep for p processors is $max_{i=1}^{p}(w_i) + max_{i=1}^{p}(h_i g) + l$. Note that it is more common for the expression to be written as $w + hg + l$ where w and h are maximums.

In our case, if we consider the intersection of m automata A_i, with a fixed symbol alphabet Σ, and the automaton outdegree equal to k (k is the number of transitions outgoing from each state), w and h are respectively $O(k^{m+1})$ and $O(mk^{m+1})$. In fact, for each common symbol σ (line 16), individual transitions are combined (line 19) to find product transitions. After that, each new product state $^{\otimes}q$ will ask for information (outgoing transitions) from each of its component states q_i for following supersteps (line 23).

6   Concluding Remarks

In this work, we have proposed and implemented a novel use of BSP (*Bulk Synchronous Parallel*) abstract computer for large NFAs intersection. The intersection problem has been widely studied, with distributed solutions using either disk-based parallel algorithms, or MapReduce programming model. Contrary to these previous and classic approaches, we don't need to process the product of the size of all inputs NFAs. In fact, we avoid producing unreachable states of the product automaton. Our Pregel algorithm is accompanied by a running example, with some execution details. Finally, we have given an analysis of state and computational complexities.

Our model would benefit from (i) being more deeply studied in order to have tighter upper bounds for product automaton size and number of required supersteps, depending on input automata, (ii) more accurate complexity and (iii) experimental validation. However, one important thing for future works is to see how to bring out relevant programming artifacts, or building blocks, for a high level distributed graph language that hides the most distributed aspects.

References

1. The Apache Software Foundation: Apache giraph. https://giraph.apache.org/
2. The Apache Software Foundation: Apache hadoop. https://hadoop.apache.org/
3. Aridhi, S., Lacomme, P., Ren, L., Vincent, B.: A mapreduce-based approach for shortest path problem in large-scale networks. Eng. Appl. Artif. Intell. **41**, 151–165 (2015)
4. Dean, J., Ghemawat, S.: Mapreduce: simplified data processing on large clusters. Commun. ACM **51**(1), 107–113 (2008)
5. Gandhi, A., Khoussainov, B., Liu, J.: On state complexity of finite word and tree languages. In: Yen, H.-C., Ibarra, O.H. (eds.) DLT 2012. LNCS, vol. 7410, pp. 392–403. Springer, Heidelberg (2012). https://doi.org/10.1007/978-3-642-31653-1_35
6. Gonzalez, J.E., Low, Y., Gu, H., Bickson, D., Guestrin, C.: Powergraph: Distributed graph-parallel computation on natural graphs. In: Thekkath, C., Vahdat, A. (eds.) 10th USENIX Symposium on Operating Systems Design and Implementation, OSDI 2012, Hollywood, CA, USA, 8–10 October, pp. 17–30 (2012)
7. Gonzalez, J.E., Xin, R.S., Dave, A., Crankshaw, D., Franklin, M.J., Stoica, I.: Graphx: graph processing in a distributed dataflow framework. In: Flinn, J., Levy, H. (eds.) 11th USENIX Symposium on Operating Systems Design and Implementation, OSDI 2014, Broomfield, CO, USA, 6–8 October, pp. 599–613 (2014)
8. Grahne, G., Harrafi, S., Hedayati, I., Moallemi, A.: DFA minimization in mapreduce. In: Afrati, F.N., Sroka, J., Hidders, J. (eds.) Proceedings of the 3rd ACM SIGMOD Workshop on Algorithms and Systems for MapReduce and Beyond, BeyondMR@SIGMOD 2016, San Francisco, CA, USA, 1 July 2016, p. 4. ACM (2016)
9. Grahne, G., Harrafi, S., Moallemi, A., Onet, A.: Computing NFA intersections in map-reduce. In: Fischer, P.M., Alonso, G., Arenas, M., Geerts, F. (eds.) Proceedings of the Workshops of the EDBT/ICDT 2015 Joint Conference (EDBT/ICDT), Brussels, Belgium, 27th March 2015. CEUR Workshop Proceedings, vol. 1330, pp. 42–45. CEUR-WS.org (2015)
10. Han, Y.-S., Salomaa, K.: State complexity of union and intersection of finite languages. In: Harju, T., Karhumäki, J., Lepistö, A. (eds.) DLT 2007. LNCS, vol. 4588, pp. 217–228. Springer, Heidelberg (2007). https://doi.org/10.1007/978-3-540-73208-2_22
11. Kozen, D.: Lower bounds for natural proof systems. In: Proceedings of the 18th Annual Symposium on Foundations of Computer Science SFCS 1977, pp. 254–266. IEEE Computer Society, USA (1977)
12. Lattanzi, S., Mirrokni, V.S.: Distributed graph algorithmics: theory and practice. In: WSDM, pp. 419–420 (2015). http://dl.acm.org/citation.cfm?id=2697043
13. Malewicz, G., et al.: Pregel: a system for large-scale graph processing. In: Proceedings of the ACM SIGMOD International Conference on Management of Data, SIGMOD 2010, Indianapolis, Indiana, USA, 6–10 June, pp. 135–146. ACM (2010)
14. Ravikumar, B., Xiong, X.: A parallel algorithm for minimization of finite automata. In: Proceedings of IPPS 1996, The 10th International Parallel Processing Symposium, 15–19 April, Honolulu, USA, pp. 187–191. IEEE Computer Society (1996)

15. Slavici, V., Kunkle, D., Cooperman, G., Linton, S.: Finding the minimal DFA of very large finite state automata with an application to token passing networks. CoRR abs/1103.5736 (2011)
16. Tewari, A., Srivastava, U., Gupta, P.: A parallel DFA minimization algorithm. In: Sahni, S., Prasanna, V.K., Shukla, U. (eds.) HiPC 2002. LNCS, vol. 2552, pp. 34–40. Springer, Heidelberg (2002). https://doi.org/10.1007/3-540-36265-7_4
17. Valiant, L.G.: A bridging model for parallel computation. Commun. ACM **33**(8), 103–111 (1990)

Tight Bound of Parallel Request Latency for Erasure-Coded Distributed Storage System

Xingshun Zou and Wei Li[✉]

Shandong University, Jinan, China
zouxingshun@foxmail.com, wli1@sdu.edu.cn

Abstract. With big data analysis and cloud storage improving at an astonishing rate in recent years, more and more parallel computing and data analysis techniques are invented and applied to accessing data from a distributed storage system, which has noticeable differences with accessing data from a separate dedicated storage system. The first and the last processes of a data analysis task in a shared filesystem may have a remarkable gap in latency. The latency of parallel file requests from these processes, however, lacks analytical model especially when erasure code is employed as storage strategy. The main contribution of this work is development of a tight bound of parallel file request latency to give a theoretical upper bound of lantency. Additionally, an exact batch file chunk request analytical model in distributed systems adopting erasure code is studied by the technique of two embedded Markov chains, performance measures such as queue length distribution, waiting time distribution, etc. are derived in closed form. Experimental evaluation verifies accuracy of the theoretical upper bound on latency.

Keywords: Parallel file requests · Distributed storage system · Erasure code · Queueing theory · Latency · Embedded Markov chain

1 Introduction

Recently, erasure code has been widely used in various distributed storage systems such as GFS, HDFS or QFS [8,9,20]. These storage systems encode files into chunks and store file chunks spreading over all of the storage nodes rather than full file replication. Comparing to traditional method, erasure code reduces space of storage by 50% [11] and achieves the tantamount reliability because of its matrix algorithm. For instance, while distributed storage system uses a (d, k) MDS code to encode a file into d equal-size file chunks and randomly store them in the system, any subset of $k(k < d)$ chunks is enough to reconstruct the original file. Gathering k distinct file chunks, though, from different storage nodes to reconstruct the file leads to a severe latency problem. Due to randomness of file chunks distribution and difference of service capabilities of various nodes, chunk request times from different nodes possibly have a considerable gap, and

© Springer Nature Switzerland AG 2020
M. Qiu (Ed.): ICA3PP 2020, LNCS 12452, pp. 355–368, 2020.
https://doi.org/10.1007/978-3-030-60245-1_25

the file reconstructing time is determined by the slowest storage node. In other words, the request of the file in erasure code distributed system has to wait even if other $k-1$ distinct chunk requests are completed. Waiting for kth chunk leads to latency. It is significant to reduce latency in distributed storage system, since Google and Amazon show that every 500 ms delay means a 1.2% user loss [18].

Latency may be tolerated for request of single file. It is, however, necessary to optimise the latency of parallel requests of multiple files from the parallel processes or threads. More and more parallel computing and data analysis techniques [4,15] are applied to domain of accessing data from distributed storage systems. These techniques manage multiple parallel processes and parallel file requests generated by these processes. The parallel file requests may be generated by various applications, such as Hadoop MapReduce applications [7,23] and MPI [5], etc. Some researchers have optimised the system load balance of parallel data access by employing the method of Opass [12]. It is noticeable, accessing files with parallel strategies [15] has less latency in traditional individual storage system than distributed storage system. To the best of our knowledge, the latency optimisation of parallel file requests in distributed system using erasure code is still an open problem.

An exposition is undertaken on the data access latency in distributed storage system using erasure code. Previous works devoted to provide bounds on mean waiting latency [6,13,22] or find upper bound of weighted latency tail probability (WLTP) for file request of distributed storage system [1,2]. A queueing architecture that leverages the coding redundancy inherent is proposed by Chen et al. [6] to improve data accessing latency performance. Xiang et al. [21,22] provided a sophisticated upper bound on the mean service delay of heterogeneous files under predetermined probability with arbitrary service time distribution and derived a Joint Latency-Cost Minimization (JLCM) to minimise latency and achieved lowest storage cost. Besides, V. Aggarwal et al. focused on investigating tail latency and expressed the tail latency bound with general file size distribution [1] and Pareto file size distribution [2] respectively. However, none of them has considered the latency analysis of the simultaneous arrival of file requests. The simultaneity of file requests arriving is vital for modelling parallel file requests. Most of the existing strategies based on quantifying and optimising single file request latency of the distributed storage system can not be directly applied to obtain parallel accessing latency of multiple files.

In this paper, an efficient and accurate analytical performance model is proposed to quantify parallel file request latency in distributed storage system employing erasure code. To describe the process of parallel file requests arriving at the distributed storage system and being served in a storage node, Fig. 1 illustrates the process of parallel requests arriving and being served at distributed storage system in which 7 storage nodes are included and 4 files are stored using $(4,2)$ erasure code. Let task T be a certain set of requests, say file requests 1 and 4. A batch file requests T arrive at the storage system in an instant. The main server node denoted as S, which is similar to "NameNode" in Hadoop (i.e., Introducer in Tahoe), responds to distribute chunk requests. For instance, chunk

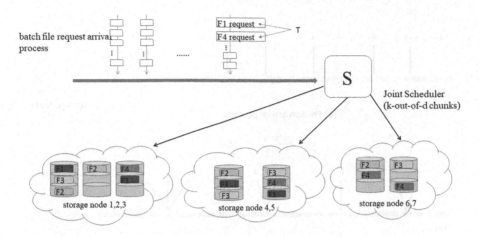

Fig. 1. Parallel file requests arrive at distributed storage system which has 7 storage nodes and stores 4 files using a $(4, 2)$ erasure code

requests of file 1 and 4 in the Fig. 1 arrive at nodes 3 and 5 at the same time, the slowest node decides the latency of task T. Therefore, batch process is considered to describe this chunk arriving process on each storage node in this distributed system. Finally, we analyse each storage node to obtain the system measures of batch arrival, such as the first two moments of chunk request time and the chunk requests queue length distribution. These measures are used to derive the upper bound of latency of parallel file requests in the distributed system.

The remaining of the paper is organised as follows. Section 2 describes the system model and the batch chunk request arriving process of a storage node in this distributed storage system. Section 3 and 4 analyses the batch queueing model and derives the tight upper bound of parallel file request latency, respectively. Section 5 presents our experimental evaluation and Sect. 6 concludes the work and proposes further work.

2   System Model

Consider a distributed storage system with m heterogeneous storage nodes, denoted by $\mathcal{M} = \{1, 2, ..., m\}$. To distributively store a set of R files, indexed by $i = 1, ..., R$, each file i is partitioned into k fixed-size chunks and then encoded with a (d, k) MDS erasure code to generate d same size distinct chunks of file i. There is r_j chunks of different file stored on storage node j, $r_j < R$. These encoded chunks are stored on and assigned to d distinct storage node-set $\mathcal{S}_i$, where $\mathcal{S}_i \subseteq \mathcal{M}$, so at most one file chunk could be attributed to the same node. Employing (d, k) MDS erasure code allows files to be reconstructed from any subset of k-out-of-d chunks. Thus, upon arrival of each file request, a batch of k distinct chunks from a group of nodes A_i, $A_i \subseteq \mathcal{S}_i$ are selected by the scheduler and retrieved to reconstruct the desired file i. Furthermore, while a batch of n file

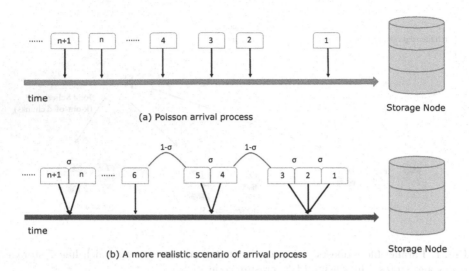

(a) Poisson arrival process

(b) A more realistic scenario of arrival process

Fig. 2. Chunk requests arrival process on the storage node

requests arrives into the distributed data storage system at the same time, the total number of $n \cdot k$ chunk requests arrives at all storage nodes simultaneously. In this distributed storage system, if a batch of n file requests including a set of file requests from task T arrives, k chunk requests of each file i of these n files appropriate a group of nodes A_i and all chunk requests from task T appropriate a group of nodes $\cup_{i \subseteq T} A_i$. Task T is served completely after all nodes belonging to the group of nodes service are completed. In other words, the parallel file request latency of task T is determined by the slowest one of the group of nodes.

To investigate parallel file request latency with arbitrarily set of file for distributed storage system with erasure code, we consider the probabilistic scheduling policy as follows: 1) client dispatches all chunk requests which correspond to a file request to appropriate a set of nodes; 2) each node buffers chunk requests in a local queue and processes them in order (FIFS). If a random file i request arrives, the probability of chunk from this file stored on node j $\frac{r_j}{R}$ since each storage manages different file chunk group. The probability (denoted by δ_j) of reconstructing file i with a chunk from node j is decided by

$$\delta_j = \frac{\binom{k-1}{d-1}}{\binom{k}{d}} \cdot \frac{r_j}{R} \tag{1}$$

where r_j is the number of file chunks stored on node j, and R is the amount of files stored all over the system, k, d are two parameters of erasure code. In order to avoid the indeterminacy of scheduling in different distributed systems and achieve a general parallel file request latency, k/d is set as the probability of scheduler chooses node j as a node of A_i in our model. Knowing the average arrival rate of file requests Λ, it is feasible to obtain arrival rate of chunk requests on node j with $\lambda_j = \Lambda \delta_j$.

Most existing works use $M/G/1$ or $M/G/1/K$ queue to model the storage node queue and then provide an evaluation on approximating $M/G/1$ using $M/M/1$ [1,2,19,22]. Figure 2 (a) shows the Poisson process of the chunk requests arriving at a storage while it does not describe the situation of multiple chunk request arriving at a storage node simultaneously. Figure 2 (b) exhibits a chunk request arrival process more suitable for the real situation, in which more than one chunk requests are allowed to arrive at a storage node at the same time. Let σ denote the probability that consecutive chunk requests belongs to the same batch. The arrival process is specified as a Poisson process under the condition of $\sigma = 0$.

On account of the situation of chunk requests arriving at storage node j simultaneously, batch Poisson arrival [17] is used to model the arrival process. The interval between batches is distributed as $a_j(t) = \sigma_j \lambda_j e^{-\sigma_j \lambda_j t}, t > 0$, and chunk batch size distribution is specified as $b_j(n) = (1 - \sigma_j)^{n-1}\sigma_j, n = 1, 2, ...$, where $\lambda_j t$ is the average number of chunk requests arriving during t and the model is stable when the file requests arrival rate Λ satisfying

$$k \cdot \Lambda = \sum_{j \in \mathcal{M}} \lambda_j < \sum_{j \in \mathcal{M}} \mu_j \tag{2}$$

where μ_j is chunk service rate of storage node j.

3   Queueing Model and Performance Analysis with Batch Poisson Arrival

With unprecedented development of parallel processing technology, part of storage nodes in the distributed system have the ability of parallel reading which can process multiple chunk requests at the same time and others still handle each chunk request serially. The storage node is modeled by batch Poisson arrival process and service process with Poisson, batch Poisson and shifted Poisson.

3.1   Queueing Model Analysis with Batch Poisson Service Processes

Batch Poisson is proposed by [17] and it is easy to know that batch Poisson is a generalisation of Poisson process. Here each storage node in distributed storage system can be considered as an $M^{Geo}/M^{Geo}/1$ queueing system, service counting function of batch Poisson process is

$$g(s,t) = \begin{cases} e^{-\tau\mu t} & s = 0 \\ \sum_{\ell=1}^{s} \binom{s-1}{\ell-1}(1-\tau)^{s-\ell}\tau^{\ell}\frac{(\tau\mu t)^{\ell}}{\ell!}e^{-\tau\mu t} & s = 1, 2... \end{cases} \tag{3}$$

Equation (3) is derived by considering the probability density function of exactly s chunk requests depart during the interval of duration t in the local queue of on a node, where μ is the service rate of the node, and τ is a geometrically distributed service chunk size parameter. The distribution of the number of chunk request waiting for service in the local queue of the node at any time, $p(n)$, is given by

$$p(0) = \frac{\tau(\mu - \lambda)}{\tau\mu + \sigma\lambda(1 - \tau)} \tag{4}$$

$$p(n) = \frac{\tau(\mu - \lambda)}{\tau\mu + \sigma\lambda(1 - \tau)} \cdot \frac{\sigma\lambda(\tau + \sigma - \tau\sigma)}{\tau\mu + \sigma\lambda(1 - \tau)} \cdot \left\{ \frac{\sigma\lambda + \tau\mu(1 - \sigma)}{\tau\mu + \sigma\lambda(1 - \tau)} \right\}^{n-1} \qquad n = 1, 2... \tag{5}$$

From (4) and (5) the mean queue length (MQL) on node j is calculated as following

$$MQL = \mathbb{E}(n) = \sum_{n=1}^{\infty} n \cdot p(n) \tag{6}$$

$$= \frac{\tau(\mu - \lambda)}{\tau\mu + \sigma\lambda(1 - \tau)} \cdot \frac{\sigma\lambda(\tau + \sigma - \tau\sigma)}{\tau\mu + \sigma\lambda(1 - \tau)} \cdot \sum_{n=1}^{\infty} n \cdot \left\{ \frac{\sigma\lambda + \tau\mu(1 - \sigma)}{\tau\mu + \sigma\lambda(1 - \tau)} \right\}^{n-1} \tag{7}$$

$$= \frac{\tau(\mu - \lambda)}{\tau\mu + \sigma\lambda(1 - \tau)} \cdot \frac{\sigma\lambda(\tau + \sigma - \tau\sigma)}{\tau\mu + \sigma\lambda(1 - \tau)} \cdot \left\{ \frac{\tau\mu + \sigma\lambda(1 - \tau)}{\tau\sigma(\mu - \lambda)} \right\}^{2} \tag{8}$$

Hence, applying Little's law to (6) yield

$$\mathbb{E}(w) = W = \frac{\mathbb{E}(n)}{\lambda} \tag{9}$$

where W is mean waiting time of a chunk request spends on the node. Similarly, by generalisation of Little's law, of course, the second moment of waiting time of a chunk request is given by

$$\mathbb{E}(w^2) = \frac{\mathbb{E}[n(n-1)]}{\lambda^2} \tag{10}$$

And

$$\mathbb{E}[n(n-1)] = \sum_{n=1}^{\infty} n(n-1) \cdot p(n) \tag{11}$$

$$= Y \cdot \left\{ \sum_{n=1}^{\infty} n^2 Z^{n-1} - \sum_{n=1}^{\infty} n Z^{n-1} \right\} \tag{12}$$

$$= 2Y \cdot \frac{Z}{(1 - Z)^3} \tag{13}$$

where Y and Z are substitutions of $\frac{\tau(\mu-\lambda)}{\tau\mu+\sigma\lambda(1-\tau)} \cdot \frac{\sigma\lambda(\tau+\sigma-\tau\sigma)}{\tau\mu+\sigma\lambda(1-\tau)}$ and $\frac{\sigma\lambda+\tau\mu(1-\sigma)}{\tau\mu+\sigma\lambda(1-\tau)}$ respectively.

3.2 Queueing Model Analysis with Poisson Service Processes

Chunk requests are assumed to be served one by one in this model. To quantify the model on these nodes, service process is described by Poisson process which $g(s,t) = \frac{(\mu t)^s}{s!}e^{-\mu t}, t > 0$. Now by Eq. (6) and (10), let $\tau = 1$, for an $M^{Geo}/M/1$ system on one node with Poisson service process, the first and second moments of w are given, respectively, by

$$\mathbb{E}(w) = \frac{1}{\sigma(\mu - \lambda)} \tag{14}$$

$$\mathbb{E}(w^2) = \frac{2\sigma(\mu - \lambda)}{\lambda\mu^2} \cdot \frac{X}{(1 - X)^3} \tag{15}$$

where $X = \frac{\sigma\lambda + \mu(1-\sigma)}{\mu}$.

3.3 Queueing Model Analysis with Shifted Poisson Service Process

Taking account of our Tahoe experiments and Amazon S3 experiments [6], The service process could be considered as Poisson process with a shifted exponential distribution, denoted by sM. Let the service time density function is specified as

$$f(x) = \begin{cases} \alpha e^{-\alpha(t-D)} & t \geq D \\ 0 & t < D \end{cases} \tag{16}$$

Equation (16) is generalization of exponential distribution. The counting function [10] of this service process derived by Eq. (16) is

$$g(s,t) = \frac{\alpha^s(t - sD)^s}{s!}e^{-\alpha(t-sD)} \qquad s = 0, 1, 2\ldots \tag{17}$$

By Eq. (17) and the solution of $GI^G/G/1$ queue using two embedded Markov Chains, the z-transform of queue length distribution of $M^{Geo}/sM/1$ is given by

$$P[z] = \frac{z(\alpha + \sigma\lambda) - \alpha e^{-\sigma\lambda D} - \sigma\lambda z + \alpha z(1 - e^{-\sigma\lambda D})}{z(\alpha + \sigma\lambda) - \alpha e^{-\sigma\lambda D} - \sigma\lambda z B[z]} \cdot p(0) \tag{18}$$

where $B[z]$ is z-transform of chunk batch size probability mass function $b(n)$. Taking $z = 1$ the probability of system being empty is specified as

$$p(0) = \frac{\alpha e^{-\sigma\lambda D} - \sigma\lambda}{\alpha e^{-\sigma\lambda D}} \tag{19}$$

Therefore, the first and second moments of w can be derived with properties of z-transform and Little's law as following

$$\mathbb{E}(w) = \frac{P'[1]}{\lambda} \tag{20}$$

$$\mathbb{E}(w^2) = \frac{P''[1]}{\lambda^2} \tag{21}$$

4 Bound of Parallel File Request Latency

Now we consider the arbitrary parallel file requests latency of task F and N is the number of file requests. All files requested by task F are encoded into file chunks with erasure code. These file chunks are stored on a set of nodes ψ_f and $|\psi_f|$ is the number of elements. Let Q_j be the sojourn time random variable of chunk requests from task F in the local queue of node j and node j is a node of ψ_f. The probability of a chunk request of task F arriving at node j is considered as

$$P = \frac{d}{|\psi_f|} \cdot \frac{\binom{k-1}{d-1}}{\binom{k}{d}} \tag{22}$$

The probability of x chunk requests arriving at an arbitrary node is $C(x) = \binom{x}{N} P^x (1-P)^{N-x}$. The average number of chunk requests arriving at node j is $N \cdot P$ and it satisfies $\sum_{j \in \psi_f} N \cdot P = N \cdot d$ to guarantee access to enough chunks for successful file retrieval. Also, because of chunks are all of the same size, group this x chunk requests into a super chunk request for obtaining the first and second moments of this super chunk request time on node j. The service times of all chunk requests are identically and independently distributed and the mean waiting time of this super chunk request can be calculated by setting service capacity of node j to $\frac{\mu_j}{x}$.

As done in [22], the expected parallel file requests latency of task F, $\bar{S}_f$ is the maximum chunk request time of ψ_f under our probability scheduling policy, expressed as

$$\bar{S}_f = \mathbb{E}_{\psi_f} \left(\max_{j \in \psi_f} \{Q_j\} \right) \tag{23}$$

where $\max_{j \in \psi_f} \{Q_j\}$ is the maximum batch chunk request time of all nodes in a certain set of ψ_f, $Q_{max} = max\{Q_j, j \in \psi_f\}$. Use the Theorem 1 and Proposition 1 proposed in [3] to obtain the bound of parallel file request latency. The upper bound of Q_{max} using arbitrary constant $z \in \mathbb{R}$ is derived as

$$Q_{max} = z + (Q_{max} - z) \leq z + \sum_{j \in \psi_f} [Q_j - z]^+ \tag{24}$$

where $[Q_j - z]^+ = Q_j - z$ if $Q_j - z > 0$, otherwise $[Q_j - o]^+ = 0$. Next, it is easy to obtain the equation as following

$$[Q_j - z]^+ = \frac{1}{2}(Q_j - z + |Q_j - z|) \tag{25}$$

Taking expectations and by the Cauchy−Schwarz inequality, it yields

$$\begin{aligned} \mathbb{E}[Q_j - z]^+ &= \frac{1}{2}(\mathbb{E}[Q_j - z] + \mathbb{E}|Q_j - z|) \\ &\leq \frac{1}{2}(\mathbb{E}[Q_j] - z + \sqrt{(\mathbb{E}[Q_j] - z)^2 + Var[Q_j]}) \end{aligned} \tag{26}$$

Hence, by Eq. (24), (25) and (26), the expected parallel file request latency of task F is bounded by

$$\bar{S}_f \leq \min_{z \in \mathbb{R}} \left\{ z + \sum_{j \in \psi_f} \sum_{x=1}^{N} \frac{C(x)}{2} (\mathbb{E}[Q_j^x] - z) \right.$$

$$\left. + \sum_{j \in \psi_f} \sum_{x=1}^{N} \frac{C(x)}{2} \left(\sqrt{(\mathbb{E}[Q_j^x] - z)^2 + Var[Q_j^x]} \right) \right\} \tag{27}$$

where Q_j^x is waiting time for all x chunk requests of task F arriving together on the local queue of node j. $\mathbb{E}[Q_j^x]$ and $Var[Q_j^x]$ are mean and variance of Q_j^x respectively, can be calculated from $\mathbb{E}(w)$ and $\mathbb{E}(w^2)$ with super chunk request in this queue system.

Furthermore, it is easy to verify the bound is tight since it is attained [3] by the distribution $Q_j = z \pm \sqrt{(\mathbb{E}[Q_j] - z)^2 + \mathbb{D}[Q_j]}$ with probabilities

$$P = \frac{1}{2} \left(1 + \frac{\mathbb{E}[Q_j] - z}{\sqrt{\mathbb{E}([Q_j] - z)^2 + Var[Q_j]}} \right) \tag{28}$$

$$1 - P = \frac{1}{2} \left(1 - \frac{\mathbb{E}[Q_j] - z}{\sqrt{\mathbb{E}([Q_j] - z)^2 + Var[Q_j]}} \right) \tag{29}$$

5   Implementation and Evaluation

To evaluate accuracy of our proposed request latency bound in a erasure-coded supporting distributed filesystem under diverse workload scenarios, we implement the model in Tahoe [16]. 10 Tahoe storage servers and 4 Tahoe clients are deployed as typical medium-sized virtual machine instances in our data center. The average round-trip time (RTT) between clients and servers measured by Tahoe statistic JSON file is 0.05 ms, which has little impact on latency in our evaluation. Each Tahoe storage server attaches a 100 GB SSD disk stores erasure-coded shares and offers storage for the external client. These servers and clients are introduced via Tahoe Introducer Node. Tahoe client can issue upload/download requests simultaneously and connect to storage server through the Tahoe-LAFS storage protocol over SSL. The workloads are generated by issuing download requests to the data center, and each client can process four download requests concurrently based on multithread. Each instance in our data center monopolizes 2 VCPUs, 2 GB of RAM, and runs Ubuntu 14.04.

In the experiment, the probability of the aiming file is stored at a particular storage node is k/d, which could be modified in the Tahoe configuration file. 8000 files are distributively stored on the 10 nodes with $(7,3)$ erasure code (file sizes range from 5 MB to 50 MB). In particular, the latency of small size file is susceptible to the phase of metadata read, and our model has not taken into account the effect of reading metadata on boundary latency. Therefore, smaller file size set in [19] is not choosen in our experiment. The average file size around 10 MB (much bigger than metadata) is consider to prevent the phase of reading metadata to affect our latency bound.

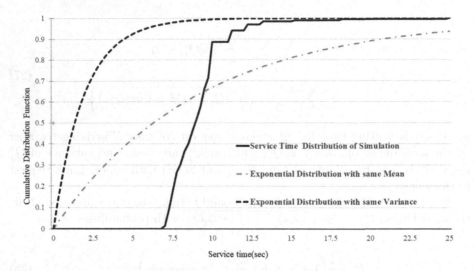

Fig. 3. The cumulative distribution function of node service time.

To obtain the genuine service process in our data center, we run an experiment to analyse service time distribution of these storage nodes and acquire each actual service rate of those 10 storage nodes. Thus, by uploading other 800 files using (7, 3) erasure code into the system and recording the chunk service time, all files are 15 MB, each chunk size is 15/3 MB. The cumulative distribution function of the chunk service time is described in Fig. 3. Moreover, using the sample mean service time and variance derived from experimental results, other two exponential distribution curves with the identical mean and variance as experimental results are depicted. Specially, the result is similar to Xiang's Tahoe experiment in [22], which uploaded 50 MB files using (7, 4) erasure code in their experiment environment. Therefore, 1) The chunk service process can not be described precisely by exponential distribution; 2) shifted exponential distribution may be used to approximate the actual service process. Consequently, the accuracy of the approximation is acceptable according to the same service process assumption in Tahoe distributed storage system [1] and our later evaluation in this section. The average service rate on each node is obtained by running the simulation.

These service rates are used in our model to calculate upper bound on the latency. For comparison purpose, two upper bounds in [14] and [22] are compared with our bound (named Zou's bound). In the [14], a modified (n, k) fork-join queue with Poisson arrival process is proposed. Further, after each request is divided and arrive at n servers, which are occupied with processing those n chunk requests, moreover, the $n - k$ chunks will be abandoned immediately if the first k chunks are served. In the reference [22], the $M/G/1$ queueing model is empirically applied to describe the storage node queue in order to obtain the latency bound. Our work uses M^{Geo} and sM to model the arrival process and

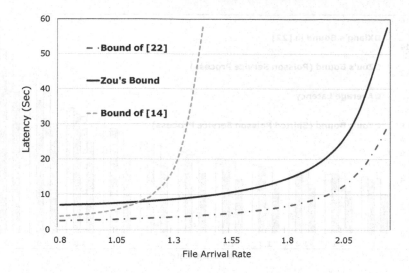

Fig. 4. Comparison of Zou's bound with Bound in [22] and [14]

service process of chunk request. For a fair comparison, $(7,3)$ erasure code is used in those three models, besides, the service capability shown in Table. The parameter σ of our model is set to 0.5 for describing the bursty arrival rate in reality. Three superlinear curves are illustrated in Fig. 4. The Xiang's bound proposed in [22] is the tightest bound under any arrival rate. Also, Xiang's bound has the similar growing trend with Zou's with λ rising. Joshi's bound proposed in [14] is outperformed our bound if chunk arrival rate is less than 1.15 but the growing slope becomes sharp after that point.

Nonetheless, that is what we have expected in Eqs. (14) and (27). The batch arrival and shifted service process are employed to preferably approximate the actual parallel arrivals. Both increase the expectation and variance of chunk request time, the Zou's bound (27) is consequently growing with these two measures.

We compare two upper bounds proposed with Poisson and shifted Poisson service process to the Xiang's bound in [22]. The statistical average latency observed in this experiment is also compared with the theoretical results. In this comparison experiment, we generate the workload in which average arrival rate increases every five minutes lasting one hour and keeps the arrival process bursty in order to approximate the actual arrival process. Figure 5 shows the two bounds derived in this paper. Both have more accurate predictions and significantly outperforms the bound in [22] under this situation. It demonstrates that using batch arrival process to describe the chunk arrival on the storage server is more accurate. On the other hand, the bound using shifted Poisson service process is closer to the real experimental data than the Poisson process at $\lambda = 1.1, 1.3, 1.4, 1.7, 1.8$ and 2.

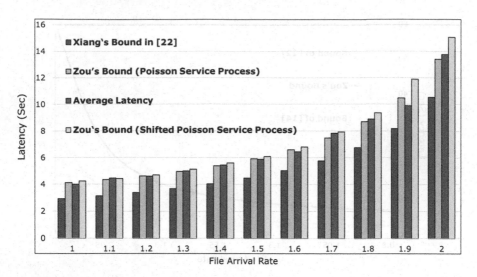

Fig. 5. Comparison and evaluation of Zou's bound with two service processes, Xiang's bound in [22] and Average latency in the experiment.

6 Conclusions and Future Work

In this paper, the problem of batch latency of parallel file requests for distributed storage system employed erasure code is studied. A mathematical model is developed to quantify and analyse the performance of storage nodes and batch latency bound in distributed storage system. In this experiment, a typical distributed storage system is set to verify the analytical model. By applying the analytical model, parallel file request latency bound can be obtained and optimised by setting model parameters on each storage node.

Most of the published work focus on individual requests to each particular file. That leads to the difficulty of dealing with multiple simultaneous file requests within one task. A novel analytical model is developed to solve the existing problem of multiple simultaneous file requests within the distributed storage system. The queueing model proposed by our previous work is employed and improved to model each storage node, assuming batch size distribution of file requests arrival is specified as geometric distribution and three kinds of service processes. For a generalisation purpose, more general arrival batch size distributions, such as GGeo (generalised Geometric distribution) is to be proposed, and more general service process can be considered. To find more general model to describe the file request process and quantify the parallel latency on the storage node of the distributed system, Besides, the processing time of metadata read is not considered in this paper due to the time constraint. To avoid this deviation, we use the 10M file in this paper, so the larger value of N is not set in this experiment. Our future work will concentrate on a more general model in erasure-coded distributed systems.

References

1. Aggarwal, V., Fan, J., Lan, T.: Taming tail latency for erasure-coded, distributee storage systems. In: IEEE INFOCOM 2017-IEEE Conference on Computer Communications. pp. 1–9. IEEE (2017)
2. Aggarwal, V., Lan, T.: Tail index for a distributed storage system with pareto file size distribution. arXiv preprint arXiv:1607.06044 (2016)
3. Bertsimas, D., Natarajan, K., Teo, C.P.: Tight bounds on expected order statistics. Probab. Eng. Inf. Sci. **20**(4), 667–686 (2006)
4. Cannataro, M., Talia, D., Srimani, P.K.: Parallel data intensive computing in scientific and commercial applications. Parallel Comput. **28**(5), 673–704 (2002)
5. Chao, L., Li, C., Liang, F., Lu, X., Xu, Z.: Accelerating apache hive with MPI for data warehouse systems. In: 2015 IEEE 35th International Conference on Distributed Computing Systems, pp. 664–673. IEEE (2015)
6. Chen, S., et al.: When queueing meets coding: optimal-latency data retrieving scheme in storage clouds. In: IEEE INFOCOM 2014-IEEE Conference on Computer Communications, pp. 1042–1050. IEEE (2014)
7. Dao, T.C., Chiba, S.: HPC-reuse: efficient process creation for running MPI and hadoop MapReduce on supercomputers. In: 2016 16th IEEE/ACM International Symposium on Cluster, Cloud and Grid Computing (CCGrid), pp. 342–345. IEEE (2016)
8. Ghemawat, S., Gobioff, H., Leung, S.T.: The google file system (2003)
9. hadoop.apache.org: https://hadoop.apache.org/docs/r3.0.3/
10. https://stats.stackexchange.com/users/10479/yves , Y.: Variance of arrival process with shifted exponential distribution. Cross Validated https://stats.stackexchange.com/q/87287. Accessed 29 Dec 2014
11. Huang, C., et al.: Erasure coding in windows azure storage. In: Presented as Part of the 2012 Annual Technical Conference, pp. 15–26 (2012)
12. Huang, D., et al.: Achieving load balance for parallel data access on distributed file systems. IEEE Trans. Comput. **67**(3), 388–402 (2018)
13. Huang, L., Pawar, S., Zhang, H., Ramchandran, K.: Codes can reduce queueing delay in data centers. In: 2012 IEEE International Symposium on Information Theory Proceedings, pp. 2766–2770. IEEE (2012)
14. Joshi, G., Liu, Y., Soljanin, E.: On the delay-storage trade-off in content download from coded distributed storage systems. IEEE J. Sel. Areas Commun. **32**(5), 989–997 (2014)
15. Kumar, V., Grama, A., Gupta, A., Karypis, G.: Introduction to Parallel Computing: Design and Analysis of Algorithms Benjamin. Cummings, Redwood City (1994)
16. tahoe lafs.org: Tahoe-lafs docs, January 2019. https://tahoe-lafs.readthedocs.io/en/tahoe-lafs-1.12.1
17. Li, W.: An investigation into batch renewal process and batch markovian arrival process and batch markovian arrival process and their performance impact on queueing models. Ph.D. thesis, University of Bradford, UK (2007)
18. Schurman, E., Brutlag, J.: The user and business impact of server delays, additional bytes, and http chunking in web search. In: Velocity Web Performance and Operations Conference (2009)
19. Su, Y., Feng, D., Hua, Y., Shi, Z.: Predicting response latency percentiles for cloud object storage systems. In: 2017 46th International Conference On Parallel Processing (ICPP), pp. 241–250. Proceedings of the International Conference on Parallel Processing (2017)

20. Weatherspoon, H., Kubiatowicz, J.D.: Erasure coding vs. replication: a quantitative comparison. In: Druschel, P., Kaashoek, F., Rowstron, A. (eds.) IPTPS 2002. LNCS, vol. 2429, pp. 328–337. Springer, Heidelberg (2002). https://doi.org/10.1007/3-540-45748-8_31

21. Xiang, Y., Lan, T., Aggarwal, V., Chen, Y.F.R.: Multi-tenant latency optimization in erasure-coded storage with differentiated services. In: 2015 IEEE 35th International Conference on Distributed Computing Systems, pp. 790–791. IEEE (2015)

22. Xiang, Y., et al.: Joint latency and cost optimization for erasure-coded data center storage. IEEE/ACM Trans. Netw. (TON) **24**(4), 2443–2457 (2016)

23. Yu, X., Li, W.: Performance modelling and analysis of MapReduce/Hadoop workloads. In: The 21st IEEE International Workshop on Local and Metropolitan Area Networks, pp. 1–6. IEEE (2015)

High-Performance Simulations on GPUs Using Adaptive Time Steps

Marcel Köster[✉], Julian Groß, and Antonio Krüger

Saarland Informatics Campus, Campus D3.2, 66123 Saarbrücken, Germany
{marcel.koester,julian.gross,antonio.krueger}@dfki.de

Abstract. Graphics Processing Units (GPUs) are widely spread nowadays due to their parallel processing capabilities. Leveraging these hardware features is particularly important for computationally expensive tasks and workloads. Prominent use cases are optimization problems and simulations that can be parallelized and tuned for these architectures. In the general domain of simulations (numerical and discrete), the overall logic is split into several components that are executed one after another. These components need step-size information which determines the number of steps (e.g. the elapsed time) they have to perform. Small step sizes are often required to ensure a valid simulation result with respect to precision and constraint correctness. Unfortunately, they are often the main bottleneck of the simulation. In this paper, we introduce a new and generic way of realizing high-performance simulations with multiple components using adaptive time steps on GPUs. Our method relies on a code-analysis phase that resolves data dependencies between different components. This knowledge is used to generate specially-tuned execution kernels that encapsulate the underlying component logic. An evaluation on our simulation benchmarks shows that we are able to considerably improve runtime performance compared to prior work.

Keywords: Simulations · Parallel computing · Adaptive time steps · Graphics processing units · GPUs

1 Introduction

There have been many distinct time-step adaption methods for completely different domain-specific problems. Such adaption techniques are particularly well investigated in the field of fluid and/or particle-based simulations. Prominent examples are SPH (Smoothed Particle Hydrodynamics) [10,21,22] and Position Based Dynamics (PBD) [23] simulations. Even sophisticated optimization problems are often modeled with the help of underlying rule-based and simulation like programs or code fragments [11–13]. Regardless of the use case and the domain, a simulation always consist of several phases that have to be evaluated

This work was supported by the SINTEG-project DESIGNETZ funded by the German Federal Ministry of Economic Affairs and Energy (BMWi) under the grant 03SIN222.

© Springer Nature Switzerland AG 2020
M. Qiu (Ed.): ICA3PP 2020, LNCS 12452, pp. 369–385, 2020.
https://doi.org/10.1007/978-3-030-60245-1_26

one after another[1]. Executing those parts in an iterative manner yields the final simulation result in the end [16,24]. An essentially required piece of information in this context is the actual time-step size which is used to execute each simulation step. Choosing this step size is very important with respect to simulation correctness and performance at the same time: Small time-step sizes often yield the best precision but are much more expensive since the simulation has to execute more steps until it reaches a final state. Since many parts of a simulation step have to be executed sequentially, the time-step size is often considered to be the primary performance-critical part.

Today, the mentioned simulations are often run on GPUs to benefit from their parallel computation capabilities. This allows to realize large-scale simulations which are even well suited for real-time applications [17,19]. In order to use these massive processing features, algorithms (as well as their surrounding applications) have to be specifically tuned for these architectures. From a high-level point-of-view, each simulation consists of a set of *components* which describe a specific module of the whole simulation logic.

In this paper, we present a new method to realize component-based simulations on GPUs using adaptive time steps. Our method is generic in respect of the application domain, as it does not rely on specific knowledge about the internal structure of the components. Instead, we use a static program analysis to determine a data-dependency graph. Based on this graph, we are able to compute the next adaptive step size. In this scope, we also allow developers to include their domain-specific knowledge with respect to time-step constraints. Our concept is based on the well-known idea to interpolate certain values at specific points in time [3]. We use this principle to simulate an intermediate component execution that has not happened in order to reduce the overall runtime of the simulation. Using our approach yields speedups between 12% (a smaller gravity-like simulation) and 22% (a larger PBD-like simulation) on our non-optimistic evaluation scenarios. Furthermore, we do not suffer from slowdowns in comparison to more conservative time-stepping approaches. This makes it a perfect extension for modern GPU-based simulation systems.

The remainder of this paper, summarizes and discusses related work from the fields of adaptive (fluid/particle) simulations to improve performance. We give a high-level introduction into the modeling of simulations in Sect. 3. This section also describes the major challenges in terms of adaptive time steps that we can solve using our new approach. Section 4 presents our generic concept and gives in-depth information about all design considerations and implementation details on GPUs. The evaluation section covers two designed benchmark scenarios. They are inspired by real-world simulation models to measure realistic performance numbers.

2   Related Work

From a theoretical point-of-view, using interpolation functions to adapt time steps is a well known concept [2,3,9,27]. There have been many different approaches from the field of numerics (solving partial differential equations, for

---

[1] A single simulation iteration is commonly referred to as a simulation step.

example). They also leverage interpolation functions to resolve intermediate values. This contribution has a different view on adaptive time steps without having explicit domain knowledge about the components. As outlined above, we focus on practical and pragmatic aspects of realizing fast and efficient simulations (see Sect. 5). To the best of our knowledge, we consider the following papers to be related to this subject.

Adams et al. [1] present an adaption approach for particle-based fluids. They define a domain-specific criterion to decide which region of the fluid simulation is more important and needs more computational resources. In contrast to our method, they are not adapting the actual time-step sizes, but focus on adaptive particle sizes to reduce the computational complexity.

Predictive-Corrective Incompressible SPH (PCISPH) [29] is a well known fluid-computation model that essentially predicts forces to enable larger time steps in order to overcome the severe time-step restrictions of previous papers. Ihmsen et al. [8] add support for adaptively computed time steps in the context of PCISPH. They use domain-specific properties of the underlying PCISPH algorithm and combine them with knowledge about maximum time-step restrictions of these simulations. Further adaptive fluid simulations are the ones by Hong et al. [6] and Zhang et al. [30]. They basically split and merge particles in order to reduce the number of elements to process. The split- and merge-criteria are based on a set of scenario-dependent properties. It is regrettable that these methods essentially modify the underlying data to be evaluated by several components. This requires domain-specific knowledge about the underlying structure of the problem and its associated optimization possibilities which cannot be easily generalized to work with black-box components.

Solenthaler et al. [28] and Horvath et al. [7] use a custom solution to enable adaptive computations in the scope of fluid simulations. They are achieving this purpose by coupling a set of differently scaled simulations. This avoids splitting and merging of particles. As before, these methods are limited to their particular field of application and cannot be applied to generic components.

Macklin et al. [18] use a PBD-based algorithm and add support for density constraints in order to model fluids. Since PBD simulations are very robust, they relax time-step restrictions of previous approaches significantly. Koester et al. [14] present an adaption scheme for these simulations. Like Macklin et al. [18], they are using an iterative constraint solver to move particles into the right positions to satisfy a fluid-density constraint. During solving, some particles can be considered less important than others. The more important a particle is, the more particle-position adjustments will be performed in upcoming iterations of the constraint solver. This significantly reduces the overall computational effort, since less important particles will not be considered in further solver iterations. This idea is similar to our high-level concept of interpolated values at certain time steps: Some values loaded from memory are not available at particular points in time and can be reconstructed using interpolation functions. In contrast to the original paper by Koester et al., we can provide intermediate values at specific time steps, whereas the related algorithm uses out-dated information without interpolation.

Garcia et al. [4] describe an adaptive time-stepping method that computes common step sizes using global reduction kernels. They estimate the time-step size in each iteration and synchronize the resolved time-step value with the CPU part of the application. We follow their approach by applying all time-step estimation methods of all components prior to the execution part of each component: We still rely on a synchronization step with the CPU side to exchange information about the computed time-step size.

Mayr et al. [20] leverage an iterative computation of the time-step size based on certain error calculations in the field of fluid-interaction solvers. From a high level point-of-view, this is closely related to our method: We compute the time-step size for each component and search for a compatible step size that is acceptable for all components. However, their adaption method relies on specific domain knowledge (which is tightly coupled to their application domain) that cannot be reused in a generic and portable way.

3 Component-Based Simulations on GPUs

This section covers a brief introduction into the general modeling of simulations on GPUs consisting of several components C_0 to C_{n-1}. Every component realizes a certain part of a simulation loop and is supposed to compute a separate piece of the overall puzzle that we want to put together. Figure 1 shows a sample simulation loop consisting of several components. Note that the overall execution order is usually determined by the domain expert/application programmer that has to take all data dependencies into account. Moreover, many simulations leverage the concept of *double buffering* to simplify data dependencies. Double buffering allows to read the same source data in each component while writing simulation updates into a separate target buffer. This ensures that each component is able to work on a consistent state without having to worry about possible changes by other components. Alternatively, some applications work on a single buffer only. This exposes all value updates to the components executed afterwards. The developers have to take special care to create a consistent logical component model that is compliant with the actual buffering approach used.

Fig. 1. A simple simulation loop with five components C_0 to C_4 in a sequential order. The back edge from C_4 to C_0 models the jump to the initial component C_0 of the next simulation step. Furthermore, it visualizes the data dependency between two steps.

Each component is applied to a set of items (e.g. particles or other data elements) from a set of source buffers. A *state* in memory thereby consists of all items in all buffers that the components can operate on. To implement this functionality on a GPU, each component C needs to have a separate GPU kernel that iterates

---

Algorithm 1. Simple application kernel algorithm for component C

```
   /* Perform a grid-stride loop over the padded value range of C    */
 1 for i := global index; i < range(C); i += grid size * group size do
      /* Initialize component by loading relevant information          */
 2    c := C.Init(i);
      /* Evaluate component                                            */
 3    c.Evaluate(Δt);
 4 end
```

---

over all items (referred to as the *range* of C). Algorithm 1 shows a straight-forward kernel implementation for an arbitrary C. It uses grid-stride loops to realize an efficient way to iterate over all items in the range of C. However, depending on the computational complexity and the performance characteristics of C, it can be beneficial to apply loop unrolling at this point [15].

Applying components in the presence of adaptive time steps means that need to retrieve information about the potential step size of each component. The estimated step sizes also include domain-specific time-step constraints to satisfy correctness/stability requirements. Figure 2 shows the high-level difference between a workflow using fixed vs. adaptively computed time steps. In the adaptive setting, all components estimate their intended time-step size they could potentially, assuming that all necessary information is available (optimistic assumption). Afterwards, the minimum time step of all potential time steps (from the first phase) can be used safely (sound assumption). Finally, the computed step size from the previous phase can be used for all components in the actual simulation step. Consequently, we need to create additional kernels that compute the possible number of steps to execute all components.

Although this approach is perfectly sound with respect to time-step constraints, it does not solve all performance issues. The time-step computation kernels require an additional iteration over all items in the state which causes an overhead compared to the fixed-step version. The latter can only be outperformed by the adaptive one if it is possible to skip many simulations steps. From now on, we consider a step size of 1 as the reference and default fixed step size that satisfies all constraints. Since we have to use the minimum common step size, the overall probability that we have to perform a simulation step of size 1 is $P = 1 - p^n$, where p is the probability that a single component needs to perform a step of 1 and n is the number of components. Even in small examples, P can easily become close to 1. This in turn can also lead to a performance degradation (see Sect. 5).

4 Our Method

Figure 3 visualizes our high-level concept to determine intermediate item values. Formally, we apply a component C_i ($i \in [0, \dots, num\ components - 1]$) to the current state S. More precisely, the component works on a part of the state S_i.

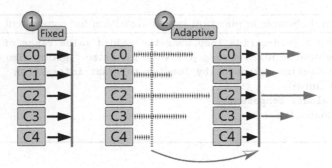

Fig. 2. A visualization of five components C_0 to C_4 and their executed step sizes (black arrows). A fixed step size that is uniform across all components (left, 1). A two-phase approach to compute a compatible step size (the minimum of all possible step sizes) across all components (right, 2). Grey arrows indicate the intended step sizes which could not be used for an actual simulation step.

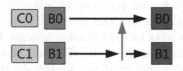

Fig. 3. Two components C_0 and C_1, where C_1 has a data dependency to the items in B_0. C_0 performs a large time step while reading from its source buffer and writing into its target buffer B_0. Our idea is to enable C_1 to do the same large time step, even if its dependencies only allow two small steps. This requires an interpolation of the values in B_0 (gray arrow) at the intermediate time step of C_1.

Taking the current simulation time T and the next time step of size Δt into account yields a new state at time step $T + \Delta t$ that is given by

$$S_i(T + \Delta t) = C_i\left(S_i(T), \Delta t\right). \tag{1}$$

We then use a problem- and domain-specific interpolation function I (see Sect. 5) to approximate the state S at time $T + \Delta k$ in the future using

$$S_i(T + \Delta k) \approx I_{\Delta k}\left(S_i(T), S_i(T + \Delta t)\right), \Delta k \in [0, \ldots \Delta t], \tag{2}$$

where Δk is a limited time offset that has to be smaller or equal to Δt. Consider a case in which a component C_j accesses interpolated information provided by another component C_i, where $i \neq j$ (for example in Fig. 3). Further, let us assume that we have to perform two small steps $\Delta l > 0$ and $\Delta o > 0$ that sum up to the intended step size $\Delta t = \Delta l + \Delta o$. This can be formally expressed via

$$S_j(T + \Delta l) = C_j\left(S_i(T), S_j(T), \Delta l\right), \tag{3}$$

$$S_j(T + \Delta t) = C_j\left(S_i(T + \Delta l), S_j(T + \Delta l), \Delta o\right) \tag{4}$$

$$= C_j\left(I_{\Delta l}\left(S_i(T), S_i(T + \Delta t)\right), S_j(T + \Delta l), \Delta o\right). \tag{5}$$

These equations demonstrate that a single use of an interpolation function can cause a simulation deviation that easily propagates to all other components. However, potential differences in terms of the simulation results depend on a huge variety of different factors. For instance, the actual interpolation being used, the number of components that can access interpolated information and even the application scenario. In comparison to related approaches, this is not a novel limitation that only applies to our method. Other methods rely on domain-specific time-step computations that also introduce simulation deviations. We support domain- and even scenario-specific time-step computations to model the required knowledge in the scope of our method (see also Sect. 4.2).

In order to apply interpolation properly to the right buffers, we have to identify locations that allow us to interpolate between items in the source and items in the target buffer. Therefore, we conceptually always rely on the idea of double buffering. However, if we execute a component an even number of times, while others have been executed an odd number of times, the items in the source and target buffers are mixed with respect to the different iteration steps. This issue is visualized in Fig. 4: Since component C_1 is executed an even number of times, its originally used source buffer contains the finally written information. To circumvent this problem, C_1's kernel has to copy the contents of the memory buffers into its intended target buffer to make C_1's updates visible to all other components. Unfortunately, the problem is even worse: If another component needs to interpolate the values written by C_1, we need its original source-buffer data. Consequently, we have to copy all items that can potentially be used for interpolation into a separate memory buffer at the beginning of the simulation step.

Fig. 4. Component C_1 from Fig. 3 performing two small steps in order to reach C_0's larger time step. First, C_1 performs its initial step and writes its computed items into its associated target buffer (1). Afterwards, C_1 is applied again and performs the next step. In this case, C_1 reads its source values from the originally intended target buffer and writes into its original source buffer (2). To ensure that the finally written values will end up in the correct target buffer, we have to copy the updated items from the source into the target after execution of the second step (3).

Finding locations to safely apply interpolation to already computed items is based on a static program analysis. It essentially resolves a data-dependency graph induced by view accesses in the program (see Fig. 5). Note that the decision on the back-edge dependency (or in other words: the first component in the schedule) is usually done by a domain expert. He or she is conceptually able to split the resolved dependency graph into semantically separate simulation steps. Although it is possible to use topological sorting of this graph to determine an execution order, it can happen that we encounter multiple possibilities to

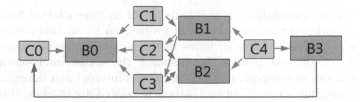

Fig. 5. A set of components C_0 to C_4 (from Fig. 1) along with some imaginary read and write dependencies to intermediate buffers B_0 to B_3. The read dependency of component C_0 from buffer B_3 is highlighted in purple, since it is also implicitly given by the back edge of the simulation loop. Note that it is possible to automatically compute a component schedule by applying topological sorting. In this example, this could yield the schedule C_0, C_1, C_2, C_3, C_4. Note further that the back-edge dependency of C_0 to buffer B_3 separates multiple simulation steps from each other, since C_0 is the first component is this schedule.

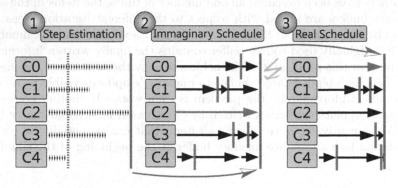

Fig. 6. Our approach to compute the next time-step size for components C_0 to C_4 from Fig. 5. Imagine that we perform an initial step estimation (1, the minimum step size is denoted by a dotted line). Consider further that we use the maximum possible step size of C_2 to be the next step size for all components (2). Although this seems to be feasible at first sight, we have to take all data dependencies into account. This reveals that we cannot interpolate an intermediate value for C_0, since it depends on buffer information from the previous iteration. Therefore, the maximum step size is computed using all components that are directly reachable via back edges (3).

schedule a component. These cases require domain knowledge to decide on the actual component order.

Once the actual schedule is available and the component dependencies have been resolved, we can focus on the adaptive time-step size computation (see Fig. 6). First, we perform a step estimation by querying all components (using *ComputeNumSteps* from Listing 1.1) that depend on immediate buffer information from a previous iteration (components that are reachable via back edges). In the case of multiple components that are reachable by following all back edges, we have to compute the minimum step size of all of them. Based on the actual domain, our approach is applied to, it may be totally fine to use the maximum possible time-step size from all components without explicit data dependencies. Note that we must always have access to the source item values at the beginning of the time step in order to interpolate between the source and target values. If the application does not use double buffering by default, we have to copy all relevant values that have to be considered during interpolation into separate global-memory buffers.

4.1 Leveraging Shared-Memory Caches

Depending on the computational load induced by an interpolation function, it can be advantageous to cache already interpolated values in shared memory. This frees up resources in terms of required bandwidth and ALUs and makes them available to the underlying component implementation. This can reduce the overhead of our method in the case of expensive interpolations (see Sect. 5). Without explicit modeling of such a cache, the GPU will have to recompute the interpolation function for each item access (see Fig. 7). Furthermore, this will trigger two additional loads from global memory for each item access. In this scope, L1 and L2 caches help to reduce the actual number of global-memory transfers automatically on modern GPUs [26] (see also Sect. 5).

Figure 8 visualizes several access patterns of an imaginary component in the scope of a single thread group. In our implementation, each thread just caches its associated item value(s) that will be accessed in the scope of the component in shared memory (1-to-1 thread-item mapping). Components in our scope are already programmed while having GPU-like architectures in mind. Therefore, they typically perform coherent memory accesses that are very close in global memory with respect to the current thread index (local access window). These cases can be covered by our shared-memory cache without the need for sophisticated program transformations. Unfortunately, we have to check that a certain access to a particular item is included in our cache (via an *if-branch*), which imposes additional runtime overhead. More advanced static program analyses can help to determine the actual access pattern(s) in order to realize more efficient caches in the future. This can even help to avoid on-the-fly cache-boundary checks that can be expensive with respect to thread divergences and register usage.

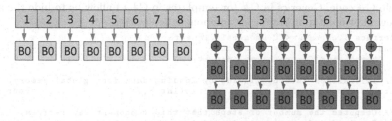

Fig. 7. An imaginary thread group of 8 threads accesses items from a buffer $B0$ in global memory (left). Using an interpolation function triggers two loads from global memory (from the source and target buffers respectively) for each access (right). Furthermore, the loaded values need to be interpolated to get the actual item value at a specific point in time.

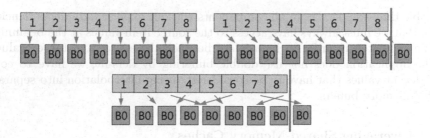

Fig. 8. Different access patterns for an imaginary thread group of 8 threads. Coherent memory accesses with respect to each thread while accessing a shared-memory cache (top). Random memory accesses that are not coherent (bottom). The vertical bars indicate the boundaries of the shared-memory cache. Accesses that cross these boundaries will be cache misses and have to be resolved via expensive global-memory loads and interpolation-function applications.

4.2   Hiding Different Access Patterns Using Views

From a practical point-of-view, every component can be modeled as a specific object-oriented *class* that implements a particular interface. An abstract pseudo-code interface definition we use for our components is shown in Listing 1.1 to get a better understanding about the general functions a component needs to provide. We propose this generic interface definition that distinguishes between *component-data* and *item-data* views. This allows us to clearly separate component-specific (uniform for each component type C) and item-specific (varies from item to item) information. Since the functions *ComputeNumSteps* and *Evaluation* work on abstract data views, it is easily possible to replace an item access with an interpolated value (that is even cached in shared memory) without touching the component implementation. Another advantage of having separate views is the improvement of static-program-analysis results with respect to potential aliases. This is due to the fact that each view has a very limited scope that is it only accessible within a single function. We do not allow storing views of any kind within member fields of a component to reduce the risk of bugs and to limit potential aliases in practice.

Listing 1.1. An abstract component interface definition used in our implementations in pseudo C# code. Generics in C# (or templates in C++) allow us to hide the actual data-view implementation from each component realization.

```
1  interface IComponent<TComponentImplementation>
2    where TComponentImplementation :
3      IComponent<TComponentImplementation>
4  {
5    // Initializes internal fields by loading data from global memory.
6    static TComponentImplementation Init(int index, ComponentView source);
7
8    // Computes the number of steps that this component can perform.
9    // Required information is loaded from the provided source data view.
10   int ComputeNumSteps<TDataView>(TDataView source);
11
12   // Evaluates this component by applying the given number of simulation
          steps.
13   // Computes results are written to the provided target data view.
14   void Evaluate<TDataView>(TDataView target, int numSteps);
15 }
```

4.3 Algorithm

The main kernel that wraps the actual functionality of a component C is shown in Algorithm 2. It is designed to be specialized by a compiler (via meta-programming techniques or code generation) to generate an individually instantiated GPU program. The first lines allocate all required shared-memory resources based on knowledge from the dependency graph. Afterwards, we perform a padded grid-stride loop (to avoid thread divergences) over the whole range of C. The body of the outer loop is another loop that performs the required small steps in the scope of a larger Δt time step. Important to mention is the group-wide computation of the next step size that will be applied to all threads in the group. This is required since all threads can access our shared-memory caches and need to have access to consistent interpolated information. If the presented shared-memory caches are not suitable for the application scenario, several parts of the algorithm are not required (like the lines 1–3 and 10). This even affects the computation of the upcoming local step-size consisting of several reductions and group barriers. These operations will be no longer required in such a case. Further optimizations (like loop unrolling, multiple components per kernel or even thread compaction) can be applied based on the actual component implementation to increase occupancy and to improve runtime performance.

4.4 Implementation Details

We have implemented our algorithm in C# using the ILGPU[2] compiler for all GPU kernels. Regarding code generation, we used a custom pre-compilation step to gather all data dependencies between all components. Afterwards, we generate all required C# kernels for each component and the whole adaptive-time-stepping driver code to execute all components in a specialized simulation loop. Each component will be wrapped in a particularly specialized time-stepping algorithm based on Algorithm 2. This includes the generation and allocation of shared-memory caches and inlining of all interpolation functions. The adaptive time-step sizes for each component are realized with the help of specialized kernels using efficient warp reductions and atomic operations [25, 26].

5 Evaluation

The evaluation section covers two different application scenarios inspired by particle-based simulations. Each scenario is described with the help of a component dependency graph to give detailed insights into the modeled simulation structure. In order to avoid hard-to-reproduce benchmarks, we use component implementations that are based on matrix-matrix multiplications to generate computational load per item. Moreover, memory-accesses to neighboring particles (that are often accessed in SPH-based simulations [5]) always consider 9 neighboring items. We used two different interpolation functions (linear and

---

[2] www.ilgpu.net.

Algorithm 2. Our adaptive time-stepping algorithm for component C

Input: maxNumSteps, source, originalSource, target
1 nextStepSize := **shared memory** int[1];
  /* Shared memory allocations for all intermediate values   */
2 sharedMemoryCache0 = **shared memory** Type0[group size];
  /* ...   */
  /* Perform a grid-stride loop over the padded value range of C */
3 **for** i := **global index**; $i < padded\_range(C)$; i += **grid size** * **group size do**
    /* Optional: synchronize group members to improve the memory access
         pattern on some GPU architectures   */
    /* **group barrier**   */
4   localSource, localTarget := source, target;
5   numPerformedIterations := 0;
6   **for** $stepIdx$:= 0; $stepIdx < maxNumSteps$; **do**
7   view := **new CachedDataView<Interpolation Function>**(
8   localSource, localTarget, originalSource, i,
9   maxNumSteps / (stepIdx + 1) **as float**,
        /* References to shared memory allocations   */
10   sharedMemoryCache0, ...);
      /* Wait for all cached values to be available and initialize the
         next maximum step size   */
11   **if** *is first thread of group* **then**
12   nextStepSize := maxNumSteps - stepIdx;
13   **end**
14   **group barrier**;
      /* Check component precondition for the current value   */
15   instanceStepSize := max(**int**);
16   C c := $\bot$;
      /* Compute next common step size for all group threads   */
17   **if** $i < range(C)$ **then**
18   c := C.Init(i);
19   instanceStepSize := c.ComputeNumSteps(view);
20   warpWideStepSize := **warp reduce min**(instanceStepSize);
21   **if** *is first lane of warp* **then**
22   **atomic min** nextStepSize, warpWideStepSize;
23   **end**
24   **end**
25   **group barrier**;
26   stepSize := nextStepSize;
      /* Apply component with the next common step size   */
27   **if** $i < range(C)$ **then**
28   c.Evaluate(view, stepSize);
29   **end**
      /* Advance step index and wait for all threads   */
30   stepIdx += stepSize;
31   numPerformedIterations++;
32   **Swap** localSource, localTarget;
33   **group barrier**;
34   **end**
35   **if** *numPerformedIterations is even* **then**
36   Perform a parallel group-wide copy operation of affected information from
      source to target buffers;
37   **end**
38 **end**

---

cubic-spline) to emulate less- and more-expensive interpolation computations. Similarly, we used a varying number of items $|R|$ (the range, particles in these scenarios) to analyze the scaling behavior of our adaptive time-stepping approach. Most important for the evaluation are the number of steps we can adaptively perform. In order to be close to application scenarios, we vary the number of steps continuously (computed using a uniform random distribution) for all components from the intervals $[1, \dots 3]$. We have not included larger intervals to show realistic performance measurements that do not assume optimistic properties of the underlying simulation. This emulates common scenarios in which we can sometimes perform larger steps, while other situations require small step sizes to satisfy all domain-specific simulation constraints.

We measured four different algorithms: *Simple* (fixed step size of 1), *Trad. Adaptive* (adaptive time stepping based on prior work, see Sect. 3), *HIP-NoCache* (our method without shared-memory caches) and *HIP* (our method with caches enabled). Every setup has been evaluated using two GPUs from NVIDIA for all benchmarks (a GeForce GTX 980 Ti and a GeForce GTX 1080 Ti). Furthermore, every performance measurement is the median execution time of 100 simulation runs, each performing a maximum number 100 simulation steps (when using a time-step size of 1). All execution times are measured in milliseconds (ms).

5.1 Gravity-Like Simulation

This evaluation scenario covers a gravity-like simulation that leverages three components $C0, C1$ and $C2$ (see Fig. 9). The first component conceptually computes particle-specific information, whereas the second one iterates over neighboring particles in memory and prepares accumulated results for C_2. The last component accesses neighboring information from B_0 and B_1 and computes item updates.

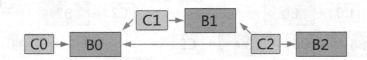

Fig. 9. Component dependency-graph of the first evaluation scenario.

Table 1 shows the performance measurements for the discussed configurations. In the presence of a small range $|R|$, the traditional method cannot improve performance significantly. It can only improve performance of about 4.5% to 7% in the case of 64k items. Using our approach allows us to increase the performance approx. 12% to 17% in comparison to the traditional adaption approach. Adding shared-memory caches does not hurt in most cases and is able to improve the runtime by approximately 2% on average using cubic-spline interpolation. However, modern GPU architectures do not seem to benefit from the explicit caching method in the case of simple linear interpolation functions.

Table 1. Performance measurements of the first evaluation scenario.

$|R|$	I	Algorithm	980 Ti	σ	1080 Ti	σ
16384	–	**Simple**	**136.36**	18.21	**83.45**	9.24
*	–	**Trad. Adaptive**	**135.49**	4.14	**82.67**	1.16
*	Linear	**HIP-NoCache**	**118.02**	5.44	**72.32**	3.62
*	Spline2	*	**121.63**	5.14	**74.02**	3.21
*	Linear	**HIP**	**116.48**	4.22	**72.16**	2.91
*	Spline2	*	**117.25**	6.60	**72.97**	2.53
65536	–	**Simple**	**399.47**	13.62	**245.88**	0.68
*	–	**Trad. Adaptive**	**381.34**	23.13	**229.36**	0.51
*	Linear	**HIP-NoCache**	**340.04**	1.74	**203.84**	0.12
*	Spline2	*	**342.30**	3.87	**209.74**	0.03
*	Linear	**HIP**	**332.88**	1.65	**204.35**	0.09
*	Spline2	*	**335.36**	4.16	**205.74**	0.12

5.2   PBD-like Simulation

This evaluation scenario is inspired by a PBD-like simulation that involves seven different components. C_0 computes basic information per particle. This can be seen as a more expensive position-prediction step derived from PBD. As before, C_1 prepares accumulated neighborhood information that is used for an imaginary collision-detection step in C_2. C_3 and C_4 use SPH-based calculations across all neighboring particles to simulate a reasonable workload per item. Afterwards, C_5 iterates over all neighbors while accessing items in B_4 and B_3. C_6 computes final item updates using B_5 (Fig. 10).

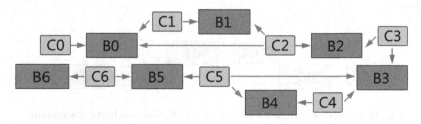

Fig. 10. Component dependency-graph of the second evaluation scenario.

Table 2 depicts the performance numbers for the discussed evaluation configurations. In all cases, the traditional adaption approach performs slower than the non-adapted version. This is due to the fact that the probability is lower than in the previous scenario to execute a larger time step (see also Sect. 3) as we are using more components. However, we are able to improve the performance approx. 16% to 22% in comparison to the non-adapted version. Again, shared-memory caches can help to improve the runtime (in this scope of up to 6%) on older GPU architectures.

Table 2. Performance measurements of the second evaluation scenario.

$|R|$	I	Algorithm	980 Ti	σ	1080 Ti	σ
16384	–	**Simple**	**327.50**	82.33	**200.95**	11.43
*	–	**Trad. Adaptive**	**346.83**	11.32	**212.75**	6.59
*	Linear	**HIP-NoCache**	**282.70**	26.43	**171.10**	3.78
*	Spline	*	**293.63**	16.05	**175.12**	13.57
*	Linear	**HIP**	**282.88**	19.57	**172.27**	10.72
*	Spline	*	**284.50**	12.92	**178.05**	18.46
65536	–	**Simple**	**1006.99**	85.99	**602.42**	2.49
*	–	**Trad. Adaptive**	**1016.65**	0.23	**615.63**	8.23
*	Linear	**HIP-NoCache**	**826.18**	16.77	**503.14**	2.94
*	Spline	*	**878.80**	18.51	**516.66**	0.50
*	Linear	**HIP**	**825.83**	18.62	**504.14**	0.19
*	Spline	*	**827.86**	19.63	**510.84**	3.24

6 Conclusion

We present a new approach to realize generic and domain-independent time-step adaptive GPU-based simulations. Our concept is based on several components that define the actual simulation loop. A static program analysis resolves data dependencies between read/write accesses of all components. The determined dependency graph is then used to generate specialized kernels that access interpolated values at intermediate time steps. In this scope, our generic component model allows to hide the actual memory-access implementation logic. This enables us to integrate our shared-memory-based caching concept and domain-specific interpolation functions touching the component implementations.

Our approach scales perfectly with the complexity of the underlying simulation model. Even in non-optimal use cases consisting of a few components, our approach was able to outperform fixed step sizes and the traditionally used conservative step-adaption method. However, our integrated caching concept is only beneficial if the simulation uses computationally expensive interpolation functions. This should not be necessary on modern GPUs when using linear or cubic-spline value-estimation methods. In case of more sophisticated simulations, we were able to significantly improve performance in comparison to prior work of up to 22% on non-optimistic benchmarks for our idea. If we assume more optimistic scenarios, we will be able to achieve impressive performance speedups in particular on large-scale simulations involving many different components. This makes our method a perfect extension to every modern GPU-based simulation that wants to benefit from adaptive time steps.

Probably the primary downside of our approach is the fact that we might introduce simulation errors with respect to adaptively adjusted time steps. As in all related papers, the actual time-step size calibration itself is tightly coupled to the domain and is usually adjusted by a domain expert. Therefore, we argue that this is not a disadvantage that has been introduced by our method. Furthermore, we have the ability to express domain-specific criteria to limit the step size.

In the future, we would like to extend the concept to support more advanced program analyses. This would allow us to simplify shared-memory-cache accesses considerably. In addition, we want to experiment with different caching concepts that even works on the warp-level in register space and in shared memory.

Acknowledgments. The authors would like to thank T. Schmeyer, A. Bosch and J. Bayer for their feedback on the paper, even in the scope of very challenging and stressful times.

References

1. Adams, B., Pauly, M., Keiser, R., Guibas, L.J.: Adaptively sampled particle fluids. ACM Trans. Graph. (2007)
2. Carey, V., Estep, D., Johansson, A., Larson, M., Tavener, S.: Blockwise adaptivity for time dependent problems based on coarse scale adjoint solutions. SIAM J. Sci. Comput. **32**, 2121–2145 (2010)
3. Gander, M., Halpern, L.: Techniques for locally adaptive time stepping developed over the last two decades. In: Bank, R., Holst, M., Widlund, O., Xu, J. (eds.) Domain Decomposition Methods in Science and Engineering XX. Lecture Notes in Computational Science and Engineering, vol. 91, pp. 377–385. Springer, Heidelberg (2013). https://doi.org/10.1007/978-3-642-35275-1_44
4. Garcia, V.M., Liberos, A., Climent, A.M., Vidal, A., Millet, J., González, A.: An adaptive step size GPU ODE solver for simulating the electric cardiac activity. In: 2011 Computing in Cardiology (2011)
5. Groß, J., Köster, M., Krüger, A.: Fast and efficient nearest neighbor search for particle simulations. In: Eurographics Proceedings (2019)
6. Hong, W., House, D.H., Keyser, J.: An adaptive sampling approach to incompressible particle-based fluid. In: Theory and Practice of Computer Graphics (2009)
7. Horvath, C.J., Solenthaler, B.: Mass preserving multi-scale SPH. Pixar Technical Memo 13–04, Pixar Animation Studios (2013)
8. Ihmsen, M., Akinci, N., Gissler, M., Teschner, M.: Boundary handling and adaptive time-stepping for PCISPH (2010)
9. Kay, D., Gresho, P., Griffiths, D., Silvester, D.: Adaptive time-stepping for incompressible flow part II: Navier-Stokes equations. SIAM J. Sci. Comput. **32**, 111–128 (2010)
10. Kelager, M.: Lagrangian fluid dynamics using smoothed particle hydrodynamics (2006)
11. Köster, M., Groß, J., Krüger, A.: Massively parallel rule-based interpreter execution on GPUs using thread Compaction. Int. J. Parallel Program. **48**, 675–691 (2020). https://doi.org/10.1007/s10766-020-00670-2
12. Köster, M., Groß, J., Krüger, A.: FANG: fast and efficient successor-state generation for heuristic optimization on GPUs. In: Wen, S., Zomaya, A., Yang, L.T. (eds.) ICA3PP 2019. LNCS, vol. 11944, pp. 223–241. Springer, Cham (2020). https://doi.org/10.1007/978-3-030-38991-8_15

13. Köster, M., Groß, J., Krüger, A.: Parallel tracking and reconstruction of states in heuristic optimization systems on GPUs. In: Parallel and Distributed Computing, Applications and Technologies (PDCAT-2019) (2019)
14. Köster, M., Krüger, A.: Adaptive position-based fluids: improving performance of fluid simulations for real-time applications. Int. J. Comput. Graph. Anim. (2016)
15. Köster, M., Leißa, R., Hack, S., Membarth, R., Slusallek, P.: Code refinement of stencil codes. Parallel Process. Lett. (PPL) **24**, 1441003 (2014)
16. Köster, M., Schmitz, M., Gehring, S.: Gravity games - a framework for interactive space physics on media facades. In: Proceedings of the International Symposium on Pervasive Displays (2015)
17. Köster, M., Krüger, A.: Screen space particle selection. In: Eurographics Proceedings (2018)
18. Macklin, M., Müller, M.: Position based fluids. ACM Trans. Graph. **32**, 1–12 (2013)
19. Macklin, M., Müller, M., Chentanez, N., Kim, T.Y.: Unified particle physics for real-time applications. ACM Trans. Graph. **33**, 1–12 (2014)
20. Mayr, M., Wall, W., Gee, M.: Adaptive time stepping for fluid-structure interaction solvers. Finite Elem. Anal. Design **141**, 55–69 (2018)
21. Monaghan, J.J.: Smoothed particle hydrodynamics. In: Annual Review of Astronomy and Astrophysics (1992)
22. Müller, M., Charypar, D., Gross, M.: Particle-based fluid simulation for interactive applications. In: Symposium on Computer Animation (2003)
23. Müller, M., Heidelberger, B., Hennix, M., Ratcliff, J.: Position based dynamics. J. Vis. Commun. Image Represent. **18**, 109–118 (2007)
24. Müller, M., Solenthaler, B., Keiser, R., Gross, M.H.: Particle-based fluid-fluid interaction. In: Symposium on Computer Animation (2005)
25. NVIDIA: Faster Parallel Reductions on Kepler (2014)
26. NVIDIA: CUDA C Programming Guide v10 (2019)
27. Pounders, J., Boffie, J.: Analysis of an adaptive time step scheme for the transient diffusion equation (2015)
28. Solenthaler, B., Gross, M.: Two-scale particle simulation. In: ACM Siggraph (2011)
29. Solenthaler, B., Pajarola, R.: Predictive-corrective incompressible SPH. In: ACM Siggraph (2009)
30. Zhang, Y., Solenthaler, B., Pajarola, R.: Adaptive sampling and rendering of fluids on the GPU. In: Eurographics Proceedings (2008)

Performance Modeling of Stencil Computation on SW26010 Processors

Yao Liu[1], Li Liu[1], Mengtao Hu[1], Wei Wang[1], Wei Xue[2], and Qingting Zhu[1(✉)]

[1] School of Data Science and Engineering, East China Normal University,
Shanghai, China
{liuyao,qtzhu}@cc.ecnu.edu.cn, bran96@163.com, izrail@163.com,
wwang@dase.ecnu.edu.cn
[2] Department of Computer Science and Technology, Tsinghua University,
Beijing, China
xuewei@tsinghua.edu.cn

Abstract. Stencil computation is a basic part in a large variety of scientific computing programs, especially for those containing partial differential equations. Due to the limited memory bandwidth, it is a challenge to improve the parallel efficiency of stencil computation on modern supercomputers. Performance modeling is the most common method of performance analysis. In this paper, we propose the generic performance model based on Sunway TaihuLight which is powered by SW26010 heterogeneous many-core processors. The generic model indicates the interaction between the programs and the computing platform from the architecture perspective, and points out the performance bottlenecks of the programs from the optimization perspective. Furthermore, we propose the specific performance model of stencil computation on SW26010 processors, and optimize the performance of stencil computation under the guidance of the model. The experimental results show that the performance models proposed in this paper are effective—the average error ratio of the predicted performance is less than 7%. Guided by the specific model, the optimized stencil computation achieves better performance than the unoptimized many-core version by 154.71% on 4096 cores.

Keywords: Stencil computation · Performance modeling · Sunway TaihuLight · Heterogeneous many-core processors

1 Introduction

Large-scale scientific computing programs such as materials science [24], atmospheric modeling [7], seismic prediction [4], molecular dynamics [9], play important roles in the national core science and technology fields. Stencil computation is an essential part of many scientific computing programs, especially in the programs containing a large number of partial differential equations [14]. Due to the particularity and importance of stencil computation, how to improve its computing efficiency has been widely concerned.

© Springer Nature Switzerland AG 2020
M. Qiu (Ed.): ICA3PP 2020, LNCS 12452, pp. 386–400, 2020.
https://doi.org/10.1007/978-3-030-60245-1_27

Considering the complexity and high time consumption of stencil computation, high-performance computing platforms are used to accelerate computing. However, with the increase of computing resources, the parallel efficiency of the programs is often unsatisfactory. One of the key reasons is that the computing capability of modern microprocessors is not fully utilized because of their complicated architecture. Therefore, improving the match between the platform and the program, analyzing and tuning the program design, and exploring the potential concurrency have become the main methods to improve the performance of the program.

Sunway TaihuLight supercomputer reached the peak performance of more than 100 PFLOPS in 2016, which provides a favorable guarantee for large-scale scientific computing [11]. Nonhydrostatic atmospheric dynamics simulation on the Sunway TaihuLight system won the Gordon Bell Award [21]. However, Sunway TaihuLight has unique architecture and programming features, it's a big challenge to make full use of its high computing capability. Generally speaking, it takes a lot of time and effort to transplant scientific computing programs to Sunway TaihuLight. The factors that affect the performance of a program include the program's computing scale, computing intensity, and programming habits. Analysis of the program performance could point out the performance bottlenecks of the programs, improve the match between the platform and the program, and achieve the purpose of performance optimization.

Performance modeling is one of the most commonly used methods for performance analysis [6]. It abstracts the various behaviors and running characteristics of the program through mathematical formulas, and can effectively identify the bottlenecks. The performance model can also be used to predict the scalability of the performance, providing the guidance for performance optimization. There have been some research on performance modeling on the GPU and Intel platforms [5,13,17,22]. How to design the performance model on Sunway TaihuLight has gradually become a research hotspot. Ao et al. [1] proposed atmospheric model performance modeling, which mainly focused on the optimization methods and the scalability of stencil computation. Zhang et al. [23] addressed 3D stencil performance modeling on GPU, which gave the best data blocking scheme on multiple levels of storage. However, these studies rarely involve quantitative modeling, and the model parameters are not rich enough to fully indicate the features of the algorithms and architecture.

In this paper, based on the architecture features of SW26010, we propose the generic performance model of scientific computing program on Sunway TaihuLight, which provides guidance for optimizing the performance of scientific computing programs. Using this generic model, we propose the performance model of stencil computation on SW26010 processors, which considers the features of architecture and program. The model quantitatively analyzes the characteristics of the program and the running status of each program part, which provides a favorable guidance for the performance optimization of the stencil computation on Sunway TaihuLight. Additionally, the model can predict the performance of the program under different blocking and optimizing schemes. In summary, this paper makes the following contributions.

- Facing the challenges of heterogeneous architecture and hybrid programing models, we propose the generic performance model on Sunway TaihuLight which provides guidance for modeling and optimizing scientific computing programs.
- We propose the stencil computation performance model on SW26010 processors. The model quantitatively analyzes the performance under different blocking schemes. The average error ratio of the predicted performance is less than 7%.
- Under the optimization guidance provided by the performance model, the stencil computation could be optimized significantly on Sunway TaihuLight. The optimized stencil computation achieves better performance than the unoptimized many-core version by 154.71% on 4096 cores.

The paper is organized as follows. Section 2 presents related work about performance modeling and stencil computation. Section 3 illustrates the details of performance modeling on SW26010 processors. Section 4 presents the experimental results and analysis. Finally, Sect. 5 concludes this paper and points out the future work.

2   Related Work

2.1   Performance Modeling

Scientific computing programs include two parts: the computation part and the communication part. For the computing part, Samuel Williams et al. [19] proposed the Roofline Model, which is a basic performance model. It is simple and easy to measure, but the performance events are coarse-grained. Barnes et al. [2] use formula to fit model. They use performance tools to directly measure the performance of parallel programs, and perform linear or nonlinear fitting. Martin et al. [3] proposed PerfExpert, to analyze the performance bottlenecks of program comprehensively.

For the communication part, Torsten et al. [12] established an analytical communication model for the communication part of the MILC program. They divided the communication behavior into point-to-point communication and collective communication. The analytical communication model is complete, but the cost is high, and the process of modeling requires domain expert knowledge. Other communication models such as PRAM, BSP, Hockney model and LogP family are classic communication models [8].

2.2   Stencil Computation

Stencil computation is a kind of computing kernels that updates the value in a certain way by steps. For example, solving the heat conduction equation in a uniform anisotropic heat-conducting medium, we usually use a regular grid to represent the temperature distributed in three-dimensional space, and the temperature of each grid point will change by steps. Then the temperature of a point at the current moment is computing from the temperature of several points around the point at the previous moment. If it depends on the temperature of 7 or 27 nearby points, it is considered as a 7-point or 27-point stencil computation problem, as shown in Fig. 1.

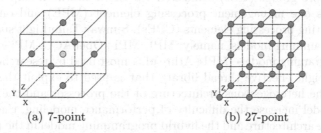

(a) 7-point   (b) 27-point

Fig. 1. Visualization of the 3D stencil.

2.3 Blocking Schemes

Due to the limitation of on-chip storage, for the GPU platform, the input data are divided into multiple small blocks, which are allocated to threads in order to optimize the stencil computation. The current work shows that there are three typical spatial blocking schemes, namely: 2D, 2.5D, and 3D blocking, as shown in Fig. 2. Vizitiu et al. [18] addressed that the effect of 3D blocking is better than the other two methods. However, the blocking schemes are flexible and should be considered in fact. Performance model proposed in this paper combines above blocking schemes, and analyzes the performance of stencil computation using different blocking schemes on Sunway TaihuLight.

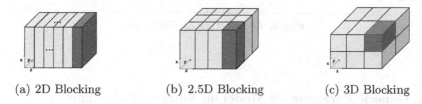

(a) 2D Blocking   (b) 2.5D Blocking   (c) 3D Blocking

Fig. 2. Blocking schemes.

3 Performance Modeling on SW26010 Processors

3.1 Overview of Sunway TaihuLight

Sunway TaihuLight system, a supercomputer independently developed by China, is the first supercomputer with a peak performance of over 125 PFlops in the world [16]. It consists of 40960 SW26010 heterogeneous many-core processors. The general architecture of the processor is shown in Fig. 3. Each processor contains 4 core groups (CGs) connected via the network on chip (NoC). Each CG includes one management processing element (MPE) and one cluster of 8×8 computing processing elements (CPEs). Sunway TaihuLight supports three major programming models, namely: MPI, MPI+OpenACC, MPI+Athread [1]. Hybrid programing model—MPI+Athread is most used by researchers. Athread is a light-weight effective thread library that is used to exploit the parallelism of CPEs. The heterogeneous architecture of the processor and the hybrid programing model increase the difficulty of performance modeling, that is, how to integrate the architecture and the hybrid programming model in the performance model.

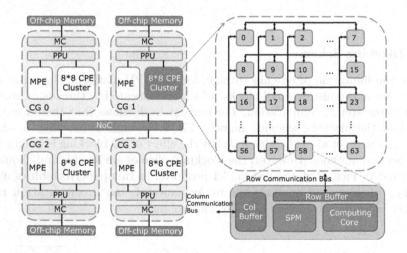

Fig. 3. Architecture of SW26010.

3.2 Generic Performance Model on Sunway TaihuLight

Most scientific computing programs use a master-slave accelerated parallel method on Sunway TaihuLight. Computing tasks are assigned to each core group. MPE (one process per MPE) is responsible for data communication between core groups, a small part of serial computation, I/O, etc., and CPEs (one thread per CPE) are responsible for computing.

We conclude from Sect. 2 that most scientific computing programs consist of two basic phases, computation phase and communication phase [8]. The computation phase consists of computations and memory accesses. The total time the program runs can be indicated as Eq. 1, where T_{comp} means the computation time, T_{mem} for the memory accesses time, T_{comm} for the communication time and T_{others} for the rest time (almost negligible) such as program initialization. Actually, the computation phase and the communication phase could be hidden from each other. Furthermore, in the computation phase, the computations and memory accesses could be hidden from each other. We use RF, RC to represent the hidden ratio, respectively.

$$T_{model} = (T_{comp} + RF \cdot T_{mem}) + RC \cdot T_{comm} + T_{others} \tag{1}$$

Computation Phase Modeling. According to the heterogeneous architecture, we extend the computation phase model. The model is indicated as Eq. 2. Each model item represents the behavior of MPE and CPEs in the computation phase.

$$\begin{aligned} T_{comp} + RF \cdot T_{mem} = T_{MPE\_comp} + RF_{MPE} \cdot T_{mem\_MPE} \\ + T_{CPE\_comp} + RF_{CPE} \cdot T_{mem\_CPE} \end{aligned} \tag{2}$$

Since the floating-point computation performance of CPEs is about 32× of MPE in one CG [11], all the computing tasks are assigned to CPEs in this section. We focus on performance modeling of CPEs computation, the performance modeling of MPE computation is the same as CPEs. The CPEs computation model is defined as Eq. 3, where $Dsize$ means the amount of computed data, P for the number of MPE, t_{comp} for minimum computation unit overhead.

$$T_{comp\_CPE} = \frac{Dsize}{P} \cdot t_{comp} \tag{3}$$

We can use timing functions to measure t_{comp} [8]. According to different types of scientific computing programs, we could choose a loop or a function as a minimum computation unit.

The speed of CPEs accessing its local device memory (LDM) is very fast, and the access delay is only 4 cycles [15]. However, due to the limited size of LDM (64 KB), CPE is required to continuously copy data from the main memory to its LDM, compute and copy the results back to the main memory finally. Although all CPEs independently complete their assigned computing tasks, the CPEs have to compete for the limited bandwidth between the main memory and LDMs. Due to the memory bound, memory access optimization is the effective method of performance optimization on SW26010 processors. Since the computing tasks are assigned in CPEs, memory access time of CPEs T_{mem_CPE} becomes an key component of the performance model. The SW26010 processor provides two ways for CPEs to access the main memory. The CPEs can directly access the main memory through the gld/gst method discretely, or can access the main memory in batches through the DMA method, and there are also mixed use methods.

T_{mem_CPE} can be indicated as Eq. 4, where $T_{gld/gst}$ means time consumed by gld/gst, T_{DMA} for time consumed by DMA.

$$T_{mem\_CPE} = T_{gld/gst} + T_{DMA} \tag{4}$$

The delay of one gld/gst is about 220 cycles [10], while the delay of one DMA is only 25 cycles [15]. We should avoid using gld/gst as much as possible in practical applications. In general, CPEs use DMA to access the main memory. When modeling performance of memory access, we mainly focus on DMA performance modeling. The modeling method of gld/gst is the same as DMA.

For one CPE, T_{DMA} can be defined as Eq. 5, where $count_{DMA}$ represents the number of DMA requests, t_{DMA} means time consumed by each DMA.

$$T_{DMA} = count_{DMA} \cdot t_{DMA} \tag{5}$$

t_{DMA} is closely related to the data size required for each DMA—$Dsize_{DMA}$ because $Dsize_{DMA}$ determines the DMA bandwidth. Therefore, we introduce the effective DMA bandwidth $effBW_{DMA}$ to represent the actual bandwidth used. t_{DMA} can be indicated as Eq. 6, where $count_{CPE}$ means the number of CPEs participating in the computation, which is another main factor that affects DMA bandwidth [10]. $effBW_{DMA}$ can be found in [15,16,20].

$$t_{DMA} = \frac{count_{CPE} \cdot Dsize_{DMA}}{effBW_{DMA}} \tag{6}$$

In the computation phase, there are cases where computations and memory accesses are hidden from each other, and how to hide them from each other as much as possible is the key to improving performance of programs. RF is related to the programmer's programming ability, the computation and memory characteristics of different problems and the system's own optimization. $RF = 0$ proves that computations and memory accesses have reached perfect hiding, and $RF = 1$ means that computations and memory accesses are not hidden from each other.

In summary, the computation phase performance model M_{comp} can be represented as Eq. 7.

$$M_{comp} = \frac{D_{size}}{P} \cdot t_{comp} + RF_{CPE} \cdot (count_{DMA} \cdot \frac{count_{CPE} \cdot Dsize_{DMA}}{effBW_{DMA}}) \tag{7}$$

Communication Phase Modeling. Programs such as partial differential equation solving usually assign grid computing tasks to each CG. When the updating of grid points require adjacent data, it needs to communicate with the surrounding grid points, so there is a need for data communication between MPEs. The communication phase can usually be divided into collective communication and point-to-point communication [8]. Communication time can be defined as Eq. 8, where T_{col} means time spent on collective communication, t_{p2p} for time consumed by point-to-point communication.

$$T_{comm} = T_{col} + T_{p2p} \tag{8}$$

We take collective communication as an example, assuming that the global reduction uses the k-tree reduction method. P means the number of MPEs, n means the number of the communication starts, $Csize$ is total communication data, t_{comm} means minimum communication unit overhead, T_{col} can be represented as Eq. 9.

$$T_{col} = n \cdot log_k(P) + \frac{Csize}{P} \cdot t_{comm} \tag{9}$$

In practical applications, we often use point-to-point communication, therefore communication model can be defined as Eq. 10, where t_{start} means communication start time. We can also measure t_{start} and t_{comm} using the timing function.

$$T_{comm} = n \cdot t_{start} + \frac{Csize}{P} \cdot t_{comm} \tag{10}$$

In general, we purpose the generic performance model on Sunway TaihuLight. The model provides a useful guide for the performance modeling of programs on Sunway TaihuLight.

3.3 Performance Modeling of Stencil Computation

When we have the generic performance model, we only need to extend the model according to the characteristics of the target programs to achieve rapid modeling.

Stencil Computation Description. Stencil computation includes computation phase and communication phase introduced in Sect. 2.2. When stencil computation runs on SW26010 processors by steps, the phases in this process include updating halo, fetching data to LDM, computing stencil, and writing data back to main memory.

Computation Phase Modeling. From the description, we found that stencil computation has a large number of memory access operations. Since LDM size is only 64 KB, it is impossible to fetch data to CPEs at once, so the memory accesses become more frequent. We use the 3D 7-point stencil computation as an example. When updating a grid point, it needs to use its top, bottom, front, back, left and right grid points as shown in Fig. 1. In this process, there is a lot of data redundancy. A lot of data are read repeatedly, so how to improve data utilization becomes the focus of the computation phase. Blocking is an effective method to improve the efficiency of memory access. Redundant memory accessing can be defined as Eq. 11, where $block_y$ means length of y-axis after blocking, $block_z$ means length of z-axis after blocking. The larger r is, the less redundant memory access.

$$r = \frac{block_y \cdot block_z}{count_{DMA}} \tag{11}$$

Since LDM size is fixed, $Dsize_{DMA}$ is inversely related to $count_{DMA}$. If the halo size is 1, $count_{DMA}$ can be defined as Eq. 12. If we take a 2.5D or 3D blocking scheme, r becomes larger, but $Dsize_{DMA}$ becomes smaller. Therefore, balancing these two indicators is the key to improving performance during the computation phase.

$$count_{DMA} = (block_y + 2) \cdot (block_z + 2) \tag{12}$$

Communication Phase Modeling. $Csize$ is a key factor affecting T_{comm}. From the description we found that the stencil computation requires swapping the halo data of the grid. For 3D stencil computation, the stencil computation intensity (5-point,7-point,27-point, etc.) determines $Csize$. $Csize$ can be defined as Eq. 13, where μ, ν are the flags that represent intensity, N for the data gird size, H for the halo size. Appropriate blocking scheme can effectively reduce $Csize$. We integrate the blocking scheme into the communication model. $Csize$ can be calculated by Eq. 14, where P_x, P_y, P_z mean the number of divisions on the x, y, z axis, and β_1, β_2, β_3 are the flags of the blocking schemes, as shown in Table 1.

Table 1. Blocking schemes

β_1	β_2	β_3	Blocking schemes
1	0	0	2D Blocking, split z-axis
1	1	0	2.5D Blocking, split y, z-axis
1	1	1	3D Blocking split x, y, z-axis

$$Csize_{3D} = 6 \cdot H \cdot N^2 + \mu \cdot 8 \cdot H^3 + \nu \cdot 12 \cdot N \cdot H^2 \tag{13}$$

$$Csize_{3D} = 2 \cdot H \cdot (\beta_1 \cdot \frac{N^2}{P_x \cdot P_y} + \beta_2 \cdot \frac{N^2}{P_x \cdot P_z} + \beta_3 \cdot \frac{N^2}{P_y \cdot P_z})$$
$$+ \mu \cdot 8 \cdot H^3 + \nu \cdot 4 \cdot (\beta_1 \cdot \frac{N}{P_x} + \beta_2 \cdot \frac{N}{P_y} + \beta_3 \cdot \frac{N}{P_z}) \cdot H^2 \tag{14}$$

For example, $Csize$ of 3D 7-point stencil computation with halo size of 1 can calculated by Eq. 15. Different process blocking schemes will make the sending data discontinuous. Therefore, we need to pack and unpack the data in the communication phase, and the amount of discontinuous sending data determines the time consumed by the data packing and unpacking. In summary, the communication model of stencil computation can be defined as Eq. 16, where $t_{packunpack}$ means time spent on data packing and unpacking.

$$Csize_{3D7p} = 2 \cdot (\beta_1 \cdot \frac{N^2}{P_x \cdot P_y} + \beta_2 \cdot \frac{N^2}{P_x \cdot P_z} + \beta_3 \cdot \frac{N^2}{P_y \cdot P_z}) \tag{15}$$

$$T_{comm} = n \cdot t_{start} + t_{packunpack}$$
$$+ 2 \cdot (\beta_1 \cdot \frac{N^2}{P_x \cdot P_y} + \beta_2 \cdot \frac{N^2}{P_x \cdot P_z} + \beta_3 \cdot \frac{N^2}{P_y \cdot P_z}) \cdot t_{comm} \tag{16}$$

4 Experiments

In this section, we evaluate the proposed performance model from the perspective of model accuracy and performance optimization.

4.1 Experimental Setup and Metrics

We use standard MPI+Athread programming model to implement the parallelization of stencil computation on Sunway TaihuLight. The data sets are named in the form of $N\text{-}STENCIL\text{-}P\text{-}STEPS$, where N means the grid size, $STENCIL$ for the computation intensity, P for the number of processes, and $STEPS$ for the iteration number of stencil. To match the grid size, different numbers of processes are used. The four data sets used in the experiments have their own features, such as different N and $STEPS$. The grid data is placed in order of x-y-z, where x is the innermost direction.

Error Ratio. The model error ratio is used to evaluate the accuracy of the model. The smaller the model error ratio, the higher the model accuracy. Let Res_e be the average model result obtained from multiple experiments, and Res_m be the model results predicted by the performance model proposed in this paper, model error ratio can be defined as

$$R_{err} = |1 - \frac{Res_m}{Res_e}| \times 100\% \tag{17}$$

Speedup. The experiments use *speedup* to evaluate the acceleration effect of model-guided optimization. Let T_{mpe} be the running time of multi-process stencil computation, and $T_{manycore}$ be the running time after model-guided optimization, then *speedup* can be defined as

$$speedup = \frac{T_{mpe}}{T_{manycore}} \tag{18}$$

4.2 Error Ratio of Models

The experiments evaluate R_{err} of the computation model, the communication model, the computation model based on the blocking schemes, and the communication model based on the blocking schemes, respectively.

For input items of the model, some of them are obtained through multiple measurements, for example, t_{comm}, t_{comp}, some are obtained by reviewing the paper [15,16,20], such as $effBW_{DMA}$, and the rest are obtained by calculation, such as $count_{DMA}$. We use the proposed performance model to predict the results, meanwhile the results of the experiment is obtained by instrumenting.

Evaluation of Computation Model. In this experiment, to utilize the DMA bandwidth, CPE take 650 double to LDM each time. Figure 4(a) shows the predicted results are basically consistent with the experimental results. With the increase of computation intensity, the proportion of T_{comp} in the total time increases. R_{err} is less than 5%, which proves the proposed computation model is effective.

Evaluation of Communication Model. The experiments use $MPI\_Sendrecv$ as the communication functions. In this experiment, we divide the grid along the z-axis, and the number of divisions is the number of processes. The results are shown in Fig. 4(b). $CSize$ is the key factor that determines pure communication time $T_{comm}(pure)$, and when $STEPS$ increases, T_{comm} also increases. The average R_{err} is satisfactory, only 6.08%.

Evaluation of Computation Model Based on Blocking Schemes. We take 512-7-16-48 as an example for the experiment. Figure 4(c) shows the blocking schemes change T_{mem} obviously, because the schemes effect $count_{DMA}$ and $effBW_{DMA}$. The improvement of $effBW_{DMA}$ plays a key role in reducing T_{mem}, and $count_{DMA}$ is inversely related to $effBW_{DMA}$. The bigger and square yz-plane, the larger r. The R_{err} is about 5%, and the performance is obviously improved when $effBW_{DMA}$ and r are both large.

Evaluation of Communication Model Based on Blocking Schemes. We take 512-7-16-48 and 768-7-64-64 as examples for the experiment. Figure 4(d) shows 2.5D and 3D blocking schemes could effectively decrease $Csize$, and $T_{comm}(pure)$ decreases accordingly. However, the advantage brought by the decrease of $Csize$ is reduced due to the increase of $t_{packunpack}$. The average R_{err} is desirable, which is 5.09%. We found that in the case of several processes, the 2D blocking scheme along the z-axis is better because $t_{packunpack}$ can be omitted. In the case of a larger number of processes, the 2.5D blocking scheme is outstanding, especially the scheme with square yz-plane and more continuous dimensions. As for the 3D blocking scheme, it is unsatisfactory because of the massive data packing and unpacking.

4.3   Model-Guided Performance Optimization

Process-Level Parallelism (PP). The computation tasks are assigned to each MPE, and each MPE is responsible for computation and communication.

Multi-level Parallelism (MP). The computation tasks on each MPE are assigned to CPEs. CPEs are responsible for computation and each MPE is responsible for communication. Mutli-level parallelism is the most common parallel version of many-core on SW26010 processors.

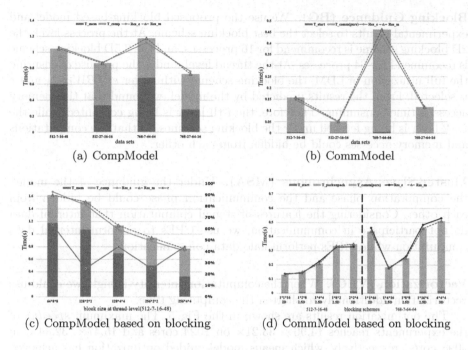

(a) CompModel

(b) CommModel

(c) CompModel based on blocking

(d) CommModel based on blocking

Fig. 4. Evaluation of the model.

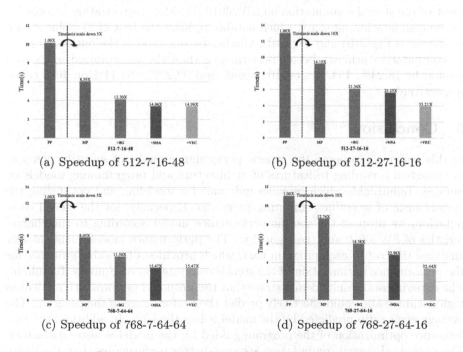

(a) Speedup of 512-7-16-48

(b) Speedup of 512-27-16-16

(c) Speedup of 768-7-64-64

(d) Speedup of 768-27-64-16

Fig. 5. Speedup by the model-guided optimization.

Blocking Guidance (BG). We use the proposed blocking-based model and experimental results to select the best blocking scheme. At the process level, the 2D blocking scheme is recommend for 16 processes, and the 2.5D blocking scheme is recommend for 64 processes. At the thread level, under the premise of ensuring the full utilization of LDM, the blocking scheme with larger $effBW_{DMA}$ and r is selected. From the results predicted by the model, we found that the memory access is time-consuming. Therefore, the i-th layer is being computed while the $i+2$ layer is being fetched under the blocking schemes, so that the computations and memory accesses could be hidden from each other.

Master-Slave Asynchronous (MSA). Under the guidance of the model, the computation phase and the communication phase could be hidden from each other. Considering the features of stencil computation that internal data do not participate in communication, we use CPEs to perform internal data computation while MPEs perform halo data communication.

Vectorization (VEC). When the computation intensity is high, we could use vectorization additionally to shorten the computing time.

The optimization results are shown in the Fig. 5. The maximum *speedup* of the experiments reaches $14.96\times$, $33.21\times$ on 1024 cores and $15.47\times$, $32.44\times$ on 4096 cores, respectively, which means model-guided optimization has achieved excellent results. Blocking guidance is significant for the performance improvement of the stencil computation on SW26010. Besides, vectorization is sensitive to the computation intensity, which notably reduces the cost of 27-point stencil as shown in Fig. 5(b) and Fig. 5(d). Guided by our model, the optimized stencil computation achieves better performance than the unoptimized many-core version by 78.24%, 134.74% on 1024 cores and 92.22%, 154.71% on 4096 cores, respectively.

5   Conclusion

In this paper, we propose the generic performance model of scientific computing program according to features of architecture and programming models on Sunway TaihuLight, which provides guidance for modeling and optimizing the performance of scientific computing programs. Especially, for the stencil computation, we propose the specific performance model according to the characteristics of SW26010 and the algorithm. The performance model quantitatively analyzes the cost of each program part, which provides a favorable guidance for the performance optimization of the stencil computation on Sunway Taihulight. The experimental results demonstrate that the proposed performance model has high accuracy and could effectively predict the performance of the program. The average error ratio predicted by the model is less than 7%. In addition, the performance optimization of the program guided by the model is also satisfactory. The optimized stencil computation achieves better performance than the unoptimized many-core version by 154.71% on 4096 cores.

In the future, we intend to conduct research on automatic performance modeling, and apply the performance model to larger scientific computing programs, which have higher computing scale and more computing kernels.

Acknowledgement. This work is supported by the Ministry of Education's University-Industry Collaborative Education Program (No. 201902146019)

References

1. Ao, Y., et al.: 26 PFLOPS stencil computations for atmospheric modeling on Sunway TaihuLight. In: 2017 IEEE International Parallel and Distributed Processing Symposium (IPDPS), pp. 535–544. IEEE (2017)
2. Barnes, B.J., Rountree, B., Lowenthal, D.K., Reeves, J., De Supinski, B., Schulz, M.: A regression-based approach to scalability prediction. In: Proceedings of the 22nd Annual International Conference on Supercomputing, pp. 368–377 (2008)
3. Burtscher, M., Kim, B.D., Diamond, J., McCalpin, J., Koesterke, L., Browne, J.: Perfexpert: an easy-to-use performance diagnosis tool for HPC applications. In: SC 2010: Proceedings of the 2010 ACM/IEEE International Conference for High Performance Computing, Networking, Storage and Analysis, pp. 1–11. IEEE (2010)
4. Chen, B., et al.: Simulating the Wenchuan earthquake with accurate surface topography on Sunway TaihuLight. In: SC18: International Conference for High Performance Computing, Networking, Storage and Analysis, pp. 517–528. IEEE (2018)
5. Chen, G., Wu, B., Li, D., Shen, X.: Porple: an extensible optimizer for portable data placement on GPU. In: 2014 47th Annual IEEE/ACM International Symposium on Microarchitecture, pp. 88–100. IEEE (2014)
6. Datta, K., Kamil, S., Williams, S., Oliker, L., Shalf, J., Yelick, K.: Optimization and performance modeling of stencil computations on modern microprocessors. SIAM Rev. **51**(1), 129–159 (2009)
7. Dennis, J.M., et al.: Cam-se: a scalable spectral element dynamical core for the community atmosphere model. Int. J. High Perform. Comput. Appl. **26**(1), 74–89 (2012)
8. Ding, N., Xu, S., Song, Z., Zhang, B., Li, J., Zheng, Z.: Using hardware counter-based performance model to diagnose scaling issues of HPC applications. Neural Comput. Appl. **31**(5), 1563–1575 (2019). https://doi.org/10.1007/s00521-018-3496-z
9. Dong, W., Li, K., Kang, L., Quan, Z., Li, K.: Implementing molecular dynamics simulation on the Sunway TaihuLight system with heterogeneous many-core processors. Concurr. Comput.: Pract. Exp. **30**(16), e4468 (2018)
10. Fu, H., et al.: Refactoring and optimizing the community atmosphere model (CAM) on the Sunway TaihuLight supercomputer. In: SC 2016: Proceedings of the International Conference for High Performance Computing, Networking, Storage and Analysis, pp. 969–980. IEEE (2016)
11. Fu, H., et al.: The Sunway TaihuLight supercomputer: system and applications. Sci. China Inf. Sci. **59**(7), 072001 (2016). https://doi.org/10.1007/s11432-016-5588-7
12. Hoefler, T., Gropp, W., Kramer, W., Snir, M.: Performance modeling for systematic performance tuning. In: SC 2011: Proceedings of 2011 International Conference for High Performance Computing, Networking, Storage and Analysis, pp. 1–12. IEEE (2011)

13. Hong, S., Kim, H.: An analytical model for a GPU architecture with memory-level and thread-level parallelism awareness. In: Proceedings of the 36th Annual International Symposium on Computer Architecture, pp. 152–163 (2009)
14. Langtangen, H.P.: Computational Partial Differential Equations: Numerical Methods and Diffpack Programming, vol. 2. Springer, Berlin (1999). https://doi.org/10.1007/978-3-662-01170-6
15. Li, L., et al.: swCaffe: a parallel framework for accelerating deep learning applications on Sunway TaihuLight. In: 2018 IEEE International Conference on Cluster Computing (CLUSTER), pp. 413–422. IEEE (2018)
16. Liu, Y., Liao, Q., Sun, J., Hu, M., Liu, L., Zheng, L.: A heterogeneous parallel genetic algorithm based on sw26010 processors. In: 2019 IEEE 21st International Conference on High Performance Computing and Communications; IEEE 17th International Conference on Smart City; IEEE 5th International Conference on Data Science and Systems (HPCC/SmartCity/DSS), pp. 54–61. IEEE (2019)
17. Shirako, J., et al.: Analytical bounds for optimal tile size selection. In: O'Boyle, Michael (ed.) CC 2012. LNCS, vol. 7210, pp. 101–121. Springer, Heidelberg (2012). https://doi.org/10.1007/978-3-642-28652-0_6
18. Vizitiu, A., Itu, L., Niţă, C., Suciu, C.: Optimized three-dimensional stencil computation on Fermi and Kepler GPUs. In: 2014 IEEE High Performance Extreme Computing Conference (HPEC), pp. 1–6. IEEE (2014)
19. Williams, S., Waterman, A., Patterson, D.: Roofline: an insightful visual performance model for multicore architectures. Commun. ACM 52(4), 65–76 (2009)
20. Xu, Z., Lin, J., Matsuoka, S.: Benchmarking sw26010 many-core processor. In: 2017 IEEE International Parallel and Distributed Processing Symposium Workshops (IPDPSW), pp. 743–752. IEEE (2017)
21. Yang, C., et al.: 10m-core scalable fully-implicit solver for nonhydrostatic atmospheric dynamics. In: SC 2016: Proceedings of the International Conference for High Performance Computing, Networking, Storage and Analysis, pp. 57–68. IEEE (2016)
22. You, Y., et al.: Accelerating the 3D elastic wave forward modeling on GPU and MIC. In: 2013 IEEE International Symposium on Parallel & Distributed Processing, Workshops and Phd Forum, pp. 1088–1096. IEEE (2013)
23. Zhang, G., Zhao, Y.: Modeling the performance of 2.5 d blocking of 3D stencil code on GPUs. In: IEEE High Performance Extreme Computing Conference, HPEC (2016)
24. Zhang, J., et al.: Extreme-scale phase field simulations of coarsening dynamics on the Sunway TaihuLight supercomputer. In: SC 2016: Proceedings of the International Conference for High Performance Computing, Networking, Storage and Analysis, pp. 34–45. IEEE (2016)

Optimizing B$^+$-Tree Searches on Coupled CPU-GPU Architectures

Han Huang and Hua Luan$^{(\boxtimes)}$

Beijing Normal University, Beijing, China
huanghan@mail.bnu.edu.cn, luanhua@bnu.edu.cn

Abstract. The B$^+$-tree is an important index in the fields of data ware-housing and database management systems. With the development of new hardware technologies, the B$^+$-tree needs to be revisited to fully take advantage of hardware resources. In this paper, we focus on opti-mization techniques to increase the searching performance of B$^+$-trees on the coupled CPU-GPU architecture. First, we propose a hierarchical searching approach on the single coupled GPU to efficiently deal with leaf nodes of B$^+$-trees. It adopts a flexible strategy to determine the number of work items in a work group to search one key in order to reduce irreg-ular memory accesses and divergent branches in the work group. Second, we present a co-processing pipeline method on the coupled architecture. The CPU and the integrated GPU process the sorting and searching tasks simultaneously to hide sorting and partial searching latencies. A distribution model is designed to support the workload balance strat-egy based on real-time performance. Our performance study shows that the hierarchical searching scheme provides an improvement up to 36% on the GPU compared to the baseline algorithm with fixed number of work items and the co-processing pipeline method further increases the throughput by a factor of 1.8. To the best of our knowledge, this paper is the first study to consider both the CPU and the coupled GPU to optimize B$^+$-trees searches.

Keywords: B$^+$-trees · The coupled architecture · Integrated GPU · Co-processing

1 Introduction

The B$^+$-tree [3] as a fundamental index is an important data structure and widely used in many applications. Due to the high throughput, large capacity and self-balance characteristics, it is efficient to use B$^+$-trees to deal with large-scale data [7]. However, with the Internet breakout and the rapidly expanding data size, data systems need to store very huge volumes of data and deal with numerous queries at the same time. The B$^+$-tree, as a traditional database index,

Supported by the National Key R&D Program of China (No. 2017YFC0804004), and a grant from the Capital Science and Technology Innovation Vouchers of China.

© Springer Nature Switzerland AG 2020
M. Qiu (Ed.): ICA3PP 2020, LNCS 12452, pp. 401–415, 2020.
https://doi.org/10.1007/978-3-030-60245-1_28

is confronted with new performance challenges. It is necessary to explore novel techniques to increase the query throughput and eliminate the search latency of B^+-trees in the big data era.

In recent years, graphics processing units (GPUs) have been used to help improve B^+-trees for better search performance. The GPU contains more compute units compared with the CPU, and is able to execute instructions in a parallel way. These features provide an opportunity to accelerate B^+-trees operations. However, in the traditional discrete GPU architecture, data are stored in the main memory and must be transferred from memory to the discrete GPU with the PCIe bus before queries are executed on the GPU, which becomes the bottlenecks of many applications and limits the benefits brought by using high-performance GPUs [13,19]. Recently a new processor architecture, known as the coupled or fused CPU-GPU architecture, has been provided by mainstream processor vendors, such as Intel and AMD. On this kind of architecture, the CPU and the coupled GPU share the same main memory and even the last level cache, thus data transmission through the PCIe bus is eliminated which could avoid transfer overhead [20].

In this paper, we aim at optimizing the search performance of B^+-trees using both the CPU and the coupled GPU in the fused architecture. To the best of our knowledge, this paper is the first effort to exploit both the CPU and the coupled GPU simultaneously to accelerate B^+-trees searches. In our study, we found that using a fixed work item number in a work group to search a key does not match with the features of GPUs and therefore the GPU is not being efficiently and properly utilized. Based on this observation, we present a hierarchical searching way to reduce irregular memory accesses and work item divergence in a work group. Compared with the coarse-grained parallel method, proposed by Daga et al. [4], with a fixed number of search keys processed by a work group, our hierarchical scheme is able to achieve a performance improvement up to 36%. Then, on the base of this optimization on a single coupled GPU, we further probe into B^+-tree search techniques using both the CPU and the coupled GPU. We present a co-processing pipeline search design which could hide sorting and partial searching latencies and adopt a workload assignment method based on real-time performance. The detailed experimental studies demonstrate that the co-processing pipeline strategy provides a throughput 40% higher than our single GPU method in the best case and outperforms the coarse-grained parallel GPU method by a factor of 1.8.

Our contributions in this work are as follows:

1. A hierarchical searching scheme is proposed to match the characteristics of the single coupled GPU. Different strategies are adopted when searching inner nodes and leaf nodes of B^+-trees.
2. The compute power of both the CPU and the coupled GPU is utilized to improve the performance of B^+-tree searches.
   (a) A pipeline searching algorithm is presented to keep the CPU and the coupled GPU busy in the searching process.

(b) Suitable tasks are assigned to the CPU and the coupled GPU based on their hardware differences and real-time performance.
3. An extensive performance evaluation is conducted under a wide range of scenarios to verify the effectiveness of the proposed methods.

In the rest of this paper, Sect. 2 provides preliminaries of B$^+$-trees, the coupled architecture and OpenCL. Section 3 explains our optimization techniques which contain a kind of search strategy on the single coupled GPU and a co-processing method for the CPU and the coupled GPU on the same die. Section 4 provides the experimental results and performance analysis. Section 5 describes the related work. We conclude our work in Sect. 6.

2 Preliminaries

In this section, we first describe the B$^+$-tree index structure. Then, we introduce the main features of the coupled CPU-GPU architecture. Finally, we describe the OpenCL programming standard, which is uesd in our implementations.

2.1 B$^+$-Tree

The B$^+$-tree is a self-balance data structure, which is widely used as an index in data warehouses, database management systems [14], file systems [12], etc. In B$^+$-trees all keys and values are stored in the last level leaf nodes. Inner nodes above leaf nodes store keys and references to the next level nodes. The maximal number of references stored in the inner node is called *fanout*. An inner node is able to store up to *fanout* child references, and *fanout* − 1 keys to identify child references. A leaf node can store *fanout* − 1 keys and their corresponding values. Some space left in a leaf node stores the reference to the next leaf node, which forms a linked list to allow range queries.

Searching a key in B$^+$-trees starts from the root and traverses the tree until a leaf node is reached. In an inner node, the process needs to compare stored keys with the key to be searched. If the maximal key which is not bigger than the key to be searched is found, the corresponding child reference to the next level node is obtained. When searching a leaf node, keys stored in the node are also compared with the search key and the corresponding value is fetched if keys are equal.

2.2 Coupled CPU-GPU Architectures

The coupled CPU and GPU architecture places a multi-core CPU and a GPU on one same chip, which forms *Heterogeneous System Architecture* and becomes increasingly common in current processors. AMD first released its coupled architecture named accelerated processing units (APUs). Intel also introduced the integrated GPU into the CPU since Ivy Bridge and Haswell processors. On this architecture, coupled GPUs are less powerful than discrete GPUs, but it shares the last level cache and physical system memory with the CPU. This feature makes the GPU not need to wait for data transfer by PCIe bus which occurs for the discrete GPU. Moreover, the emergence of the shared virtual memory (SVM) technology allows sharing virtual address space between the CPU and the coupled GPU, which makes it much easier to effectively leverage the power of both CPUs and coupled GPUs.

2.3 OpenCL

The OpenCL [17] is an open standard for parallel programming on heterogeneous platforms, proposed by Khronos Group. The OpenCL describes a framework that allows applications to use devices across various compute platforms by the same portable C-like language. Many hardware vendors provide drivers and programming tools to support OpenCL implementations. We used C and OpenCL in our study. Processors on the same chip can share the global memory in the same context in OpenCL program model, which helps us to implement the co-processing method in integrated CPU-GPU architectures.

3   Optimization Techniques

In this section, we first optimize the B^+-tree search operation on the single coupled GPU by proposing a hierarchical scheme which distinguishes inner nodes and leaf nodes of B^+-trees. Then, in order to further improve B^+-tree search performance in the coupled architecture, we present a pipeline searching method to use the computing power of both processors.

3.1   Hierarchical Searching on the Coupled GPU

We first analyze the GPU parallel execution behavior. On GPUs, kernel instances are organized into work groups, and each work group contains the same number of work items. A work group is assigned to run on a single compute unit. All the work items in a work group are executed in a single instruction multiple thread (SIMT) manner. It is efficient for work items to access coalesced memory since the data may be fetched in one operation. In addition, if there exist branch instructions in a work group, that is, work items execute different branches, some work items will wait until the others finish their jobs. When these work items begin to work, the others stop again. Thus, keeping all work items executing the same instructions helps to efficiently utilize the parallel ability of GPUs.

In order to match with this feature, previous studies [4,18] proposed the solution way that one key is searched by one or several work items in a work group. If the data was not ordered, there would exist a number of divergent branches as well as irregular memory accesses, which are the main factors damaging the overall performance. Thus keys should be sorted in advance of the searching phase to increase the probability that keys in the charge of one work group are located in fewer nodes. However, in B^+-trees, the key range in a node of different

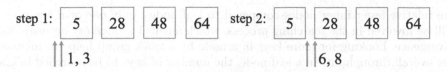

Fig. 1. In the root, stored keys that are not bigger than keys to be searched are located at the first and second positions. Work items in the work group only need to access two keys at most to fetch the child references to go to the next level. The total step is 2.

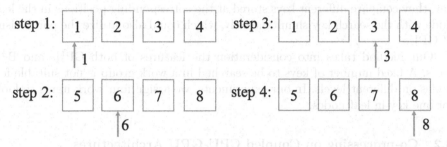

Fig. 2. In the leaf, ordered search keys are located at two leaf nodes with different positions. The work group visits two different leaf nodes. Each work item had different searching steps. The total step of is 4 and all the other work items should wait for the fourth work item to finish.

levels is not the same. The number of nodes in the lower level becomes larger. Sorted keys belonging to a work group may also be scattered among multiple nodes, which happens more often for lower level nodes, especially leaf nodes. Thus, it may bring more benefits to distinguish inner nodes and leaf nodes in the searching design.

Figures 1 and 2 show an example where a work group with 4 work items searches 4 sorted keys: 1, 3, 6 and 8 in a two-level B$^+$-tree. Every work item in the work group is responsible for one key. As Fig. 1 shows, in the root, the total step number of this work group is two, because at most work items need to access the first and second keys to fetch child references. But, in the leaf nodes in Fig. 2, search keys are distributed among different nodes. The work group needs to visit two leaf nodes and each work item in the work group has different steps to visit the keys. Until the work item has found the key 8, the whole work group finishes its job. During this period, the other three work items in the same work group need to wait for the fourth work item to find its key. Even if search keys have been sorted, there still exist irregular memory accesses and work item divergence problems.

We present a hierarchical searching method to optimize the search performance. We separate the B$^+$-tree into two parts, inner nodes and leaf notes. In an inner node, one work item is in charge of searching one key, and the work group size is set to the maximal number of keys that could be stored in a node, i.e., $fanout - 1$. This will lead to the high parallelism in the node. Because there are relative fewer inner nodes compared to leaf nodes, only one or a few nodes will be involved in the searching process, which means less memory accesses and divergence. Looking for more keys in a node by a work group helps to increase the overall throughput. In a leaf node, the number of keys to be searched in one work group is reduced and several work items are used to find the same search key. The reason is that we want to decrease the possibility that one work group will visit multiple different leaf nodes. In the work group, work items are separated into several subgroups. When work items in each subgroup process one key, they compare different keys stored at the corresponding positions in the leaf node with the search key simultaneously, which could also utilize the parallelism of GPUs.

Our method takes into consideration the features of both GPUs and B$^+$-trees. A fixed number of keys to be searched in a work group is not suitable for nodes on different levels. In our experiments, we assign four work items to look for one key in leaf nodes.

3.2   Co-processing on Coupled CPU-GPU Architectures

To further optimize B$^+$-tree searches on coupled CPU-GPU architectures, we propose a co-processing pipeline method to make use of the CPU and the integrated GPU. The principle is to keep both processors always busy and produce a higher throughput.

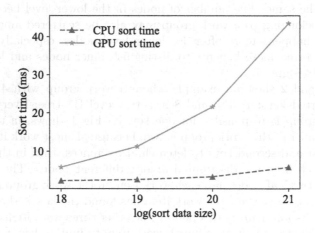

Fig. 3. The radix sorting time on CPU and GPU with various data sizes.

CPU-GPU Co-processing Pipeline. The architectures of CPUs and coupled GPUs are different. GPUs have more computing cores compared with CPUs, which can hide the memory latency by setting a wide range of work group size. While CPUs have relatively large caches, and are more suitable on programs with good locality [20]. Thus, it is necessary to carefully consider and assign appropriate tasks for CPUs and the coupled GPUs to make full use of their advantages.

Figure 3 shows the performance comparison of radix sorting on CPUs and coupled GPUs with different data sizes. Our implementation is based on an OpenCL radix-sort provided by Helluy et al. [10]. It could be found that the gap between the sorting times on the CPU and the coupled GPU is huge, and the CPU is much faster than the coupled GPU with all data sizes. So, we assign the task to sort the search keys using radix sorting to the CPU. As for as the GPU, the hierarchical searching method described in the previous subsection is exploited to retrieve keys. This forms an initial co-processing pipeline way, in which all the search keys are divided into batches. The CPU sorts the first batch of keys. Then the coupled GPU begins to search keys in the batch in the B$^+$-tree, and in the meantime the CPU starts to sort the second batch prepared for the GPU. Within OpenCL programming standard, the CPU and the coupled GPU have their own task queues, and processors do their individual jobs in separate ways. We use $clWaitForEvent()$ to synchronize the sort task and the search task on the CPU and the GPU in a batch size. Data do not need to be transferred from the CPU to the GPU with PCIe bus, which reduces the delay of data transmission. Therefore, the sorting time occupied by the CPU can be efficiently hidden by the GPU searching phase.

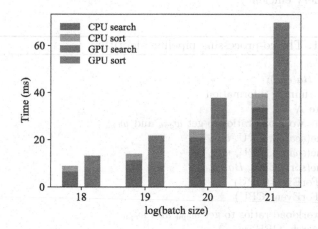

Fig. 4. The time breakdown on CPU and GPU with different batch sizes, when the tree size is fixed at 2^{26}. The total time is divided into the sorting time and the search time.

We further explore whether there is optimization room in the above simple pipeline strategy. We compare the performance of a single CPU and a coupled GPU when each processor sorts and searches the same keys. The results are shown in Fig. 4. It could be found that the sorting time required by the CPU is much less than the searching time occupied by the CPU or the coupled GPU.

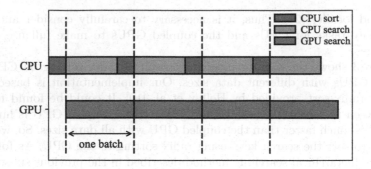

Fig. 5. The co-processing pipeline method.

Thus, when the CPU finishes the sorting task, it becomes idle to wait for the GPU. The performance could be increased further if the CPU always stays busy. Therefore, we propose a novel pipeline algorithm where part of the searching task is also distributed to the CPU besides the sorting task. A work item is responsible for searching one key on the CPU since its execution model is not similar to that of the GPU with the SIMT paradigm. The pipeline process is illustrated in Fig. 5. In most cases, the sorting and searching operations performed by the CPU are overlapped with the searching task belonging to the GPU, which will improve the overall query efficiency.

Algorithm 1. The co-processing pipeline algorithm

Input: $Batch_{0,1,...,N-1}$

1: sort(CPU, $Batch_0$);
2: collect the running information
3: **for** i = 0 to $N - 2$ **do**
4:     compute workload ratios to get w_{gpu} and w_{cpu}
5:     enqueue(search, GPU, w_{gpu})
6:     enqueue(search, CPU, w_{cpu})
7:     enqueue(sort, CPU, $Batch_{i+1}$)
8:     clWaitForEvent(GPU)
9:     clWaitForEvent(CPU)
10: compute workload ratios to get w_{gpu} and w_{cpu}
11: enqueue(search, GPU, w_{gpu})
12: enqueue(search, CPU, w_{cpu})
13: clWaitForEvent(GPU)
14: clWaitForEvent(CPU)

Algorithm 1 gives the implementation framework. In the algorithm, we first sort a batch of keys by the CPU (Line 1), and in the meantime start to collect the running times (Line 2) which will be explained in the next subsection. Then the algorithm iterates through $N - 1$ batches (Line 3), where the parameter

N refers to the number of batches needed to be searched. For each batch, the workload assigned to the GPU w_{gpu} and the CPU w_{cpu} is obtained (Line 4) and the searching tasks are enqueued into the GPU and the CPU (Line 5 and Line 6). The sorting task for the next batch is also enqueued into the CPU (Line 7). In order to synchronize CPU tasks and GPU tasks, $clWaitForEvent(GPU)$ and $clWaitForEvent(CPU)$ are used sequentially in the end of each loop (Line 8 and Line 9). Lines 10–14 process the last batch of keys and finish the searching algorithm.

Workload Assignment Based on Real-Time Performance. In order to assign appropriate workload between CPUs and coupled GPUs, we adopt a workload balance method based on real-time performance. Batches are further divided into blocks. The first batch and the first few blocks are processed respectively by the single CPU and GPU in order to collect the running time of the CPU sorting, CPU searching and GPU searching. Based on the information, we design a distribution model to partition the workload between processors. The distribution model is described as follows.

$$workload_{GPU} = min\left(\frac{search_{CPU}}{search_{GPU}+search_{CPU}} * \left(total + \frac{test\_block}{search_{CPU}} * sort\right), total\right)$$
$$workload_{CPU} = total - workload_{GPU}$$

$total$: the whole searching workload to be assigned
$workload_{GPU/CPU}$: the workload ratio assigned to GPU/CPU.
$search_{GPU/CPU}$: the average GPU/CPU searching time for one block
$test\_block$: the block size
$sort$: the CPU sorting time for one batch

During the phase of information collection, the CPU and the coupled GPU deal with a few small blocks in parallel to simulate the real running mode and save the profiling cost. The statistics are used to determine the distribution of the left workload. By use of the real-time performance, this method could avoid the estimated error caused by the data distribution, data sizes and hardware processors. In our study, we divide a batch into 128 blocks and use four blocks to measure the execution time. The effects will be verified by testing ratios ranging from 0% to 100% in the experiments.

4   Performance Evaluation

In this section, we first explain the experiment environment. Then we show the performance of the hierarchical searching method compared with the coarse-grained parallelism searching method. Lastly, the overall evaluation related to the co-processing pipeline strategy is conducted.

4.1 Experimental Setup

Intel i7-8700, an Intel Coffee Lake series processor, is used in our experiments. The CPU on this architecture contains 6 cores, running at a base clock of 3.2 GHz. A coupled GPU named Intel UHD Graphics 630 is supported, which has 24 execution units and runs at a base clock of 350 MHz. The machine we used for our experiments is equipped with a 16 GB DDR4 memory and using 64-bit Windows 10 as the operating system. An overview of hardware parameters is presented in Table 1. Our codes were implemented by C and OpenCL 2.0 which supports share virtual memory. All the methods are executed 3 times and average values are used.

In the experiments, we use 32-bit integer keys and values. All the data is generated by the Mersenne Twister pseudo-random number generator using the uniform distribution [4]. We create B^+-trees with different sizes and the number of keys in trees is 2^{23}, 2^{24}, 2^{25} and 2^{26}. B^+-trees are built by the CPU in the memory before the searching process. The keys to be searched are divided into eight batches and every batch size is varied and set to 2^{18}, 2^{19}, 2^{20} and 2^{21}. Nodes in B^+-trees contain 32 keys at most.

Table 1. Hardware parameters

Processor	Intel i7-8700	
Type	CPU	GPU
Name	N/A	Intel UHD Graphics 630
Basic units	Core	EU
Number of units	6	24
Base Clock Rate	3.2 GHz	350 MHz
TDP	65 W	
LLC	12 MB	

4.2 Performance of the Hierarchical Searching Scheme

Figure 6 and Fig. 7 show the performance of our hierarchical searching method on a single coupled GPU when batch sizes and tree sizes are varied. The baseline approach uses a coarse-grained parallelism searching strategy, in which leaf nodes are treated as inner nodes and a work item searches one key. The keys to be searched are sorted in advance, while the sorting time is not included in the experimental results since two methods could use the same sorting strategy.

We can find that the hierarchical method always outperforms the coarse-grained parallelism searching with various batch sizes and tree sizes. Both methods occupy more time if the batch size or the tree size is increased. In Fig. 6 where the tree size is set to 2^{26}, when the batch size is increased to 2^{21}, the

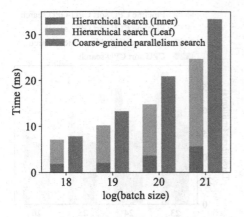
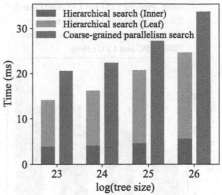

Fig. 6. The searching time with different search batch sizes, when the tree size is fixed at 2^{26}.

Fig. 7. The searching time with different tree sizes, when the search batch size is fixed at 2^{21}.

hierarchical searching method runs 27% faster than the baseline method. There also exists 7% performance improvement if the batch size is 2^{18}. The advantage of our strategy is more obvious on larger batches. When the batch size is fixed at 2^{21} and the tree size is increased from 2^{23} to 2^{26} as shown in Fig. 7, the hierarchical searching scheme can achieve 23%–36% performance improvement compared with the coarse-grained parallelism searching method.

The total searching time is broken down into two parts: the inner node searching time and the leaf node searching time. From two figures, we can find that accessing leaf nodes is more time-consuming, which is about 2X–3X as fast as traversing inner nodes. It is meaningful to carefully design the searching scheme in the leaf level. The experimental results also demonstrate that the overall performance benefits from the special treatment of leaf nodes in our method.

4.3 Performance of the Co-processing Pipeline Searching

Figure 8 and Fig. 9 show the throughput of four searching methods with the tree size set at 2^{26} and the batch size 2^{21} respectively. We use "Pipeline CPU sort CPU-GPU search" to represent our co-processing pipeline method. "Pipeline CPU sort GPU search" means the initial pipeline strategy that we take into account. "CPU sort GPU Hierarchical search" and "CPU sort CPU search" denote two non-pipeline solutions. The former refers to the proposed hierarchical searching strategy on the coupled GPU plus the CPU sorting operation, while the latter is the complete CPU version for comparison.

On the whole, our novel co-processing pipeline searching algorithm provides the highest throughput in all cases. This pipeline scheme can hide the sorting and partial searching time contributed by the CPU, thus an expected performance improvement is achieved. The initial pipeline method where the CPU sorting is overlapped with the GPU searching outperforms the other two non-pipeline

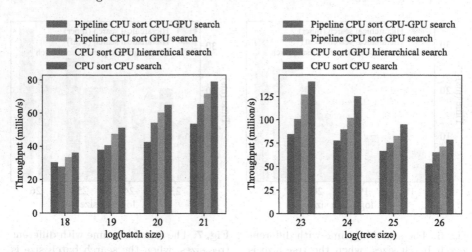

Fig. 8. The throughput with different search batch sizes, when the tree size is fixed at 2^{26}.

Fig. 9. The throughput with different tree sizes, when the search batch size is fixed at 2^{21}.

approaches. When the tree size is set to 2^{23} and the batch size 2^{21}, the optimized pipeline scheme could complete 141.4 million queries per second (MQPS) and 127 MQPS for the initial pipeline solution. Our hierarchical searching method on a coupled GPU shows better performance than the CPU algorithm, while the pipeline method further enhances the performance by 20%–40%. If compared with the baseline approach which has been studied in the previous subsection, our optimized pipeline method could increase the throughput by a factor of 1.8. All the four methods can introduce a larger throughput when the batch becomes bigger in size. Although it needs more time to finish the searching operation for larger batches, the overall throughput is increased because more keys are processed. The throughput is better when the tree size is small, which is reasonable since a larger tree means more nodes are involved leading to more memory accesses.

We examine the performance of the co-processing pipeline searching method when the GPU workload ratio α is varied from 0% to 100% and the remaining workload radio $(1 - \alpha)$ is assigned to the CPU. From Fig. 10 and Fig. 11, two conclusions could be drawn. First, the optimal workload ratio is not fixed, which should be determined dynamically. The optimal ratio α falls between 0.6 and 0.8, that is, the coupled GPU should be assigned more tasks in terms of searching jobs. Second, the used ratios which are calculated based on the real-time performance are close to the optimal ones and the error is below 5%, which shows that the workload distribution strategy is efficient and works well with different tree sizes and batch sizes.

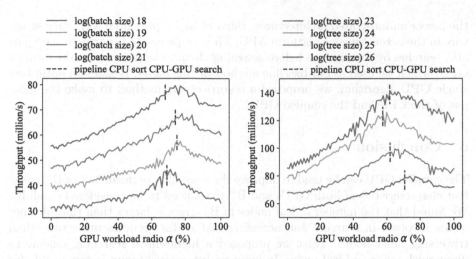

Fig. 10. The searching performance under various GPU workload ratios with different batch sizes, when the tree size is fixed at 2^{26}.

Fig. 11. The searching performance under various GPU workload ratios with different tree sizes, when the batch size is fixed at 2^{21}.

5 Related Work

There are many research studies focusing on multi-core CPUs, discrete GPUs or both processors. Sewall et al. [15] presented an architecture-friendly latch-free B$^+$-tree called PALM. Multiple concurrent queries and modifies in the multi-core CPU were performed by using the bulk synchronous parallel model. Fix et al. [6] used CUDA in a discrete GPU to speed up braided B$^+$-trees searches with GPU parallelism. Kaczmarski [11] created a B$^+$-tree on the discrete GPU which could benefit from GPU compute power to search and insert keys quickly. Yan et al. [18] aimed to bridge the gap between B$^+$-trees and discrete GPUs. They designed a high-performance B$^+$-tree structure called Harmonia which separates the node into key region and child region. The key region stores nodes in a breadth-first order, while the child region stores the first child index in the prefix-sum array. Awad et al. [1] considered the implementation of high-performance concurrent insertions, searches and deletions under the traditional B-tree data structure. They designed a GPU B-Tree for batch search and update performance with a warp cooperative work-sharing strategy. Shahvarani et al. [16] proposed HB$^+$-tree to jointly leverage the heterogeneous computing power of the CPU and the discrete GPU and utilize both of their memories to achieve a higher bandwidth over a high volumes of data.

There have existed some studies on the coupled CPU-GPU architecture or integrated GPUs. Luan et al. [13] conducted an extensive experimental study to evaluate analytical queries on CPUs and coupled GPUs. Hash joins [8,9], MapReduce programs [2] and breadth-first searches [5] were studied to leverage the power on integrated architectures. Daga et al. [4] presented a B^+-tree structure in the heterogeneous platform APU. They implemented coarse-grained parallel searches by sorting keys before assigning them to work-items on the coupled GPU, which we used as our baseline method in this paper. Besides an optimized single GPU algorithm, we proposed a co-processing method to make the most use of the CPU and the coupled GPU.

6 Conclusion

The coupled GPU can be used to improve the searching performance of B^+-trees. But characteristics of both GPUs and B^+-trees need to be considered carefully. We found that the number of leaf nodes in B^+-trees is larger than that of inner nodes involved in searches and accessing leaf nodes occupies more time than traversing inner nodes. Thus, we proposed a hierarchical searching scheme to distinguish inner and leaf nodes. In inner nodes, a work group is responsible for a relatively large number of keys to leverage the advantage of the instruction parallelism in GPUs, while in leaf nodes, the number of keys in a work group is reduced to lower irregular memory accesses and divergent branches in a work group. Our hierarchical scheme is able to achieve a performance improvement up to 36% compared with the method with a fixed number of search keys in all levels.

Furthermore, we investigated how to use the computing resources of both the CPU and the coupled GPU to search keys in B^+-trees. A co-processing pipeline method was presented which assigns the sorting task to the CPU and the searching task to the CPU and the coupled GPU. Real-time performance including the sorting time and the searching time was utilized to distribute the workload and a distribution model was proposed to calculate the workload ratios belonging to the processors. A performance evaluation was conducted to demonstrate the effectiveness of the optimization techniques. In the future, we would like to accelerate more operations on B^+-trees like insertions, updates and deletions.

References

1. Awad, M.A., Ashkiani, S., Johnson, R., Farach-Colton, M., Owens, J.D.: Engineering a high-performance GPU B-Tree. In: Proceedings of the 24th Symposium on Principles and Practice of Parallel Programming, pp. 145–157. ACM (2019)
2. Chen, L., Huo, X., Agrawal, G.: Accelerating mapreduce on a coupled CPU-GPU architecture. In: Proceedings of the International Conference on High Performance Computing, Networking, Storage and Analysis, pp. 25:1–25:11. IEEE (2012)
3. Comer, D.: The ubiquitous B-tree. ACM Comput. Surv. **11**(2), 121–137 (1979)
4. Daga, M., Nutter, M.: Exploiting coarse-grained parallelism in B+ tree searches on an APU. In: 2012 SC Companion: High Performance Computing, Networking Storage and Analysis, pp. 240–247. IEEE (2012)

5. Daga, M., Nutter, M., Meswani, M.: Efficient breadth-first search on a heterogeneous processor. In: 2014 IEEE International Conference on Big Data, pp. 373–382. IEEE (2015)
6. Fix, J., Wilkes, A., Skadron, K.: Accelerating braided B+ tree searches on a GPU with CUDA. In: Proceedings of the 2nd Workshop on Applications for Multi and Many Core Processors: Analysis, Implementation, and Performance (2011)
7. Graefe, G., Kuno, H.: Modern B-tree techniques. In: 2011 IEEE 27th International Conference on Data Engineering, pp. 1370–1373. IEEE (2011)
8. He, J., Lu, M., He, B.: Revisiting co-processing for hash joins on the coupled CPU-GPU architecture. Proc. VLDB Endow. **6**(10), 889–900 (2013)
9. He, J., Zhang, S., He, B.: In-cache query co-processing on coupled CPU-GPU architectures. Proc. VLDB Endow. **8**(4), 329–340 (2014)
10. Helluy, P.: A portable implementation of the radix sort algorithm in OpenCL (2011). https://hal.archives-ouvertes.fr/hal-00596730
11. Kaczmarski, K.: Experimental B+-tree for GPU. In: Proceedings II of the 15th East-European Conference on Advances in Databases and Information Systems, pp. 232–241 (2011)
12. Levandoski, J.J., Lomet, D.B., Sengupta, S.: The Bw-tree: a B-tree for new hardware platforms. In: 2013 IEEE 29th International Conference on Data Engineering, pp. 302–313. IEEE (2013)
13. Luan, H., Chang, L.: An evaluation of analytical queries on CPUs and coupled GPUs. Concurr. Comput.: Pract. Exp. **29**(5), e3982 (2017)
14. Ramakrishnan, R., Gehrke, J.: Database Management Systems, 3rd edn. McGraw-Hill, London (2002)
15. Sewall, J., Chhugani, J., Kim, C., Satish, N., Dubey, P.: PALM: parallel architecture-friendly latch-free modifications to B+ trees on many-core processors. Proc. VLDB Endow. **4**(11), 795–806 (2011)
16. Shahvarani, A., Jacobsen, H.A.: A hybrid B+-tree as solution for in-memory indexing on CPU-GPU heterogeneous computing platforms. In: Proceedings of the 2016 International Conference on Management of Data, pp. 1523–1538. ACM (2016)
17. Stone, J.E., Gohara, D., Shi, G.: OpenCL: a parallel programming standard for heterogeneous computing systems. Comput. Sci. Eng. **12**(3), 66–73 (2010)
18. Yan, Z., Lin, Y., Peng, L., Zhang, W.: Harmonia: a high throughput B+tree for GPUs. In: Proceedings of the 24th Symposium on Principles and Practice of Parallel Programming, pp. 133–144. ACM (2019)
19. Yuan, Y., Lee, R., Zhang, X.: The yin and yang of processing data warehousing queries on GPU devices. Proc. VLDB Endow. **6**(10), 817–828 (2013)
20. Zhang, F., Zhai, J., He, B., Zhang, S., Chen, W.: Understanding co-running behaviors on integrated CPU/GPU architectures. IEEE Trans. Parallel Distrib. Syst. **28**(3), 905–918 (2017)

OCVM: Optimizing the Isolation of Virtual Machines with Open-Channel SSDs

Zhe Liu, Xiaojian Liao, Fei Li, Zhe Yang, Youyou Lu, and Jiwu Shu$^{(\boxtimes)}$

Tsinghua University, Beijing, China
shujw@tsinghua.edu.cn

Abstract. A longstanding goal of virtual machines (VMs) isolation running on commercial Solid State Drives (SSDs) is to avoid performance degradation which could be quite severe in certain cases. However, it has been a challenge due to the limitations of the traditional flash translation layer (FTL). We propose OCVM, a novel storage stack that optimizes the isolation of VMs in a multiple VMs environment. By providing channel-granular and block-granular isolation for VMs, OCVM significantly reduces storage internal resource conflicts and eases the pressure of garbage collection (GC). To achieve good VM isolation as well as hardware utilization, OCVM applies a dynamic allocation mechanism for the underlying storage resources of the open-channel SSD (OCSSD). The evaluation results demonstrate that the average execution time of data-intensive applications on OCVM shortens by 28%, compared to those on a baseline system. In addition, OCVM achieves better GC efficiency by reducing the frequency of data migration.

Keywords: Open-Channel SSD · Isolation · Virtual machine · Flash channel · Flash memory

1 Introduction

The virtual machine (VM) has been the infrastructure of cloud service for recent years. Since it provides CPU and memory isolation for multiple tenants, it is used to increase the computing utilization by sharing resources. Nevertheless, it does not efficiently isolate underlying storage due to the complex management of the modern SSD's controller [1]. However, there are several issues to achieve strong isolation and efficient resource sharing at the same time when multiple VMs share the same SSD.

This work is supported by Special Topics of Major Scientific and Technological Projects in Sichuan Province (Grant No. 2018GZDZX0049), Research and Development Plan in Key field of Guangdong Province (Grant No. 2018B010109002), the National Natural Science Foundation of China (Grant No. 61832011), and Research Program of ZTE Corporation Limited (Grant No. 20182002008).

© Springer Nature Switzerland AG 2020
M. Qiu (Ed.): ICA3PP 2020, LNCS 12452, pp. 416–432, 2020.
https://doi.org/10.1007/978-3-030-60245-1_29

Firstly, multiple VMs introduce noisy neighbor problems. To enhance performance, SSDs exploit internal parallelism [2,3]. However, VMs cannot directly leverage flash parallelism for isolation, because FTLs hide the parallelism. Since multiple VMs and host share one SSD, the interference among VMs I/O executions introduces many storage recourse conflicts [1,4]. For example, the bandwidth-aggressive VMs may occupy more flash channels and block the I/O requests of neighbors. Since in flash memory the write latency is about several times of the read latency, the latency-sensitive VMs read performance would suffer from the bandwidth-aggressive VMs writes.

Secondly, the garbage collection (GC) of SSD impedes the I/O requests of VMs. Due to the no-overwrite property of flash memory, SSD leverages garbage collection operations to reclaim free space for incoming data pages. Unfortunately, because of the application-unawareness of commercial SSDs in the multiple VMs environment [5], data pages (e.g., 4 KB) of different VMs requests may belong to the same flash block (e.g., 2 MB). Since it is necessary to migrate the valid data before erasing the flash memory block, the flash memory blocks which involve data pages of different VMs' requests slow down the GC, and further impede the following VMs' requests.

Thirdly, strong isolation contradicts with the demands of high resource utilization. A straightforward approach to achieve high isolation is running VMs atop dedicated flash channels, as the partition method proposed by Huang et al. [1] and Kwon et al. [4]. However, these approaches run at a sacrifice of flexibility and resource utilization. As flexibility and resource utilization are vital considerations of the cloud environment, the cloud storage stack should achieve high resource utilization while providing tenant-desired isolation.

Today, a new architecture of SSD, i.e., the Open-Channel SSD (OCSSD), has been introduced to bridge the semantic gap between the host and flash devices. It exports the device geometry and further allows the host side software to directly manage the storage for better performance [1,6].

To address the aforementioned challenges, we propose OCVM, an OCSSD-based VM storage stack for multi-VMs systems. The key idea of OCVM is combining the channel-granular and block-granular isolation, and dynamically adjusting storage resource at runtime.

Specifically, OCVM divides the data space into multiple regions. Each VM has a private set of regions and a set of pool regions is shared by throughput-intensive VMs. The channel-granular isolation assigns a set of dedicated flash channels to each private set of regions under the "pay-as-you-go" [7] model in the cloud. VMs should specify the desired bandwidth. The block-granular isolation groups the data blocks of a VM to dedicated flash memory blocks. OCVM leverages the hybrid isolation scheme to allocate private sets of channels at startup and dynamically re-balances the shared flash channels at runtime.

We have implemented the prototype with a set of modifications to F2FS [8] in a real OCSSD. We demonstrate the effectiveness of our proposed mechanism through a variety of workloads against the baseline (i.e., Qemu [9] atop F2FS [8]) and a recently proposed file system designed for OCSSD (i.e., Qemu atop ParaFS [6]). The evaluation results demonstrate that data-intensive applications on OCVM exhibit 28% shorter execution time compared to the baseline. The multiple VMs achieve much more stable performance on OCVM and OCVM improves endurance and utilization of the SSD.

The rest of the paper is organized as follows. Section 2 introduces the background and motivation of the research work. Section 3 describes the design of OCVM. Section 4 presents our evaluation results. Section 5 discusses related work. Section 6 concludes this paper.

2   Background

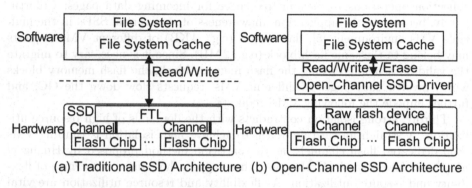

(a) Traditional SSD Architecture   (b) Open-Channel SSD Architecture

Fig. 1. The difference between traditional SSD architecture and Open-Channel SSD architecture. [6,10]

Open-Channel SSD Architecture. Commercial SSDs usually use FTL [11] to handle data accesses [12], as Fig. 1(a) shows. Since the flash chip, which contains many flash memory blocks, has the ability to execute I/O operations independently, FTL implements multi-channel technology to increase the data transmission speed [2]. Although the file system could read and write data using the interfaces provided by FTL, the location of the data in the underlying flash device is determined by FTL. What's more. since flash blocks must be erased before they can be reused for new data [13], FTL applies out-place-update operations. Because of the semantic gap between the file system and the SSD, data with different update frequencies may be stored in the same flash block. The fragmented underlying flash media cause write amplification when the FTL tries to reclaim free space to store new data. Although researchers have improved data location performance by modifying the FTL [5,14], it does not have desired VM compatibility.

OCSSD architecture [10,15] is a recently proposed architecture that software directly manages underlying flash memory, as shown in Fig. 1(b). It redesigns the internal management mechanism of the SSD. The OCSSD driver provides read/write/erase interfaces and the device geometry information to the host-side systems, such as file systems, database systems. The file system can flexibly control the data access process and directly manage the data layout of the underlying flash device. On the basis of it, we propose OCVM to optimize the isolation performance when VMs share the same OCSSD.

Storage Isolation of VMs. VMs require storage isolation to guarantee the storage service level agreements (SLAs) in the multiple VMs environment [16]. However, the commercial SSDs could not provide excellent storage isolation for VMs. Because of the standard block device interface and internal mechanism of the commercial SSDs, the semantics of VMs isolation in the host file system cannot be directly conveyed to SSDs [6]. Since FTLs manage the data layout and does not distinguish I/O requests from different VMs, different VMs may access the same physical storage resource (e.g. flash channel) in the meantime, which causes I/O collisions on the physical media and degrades the I/O performance of VMs. The interference among multiple VMs in a single SSD makes the VMs I/O operation latencies unpredictable. Furthermore, If the GC is triggered by a VM, it causes more severe degradation to the storage performance of other VMs [17]. These issues need to be addressed.

Under the management of VM monitor (e.g. KVM [18]), the data accesses of VMs can be converted to the I/O requests of VM image files by the VM emulator. Therefore, we can provide a storage isolation mechanism for VMs at the host file system level [19].

3   Design

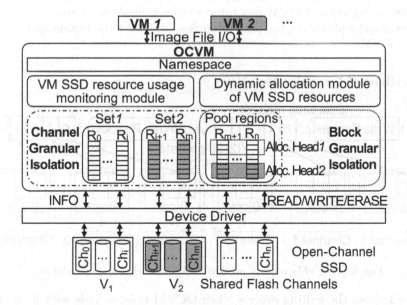

Fig. 2. The OCVM architecture.

The OCVM Architecture. OCVM aims to improve both the storage isolation of VMs and the utilization of SSD in the multiple VMs environment. It features three key techniques: channel-granular isolation, block-granular isolation, and dynamic allocation of pool resources. As shown in Fig. 2, with the geometry information provided by the OCSSD driver, OCVM can allocate and recycle space that aligned to the underlying flash memory device. We divide the data space into regions that match flash channels one to one. A region consists of many data segments. The address and size of a data segment are aligned to a flash memory block. Each data segment contains 512 data blocks which are aligned to the flash pages. OCVM maintains a data group for each VM which is assigned numerous free data segments. An allocator head pointing to the VM data group allocates free data blocks to the VM which generates write requests.

OCVM writes VM data pages in a log-structured way because of out-place-update feature of the flash memory. Since the entire VM is encapsulated in a VM image file, OCVM manages data allocation and I/O requests scheduling of multiple VMs on the host file system level based on the inode information of the VM image files. The VM SSD resource monitoring module handles the data write of each VM, and the dynamic allocation module allocates regions to VMs. To provide channel-granular isolation of VMs, OCVM assigns a private set of regions (e.g. set1 or set2 in Fig. 2) for each VM to avoid the interference of VM I/O operations. Since the internal parallelism provided by each private set of regions is lower than that of the whole device, we reserve a set of regions as pool regions sharing by throughput-intensive VMs to guarantee the QoS in terms of throughput. OCVM leverages channel-granular and block-granular isolation of VMs to reduce interference and improve data locality. According to the VMs runtime information, OCVM can improve the utilization of storage resources by dynamically allocating the pool resources, and meet the throughput requirements of VMs.

3.1 Channel-Granular Isolation

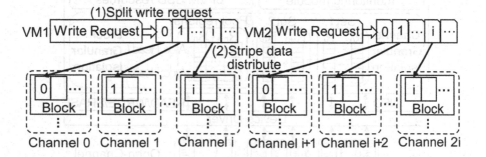

Fig. 3. Two VMs writing process with channel-granular isolation.

Figure 3 shows the writing process when OCVM isolates VMs with flash channel granularity. To provide I/O isolation of VMs in the multiple VMs environment, OCVM issues I/O requests of each VM to specific flash channels in the device. Since the flash controller connects flash chips through flash channels in the underlying device, OCVM can better utilize the flash internal parallelism by distributing VM I/O requests to different flash channels. Specifically, as in Fig. 3, OCVM first splits the write requests of VMs into 4 KB pages, and then stripes these pages over the exclusive regions allocated to VMs. Since each region in OCVM is an abstraction of a specific physical flash channel in OCSSD. the striped write requests of VMs are sent to a dedicated set of flash channels for parallel execution. In this way, OCVM isolates the data access paths of VMs on the channel-level.

The channel-granular isolation also avoids the interference caused by VMs' valid data pages migration during garbage collection. The valid data pages of a VM are only migrated to the flash channels allocated to the VM during garbage collection. As a result, the valid data pages migration of VMs does not conflict with the normal read/write operations of other VMs.

By executing I/O operations of one VM in exclusive physical flash channels, channel-granular isolation provides stable I/O performance for VMs. Because I/O requests that emanate from multiple VMs would not collide when reaching the physical flash media. In this view, OCVM provides predictable delay and bandwidth for VMs by reducing the interference of neighbors.

3.2   Block-Granular Isolation

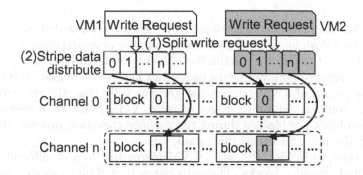

Fig. 4. The VMs writing process with block-granular isolation.

For VMs that can tolerate some interference, such as the VMs running the non-premium cloud database offerings [1], we propose a more fine-grained VM storage isolation mechanism based on the OCSSD architecture. As Fig. 4 shows, when multiple VMs share flash channels, we store the data from different VMs in separate physical flash memory blocks to isolate VMs with block granularity.

Considering that multiple VMs share multiple flash channels, to improve the degree of internal parallelism, OCVM divides the write requests of each VM into data pages, and then stripes them over assigned regions. As mentioned above, OCVM writes data pages in a log-structured way. When a VM write request is sent to a region, the VM specific allocator head allocates a free page from a

422   Z. Liu et al.

log-structured segment to the write request. Since the data segments are aligned to the underlying flash memory blocks, the write requests from different VMs are stored in separate flash memory blocks.

With block-granular isolation, OCVM reduces the write amplification caused by the mixed write of VMs. In general, the data access patterns of different VMs vary. For examples, some VMs' workloads may be update-intensive, and others may be read-intensive. Being unaware of the access patterns of VMs, the FTL of commercial SSDs stores different VMs' data in the same flash memory block (see Fig. 5(a)), which damages the data locality.

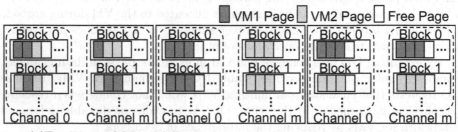

(a)Traditional SSD   (b) Flash channel isolation (c) Flash block isolation

Fig. 5. Data layout of two VMs in SSD.

The VMs data layout of traditional SSDs leads to two undesired results: on one hand, each flash block contains a similar number of invalid data pages, which increases the cost in selecting victim flash blocks. On the other hand, the victim flash block contains more valid data pages that need to be migrated. If the data update operations of a VM triggers the garbage collection, the selected victim flash block contains valid data pages of other VMs. Since all accesses to this victim flash block are blocked during the garbage collection process, there is a long delay if other VMs access these data.

In contrast, as in Fig. 5(b), (c), OCVM places the data of different VMs in separate flash memory blocks. The data pages in a flash memory block have similar hotness. OCVM facilitates the selection of victim blocks and reduces data migration during garbage collection, because the victim flash blocks only contain the data pages belonging to one VM. By relieving the costs of garbage collection and reducing the adverse impact of garbage collection triggered by other VMs, OCVM prevents the storage performance of VMs from degradation caused by garbage collection to some extent.

3.3 Dynamic Allocation of Pool Resources

To improve the utilization of SSD resources and meet the storage performance requirements of throughput-intensive VMs, OCVM adopts a dynamic allocation method for the OCSSD. OCVM reserve several regions as pool regions shared by throughput-intensive VMs. Since a region in OCVM matches a flash channel, there are several corresponding flash channels in flash memory as pool resources shared by throughput-intensive VMs. To provide more stable write bandwidth and improve the data locality, OCVM provides both channel-granular and block-granular isolation for VMs in the pool regions.

We design a VM resource monitoring module and a dynamic allocation module in OCVM, as Fig. 6 shows. The monitoring module monitors the amount of data written by the throughput-intensive VMs periodically. The dynamic allocation module uses the information from the monitor module to predict the amount of data written by the VM and adjusts the pool flash channels allocation in the next monitoring period. The period length can be set according to the characteristics of the VMs' applications.

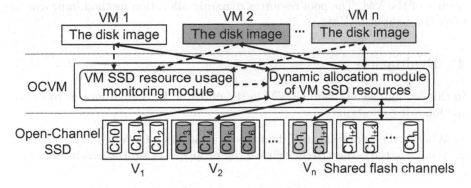

Fig. 6. Dynamically allocate pool resources of VMs.

Our hypothesis is then that the tenants know storage requirements in advance. One flash channel represents a unit of SSD resources. We use L_i to denote the minimum SSD resources required by each VM, and m to indicate the number of throughput-intensive VMs. Assume that there are n flash channels in the OCSSD. These variables should satisfy the constraints of formula (1).

$$\sum_{k=0}^{n} L_i \leq n \tag{1}$$

If the number of flash channels is less than the minimum storage resource requirements of all VMs, we should reduce the number of VM instances or increase the flash memory hardware devices.

The dynamic allocation method is as follows:

1. Pre-allocate L_i flash channels for each throughput-intensive VM.
2. The monitoring module records the amount of data written by each VM in a period of time t, which we use B_i^t to denote.

$$C_i^{t+1} = L_i + \lfloor (n - \sum_{k=0}^{m} L_i) \times \frac{B_i^t}{\sum_{i=0}^{m} B_i^t} \rfloor \tag{2}$$

3. Based on the formula (2), the dynamic allocation module allocates C_i^{t+1} flash channels to the VM in the period of time $t+1$, which contain two parts: the pre-allocated part, and the part allocated from the pool resources.
4. Repeat steps 2 and 3.

Since the proportion of data written in the whole device by a VM in a time period represents the degree of demand for resources of this VM, OCVM first monitors and counts the proportion of data writes of a VM in the whole device in a time period and then allocates the same proportion of pool resources to the VM in the next time period.

By reallocating pool resources to the VMs, which are in urgent need of resources, OCVM improves the overall utilization of OCSSD in a more flexible way. The pre-allocated flash channels guarantee the necessary storage service quality of the VM. The pool resources dynamic allocation method improves the peak throughput of VMs.

4   Evaluation

In this section, we evaluate OCVM in the multiple VMs environment to answer the following questions:

1. What is the performance of data-intensive VMs on OCVM?
2. What are the respective benefits of the proposed optimizations in OCVM?

4.1   Experimental Setup

We prepare a real customized Open-Channel SSD in an X86 architecture server. Table 1 lists the parameters of the VM environment. In this evaluation, we use QEMU [9] to generate the VM image files and run the operating system of the VM. We take advantage of KVM to accelerate VMs and choose the writethrough cache mode to ensure data consistency.

Table 1. Important parameters of our system and device.

Host		Guest	
CPU	Intel Xeon E5-2620 2.10 GHz	QEMU	2.1.0
Main memory	16G	Virtual memory	2G
Linux Kernel	2.6.32	OS	Ubuntu 16.04
Open-Channel SSD			
Number of flash channel		32	
Host interface	PCIe 2.0x8	NAND type	25 nm MLC

Tools and Workloads. We have selected 4 tools and workloads: (1) Flexible I/O Tester (FIO v2.0.7) [20], which is a tool that does a particular type of I/O action as specified by the user. We use FIO to simulate the different read and write operations. We set a O_DIRECT flag for all evaluations to bypass page caches. (2) Fileserver [21], which is one of the predefined workloads in Filebench (v1.5-alpha3), from which we can emulate the I/O behavior of the VM which running file services. (3) ab [22], which is a tool for benchmarking Apache HTTP server. We used ab to test the latency of accessing local web pages when the VM runs the Apache service. (4) three workloads from the Yahoo Cloud Serving Benchmarks (YCSB) [23]. YCSB-A, YCSB-B, and YCSB-C are used for the evaluation. Their I/O patterns are 50%/50% read/updata, 95%/5% read/updata, and 100% read respectively. The workloads run directly on MySQL database in the VM system. There are 12 million records in the database. We set the operation count of each workload to 10 thousand.

As comparison objects of OCVM, we run F2FS [8] and ParaFS [6] as the host file system respectively. F2FS is a flash friendly log-structure file system that requires the support of FTL. ParaFS is a host file system based on OCSSD architecture, which exploits internal parallelism to improve performance.

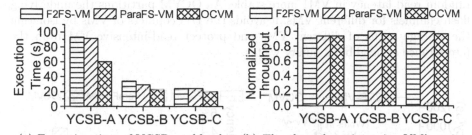

(a) Execution time of YCSB workloads. (b) The throughput-intensive VM's write throughput when VMs running together.

Fig. 7. Overall performance analysis.

4.2 Overall Performance

In this section, we test the overall performance of VMs on OCVM. In our evaluations, we co-run a data-intensive VM and a throughput-intensive VM together to simulate a real-world multiple VMs environment. We start two VMs on OCSSD, F2FS, ParaFS, respectively. VM1 is a data-intensive VM that runs the YCSB workloads, and VM2 is a throughput-intensive VM that uses FIO to emulate bursty writes. On OCVM, we pre-allocate 13 flash channels for each VM, and the rest 6 flash channels are as pool resources sharing by throughput-intensive VMs. Figure 7(a) shows the execution time by co-running YCSB workloads with bursty writes in VM2. Compared to the execution times on F2FS, OCVM shortens the execution times of YCSB-A, YCSB-B, and YCSB-C by 35%, 33%, and

18%, respectively. Since write requests from VM2 are sent to every flash channel in F2FS and ParaFS, these requests compete with the requests of YCSB
workloads and degrade the performance. Since OCVM allocates I/O requests
from different VMs to different flash channels for execution, which avoids internal resource conflicts, severe performance degradation of VM1 on OCVM is not
occurring.

Even though the number of channels that VM2 could use on OCVM would be
lower than that of F2FS, the write throughput of VM2 show negligible difference
(see Fig. 7(b)). It is because VM2 could use the pool resources, and VM2's writes
can be buffered by the host-side memory.

4.3   The Impact of Channel-Granular Isolation on VMs

Eliminating Disturbing Neighbors. We first perform the evaluation by the
scenario of Facebook-like workloads [24] (sustained reads with bursty writes). We
enable FIO applications in three VMs and issue sustained random reads in VM1.
To show the level of I/O interference between VMs, we enable bursty writes in
the other two VMs for every 30 s. Figure 8 shows the level of I/O determinism
in terms of random read latency. Compared with the F2FS, OCVM makes the
random read latency in VM1 more stable. As OCVM partitions the underlying
flash channels for different VMs, it avoids the I/O conflicts of VMs by isolating
the storage paths of different VMs and protect read-intensive VM from the
interference of neighbor's bursty writes.

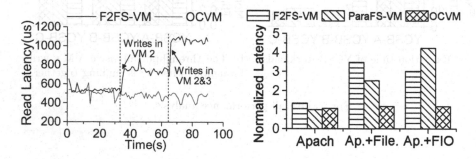

Fig. 8. Read latency with bursty writes.   **Fig. 9.** Latency of Apache service.

Then, we test the quality of service improvement when the VM runs read-
latency sensitive services. We run the Apache service in VM1 and the fileserver
service in VM2. Figure 9 shows the time it takes the Apache service to read
a local web page when the two services are co-running. Compared with F2FS
and ParaFS, OCVM ensures that the Apache service is not disturbed by the
applications of the other VMs, thereby providing better storage isolation for
VMs. We also observed similar experimental results when using FIO to issue
heavy writes in VM2.

Isolation Effect of VM Image File I/O Requests. We compare OCVM with F2FS and ParaFS to test the read-write interference of two VM image files on the host file system level. We store two VM image files on OCSSD. Each of them uses half of the flash channels of OCSSD on OCVM, respectively. We use FIO to read the VM1 image file and use another FIO to write the VM2 image file in the meantime.

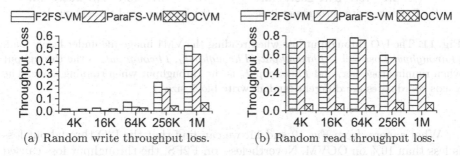

(a) Random write throughput loss.   (b) Random read throughput loss.

Fig. 10. The I/O throughput loss when writing to the VM2 image file under heavy load. ($Throughput\ Loss = 1 - Throughput_t/Throughput_a$, $Throughput_t$ is the throughput when running together, and $Throughput_a$ is the throughput when running alone. The x-axis coordinates are different random read block sizes.)

We first test the isolation effect of reads with heavy writes. We set the random write block size to 2 MB to simulate heavy write behavior, and change the random read block size (from 4 KB to 1 MB) to emulate different read behavior. We record the throughput when reading and writing at the same time, and compare it with the throughput of each separate operation. Figure 10 shows the throughput losses. As we can see in Fig. 10(b), on F2FS and ParaFS, when the read load is light, the read throughput loss is large. In the flash device, a typical write operation takes more time than a read operation. Once the write request in the flash channel executes before the read request, the read request has a long delay.

Then we test the isolation effect of writes with heavy reads. We set the random read block size to 2 MB to emulate heavy read behavior, and change the random write block size to emulate different write behavior. From Fig. 11, we find that, on F2FS and ParaFS, under the heavy random read, the random write throughput loss is higher, while the random read throughput loss is lower. It shows that the read operation on the flash memory has more advantages than the write operation. The execution of read requests is completed faster and preempts the execution resources of write requests. It is also a manifestation of the uneven reading and writing of flash memory.

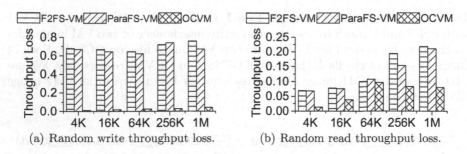

(a) Random write throughput loss.   (b) Random read throughput loss.

Fig. 11. The I/O throughput loss when reading the VM1 image file under heavy load. ($Throughput\ Loss = 1 - Throughput_t/Throughput_a$, $Throughput_t$ is the throughput when running together, and $Throughput_a$ is the throughput when running alone. The x-axis coordinates are different random write block sizes.)

What's more, from Fig. 10 and 11, we can find that the I/O throughput loss is less than 10% on OCVM. Nevertheless, on F2FS, the throughput loss caused by mutual interference reaches 70% in the worst case. This shows that channel-granular isolation can avoid the interference between I/O requests to different VM image files.

4.4   The Impact of Block-Granular Isolation on Garbage Collection

We perform an experiment to study the impact of block-granular isolation on flash garbage collection. We set the OCSSD capacity to 40 GB to trigger garbage collection faster and create two VMs (VM1 and VM2) that are stored in the clean OCSSD. Each VM image file size is 15 GB. We use FIO to write 5GB data in each VM concurrently and then write 5GB data in turn.

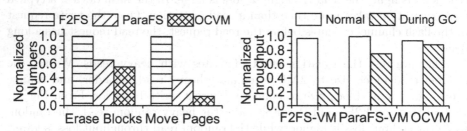

(a) Normalized number of erase blocks and (b) Impact of the GC on VM throughput.
move pages.

Fig. 12. Improvement of SSD garbage collection.

Figure 12(a) shows the number of flash block erases and the number of flash page migrations in the device when using different host file systems. OCVM has the lowest garbage collection overheads. In our experimental scenario, the number of recycled blocks and migrated pages decrease by 44% and 86%, respectively, compared with those on F2FS. Although both OCVM and ParaFS complete the data migration during garbage collection on the host file system level, the data locality of VMs on OCVM is better, which improves the garbage collection efficiency. The data of different VMs have diverse update frequencies. Because OCVM isolates the VMs' data with flash block granularity, hot data are gathered in a few underlying flash blocks. It reduces the pressure of the garbage collection from the erase of flash memory blocks and the migration of valid pages. By alleviating the write amplification problem, OCVM extends the lifetime of SSD running in the VM environment.

Besides, the benefits of the GC improvement also result in higher performance of VMs. Figure 12(b) represents the average write throughput of the VM when the last 5 GB file is written. Since OCVM reduces data migration, the device can focus on performing I/O operation requests of VMs. It is beneficial to both the performance of VMs and the durability of SSD.

4.5   The Effect of Dynamically Allocating Pool Resources

In this section, we study the effect of dynamically allocating pool resources through experiments. We enable FIO applications in three VMs and issue random writes to emulate throughput-intensive applications. To distinguish between VMs, we set the minimum storage resources required by the three VMs to 8, 4, and 4. OCSSD includes 32 flash channels and half of them compose the pool resources. To observe the changes of VMs throughput, we set the running time of FIO in three VMs to 100 s, 120 s, and 180 s and dynamical reallocate the pool resources in every 5 s.

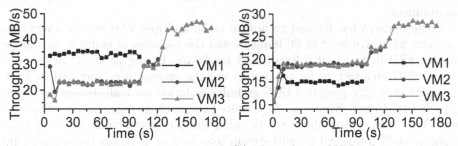

(a) Throughput of VMs when running the same heavy workload in each VM.

(b) Throughput of VMs when running the different heavy workloads in each VM.

Fig. 13. Throughput of three VMs during the experiment.

We first set the write block size to 2 MB in each VM to observe the write bandwidth that each VM can obtain. Figure 13(a) shows that the throughput of VM1 is higher than that of the other two VMs. It is due to two reasons. First, the flash channel resources pre-allocated to VM1 are more than those pre-allocated to the other VMs. Second, under the same heavy workload, the amount of data that VM1 can write in a monitoring period is more than that of the other VMs so that it gets more flash channels from the pool resources in the next period. The tenants that require more SSD resources need to pay higher fees. Since the tenant of VM1 pays more, it has advantages in allocating pool resources. After the workload of VM1 stops working, other VMs occupy the resources originally allocated to VM1. It improves the utilization of the device.

Then we set the write block size to 256 KB in VM1 and 512 KB in other VMs to emulate different heavy workloads in each VM. As we can see in Fig. 13(b), Since the workloads of VM2 and VM3 is heavier than that of VM1, OCVM allocates more flash channels to VM2 and VM3 from the pool resources. In the meantime, OCVM reduces the number of flash channels allocated to VM1. After a period of adjustment, the allocation of pool resources is balanced among the VMs. The write throughput of each VM is stable. It proves that the dynamic allocation mechanism can improve the utilization of SSD when VMs are isolated by channel-granular, and it is the balance between the isolation of VMs and the utilization of SSD.

5   Related Work

With more researches on SSD in recent years, the problems brought by FTL are exposed. Lu et al. [10] proposed OCSSD architecture to eliminate functional redundancy. Based on OCSSD architecture, Zhang et al. [6] implemented a file system to exploit the internal parallelism of flash devices. Lightnvm [15] implemented OCSSD in Linux kernel. Based on it, applications get an abstraction of OCSSD in user space and access data directly. These studies show the flexibility of OCSSD and provide new ideas for the use of flash devices in the multiple VMs environment.

Some research has focused on how to make multiple VMs or containers better share SSDs. Kwon et al. [4] implemented the isolation of multiple containers sharing SSD in terms of software and hardware. Huang et al. [1] designed a virtual SSD architecture for tenants to use the storage area specified in SSD. Gonzalez et al. [25] proposed LUN-granular isolation for multi-tenant by modifying kernel drivers. Kim et al. [26] reduced the garbage collection overheads to meet VMs service level objectives by using the over-provisioning space of SSD. Kang et al. [5] proposed a multi-stream interface to improves throughput and latency QoS by modifying the FTL. In similar, FStream [14] reduces garbage collection overheads by storing metadata and journal of the file system in different streams. Kim et al. [27] proposed a method to improve the utilization of SSDs for different tenants by using the parallelism of chip level in SSDs. S-CAVE [28] manages a shared SSD cache based on VMs and SSD runtime information to improve the utilization of the device.

The difference between our work and the above works is that we focus on the OCSSD host file system to optimize the performance of VM isolation and dynamic allocate storage resources. Without modifying the VM operating system and VM emulator, it provides better compatibility and flexibility for the multiple VMs environment.

6 Conclusion

Based on OCSSD architecture, we propose OCVM to improve storage isolation for the VMs sharing the same SSD. OCVM is an OCSSD-based VM storage stack and does not require any modification to VM emulator or guest OS. Experiments show that OCVM controls the performance loss caused by the interference between VMs within 10% and shortens the data-intensive applications by 28%, on average, compared to those on a baseline host file system. What's more, by reducing garbage collection overhead and dynamically allocating pool resources, it enhances the storage lifetime and improves the utilization of the device in the multiple VMs environment. In the future, we wish to expand OCVM in two directions. First, we would like to investigate how to integrate with VM emulator to improve VMs storage performance. Second, we would like to explore the integration method with the distributed data center to guarantee cloud storage SLAs for tenants.

References

1. Huang, J., et al.: Flashblox: achieving both performance isolation and uniform lifetime for virtualized SSDs. In: 15th USENIX Conference on File and Storage Technologies (FAST 17), vol. 375 (2017)
2. Park, S., Seo, E., Shin, J., Maeng, S., Lee, J.: Exploiting internal parallelism of flash-based SSDs. IEEE Comput. Archit. Lett. 9, 9–12 (2010)
3. Hu, Y., Jiang, H., Feng, D., Tian, L., Luo, H., Ren, C.: Exploring and exploiting the multilevel parallelism inside SSDs for improved performance and endurance. IEEE Trans. Comput. 62, 1141–1155 (2012)
4. Kwon, M., Gouk, D., Lee, C., Kim, B., Hwang, J., Jung, M.: DC-store: eliminating noisy neighbor containers using deterministic I/O performance and resource isolation. In: 18th USENIX Conference on File and Storage Technologies (FAST 20), vol. 183 (2020)
5. Kang, J., Hyun, J., Maeng, H., Cho, S.: The multi-streamed solid-state drive. In: 6th USENIX Workshop on Hot Topics in Storage and File Systems (HotStorage 14) (2014)
6. Zhang, J., Shu, J., Lu, Y.: ParaFS: a log-structured file system to exploit the internal parallelism of flash devices. In: 2016 USENIX Annual Technical Conference (USENIXATC 16), vol. 87 (2016)
7. PAYG cloud computing. https://searchstorage.techtarget.com/definition/pay-as-you-go-cloud-computing-PAYG-cloud-computing. Accessed 30 Mar 2015
8. Lee, C., Sim, D., Hwang, J., Cho, S.: F2FS: a new file system for flash storage. In: 13th USENIX Conference on File and Storage Technologies (FAST 15), vol. 273 (2015)
9. QEMU. https://www.qemu.org. Accessed 28 Apr 2020

10. Lu, Y., Shu, J., Zheng, W.: Extending the lifetime of flash-based storage through reducing write amplification from file systems. In: Presented as part of the 11th USENIX Conference on File and Storage Technologies (FAST 13), vol. 257 (2013)
11. Shin, J., et al.: FTL design exploration in reconfigurable high-performance SSD for server applications. In: Proceedings of the 23rd International Conference on Supercomputing, vol. 338 (2009)
12. Zhe, Y., Lu, Y., Xu, E., Shu, J.: CoinPurse: a device-assisted file system with dual interfaces. In: the 56th Annual Design Automation Conference (DAC) (2020)
13. Agrawal, N., Prabhakaran, V., Wobber, T., Davis, J.D., Manasse, M.S., Panigrahy, R.: Design tradeoffs for SSD performance. In: USENIX Annual Technical Conference, vol. 8, p. 57 (2008)
14. Rho, E., et al.: FStream: managing flash streams in the file system. In: 16th USENIX Conference on File and Storage Technologies (FAST 18), vol. 257 (2018)
15. Bjørling, M., González, J., Bonnet, P.: Lightnvm: the linux open-channel SSD subsystem. In: 15th USENIX Conference on File and Storage Technologies (FAST 17), vol. 359 (2017)
16. Park, H., Yoo, S., Hong, C.-H., Yoo, C.: Storage SLA guarantee with novel SSD I/O scheduler in virtualized data centers. IEEE Trans. Parallel Distrib. Syst. **27**, 2422–2434 (2015)
17. Kim, J., Lee, D., Noh, S.H.: Towards SLO complying SSDs through OPS isolation. In: 13th USENIX Conference on File and Storage Technologies (FAST 15), pp. 183–189 (2015)
18. Kernel Virtual Machine. https://linux-kvm.org. Accessed 30 Apr 2020
19. Li, S., Lu, Y., Shu, J., Hu, Y., Li, T.: Locofs: a loosely-coupled metadata service for distributed file systems. In: Proceedings of the International Conference for High Performance Computing, Networking, Storage and Analysis, pp. 1–12 (2017)
20. Fio: Flexible I/O Tester. https://github.com/axboe/fio. Accessed 30 Apr 2020
21. Filebench. https://github.com/filebench/filebench/eiki. Accessed 30 Apr 2020
22. Apache Bench. https://httpd.apache.org/docs/2.4/programs/ab.html. Accessed 30 Apr 2020
23. Cooper, B.F., Silberstein, A., Tam, E., Ramakrishnan, R., Sears, R.: Benchmarking cloud serving systems with YCSB. In: Proceedings of the 1st ACM Symposium on Cloud Computing, vol. 143 (2010)
24. Petersen, C., Zhang, W., Naberezhnov, A.: Enabling NVMe I/O determinism at scale. In: The 12th annual Flash Memory Summit (2018)
25. González, J., Bjørling, M.: Multi-tenant I/O isolation with open-channel SSDs. In: Nonvolatile Memory Workshop (NVMW) (2017)
26. Kim, J., Lee, D., Noh, S.H.: Towards SLO complying SSDs through OPS isolation. In: 13th USENIX Conference on File and Storage Technologies (FAST 15), vol. 183 (2015)
27. Kim, B.S.: Utilitarian performance isolation in shared SSDs. In: 10th USENIX Workshop on Hot Topics in Storage and File Systems (HotStorage 18) (2018)
28. Luo, T., Ma, S., Lee, R., Zhang, X., Liu, D., Zhou, L.: S-cave: effective SSD caching to improve virtual machine storage performance. In: Proceedings of the 22nd International Conference on Parallel Architectures and Compilation Techniques, vol. 103. IEEE (2013)

CANRT: A Client-Active NVM-Based Radix Tree for Fast Remote Access

Yaoyao Ying[1], Kaixin Huang[1], Shengan Zheng[2], Yaofeng Tu[3],
and Linpeng Huang[1(✉)]

[1] Shanghai Jiaotong University, Shanghai, China
{yingyy77,Kaixinhuang,Lphuang}@sjtu.edu.cn
[2] Tsinghua University, Beijing, China
venero@tsinghua.edu.cn
[3] ZTE Corporation, Nanjing, China
tu.yaofeng@zte.com.cn

Abstract. This paper presents the first study of building a remote-accessible persistent radix tree, named CANRT. Unlike prior works that only focus on designing single-node tree structure for non-volatile memory, we focus on optimizing remote access performance for a persistent radix tree while minimizing the persistence overhead. Simply adopting server-reply paradigm will incur heavy server CPU consumption and hence lead to high operation latency under concurrent workloads. Therefore, we design a low-latency *node-oriented read* mechanism and a *fine-grained lock-based write* mechanism to minimize the server CPU involvement in the critical path. We also devise a *non-blocking resizing* scheme in CANRT. The extensive experimental results on commercial Intel Optane DC Persistent Memory platform show that CANRT outperforms the state-of-art server-centric persistent radix trees by 1.19x–1.22x and 1.67x–1.72x in read and write latency, respectively. CANRT also gains improvement of 7.44x–11.15x in terms of concurrent throughput under YCSB workloads.

Keywords: Radix tree · Non-volatile memory · RDMA · Data consistency · Concurrent access

1 Introduction

Emerging storage and networking technologies such as non-volatile memory(NVM) and Remote Direct Memory Access (RDMA) technology are poised to reshape conventional data storage and memory systems in data centers, bringing new opportunities to achieve efficient remote data storage and access. Non-volatile memory technologies such as PCM [17], STT-RAM [16] and the recently released Intel Optane DC Persistent Memory [8] promise non-volatility, byte-addressability and DRAM-comparable latency. RDMA technology makes it possible for clients to access remote memory region while bypassing server CPU and kernel, which contributes to high bandwidth and low latency.

© Springer Nature Switzerland AG 2020
M. Qiu (Ed.): ICA3PP 2020, LNCS 12452, pp. 433–447, 2020.
https://doi.org/10.1007/978-3-030-60245-1_30

To further exploit the advantages of NVM and RDMA technology, indexing data structures and algorithms are required to be carefully redesigned to promise both data consistency and data concurrency for higher scalability. In recent years, a large number of tree-based indexing structures are proposed for non-volatile memory. Most of these works focus on the design of persistent B+-tree, [1,7,14,19]. They achieve considerable performance by reducing the number of expensive memory fencing and cache line flush operations.

Lee et al. [12] propose three radix tree variants (i.e., WORT, WOART, ART+CoW), and demonstrate that radix tree is more appropriate for persistent memory than B+-tree variants due to its unique features. For instance, the radix tree does not demand tree rebalancing operations and each insertion or deletion only results in a single atomic update operation to the tree, which is perfect for NVM [12]. However, whether the B+-tree variants or the radix tree variants in existing papers are designed for NVM in a single machine environment instead of distributed systems. Therefore, when deployed in distributed system environment such as data centers, they may suffer from high access latency and poor throughput scalability. Even though they can be equipped with RDMA server-reply paradigm [2,5,9,13], where both reads and writes are processed by the server and requires replies to the clients, the limited server CPU resource will become the bottleneck for highly-concurrent remote requests and impair the overall performance. The server-bypass merits of RDMA cannot be exploited with such designs.

In this paper, we propose a client-active NVM-based radix tree for fast remote access with server-bypass paradigm using RDMA. In the server side, we decouple the radix tree into two parts: data nodes that store actual key-value pairs with the same prefix, and prefix nodes that locate the target data nodes. Each data node contains an 8-byte metadata, which is used to resolve concurrent write collisions. Specifically, prefix nodes are also stored in all clients to support server-bypass remote data access. Our contributions can be summarized as follows.

- We propose CANRT, a client-active and NVM-optimized radix tree that leverages RDMA technology to speed up remote data access. To the best of our knowledge, CANRT is the first persistent radix tree that is optimized for fast remote access.
- We decouple the radix tree nodes into prefix nodes and data nodes to enable server-bypass remote data access and utilize a bitmap-assisted strategy for all types of writes to efficiently guarantee data consistency while resolving concurrent write collisions.
- We design a node-oriented remote read mechanism and a fine-grained lock-based remote write mechanism to support low-latency and high-concurrency remote access to the persistent radix tree. We also propose a non-blocking resizing scheme for CANRT to boost its performance during tree reconstruction.
- We conduct extensive experiments for CANRT. The results show that CANRT outperforms the server-centric persistent radix trees by 1.19x-1.22x for remote reads, 1.67x–1.72x for remote writes and 3.10x–3.83x for range

query in terms of latency. For concurrent throughput, CANRT outperforms the counterparts by up to 11.15x under write-intensive workloads.

Organization. The rest of this paper is organized as follows. Section 2 introduces the background and motivation of our work. Section 3 describes the design and implementation of CANRT. Section 4 presents experimental results and we conclude this paper in Sect. 5.

2   Background and Motivation

2.1   Non-Volatile Memory

Emerging memory technologies such as phase change memory(PCM) [17], spin-transfer torque RAM (STT-RAM) [16] and the recently released Intel Optane DC Persistent Memory [8], can be directly accessed through the memory bus via processor loads and stores, delivering DRAM-like performance while providing persistent data storage [3,15]. NVM provides potential to design efficient persistent index structures [1,7,12,14,19], storage systems [2,3,11,15,18] and etc, with leveraging its non-volatility, byte-addressability and DRAM-comparable latency. Data consistency is a significant issue for NVM as there may be partial writes and reordering writes during system crash, which will lead to inconsistent data state and unexpected system error. The mainstream solutions are using logging or copy-on-write techniques to guarantee data consistency [3,11,15]. The persistence ordering is maintained by using $clflush/clwb/clflushopt$ and $mfence$ instructions.

2.2   Remote Direct Memory Access

Remote direct memory access (RDMA) is a new networking technology that supports low-latency, high-bandwidth, zero-copy and kernel-bypass access to remote memory region. Since Reliable Connection (RC) provides full support for RDMA semantics, in this paper, we mainly focus on the RC mode of using RDMA.

Normally, there are two types of RDMA semantics: message semantics and memory semantics. Message semantics employ SEND/RECV verbs for user-level message exchange. Due to the requirement of the matching RECV request posted at the server side, these verbs are generally referred as two-sided RDMA operations. Memory semantics, on the other hand, are one-sided semantics. These semantics utilize READ/WRITE verbs and are proved to achieve higher throughput and lower latency than message semantics [10]. Adopting these semantics is attractive in that they free server from sending requested data to clients, replying acknowledge information or coordinate collisions, as well as continuously polling for the incoming requests [5,6,9,13].

Apart from above semantics, RDMA also support one-sided atomic operations, including FETCH_AND_ADD (FAA) and COMPARE_AND_SWAP (CAS). Both of these verbs operate on data with 64 bits and are atomic operations relative to other operations on the same NIC. This paper utilizes FAA to support concurrent remote write operations to the radix tree.

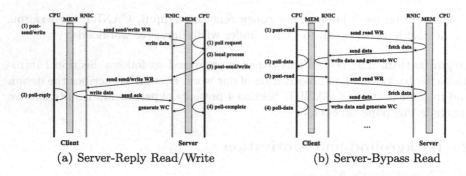

(a) Server-Reply Read/Write   (b) Server-Bypass Read

Fig. 1. Remote access mechanisms for current tree-based indexing structures

2.3   RDMA-enabled Index Structure

A few recent researches have used RDMA technology in key-value stores to accelerate the remote data access [2,5,6,9,13]. However, in these designs, the fundamental indexing structures are all hash tables. In fact, few researches have been proposed to optimize the current tree-based indexing structures using RDMA to fit in distributed environment for efficient remote data access. This can be attributed to three challenges.

First, *unawareness of data location.* Since traversing through an indexing tree means the retrieval of data at multiple memory addresses, it is relatively harder to locate the target data than hash tables. Second, *data consistency.* The employment of NVM raises consistency issues. While tree structure is often more complicated compared to hash table, it demands more careful design to avoid such problems. Third, *concurrent write collision.* Since the write operations to a tree tend to affect multiple nodes, concurrent writes are easy to conflict, which as a result can harm the concurrent performance.

2.4   Persistent Radix Tree

Most of the previous proposed persistent trees are variants of B+-tree, such as NVTree [19], wB+tree [1], FPTree [14] and FAST&FAIR [7]. Lee et al. [12] proposed three variants of a radix tree: WORT, WOART and ART+CoW. It is pointed out that radix tree is actually more appropriate for NVM compared to B+-tree variants. For instance, key comparisons in a radix tree are no longer required since the radix tree structure is decided only by each character of the inserted keys. Furthermore, updates in node granularity and expensive tree rebalancing operations are also avoided. The experimental results of [12] show that the radix tree variants outperform most of the current B+-tree variants(i.e., NVTree, wB+Tree and FPTree).

Although WORT has already improved the indexing performance on NVM, it cannot be directly deployed in an RDMA-enabled distributed system environment since it is designed for a single machine environment. Fortunately, similar to previous hashtable-based key-value stores [5,9,13], we can apply the server-reply paradigm as a general RPC mechanism for WORT to support remote read and write requests. However, there are two performance limitations: 1) high server CPU overhead/high latency. Since data access through a radix tree means the traversing of the tree, clients require the server to coordinate and process requests to access target data. Taking read request as an example, there are two methods to execute it which are illustrated in Fig. 1. One method is to let the server collect all the requested keys and send back to the client (shown in Fig. 1(a)). Although it can be performed in 1 RTT, the server CPU overhead will be quite high for concurrent access and thus incurs high operation latency. The other is to allow the server to only reply addresses to the client and it is the client's responsibility to complete the reads using one-sided READ verb (shown in Fig. 1(b)). It is almost impossible for clients to directly access data in the server side within 2 RTTs, which also leads to high operation latency and heavy bandwidth consumption. As a result, either significant server CPU overhead or high operation latency will be incurred. 2) poor concurrency. Concurrent access is demanded in most distributed database systems or key-value store systems. Unfortunately, WORT does not support concurrent access requests, which is harmful to both system throughput and latency performance.

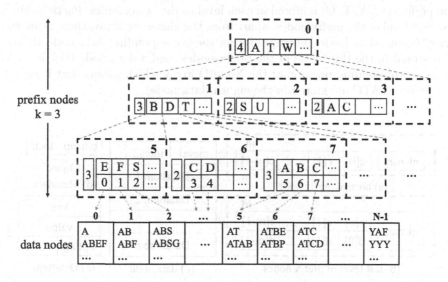

Fig. 2. CANRT storage architecture

3   CANRT

3.1   CANRT Data Structures

Figure 2 depicts the server-side storage architecture of CANRT. Basically, CANRT is composed of two major parts:

Prefix nodes, which exist in both the server and clients, are organized in a pointer-less format. In CANRT, all the prefix nodes are pre-allocated in a consecutive memory space in the server, and will be sent to each client when the tree is reorganized. Figure 3 (a) and Fig. 3 (b) illustrate the layouts of prefix nodes. CANRT does not store explicit keys in its prefix nodes. Instead, a prefix node consists of an array of characters, each of which represents one character of the inserted keys. Additionally, each prefix node contain an *nChars* field, which records the number of valid characters stored in that node.

Data nodes, which exist only in the server and contain actual key-value pairs, can be accessed through the node ID stored in the last level of the prefix nodes. Each key-value pair in the data node is encapsulated in a DataItem entry and the DataItems in each data node are kept unsorted. As depicted in Fig. 3(c) and Fig. 3(d), to both accelerate data access and enable high write concurrency, an 8-byte metadata structure is adopted in the data node. The metadata contains a bitmap and a write lock.

In CANRT, only the first k characters of the inserted keys will be stored in the prefix nodes, which we call the prefix of the key. Here, k also denotes the height of the prefix nodes. Figure 2 shows an example of a CANRT with $k = 3$. Taking the key ATCD as an example, the prefix is ATC, and each character of the prefix (i.e., A, T, C) is stored in each level of the prefix nodes. Particularly, in the last level of the prefix nodes, apart from the character array, there is another array (denoted as NodeID array) storing the corresponding data node ID (i.e., C is stored in the third level of the prefix nodes, and a data node ID 7 is stored in the corresponding position of the NodeID array which means that keys with the prefix of ATC are stored in the eighth data node).

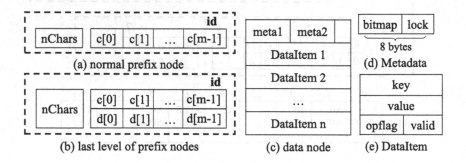

Fig. 3. CANRT node layout

3.2 Locating Target Data Node

Both read and write operations in CANRT need to find the target data node first. In CANRT, the position of each prefix node is fixed and the node ID of each prefix node increases sequentially from zero. Hence it can be used to locate the target prefix node in each level and finally find the target data node ID.

Generally, the i_{th} $(i < k)$ character of the search key will be used to probe the target node in $(i + 1)_{th}$ level, and the k_{th} character will finally locate the data node which contains the search key. For instance, with each prefix node contains m characters, the i_{th} level of prefix nodes contains m^{i-1} nodes and the base node ID in $(i + 1)_{th}$ level (denoted as b_{i+1}) can be calculated by

$$b_{i+1} = b_i + m^{i-1} \tag{1}$$

where b_i is the base node ID in i_{th} level. Based on the recursive relations in (1), we can calculate the value of b_i by the following equation:

$$b_i = \frac{m^{i-1} - 1}{m - 1} \tag{2}$$

Now we can locate the target node ID in $(i+1)_{th}$ level (denoted as I_{target}) using the equation below:

$$I_{target} = b_{i+1} + (I_{cur} - b_i) * m + p \tag{3}$$

where I_{cur} denotes the node ID of the current node in i_{th} level and p is the position of the i_{th} character of the search key in the character array of the current node. When $i = k$, the current node is in the last level of prefix nodes, and the data node ID can be directly obtained from the NodeID array of current node with the position p.

In the cases when the i_{th} character of the request key does not exist in the corresponding character array, the search process will be terminated and the target data node ID will be set as the node ID of the last data node. Once the target data node ID is obtained, the address of the target data node can be calculated based on the memory address of the first data node and the size of each data node.

3.3 Fine-Grained Lock-Based Remote Write

To support high-concurrency remote writes, we design a fine-grained lock-based remote write mechanism. It employs server-bypass paradigm using one-sided RDMA verbs and makes append-only updates to the server's persistent region in order to avoid any data corruption. There are three phases in remote write procedure, all of which are client-active: locking phase, checking phase and writing phase. The whole remote write procedure is provided in Algorithm 1.

In the locking phase (lines 1 to 9), the client first calculates the target data node ID with locally-buffered prefix nodes. Then it uses a FETCH_AND_ADD verb to obtain the 8-byte metadata of the target data node from the server, while atomically adding 1 to the *lock* field of metadata. If the fetched *lock* value is greater than 0, it means that the target data node has been locked by another client or the server itself. In such cases, the client will wait for random backoff

---

Algorithm 1. remote_write(key, value)

1: /* locking phase: fetch the metadata of target data node */
2: node_id ← findDataNode(key);
3: raddr ← data_base + datanode_size * node_id + shadow_meta_off;
4: len ← 8;
5: post_rdma_fetch_and_add(raddr, len, laddr)
6: meta ← laddr;
7: **if** meta.lock > 0 **then**
8:    retry from step 5 after random backoff;
9: **end if**
10: /* checking phase: find a free bit in the bitmap */
11: **for** bit in meta.bitmap **do**
12:    **if** bit == 0 **then**
13:       /* writing phase: write a new DataItem to the target node */
14:       generateDataItem(OPFLAG, key, value, 1);
15:       raddr ← raddr+meta_size+bit_index * dataitem_size;
16:       len ← dataitem_size; imm ← (node_id << 8 | bit_index);
17:       post_rdma_write_with_imm(raddr, len, laddr, imm);
18:       **return**
19:    **end if**
20: **end for**
21: imm ← (node_id << 8 | RESIZE_FLAG);
22: post_send_with_imm(imm);
23: retry from step 1 after being notified a resizing accomplishment;

---

time (e.g., 1 us), after which the lock operation will be replayed. Otherwise, current client successfully locks target node and enters the checking phase.

In the checking phase (lines 10 to 12 and 19 to 23), the client will verify if there are free bits (i.e., 0-bit) in the *bitmap* field. If there is one or more free bits, the client proceeds to the writing phase. Otherwise, it indicates that the target node is full (i.e., there is no free DataItem slot), the client will inform the server to resize the prefix nodes (see Sect. 3.5) and resending the request after the server informs the accomplishment of resizing.

In the writing phase (lines 13 to 18), the DataItem location corresponding to the first free bit in the *bitmap* will be regarded as the target location to append a new write. For all types of remote write operations, including insert, delete and update, the write steps in the client side are almost the same, except the *opflag* value in each packed DataItem to be appended. The *opflag* indicates the specific operation type. For example, insert, delete and update can be represented by 1, 2 and 3, respectively. The client then immediately transmits the newly-generated DataItem to the target location in the server using a WRITE_WITH_IMM verb. The carried 32-byte immediate informs the server of both the node ID of the target data node and the specific location of the newly inserted DataItem in the data node. From the perspective of clients, a remote write operation is completed once the WRITE_WITH_IMM request has been successfully posted, which incurs no server CPU involvement in the critical path.

Algorithm 2. remote_read(key, find_value)

```
1: /* fetch both metadata and data*/
2: node_id ← findDataNode(key);
3: raddr ← data_base + datanode_size * node_id;
4: len ← datanode_size;
5: post_rdma_read(raddr, len, laddr);
6: datanode ← laddr;
7: (meta, data) ← datanode;
8: /* local search */
9: find_value ← NULL ;
10: for i ← 0, i < dataitem_num, i++ do
11:    if meta.bitmap[i] == 1 && data[i].key == key then
12:       find_value ← data[i].value;
13:       return find_value;
14:    end if
15: end for
16: return NULL;
```

In the server side, a background thread will process the appending messages from clients, which is out of the critical path of client request execution. When polling a work completion (WC), it will first locate the newly inserted DataItem according to the immediate data, and then checks the *opflag* field to determine the write operation type. For insert or delete operation, the server simply atomically modifies the metadata of the data node by changing the corresponding bit (i.e., 0 to 1 for insert and 1 to 0 for delete) in the *bitmap* and resetting the *lock* field to 0. For update operation, the server will mark the old DataItem as invalid and validate the newly inserted DataItem at the same time with two bit flips. For all types of write operations, the server will set the *valid* field in the DataItem to 0 once the metadata update completes. To maintain persistence ordering for data consistency, we utilize *clwb* and *mfence* primitives accordingly.

Notice that a remote write operation is only visible after the corresponding metadata in the server is modified, which avoids data inconsistency during system crash. Since the modification of the metadata is an 8-byte atomic write operation, the expensive logging or copy-on-write is unnecessary. The fine-grained lock-based remote write mechanism of CANRT has two merits: 1) it enables clients to write data without the involvement of the server CPU, which considerably reduces the server processing overhead in the critical path; 2) it allows DataItems in each data node to be unsorted and no data shift is required, hence reducing extra writes to NVM.

3.4 Node-Oriented Remote Read

Instead of simply accessing one DataItem, CANRT fetches an entire node from the server-side for each single read. Remote Read operations also start with locating the target data node. As introduced in Sect. 3.2, the target node locating can be conducted in the client directly. After obtaining the target data node ID,

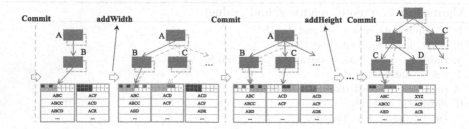

Fig. 4. Non-blocking resizing

the client posts a READ request to fetch the target data node in the server radix tree. Algorithm 2 illustrates the remote read mechanism of CANRT. Since only one RDMA operation is posted, the total network overhead is only 1 RTT. After fetching the required data node to local buffer, the client then checks the *bitmap* entry in the metadata field and scans the valid DataItems to retrieve the matched DataItem.

Although the unsorted DataItems increase the searching time inside the data node, the searching performance is still acceptable since the number of DataItems inside a data node is limited (see Sect. 4). Thanks to the bitmap-assisted strategy, the read operations will never be blocked by any concurrent write operation since the search process will only check the DataItems whose bit value is one in the *bitmap*.

3.5   Non-blocking Resizing

As the size of each data node is fixed, when one data node is full, the server need to allocate more space to hold more keys. As such, existing keys in the tree may be reorganized and migrated to newly-generated data nodes. This procedure is called radix tree resizing. The resizing can be triggered either by the server itself or the notification of a client, as lines 21–22 in Algorithm 1 indicate.

We propose a non-blocking resizing scheme for CANRT. That is, even during data migration, the radix tree in the server can still be accessed by remote clients. The key idea is using a tick-tock design (see Fig. 4) inspired by Redis dictionary [20] to avoid any modifications to current tree structure during data migration. Specifically, CANRT maintains two versions of radix tree, one version as the valid tree and the other as the shadow tree. Any read or write operation is performed on the valid tree and the shadow tree is exclusively designed for resizing.

In CANRT, there are two types of resizing: addWidth and addHeight. When the last data node is full, addWidth resizing is performed to store more characters in each level of prefix nodes. Otherwise, the resizing procedure is launched in the addHeight manner. Both addWidth and addHeight will not affect the old data nodes, the newly generated data nodes will simply be appended to the end of old data nodes and the DataItem migration will be reflected by the metadata.

With the shadow tree, the resizing scheme in CANRT is non-blocking for read operations, and only a portion of write operations will be blocked due to locking failure to the full data nodes. Compared to read and write operations, the resizing procedure seems time-consuming in large CANRT, but the frequency of resizing can be significantly dropped with more and more prefixes inserted into the prefix nodes (6.20% when inserting one thousand keys and 1.47% when inserting one million keys). We believe that such overhead is acceptable.

3.6   Crash Recovery

Inconsistency may occur when system crashes during remote write operations or resizing procedure and thus recovery is required. The recover procedure starts with detecting the locked data nodes and then checks the *valid* field of each DataItem in those data nodes. The *valid* field is the last bit of each DataItem and indicates both the integrity of the DataItem and whether the server side process for that DataItem is completed. If the *valid* value is 1, which means the DataItem is not processed, the server will simply execute server side metadata update, as discussed in Sect. 3.3. If the valid value is 0, it means current DataItem 1) has already been processed by the server or 2) is an incomplete write from the client. In either case, the server-side data state is consistent.

After fixing the remote write inconsistency, the server will then check the resizing consistency. We use two state flags (i.e., *pxflag* and *dtflag*) to mark the completion of prefix nodes reconstruction and data migration, respectively. Only when the two flags are both marked, the system crash does not occur during resizing procedure. Otherwise, the recover procedure should be performed. When *pxflag* is unmarked, it means that the resize procedure just starts, and the server will simply replay the whole resizing procedure. When *pxflag* is marked and *dtflag* is unmarked, it means that system crashes after the reconstruction of the prefix nodes but before the completion of data migration. In such cases, the server will perform the data migration recovery. The procedure ends with atomically marking the *dtflag* which transforms the shadow tree to the new valid tree.

4   Evaluation

4.1   Experimental Setup

Different from previous works that leverage a DRAM-based PM performance emulator (e.g., Intel PMEP/Quartz) to emulate PM latency [1,7,12,14], we conduct our experiments on a small cluster of three machines, which are all equipped with commercial Intel Optane DC Persistent Memory, and Intel Xeon(R) 2.60 GHz CPU (32 KB/1 MB/24 MB L1/L2/L3 cache). Our experiments are performed in the AppDirect mode of AEP, which exposes a separate

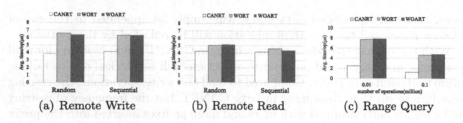

Fig. 5. Remote read and write performance comparisons

persistent memory device. As for RDMA hardware configuration, all of these machines are equipped with Mellanox ConnectX-5 InfiniBand NIC and run on Ubuntu 18.04 (kernel version 4.18.04). For each experiment, we use one machine as server, while the remaining two machines work as clients. Since 16-byte key has been broadly adopted in current key-value stores [18], we set the key and value in CANRT to be 16 bytes and 15 bytes, respectively.

For comparison, we implement both WORT and WOART by adding RDMA mechanism to it. To accelerate remote access, similar to previous RDMA-enabled hash works [5,9], we design a ring buffer mechanism for both WORT and WOART. That is, for each client, the server keeps a request ring buffer and for both remote read and write procedure, clients only need to write requests to the ring buffer using a WRITE_WITH_IMM verb. The server dedicatedly checks the corresponding client ring buffer and makes processing once a request is found. Afterwards, the server will inform the client using a SEND verb. The procedure ends once the client receives the server's reply message. We evaluate the operation latency with random String workload and test throughput with the widely-used YCSB [4] benchmark. Each result is averaged for 10 runs.

4.2 Remote Write Latency

Figure 5(a) shows the comparison of remote write performance among CANRT and the other two radix tree variants under random and sequential workloads. For random workloads, CANRT is roughly 1.7x faster than the other two trees. Although it seems that CANRT requires more RTTs (i.e., about 1.5 RTTs) in one remote write operation, it is steady and little affected by the server's CPU usage. For WORT and WOART, apart from 1 RTT (i.e., 0.5 RTT for client's WRITE and 0.5 RTT for server's SEND), waiting for the server's reply also incurs heavy time consumption, including request polling, key locating, data persistence, and posting reply message. All these processing steps are in the critical path of request execution. Since each operation requires interaction between the client and the server, the latency is more unstable compared with CANRT. Although WOART has better local write performance than WORT, the relatively much higher network overhead makes this difference no longer noticeable. For sequential workloads, the performance of CANRT is dropped by 7.94% compared with random workloads. This is because sequential workloads contribute to a higher probability of data node access conflict.

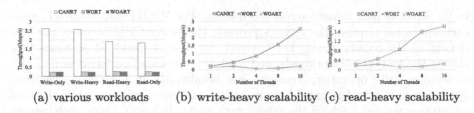

(a) various workloads   (b) write-heavy scalability (c) read-heavy scalability

Fig. 6. Throughput and scalability comparisons under YCSB workloads

4.3 Remote Read Latency

Figure 5(b) compares the remote write latency of three radix trees. We observe that CANRT outperforms its counterparts by 1.19x-1.22x. It is also noticed that the read latency of CANRT is slightly higher than the write latency shown in Fig. 5(a). Although each remote read operation of CANRT only consumes one RTT, the size of the data transferred per operation is larger compared with write operations. For WORT and WOART, the server-reply mechanism incurs higher latency than CANRT due to the processing overhead in the critical path. For sequential workloads, the latency of WORT and WOART is lower than the latency under random workloads. The reduction of server side cache misses under sequential workloads contributes to the improvement of WORT and WOART. Although there is almost no performance improvement for CANRT under sequential workloads, the total latency of CANRT remote read operations is still lower than WORT and WOART.

4.4 Range Query Latency

Since DataItems with same prefix are stored in the same data node in CANRT, we integrate consecutive requests for the same data node into one request, which significantly improves the range query performance. For WORT and WOART, due to the structure limitation, the remote range query functions are simply implemented by calling a remote read request for each key. We collect the range query performance by querying 10000 and 100000 keys respectively, and the results shown in Fig. 5(c) demonstrate that CANRT achieves much better performance for range query. When querying 10000 keys, CANRT is roughly 3.1x faster compared with WORT or WOART. When querying 100000 keys, CANRT is 3.71x faster than WORT and 3.82x faster than WOART.

4.5 Concurrent Throughput

In the experiments shown in Fig. 6(a), we evaluate the concurrent throughput under four types of YCSB workloads: Write-Only (100% update), Write-Heavy (50% read + 50% update), Read-Heavy (95% read + 5% update) and Read-Only (100% read). The experiments are conducted using 16 threads. We can observe that CANRT consistently outperforms the other two radix trees due to its server-bypass remote access mechanism. For WORT and WOART, since each operation requires the server's process, the server CPU becomes the bottleneck of the remote access. For CANRT, write-only workload performs best among the four YCSB workloads, and read-only workload performs worst. This is because write operations consume less bandwidth. In CANRT, each write operation only transfers one 8-byte metadata and one DataItem while the read operations require fetching the whole data node. However, since each remote read operation only requires one RTT and the number of DataItems in one data node is limited, the remote read throughput is still acceptable.

Figure 6(b) and Fig. 6(c) show the scalability performance of three radix trees under write-heavy and read-heavy workloads, respectively. Due to the server CPU limitation, WORT and WOART exhibit poor scalability and they are unable to saturate the network bandwidth. By contrast, the throughput of CANRT increases with the number of client threads. For Write-Heavy workloads, Fig. 6(b) shows that when the number of threads increases from 2 to 16, the throughput of CANRT is increased by 10.88x, which almost increases proportionally to the number of threads. However, for read-heavy workload, the experimental results in Fig. 6(c) show that the performance of CANRT is increased by a factor of 7.93, when the number of threads increases from 2 to 16, which increases proportionally to the number of threads only when the number of threads is less than 8. The reason is that when the number of client threads reaches 16, the network bandwidth becomes the bottleneck of remote access. Notice that remote read operations carries larger packet each time than remote write and hence the read throughput bottleneck is reached earlier.

5   Conclusion

In this paper, we propose CANRT, a client-active NVM-based radix tree for fast remote access using RDMA technology. We decouple the radix tree nodes into data nodes and prefix nodes to exert the full potential of both NVM and RDMA. We design a node-oriented read mechanism for low-latency remote reads and a fine-grained lock-based write mechanism for fast remote writes. The experimental results show that CANRT performs better than server-centric persistent radix trees in terms of both remote access latency and throughput.

Acknowledgements. This work is supported by the National Key Research and Development Program of China (No. 2018YFB1003302), SJTU-Huawei Innovation Research Lab Funding, and the China Scholarship Council (No. 201906230180).

References

1. Chen, S., Jin, Q.: Persistent b+-trees in non-volatile main memory. Proc. VLDB Endow. **8**(7), 786–797 (2015)
2. Chen, Y., Lu, Y., Yang, F., Wang, Q., Wang, Y., Shu, J.: Flatstore: an efficient log-structured key-value storage engine for persistent memory. In: Proceedings of the Twenty-Fifth International Conference on Architectural Support for Programming Languages and Operating Systems, pp. 1077–1091 (2020)

3. Coburn, J., et al.: Nv-heaps: making persistent objects fast and safe with next-generation, non-volatile memories. ACM SIGARCH Comput. Archit. News **39**(1), 105–118 (2011)
4. Cooper, B.F., Silberstein, A., Tam, E., Ramakrishnan, R., Sears, R.: Benchmarking cloud serving systems with YCSB. In: Proceedings of the 1st ACM symposium on Cloud computing, pp. 143–154 (2010)
5. Dragojević, A., Narayanan, D., Castro, M., Hodson, O.: Farm: fast remote memory. In: 11th {USENIX} Symposium on Networked Systems Design and Implementation ({NSDI} 14), pp. 401–414 (2014)
6. Huang, H., Huang, K., You, L., Huang, L.: Forca: fast and atomic remote direct access to persistent memory, pp. 246–249 (2018). https://doi.org/10.1109/ICCD.2018.00045
7. Hwang, D., Kim, W.H., Won, Y., Nam, B.: Endurable transient inconsistency in byte-addressable persistent b+-tree. In: 16th {USENIX} Conference on File and Storage Technologies ({FAST} 18), pp. 187–200 (2018)
8. Intel: Intel optane dc persistent memory (2019). https://newsroom.intel.com/news-releases/intel-data-centric-launch/
9. Kalia, A., Kaminsky, M., Andersen, D.G.: Using rdma efficiently for key-value services. In: ACM SIGCOMM Computer Communication Review, vol. 44, pp. 295–306. ACM (2014)
10. Kalia, A., Kaminsky, M., Andersen, D.G.: Design guidelines for high performance {RDMA} systems. In: 2016 {USENIX} Annual Technical Conference ({USENIX}{ATC} 16), pp. 437–450 (2016)
11. Kim, W.H., Kim, J., Baek, W., Nam, B., Won, Y.: NVWAL: Exploiting NVRAM in write-ahead logging. ACM SIGPLAN Not. **51**(4), 385–398 (2016)
12. Lee, S.K., Lim, K.H., Song, H., Nam, B., Noh, S.H.: {WORT}: Write optimal radix tree for persistent memory storage systems. In: 15th {USENIX} Conference on File and Storage Technologies ({FAST} 17), pp. 257–270 (2017)
13. Mitchell, C., Geng, Y., Li, J.: Using one-sided {RDMA} reads to build a fast, CPU-efficient key-value store. In: Presented as part of the 2013 {USENIX} Annual Technical Conference ({USENIX}{ATC} 13), pp. 103–114 (2013)
14. Oukid, I., Lasperas, J., Nica, A., Willhalm, T., Lehner, W.: FPTree: a hybrid SCM-dram persistent and concurrent b-tree for storage class memory. In: Proceedings of the 2016 International Conference on Management of Data, pp. 371–386. ACM (2016)
15. Volos, H., Tack, A.J., Swift, M.M.: Mnemosyne: lightweight persistent memory. ACM SIGARCH Comput. Archit. News **39**(1), 91–104 (2011)
16. Wang, K., Alzate, J., Amiri, P.K.: Low-power non-volatile spintronic memory: STT-RAM and beyond. J. Phys. D Appl. Phys. **46**(7), 074003 (2013)
17. Wong, H.S.P., et al.: Phase change memory. Proc. IEEE **98**(12), 2201–2227 (2010)
18. Xia, F., Jiang, D., Xiong, J., Sun, N.: Hikv: a hybrid index key-value store for dram-nvm memory systems. In: 2017 {USENIX} Annual Technical Conference ({USENIX}{ATC} 17), pp. 349–362 (2017)
19. Yang, J., Wei, Q., Chen, C., Wang, C., Yong, K.L., He, B.: Nv-tree: reducing consistency cost for nvm-based single level systems. In: 13th {USENIX} Conference on File and Storage Technologies ({FAST} 15), pp. 167–181 (2015)
20. Zawodny, J.: Redis: lightweight key/value store that goes the extra mile. Linux Mag. **79**(8), 1–10 (2009)

Distributed and Parallel Ensemble Classification for Big Data Based on Kullback-Leibler Random Sample Partition

Chenghao Wei[1], Jiyong Zhang[2], Timur Valiullin[1], Weipeng Cao[1(✉)],
Qiang Wang[3(✉)], and Hao Long[1]

[1] Big Data Institute, College of Computer Science and Software Engineering,
Shenzhen Univresity, Shenzhen 518000, China
caoweipeng@szu.edu.cn
[2] School of Automation, Hangzhou Dianzi University, Hangzhou 311305, China
[3] SUSTech Academy for Advanced Interdisciplinary Studies,
Southern University of Science and Technology, Shenzhen 518055, China
wangq8@sustech.edu.cn

Abstract. In this article, we use a Kullback-Leibler random sample partition data model to generate a set of disjoint data blocks, where each block is a good representation of the entire data set. Every random sample partition (RSP) block has a sample distribution function similar to the entire data set. To obtain the statistical measure between them, Kernel Density Estimation (KDE) with a dual-tree recursion data structure is firstly applied to fast estimate the probability density of each block. Then, based on the Kullback-Leibler (KL) divergence measure, we can obtain the statistical similarity between a randomly selected RSP data block and other RSP data blocks. We rank the RSP data blocks according to their divergence values in descending order and choose the first ten for an ensemble classification learning. The classification models are established in parallel for the selected RSP data blocks and the final ensemble classification model is obtained by the weighted voting ensemble strategy. The experiments were conducted by building XGboost model based on those ten blocks in parallel, and we incrementally ensemble them according to their KL values. The testing classification results show that our method can increase the generalization capability of the ensemble classification model. It could reduce the model building time in parallel computation environment by using less than 15% of the entire data, which could also solve the memory constraints of big data analysis.

Keywords: Big data analysis · Approximate computing · Random sample partition · Ensemble classification · Parallel distributed computing

This work was supported by National Natural Science Foundation of China (61836005), and the Opening Project of Shanghai Trusted Industrial Control Platform (TICPSH202003008-ZC).
C. Wei and J. Zhang—Joint first authors.

M. Qiu (Ed.): ICA3PP 2020, LNCS 12452, pp. 448–464, 2020.
https://doi.org/10.1007/978-3-030-60245-1_31

1   Introduction

One of the biggest challenges of big data analysis is how to perform complex computation tasks under certain computing resources and within an acceptable time range. Divide and conquer is the main strategy to deal with big data computing, that is to divide big data into several small data partitions and store them on cluster nodes for analysis [1]. The Hadoop Distributed File System (HDFS) is a distributed file system designed to run on distributed computation nodes, which is highly fault-tolerant and can be deployed on low-cost hardware [2]. For big data computing, MapReduce [3] is a programming model, where users can specify a map function that processes a key/value pair to generate a set of intermediate key/value pairs, and a reduce function that merges all intermediate values [4]. However, MapReduce programs will result in high I/O cost, it reads input data from disk, maps a function across the data, reduces the results of the map function, and stores reduction results on disk. Spark [5] and its RDDs were developed in 2012 in response to limitations of the MapReduce. The RDDs function as a working set for distributed programs offers a restricted form of distributed shared memory. In such a way, memory structure can be used to read big data into node memory for calculation, which avoids the repeated read and write disk operations in MapReduce. Although Spark provides many optimized upper-level libraries, e.g., Sparks MLlib for machine learning, GraphX for graph analysis, Spark Streaming for stream processing and Spark SQL for structured data processing [6], the execution efficiency of spark algorithm will be greatly reduced by iterative operation, especially when the size of the data is heavily over the memory of cluster [7]. As a result, memory resources become a big data analysis problem.

Can we use sampling techniques for big data analysis in order to avoid the bottleneck of memory resources? In the application of data mining, it is unnecessary to use the entire big dataset for analysis. A big dataset is an N * d matrix that contains N observations with d variables. N is extremely high, while d may be high or not, depending on the context from which the big data originate. A good sampling technique needs to be designed for the big data analysis with consideration of cluster storage, I/O costs and manageable size for processing [8]. Current mainstream cluster computing frameworks and engines, such as Apache Hadoop[1], Apache Spark[2] and Apache MapReduce[3], were implemented by a shared-nothing architecture, where each node is independent in terms of both data and resources. The MapReduce and Spark computing models are adapted to the distributed HDFS data blocks. With such architecture, Record-Level Sampling (RLS) from an HDFS file becomes time-consuming because selecting records with equal probability requires scanning the entire data. Instead of sampling the big data records, Salman proposed the random sample partition (RSP) storage strategy as an alternative, which enables block-level sampling to

---

[1]   Available: http://hadoop.apache.org/.

[2]   Available: https://spark.apache.org/.

[3]   Available: http://hadoop.apache.org/docs/current/hadoop-mapreduce-client.

speed up big data analysis and it can be also used to estimate the statistical properties of the entire big data [9].

By increasing the diversity of base classifiers in terms of data disturbance and attribute disturbance, ensemble learning [10] is to combine several weak training models to get a better and more comprehensive model for data analysis. Based on the idea of "good but different", the generalization performance of the combined learner is greatly improved. Several research works of ensemble learning have been done for big data analysis. Ensemble composed of the decision tree, the gradient boosted trees and the random forest has been proposed for multi-step forecasting of big data time series [11]. An ensemble learning based approach towards big data Bayesian network was provided in [12]. A label-aware distributed ensemble learning strategy was proposed for large scale dataset in [13]. A low-latency multi-threaded ensemble learning method was proposed for dynamic big data streams [14]. However, most of them are based on the implementation of the ensemble algorithm in parallel and few of them consider the reduction of the computational costs. From the view of big data computation, neither Bagging nor Boosting can avoid the bottleneck of memory resources due to the inevitable usage of the entire big dataset. Although an asymptotic ensemble learning framework based on RSP block for big data analysis was proposed for reduction of processing data [15], the learning framework lacked consideration of the statistical difference between the generated partitions. Meanwhile, it did not answer the question of how many blocks should we use in the analysis task. Selective ensemble learning is mainly based on some criteria, and it selects some of the existing base classifiers to speed up the prediction and classification as well as reduce the storage space requirements. It is better to ensemble well selected base learners instead of all of the learners at hand [16] that could decrease the demands of computer memories and I/O costs. The selective ensemble learning algorithm can be classified into three types, which are clustering based, ranking based and optimized based algorithms. Clustering based ensemble learning algorithms usually define a distance to measure the similarity between based learners for searching those learners with consistent results [17]. In ranking ensemble learning algorithms, base classifiers are sorted based on a certain index, e.g., weighted training error [18], attribute subset [19], degree of the signature vector [20] and out-of-bag degree [21], a certain number of base classifiers are then selected by appropriate stopping criteria. The ensemble learning algorithm based on optimization strategy gives weight to based classifiers in the process of base learners merging. It selects the best classifier by sparsity constraint and optimization algorithm [16,22]. However, most of those selective ensemble strategies are based on iterative calculation and unavoidable to access the entire data repeatedly. Compared to the high utilization of memory computing resources, additional usage of hard disk storage resources can be a trade-off strategy for solving memory bottleneck. Random sample partition data block [9], which has the same sample distribution of the entire dataset, were used for analysis. In this paper, we first use the Kernel Density Estimation (KDE) method to estimate the probability density function of the generated RSP block

[9], and then Kullback–Leibler (KL) divergence is applied to the measure similarity between RSP blocks, which is a key factor for choosing the data blocks and ensure the data diversity during ensemble learning.

In Sect. 2, data storage in HDFS and Spark framework are introduced additionally, sampling method for big data is presented. Then, KDE and KL divergence measures are discussed. Section 3 contains the mathematical definition of KL-RSP data storage model, and relevant proof of distribution consistency is provided. Section 4, we introduce the ensemble classification for KL-RSP learning model for approximate big data analysis and discuss its performance and extensions. In Sect. 5, we provide the our experiment results. The final conclusion is presented in the Sect. 6.

2 Preliminaries

2.1 Data Storage in HDFS and Spark

A big data set is sequentially chunked into small data blocks, they are distributed saved on the nodes of a cluster by HDFS [2]. For big data analysis, all the data blocks belonging to the same big data file will be processed in parallel on the cluster. By integrating the intermediate results from the local data blocks processed on local nodes, the final analysis results are obtained for the entire data set. An HDFS file has a fixed size for each small data block, e.g., 64 MB or 128 MB. For data safety, three copies of the same data block are usually stored on three different nodes. The resilient distributed data set (RDD) [6] is introduced by Apache Spark for providing processing-level distributed data abstractions. It is held in memory to facilitate the data processing and analysis. The RDDs APIs are provided to import HDFS files or other data files to Spark RDD, it controls RDD partitioning and manipulates RDD using a rich set of operators. Different partitioning methods such as hash partitioner can be used for the obtained RDD. However, both HDFS data blocks and Spark RDD do not well cover the statistical characteristics of the entire big data. So, the data blocks in HDFS files and Spark RDDs cannot directly use in approximate analysis due to their statistical deviation.

2.2 Sampling Method for Big Data

Sampling is the basis of statistics theory, which extracts one part from all the respondents according to a certain procedure and do estimation for the original data. The evaluation of the obtained samples is mainly carried out in two aspects. The first one is the value coverage of the sample. We need to obtain the value range of each attribute in the sample, and then compare it with the corresponding attribute value range in the original data. According to the comparison results, we can get the coverage rate of the obtained sample data. Secondly, the distribution of sample data, which is quite important for machine learning model. If the strategy is random sampling, it is expected that the distribution

Table 1. Common sampling methods

Bernoulli sampling [23,24]	Extract data item with equal probability and random sample size
Simple random sampling [25,26]	Extract data item with equal probability and fix sample size
Stratified random sampling [27]	Population was divided into several layers and samples were extracted from each layer by simple random sampling
Boostrapping [28]	Extract data items with replacement

of sample data is similar to that of population data. Table 1 summarizes all the common sampling methods. Bernoulli sampling method [23] operates without replacement and the sample size is random and not fixed, which is difficult to estimate the processing latency. Simple random sampling [25] performed with replacement, which allows each data item to appear one time at most. However, simple random sampling does not ensure that each group in the original data is considered fairly in the sample. Compared to simple random sampling, stratified sampling [27] provides higher statistical precision and reduces the sampling error. Boostrap [28] uses multiple samples with replacement from the observed dataset, but it requires high computational and I/O costs as it depends on repeatedly drawing samples of sizes comparable to the original dataset. Unlike those sampling strategies without consideration of cluster computation capability, the RSP model uses additional storage for saving big data with HDFS blocks, whose distribution is quite similar to that of the entire data set. Figure 1 shows the distribution of each block after RSP data transformation process. There are two advantages by using RSP blocks. Firstly, the distribution of the big data itself is hard to obtained and will cost an extra high amount of computation. RSP blocks can be used to approximately estimate the density distribution of the big data and build ensemble models without computing the entire data set. Secondly, block based sampling can avoid repeatedly scanning big data process caused by record sampling. Obviously, there exist statistical differences between data blocks, without consideration of such issues, low generalization of the ensemble model could be caused indeed.

2.3 Kernel Density Estimation and Kullback−Leibler Measure

In this section, KDE is introduced for RSP block density estimation. KDE was proposed by Murray Rosenblatt [29] and Emanuel Parzen [30]. It does not introduce a priori assumption of data distribution, only obtains the data distribution characteristics from the sample itself. It can be used to estimate the density function of arbitrary shape. It has obvious advantages over the traditional parameter estimation method and semi-parameter estimation method.

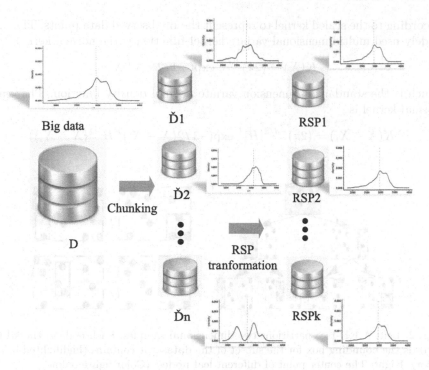

Fig. 1. Random sample partition storage model [9]

Let $D = x_1, x_2, \ldots, x_n$ be a univariate independent and identically distributed sample drawn from a distribution with an unknown density F. The kernel density estimator can be achieved by using the followers.

$$\widetilde{F}(x) = \frac{1}{n} \sum_{i=1}^{n} K_h(x - x_i) = \frac{1}{nh} \sum_{i=1}^{n} K(\frac{x - x_i}{h}) \tag{1}$$

where h is the window width. The kernel function should always satisfy the following condition:

$$\int_{-\infty}^{+\infty} K(u)du = 1 \tag{2}$$

Let $D = X_1, X_2, \ldots, X_n$, and X_i is a d-dimensional random variate. The kernel density estimator of multi-dimension variate can be achieved by using the followers.

$$\widetilde{F}(x; H) = \frac{1}{n} \sum_{i=1}^{n} K_H(X - X_i) \tag{3}$$

The crucial tuning parameter is the bandwidth H, also called the window width matrix. It is a symmetric, positive definite, d ∗ d matrix of smoothing parameters. It is a local weighted averaging estimator with a given observer point X, and for a data point X_i, the probability mass is smoothed in the local neighborhood according to the scaled kernel to represent the unobserved data points. The most widely used multidimensional variate kernel function is the normal kernel.

$$K(X) = (2\pi)^{-d/2} \exp(-1/2 * X^T X) \tag{4}$$

which is the standard d-dimension variate normal density function. The scaled normal kernel is

$$K(X - X_i) = (2\pi)^{-d/2} |H|^T \exp(-1/2(X - X_i)^T H^{-1}(X - X_i)) \tag{5}$$

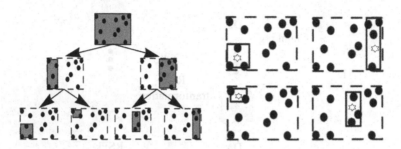

Fig. 2. Left: A KD-tree partitions two dimensional samples. Each node in the kd-tree records the bounding box for the subset of the dataset it contains (highlighted in blue color). Right: The center point of different leaf nodes. (Color figure online)

But there are two problems during KDE application for big data computation. Firstly, using the data indiscriminately and intensively makes the algorithm have a high time complexity. Selecting the optimal window width needs extra computing overhead. If the data contains m test samples and n observe samples, the time complexity of KDE calculation is $O(mn)$, and it will increase with the data dimension d. Secondly, for KDE calculation of any test sample subject to the same distribution, all training data need to reside in memory, so that the spatial complexity of the algorithm is proportional to the capacity of the training set. When the training set contains massive data, e.g., TB-level data, the task of reading the data is not feasible. In order to solve this problem, the orthogonal forward regression was applied to KDE calculation, which has a fast calculation speed and good calculation accuracy [31]. We applied the Dual-tree recursion to reduce the computation cost using Eq. (5). We first use a variant of KD-trees [32] to form hierarchical groupings of points based on their locations using the recursive procedure shown in the following Algorithm 1. After a recursive call of the dual-tree implementation, it has the following task, for $x_q \in Qnode$ compute the contribution to x_q's summed weights that are due to all points in Dnode. We use simple rectangle geometry to compute the shortest and furthest possible distances between any (x_q, x_d) pair. This bounds the minimum and maximum possible values of $Kw(x_q$-$x_d)$. If these bounds are tight enough, we prune by simply distributing the midpoint weight to all the points in Qnode as shown in the above figure's right part [33] (Fig. 2).

Algorithm 1. KD-BUILD algorithm

Input: Dataset:$\mathbb{D} = \{x_1, x_2, \cdots, x_N\}$, $x_i \in R^D$
Output: Sparse tree: T
1: Generate a tree with empty node.
2: Obtain the number of samples $Tree_N.P$
3: Set the left child Node $Tree_N{}^L$ as null.
4: Set the right child Node $Tree_N{}^R$ as null.
5: **for** $d \in [1, D]$ **do** :
6:      Calculate the $N.b[d].u$ by $\min\limits_{x \in p} x[d]$
7:      Calculate the $N.b[d].l$ by $\max\limits_{x \in p} x[d]$
8: **end for**
9: **if** $|T|$ is above the leaf threshold **then**
10:      Obtained the $N.sd$ by $\arg\max\limits_{1<d<D} N.b[d].u - N.b[d].l$
11:      Obtained the $N.sc$ by $\frac{N.b[N.sd].l - N.b[N.sd].u}{2}$
12:      Obtained the P^L by $x \in \{x \in P \,|\, x[N.sd] \leq N.sc\}$
13:      Obtained the P^R by $x \in \{x \in P \,|\, x[N.sd] \geq N.sc\}$
14:      Obtained the T^L by KD-BUILD(P^L)
15:      Obtained the T^R by KD-BUILD(P^R)
16: **end if**
17: Return T

Kullback–Leibler divergence, also called relative entropy, is the difference measure between two probability distributions [34]. Applications include characterizing the relative Shannon entropy in information systems, randomness in continuous time-series, and information gain when comparing statistical models of inference. For discrete probability distributions U and V defined on the same probability space, χ, the Kullback–Leibler divergence from U to V is defined to be:

$$D_{KL}(U||V) = \sum_{x \in \chi} U(x) \log(\frac{U(x)}{V(x)}) \tag{6}$$

If U and V are distribution of a continuous random variable, the KL divergence is defined to be the integral form.

$$D_{KL}(U||V) = \int_{-\infty}^{+\infty} u(x) \log(\frac{u(x)}{v(x)}) dx \tag{7}$$

where the $u(x)$ and $v(x)$ denote the obtained RSP probability densities of U and V. Based on the KDE results, we can use KL measurements to compare the similarity of distribution between RSP blocks.

3   KL-RSP Data Storage Model

To enable HDFS distributed data blocks to be used as random samples for estimation and analysis of the entire big data set, we use the KL-RSP model to generate a data block for representing big data, which ensures that each data block is a random sample of the big data set. The main properties of the RSP model are given in the followers. Let $\mathbb{D} = \{x_1, x_2, \cdots, x_N\}$ be a data set containing N objects. Let $\mathbb{OP}$ be an operation which divides $\mathbb{D}$ into a family of subsets $T = \{D_1, D_2, \cdots, D_K\}$, where T is called a partition of data set $\mathbb{D}$. It should satisfy the following two conditions. The first one is all the union of all sub data sets should be equal to the original big data, which means $\bigcup_{k=1}^{K} D_k = \mathbb{D}$. The second one is, for any given two blocks, there is no joint data set between blocks, which is $D_i \cap D_j = \emptyset$, when $i, j \in \{1, 2, \cdots, K\}$ and $i \neq j$. An HDFS file is a partition of data set $\mathbb{D}$ where data blocks $\{D_1, D_2, \cdots, D_K\}$ are generated by sequentially cutting the big data. Unfortunately, these data blocks in HDFS files do not have similar distribution properties as the input big data file. The data blocks in the RSP as defined below can be used as random samples of the big data set.

Definition 1 (Kullback-Leibler Random Sample Partition): Suppose $\mathbb{D} = \{x_1, x_2, \cdots, x_N\}$ be a big data set, which is a random sample of a population. $F(x)$ to be the sample distribution function $(s.d.f.)$ of $\mathbb{D}$. Let $\mathbb{OP}$ be a partition operation on $\mathbb{D}$ and $D = \left\{ P_1^{KL_0}, P_2^{KL_1}, \cdots, P_Q^{KL_{Q-1}} \right\}$ be a partition of data set $\mathbb{D}$ accordingly. $P_Q^{KL_{Q-1}}$ is called a *Kullback−Leibler Random Sample Partition* of $\mathbb{D}$ if

$$E[\tilde{F}(P_q^{KL_q}(x))] = F(x), q = 1, 2, \cdots, Q, \text{and}, KL_1 > KL_2 > .... > KL_{Q-1} \quad (8)$$

where $\tilde{F}(P_q^{KL_q}(x))$ denotes the sample distribution function of P_q and $E[\tilde{F}_q(x)]$ denotes its expectation.

Corollary 1. Let $\{D_1, D_2, \cdots, D_Q\}$ denote the above Q small blocks. Each D_k is an RSP data block of $\mathbb{D}$.

Proof: Set $D_k = \left\{ x_1^{(k)}, x_2^{(k)}, \cdots, x_{N_k}^{(k)} \right\}$, $k \in \{1, 2, \cdots, Q\}$ and N_k is the number of objects in D_k. It is obvious that

$$P\left\{ D_k = \left\{ x_{s_1}, x_{s_2}, \cdots, x_{s_{N_k}} \right\} \right\} = \frac{1}{C_N^{N_k}}, \quad (9)$$

where $x_{s_1}, x_{s_2}, \cdots, x_{s_{N_k}}$ are s_{N_k} objects selected arbitrarily from $\mathbb{D}$. Assume $F(x)$ is the $s.d.f.$ of $\mathbb{D}$. On one hand, for each real number $x \in R^1$, the number of samples whose values are not greater than x is $F(x) \cdot N$. On the other hand, for each data block D_k and object $x_i \in \mathbb{D}$, $P\{x_i \in D_k\} = C_{N-1}^{N_k-1} \cdot \frac{1}{C_N^{N_k}} = \frac{N_k}{N}$. Thus the number of objects in D_k whose values are not greater than x is $F(x) \cdot N \cdot \frac{N_k}{N} = F(x) \cdot N_k$. We therefore obtain that the expectation of the $s.d.f.$ of D_k is $\frac{1}{N_k} \cdot F(x) \cdot N_k = F(x)$. According to the definition, D_k is an RSP data block of $\mathbb{D}$.

Theorem 1. *Let* $\mathbb{D} = \{\mathbb{D}_1, \mathbb{D}_2, \cdots, \mathbb{D}_K\}$. *Suppose* $K = 2$, *then* $\mathbb{D}_1$ *and* $\mathbb{D}_2$ *be two big data sets of* $\mathbb{D}$ *with* N_1 *and* N_2 *objects respectively. By* $\mathbb{OP}$ *operation to* $\mathbb{D} = \{P_1, P_2, \cdots P_i, P_{i+1}, \cdots P_j, P_{j+1}, \cdots P_Q\}, i < j < Q$. *Assume that* P_i *with* n_i *objects is an RSP data block of* $\mathbb{D}_1$ *and* P_j *with* n_j *objects is an RSP data block of* $\mathbb{D}_2$. *Then,* $P_i \bigcup P_j$ *is an RSP data block of* $\mathbb{D}_1 \bigcup \mathbb{D}_2$ *under the condition that* $\frac{n_i}{n_j} = \frac{N_1}{N_2}$.

Proof: Let $F_1(x)$ and $F_2(x)$ denote the *s.d.f.s* of $\mathbb{D}_1$ and $\mathbb{D}_2$ respectively. Assume that the *s.d.f.s* of P_i and P_j are $\tilde{F}_i(x)$ and $\tilde{F}_j(x)$, respectively. According to Definition 2, we have $E[\tilde{F}_1(x)] = F_1(x), E[\tilde{F}_2(x)] = F_2(x)$. For any real number x, the number of objects in $P_i \bigcup P_j$ whose values are not greater than x is $n_i \tilde{F}_i(x) + n_j \tilde{F}_j(x)$. Therefore, the *s.d.f.* of $D_i \bigcup D_j$ is:

$$\tilde{F}(x) = \frac{n_i \tilde{F}_i(x) + n_j \tilde{F}_j(x)}{n_i + n_j}.$$

Similarly, the *s.d.f.* of $\mathbb{D}_1 \bigcup \mathbb{D}_2$ is:

$$F(x) = \frac{N_1 F_1(x) + N_2 F_2(x)}{N_1 + N_2}.$$

The expectation of $\tilde{F}(x)$ is

$$E[\tilde{F}(x)] = E\left[\frac{n_i \tilde{F}_i(x) + n_j \tilde{F}_j(x)}{n_i + n_j}\right] = \frac{n_i E[\tilde{F}_i(x)] + n_j E[\tilde{F}_j(x)]}{n_i + n_j}$$
$$= \frac{N_1 F_1(x) + N_2 F_2(x)}{N_1 + N_2}$$
$$= F(x).$$

Remark: With a subtle modification, the proof of **Corollary** 1 can be extended to data in multiple dimensions and the proof of **Theorem** 1 can also be extended to the multiple dimensions and more than two data sets.

Let $\mathbb{D} = \{x_1, x_2, \cdots, x_N\}$ be a data set with N objects and M features. To generate RSP data blocks (i.e., random samples) from $\mathbb{D}$, if N is not big, we can easily use the following algorithm to convert $\mathbb{D}$ into Q RSP data blocks.

4   Ensemble Classification for KL-RSP Data Model

Bagging often considers homogeneous weak learners, learns them independently from each other in parallel and combines them following a deterministic averaging process. In order to fit several independent models and average their predictions for obtaining a model with a lower variance, users need to create multiple bootstrap samples so that each new bootstrap sample would act as another possible independent data set drawn from the true distribution. This method is

Algorithm 2. KL-RSP algorithm

Input: Dataset:$\mathbb{D} = \{x_1, x_2, \cdots, x_N\}$

Output: Dataset:$\mathbb{D}_p = \left\{P_1^{KL_0}, P_2^{KL_1}, \cdots, P_Q^{KL_{Q-1}}\right\}$

1: Generate N unique random integer numbers for N objects from a uniform distribution;
2: Sort N objects on the random numbers to reorganize $\mathbb{D}$;
3: Sequentially cut N reordered objects into Q small data blocks, each with N/Q objects.
4: Obtain random sample partition blocks $\mathbb{D}_p = \{P_1^+, P_2^+, \cdots, P_Q^+\}$
5: Calculate the density estimation function with kernel density estimation for generating $\tilde{F}(P_1(x)), \tilde{F}(P_2(x)), \ldots, \tilde{F}(P_q(x))$
6: Random selected data block with estimated function $\tilde{F}(P_q(x))$
7: Calculate the KL measures between the selected data block and the rest block $\tilde{F}(P_q(x))$ to generated $KL_1, KL_2, \ldots, KL_{Q-1}$

---

based on the assumption of the data itself is relevant small size. However, in practice of big data, fit fully independent models is not easy because the data set itself is big enough. As an offline operation, RSP model generates ready-to-use disjoint data blocks, which aims to achieve Write-Once-Use-Many-Times (WOUM) strategy. By considering the statistical difference of RSP blocks, we proposed a weighted ensemble strategy for classification by using the obtained the KL measure vector $KL = [KL_1, KL_2, \ldots, KL_{n-1}]$, where n is the number of RSP block. The final ensemble output Π is given by the following equations.

$$\Pi = \text{Sigmoid}(\frac{1}{q} \sum_{i=1}^{q} (\Pi_0 + \frac{KL_i}{\sum_{k=1}^{i} KL_k} * \Pi_i) - 0.5), \text{where}, q \leq n, \qquad (10)$$

where Π_0 indicates the probability output of the selected base model, KL_i is the i-th KL values and Π_i is the i-th output of model. A detail description is shown in Fig. 3, the first stage is partition of big file to number of small blocks using random sample partition, and then the RSP block is applied for calculation of KL parameters. Then, the above equation is used for obtaining the final output with outputs of selected base models.

5   Experiment Results

A Spark cluster of eight computational nodes was used for the evaluation of the proposed method. In the cluster, each node has 24 cores, 128 GB RAM, 20 TB total disk storage, 1 TB total memory, 160 execution containers, Hadoop version CDH 5.13.3 and Spark version 2.4.0. Firstly, we tested the RSP generation algorithm performance in contrast to Bootstrap algorithm. Then, we discussed about the performance of our ensemble classification strategy using RSP blocks.

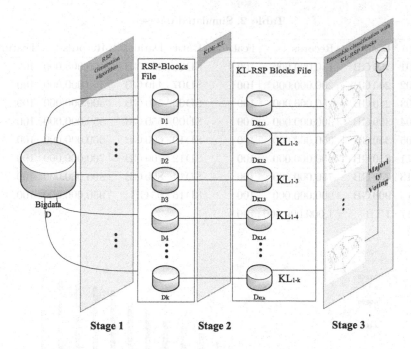

Fig. 3. The procedure of the proposed method

5.1 Sampling Time Efficiency Comparison

In this section, we have done sampling experiments on datasets in Table 2. Figure 4 shows that the partitioning time of RSP transformation increases almost linearly with the increase of the data size. In this example, the number of records in each block is 100000. Compared with records level sampling strategy, RSP block level sampling strategy greatly saves computational time. On a computing cluster, an RSP operation ℗ is saved as an RSP file with metadata, which contains the storing RSP block information including the size, location and reference. Selecting the distributed RSP blocks directly as random samples save a lot of sampling time, especially when many samples are required. For instance, it takes 10 s−20 s on average to select 10 RSP blocks from 150 GB data and load them locally using Apache Spark. The Write-Once-Use-Many-Times (WOUM) strategy enables the success of RSP. To get a record-level sample from an HDFS file, all the data should be loaded for the selection of records with equal probability. We show the sampling time of 100000 record samples which is the size of RSP block without replacement. It is normally done by Boostrap sampling. These results were produced with Apache Spark by repeating the sample transformation function. When the data size is small, the time costs are generally the same. However, when the data records increased, the RLS time will be further increased dramatically. The results illustrate the computational advantage of generating an RSP from big data and using RSP blocks as random samples.

Table 2. Simulated dataset

Data	Data size	Records	Features	Data	Data size	Records	Features
SD01	150 GB	150,000,000	100	SD06	400 GB	400,000,000	100
SD02	200 GB	200,000,000	100	SD07	440 GB	450,000,000	100
SD03	250 GB	250,000,000	100	SD08	500 GB	500,000,000	100
SD04	300 GB	300,000,000	100	SD09	550 GB	550,000,000	100
SD05	350 GB	350,000,000	100	SD10	600 GB	600,000,000	100
SD11	650 GB	650,000,000	100	SD12	700 GB	700,000,000	100
SD13	750 GB	750,000,000	100	SD14	800 GB	800,000,000	100
SD15	900 GB	900,000,000	100	SD16	95 GB	950,000,000	100
SD17	1 TB	1,000,000,000	100				

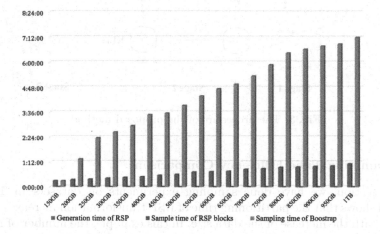

Fig. 4. RSP transformation algorithm execution time with different data size

5.2 Ensemble Learning Results

Table 3 illustrates three storage format of HIGGS[4]. Sixty blocks are used for saving HIGGS dataset by HDFS. HIGGS RSP1 and HIGGS RSP2 were obtained by the RSP transformation algorithm with the default HIGGS HDFS file. We randomly sample 20 blocks from HIGGS default, HIGGS RSP1 and HIGGS RSP2 for the ensemble classification, dual-tree KDE algorithm was applied to 20 of them for density estimation and then KL measurement is applied to them. The first 10 blocks with large KL values were used for training strong base classifiers with XGboost[5] algorithm, 5 RSP blocks data were used for validation and the rest 5 RSP blocks data were applied to testing. We set the parameter *objective* as *binaryLogistic* for bi-classification during training. Other parameters of XGboost model are obtained

---

[4] Available: http://archive.ics.uci.edu/ml/datasets/HIGGS.
[5] Available: https://xgboost.readthedocs.io/en/latest/.

Table 3. HIGGS dataset

Data	Records per each block	Block number
HIGGS HDFS	180000	60
HIGGS RSP1	55000	200
HIGGS RSP2	27500	400

by grid searching method during training. We set the range of *maxDepth* as [3, 10] and *eta* as [0.1, 0.3]. According to the proposed ensemble strategy, we increase the base model for calculating the final classification accuracy.

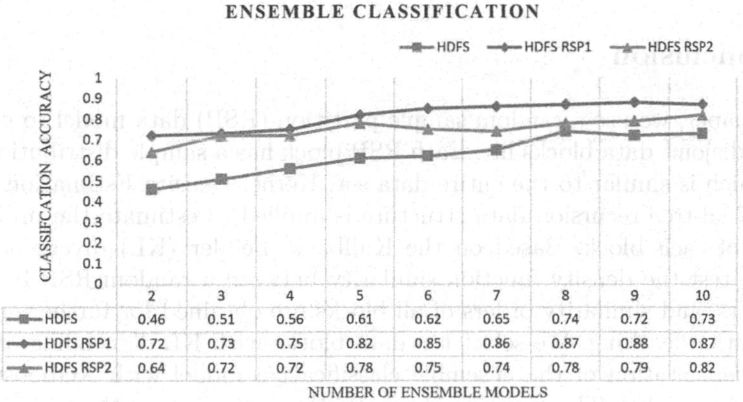
ENSEMBLE CLASSIFICATION
HDFS
HDFS RSP1
HDFS RSP2
CLASSIFICATION ACCURACY
1
0.9
0.8
0.7
0.6
0.5
0.4
0.3
0.2
0.1
0
2 3 4 5 6 7 8 9 10
HDFS 0.46 0.51 0.56 0.61 0.62 0.65 0.74 0.72 0.73
HDFS RSP1 0.72 0.73 0.75 0.82 0.85 0.86 0.87 0.88 0.87
HDFS RSP2 0.64 0.72 0.72 0.78 0.75 0.74 0.78 0.79 0.82
NUMBER OF ENSEMBLE MODELS

Fig. 5. Ensemble classification results with different number of ensemble models

Figure 5 shows the change of the classification accuracy over the number of ensemble models. The figure shows that all the classification accuracy increases as the increase of the ensemble models. But, the model built on traditional HDFS blocks shows a decrease when the number of ensemble model reaches to eight. This is due to the fact that each HDFS block cannot represent the entire big data. By using such data block, the base learning model will cause low generalization. Compared with model using the HDFS RSP1, HDFS RSP2 shows a high classification accuracy. And less than 15% of the entire data is enough for building a stable model. We agree that the bigger number of records in each RSP block will lead to a high classification results finally. Figure 6 shows the time costs of kernel density estimation and model training. The time costs include three main parts, which are the Spark task computation time, job scheduling time and job communication time. As can be seen from the figure, we can conclude that the original HDFS blocks still need to chunk into small ones under the iterative computation task. Due to fewer records, the HDFS RSP2 shows a low computation cost compared to that of HDFS RSP1.

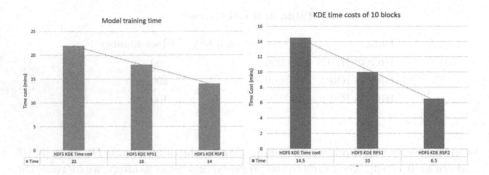

Fig. 6. Left: Time cost of XGboost training in parallel. Right: Time cost of Duel-Tree KDE in parallel

6 Conclusion

In this paper, we use a random sample partition (RSP) data model to generate a set of disjoint data blocks file. Each RSP block has a sample distribution function, which is similar to the entire data set. Kernel Density Estimation (KDE) with a dual-tree recursion data structure is applied to estimate the probability density of each block. Based on the Kullback−Leibler (KL) divergence measure, we test the density function similarity between a random RSP block and the others, and similarity orders of all blocks are obtained for further ensemble classification learning. We select ten data blocks with KL descending order for the implementation of the ensemble classification model with XGboost based learners in parallel. The experiment results show that our method can increase the generalization capability of the ensemble classification model. By using less than 15% of the entire data, the model building time in parallel computation environment is reduced. With small data blocks, the memory constraints of big data analysis are solved.

References

1. Chen, B.W., Wen, J., Seungmin, R.: Divide-and-conquer signal processing, feature extraction, and machine learning for big data. Neurocomputing **174**, 383 (2016)
2. Shvachko, K., Kuang, H., Radia, S., Chansler, R.: The hadoop distributed file system. In: IEEE 26th Symposium on Mass Storage Systems and Technologies, pp. 1–10 (2010)
3. Dean, J., Ghemawat, S.: MapReduce: simplified data processing on large clusters. Commun. ACM. **51**(1), 107–13 (2008)
4. Elteir M., Lin H., Feng W.C.: Enhancing mapreduce via asynchronous data processing. In: IEEE International Conference on Parallel and Distributed Systems, pp. 397–405 (2010)
5. Zaharia M., Chowdhury M., Franklin M.J., Shenker S., Stoica I.: Spark: cluster computing with working sets. In: Proceedings of the 2nd USENIX Conference on Hot Topics in Cloud Computing, pp. 10 (2010)

6. Salloum, S., Dautov, R., Chen, X., Peng, P.X., Huang, J.Z.: Big data analytics on apache spark. Int. J. Data Sci. Anal. **1**, 145–164 (2016)
7. Lei G., Huan L.: Memory or time: performance evaluation for iterative operation on hadoop and spark. In: IEEE 10th International Conference on High Performance Computing and Communications, pp. 721–727 (2013)
8. Wu, X., Zhu, X., Wu, G.Q., Ding, W.: Data mining with big data. IEEE Trans. Knowl. Data Eng. **26**(1), 97–107 (2014)
9. Salloum, S., Huang, J.Z., He, Y.L.: Random sample partition: a distributed data model for big data analysis. IEEE Trans. Ind. Inform. **15**(11), 5846–5854 (2019)
10. Dong, X., Yu, Z., Cao, W., Shi, Y., Ma, Q.: A survey on ensemble learning. Front. Comput. Sci. **14**(2), 241–258 (2020)
11. Galicia, A., Talavera, L.R., Troncoso, A., Koprinska, I., Martnez, A.F.: Multi-step forecasting for big data time series based on ensemble learning. Knowl. Base Syst. **163**, 830–841 (2018)
12. Tang Y., Wang Y., Cooper K.M.L., Li L.: Towards big data Bayesian network learning - an ensemble learning based approach. In: IEEE International Congress on Big Data, pp. 355–357 (2014)
13. Shadi, K., Patrick, M., Rebecca, Y.: Label-aware distributed ensemble learning: a simplified distributed classifier training model for big data. Big Data Res. **15**, 1–11 (2019)
14. Diego, M., Eduard, A., Jose, R. Herrero, R.J., Bifet, A.: Low-latency multi-threaded ensemble learning for dynamic big data streams. In: IEEE International Conference on Big Data, pp. 223–232 (2017)
15. Salman, S., Joshua, Z.X.H., He, Y.L., Chen, X.J.: An asymptotic ensemble learning framework for big data analysis. IEEE Access **7**, 3675–3693 (2019)
16. Zhou, Z.H., Wu, J.X., Tang, W.: Ensembling neural networks: many could be better than all. AI **137**(1–2), 239–263 (2002)
17. Giancinto, G., Roli, F.: An approach to the automatic design of multiple classifier ensembles. Pattern Recogn. Lett. **22**(1), 25–33 (2001)
18. Cheng X.Y., Guo H.L.: The technology of selective multiple classifiers ensemble based on kernel clustering. In: International Symposium on Intelligent Information Technology Application, pp. 146–150 (2008)
19. Martinez, M.G., Suarez, A.: Using boosting to prune bagging ensembles. Pattern Recogn. Lett. **28**(1), 156–165 (2007)
20. Martinez M.G., Suarez A.: Pruning in ordered bagging ensembles. In: Proceedings of the 23rd International Conference on Machine Learning, pp. 609–368 (2006)
21. Breiman, L.: Out-of-bag estimation. Statistics deparment in university of California, Technical Report (1996)
22. Zhang, L., Zhou, W.D.: Sparse ensembles using weighted combination methods based on linear programming. Pattern Recogn. Lett. **44**(1), 97–106 (2011)
23. Fan, C.T., Muller, M.E., Rezucha, I.: Development of sampling plans by using sequential (item by item) selection techniques and digital computers. J. Am. Stat. Assoc. **57**(298), 387–402 (1962)
24. Haas, P.J.: Data-stream sampling: basic techniques and results. Data Stream Management. DSA, pp. 13–44. Springer, Heidelberg (2016). https://doi.org/10.1007/978-3-540-28608-0_2
25. Oliphant, T.E.: SciPy: open source scientific tools for python. Comput. Sci. Eng. **9**(3), 10–20 (2007)
26. Swami, A., Jain, R.: Scikit-learn: machine learning in python. J. Mach. Learn. Res. **12**(10), 2825–2830 (2013)

27. Podgurski, A., Yang, C.: Partition testing, stratified sampling, and cluster analysis. ACM SIGSOFT Softw. Eng. Notes **18**(5), 169–181 (1993)
28. Kleiner A., Talwalkar A., Sarkar P., Jordan M.I.: The big data bootstrap. In: Proceedings of the 29th International Conference on Machine Learning, pp. 1787–1794 (2012)
29. Rosenblatt, M.: Remarks on some nonparametric estimates of a density function. Ann. Math. Stat. **27**, 832–837 (1956)
30. Parzen, E.: On the estimation of probability density functions and mode. Ann. Math. Stat. **33**, 1065–1076 (1962)
31. Chen, S., Hong, X., Harris, C.J.: Sparse kernel density construction using orthogonal forward regression with leave-one-out test score and local regularization. IEEE Trans. Syst. Man Cybern. Part B Cybern. **34**(4), 1708–1717 (2004)
32. Friedman, J.H., Bentley, J.L., Finkel, R.A.: An algorithm for finding best matches in logarithmic expected time. ACM Trans. Math. Softw. **3**(3), 209–226 (1977)
33. Gray, A.G., Moore, A.W.: 'N-Body' problems in statistical learning. In: Advances in Neural Information Processing Systems, vol. 4, no. 1, pp. 521–527 (2001)
34. Kullback, S., Leibler, R.A.: On information and sufficiency. Ann. Math. Stat. **22**(1), 79–86 (1951)

SWAF: A Distributed Solar WSN Adaptive Framework

Yuekun Hu[1], Dongchao Ma[2($\boxtimes$)], Xiaofu Huang[2], Xinlu Du[2], and Ailing Xiao[2]

[1] Beijing University of Posts and Telecommunications, Beijing, China
16151010124@mail.ncut.edu.cn
[2] North China University of Technology, Beijing, China
madongchao1980@wo.cn

Abstract. This article comes from a solar WSN air monitoring system deployed outdoors. We find that two issues have not been properly resolved: 1, The actual deployment environment of the node has a part time shadow, resulting in a significant reduction in the accuracy of solar prediction algorithms. 2, The length of solar prediction and the topology adjustment period have a great influence on the network lifetime. According to the above, this paper proposes a distributed Solar WSN Adaptive Framework (SWAF), designs a distributed method to distinguish the shadow time of nodes and a dynamic method to select the charging prediction period, which can effectively integrate the existing charging prediction algorithm and energy aware routing algorithm. The experimental results show that SWAF can reduce node mortality, and thus improve network lifetime. Compared with the case of simply using the existing prediction model and the routing algorithm, the SWAF can increase the network lifetime by 5%–27%.

Keywords: Energy harvesting wireless sensor networks · Maximum lifetime · Charging prediction algorithm · Energy aware routing · Micro-solar power system · Prediction period

1 Introduction

Considering the cost of photovoltaic panels, lithium-ion capacitors and other charging and energy storage devices, the application of solar WSN needs effective solar charging prediction (SCP) algorithm [1, 2, 6, 11, 12] and energy aware routing strategy [3, 4, 13]. Although some "permanent" experimental systems are proposed [5], research in this field is actually not sufficient. Based on such existing research, we have tried to make a WSN prototype system (based on solar energy) for air monitoring in the education and scientific research project, which is used to collect the basic air indexes such as CO_2, nitrogen oxide, temperature and humidity in Shijingshan District (Shougang old Industrial Zone) in Beijing. However, during the trial operation of the system, we found that its working status was unsatisfactory. For example, the node mortality rate was high, and the network lifetime and other indicators were lower than expected. After a simple analysis of the data, we find the following two reasons: First, the accuracy of the SCP is

© Springer Nature Switzerland AG 2020
M. Qiu (Ed.): ICA3PP 2020, LNCS 12452, pp. 465–479, 2020.
https://doi.org/10.1007/978-3-030-60245-1_32

significantly lower than expected [1, 2, 11]. Secondly, it is difficult to set the charging prediction period. The following is a brief analysis of these two issues.

First of all, the existing SCP can indeed achieve a high accuracy rate in the open area and fixed orientation scenes [1, 2, 6, 11, 12]. But in the actual deployment practice in the city, the size of the node is very small, and the deployment environment is also different. During the effective lighting time (for example, 8AM–5PM in winter), some nodes are seriously shaded, and may even reach more than half of the time. In these shadow scenes, the accuracy of SCPs is greatly affected. In this case, the prediction result is used as an input for energy aware routing, which causes some hotspot nodes to run out of power. Secondly, there are some difficulties in setting the prediction period. If the setting is too large, it may lead to the repeated death and resurrection of the node; if the setting is too small and ignores long-term weather changes, the short-sighted adjustment of routing will also lead to the early end of network lifetime.

The above two aspects are the reasons for the poor operation of the environmental collection system we deployed. In conclusion, we believe that the accuracy of the existing algorithms and the optimal effect of the routing algorithm basically meet the requirements. However, how to effectively integrate these new technologies and apply them to "field system" needs to solve the following two problems: 1. The shadow in the complex deployment environment of small nodes. 2. When to predict the solar energy and adjust the network topology.

The main motivation of this paper is to propose a distributed Solar WSN Adaptive Framework for the above problems, and integrate some recent research programs to focus on solving the adaptation problems. See Sect. 3 for details.

This paper makes the following contributions:

1. For the first time, a set of Solar WSN Adaptive Framework (SWAF) is proposed, including three main levels of charging prediction, individual feature adaptation, energy aware routing and scheduling.
2. Aiming at the adaptation problem between the SCP and the individual environment, a distributed node shadow judgement algorithm is given in the SWAF adaptation layer, which effectively improves the accuracy of the SCP in complex deployment environments and effectively reduce computing costs.
3. Aiming at the problem of prediction and topology adjustment period selection, a dynamic selection algorithm is presented. Prediction can be performed only when necessary, and the node mortality is reduced with low computational cost.

The rest of the paper is organized as follows: Sect. 2 analyzes the research status of the SCP and energy aware routing, and shows the research motivation of the SWAF; Sect. 3 gives the distributed Solar WSN Adaptive Algorithm (SWA). Section 4 analyzes the optimization effect and the required cost through experiments.

2 Related Work and Problem Analysis

The following describes the relevant research and problem analysis from three aspects, including the applicability of the SCP in the complex individual environment where small nodes are located, the impact of the prediction period on network lifetime and node mortality, and energy aware routing algorithms.

2.1 SCP and Applicability

The routing adjustment considered solar prediction is more profitable for the WSN lifetime than that based only on the remaining power [9]. SCPs can generally be divided into machine learning-based and statistics-based.

In machine learning, Bao Y [1] proposed a combined prediction model ANN-Linear combining artificial neural network and linear model. The algorithm can select different models for prediction according to the length of the period. Rodriguez [11] and others proposed an artificial neural network model (ANN) to predict the amount of solar power generated by photovoltaic generators in the next 10 min and verified the correctness of the results. The difference between the predicted energy and the actual energy produced is about 0.5–9%, indicating that the model can be applied to systems with integrated solar generators.

To predict energy, the statistical model uses different statistical data, such as average, moving average, standard deviation, and variance. Muhammad [2] proposed a statistical-based prediction model Ipro-Energy, which has a significant improvement in short-term and medium-term prediction accuracy. In addition, Ahmed et al. [12] proposed a lightweight linear energy prediction model LINE-P for sensor network nodes based on abstract approximation theory, and achieved high prediction accuracy for both solar and wind energy.

Whether it is a SCP based on machine learning or statistics, it already has more than 80% accuracy, and has actual conditions of use. However, due to the differences in the environment of large-scale photovoltaic power plants and the small rechargeable nodes, the SCP cannot be directly copied. Existing SCPs ignore the features of small nodes that are sensitive to precise locations (such as buildings, tree shadow, orientation, etc.), and mostly only use atmospheric temperature, wind speed, cloudiness, and historical power generation data of sensor nodes as input variables of the model to make predictions. This is not suitable for nodes that are actually deployed outdoors rather than ideally placed. We find that the light intensity under shadow has a certain rule through actual measurements. When the node is inside the shadow, its charging power is approximately constant (about 1.14 W/m^2). Therefore, it is proposed to introduce a shadow judgement method based on geographical and geometric knowledge to calculate the time of nodes in the shadow. Integrated with the existing high-precision SCP, it can effectively solve the charging prediction problem of smaller nodes in the messy individual environment.

2.2 Influence of Prediction Period on Network Lifetime

The data collection system of the WSN we made is used to collect the main indicators in the air such as CO_2, nitrogen oxides and temperature of Shijingshan District (Shougang Old Industrial Zone). After a period of actual operation, we found that the mortality rate of the node was high. And no matter whether the prediction period is increased or decreased, the problem still exists. This problem will be more obvious in the period when the data collection frequency is high or the solar condition is bad. By carefully analyzing all the records of the system, we found that the length of the prediction period will significantly affect the lifetime of the network.

The essential reason is that considering the calculation cost, the granularity of the prediction period cannot be too small, the fixed-period approach may encounter two situations: some nodes have less remaining energy, but subsequent charging is fast; some nodes have sufficient remaining energy, but subsequent charging is slow. Therefore, the prediction period needs to be dynamically adjusted according to the weather and network status in order to reduce the risk of node death and extend the lifetime of the network.

2.3 Energy-Aware Routing Strategies

According to its technical characteristics, the existing energy aware routing algorithms can be divided into two categories: energy-efficient routing schedule and data compression transmission.

The technical characteristic of energy aware routing scheduling is to schedule some nodes to enter the sleep state, and to distribute the data traffic to the nodes with sufficient residual power depending on the routing strategy. Lu [3] proposed an energy-efficient data sensing and routing scheme (EEDSRS), so that the EH-WSN can use the harvested energy wisely for data sensing and transmission according to current available energy to maximize network utility and route all the collected data to the sink along energy-efficient paths. Lv [4] proposed a sparsity feedback-based compressive data gathering algorithm. The algorithm first estimates the sparsity accurately, then adaptively adjusts the number of measurements for accurate reconstruction. The algorithm solves the problem of balancing energy balance amongst sensor nodes. Zhang [13] studied the problem of allocating sensor energy, sensor data rates, and data routing in an energy harvesting sensor network for a given monitoring period T, such that the utility sum of temporally correlated sensing data collected during the period T is max- imized.

The research of data compression transmission is to reduce the times of data transmission by means of cache forwarding or compression in the data collection process. Tan et al. [14] gave an upper bound on WSN lifetime optimization and proposed a method for generating a distributed data aggregation tree that increases the lifetime of the network. This method can support continuous optimization in the process of nodes continuously joining and exiting. Similar research also includes the heuristic algorithm recently proposed by Lin [15] et al. Its characteristic is the introduction of transmitting power parameters that can be adjusted with the communication distance.

3 Solar WSN Adaptive Algorithm (SWA)

This section proposes a distributed Solar WSN Adaptive Framework and integrates some recent research programs to focus on solving the adaptation problems.

As Fig. 1, the top-level SCP can obtain important parameters for network operation, which is the consensus of such research [1, 2, 11]. Existing SCP research is relatively sufficient, and it is considered to be introduced into this framework in an integrated mode. Later in the experimental part, we try to introduce a variety of SCPs. In addition, the underlying energy-aware routing algorithm also has a wealth of options for integration into the framework. For example, the EEDSRS [3] model integrates node scheduling and routing algorithms, attaches importance to predicted charge as an input parameter, and is suitable for the application scenario of this article. For another example, the DLEX-DAG [4] model can compress the data according to the sparseness of the collected data, and then balance the remaining energy of each node, which is suitable for a network where nodes are densely deployed. The EEDSRS model is integrated into the system of this paper as representative of energy-aware routing algorithm.

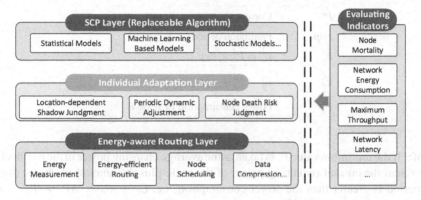

Fig. 1. Solar WSN adaptive framework (SWAF)

As shown in the middle layer of Fig. 1, this paper mainly focuses on the applicability of the SCP in the complex environment (shadow occlusion) of small nodes, described in Sect. 2.1, and the reasonable selection of the prediction period described in Sect. 2.2, adding the individual node feature adaptation layer (Fig. 1). These include shadow judgement algorithms, dynamic period adjustment strategies, and node death risk judgement methods, which specifically solve several key adaptive problems in the integration of existing technologies, thereby improving network lifetime. In addition, considering the scalability problem of the centralized computing burden as the node scale increases, the distributed computing under the "full power" state of the node is introduced to reduce the server's computing burden.

In this section, we first introduce the node death risk judgment method and shadow judgment method, and give the SWA algorithm in Sect. 3.3. Specific parameters are shown in Table 1:

3.1 Judging the Risk of Node Death

In order to facilitate the description, this section first gives a method for judging whether a node has a risk of death in the case of accurately predicting the future charging.

Theorem 1. If (i). The current charging power P_c of the node is equal to the minimum value of the charging power in the next n collection cycles, and (ii). During this nT time, there is $\frac{E_r + E_c}{nT} > P_e$. These two conditions ensure that the node has no risk of death during this nT time. ($E_c < E_f$)

Table 1. Variable description

Parameters	Meaning
V	Set of nodes
T	Collection cycle
E_r^i	Current remaining capacity of node i
P_f^i	Charging power prediction of node i
P_c^i	Current charging power of node i
step	Prediction period
E_c^i	$E_c^i = P_c^i \times step$
E_f^i	$E_f^i = P_f^i \times step$
P_e^i	Power consumption of node i

Proof: If the minimum value P_c is taken as the average charging power in the n collection cycles, and the current remaining power of the node plus the amount of charge during this period is greater than the power consumption, i.e. $E_r + P_c \times nT > P_e \times nT$, equivalent to $\frac{E_r + E_c}{nT} > P_e$, therefore, the node can be guaranteed to survive in this nT time.

Theorem 2. If (i). The current charging power of the node is not less than the average charging power in the next n collection cycles, i.e. $E_c \geq E_f$, (ii). Let P_i denote the average charging power of the ith collection cycle in this nT time, then there is $x \in [0, n]$, which satisfies $P_1, P_2, P_3, \ldots, P_x \geq P_f, P_{x+1}, P_{x+2}, \ldots, Pn \leq P_f$, (iii). Meet $\frac{E_r + E_f}{nT} > P_e$ in the future nT time. These three conditions ensure that the node has no risk of death during this nT time.

Proof: It is easy to obtain by analysis, and the condition for the node to survive in the ith collection cycle is:

$$E_r + \sum_{j=1}^{i} P_j \times T \geq i \times P_e \times T \tag{1}$$

If $m \in [1, x]$, by $P_1, P_2, P_3, \ldots, P_m \geq P_f$, there is $P_1 + P_2 + \ldots + P_m \geq mP_f$, $E_r + \sum_{j=1}^{m} P_j \times T \geq E_r + P_f \times mT$. According to condition c, $E_r + P_f \times nT > P_e \times nT$, then $\frac{m}{n} E_r + P_f \times mT > P_e \times mT$.

By $m < n$, $E_r + P_f \times mT > \frac{m}{n} E_r + P_f \times mT$, i.e. $E_r + \sum_{j=1}^{m} P_j \times T > P_e \times mT$.

Therefore, The node has no risk of death within $[0, xT]$.

If $m \in [x + 1, n]$, then

$$
\begin{aligned}
E_r + \sum_{j=1}^{m} P_j \times T &= E_r + P_f \times nT - P_n - P_{n-1} - \ldots - P_{m+1} \\
&= E_r + P_f \times mT + P_f \times (n - m)T - P_n - P_{n-1} - \ldots - P_{m+1} \\
&\geq E_r + P_f \times mT > \tfrac{m}{n}E_r + P_f \times mT > P_e \times mT
\end{aligned}
\tag{2}
$$

Therefore, the node also has no risk of death in $[(x + 1)T, nT]$, i.e. the node can be guaranteed to survive in this nT time.

In summary, when the prediction period is '*step*', if certain conditions are met and

$$
\frac{E_r + \min(E_c, E_f)}{step} > P_e
\tag{3}
$$

The node can be considered to have no risk of death.

3.2 Judgement Method of Shadow

This section presents a simple way to calculate whether a building shadow will occlude a node at a particular moment. The purpose of this method is to improve the accuracy of the SCPs and reduce the prediction overhead. We have learned through a lot of experiments that when a node is blocked by the shadow of a nearby building, its charging power tends to approach a fixed value P_{in_shade} (about 1.14 W/m^2), which has little to do with weather and other factors. Therefore, if it is known that the node will be shadowed by nearby buildings in a certain period in the future, the charging power of the node in this period can be directly set to the constant P_{in_shade}. This significantly reduces the number of times the SCP is run and reduces the computational burden. Many literatures have given low-cost three-dimensional map collection methods [7]. Therefore, this paper uses these three-dimensional maps and the basic knowledge of geography and geometry, only taking the coordinates of the nodes as input, can simply calculate the time range of nodes being shaded.

As Fig. 2a, a three-dimensional Cartesian coordinate system is established, and the rectangular parallelepiped A-H represents the vertices of the building. Figure 2b shows the projection of the building on the ground plane. It is assumed that the plane composed of xOy is the ground plane, the node is at the origin O, the y-axis direction is the north direction, the building is a rectangular parallelepiped, and its bottom surface is a rectangle ABCD. The regulations are as follows:

1. In the figure at the bottom of the building, the point closest to the southwest is point A, and its coordinate is (a, b).
2. The solar elevation angle is H_A and the azimuth is A [8].
3. The angle between the extension of the line segment AD and the east-west direction (x-axis) is θ ($0° \leq \theta < 90°$).
4. The building length AD is l, the building width AB is d, and the building height AE is h.

Fig. 2. Building and building floor.

When the sun's rays hit the building from different angles, one of the following 4 cases will occur: Case 1. Cut to the edge FE and edge HE; Case 2. Cut to the edge EF and edge GF; Case 3. Cut to the edge FG and edge HG; Case 4. Cut to the side EH and edge GH. The conditions for determining the node being obscured by the shadow of the building in Case 1 are given below.

$$- \cot A(a + d \sin \theta) + b + d \cos \theta \geq 0 \text{ and } -a \cot A + b \leq 0 \text{ and}$$

$$\cot \theta \left(-a + \frac{h \sin A}{\tan H_A} \right) + b - \frac{h \cos A}{\tan H_A} \geq 0 \text{ and } -a \cot \theta + b \leq 0$$

or

$$-a \cot A + b \geq 0 \text{ and } -\cot A(a + l \cos \theta) + b - l \sin \theta \leq 0 \text{ and}$$

$$a \tan \theta + b \geq 0 \text{ and } \tan \theta \left(a - \frac{h \sin A}{\tan H_A} \right) + b - \frac{h \cos A}{\tan H_A} \leq 0 \qquad (4)$$

The shadow judgement process in the other three cases is similar to that in case 1, so it is omitted here.

3.3 SWA Algorithm

This section gives the SWA algorithm. The main purpose of the algorithm is to solve the applicability of the SCP in the complex environment (shadow occlusion) of small nodes and the selection of the prediction period. In addition, considering the scalability problem of the centralized computing burden as the node size increases, the distributed computing under the "full power" state of the node is introduced to reduce the computing burden on the server. Details are as follows: if the server is used to centrally calculate the shadow judgement algorithm, when the number of nodes is large, the computational cost will increase approximately linearly. According to an actual measurement, when the server undertakes 20,000 nodes, it takes about 10 min to run the shadow judgement algorithm each time. In the morning or afternoon, due to the low sun height, the nodes are easily blocked by building shadows. The SWA algorithm may be triggered multiple times in a short period of time, significantly increasing the server's computational burden. Considering that distributed computing has good applications in many aspects [16–20], this article introduces it to reduce the computing burden of the server. After actual measurement, the running time of the shadow judgement algorithm on the node with capacitance 124 J only takes 2 ms, and consumes about 0.6 mJ (accounting for 0.00048% of the node capacitance). In terms of computing energy consumption, considering that the node is often in a "full power" state for some time, allowing the node in the full power state to run this algorithm can effectively reduce the burden on the server.

Figure 3 shows the main process of the algorithm. Among them, the reason for performing the second step 'Adjusting the '*step*' by the shadow' is as follows: As shown in Fig. 4a and Fig. 4b, the node charging power does not satisfy the conditions described in **Theorem 1** (i) and **Theorem 2** (ii) due to the decrease and rise. The main reason is that the node is occluded. Therefore, the prediction period should be properly adjusted according to the shaded time of the nodes to ensure that the node charging power satisfies the two theorems in the prediction period.

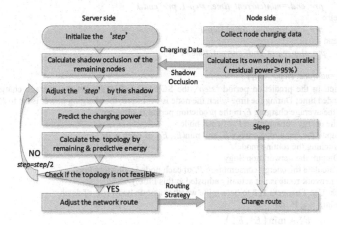

Fig. 3. The basic flow of SWA algorithm.

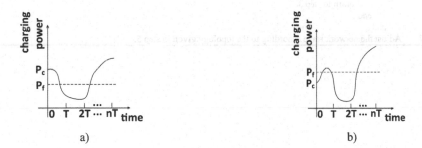

a)   b)

Fig. 4. Case 1 and case 2.

The SWA algorithm is as follows:

Algorithm 1. Solar WSN Adaptive Algorithm (SWA)

---

1. Initialize the '*step*'. Refer to the test results of the predictive model, a period '*step*' is selected as long as possible under the condition that the prediction accuracy allows.

2. Calculate the shaded time of the node. According to the shadow judgement algorithm, calculate the shaded time of each node in the prediction period '*step*'. The nodes whose current capacity is greater than 95% of the capacitance are calculated in parallel, and the remaining nodes are calculated by the server.

3. Adjust the '*step*' according to the shaded time of the node.(t is the time at which the node v is first occluded by the building shadow in the prediction period '*step*'; pre_end_v is the last occluded time of node v in the continuous shaded time starting from time t.)

   while $v \in V$ do
       if (In the period '*step*', the node v has an occluded time)
           pre_end_v=min(current_time+step-1, pre_end_v)
       else
           pre_end_v=current_time+step-1
       end
           pre_step_v= pre_end_v-current_time+1
   end
   step=min(pre_step_v), $v \in V$

4. Predict. In the prediction period '*step*', the SCP is used to predict the average charging of the unshaded time; During the time when the node is occluded, the charging power is set to P_{in_shade}, and then the average charging E_f in the prediction period is obtained.

5. Get the energy consumption P_e of each node.
   Input: the remaining energy E_r and $min(E_c, E_f)$ of each node
   Running the routing model.
   Output: the network topology
   Calculate the energy consumption P_e of each node in this topology.
   (The network route is not actually adjusted at this time.)

6. Check. Use **Theorem 1, 2** to determine whether a node has a risk of death:
   while $v \in V$ do

       if $\dfrac{E_r^v + min\left(E_c^v, E_f^v\right)}{step} < P_e^v$

       step=step/2
       return to step 3
       end
   end

7. Adjust the network route according to the topology given in step 5.

Table 2. Parameter settings

Configuration	Value
Number of nodes	20~200
Communication radius	8 m
Initial power of the node	40 mWh
Deployment density (even)	$0.625/m^2$
Default collecting cycle	1/s
Sink location	Center of the circle
Photovoltaic panel area	$110 * 80 \ (mm^2)$
Photovoltaic panel conversion efficiency	18%
Energy consumption (sending)	0.268 J/time
Energy consumption (receiving)	0.161 J/time
Unit time	5 min

4 Experimental

This section simulates the SWA algorithm for the following purposes. 1) According to the SWA algorithm in Sect. 3, verify the effectiveness of the existing SCPs, including ANN linear [1], Ipro energy [2], and a routing algorithm named EEDSRS [3] combined with the SWA algorithm, and integrating them into SWAF. The main purpose is to verify the reduction of power consumption and the improvement of network lifetime compared with the original algorithms without SWA algorithm (NO-SWA). 2) Analyze the cost of topological reorganizations and time complexity in the SWA algorithm. Detailed parameters are shown in Table 2. Among them, the conversion efficiency of photovoltaic panels is determined by actual measurements. The energy consumption of the data transmission is used the high-accuracy power monitor (Monsoon FTA22) on the TI CC 2538 node.

The solar data required for the experiment came from the actual data of Denver, USA in January 2018, published by the US Renewable Energy Laboratory. The ANN-linear, Ipro-energy, EEDSRS and SWA algorithm are implemented, the resulting of prediction period length and harvested energy are used as input to the EEDSRS. The EEDSRS is implemented in PyCharm, the network energy consumption and node residual energy are transmitted back to the SWA algorithm.

First, the Denver solar irradiance data from January 2 to 31, 2019 was used to verify the prediction accuracy improvement and prediction cost reduction brought by the judgement method of shadow. Three prediction models, ANN-linear [1], Ipro-energy [2], and ANN [11], are tested here. As described in Sect. 3.2, it is assumed here that the angle θ is 0, the building is 50 m long, 30 m wide, and 100 m high, and the point A coordinate is $(0, -40)$. Therefore, by running the shadow judgement algorithm, it can be obtained that the node is blocked by the building shadow every day in the time period of about 7:30–12:00. Figure 5 shows the solar irradiance prediction error of the prediction period of 1 h. Since it is impossible to train each node one by one in actual use, the training set uses solar irradiance data without shadow. We use two commonly used error evaluation indicators, MAE and RMSE, focusing on the prediction error value. From Fig. 5, after adding the shadow judgement algorithm, the prediction error of the prediction algorithm under the two prediction error indicators has decreased significantly. It can be seen that the shadow judgement is of great practical significance. In addition, there is no need to use a more complicated SCP for prediction during the time when the node is shaded, which also reduces the computational cost.

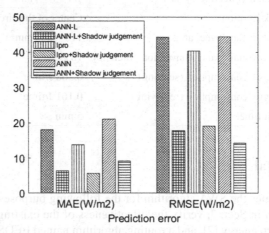

Fig. 5. Comparison of forecast errors under 1 h prediction period.

It can be seen from Fig. 6a that under the LST1, in the case of ANN-linear and Ipro-energy using SWA algorithm, nodes harvest 7~25 J and 2~16 J of energy in a time unit. In the case of NO-SWA (120 min), the nodes consume an average of 60–103 J and 65–95 J of energy in a time unit. This is because the core idea of the EEDSRS is to reduce the workload of nodes with less residual energy at the expense of increasing hops. In this situation, the total energy consumption of the network becomes greater, and we call this "Expand Hops". Compared with the NO-SWA algorithm, the SWA can capture some nodes with less residual energy at the moment but more harvest energy in the future. Since there are more solar charges for such nodes subsequently, they will not die. Therefore, letting such nodes continue to work can effectively avoid the phenomenon of "Expand Hops" and significantly reduce the average energy consumption of the entire network. From Fig. 6b, under the life standard of death of 50% node death (hereinafter referred to LST2) [10], the nodes using SWA have more energy harvested in a time unit than LST1, reaching 6 to 40 and 9 to 28 J. However, based on the NO-SWA, the nodes consume 27 to 98 and 34 to 58 J in a time unit. This is because the LST2 has a looser requirement for network death, the SWA effect is more significant, and the average energy consumption of the node is lower than that under the LST1.

From Fig. 7a, under the LST1, the ANN-linear and the Ipro-energy based on SWA have a lifetime increase of 5% to 7% compared to the NO-SWAalgorithm. Because LST1 defines the network death more severely, SWA has less advantages. From Fig. 7b, under

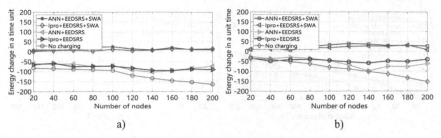

Fig. 6. Average energy change of nodes in a unit time.

the LST2, the SWA has a network life increase of 23% to 27% compared to NO-SWA. Figure 8a and Fig. 8b show the network lifetimes under LST1 and LST2 in the form of CDF indicators.

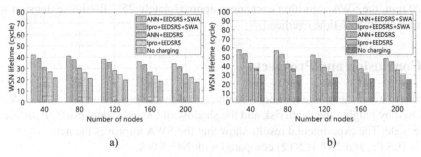

Fig. 7. Network lifetime.

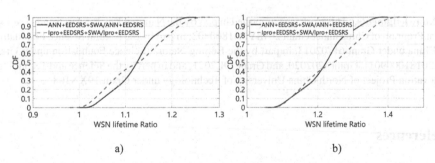

Fig. 8. Network lifetime CDF.

In this simulation experiment, we deployed the nodes in a complex environment, including buildings, trees, shadows, and cloudy conditions. Therefore, the frequency of SWA is faster. It can be seen from Fig. 9a that the number of network topology oscillations based on the SWA is 1.5 to 1.8 times of NO-SWA. In the case of frequently changing charging conditions, we extend network life at the cost of increasing the number of topological oscillations.

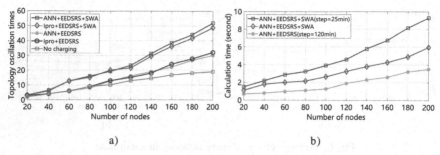

Fig. 9. Topology oscillation and calculation times.

From Fig. 9b, the time complexity of the SWA algorithm is between the two NO-SWAs (25 min, 120 min). This is because the longer the prediction period, the fewer the number of predictions and vice versa. Compared to NO-SWA with a prediction period of 120 min, the SWA algorithm exchanges approximately 25% lifetime extension with approximately 46% time overhead.

5   Conclusion and Prospect

In order to avoid the problem of node death caused by the prediction period, the SWA is proposed by judging the death risk and the shadow of the individual position difference of the node. The experimental results show that the SWA improves the network lifetime by 7% (LST1) and 25% (LST2) compared with NO-SWA.

We intend to introduce a variety of environmental characteristics (humidity, temperature, wind speed) as parameters, and further study the SCPs to improve accuracy.

Acknowledgments. This work was supported in part by the National Key Research and Development Program of China under Grant 2018YFB1800302, in part by the Natural Science Foundation of China under Grant 61702013, in part by the Beijing Natural Science Foundation under Grant KZ201810009011, Grant 4202020, and Grant 19L2021, and in part by the Science and Technology Innovation Project of North China University of Technology under Grant 19XN108.

References

1. Bao, Y., et al.: Solar radiation prediction and energy allocation for energy harvesting base stations. In: ICC 2014 - 2014 IEEE International Conference on Communications. IEEE (2014)
2. Muhammad, K.Q.H., Umber, S., et al.: Harvested energy prediction schemes for wireless sensor networks: performance evaluation and enhancements. Wirel. Commun. Mob. Comput. **2017**, 1–14 (2017)
3. Lu, T., Liu, G., Chang, S.: Energy-efficient data sensing and routing in unreliable energy-harvesting wireless sensor network. Wirel. Netw. **24**(3), 1–15 (2016)
4. Cuicui, L., et al.: A sparsity feedback-based data gathering algorithm for wireless sensor networks. Comput. Netw. **141**, 145–156 (2018)

5. Xin, J., Liu, W., Zhang, C.: An energy conserving and transmission radius adaptive scheme to optimize performance of energy harvesting sensor networks. Sensors **18**, 2885 (2018)
6. Lv, C., Zhang, T., Ma, F., Yue, D.: A very short-term online forecasting model for photovoltaic power based on two-stage resource allocation network. In: 2018 International Joint Conference on Neural Networks (IJCNN 2018) (2018)
7. He, Y., Zheng, Y.: Making series of achievements for building entity measured by terrestrial laser scanning. In: 2016 4th International Workshop on Earth Observation and Remote Sensing Applications (EORSA). IEEE (2016)
8. Walraven, R.: Calculating the position of the sun. Solar Energy **20**(5), 393–397 (1978)
9. Adu-Manu, K.S., Adam, N.: Energy-harvesting wireless sensor networks (EH-WSNs): a review. ACM Trans. Sensor Netw. **14**(2), 1–50 (2018). Article 10
10. Isabel, D., Falko, D.: On the lifetime of wireless sensor networks. ACM Trans. Sensor Netw. **5**(1), 1–39 (2009)
11. Rodriguez, F., et al.: Predicting solar energy generation through artificial neural networks using weather forecasts for microgrid control. Renew. Energy **126**, 855–864 (2018)
12. Ahmed, F., Tamberg, G., Le Moullec, Y., Annus, P.: Dual-source linear energy prediction (LINE-P) model in the context of WSNs. Sensors **7**, 1666 (2017)
13. Zhang, R., et al.: Utility maximization of temporally correlated sensing data in energy harvesting sensor networks. IEEE Internet Things J. **6**(3), 5411–5422 (2019)
14. Tan, H.O., Korpeoglu, I., Stojmenovi, I.: Computing localized power-efficient data aggregation trees for sensor networks. IEEE Trans. Parallel Distrib. Syst. **22**(3), 489–500 (2011)
15. Lin, H.C., Chen, W.Y.: An approximation algorithm for the maximum-lifetime data aggregation tree problem in wireless sensor networks. IEEE Trans. Wirel. Commun. **16**(6), 2017, PP (99) 1–1 (2017)
16. Qiu, H., et al.: A user-centric data protection method for cloud storage based on invertible DWT. IEEE Trans. Cloud Comput. 1 (2019)
17. Qiu, M., Ming, Z., Li, J., Liu, S., Wang, B., Lu, Z.: Three-phase time-aware energy minimization with DVFS and unrolling for chip multiprocessors. J. Syst. Archit. **58**(10), 439–445 (2012)
18. Tian, Z., et al.: Block-DEF: a secure digital evidence framework using blockchain. Inf. Sci. **491**, 151–165 (2019)
19. Qiu, M., Dai, W., Vasilakos, A.V.: Loop parallelism maximization for multimedia data processing in mobile vehicular clouds. Cloud Comput. IEEE Trans. **7**(1), 250–258 (2019)
20. Qiu, M., et al.: Online optimization for scheduling preemptable tasks on IaaS cloud systems. J. Parallel Distrib. Comput. **72**(5), 666–677 (2012)

Formalizing and Verifying Decentralized Systems with Extended Concurrent Separation Logic

Yepeng Ding$^{(\boxtimes)}$ and Hiroyuki Sato

The University of Tokyo, Tokyo, Japan
{youhoutei,schuko}@satolab.itc.u-tokyo.ac.jp

Abstract. Decentralized techniques are becoming crucial and ubiquitous with the rapid advancement of distributed ledger technologies such as the blockchain. Numerous decentralized systems have been developed to address security and privacy issues with great dependability and reliability via these techniques. Meanwhile, formalization and verification of the decentralized systems is the key to ensuring correctness of the design and security properties of the implementation. In this paper, we propose a novel method of formalizing and verifying decentralized systems with a kind of extended concurrent separation logic. Our logic extends the standard concurrent separation logic with new features including communication encapsulation, environment perception, and node-level reasoning, which enhances modularity and expressiveness. Besides, we develop our logic with unitarity and compatibility to facilitate implementation. Furthermore, we demonstrate the effectiveness and versatility of our method by applying our logic to formalize and verify critical techniques in decentralized systems including the consensus mechanism and the smart contract.

Keywords: Decentralized system · Extended concurrent separation logic · Formal methods · Consensus · Smart contract

1 Introduction

Nowadays, decentralized technology has evolved into a new stage with the advancement of many kinds of decentralized techniques such as the consensus mechanism, smart contract. Based on these decentralized techniques, numerous solutions have been proposed to circumvent security and privacy issues widely existing in centralized systems such as unauthorized disclosure, deception, disruption, and usurpation, which is attracting a huge amount of attention from both academy and industry. These solutions have been developed as decentralized systems and applied to a wide range of fields such as economics [7], the

This research was partially supported by KAKENHI (Grants-in-Aid for Scientific Research) (C) 19K11958 and (B) 19H04098.

Internet of things [11], smart health [13], and infrastructures [12]. However, it is challenging to ensure the correctness and properties of decentralized techniques and systems due to the high complexity of the design and intricate factors.

Formal methods play an important role in verifying complex systems by specifying systems with rigorous mathematical syntax and semantics to eliminate imprecision and ambiguity of the design and implementation. Significant contributions have been made to verify distributed systems with two mainstream techniques of formal methods including model checking and automated theorem proving. Model-checking techniques [15,22,33] prove to be effective to check safety and liveness properties of distributed system design and algorithms. However, it is noteworthy that these model checking techniques suffer from the state-space explosion problem, which leads to the ineffectiveness of complicated realistic systems, though many model-checking tools such as SPIN [19], NuSMV [9], Cubicle [10] provide space-efficient and on-the-fly algorithm to optimize the methods. Meanwhile, theorem proving techniques [17,27,31] provide a deductive method to prove the correctness and properties of distributed systems in a mathematical style without the requirement of the exploration of the state space. To date, the proof assistant is the main tool for the development of formal proofs with the collaboration of humans and machines such as Coq [35] and Isabelle [20]. These formal methods make it possible to locate design flaws and implementation pitfalls and ensure correctness and properties. Particularly, sound logic is imperative in formal reasoning about program correctness. Concurrent separation logic (CSL) [25] has been widely used to reason about concurrent systems such as cryptographic implementation [2], the concurrent operating system kernels [38]. As the extension of separation logic [21,23], CSL enhances the modularity to reason about resources locally, which makes it capable of formalizing the disjoint concurrency and inter-process interactions.

Admittedly, it is natural that decentralized technology highly requires the guarantee of correctness and system properties since there is no central entity to supervise and monitor network and system behaviors, especially in untrustworthy environments. Nevertheless, formal reasoning about decentralized systems with the standard CSL is restrictive.

- CSL lacks encapsulated components to reason about the communication that is the basic and non-negligible component in the specification of decentralized systems to formalize interactions.
- CSL has restrictive expressiveness of reasoning about the environment factor and temporal conditions that can facilitate formalizing complex protocols of decentralized systems.
- CSL focuses on reasoning about low-level programs such as memory management and resource relationship. It is restrictive for CSL to reason about high-level systems such as nodes in decentralized systems.

Hence, it is still a significant challenge to reason about decentralized systems with an effective logic with rich expressiveness and high modularity.

In this paper, we propose a method of formalizing and verifying decentralized systems with an extended concurrent separation logic with three novel features:

communication encapsulation, environment perception, and node-level reasoning. These features are developed to bridge the gulf of reasoning about decentralized systems with CSL. We encapsulate the formalization of communications as the basic specification and proof component. Besides, our logic enriches the expressiveness of CSL with the capability of environment perception. The environment perception also extends CSL into two-dimension reasoning including both spatial and temporal reasoning. Moreover, we introduce the node-level reasoning to enable our logic to formalize high-level decentralized systems, which enhances the modularity to specify and verify systems at different levels.

We summarize our main contributions as follows:

1. We propose a novel method of formalizing and verifying decentralized systems with a kind of extended concurrent separation logic. Our logic addresses the issues of CSL while reasoning about decentralized systems with novel features including communication encapsulation, environment perception, and node-level reasoning.
2. We formalize a consensus mechanism with our method to prove the effectiveness of reasoning about complex protocols in decentralized systems with our logic. Our logic simplifies high-level reasoning with great modularity and rich expressiveness. It also presents unitarity while reasoning about systems at different abstraction levels.
3. We also demonstrate our work by applying our logic to the specification and verification of smart contracts. We locate the design flaw of a typical smart contract with vulnerability to prove the effectiveness of formalizing and verifying decentralized applications.

The remainder of this paper is organized as follows. We give a short introduction to the related work in Sect. 2. We then illustrate our logic that is the core of our method in Sect. 3. Subsequently, we describe the method of the application of our logic in a consensus algorithm and a smart contract in Sect. 4. We discuss our work in Sect. 5, followed by the conclusion in Sect. 6.

2 Related Work

Prior work has made significant contributions to the formalization and verification of distributed systems such as [1,15,31,33]. Model checking and theorem proving are two main methods in these works with the support of typical tools such as HOL4 [16], TLC model checker [39], NuSMV [9]. In addition, reasoning about distributed systems with Hoare-style logic has also proved effective. Iron-Fleet [17] was proposed to build practical and provably correct distributed systems based on the blend of TLA-style state-machine refinement and Hoare-logic verification. DISEL [32] provided a framework for implementation and compositional verification of distributed systems and clients based on the distributed separation logic. Particularly, our logic improves the work [14] to make it practical to formalize and verify the decentralized systems.

In the meanwhile, the decentralized technology has been rapidly developed to address the common issues in centralized distributed systems since Bitcoin [29]. Due to the lack of supervision from central entities, ensuring the correctness and properties of decentralized systems during the development is imperative. Formal verification provides a mathematical approach to analyze the decentralized system in a rigorous manner. However, the specific research of the formalization and verification of decentralized systems just gets started with the boost of decentralized technology.

As one of the critical techniques in decentralized systems, the consensus algorithm has been applied with formal methods to ensure the correctness in the distributed environment. The agreement safety property of the PBFT [8] was proved in [28]. Besides, the Raft [24] state replication library was formally verified by Verdi [37], a framework for formal verification of distributed systems implemented in Coq.

In recent years, formal verification of smart contracts has been an attractive topic since TheDAO attack [3] that brought great damage to the cryptocurrency market and successfully transferred about $50M worth of Ether into the control of the attacker by exploiting the reentrancy vulnerability. In the prompt work [6], they proposed a method to translate smart contracts implemented in Solidity to F* [34] and decompile Ethereum virtual machine (EVM) bytecode into F* for formal verification. Later, model-checking-based approaches were proposed for formal verification of smart contracts. In the work [4], SPIN has been used to verify the correctness and properties of a smart contract template to reduce the potential errors.

The EVM that supports the execution of smart contracts has also been formally specified and verified. In [18], the formal specification of the EVM bytecode named KEVM was developed with $\mathbb{K}$ framework [30], which provides the foundations for the verification. A deductive verifier was constructed in [26] to precisely reason about possible behaviors of the EVM bytecode with KEVM.

3 Our Logic

The standard CSL has great significance in reasoning about concurrent programs with preeminent expressiveness. It is noteworthy that the standard CSL and typical variants can well support the formalization of parallel systems at the thread level or process level about memory management and resources. Our logic is compatible with CSL and inherits the capability of formalizing low-level systems. Based on that, we extend the standard CSL into formalizing high-level decentralized systems with better modularity and richer expressiveness.

3.1 Communication Encapsulation

In a decentralized system, communications among nodes are indispensable. It is hard for CSL to formalize the complex interactions among programs in an elegant manner. Hence, we simplify the formalization by encapsulating communications as the basic component.

Our logic defines a minimal program unit over (Var, Ch) in (1).

$$\mathfrak{P} \triangleq (L, A, \mathcal{E}, \hookrightarrow, L_0, g_0) \tag{1}$$

Here, Var is a set of typed variables and Ch is a set of channels. L is a set of locations and A is a set of actions. $\mathcal{E}$ denotes the effect function $A \times [\![Var]\!] \mapsto [\![Var]\!]$. The notation $\hookrightarrow \subseteq L \times \|Var\| \times A \times L$ represents the conditional transition relation. $L_0 \subseteq L$ and $g_0 \in \|Var\|$ denotes a set of initial locations and the initial condition respectively. $[\![Var]\!]$ denotes the set of variable evaluations. $\|Var\|$ denotes the set of Boolean conditions over Var.

For convenience, we use the notation $l \xrightarrow{g:\alpha} l'$ as shorthand for $(l, g, \alpha, l') \in \hookrightarrow$ where $l \in L$ and $\alpha \in A$, meaning that the program $\mathfrak{P}$ goes from location l to l' when the current variable evaluation $\eta \models g$. We connect the program unit with CSL by specifying $l \xrightarrow{g:\alpha} l'$ as $\{g\} \alpha \{g'\}$, where $\mathcal{E}(\alpha, \eta) \models g'$.

Let $c!s$ denote sending signal s via channel c and $c?v$ denote receiving a signal from channel c and assign the signal to variable v. A communication $\pi \in \Pi = \{c!s, c?v\}$ is an action where $c \in Ch, s \in Dom(c), v \in Var$ with $Dom(v) \supseteq Dom(c)$.

In a practical communication scenario, the finite asynchronous channel is commonly used. The essence of a finite asynchronous channel is a buffer with a capacity $Cap(c) \in \mathbb{N}^+$ and a domain $Dom(c)$. In this manner, a communication $c!s$ produces signal s into the buffer whereas a communication $c?v$ consumes a signal from the buffer while assigning it to variable v.

For example, two parallel programs $c!s$ and $c?v$ can be specified at a high specification level in (2).

$$\{s \mapsto -\} \ c!s \ \| \ c?v \ \{[\![v]\!] = [\![s]\!]\} \tag{2}$$

We can give a proof outline of this parallel system containing the communication between two parallel programs in Fig. 1.

$$\{s \mapsto -\}$$
$$\{s \mapsto - * \top\}$$
$$\{s \mapsto -\} \qquad \{\top\}$$
$$c!s \quad \| \quad c?v$$
$$\{\top\} \qquad \{[\![v]\!] = [\![s]\!]\}$$
$$\{\top * [\![v]\!] = [\![s]\!]\}$$
$$\{[\![v]\!] = [\![s]\!]\}$$

Fig. 1. Proof outline of $\{s \mapsto -\} \ c!s \ \| \ c?v \ \{[\![v]\!] = [\![s]\!]\}$.

Since we encapsulate the formalization of communications as the basic component, we have $\Pi \subseteq A$. The specification of communications can be automatically verified in the manner of Fig. 1.

3.2   Environment Perception

Our logic has the capability of perceiving the environment factors including foreign factor and native factor, which means that we can formulate specifications with both internal (native) and external (foreign) temporal conditions.

In our logic, we follow the style of Hoare triples as (3).

$$\{\Gamma, \gamma \wedge P\} \; \alpha \; \{\Gamma, \gamma' \wedge P'\} \tag{3}$$

Here, Γ is used to specify the foreign pre-conditions and post-conditions while γ and γ' are used to specify the native pre-conditions and post-conditions. P and P' are assertions with the same semantics of the standard CSL. $\alpha \in A$ is the action to change the state of programs.

Before illustrating the structure of the environment factor, we firstly introduce a partially ordered relation $\lhd$ defined in (4).

$$a \lhd a' \iff a = Pred(a') \tag{4}$$

Here, $a, a' \in \acute{A}$ and $Pred(a)$ denotes the predecessor action set of a. $\acute{A}$ is the set of occurred actions derived from a partial function $A \rightharpoonup \acute{A}$.

We use the notation $a \lhd a'$ as shorthand for $(a, a') \in \lhd$. Intuitively, $a \lhd a'$ means that action a happens before action a'. Furthermore, the relation $\lhd$ has transitivity that is $a \lhd a' \lhd a'' \implies a \lhd a''$. With this ordered relation, we define a finite action path ϱ as a finite action sequence $a_0 a_1 ... a_n$ such that $\forall i \in [0, n) :$ $(a_i, a_{i+1}) \in \lhd$, where $n \geq 1$ if the length of the sequence is greater than 1.

Here, we use the finite action path as the atomic proposition that can be formulated as the environment factor to express that the occurrence of actions in the path must be true. In fact, the environment factor can be formulated in any temporal logic such as linear temporal logic (LTL) to formalize temporal properties as conditions.

We consider a simple network consisting of two parallel programs. One program sends a signal s through channel c while another program receives a signal from channel c.

The sending program specified in (5) does not need to perceive the environment factor, meaning that it can send s at any time.

$$\begin{array}{c} \{\top, \top \wedge s \mapsto -\} \\ c!s \\ \{\top, c!s\} \end{array} \tag{5}$$

The receiving program can only execute the receiving action after perceiving that the signal has been sent, which is specified in (6).

$$\{\top, v \mapsto -\}$$
$$\{c!s, v \mapsto -\}$$
$$c?v$$ (6)
$$\{c!s, c?v \wedge \llbracket v \rrbracket = \llbracket s \rrbracket\}$$

To give the proof for environment extension, we introduce **Environment Composition Rule** in (7).

$$\frac{\{\Gamma_0, \Upsilon_0\} \; \alpha_0 \; \{\Gamma_0, \Upsilon_0'\} \; ... \; \{\Gamma_n, \Upsilon_n\} \; \alpha_n \; \{\Gamma_n, \Upsilon_n'\}}{\{\circledast_{i=0}^n \Upsilon_i\} \; \alpha_0 \; \| \; ... \; \| \; \alpha_n \; \{\circledast_{i=0}^n \Upsilon_i'\}}$$ (7)

(Environment Composition Rule)

For brevity, we use Υ to denote the conjunction of γ and P. The big star notation $\circledast_i^n$ denotes consecutive separating conjunction from index i to n.

It is notable that the inference in **Environment Composition Rule** eliminates the foreign environment naturally if we regard the proved parallel system as the highest level of specification.

In the example above, we can specify two programs locally and combine the local specifications to produce a high-level specification in Fig. 2. The verification can be done with **Environment Composition Rule**.

$$\{\top, s \mapsto - * v \mapsto -\}$$
$$c!s \qquad \| \qquad c?v$$
$$\{\top, (c!s \wedge \top) * (c?v \wedge \llbracket v \rrbracket = \llbracket s \rrbracket)\}$$

Fig. 2. Specification of the network in Sect. 3.2.

The capability of the environment perception enhances the expressiveness and extends the standard CSL into temporal formalization. The new proof rule also preserves the modularity to formalize a complex system with environment factors by permitting reasoning about programs locally.

3.3 Node-Level Reasoning

In decentralized systems, nodes are distributed in the network and interact with each other under specific protocols including communication protocols and execution protocols. Each node in the network perceives states of others by communication protocols regularizing the way of passing information in the network. With the information perceived from communication protocols and local states, a node enforces execution protocols with corresponding parameters. According to the scope of communication and execution protocols, it is reasonable to regard the enforcement of communication protocols as the foreign environment while the enforcement of execution protocols being the native environment.

In node-level reasoning, communications from programs on the same node and on other nodes can be distinguished with the introduction of node-level parallelism. Besides, the temporal conditions imposed by other nodes can be used as the environment factors while specifying a parallel system on a node. For instance, to do action $c!s_1$, the sending program on node N needs to satisfy that another program on node N has received a signal s_0 from another node N' through channel c and assigned s_0 to variable v, which can be specified as $\{c!s_0@N' \lhd c?v@N, s_1 \mapsto -\}$ $c!s_1$ $\{\top, c!s_1\}$.

To facilitate formalizing and verifying the node and the interactions among nodes, we extend the CSL with the support of the node parallel. We introduce **Node Environment Composition Rule** in (8) and **Node Composition Rule** in (9).

$$\frac{\{\Gamma_0, \Upsilon_0\} \; \alpha_0@N \; \{\Gamma_0, \Upsilon_0'\} \; ... \; \{\Gamma_n, \Upsilon_n\} \; \alpha_n@N \; \{\Gamma_n, \Upsilon_n'\}}{\{\hat{\Gamma}, \circledast_{i=0}^n \Upsilon_i\} \; \alpha_0@N \; \| \; ... \; \| \; \alpha_n@N \; \{\hat{\Gamma}, \circledast_{i=0}^n \Upsilon_i'\}} \tag{8}$$

(Node Environment Composition Rule)

$$\frac{\{\Gamma_0, \Upsilon_0\} \; \alpha_0@N_0 \; \{\Gamma_0, \Upsilon_0'\} ... \{\Gamma_n, \Upsilon_n\} \; \alpha_n@N_n \; \{\Gamma_n, \Upsilon_n'\}}{\{\circledast_{i=0}^n \Upsilon_i\} \; \alpha_0@N_0 \; \|_N \; ... \; \|_N \; \alpha_n@N_n \; \{\circledast_{i=0}^n \Upsilon_i'\}} \tag{9}$$

(Node Composition Rule)

Here, $\mathfrak{N}$ denote a set of nodes. We have $N_0, .., N_n \in \mathfrak{N}$. @ is the ownership relation between action and node. $(\alpha, N) \in @$ denotes action α happens at node N, meaning that node N has the ownership of action α. We use the notation $\alpha@N$ as shorthand for $(\alpha, N) \in @$. $\|_N$ is the notation for node-level parallel to distinguish with program-level parallel $\|$. For a node N, the foreign environment Γ includes the temporal conditions imposed by other programs on N and the temporal conditions imposed by other nodes. $\hat{\Gamma}$ denotes the foreign environment without the temporal conditions imposed by other programs on N.

In **Node Environment Composition Rule**, the foreign environment factors on node N are composite in the inference with the persistence of the temporal conditions imposed by other nodes and the elimination of conditions imposed by other programs on N. Local specifications on different nodes can be combined to make a high-level specification with **Node Composition Rule**.

4 Application

In this section, we present two practical applications to show how our method works with our logic including a consensus mechanism and a smart contract. These are two critical decentralized techniques widely used in decentralized systems. By formalizing and verifying them, it proves the effectiveness of our method applied in decentralized systems.

4.1 Consensus Mechanism

We use our logic to formalize the consensus mechanism, a critical role in decentralized techniques. Hashgraph [5] is a recently developed consensus algorithm, which adopts a directed acyclic graph (DAG) structure and proves effective in permissioned blockchain.

Let us recall the data structures in Hashgraph. An event e is a tuple defined in (10).

$$e \triangleq \langle TS, TX, SH, OH \rangle \tag{10}$$

Here, TS denotes a timestamp signed by the creator. TX is a set of transactions embedded into the event. SH and OH denote pointers pointing to a self-parent and an other-parent respectively. In this manner, all events associated with a set of transactions compose a DAG.

For a node N, we consider an immutable transaction list $\mathbf{T}$ to persist transactions accepted in the network. $\mathbf{T}$ is used to simplify the DAG structure. The acceptance is the alternative of *round received*. There is a set of events E associated with a set of transactions TX on N. Each event is either accepted by N or rejected by N. For brevity, we introduce two notations $\ominus$ and $\oplus$ to denote acceptance and rejection. Formally, we have definitions in (11).

$$
\begin{aligned}
\ominus e &\triangleq t \in e.TX \implies t \notin \mathbf{T} \\
\oplus e &\triangleq t \in e.TX \implies t \in \mathbf{T} \\
\ominus E &\triangleq \forall e \in E : t \in e.TX \implies t \notin \mathbf{T} \\
\oplus E &\triangleq \exists e \in E : t \in e.TX \implies t \in \mathbf{T}
\end{aligned}
\tag{11}
$$

In this paper, we consider a small network to illustrate our methodology of reasoning about the Hashgraph mechanism in an outline. With the support of modular reasoning, our logic can reason about nodes and programs on nodes separately.

There are four nodes N_0, N_1, N_2, N_3 in the network. Each node is deployed with programs enforcing the Hashgraph consensus mechanism. Generally, there must be at least one program for broadcasting events to other nodes and one program for receiving events from other nodes. All programs run in a concurrent manner.

We specify the outline of the consensus mechanism for a set of transactions TX in Fig. 3, where $\#$ denotes the occurred action that receives transactions from a client. E^i is a set of events created on node N_i. Let e_j^i denotes the jth event on node N_i where $j \in \mathbb{N}$, we have $\forall e_j^i \in E^i \setminus \{e_0^i\} : e_j^i.SH = \mathcal{H}(e_{j-1}^i)$ where $\mathcal{H}$ is the hash function. $P(N_i)'$ is the post-condition of node N_i, which specifies the termination condition of the consensus on TX. For instance, we have $P(N_0)'$ as follows:

$$P(N_0)' = \{\Diamond RR(E^0)@N_1 \wedge \Diamond RR(E^0)@N_2 \wedge \Diamond RR(E^0)@N_3), \Diamond RR(E^0)@N_0 \wedge [\![E^0]\!] = \oplus E^0\},$$

where $RR(E)$ denotes the action *round received*, meaning that an event $e \in E$ is accepted in some round. In this application, we use LTL to formulate the environment factor. Here, $\Diamond$ is a temporal modality denoting *eventually*.

$$\{\top, emp\}$$
$$\{\#, [\![E^0]\!] = \ominus E^0\} \quad \{\#, [\![E^1]\!] = \ominus E^1\} \quad \{\#, [\![E^2]\!] = \ominus E^2\} \quad \{\#, [\![E^3]\!] = \ominus E^3\}$$
$$\mathfrak{P}@N_0 \qquad \|_N \qquad \mathfrak{P}@N_1 \qquad \|_N \qquad \mathfrak{P}@N_2 \qquad \|_N \qquad \mathfrak{P}@N_3$$
$$P(N_0)' \qquad\qquad P(N_1)' \qquad\qquad P(N_2)' \qquad\qquad P(N_3)'$$
$$\{\top, [\![E^0]\!] = \oplus E^0 * [\![E^1]\!] = \oplus E^1 * [\![E^2]\!] = \oplus E^2 * [\![E^3]\!] = \oplus E^3\}$$

Fig. 3. Specification of the event consensus in Hashgraph network.

This is a high-level specification of the whole network to verify that all nodes will eventually get the consistent $\top$. With the support of **Node Composition Rule**, we can break down the specification and focus on specifying and verifying programs on nodes separately. On each node, there is a parallel system containing a set of programs, which can also be formalized separately.

Besides, We can combine several highly coupled programs as a module. For instance, *famous witness election* is a critical module to make the Hashgraph algorithm converge. Recall the concept of *witness* in Hashgraph that is the first event a node creates in each round. Each node will decide its witness events after executing action *round created*. The election is for witnesses to vote on a witness before the current round to determine whether it is famous. In fact, the vote is virtual, which means that there is not an actual vote based on communications formed in the network. The voting process is enforced on the local environment. Here, for a vote function, we can specify it as (12).

$$\{\top, \Diamond See(e, e') \wedge IsWitness(e) \wedge IsWitness(e')\} \ Vote(e, e') \ \{\top, IsFamous(e, e')\} \tag{12}$$

Here, the specification has the meaning that if a witness e can *see* another witness e', then e votes e' as a famous witness and the ballot is signed with e.

Besides, vote collection is also enforced on the local environment in the round after voting. Each node can make the election by itself without synchronous interactions with others based on a directed acyclic graph that is a data structure that can be trivially formalized in the standard CSL.

In this manner, our logic can effectively formalize such a module with modular reasoning and other critical modules such as *see* and *strongly see* with the support of environment perception.

Listing 1.1. Pseudocode of Malicious contract

```
1  contract Malicious {
2     ...
3     function Drain() {
4         myBank.Deposit.value(amount);
5         myBank.Withdraw.value(amount);
6     }
7
8     function () {
9         myBank.Withdraw.value(amount);
10    }
11    ...
12 }
```

4.2 Smart Contract

We follow the unitarity principle to unify the specification and verification of decentralized systems at different abstraction levels with our logic. Besides complex protocols such as consensus mechanisms, our logic can also specify and verify the decentralized applications developed by smart contracts.

We take a simple example of the smart contract slightly modified from [6], which is similar to the behavior of TheDAO attack. It defines a contract named *MyBank* with three functions *Deposit*() for depositing money, *Withdraw*(amount::Integer) for withdrawing money, and *Balance*() for looking up the balance. Another contract named *Malicious* behaves maliciously with two evil functions *Drain*() to trigger the malicious behavior and a fallback method, which is shown in Listing 1.1.

With our logic, we can specify the contract *MyBank* in Listing 1.2.

Then, we consider a concurrent specification and introduce a resource invariant RI_b where resource b is the balance. We have $RI_b \triangleq b \geq 0$. The function *Withdraw* is specified in (13) from a high-level while *Deposit* is specified in the same manner.

$$\{c?msg \wedge c?a, RI_b \wedge [\![b]\!] = n\}$$
$$Withdraw(a)' \qquad (13)$$
$$\{\top, RI_b \wedge [\![b]\!] = n - a\}$$

The function *msg.sender.call.value* in Listing 1.2 transfers control back to contract *Malicious* by invoking the fallback method. In the fallback method, a reentrant call *Withdraw*() is triggered in another thread or process of *MyBank'* as the shadow of *MyBank*, which can be specified in Fig. 4.

Listing 1.2. Pseudocode of MyBank contract

```
1   contract MyBank {
2       ...
3       {c?msg, balances[msg.sender] ↦ −}
4       function Deposit() {
5           {⊤, [[balances[msg.sender]]] = n}
6           balances [msg.sender] += msg.value;
7           {⊤, [[balances[msg.sender]]] = n + msg.value}
8       }
9       {⊤, [[balances[msg.sender]]] = n + msg.value}
10
11      {c?msg ∧ c?amount, balances[msg.sender] ↦ −}
12      function Withdraw(amount :: Integer) {
13          {⊤, [[balances[msg.sender]]] = n ∧ n ≥ amount}
14          if (balances [msg.sender] ≥ amount) {
15              msg.sender.call.value(amount);
16              balances [msg.sender] −= amount;
17          }
18          {⊤, [[balances[msg.sender]]] = n − amount]}
19      }
20      {⊤, [[balances[msg.sender]]] = n − amount]}
21      ...
22  }
```

$$\{⊤, emp * emp\}$$

$$\{c?msg, RI_b ∧ [[b]] = n\}$$
$$Deposit()$$
$$\{c?msg ∧ c?a, RI_b ∧ [[b]] = n + a\} \qquad\qquad \{fallback() ⊲ c'?msg ∧ c'?a, RI_b ∧ [[b]] = n\}$$
$$Withdraw(a)' \qquad\qquad\qquad ‖ \qquad\qquad\qquad Withdraw(a)'$$
$$\{⊤, RI_b ∧ [[b]] = n\} \qquad\qquad\qquad\qquad \{⊤, RI_b ∧ [[b]] = n − a\}$$
$$\{⊤, RI_b ∧ [[b]] = n\}$$

Fig. 4. Specification of the parallel smart contract system.

We can obtain that the specification of this parallel smart contract system cannot be proved in any inference rules while preserving the resource invariant and satisfying the post-condition if both *MyBank* contract program and its shadow program *MyBank'* are verified separately. Therefore, the design flaw is located. In this manner, our logic can prove the correctness of smart contracts and assist in finding design flaws in decentralized applications.

5   Discussion

The core of our method of formalizing and verifying decentralized systems is our extended concurrent separation logic. Our logic presents rich expressiveness with the encapsulation of communication formalization and the capability of environment perception. It also has great modularity with the node-level reasoning feature to enable formalizing decentralized systems from memory level to node level. Furthermore, the formalization and application at different abstraction levels present unitarity with unified forms.

In the meanwhile, our logic has compatibility, which allows the interpretation from our logic to the standard CSL. We have proved the soundness of our logic by the general structure of [36]. We have also mechanized our logic by an interpreter together with a mechanized CSL prover implemented in the Coq proof assistant to implement the applications in Sect. 4. However, it is not dependable to reduce our logic to the CSL for automated reasoning by the interpreter. We plan to mechanize our logic directly with the proof assistant.

We do not address formulating temporal properties as temporal conditions in the environment factors in this paper. However, it is straightforward to formulate them in temporal logic. For instance, we can express temporal properties with LTL formulae such as $\{\phi, \top\}\, \alpha\, \{\top, \top\}$ to add the constraint for the occurrence of α with the LTL formula ϕ.

6 Conclusion

Reasoning about decentralized systems during the development is indispensable to ensure the correctness and system properties since it is a significant challenge to establish trust without central entities in untrustworthy distributed environments. In this paper, we have proposed a novel method of formalizing and verifying decentralized systems with an extended concurrent separation logic enhanced by three novel features: communication encapsulation, environment perception, and node-level reasoning. Besides, we have applied our method to reason about the consensus mechanism to prove the effectiveness of reasoning about complex protocols with the support of new features. Furthermore, we have demonstrated the capability of formalizing and verifying smart contracts with an instance to locate the design flaw. Particularly, we follow the unitarity principle and compatibility principle to facilitate the implementation. We plan to optimize the mechanization of our logic directly with the proof assistant. Formalizing and verifying a more complicated decentralized system such as a blockchain platform is also on our schedule.

References

1. Aminof, B., Rubin, S., Stoilkovska, I., Widder, J., Zuleger, F.: Parameterized model checking of synchronous distributed algorithms by abstraction. VMCAI 2018. LNCS, vol. 10747, pp. 1–24. Springer, Cham (2018). https://doi.org/10.1007/978-3-319-73721-8_1
2. Appel, A.W.: Verification of a cryptographic primitive: SHA-256. ACM Trans. Program. Lang. Syst. (TOPLAS) 37(2), 1–31 (2015). ISBN 0164-0925 Publisher: ACM New York, NY, USA
3. Atzei, N., Bartoletti, M., Cimoli, T.: A Survey of attacks on ethereum smart contracts (SoK). In: Maffei, M., Ryan, M. (eds.) POST 2017. LNCS, vol. 10204, pp. 164–186. Springer, Heidelberg (2017). https://doi.org/10.1007/978-3-662-54455-6_8
4. Bai, X., Cheng, Z., Duan, Z., Hu, K.: Formal modeling and verification of smart contracts. In: Proceedings of the 2018 7th International Conference on Software and Computer Applications, pp. 322–326 (2018)

5. Baird, L.: The swirlds hashgraph consensus algorithm: Fair, fast, byzantine fault tolerance. Swirlds, Inc., Technical Report SWIRLDS-TR-2016 1 (2016)
6. Bhargavan, K., et al.: Formal verification of smart contracts: short paper. In: Proceedings of the 2016 ACM Workshop on Programming Languages and Analysis for Security, pp. 91–96 (2016)
7. Böhme, R., Christin, N., Edelman, B., Moore, T.: Bitcoin: Economics, technology, and governance. J. Econ. Perspect. 29(2), 213–38 (2015). ISBN 0895-3309
8. Castro, M., Liskov, B.: Practical Byzantine fault tolerance. In: OSDI, vol. 99, no. 1999, pp. 173–186 (1999)
9. Cimatti, A., Clarke, E., Giunchiglia, E., Giunchiglia, F., Pistore, M., Roveri, M., Sebastiani, R., Tacchella, A.: NuSMV 2: an opensource tool for symbolic model checking. In: Brinksma, E., Larsen, K.G. (eds.) CAV 2002. LNCS, vol. 2404, pp. 359–364. Springer, Heidelberg (2002). https://doi.org/10.1007/3-540-45657-0_29
10. Conchon, S., Goel, A., Krstić, S., Mebsout, A., Zaïdi, F.: Cubicle: a parallel SMT-based model checker for parameterized systems. In: Madhusudan, P., Seshia, S.A. (eds.) CAV 2012. LNCS, vol. 7358, pp. 718–724. Springer, Heidelberg (2012). https://doi.org/10.1007/978-3-642-31424-7_55
11. Ding, Y., Sato, H.: Bloccess: towards fine-grained access control using blockchain in a distributed untrustworthy environment. In: 2020 8th IEEE International Conference on Mobile Cloud Computing, Services, and Engineering (MobileCloud), pp. 17–22. IEEE (2020)
12. Ding, Y., Sato, H.: Dagbase: a decentralized database platform Using DAG-based consensus. In: 2020 IEEE 44rd Annual Computer Software and Applications Conference (COMPSAC). IEEE (2020). To appear
13. Ding, Y., Sato, H.: Derepo: a distributed privacy-preserving data repository with decentralized access control for smart health. In: 2020 7th IEEE International Conference on Cyber Security and Cloud Computing (CSCloud)/2020 6th IEEE International Conference on Edge Computing and Scalable Cloud (EdgeCom), pp. 29–35. IEEE (2020)
14. Ding, Y., Sato, H.: Extending concurrent separation logic to enhance modular formalization. arXiv preprint arXiv:2007.13685 (2020)
15. Fatkina, A., Iakushkin, O., Selivanov, D., Korkhov, V.: Methods of formal software verification in the context of distributed systems. In: Misra, S., et al. (eds.) ICCSA 2019. LNCS, vol. 11620, pp. 546–555. Springer, Cham (2019). https://doi.org/10.1007/978-3-030-24296-1_43
16. Gordon, M.J., Melham, T.F.: Introduction to HOL: A Theorem Proving Environment for Higher Order Logic. Cambridge University Press, Cambridge (1993)
17. Hawblitzel, C., et al.: IronFleet: proving practical distributed systems correct. In: Proceedings of the 25th Symposium on Operating Systems Principles, pp. 1–17 (2015)
18. Hildenbrandt, E., et al.: Kevm: A complete formal semantics of the ethereum virtual machine. In: 2018 IEEE 31st Computer Security Foundations Symposium (CSF), pp. 204–217. IEEE (2018)
19. Holzmann, G.J.: The model checker SPIN. IEEE Trans. Softw. Eng. 23(5), 279–295 (1997). ISBN 0098-5589 Publisher: IEEE
20. Isabelle. https://isabelle.in.tum.de/
21. Ishtiaq, S.S., O'hearn, P.W.: BI as an assertion language for mutable data structures. In: Proceedings of the 28th ACM SIGPLAN-SIGACT Symposium on Principles of Programming Languages, pp. 14–26 (2001)
22. Konnov, I., Veith, H., Widder, J.: On the completeness of bounded model checking for threshold-based distributed algorithms: Reachability. Inf. Comput. 252, 95–109 (2017). ISBN 0890-5401 Publisher: Elsevier

23. O'Hearn, P.W., Pym, D.J.: The logic of bunched implications. Bull. Symbol. Logic **5**(2), 215–244 (1999). ISBN 1079-8986 Publisher: Cambridge University Press
24. Ongaro, D., Ousterhout, J.: In search of an understandable consensus algorithm. In: Proceedings of the 2014 USENIX Conference on USENIX Annual Technical Conference, pp. 305–320 (2014)
25. O'hearn, P.W.: Resources, concurrency, and local reasoning. Theor. Comput. Sci. **375**(1–3), 271–307 (2007). ISBN 0304-3975 Publisher: Elsevier
26. Park, D., Zhang, Y., Saxena, M., Daian, P., Roşu, G.: A formal verification tool for ethereum VM bytecode. In: Proceedings of the 2018 26th ACM Joint Meeting on European Software Engineering Conference and Symposium on the Foundations of Software Engineering, pp. 912–915 (2018)
27. Rahli, V., Guaspari, D., Bickford, M., Constable, R.L.: Formal specification, verification, and implementation of fault-tolerant systems using EventML. Electron. Commun. EASST **72** (2015)
28. Rahli, V., Vukotic, I., Völp, M., Esteves-Verissimo, P.: Velisarios: byzantine fault-tolerant protocols powered by Coq. In: Ahmed, A. (ed.) ESOP 2018. LNCS, vol. 10801, pp. 619–650. Springer, Cham (2018). https://doi.org/10.1007/978-3-319-89884-1_22
29. Raval, S.: Decentralized Applications: Harnessing Bitcoin's Blockchain Technology. O'Reilly Media Inc, Newton (2016)
30. Rosu, G., Serbănută, T.F.: An overview of the K semantic framework. J. Logic Algebraic Program. **79**(6), 397–434 (2010). ISBN 1567-8326 Publisher: Elsevier
31. Sardar, M.U., Hasan, O., Shafique, M., Henkel, J.: Theorem proving based formal verification of distributed dynamic thermal management schemes. J. Parallel Distrib. Comput. **100**, 157–171 (2017). ISBN 0743-7315 Publisher: Elsevier
32. Sergey, I., Wilcox, J.R., Tatlock, Z.: Programming and proving with distributed protocols. Proc. ACM Program. Lang. **2**(POPL), 1–30 (2017). ISBN 2475-1421 Publisher: ACM New York, NY, USA
33. Souri, A., Rahmani, A.M., Navimipour, N.J., Rezaei, R.: A symbolic model checking approach in formal verification of distributed systems. Hum.-Centric Comput. Inf. Sci. **9**(1), 4 (2019). https://doi.org/10.1186/s13673-019-0165-x, ISBN 2192-1962 Publisher: Springer
34. Swamy, N., et al.: Dependent types and multi-monadic effects in F. In: Proceedings of the 43rd Annual ACM SIGPLAN-SIGACT Symposium on Principles of Programming Languages, pp. 256–270 (2016)
35. Welcome! — The Coq Proof Assistant. https://coq.inria.fr/
36. Vafeiadis, V.: Concurrent separation logic and operational semantics. Electron. Notes Theor. Comput. Sci. **276**, 335–351 (2011). ISBN 1571-0661 Publisher: Elsevier
37. Wilcox, J.R., et al.: Verdi: a framework for implementing and formally verifying distributed systems. In: Proceedings of the 36th ACM SIGPLAN Conference on Programming Language Design and Implementation. pp. 357–368 (2015)
38. Xu, F., Fu, M., Feng, X., Zhang, X., Zhang, H., Li, Z.: A practical verification framework for preemptive OS kernels. In: Chaudhuri, S., Farzan, A. (eds.) CAV 2016. LNCS, vol. 9780, pp. 59–79. Springer, Cham (2016). https://doi.org/10.1007/978-3-319-41540-6_4
39. Yu, Y., Manolios, P., Lamport, L.: Model checking TLA$^+$ specifications. In: Pierre, L., Kropf, T. (eds.) CHARME 1999. LNCS, vol. 1703, pp. 54–66. Springer, Heidelberg (1999). https://doi.org/10.1007/3-540-48153-2_6

PRIAG: Proximal Reweighted Incremental Aggregated Gradient Algorithm for Distributed Optimizations

Xiaoge Deng[iD], Tao Sun[(✉)], Feng Liu, and Feng Huang

National Laboratory for Parallel and Distributed Processing (PDL), College
of Computer, National University of Defense Technology, Changsha 410073, China
{nudtxgdeng,nudtsuntao}@163.com,
{richardlf,fhuang}@nudt.edu.cn

Abstract. Large-scale machine learning problems are nowadays tack-
led by distributed optimization algorithms, i.e., algorithms that leverage multiple workers for training. However, collecting the information
from all workers in every iteration is sometimes expensive or even prohibitive. In this paper, we propose an iterative algorithm called proximal reweighted incremental aggregated gradient (PRIAG) for solving a
class of nonconvex and nonsmooth problems, which are ubiquitous in
machine learning tasks and distributed optimization problems. In each
iteration, this algorithm just needs the information from one worker due
to the incremental aggregated method. Combined with the reweighted
technique, we only require an easy-to-calculate proximal operator to deal
with the nonconvex and nonsmooth properties. Using the Lyapunov function analysis method, we prove that the PRIAG algorithm is convergent
under some mild assumptions. We apply this approach to nonconvex nonsmooth problems and distributed optimization tasks. Numerical experiments on both synthetic and real data sets show that our algorithm
can achieve comparative learning performance, but more efficiently, compared with previous nonconvex solvers.

Keywords: Distributed optimization · Incremental method · Proximal
operator · Nonconvex optimization · Iteratively reweighted algorithm

1 Introduction

In recent years, nonconvex and nonsmooth models have been attracting increasing attention for its efficiency in distributed data processing and machine learning research. Among them, a great class can be presented as

$$\min_{x \in \mathbb{R}^n} \left\{ \Phi(x) := F(x) + H(x) \right\} \tag{1}$$

where $F(x) := \sum_{i=1}^{M} f_i(x)$ usually represents the loss function from M workers
and $H(x)$ models the regularization term. Additionally, we assume each compo-
nent function $f_i : \mathbb{R}^n \to \mathbb{R}$ is continuously differentiable, while the regularization

© Springer Nature Switzerland AG 2020
M. Qiu (Ed.): ICA3PP 2020, LNCS 12452, pp. 495–511, 2020.
https://doi.org/10.1007/978-3-030-60245-1_34

function $H : \mathbb{R}^n \to \mathbb{R}$ is not necessarily differentiable. Notable examples include constrained and regularized least squares problems that arise in various machine learning applications [10,29], distributed optimization problems that arise in wireless sensor network [12,23,31] as well as smart grid applications [15,17], constrained optimization of separable problems [3,25], communication systems and signal processing [6,11,13].

1.1 Related Work

The well-known method to solve (1) is the proximal gradient (PG) method proposed in [8], which uses the functions F and H separately at every iteration. When ∇F is Lipschitz continuous and function Φ is convex, this method can be shown to converge with the sublinear rate with a proper step-size. The PG method is also proved convergent for the nonconvex case but with a smaller step-size than the convex one [1]. However, the PG method requires a self-explanatory assumption that the proximal map of H (see detailed definition in Sect. 3) is easy to calculate, which may be broken in various nonconvex tasks. Typical examples in machine learning and computer vision [2,16,18,39] can be formulated as a class of nonconvex structured composite minimizations, in which $H(x) = h(g(x))$ with $h(y)$ being a nonnegative concave function while $g(x)$ being a nonnegative coordinate-wise convex mapping.

To solve the previously mentioned structured nonconvex composite minimization problems, the proximal iteratively reweighted (PIRE) algorithm is developed and studied in [20,33], which uses the linearization techniques to both losses and regularized parts. The PIRE is also extended to the matrix form [21,22,32] and constrained minimizations [34,35].

Besides the regularized part, the loss part also poses difficulties in using PIRE or PG. Two main reasons are : (a). if the number of workers M is huge, calculating the full gradient of F is sometimes expensive or even prohibitive; (b). when $(f_i)_{1 \le i \le M}$ is stored in different machines (like a network), we cannot read any f_i as wished. Consequently, it is impossible to get the full gradient of F in some iteration. These two reasons motivate us to employ the incremental method proposed in [4], which exploits the additive structure of the problem and updates the decision vector using one component function at a time. Instead of calculating the full gradient for the loss function F, the proximal incremental aggregated gradient (PIAG) method proposed in [36,37] uses an aggregated gradient which calculates the gradient of one component function in one iteration. Thus, we consider recruiting the gradient updating strategy to design a novel iteratively reweighted algorithm.

1.2 Our Contributions

This paper proposes an iterative algorithm called proximal reweighted incremental aggregated gradient (PRIAG) algorithm, which selects a component function to update the gradient of F and reweight the nonconvex nonsmooth part H to get a more appropriate estimation of the surrogate functions. Our algorithm only needs to calculate the gradient of one component function and an easy-to-calculate proximal operator for each iteration. Using the Lyapunov function analysis, we prove that the PRIAG algorithm is convergent, provided that the stepsize keeps smaller than some positive constant. This method is applied to solve nonconvex nonsmooth problems and distributed optimization tasks. Numerical experiments on both synthetic and real data sets demonstrate that our methods outperform previous nonconvex solvers.

2 Set Up

In this section, We will describe the optimization model in detail and give some examples. Then we will show how to solve the problem (2) by our PRIAG algorithm.

2.1 Nonconvex and Nonsmooth Model

This paper is devoted to the following nonconvex and nonsmooth minimizations

$$\min_x \left\{ \Phi(x) = F(x) + \sum_{j=1}^{n} h\left(g\left(x_j\right)\right) \right\} \tag{2}$$

where $F(x) := \sum_{i=1}^{M} f_i(x)$, $x = (x_1, x_2, \ldots, x_n)^\top \in \mathbb{R}^n$, and functions f_i, h and g satisfy the following assumptions:

(a) Function $f_i : \mathbb{R}^n \to \mathbb{R}$ is differentiable and the gradient function of f_i, namely, $\nabla f_i : \mathbb{R}^n \to \mathbb{R}^n$, is Lipschitz continuous with Lipschitz constant $L_i > 0$, i.e.

$$\|\nabla f_i(w) - \nabla f_i(\overline{w})\|_2 \le L_i \|w - \overline{w}\|_2 \tag{3}$$

where $\{w, \overline{w}\} \subset \mathbb{R}^n$.
(b) Function $h : \text{Im}(g) \to \mathbb{R}$ is a differentiable concave function.
(c) Function $g : \mathbb{R} \to \mathbb{R}$ is a convex function.

We highlight that in model (2), function F can be nonconvex. Although both F and h are differentiable, Φ may still be a nonsmooth and nonconvex function. Functions g and h are usually determined by the penalties that model the priors known about the desired solution (such as sparsity). Although with several assumptions on the structure and functions involved, model (2) is quite common. To demonstrate this claim, three application examples are presented in the next subsection.

2.2   Some Examples

This section contains several examples of model (2), in which the Assumptions (a), (b), and (c) can all be satisfied.

Example 1 (ℓ_q-norm linear regression). This example considers the classical ℓ_q-norm linear regression, which is widely studied in the sparse coding community [5, 7].

$$\min_{x} \left\{ \frac{1}{2} \sum_{i=1}^{M} \|A_i x - b_i\|_2^2 + \lambda \sum_{j=1}^{n} (|x_j| + \varepsilon)^q \right\} \tag{4}$$

where $A_i \in \mathbb{R}^{m \times n}, b_i \in \mathbb{R}^m, x \in \mathbb{R}^n, \varepsilon, \lambda > 0$, and $0 < q < 1$. In this example, $f_i(x) = \frac{1}{2}\|A_i x - b_i\|_2^2$, $h(t) = \lambda(t + \varepsilon)^q$, and $g(t) = |t|$. Notice that $\mathrm{Im}(g) = \mathbb{R}^+$, and $\lambda(t + \varepsilon)^q$ is concave on $\mathbb{R}^+$.

Example 2 (Smoothed ℓ_q-norm linear regression). This example aims to solve the smoothed ℓ_q-norm linear regression [19]

$$\min_{x} \left\{ \frac{1}{2} \sum_{i=1}^{M} \|A_i x - b_i\|_2^2 + \lambda \sum_{j=1}^{n} \left(x_j^2 + \varepsilon\right)^{\frac{q}{2}} \right\} \tag{5}$$

where $A_i \in \mathbb{R}^{m \times n}, b_i \in \mathbb{R}^m, x \in \mathbb{R}^n, \varepsilon, \lambda > 0$, and $0 < q < 1$. To fit the general model, we can use $f_i(x) = \frac{1}{2}\|A_i x - b_i\|_2^2$, $h(t) = \lambda(t + \varepsilon)^{\frac{q}{2}}$, and $g(t) = t^2$. We can see that $\lambda(t + \varepsilon)^{\frac{q}{2}}$ is concave when constrained on $\mathrm{Im}(g) = \mathbb{R}^+$.

Example 3 (Logistic regularized linear regression). This example is devoted to the minimization of the following problem [14, 38]

$$\min_{x} \left\{ \frac{1}{2} \sum_{i=1}^{M} \|A_i x - b_i\|_2^2 + \lambda \sum_{j=1}^{n} \log \left(\frac{|x_i| + \varepsilon}{\varepsilon} \right) \right\} \tag{6}$$

where $A_i \in \mathbb{R}^{m \times n}, b_i \in \mathbb{R}^m, x \in \mathbb{R}^n, \varepsilon, \lambda > 0$, and $0 < q < 1$. We set that $f_i(x) = \frac{1}{2}\|A_i x - b_i\|_2^2$, $h(t) = \lambda \log \left(\frac{t + \varepsilon}{\varepsilon} \right)$, and $g(t) = |t|$.

2.3   The PRIAG Algorithm

Let x^k denote the k-th iteration. The proximal reweighted incremental aggregated gradient (PRIAG) algorithm performs as

$$\begin{cases} v^k = \sum_{i=1}^{M} \nabla f_i(x^{k - \tau_{i,k}}), \\ x^{k+1} = \mathbf{prox}_{\gamma w^k \cdot \tilde{g}}(x^k - \gamma v^k), \end{cases} \tag{7}$$

where $\tau_{i,k}$ is the delay associated with f_i in the k-th iteration, γ is a parameter, $w^k := (h'(g(x_1^k)), \dots, h'(g(x_n^k)))$ is a reweighted vector, $\tilde{g} = (g(x_1), \dots, g(x_n))^\top$, and $w^k \cdot \tilde{g} = \sum_{i=1}^{n} h'(g(x_i^k))g(x_i)$.

The first equation in (7) indicates that v^k is an approximation of the full gradient $\nabla F(x^k)$. The delays $\tau_{i,k}$ come from the data selection style to implement the algorithm. In this paper, we assume it is bounded, which means all ∇f_i will be visited in a window. This is quite reasonable otherwise some information must be missed, and the iteration will not find the ground truth. A typical form is the cyclic selection method, in which

$$\tau_{i,k+1} = \begin{cases} 0, & \text{if } i \equiv k \bmod M, \\ \tau_{i,k} + 1, & \text{if else.} \end{cases} \tag{8}$$

Implementing PRIAG under cyclic selection rule is also very easy: we use a space to storage the gradients $\{\phi^1, \phi^2, \ldots, \phi^M\}$. In the k-th iteration, selecting $i = i_k = (k \bmod M)$, computing $\nabla f_i(x^k)$ and update the data as

$$\begin{cases} v \leftarrow v - \phi^i + \nabla f_i(x^k), \\ \phi^i \leftarrow \nabla f_i(x^k). \end{cases} \tag{9}$$

The iteration is completed by using $v^k \leftarrow v$ in the second iteration of (7). Due to that $w^k \cdot \tilde{g}$ is splittable, the updating of x^{k+1} is also

$$x_i^{k+1} = \mathbf{prox}_{\gamma h'(g(x_i^k))g}(x_i^k - \gamma v_i), \quad i \in \{1, 2, \ldots, n\}. \tag{10}$$

The iteration is fast if the proximal operator of g is readily available. Luckily, in various applications (like the examples), $g = |\cdot|$, and calculating this proximal operator is very easy. In each iteration, PRIAG requires much fewer computations than the full gradient as it calculates the gradient of one function rather than all. Thus, it is possible to implement in the case that M is huge. On the other hand, the PRIAG algorithm can solve several distributed optimization problems. Assume that data $\{f_i\}_{i=1}^M$ are stored in different nodes connected by a graph but do not communicate with each other for privacy. We first fix a routine across all nodes (some nodes may be visited more than once) and then employ a token that contains $\{\phi^1, \phi^2, \ldots, \phi^M\}$. After the token reaches node i, it will send x^k to the node and the node tells the token $\nabla f_i(x^k)$. After receiving $\nabla f_i(x^k)$, the token performs PRIAG. In this way, we still solve the distributed leaning problem with keeping privacy.

In the following, we further elaborate on the principle of the PRIAG algorithm. Use the definition (16), we have

$$\begin{aligned} x^{k+1} &= \operatorname*{argmin}_{x \in \mathbb{R}^n} \left\{ \gamma w^k \cdot \tilde{g}(x) + \frac{1}{2} \|x - (x^k - \gamma v^k)\|_2^2 \right\} \\ &= \operatorname*{argmin}_{x \in \mathbb{R}^n} \left\{ \frac{1}{2\gamma} \|(x - x^k) + \gamma v^k)\|_2^2 + w^k \cdot \tilde{g}(x) \right\} \\ &= \operatorname*{argmin}_{x \in \mathbb{R}^n} \left\{ \frac{\gamma}{2} \|v^k\|_2^2 + \langle v^k, x - x^k \rangle + \frac{1}{2\gamma} \|x - x^k\|_2^2 + w^k \cdot \tilde{g}(x) \right\} \tag{11} \\ &= \operatorname*{argmin}_{x \in \mathbb{R}^n} \Big\{ H(x^k) + w^k \cdot (\tilde{g}(x) - \tilde{g}(x^k)) + F(x^k) + \langle v^k, x - x^k \rangle \\ &\qquad\qquad + \frac{1}{2\gamma} \|x - x^k\|_2^2 \Big\}, \end{aligned}$$

Algorithm 1. PRIAG Algorithm.

Input: MaxIter: K; Step-size: γ. Initialize $x^1 = \mathbf{0}$.

1: **for** $k = 1, 2, \cdots, K$ **do**

2:     Choose the delay $\tau_{i,k}$, (e.g., (8));

3:     Update $v^k = \sum_{i=1}^M \nabla f_i(x^{k-\tau_{i,k}})$;

4:     Update $x^{k+1} = \mathbf{prox}_{\gamma w^k \cdot \tilde{g}}(x^k - \gamma v^k)$;

5:     **if** Convergence is achieved **then**

6:         Break;

7:     **end if**

8: **end for**

Output: x

---

where $H(x) := \sum_{j=1}^n h(g(x_j))$. In the last step, we change the item $\frac{\gamma}{2}\|v^k\|_2^2$ to $F(x^k)$ and add $H(x^k) - w^k \cdot \tilde{g}(x^k)$, because these items are independent of the argument x. On the other hand, noticing that $h(y)$ is concave, we have

$$H(x) \leq H(x^k) + w^k \cdot (\tilde{g}(x) - \tilde{g}(x^k)). \qquad (12)$$

The loss function $F(x)$, which has L-Lipschitz gradient, enjoys the following property

$$F(x) \leq F(x^k) + \langle \nabla F(x^k), x - x^k \rangle + \frac{L}{2}\|x - x^k\|_2^2. \qquad (13)$$

That is, instead of minimizing $\Phi(x)$ in (1) directly, the PRIAG algorithm update x^{k+1} by minimizing the sum of two surrogate functions (12) and (13), which correspond to two terms of $\Phi(x)$, respectively. The PRIAG algorithm is summarized in Algorithm 1.

3   Convergence Analysis

This section will present some preliminaries, including several definitions, as well as useful properties for analysis. In the following, we will show the key lemma and the main result.

3.1   Preliminaries

Definition 1 (Subdifferentials). *Let $J : \mathbb{R}^n \to (-\infty, +\infty]$ be a proper and lower semicontinuous function.*

(a) For a given $x \in dom(J)$, the Fréchet subdifferential of J at x, written $\hat{\partial}J(x)$, is the set of all vectors $u \in \mathbb{R}^n$ that satisfy

$$\liminf_{\substack{y \neq x \ y \to x}} \left\{ \frac{J(y) - J(x) - \langle u, y - x \rangle}{\|y - x\|_2} \geq 0 \right\}, \qquad (14)$$

where $\mathrm{dom}(J) := \{x \in \mathbb{R}^n : J(x) < +\infty\}$, *when* $x \notin \mathrm{dom}(J)$, *we set* $\hat{\partial}J(x) = \emptyset$.

(b) *The (limiting) subdifferential, or simply the subdifferential, of J at $x \in \mathbb{R}^n$, written $\partial J(x)$, is defined through the following closure process:*

$$\partial J(x) := \left\{ u \in \mathbb{R}^n : \exists x^k \to x, J(x^k) \to J(x) \, and \, u^k \in \hat{\partial}J(x^k) \to u \, as \, k \to \infty \right\} \tag{15}$$

The definition of subdifferential plays a central role in convex optimization. The details can be found in [24,30].

Definition 2 (Proximal Operator). *The proximal operator* $\mathbf{prox}_f : \mathbb{R}^n \to \mathbb{R}^n$ *of f is defined by*

$$\mathbf{prox}_f(v) = \underset{x \in \mathbb{R}^n}{\mathrm{argmin}} \left\{ f(x) + \frac{1}{2}\|x - v\|_2^2 \right\} \tag{16}$$

where $\| \cdot \|_2$ *is the usual Euclidean norm.*

The proximal operator is an essential tool that well-suited to solve many modern optimization problems, particularly those involving nonsmooth regularization terms, modern computing, and distributed computing frameworks [26].

Our convergence analysis is heavily based on the following Lyapunov function, which is constructed by a delicate embedding of growth-type conditions into descent-type lemmas.

$$\Gamma_k(\xi) := \frac{L}{2\xi} \sum_{d=k-\tau}^{k-1} (d - (k - \tau) + 1) \left\|x^{d+1} - x^d\right\|_2^2 + \Phi\left(x^k\right) - \min \Phi \tag{17}$$

where ξ is parameter to be determined by step-size γ and τ (the bound for $\tau_{i,k}$). We highlight that in definition (17), $\Gamma_k(\xi) > 0$ for any $k \in \mathbb{N}$.

3.2 Key Result

Lemma 1. *Let F be a function (may be nonconvex) with L-Lipschitz gradient, h and g are concave and convex functions, respectively. Let $\{x^k\}_{k=1,2,\ldots}$ be the sequence generated by PRIAG algorithm, and $\max_{i,k}\{\tau_{i,k}\} \leq \tau < +\infty$. Choose the step-size*

$$\gamma_k \equiv \gamma = \frac{2c}{(2\tau + 1)L}, \tag{18}$$

for arbitrary fixed $0 < c < 1$. Then we can choose $\xi > 0$ to obtain

$$\Gamma_k(\xi) - \Gamma_{k+1}(\xi) \geq \frac{1}{4}\left(\frac{1}{\gamma} - \frac{L}{2} - \tau L\right)\left\|x^{k+1} - x^k\right\|_2^2 \tag{19}$$

Consequently, we have

$$\lim_{k \to \infty} \left\|x^{k+1} - x^k\right\|_2 = 0 \tag{20}$$

and

$$\min_{1 \leq i \leq k} \left\| x^{i+1} - x^i \right\|_2 = o \left(\frac{1}{\sqrt{k}} \right). \tag{21}$$

The proof is given in the appendix. Lemma 1 provides the sufficient descent of the Lyapunov function. Here, $\tau < +\infty$ is necessary; otherwise, we get $\gamma = 0$. For the cyclic rule, it is easy to see $\tau = M$, and the step-size requirement is then smaller than $\frac{2}{(2M+1)L}$. With the Lipschitz continuity of F, we are prepared to present the main result.

Theorem 1. *Assume the conditions of Lemma 1 hold and $\{x^k\}_{k=1,2,\ldots}$ is generated by PRIAG, then we have the following results:*

$$\lim_{k \to \infty} \text{dist} \left(\mathbf{0}, \partial \Phi \left(x^k \right) \right) = 0 \tag{22}$$

and

$$\min_{1 \leq i \leq k} \left\{ \text{dist} \left(\mathbf{0}, \partial \Phi \left(x^i \right) \right) \right\} = o \left(\frac{1}{\sqrt{k}} \right). \tag{23}$$

Assume x^* is any stationary point of $\{x^k\}_{k=0,1,2,\ldots}$. The results of Theorem 1 directly indicate that $\text{dist} \left(\mathbf{0}, \partial \Phi \left(x^* \right) \right) = 0$, which is actually $\mathbf{0} \in \partial \Phi(x^*)$. That is to say any stationary point of $\{x^k\}_{k=0,1,2,\ldots}$ is a critical point of Φ.

4   Applications

This section will contain the applications as well as some numerical experiments to demonstrate the effectiveness of the proposed PRIAG algorithm. All the numerical experiments are implemented by Matlab on a PC with 16 GB of RAM and Intel Core i7-7700.

4.1   Application to the Distributed Optimization

To illustrate that our algorithm can be used in distributed optimization problems, we first describe the implementation process of the algorithm in detail. Given the initial point x^1, we first compute $v^1 = \sum_{i=1}^{M} \nabla f_i(x^1)$. Then for $k \geq 1$, do

$$x^{k+1} = \mathbf{prox}_{\gamma w^k \cdot \tilde{g}}(x^k - \gamma v^k),$$

$$v^{k+1} = \begin{cases} v^k + \nabla f_{k+1}(x^{k+1}) - \nabla f_{k+1}(x^1), & k < M \\ v^k + \nabla f_{(k+1)_M}(x^{k+1}) - \nabla f_{(k+1)_M}(x^{k+1-M}), & k \geq M \end{cases} \tag{24}$$

where $(k)_M$ represents the modulo operation of k to M.

The key feature of the PRIAG method that makes it suitable for distributed optimization such as wireless sensor network applications is that it can be implemented in a distributed manner. Consider a distributed system of M processors enumerated over $1, 2, \cdots, M$, each of which can be access to one of the functions

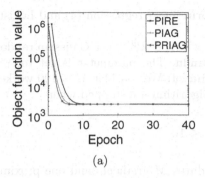
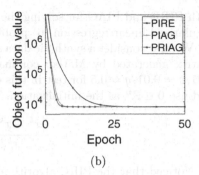

Fig. 1. Epoch v.s. Objective function value on the synthetic data. (a): ℓ_q minimization, (b): Logistic minimization.

$f_i(x)$. The calculation process (24) (also called as token) begins with x_1 at processor 1. Then, processor 1 sets $v^1 = \sum_{i=1}^{M} \nabla f_i(x^1)$ and transmits x^1 and v^1 to processor 2. Upon receiving x^{k-1} and v^{k-1} from processor $k-1$, processor k calculates x^k and v^k according to (24) and transmits them to processor $k+1$. Therefore, after $M-1$ iterations we obtain $x^1, \cdots, x^M$ and $v^M = \sum_{i=1}^{M} \nabla f_i(x^i)$. In that way, the algorithm progresses in a cyclic manner. Upon receiving x^{k-1} and v^{k-1} from processor $(k-1)_M$, processor $(k)_M$ computes x^k and v^k according to (24), and transmits them to processor $(k+1)_M$. Note that $\nabla f_{(k)_M}(x^{k-M})$ is available at processor $(k)_M$, since it was the last gradient computed at that processor. Therefore, the only gradient computation at processor $(k)_M$ is $\nabla f_{(k)_M}(x^k)$. At no phase of the algorithm do the processors share information regarding the component function $f_i(x)$ or its gradient $\nabla f_i(x)$.

There are two motivations to use the PRIAG method: (a) reduced computational burden due to the evaluation of a single gradient per iteration compared to M gradients required for the PIRE method, and (b) the possibility of a distributed implementation of the method in which each component has access to one of the functions $f_i(x)$. The second item is very useful in the literature of wireless sensor networks [27,28]. Wireless sensor networks provide a means for efficient large scale monitoring of large areas. Often the ultimate goal is to estimate certain parameters based on measurements that the sensors collect, giving rise to an optimization problem. If measurements from distinct sensors are modeled as statistically independent, the estimation problem takes the form of (1), where $f_i(x)$ is indexed by the measurements available at the sensor i. When transmitting the complete set of data to a central processor is impractical due to bandwidth and power constraints, the PRIAG method can be implemented in a distributed manner as described above.

4.2   Application to the Nonconvex and Nonsmooth Problem

This section applies our algorithm to solve the optimization problem of nonconvex nonsmooth object functions. We compare our proposed PRIAG algorithm with PIRE and PIAG for solving the ℓ_q-norm linear regression (4) and Logistic regularized linear regression (6) problems.

We first consider a synthetic data set. Each $A_i \in \mathbb{R}^{m \times n}$ is a Gaussian random matrix generated by Matlab command `randn`. The parameter is set to $\lambda = 0.01, \varepsilon = 0.01, q = 0.5$ for test in this experiment. We consider $M = 100$ workers and use $\mathbf{0} \in \mathbb{R}^n$ as the initialization. The algorithm are stopped when

$$\frac{\left\|x^k - x^{k+1}\right\|_2}{\left\|x^k\right\|_2} \leq 10^{-8}. \tag{25}$$

Noticed that the PIRE algorithm calculates M gradients and one proximal operator for each iteration, while the PRIAG and PIAG algorithms only need to calculate one gradient and one proximal operator once per iteration. Moreover, the numerical results show that the time to calculate a proximal operator is much less than the time to calculate a gradient. Therefore, we can consider the M iterations of the PRIAG and PIAG algorithms as an epoch. We plot the epoch v.s. objective function value on the synthetic data sets in Fig. 1, from which we can see that our algorithm can get the same solution as PIRE and PIAG, but with fewer epochs. This fact means that our algorithm is more efficient.

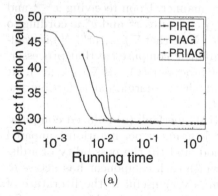 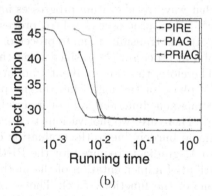

(a)   (b)

Fig. 2. Running time v.s. Objective function value on MNIST data sets. (a): ℓ_q minimization, (b): Logistic minimization.

We also plot the running time v.s. objective function value on MNIST data in Fig. 2 for minimizing the ℓ_q and logistic problems. We record the time required for each iteration and the value of the objective function to plot these figures. It can be seen that our algorithm completes the first iteration faster than PIRE and PIAG algorithm. In fact, the numerical results indicate that the calculation of our algorithm is faster in every iteration. In a nutshell, our algorithm decreases the objective function value faster when we get the same result.

The next experiment examined the recovery performance of sparse signals by our PRIAG method. The sparse signal x is generated by Matlab command `sprandn(n, 1, 0.1)`. The response $b = Ax + 0.001e$, where e is a Gaussian random vector. Given A and b, we can get the signal x^* by minimizing the ℓ_q problem by different methods. We use the recovery error $\|x^* - x\|_2/\|x\|_2$ to measure the recovery performance. As can be seen in Fig. 3 our method can achieve better recovery performance.

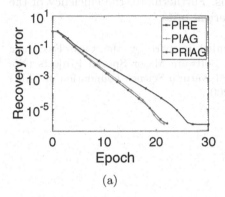
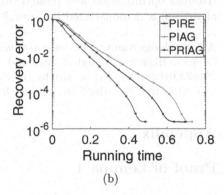

(a)   (b)

Fig. 3. ℓ_q minimization. (a): Epoch v.s. Recovery error, (b): Running time v.s. Objective function value.

The experiments on both synthetic and real data sets demonstrate that our methods achieve comparative performance but more efficient than previous nonconvex solvers. Next, we will compare and analyze these algorithms in detail.

Compared with the PIAG algorithm, numerical experiments have shown that our algorithm converges faster. More importantly, in some nonconvex and nonsmooth optimization models, the proximal operator of function H is difficult to calculate, and usually without explicit solutions. For this case, it takes more time to calculate the proximal operator in PIAG. In our algorithm, we calculate the proximal operator of g rather than H, which has a more straightforward form. In this way, we simplify calculations and programming to speed up the time required for optimization.

Compared with the PIRE algorithm, our algorithm only needs to calculate the gradient from one worker in each iteration. When dealing with the large-scale distributed tasks, it is difficult to calculate the gradient from all workers in every iteration, and then our algorithm will be more advantageous. Furthermore, we can flexibly select a component function to update the gradient in each iteration, and then apply our algorithm to distributed optimization.

5 Conclusion

In this paper, we propose the proximal reweighted incremental aggregated gradient algorithm for a class of nonconvex and nonsmooth problems. The algorithm only requires a single gradient evaluation and an easy-to-calculate proximal operator per iteration. We describe in detail how to apply our algorithm to the distributed optimization and related problems. Furthermore, the efficiency of the algorithm is demonstrated by several numerical tests.

Acknowledgement. This research was funded in part by the Core Electronic Devices, High-end Generic Chips, and Basic Software Major Special Projects (No. 2018ZX01028101), and in part by the National Natural Science Foundation of China (No. 61907034, No. 61932001, and No. 61906200).

Appendix

Proof of Lemma 1

Note that x^{k+1} is obtained by (7), combining with the Definition 1 and 2 we have

$$\frac{x^k - x^{k+1}}{\gamma} - v^k \in \partial w^k \cdot \tilde{g}\left(x^{k+1}\right). \tag{26}$$

With the convexity of g, we have

$$w^k \cdot \left(\tilde{g}\left(x^{k+1}\right) - \tilde{g}\left(x^k\right)\right) \le \left\langle \frac{x^k - x^{k+1}}{\gamma} - v^k, x^{k+1} - x^k \right\rangle. \tag{27}$$

The assumption (b) implies that the sum function F is differentiable with L-continues gradient, i.e.,

$$\|\nabla F(w) - \nabla F(\overline{w})\|_2 \le L\|w - \overline{w}\|_2, \tag{28}$$

where $L = \sum_{i=1}^m L_i$. We can further get

$$F\left(x^{k+1}\right) - F\left(x^k\right) \le \left\langle \nabla F\left(x^k\right), x^{k+1} - x^k \right\rangle + \frac{L}{2}\left\|x^{k+1} - x^k\right\|_2^2. \tag{29}$$

Then we have

$$\Phi(x^{k+1}) - \Phi(x^k) = F\left(x^{k+1}\right) - F\left(x^k\right) + \sum_{j=1}^{n} h\left(g\left(x_j^{k+1}\right)\right) - h\left(g\left(x_j^k\right)\right)$$

$$\leq \left\langle \nabla F\left(x^k\right), x^{k+1} - x^k \right\rangle + \frac{L}{2}\left\|x^{k+1} - x^k\right\|_2^2 + \sum_{j=1}^{n} h\left(g\left(x_j^{k+1}\right)\right) - h\left(g\left(x_j^k\right)\right)$$

$$\leq \left\langle \nabla F\left(x^k\right), x^{k+1} - x^k \right\rangle + \frac{L}{2}\left\|x^{k+1} - x^k\right\|_2^2 + \sum_{j=1}^{n} w_j^k\left(g\left(x_j^{k+1}\right) - g\left(x_j^k\right)\right)$$

$$= \left\langle \nabla F\left(x^k\right), x^{k+1} - x^k \right\rangle + \frac{L}{2}\left\|x^{k+1} - x^k\right\|_2^2 + w^k \cdot \left(\tilde{g}\left(x^{k+1}\right) - \tilde{g}\left(x^k\right)\right)$$

$$\leq \left\langle \nabla F\left(x^k\right), x^{k+1} - x^k \right\rangle + \frac{L}{2}\left\|x^{k+1} - x^k\right\|_2^2 + \left\langle \frac{x^{k+1} - x^k}{\gamma} + v^k, -x^{k+1} - x^k \right\rangle$$

$$= \underbrace{\left\langle \nabla F\left(x^k\right) - v^k, x^{k+1} - x^k \right\rangle}_{I} + \left(\frac{L}{2} - \frac{1}{\gamma}\right)\left\|x^{k+1} - x^k\right\|_2^2,$$

$$(30)$$

where $w_j^k := (h'(g(x_j^k)))$ and $\tilde{g} = (g(x_1), g(x_2), \ldots, g(x_N))^\top$. The first inequality uses (29). The second inequality uses the concavity of h. The third inequality uses (27). In the following, we will give the bound of I. First, notice that $\max_{i,k}\{\tau_{i,k}\} \leq \tau$, then

$$\left\|x^k - x^{k-\tau_{i,k}}\right\|_2 \leq \sum_{d=k-\tau_{i,k}}^{k-1} \left\|x^{d+1} - x^d\right\|_2 \leq \sum_{d=k-\tau}^{k-1} \left\|x^{d+1} - x^d\right\|_2. \quad (31)$$

Combining with (7), we have

$$I = \left\langle \nabla F\left(x^k\right) - v^k, x^{k+1} - x^k \right\rangle$$

$$= \left\langle \sum_{i=1}^{m} \left(\nabla f_i\left(x^k\right) - \nabla f_i\left(x^{k-\tau_{i,k}}\right)\right), x^{k+1} - x^k \right\rangle$$

$$\leq \sum_{i=1}^{m} L_i \left\|x^k - x^{k-\tau_{i,k}}\right\|_2 \cdot \left\|x^{k+1} - x^k\right\|_2 \quad (32)$$

$$\leq \sum_{i=1}^{m} L_i \left(\sum_{d=k-\tau}^{k-1} \left\|x^{d+1} - x^d\right\|_2\right) \cdot \left\|x^{k+1} - x^k\right\|_2$$

$$= L \sum_{d=k-\tau}^{k-1} \left\|x^{d+1} - x^d\right\|_2 \cdot \left\|x^{k+1} - x^k\right\|_2.$$

The first inequality uses the Lipschitz continuity of ∇f_i. The second inequality uses (31). Meanwhile, for any $\xi > 0$, we have the following Cauchy's inequality

$$\left\|x^{d+1} - x^d\right\|_2 \cdot \left\|x^{k+1} - x^k\right\|_2 \leq \frac{1}{2\xi}\left\|x^{d+1} - x^d\right\|_2^2 + \frac{\xi}{2}\left\|x^{k+1} - x^k\right\|_2^2. \quad (33)$$

Then we have

$$I \le \frac{L}{2\xi} \sum_{d=k-\tau}^{k-1} \left\| x^{d+1} - x^d \right\|_2^2 + \frac{\tau\xi L}{2} \left\| x^{k+1} - x^k \right\|_2^2. \tag{34}$$

Combining (30), (34), we have

$$\Phi(x^{k+1}) - \Phi(x^k) \le \frac{L}{2\xi} \sum_{d=k-\tau}^{k-1} \left\| x^{d+1} - x^d \right\|_2^2 + \left[\frac{(\tau\xi+1)L}{2} - \frac{1}{\gamma} \right] \left\| x^{k+1} - x^k \right\|_2^2. \tag{35}$$

If $0 < \gamma < \frac{2}{(2\tau+1)L}$, we can choose $\xi > 0$, such that

$$\xi + \frac{1}{\xi} = 1 + \frac{1}{\tau} \left(\frac{1}{\gamma L} - \frac{1}{2} \right). \tag{36}$$

Then, with direct calculations and substitutions, we have

$$\begin{aligned}
&\Gamma_k(\xi) - \Gamma_{k+1}(\xi) \\
&= \Phi\left(x^k\right) - \Phi\left(x^{k+1}\right) + \frac{L}{2\xi} \sum_{d=k-\tau}^{k-1} (d - (k-\tau) + 1) \left\| x^{d+1} - x^d \right\|_2^2 \\
&\quad - \frac{L}{2\xi} \sum_{d=k+1-\tau}^{k} (d - (k-\tau)) \left\| x^{d+1} - x^d \right\|_2^2 \\
&= \Phi\left(x^k\right) - \Phi\left(x^{k+1}\right) + \frac{L}{2\xi} \sum_{d=k-\tau}^{k-1} (d - (k-\tau) + 1) \left\| x^{d+1} - x^d \right\|_2^2 \\
&\quad - \frac{L}{2\xi} \sum_{d=k-\tau}^{k-1} (d - (k-\tau)) \left\| x^{d+1} - x^d \right\|_2^2 - \frac{L}{2\xi}\tau \left\| x^{k+1} - x^k \right\|_2^2 \\
&= \Phi\left(x^k\right) - \Phi\left(x^{k+1}\right) + \frac{L}{2\xi} \sum_{d=k-\tau}^{k-1} \left\| x^{d+1} - x^d \right\|_2^2 - \frac{L}{2\xi}\tau \left\| x^{k+1} - x^k \right\|_2^2 \\
&\ge (\frac{1}{\gamma} - \frac{(\tau\xi+1)L}{2} - \frac{L}{2\xi}\tau) \left\| x^{k+1} - x^k \right\|_2^2 \\
&= \frac{1}{4} \left(\frac{1}{\gamma} - \frac{L}{2} - \tau L \right) \left\| x^{k+1} - x^k \right\|_2^2.
\end{aligned} \tag{37}$$

The first inequality uses (35). The last equation uses (36). We then prove the first result. By summing the inequality (37), we have:

$$\sum_{k=1}^{\infty} \left\| x^{k+1} - x^k \right\|_2^2 < \infty. \tag{38}$$

The second then obviously holds. Using [Lemma 3, [9]], we are then led to

$$\min_{1 \leq i \leq k} \left\| x^{i+1} - x^i \right\|_2^2 = o\left(\frac{1}{k}\right), \tag{39}$$

which directly derives the third one.

Proof of Theorem 1

By the definition of subdifferential, we have

$$\frac{x^k - x^{k+1}}{\gamma} - v^k \in \partial w^k \tilde{g}\left(x^{k+1}\right). \tag{40}$$

That means

$$\frac{x^k - x^{k+1}}{\gamma} + \nabla F\left(x^{k+1}\right) - v^k \in \nabla F\left(x^{k+1}\right) + \partial H\left(x^{k+1}\right) = \partial \Phi\left(x^{k+1}\right), \tag{41}$$

where $H(x) := \sum_{j=1}^{n} h\left(g\left(x_j\right)\right)$. Thus we have

$$\text{dist}^2\left(0, \partial \Phi\left(x^{k+1}\right)\right) = \left\| \frac{x^k - x^{k+1}}{\gamma} + \nabla F\left(x^{k+1}\right) - v^k \right\|_2^2$$
$$\leq \frac{2\left\| x^{k+1} - x^k \right\|_2^2}{\gamma^2} + 2L^2\tau \sum_{d=k-\tau}^{k} \left\| x^{d+1} - x^d \right\|_2^2. \tag{42}$$

Combining with Lemma 1,

$$\sum_k \text{dist}^2\left(0, \partial \Phi\left(x^{k+1}\right)\right) < +\infty. \tag{43}$$

Still using [Lemma 3, [9]], the result can be proved.

References

1. Attouch, H., Bolte, J., Svaiter, B.F.: Convergence of descent methods for semi-algebraic and tame problems: proximal algorithms, forward-backward splitting, and regularized Gauss-Seidel methods. Math. Program. 137(1–2), 91–129 (2013)
2. Beck, A., Teboulle, M.: A fast iterative shrinkage-thresholding algorithm for linear inverse problems. SIAM J. Imaging Sci. 2(1), 183–202 (2009)
3. Bertsekas, D.P.: Incremental gradient, subgradient, and proximal methods for convex optimization: a survey. Optim. Mach. Learn. 2010(1–38), 3 (2011)
4. Blatt, D., Hero, A.O., Gauchman, H.: A convergent incremental gradient method with a constant step size. SIAM J. Optim. 18(1), 29–51 (2007)
5. Chartrand, R., Yin, W.: Iteratively reweighted algorithms for compressive sensing. In: 2008 IEEE International Conference on Acoustics, Speech and Signal Processing, pp. 3869–3872. IEEE (2008)

6. Chen, S.S., Donoho, D.L., Saunders, M.A.: Atomic decomposition by basis pursuit. SIAM Rev. **43**(1), 129–159 (2001)
7. Chen, X., Ng, M.K., Zhang, C.: Non-lipschitz ℓ_p-regularization and box constrained model for image restoration. IEEE Trans. Image Process. **21**(12), 4709–4721 (2012)
8. Combettes, P.L., Wajs, V.R.: Signal recovery by proximal forward-backward splitting. Multiscale Model. Simul. **4**(4), 1168–1200 (2005)
9. Davis, D., Yin, W.: Convergence rate analysis of several splitting schemes. In: Glowinski, R., Osher, S.J., Yin, W. (eds.) Splitting Methods in Communication, Imaging, Science, and Engineering. SC, pp. 115–163. Springer, Cham (2016). https://doi.org/10.1007/978-3-319-41589-5_4
10. Defazio, A., Bach, F., Lacoste-Julien, S.: SAGA: a fast incremental gradient method with support for non-strongly convex composite objectives. In: Advances in Neural Information Processing Systems, pp. 1646–1654 (2014)
11. Donoho, D.L.: Compressed sensing. IEEE Trans. Inf. Theory **52**(4), 1289–1306 (2006)
12. Duchi, J.C., Agarwal, A., Wainwright, M.J.: Dual averaging for distributed optimization: convergence analysis and network scaling. IEEE Trans. Autom. control **57**(3), 592–606 (2011)
13. Figueiredo, M.A., Nowak, R.D., Wright, S.J.: Gradient projection for sparse reconstruction: application to compressed sensing and other inverse problems. IEEE J. Sel. Top. Signal Process. **1**(4), 586–597 (2007)
14. Gasso, G., Rakotomamonjy, A., Canu, S.: Recovering sparse signals with a certain family of nonconvex penalties and DC programming. IEEE Trans. Signal Process. **57**(12), 4686–4698 (2009)
15. Giannakis, G.B., Kekatos, V., Gatsis, N., Kim, S.J., Zhu, H., Wollenberg, B.F.: Monitoring and optimization for power grids: a signal processing perspective. IEEE Signal Process. Mag. **30**(5), 107–128 (2013)
16. Gong, P., Ye, J., Zhang, C.: Robust multi-task feature learning. In: Proceedings of the 18th ACM SIGKDD International Conference on Knowledge Discovery and Data Mining, pp. 895–903 (2012)
17. Guo, F., Wen, C., Mao, J., Song, Y.D.: Distributed economic dispatch for smart grids with random wind power. IEEE Trans. Smart Grid **7**(3), 1572–1583 (2015)
18. Jacob, L., Obozinski, G., Vert, J.P.: Group lasso with overlap and graph lasso. In: Proceedings of the 26th Annual International Conference on Machine Learning, pp. 433–440 (2009)
19. Lai, M.J., Xu, Y., Yin, W.: Improved iteratively reweighted least squares for unconstrained smoothed ℓ_q minimization. SIAM J. Numeric. Anal. **51**(2), 927–957 (2013)
20. Lu, C., Wei, Y., Lin, Z., Yan, S.: Proximal iteratively reweighted algorithm with multiple splitting for nonconvex sparsity optimization. In: Twenty-Eighth AAAI Conference on Artificial Intelligence (2014)
21. Lu, C., Zhu, C., Xu, C., Yan, S., Lin, Z.: Generalized singular value thresholding. In: Twenty-Ninth AAAI Conference on Artificial Intelligence (2015)
22. Lu, Z., Zhang, Y.: Schatten-p quasi-norm regularized matrix optimization via iterative reweighted singular value minimization. arXiv preprint arXiv:1401.0869 (2015)
23. Mateos, G., Bazerque, J.A., Giannakis, G.B.: Distributed sparse linear regression. IEEE Trans. Signal Process. **58**(10), 5262–5276 (2010)
24. Mordukhovich, B.S.: Variational Analysis and Generalized Differentiation I: Basic Theory, vol. 330. Springer, Heidelberg (2006)
25. Padakandla, A., Sundaresan, R.: Separable convex optimization problems with linear ascending constraints. SIAM J. Optim. **20**(3), 1185–1204 (2010)

26. Parikh, N., Boyd, S., et al.: Proximal algorithms. Found. Trends® Optim. **1**(3), 127–239 (2014)
27. Rabbat, M., Nowak, R.: Distributed optimization in sensor networks. In: Proceedings of the 3rd International Symposium on Information Processing in Sensor Networks, pp. 20–27 (2004)
28. Rabbat, M.G., Nowak, R.D.: Decentralized source localization and tracking [wireless sensor networks]. In: 2004 IEEE International Conference on Acoustics, Speech, and Signal Processing. vol. 3, pp. iii–921. IEEE (2004)
29. Recht, B., Re, C., Wright, S., Niu, F.: Hogwild: a lock-free approach to parallelizing stochastic gradient descent. In: Advances in Neural Information Processing Systems, pp. 693–701 (2011)
30. Rockafellar, R.T., Wets, R.J.B.: Variational Analysis, vol. 317. Springer, Heidelberg (2009)
31. Shi, W., Ling, Q., Wu, G., Yin, W.: A proximal gradient algorithm for decentralized composite optimization. IEEE Trans. Signal Process. **63**(22), 6013–6023 (2015)
32. Sun, T., Jiang, H., Cheng, L.: Convergence of proximal iteratively reweighted nuclear norm algorithm for image processing. IEEE Trans. Image Process. **26**(12), 5632–5644 (2017)
33. Sun, T., Jiang, H., Cheng, L.: Global convergence of proximal iteratively reweighted algorithm. J. Global Optim. **68**(4), 815–826 (2017). https://doi.org/10.1007/s10898-017-0507-z
34. Sun, T., Jiang, H., Cheng, L., Zhu, W.: Iteratively linearized reweighted alternating direction method of multipliers for a class of nonconvex problems. IEEE Trans. Signal Process. **66**(20), 5380–5391 (2018)
35. Sun, T., Li, D., Jiang, H., Quan, Z.: Iteratively reweighted penalty alternating minimization methods with continuation for image deblurring. In: ICASSP 2019–2019 IEEE International Conference on Acoustics, Speech and Signal Processing (ICASSP), pp. 3757–3761. IEEE (2019)
36. Sun, T., Sun, Y., Li, D., Liao, Q.: General proximal incremental aggregated gradient algorithms: Better and novel results under general scheme. In: Advances in Neural Information Processing Systems, pp. 994–1004 (2019)
37. Vanli, N.D., Gurbuzbalaban, M., Ozdaglar, A.: Global convergence rate of proximal incremental aggregated gradient methods. SIAM J. Optim. **28**(2), 1282–1300 (2018)
38. Weston, J., Elisseeff, A., Schölkopf, B., Tipping, M.: Use of the zero-norm with linear models and kernel methods. J. Mach. Learn. Res. **3**(Mar), 1439–1461 (2003)
39. Wright, J., Yang, A.Y., Ganesh, A., Sastry, S.S., Ma, Y.: Robust face recognition via sparse representation. IEEE Trans. Pattern Anal. Mach. Intell. **31**(2), 210–227 (2008)

Decentralized Expectation Maximization Algorithm

Honghe Jin[1], Xiaoxiao Sun[2($\boxtimes$)], and Liwen Xu[3]

[1] University of Georgia, Athens, GA 30605, USA
hj49351@uga.edu
[2] University of Arizona, Tucson, AZ 85724, USA
xiaosun@arizona.edu
[3] North China University of Technology, Beijing 100144, China
xulw@ncut.edu.cn

Abstract. As a promising paradigm that does not require a central node, decentralized computing provides better network load balance and has advantages over centralized computing in terms of data protection. Although decentralized algorithms such as decentralized gradient descent algorithms have been extensively studied, there is no such research on the expectation maximization (EM) algorithm, which includes the expectation step (E-step) and the maximization step (M-step) and is a popular iterative method for missing data studies and latent variable models. In this paper, we propose decentralized EM algorithms under different communication and network topology settings such as synchronous communication and dynamic networks. Convergence analysis of the proposed algorithms is provided in the synchronous scenario. Empirical studies show that our proposed algorithms have numerical advantages over the EM algorithms based on local data or full data only, especially when there is no closed-form maximization in the M-step.

Keywords: Decentralized computing · EM algorithm · Asynchronous optimization · Dynamic network · Mixture model

1 Introduction

Decentralized optimization over a network, a paradigm to obtain optimal distributed decisions without a fusion center, has been implemented in a broad range of applications. Examples include distributed agreement problems [28], online learning [11,12,26], matrix completion [16,32], decentralized control [5,6], and distributed learning [10,17]. Compared to centralized computing, decentralized computing does not require a central server, i.e., fusion center, and thus can keep data safe and private [15,20,29]. The difference between centralized and decentralized topology is illustrated in Fig. 1. Existing work on decentralized optimization usually focuses on consensus optimization problems [1,4], aiming to obtain a global solution. For such a problem, decentralized gradient-based

© Springer Nature Switzerland AG 2020
M. Qiu (Ed.): ICA3PP 2020, LNCS 12452, pp. 512–527, 2020.
https://doi.org/10.1007/978-3-030-60245-1_35

algorithms and the subgradient variants have been proposed [9,22,31]. These algorithms assume that the problem is convex or strongly convex. The local solution to their problems converges to the exact global solution or the neighborhood of the global solution.

While decentralized gradient-based algorithms have been studied extensively, the expectation maximization (EM) algorithms in the context of decentralized computing are shown in the control community. Most of the methods for the distributed EM are designed for the centralized computing framework [23,30]. These methods require a central server communicating with all nodes. Therefore, the central server has information stored in each node. However, such a network topology is not feasible in some scenarios where data are stored/collected by each local node and are only shared with neighboring nodes due to the privacy concern and communication bandwidth constraint [33].

The EM algorithm sequentially iterates two steps: the expectation step (E-step), which calculates the expectation of the log-likelihood given current parameter estimates, and the maximization step (M-step), which updates parameters based on the expected log-likelihood. Due to the complexity of the problem, there are several difficulties associated with the theoretical analysis and implementation of the decentralized EM algorithm under the consensus optimiza-

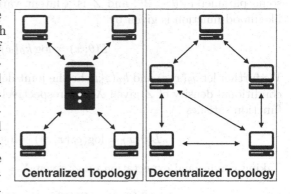

Fig. 1. A centralized topology versus a decentralized topology.

tion framework. First, the negative log-likelihood function usually is not convex. Most of the works on consensus optimization focus on the convex or strictly convex problems. Thus, it is difficult to adapt the existing theoretical results to our problem. Second, in many situations, the closed-form maximization cannot be obtained in the M-step, which makes the direct calculation of local estimate averages challenging [13,25].

In this paper, we develop decentralized EM algorithms under different network topology settings. In particular, our main contributions are:

1. We propose a general framework of decentralized EM algorithms which can be implemented in applications without a closed-form M-step.
2. We propose an asynchronous version of the decentralized EM algorithm and a decentralized EM algorithm in dynamic networks.
3. The convergence rate of our algorithm is shown as $O\left(\frac{1}{T} + \frac{1}{\sqrt{MT}}\right)$, where T is the number of iteration, and M is the number of nodes in a network.

The rest of the paper is organized as follows. Section 2 presents the background of EM algorithm and the decentralized optimization. Details of the proposed algorithms and the theoretical results are shown in Sect. 3. In Sect. 4, we present the experimental evaluation of the proposed algorithms.

2   Background

2.1   The EM Algorithm

As a popular statistical tool, the EM algorithm [8] has broad applicability in many problems, such as clustering [18,21], hidden Markov models [19], and topic models [3]. In the context of latent variable models, we consider a pair of random variables (X, Z), where X is the observed part with a density function $k_\theta(x)$ for some parameters $\theta \in \mathbb{R}^p$, and Z is a latent variable. Then the stochastic log-likelihood function is given by

$$L(\theta; x) = \log k_\theta(x).$$

We further let $g_\theta(x, z)$ and $h_\theta(z|x)$ be the joint density function of (X, Z) and the conditional density of Z given $X = x$ respectively. The stochastic log-likelihood function satisfies

$$L(\theta; x) = \log(g_\theta(x, z)) - \log(h_\theta(z|x)).$$

Taking conditional expectation on Z given $X = x$ at θ' on both sides, we have

$$L(\theta; x) = Q(\theta|\theta'; x) - H(\theta|\theta'; x),$$

where the stochastic auxiliary function $Q(\theta|\theta'; x) := \mathbb{E}_{\theta'}[\log g_\theta(x, Z)|x]$ and $H(\theta|\theta'; x) := \mathbb{E}_{\theta'}[\log h_\theta(Z|x)|x]$. In practice, we can only compute the sample auxiliary function

$$Q(\theta|\theta'; \{x_k\}) := \frac{1}{n} \sum_{k=1}^{n} Q(\theta|\theta'; x_k),$$

where $\{x_k\}_{k=1}^{n}$ are realizations of X. The sample auxiliary function is the empirical mean of the stochastic auxiliary function.

In a standard EM algorithm, we compute the sample auxiliary function in the E-step and maximize the function $Q(\theta|\theta'; \{x_k\})$ in the M-step. More specifically, on the $(t + 1)$th iteration, E- and M-steps are written as,

- E-step: Compute the sample auxiliary function $Q(\theta|\theta^{(t)}; \{x_k\})$.
- M-step: Choose $\theta^{(t+1)}$ that maximizes $Q(\theta|\theta^{(t)}; \{x_k\})$, i.e.,

$$\theta^{(t+1)} = \arg\max_{\theta \in \mathbb{R}^p} Q(\theta|\theta^{(t)}; \{x_k\}).$$

When the data set is large, that is, n is large, maximizing the auxiliary may be too expensive. Instead of maximizing the auxiliary function, the first-order EM algorithm operates by taking a gradient step in the M-step. For this algorithm, on the $(t+1)$th iteration, parameters are updated by: $\theta^{(t+1)} = \theta^{(t)} + \alpha \nabla Q(\theta|\theta^{(t)}; \{x_k\})$, where $\alpha > 0$ is an appropriately chosen step size. To further reduce the computational cost, one random sample can be selected to calculate the stochastic gradient. For the algorithm, the stochastic auxiliary functional, $Q(\theta|\theta'; X) := \mathbb{E}_{\theta'}[\log g_\theta(X, Z)|X]$ is needed to study its theoretical properties [27]. The stochastic log-likelihood functional $L(\theta; X)$ can be similarly defined.

2.2   The Decentralized Algorithm

In the decentralized algorithm, we consider a decentralized network with M nodes, each of which contains a local data set $\{x_{k,m}\}_{k=1}^{n_m}$, $m = 1, ..., M$, where $x_{k,m} \in \mathbb{R}^d$. Then, the consensus optimization problem is given by

$$\max_{\theta \in \mathbb{R}^p} \quad L(\theta) = \frac{1}{M} \sum_{m=1}^{M} L_m(\theta), \tag{1}$$

where $L_m(\theta) := \mathbb{E}L_m(\theta; \{X_{k,m}\})$ is the population log-likelihood function in the mth node, and $L_m(\theta; \{X_{k,m}\})$ is the empirical log-likelihood functional based on the stochastic log-likelihood functional $L(\theta; X)$. Without loss of generality, we select n random samples, $\{X_{k,m}\}_{k=1}^n$, from local data in each node at each iteration. If we set $n = 1$, the stochastic gradient algorithm is implemented. Particularly, the following updates are considered in the mth node,

$$\theta_m^{(t+1)} = \sum_{j=1}^{M} w_{mj}\theta_j^{(t)} + \alpha \nabla L_m(\theta_m^{(t)}; \{X_{k,m}\}),$$

where w_{mj}'s are elements of a doubly stochastic mixing matrix $W = (w_{mj}) \in \mathbb{R}^{M \times M}$. The mixing matrix models the connectivity of the network. For instance, $w_{mj} \neq 0$ if and only if the nodes m and j are directly connected. Since W is a doubly stochastic matrix, each row and column of W sums to one. In addition, all of its eigenvalues are real, and $1 = \lambda_1(W) \geq \cdots \geq \lambda_M(W) \geq -1$, where $\lambda_i(W)$ denotes the ith largest eigenvalue of W. Let $\rho := \max\{|\lambda_2(W)|, |\lambda_M(W)|\}$ be the second largest eigenvalues of W. $1 - \rho$ is referred to as the spectral gap measuring the level of connectivity in a network. For instance, zero spectral gaps indicate that there is no communication among nodes.

3   The Decentralized EM Algorithm

3.1   Algorithm Description

The Synchronous Algorithm. Throughout the paper, we focus on one scenario, in which all data are partitioned into M nodes, and the local data follow

---

Algorithm 1. Decentralized EM algorithm

input: Initial values of parameters $\theta_1^{(0)} = ... = \theta_M^{(0)}$, tolerance ϵ, and a doubly stochastic matrix $W \in \mathbb{R}^{M \times M}$.

for $t = 1, ..., T$, **do:**

   1. E-step: Compute the local auxiliary function for each node m:

$$Q_m(\theta|\theta_m^{(t)}; \{X_{k,m}\}) =$$

$$\frac{1}{n_m} \sum_{k=1}^{n_m} \mathbb{E}_{\theta_m^{(t)}}[\log g_\theta(X_{k,m}, Z)|X_{k,m}]$$

   2. M-step: Update parameters in the mth node by

$$\theta_m^{(t+1)} = \sum_{j=1}^{M} w_{mj}\theta_j^{(t)} +$$

$$\alpha \nabla Q_m(\theta_m^{(t)}|\theta_m^{(t)}; \{X_{k,m}\})$$

  if $\|\theta_m^{(t+1)} - \theta_m^{(t)}\| < \epsilon$ for any m, **break**

end for
output: $\theta_m^{(t+1)}, m = 1, ..., M$.

---

the predefined identical distribution. Due to the additivity of expectation, the E-step operates locally. Thus, in the E-step, we compute the local auxiliary function for each node individually. Based on four different communication and network settings, the M-step is modified to update local parameters. In Algorithm 1, we incorporate the decentralized gradient descent method with the EM algorithm and perform the one-step decentralized gradient descent method in the M-step using the fixed step size.

If one random sample is selected to compute the gradient in the M-step, the decentralized stochastic gradient descent is implemented. This decentralized algorithm updates local parameters of each node to the weighted average of parameter estimates from its neighbors. As the input of Algorithm 1, the mixing matrix W provides weights for the update. More details about the decentralized stochastic gradient descent algorithm are shown in [15]. In the M-step of Algorithm 1, we can also find a consensus maximizer of $Q(\theta|\theta^{(t)}) = \frac{1}{M} \sum_{m=1}^{M} Q_m(\theta|\theta^{(t)}; \{x_{k,m}\})$ through the decentralized gradient descent method in [31]. All observations of each node are utilized in the M-step for the decentralized gradient descent method.

The Asynchronous Algorithm. In some networks, the computing power of nodes may be different. Therefore, it is beneficial to decide when to perform updates for each agent independently. Such an update scheme is the asynchronous update, which may require much less communication cost and synchronization than its synchronous counterpart. Based on the workload and availability of the network, we propose an asynchronous version of the decentralized EM algorithm. The major difference between the asynchronous and synchronous algorithms is that the former updates parameters independently in each node without waiting for its neighbors to finish the previous updates. In particular, the parameters in the mth node for the M-step are updated by

$$\theta_m^{(t_m+1)} = \sum_{j=1}^{M} w_{mj}\theta_j^{(t_m)} +$$

$$\alpha \nabla Q_m(\theta_m^{(t_m)}|\theta_m^{(t_m)}; \{X_{k,m}\}).$$

The Algorithm in Dynamic Networks. For a dynamic network, the connectivity of the network is changing over time. The network topology is determined by its current mixing matrix $W^{(t)}$. In the proposed algorithm for dynamic networks, the agents only broadcast local data to their current neighbors based on the mixing matrix $W^{(t)}$ in the M-step. The same E-step and stopping rule, as shown in Algorithm 1, shall be applied to the dynamic decentralized EM algorithm.

3.2 Convergence Analysis

In this paper, we focus on analyzing the convergence rate of the decentralized first-order EM algorithm in Algorithm 1. The following assumptions are required for our convergence analysis.

1. All functions $L_m(\cdot)$'s are with ℓ-Lipschitz continuous gradients.
2. Given the symmetric doubly stochastic matrix W, $\rho^2 < 1$.
3. There exist constants σ and τ such that

$$\mathbb{E}\left\|\nabla L_m(\theta; \{X_{k,m}\}) - \nabla L_m(\theta)\right\|^2 \leq \sigma^2,$$

for any m and θ, and

$$\mathbb{E}\left\|\nabla L_m(\theta) - \nabla L(\theta)\right\|^2 \leq \tau^2,$$

for any θ, where $\nabla L(\cdot)$ denotes the gradient of a function L, and m is uniformly sampled from $\{1, \cdots, M\}$.

Assumption 1 is essential to ensure the convergence of many gradient-based algorithms. Note that this assumption does not assume the convexity of L_m. If Assumption 2 is violated, there is no communication among nodes in a network, which indicates the mixing matrix is an identity matrix. In Assumption 3, we assume that the variance of the stochastic gradient is bounded.

To show our convergence results, we define the concatenation of all local parameters $\boldsymbol{\theta}^{(t)} := (\theta_1^{(t)}, \cdots, \theta_M^{(t)})$. Without loss of generality, we further assume that $\boldsymbol{\theta}^{(0)} = 0$ and $n = 1$. Under these assumptions, we have the following convergence rate for Algorithm 1.

Theorem 1. *If the step size α is set to $1/(2\ell + \sigma\sqrt{T/M})$, we have the following convergence rate*

$$\frac{\sum_{t=0}^{T-1} \mathbb{E}\left\|\nabla L\left(\frac{\theta^{(t)}\mathbf{1}_M}{M}\right)\right\|^2}{T} \leq$$

$$\frac{8\ell(L(0) - L^*)}{T} + \frac{(4L(0) - 4L^* + 2\ell)\sigma}{\sqrt{TM}},$$

for sufficient large T, where $\mathbf{1}_M$ is the column vector with all elements equal to one, and L^* denotes the optimal value of (1).

The proof of Theorem 1 can be found in the Appendix. Theorem 1 implies that the convergence rate of Algorithm 1 is $O\left(\frac{1}{T} + \frac{1}{\sqrt{TM}}\right)$. Same convergence rate for the decentralized stochastic gradient descent algorithm can be found in [15]. Our proof can be viewed as extending this proof for one-stage decentralized algorithms to the two-stage ones. We also conclude that the decentralized first-order EM algorithm can achieve an ϵ-approximation solution, which is defined as

$$\frac{\sum_{t=0}^{T-1} \mathbb{E}\left\|\nabla L\left(\frac{\theta^{(t)}\mathbf{1}_M}{M}\right)\right\|^2}{T} \leq \epsilon.$$

If T is large enough, the second term, $\frac{1}{\sqrt{TM}}$, in $O\left(\frac{1}{T} + \frac{1}{\sqrt{TM}}\right)$ will dominate. The convergence rate becomes $O\left(\frac{1}{\sqrt{TM}}\right)$ for Algorithm 1. Then, each node shares the computational complexity of $O\left(\frac{1}{M\epsilon^2}\right)$, which is referred to as the linear speedup.

4   Experiments

4.1   Mixture Gamma Distribution

We considered a connected network with $M = 15$ nodes and 20 edges. The data in each node were generated from a mixture Gamma distribution with the density function $f(x) = \sum_{k=1}^{K} \pi_k f_k(x)$, where $K = 2$ was the number of clusters, $\pi_k > 0$ was the mixture weight and $\sum \pi_k = 1$, and f_k was the density of the kth cluster with the shape parameter γ_k and the scale parameter β_k. The density function

$$f_k(x) = \frac{x^{\gamma_k - 1} e^{-x/\beta_k}}{\Gamma(\gamma_k)\beta_k^{\gamma_k}},$$

where $\Gamma(\cdot)$ is the Gamma function. There is no closed-form maximization in the M-step. In our experiments, we set $\pi_1 = 0.3, \pi_2 = 0.7$ for two scenarios below,

1. $\gamma_1 = 10, \gamma_2 = 15, \beta_1 = 5, \beta_2 = 10$.
2. $\gamma_1 = 8, \gamma_2 = 9, \beta_1 = 10, \beta_2 = 12$.

In Scenario 1, two clusters were well separated, whereas they were mixed in Scenario 2. In our setting, each node was able to perform local computation and broadcast local information only to its neighbors based on the mixing matrix.

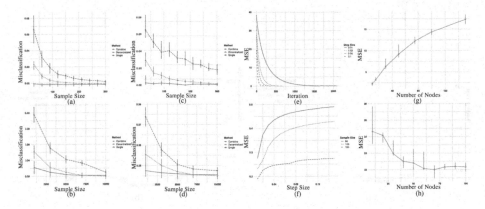

Fig. 2. Misclassification rates in balanced networks for Scenario 1 (a) and Scenario 2 (b). Misclassification rates in imbalanced networks for Scenario 1 (c) and Scenario 2 (d). The MSE between true parameters and estimated ones versus (e) iteration steps for different step sizes; (f) step sizes for different sample sizes; (g) the number of nodes with fixed total sample size; (h) the number of nodes with fixed sample sizes in each node.

We compared the performance of Algorithm 1, denoted as *Decentralized*, with that of an EM algorithm performed locally, denoted as *Single*. We also applied the EM algorithm to the full data, denoted as *Combine*, as the benchmark. Each setting was repeated 50 times. In addition, 1,000 samples were generated as the testing data set to evaluate the misclassification rate of clustering results. In Fig. 2, we showed misclassification rates of *Decentralized*, *Single*, and *Combine* for Scenario 1 in panel (a) and for Scenario 2 in panel (b) respectively. The step size was set to 0.02, and the number of observations in each node increased from 30 to 300 for Scenario 1 and increased from 1,000 to 10,000 for Scenario 2. In addition, we tested the performance of the proposed algorithms for an imbalanced network, in which the sample sizes on different nodes were varied. The local sample sizes for nine nodes were the same as the balanced ones, while the other six nodes contained twice as many. The results were shown in panels (c) and (d) of Fig. 2. All methods in comparison have better performance as the number of observations in each node increases. The proposed algorithms usually perform better than *Single* since our algorithms borrow information from neighbors. When there are enough observations in each node, the misclassification rate of the decentralized EM algorithms approximately achieves that of the centralized EM algorithm though the latter is not feasible in the decentralized setting.

To evaluate how the step size, α, affected the performance of the decentralized EM algorithm, we presented mean square errors (MSE) between true parameters and estimated ones for different step and sample sizes in panels (e) and (f) of Fig. 2. For results shown in the panel (e) of Fig. 2, the sample size was fixed to 100 for each node, and we let the step size change from 0.1 to 0.01. A smaller

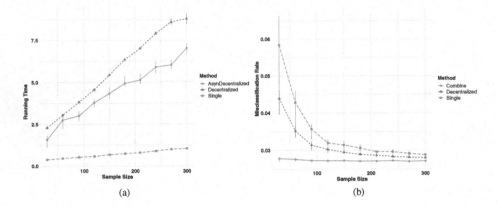

Fig. 3. (a) Running time comparison between synchronous and asynchronous algorithms. (b) Misclassification rates of the decentralized EM algorithm for a dynamic network under varying sample sizes.

step size α tends to reduce the convergence speed of the proposed algorithms, while it usually provides better numerical performance in terms of MSE. For the varying step sizes, the more samples in each node, the better performance we obtain, as is shown in panel (f) of Fig. 2. Under Scenario 2, we studied how the number of nodes affected the performance of our algorithm in two different ways. First, we fixed the total sample size to 15,000 and varied the number of nodes from 10 to 150. The results were shown in panel (g) of Fig. 2. As the number of nodes increases, the number of observations in each node decreases. Thus, the MSE increases as the number of nodes grows. Second, we showed the MSE for different numbers of nodes with 1,000 samples in each node in panel (h) of Fig. 2. The MSE decreases as the number of nodes increases, which has been proved in Theorem 1.

We also conducted numerical studies to compare synchronous algorithms with asynchronous ones in terms of running time. In this study, we generated local data sets with different sample sizes using settings in Scenario 1. Ten nodes contained 30 to 300 observations, while five nodes had 60 to 600 observations. The step size was set to 0.02. The misclassification rate of the asynchronous version is similar to that of the synchronous counterpart, with the misclassification rate $\frac{|\text{Decentralized}-\text{AsynDecentralized}|}{\max(\text{Decentralized},\text{AsynDecentralized})} < 0.1$ for all settings. However, in terms of running time, the asynchronous algorithm yielded a fast convergence rate by avoiding synchronization, as was reported in panel (a) of Fig. 3. To verify the performance of the decentralized EM algorithm in a dynamic network, we built a network with $M = 16$ nodes, which changed the position over time in a two-dimensional Cartesian coordinate system. In particular, the initial locations of the 16 agents were set on $(0,0), (0,1), ..., (3,3)$. Suppose all the agents moved to a random direction with the moving length following random uniform distribution $U(-0.5, 0.5)$. We let every edge (i, j), with $j \neq i$, be a part of the communication graph if the distance between two agents was smaller than 2. The network made such a move randomly at each iteration. Based on the setting in Scenario 1, we generated 50 to 500 observations for each agent. The misclassification rates were presented in the panel (b) of Fig. 3, in which we denoted the decentralized EM algorithm for a dynamic network topology as *DynDecentralized*. The proposed algorithm *DynDecentralized* has a lower misclassification rate than *Single* and gradually achieves the performance of *Combine* as the sample size in each node increases.

4.2 Hidden Markov Models with Mixtures as Emission Distribution

Hidden Markov model (HMM) is widely used in speech recognition [19], computational biology [14], data compression [7], and pattern recognition [2]. In HMM, the neighborhood structure is modeled via a Markov dependency among unobserved labels, whereas the distribution of observations is ruled by an emission distribution. In many applications, the emission distribution is pre-specified by a given distribution, such as Gaussian and Gamma distributions, or a mixture distribution. We apply the proposed decentralized EM algorithms in HMM with a mixture emission distribution and utilize the same notations as those in [24]. Denote $\{S_i\}_{i=1}^n$ as a D-state homogeneous Markov chain with the transition matrix Π. Assume that X_i follows an emission distribution $\psi_d, d = 1, ..., D$ conditional on the hidden state S_i,

$$X_i|S_i = d \sim \psi_d.$$

Then, an EM algorithm is developed to estimate probabilities and parameters of each hidden state.

In the experiment, we considered a four-state HMM. For different transition matrices, we assumed $P(S_i = d|S_{i-1} = d) = a$ with $a = 0.25, 0.5, 0.75, 0.9$, and $P(S_i = d'|S_{i-1} = d) = (1 - a)/3$ when $d' \neq d$. We used the same static network topology as that used in mixture Gamma distribution setting, with the sample size in each node varying from 100 to 1,000. Each setting was repeated 50 times. The criterion to evaluate the performance was the MSE between the true τ_{id} and the estimated one, where $\tau_{id} = P(S_i = d|X)$. We compared the performance of *Single*, *Combine*, and *Decentralized*. Results for different transition matrices, i.e., different a's, were shown in four subplots of Fig. 4 respectively. Since a larger a means a data point is more likely to keep its current state, all methods have a lower MSE for a larger a. For any fixed a, the proposed methods have better performance than *Single* and gradually achieve the performance of *Combine* as the sample size increases.

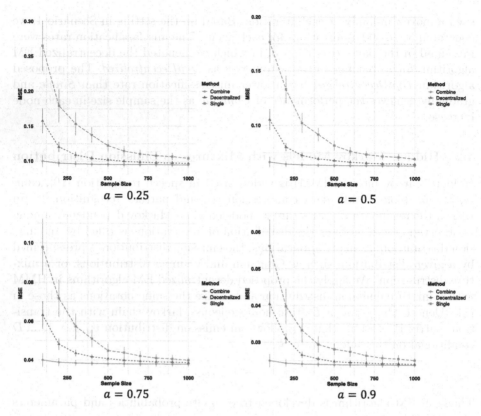

Fig. 4. The MSE between true parameters and the estimated ones for four different transition matrices.

5   Conclusion

In this paper, we propose a framework for the decentralized EM algorithm. A convergence rate has been studied for the synchronous version. Extensive numerical studies reveal that the proposed algorithms outperform the algorithms in comparison under different communication and network settings. The potential limitation of the proposed algorithms is a lack of global convergence analysis. We will leave this direction for future work.

Appendix

Proof of Theorem 3.1

To facilitate our proof, we define the following notations,

$$\boldsymbol{\theta}^{(t)} := \left(\theta_1^{(t)}, \cdots, \theta_M^{(t)} \right),$$

$$\mathbf{X}^{(t)} := \left(\{X_{k,1}^{(t)}\}, \cdots, \{X_{k,M}^{(t)}\} \right),$$

$$\nabla \mathbf{Q}(\boldsymbol{\theta}^{(t)}|\boldsymbol{\theta}^{(t)}; \mathbf{X}^{(t)}) := \left(\nabla Q_1(\theta_1^{(t)}|\theta_1^{(t)}; \{X_{k,1}^{(t)}\}), \cdots, \nabla Q_M(\theta_M^{(t)}|\theta_M^{(t)}; \{X_{k,M}^{(t)}\}) \right),$$

$$\nabla \mathbf{L}(\boldsymbol{\theta}^{(t)}; \mathbf{X}^{(t)}) := \left(\nabla L_1(\theta_1^{(t)}; \{X_{k,1}^{(t)}\}), \cdots, \nabla L_M(\theta_M^{(t)}; \{X_{k,M}^{(t)}\}) \right), \text{ and}$$

$$\nabla \mathbf{L}(\boldsymbol{\theta}^{(t)}) := \left(\nabla L_1(\theta_1^{(t)}), \cdots, \nabla L_M(\theta_M^{(t)}) \right).$$

In our proof, we minimize the negative log-likelihood function. Throughout the proof, L_m and Q_m are negative log-likelihood function and negative auxiliary function respectively. Then, the $(t+1)$th iteration of Algorithm 1 can be rewritten as

$$\theta_m^{(t+1)} = \sum_{j=1}^{M} w_{mj}\theta_j^{(t)} - \alpha \nabla Q_m(\theta_m^{(t)}|\theta_m^{(t)}; \{X_{k,m}\}) = \sum_{j=1}^{M} w_{mj}\theta_j^{(t)} - \alpha \nabla L_m(\theta_m^{(t)}; \{X_{k,m}\}), \quad (2)$$

where the last equation comes from the fact that

$$[\nabla H(\theta_m|\theta_m^{(t)}; \{X_{k,m}\})]_{\theta_m=\theta_m^{(t)}} = 0. \quad (3)$$

To see this, we show that for any θ_m,

$$H(\theta_m|\theta_m^{(t)}; \{X_{k,m}\}) - H(\theta_m^{(t)}|\theta_m^{(t)}; \{X_{k,m}\}) = \mathbb{E}_{\theta_m^{(t)}}[\log\{h_{\theta_m}(Z|X_k)/h_{\theta_m^{(t)}}(Z|X_k)\}|X_k]$$

$$\leq \log[\mathbb{E}_{\theta_m^{(t)}}\{h_{\theta_m}(Z|X_k)/h_{\theta_m^{(t)}}(Z|X_k)\}|X_k]$$

$$= 0,$$

where the inequality, which implies (3), is a consequence of Jensen's inequality. Then, (2) can be viewed as the following update

$$\boldsymbol{\theta}^{(t+1)} = \boldsymbol{\theta}^{(t)} W - \alpha \nabla \mathbf{Q}(\boldsymbol{\theta}^{(t)}|\boldsymbol{\theta}^{(t)}; \mathbf{X}^{(t)})$$

$$= \boldsymbol{\theta}^{(t)} W - \alpha \nabla \mathbf{L}(\boldsymbol{\theta}^{(t)}; \mathbf{X}^{(t)}).$$

For the function L satisfying the Lipschitzian gradient condition, we have the following inequality

$$L(y) \leq L(x) + \nabla L(x)^T (y - x) + \frac{\ell}{2} \|y - x\|^2.$$

Based on the inequality, we have

$$
\begin{aligned}
\mathbb{E}L\Big(\frac{\boldsymbol{\theta}^{(t+1)}\mathbf{1}_M}{M}\Big) &= \mathbb{E}L\Big(\frac{\boldsymbol{\theta}^{(t)}W\mathbf{1}_M}{M} - \alpha\frac{\nabla\mathbf{L}(\boldsymbol{\theta}^{(t)};\mathbf{X}^{(t)})\mathbf{1}_M}{M}\Big) \\
&= \mathbb{E}L\Big(\frac{\boldsymbol{\theta}^{(t)}\mathbf{1}_M}{M} - \alpha\frac{\nabla\mathbf{L}(\boldsymbol{\theta}^{(t)};\mathbf{X}^{(t)})\mathbf{1}_M}{M}\Big) \\
&\le \mathbb{E}L\Big(\frac{\boldsymbol{\theta}^{(t)}\mathbf{1}_M}{M}\Big) - \alpha\mathbb{E}\Big\langle\nabla L\Big(\frac{\boldsymbol{\theta}^{(t)}\mathbf{1}_M}{M}\Big),\frac{\nabla\mathbf{L}(\boldsymbol{\theta}^{(t)})\mathbf{1}_M}{M}\Big\rangle \qquad (4)\\
&\quad + \frac{\alpha^2\ell}{2}\mathbb{E}\Big\|\frac{\sum_{m=1}^M \nabla L_m(\theta_m^{(t)};\{X_{k,m}^{(t)}\})}{M}\Big\|^2.
\end{aligned}
$$

It is easy to see that

$$
\mathbb{E}\Big\|\frac{\sum_{m=1}^M \nabla L_m(\theta_m^{(t)};\{X_{k,m}^{(t)}\})}{M}\Big\|^2 = \mathbb{E}\Big\|\frac{\sum_{m=1}^M \nabla L_m(\theta_m^{(t)};\{X_{k,m}^{(t)}\}) - \sum_{m=1}^M \nabla L_m(\theta_m^{(t)})}{M}\Big\|^2
$$
$$
+ \mathbb{E}\Big\|\frac{\sum_{m=1}^M \nabla L_m(\theta_m^{(t)})}{M}\Big\|^2.
$$

Then, it follows from (4)

$$
\begin{aligned}
\mathbb{E}L\Big(\frac{\boldsymbol{\theta}^{(t+1)}\mathbf{1}_M}{M}\Big) &\le \mathbb{E}L\Big(\frac{\boldsymbol{\theta}^{(t)}\mathbf{1}_M}{M}\Big) - \alpha\mathbb{E}\Big\langle\nabla L\Big(\frac{\boldsymbol{\theta}^{(t)}\mathbf{1}_M}{M}\Big),\frac{\nabla\mathbf{L}(\boldsymbol{\theta}^{(t)})\mathbf{1}_M}{M}\Big\rangle \\
&\quad + \frac{\alpha^2\ell}{2}\mathbb{E}\Big\|\frac{\sum_{m=1}^M \nabla L_m(\theta_m^{(t)};\{X_{k,m}^{(t)}\}) - \sum_{m=1}^M \nabla L_m(\theta_m^{(t)})}{M}\Big\|^2 \\
&\quad + \frac{\alpha^2\ell}{2}\mathbb{E}\Big\|\frac{\sum_{m=1}^M \nabla L_m(\theta_m^{(t)})}{M}\Big\|^2 \\
&= \mathbb{E}L\Big(\frac{\boldsymbol{\theta}^{(t)}\mathbf{1}_M}{M}\Big) - \frac{\alpha-\alpha^2\ell}{2}\mathbb{E}\Big\|\frac{\nabla\mathbf{L}(\boldsymbol{\theta}^{(t)})\mathbf{1}_M}{M}\Big\|^2 - \frac{\alpha}{2}\mathbb{E}\Big\|\nabla L\Big(\frac{\boldsymbol{\theta}^{(t)}\mathbf{1}_M}{M}\Big)\Big\|^2 \\
&\quad + \underbrace{\frac{\alpha}{2}\mathbb{E}\Big\|\nabla L\Big(\frac{\boldsymbol{\theta}^{(t)}\mathbf{1}_M}{M}\Big) - \frac{\nabla\mathbf{L}(\boldsymbol{\theta}^{(t)})\mathbf{1}_M}{M}\Big\|^2}_{:=T_1} \\
&\quad + \underbrace{\frac{\alpha^2\ell}{2}\mathbb{E}\Big\|\frac{\sum_{m=1}^M \nabla L_m(\theta_m^{(t)};\{X_{k,m}^{(t)}\}) - \sum_{m=1}^M \nabla L_m(\theta_m^{(t)})}{M}\Big\|^2}_{:=T_2},
\end{aligned}
$$

where the last equation is resulting from $2\langle a,b\rangle = \|a\|^2 + \|b\|^2 - \|a-b\|^2$. Following Assumption 1,

$$
T_1 \le \frac{\ell^2}{M}\sum_{m=1}^M \mathbb{E}\underbrace{\Big\|\frac{\sum_{m'=1}^M \theta_{m'}^{(t)}}{M} - \theta_m^{(t)}\Big\|^2}_{D_m^{(t)}}.
$$

Based on Theorem 1 in [15], we have

$$\mathbb{E}T_1 \leq \ell^2 \mathbb{E}M_k,$$

where $\mathbb{E}M_k := (\mathbb{E}\sum_{m=1}^{M} D_m^{(t)})/M$. Following Assumption 3,

$$T_2 \leq \frac{\sigma^2}{M}.$$

Then,

$$\mathbb{E}L\Big(\frac{\theta^{(t+1)}\mathbf{1}_M}{M}\Big) \leq \mathbb{E}L\Big(\frac{\theta^{(t)}\mathbf{1}_M}{M}\Big) - \frac{\alpha - \alpha^2\ell}{2}\mathbb{E}\left\|\frac{\nabla \mathbf{L}(\theta^{(t)})\mathbf{1}_M}{M}\right\|^2 - \frac{\alpha}{2}\mathbb{E}\left\|\nabla L\Big(\frac{\theta^{(t)}\mathbf{1}_M}{M}\Big)\right\|^2$$
$$+ \frac{\alpha}{2}\ell^2\mathbb{E}M_k + \frac{\alpha^2\ell}{2M}\sigma^2.$$

We set step-size to $1/(2\ell + \sigma\sqrt{T/M})$ and sum the above inequality from $t = 0$ to $t = T - 1$. Given Corollary 2 in [15] and rearranging the inequality, we have

$$\frac{\sum_{t=1}^{T-1}\mathbb{E}\left\|\nabla L\Big(\frac{\theta^{(t)}\mathbf{1}_M}{M}\Big)\right\|^2}{4T} \leq \frac{2(L(0) - L^*)\ell}{T} + \frac{(L(0) - L^* + \ell/2)\sigma}{\sqrt{TM}} \tag{5}$$
$$+ \frac{2\ell^2 M}{(\sigma\sqrt{T/M})^2}\Big(\frac{\sigma^2}{1 - \rho} + \frac{9\tau^2}{(1 - \sqrt{\rho})^2}\Big).$$

For sufficiently large T, i.e.,

$$T \geq \frac{4\ell^4 M^5}{\sigma^6(L(0) - L^* + \ell)^2}\Big(\frac{\sigma^2}{1 - \rho} + \frac{9\tau^2}{(1 - \sqrt{\rho})^2}\Big),$$

the last term in (5) is bounded by the second term.

References

1. Aysal, T.C., Yildiz, M.E., Sarwate, A.D., Scaglione, A.: Broadcast gossip algorithms for consensus. IEEE Trans. Signal Process. **57**(7), 2748–2761 (2009)
2. Bishop, C.M.: Pattern Recognition and Machine Learning. Springer, Heidelberg (2006)
3. Blei, D.M., Ng, A.Y., Jordan, M.I.: Latent Dirichlet allocation. J. Mach. Learn. Res. **3**(Jan), 993–1022 (2003)
4. Boyd, S., Ghosh, A., Prabhakar, B., Shah, D.: Gossip algorithms: design, analysis and applications. In: Proceedings IEEE 24th Annual Joint Conference of the IEEE Computer and Communications Societies, vol. 3, pp. 1653–1664. IEEE (2005)
5. Bullo, F., Cortes, J., Martinez, S.: Distributed control of robotic networks: a mathematical approach to motion coordination algorithms, vol. 27. Princeton University Press (2009)
6. Cao, Y., Yu, W., Ren, W., Chen, G.: An overview of recent progress in the study of distributed multi-agent coordination. IEEE Trans. Ind. Inform. **9**(1), 427–438 (2012)

7. Crouse, M., Nowak, R.D., Baraniuk, R.G.: Wavelet-based statistical signal processing using hidden Markov models. IEEE Trans. Signal Process. **46**(4), 886–902 (1998)
8. Dempster, A.P., Laird, N.M., Rubin, D.B.: Maximum likelihood from incomplete data via the EM algorithm. J. Roy. Stat. Soc. Series B 1–38 (1977)
9. Eisen, M., Mokhtari, A., Ribeiro, A.: Decentralized quasi-newton methods. IEEE Trans. Signal Process. **65**(10), 2613–2628 (2017)
10. Forero, P.A., Cano, A., Giannakis, G.B.: Consensus-based distributed support vector machines. J. Mach. Learn. Res. **11**(May), 1663–1707 (2010)
11. Gai, Y., Krishnamachari, B.: Decentralized online learning algorithms for opportunistic spectrum access. In: 2011 IEEE Global Telecommunications Conference-GLOBECOM 2011, pp. 1–6. IEEE (2011)
12. Koppel, A., Paternain, S., Richard, C., Ribeiro, A.: Decentralized online learning with kernels. IEEE Trans. Signal Process. **66**(12), 3240–3255 (2018)
13. Kowalczyk, W., Vlassis, N.: Newscast EM. In: Advances in Neural Information Processing Systems, pp. 713–720 (2005)
14. Krogh, A., Brown, M., Mian, I.S., Sjölander, K., Haussler, D.: Hidden Markov models in computational biology: applications to protein modeling. J. Mol. Biol. **235**(5), 1501–1531 (1994)
15. Lian, X., Zhang, C., Zhang, H., Hsieh, C.J., Zhang, W., Liu, J.: Can decentralized algorithms outperform centralized algorithms? A case study for decentralized parallel stochastic gradient descent. In: Advances in Neural Information Processing Systems, pp. 5330–5340 (2017)
16. Ling, Q., Xu, Y., Yin, W., Wen, Z.: Decentralized low-rank matrix completion. In: 2012 IEEE International Conference on Acoustics, Speech and Signal Processing (ICASSP), pp. 2925–2928. IEEE (2012)
17. Liu, K., Zhao, Q.: Distributed learning in multi-armed bandit with multiple players. IEEE Trans. Signal Process. **58**(11), 5667–5681 (2010)
18. McLachlan, G.J., Peel, D.: Finite Mixture Models. Wiley, New York (2000)
19. Rabiner, L.R.: A tutorial on hidden Markov models and selected applications in speech recognition. Proc. IEEE **77**(2), 257–286 (1989)
20. Ram, S.S., Nedić, A., Veeravalli, V.V.: Asynchronous gossip algorithms for stochastic optimization. In: Proceedings of the 48h IEEE Conference on Decision and Control (CDC) held jointly with 2009 28th Chinese Control Conference, pp. 3581–3586. IEEE (2009)
21. Redner, R.A., Walker, H.F.: Mixture densities, maximum likelihood and the EM algorithm. SIAM Rev. **26**(2), 195–239 (1984)
22. Shi, W., Ling, Q., Wu, G., Yin, W.: Extra: an exact first-order algorithm for decentralized consensus optimization. SIAM J. Optim. **25**(2), 944–966 (2015)
23. Srivastava, S., DePalma, G., Liu, C.: An asynchronous distributed expectation maximization algorithm for massive data: the DEM algorithm. J. Comput. Graph. Stat. **28**(2), 233–243 (2019)
24. Volant, S., Bérard, C., Martin-Magniette, M.L., Robin, S.: Hidden Markov models with mixtures as emission distributions. Stat. Comput. **24**(4), 493–504 (2014)
25. Whipps, G., Ertin, E., Moses, R.: A consensus-based decentralized EM for a mixture of factor analyzers. In: 24th IEEE International Workshop on Machine Learning for Signal Processing (MLSP) (2014)
26. Wiley, D.A., Edwards, E.K.: Online self-organizing social systems: the decentralized future of online learning. Q. Rev. Distance Educ. **3**(1), 33–46 (2002)
27. Wu, C., Yang, C., Zhao, H., Zhu, J.: On the convergence of the EM algorithm: a data-adaptive analysis. arXiv preprint arXiv:1611.00519 (2016)

28. Xiao, L., Boyd, S., Kim, S.J.: Distributed average consensus with least-mean-square deviation. J. Parallel Distrib. Comput. **67**(1), 33–46 (2007)
29. Yan, F., Sundaram, S., Vishwanathan, S., Qi, Y.: Distributed autonomous online learning: regrets and intrinsic privacy-preserving properties. IEEE Trans. Knowl. Data Eng. **25**(11), 2483–2493 (2012)
30. Yin, J., Zhang, Y., Gao, L.: Accelerating distributed expectation-maximization algorithms with frequent updates. J. Parallel Distrib. Comput. **111**, 65–75 (2018)
31. Yuan, K., Ling, Q., Yin, W.: On the convergence of decentralized gradient descent. SIAM J. Optim. **26**(3), 1835–1854 (2016)
32. Yun, H., Yu, H.F., Hsieh, C.J., Vishwanathan, S., Dhillon, I.: Nomad: Non-locking, stochastic multi-machine algorithm for asynchronous and decentralized matrix completion. P. VLDB Endowment **7**(11), 975–986 (2014)
33. Zhao, L., Song, W.Z., Shi, L., Ye, X.: Decentralised seismic tomography computing in cyber-physical sensor systems. Cyber-Phys. Syst. **1**(2–4), 91–112 (2015)

Towards a Deep-Pipelined Architecture for Accelerating Deep GCN on a Multi-FPGA Platform

Qixuan Cheng, Mei Wen[✉], Junzhong Shen, Deguang Wang, and Chunyuan Zhang

School of Computer Science, National University of Defense Technology, Changsha, Hunan, China
wenmei@nudt.edu.cn
https://www.nudt.edu.cn/

Abstract. CNN (convolutional neural networks) have achieved great success in learning features from Euclidean-structured data. While lots of learning tasks require dealing with graph data. In these application scenarios where CNN cannot operate, GCN (graph neural networks) have shown appealing performance and increasing attention in recent years. However, according to our research, the computational complexity and storage overhead of the network also increase, making it a challenge to accelerate on a single FPGA. Accordingly, in this work, we focus on accelerating a deep GCN (DAGCN) on a CPU-multi FPGA platform by proposing a deep-pipelined acceleration scheme. To fully explore the parallelism that exists in DAGCN, we propose a graph convolutional neural accelerator (GCNAR) characterized by integration of a multiple 1-D systolic array. In addition, we also adopt an existing CSR algorithm-based partitioning scheme for large-scale matrix-vector multiplication in the design of our GCNAR, which effectively improves the computational efficiency of GCNAR. Moreover, we develop performance and resource evaluation models to help us determine the optimal design parameters for maximizing the accelerator throughput. Evaluation on real-world graph datasets demonstrates that our FPGA-based solution can achieve comparable performance to state-of-the-art GCN accelerations. In addition, compared to CPU and GPU solutions, our accelerator can achieve 196 times and 115 times the improvement for graph classification respectively in terms of processing latency.

Keywords: Deep Graph Neural Network (GCN) · CPU-multi-FPGA · Systolic array

Supported by Supported by organization National Natural Science Foundation of China (NSFC) project 61802420 and National Program on Key Basic Research Project 2016YFB1000401 and 2016YFB1000403.

M. Qiu (Ed.): ICA3PP 2020, LNCS 12452, pp. 528–547, 2020.
https://doi.org/10.1007/978-3-030-60245-1_36

1   Introduction

Due to the rapid development of CNNs in the applications of deep learning and data mining, increasing attention has been paid to the related work, and good performance in voice, image, video and network information retrieval has been achieved. However, non-Euclidean structure data mainly refers to graph data capable of representing most applications in the real world, which has a number of advantages. Compared with Euclidean structure data, graphic data can express richer semantics. This greatly enriches the expression of nodes, and thus enables applications such as chemical equations, bibliographies, and social networks to be represented and computed by convolutional neural networks. In addition, graph convolutional neural networks can also be used for a variety of applications including, graph classification, edge classification and node classification.

Graph convolutional neural network algorithms are one of the most popular algorithms at present. While there is a lot of related research work emerging, only a small proportion of it is hardware-related. Current accelerators work for shallow, simple connected networks. However, at present, there are some new trends developing in the sphere of graph convolutional neural networks. The algorithm begins to evolve from simple to complex, and from the initial shallow connection network to the deep network. Although most existing GCNs have two or three layers, the current trend shows that GCNs are becoming deeper (as occurred with CNNs). A GCN network with 152 layers has recently been proposed, and its efficiency in cloud semantic segmentation tasks has been demonstrated [1]. Our research shows that the training accuracy increases as the number of network layers also increases. At the same time, network computing and memory complexity are far more than the computing power of a single FPGA chip, so multi-chip acceleration is needed. In addition, similar to [3], we find that A is quite sparse (sparsity $\geq 99\%$). For the most of datasets, the distribution of input features for the first layer X1 is also very sparse (sparsity $\geq 90\%$).

To address these issues, this paper promotes accelerations for the inference of deep GCN on a CPU-multi-FPGA platform. Our design goal is to fully utilize the computational power of the multi-FPGA platform and explore the parallelism in DAGCN. The two distinct advantages of this approach are as follows:

(1) We propose a graph convolutional neural accelerator (GCNAR) characterized by the integration of a multiple 1-D systolic array. With the help of GCNAR, we are able to build accelerators for GCN in only a short time.
(2) In order to make good use of the system properties of the matrix, we also incorporate an existing CSR algorithm-based partitioning scheme for large-scale matrix-vector multiplication into the design of GCNAR, which effectively improves the computational efficiency of our proposed model.
(3) Based on the on-chip resources, we designed the model and calculated the optimal design parameters in order to maximize the throughput of the accelerator. Comparing the experimental data reveals that, our FPGA-based solution can achieve performance comparable to the most advanced GCN

acceleration. The test results also show that the throughput reaches 196 times of CPU and 115 times of GPU.

2   Related Works

GCN accelerator. To accelerate GCN training, the work in [3] proposes parallelization techniques for a multi-core platform. It partitions features to increase the cache-hit of each core. This proposes a GCN acceleration platform based on FPGA, which is mainly aimed at load balancing. [13] proposes an acceleration algorithm based on a subgraph, which results in partial information loss. [11] propose a hardware design with two efficient processing engines to alleviate the irregularity of the Aggregation phase and leverage the regularity of the Combination phase. These accelerators are mainly aimed at shallow GCN, and they use a single chip to accelerate the computation process, which is unable to realize the acceleration of deep GCN.

Scale-out accelerations for DNNs. Various FPGA-based accelerators [4–6,9,14] have been proposed to train CNNs. The work in [14] proposes a modular design based on layer operation types, and improves performance via reconfiguration. The work in [4] proposes a scalable framework for training CNNs on multi-FPGA clusters. Its partitioning and mapping strategy ensures load-balance. The works in [5,6] accelerate training by means of model compression. The reduced model size alleviates the burden on BRAM and thus improves resource utilization. Although GCNs are an extension of CNNs to graphs, the challenges associates with accelerating GCNs are significantly different. GCNs require both sparse and dense matrix operations. In addition to intensive computation, GCN accelerators need to address issues including irregular memory access and load-balance.

On the other hand, the development of the graph convolutional neural network algorithm has led to the emergence of some variant algorithms, which has in turn led to increases in the computational and storage complexity of the algorithm. Due to the large amount of computation involved, FPGA has a natural advantage in this field; accordingly, we try to apply distributed parallel acceleration FPGA to solve this problem. At present, [3,13] uses single-node FPGA to calculate GCN. [3] propose an architecture design called Ultra-Workload-Balanced-GCN (UWB-GCN) to accelerate the graph convolutional network inference. They tackle the major performance bottleneck of workload imbalance by proposing two techniques: dynamic local sharing and dynamic remote switching, both of which rely on hardware flexibility to achieve performance auto-tuning with negligible area or delay overhead. [13] design a novel accelerator for training GCNs on CPU-FPGA heterogeneous systems, which operates by incorporating multiple algorithm-architecture co-optimizations. They first analyze the computation and communication characteristics of various GCN training algorithms, then select a subgraph-based algorithm that is well suited for hardware execution. Moreover, to accelerate the weight update in GCN layers, these authors propose a systolic array-based design for efficient parallelization.

All of these works focus only on traditional GCN with a small number of layers (2–3 layers in general) on a single FPGA- or ASIC-based platform; therefore, their designs are unable to meet the acceleration requirements of deep GCN. To the best of our knowledge, we are the first article to propose hardware parallel acceleration GCN.

3 Background

In this section, we will introduce the DAGCN algorithm, then analyze the sparsity of GCN's commonly-used network matrix, along with the corresponding compression strategies and methods.

3.1 Traditional GCN Algorithm

In this paper, we focus on spectral-based graph convolutional networks, as these are among the most fundamental and widely used GCN structures. A GCN algorithm is generally a multi-layer graph convolutional neural network, each of which transmits and processes the eigenvalues of neighboring nodes among the nodes. By superimposing several convolutional layers, information transfer between nodes in the distance can be realized. Information transfer shown as the Eq. (1):

$$H(l+1) = \sigma \left(\widetilde{D}^{-\frac{1}{2}} \widetilde{A} \widetilde{D}^{-\frac{1}{2}} H^{(l)} W^{(l)} \right) \tag{1}$$

In Eq. (1), σ is the activation function, A is the adjacent matrix, and D is the degree matrix; that is, the diagonal matrix whose diagonal elements are the sum of one row of the matrix A. H denotes the matrix of features, where $H0 = X$. Finally, W is the weight matrix. As mentioned earlier, since each iteration of the traditional GCN algorithm can only obtain the key information from the previous iteration, over-smoothing will occur after many iterations, and many feature information of the graph will be lost. To address this issue, several GCN modules are developed, e.g. DAGCN and DSP-GCN [12].

3.2 Deep GCN Algorithm

In this section, we will introduce and analyze the advantages of DAGCN.

Overview of DAGCN. In Fig. 1, the basic process of DAGCN is introduced. The result of each convolutional layer is added to the graph input as the input of the next convolutional layer, after which the final result is weighted and output. Unlike the traditional graph convolution layer, the results of K hops in the AGC layer are aggregated according to specific weight relations, instead of just carrying out simple linear transmission. Such connection relations ensure that the final result covers more information for each hop, so that a better feature extraction effect can be attained.

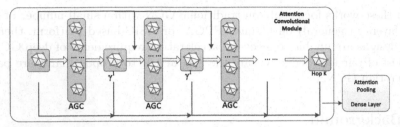

(a) The architecture of the dual attention graph convolution network(DAGCN)

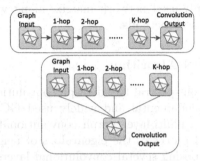

(b) Traditional Graph Convolution Layer(up section). Attention Graph Convolution Layer(down section)

Fig. 1. DAGCN algorithm process

Analysis of DAGCN. We carried out several experiments to determine the characteristics of DAGCN. Note that the datasets used in the experiments is Cora and the results are obtained on a Nvidia GTX1080 GPU; moreover, the number of AGC layers in DAGCN is 1 for all experiments and the number of hops is 12.

Table 1. Time distribution of the computation layers in DAGCN

Layer	AGC	FC	Pooling
Proportion	96.85%	1.42%	1.73%

Firstly, Table 1 presents the timing distribution of AGC, FC and the pooling layers. It can be observed that the AGC layers account for the highest computation cost (up to 90%), while the FC and pooling layers only account for 1.4% and 1.7%. Therefore, we focus on the acceleration of AGC layer in this work.

We further carried out experiments to test the relationship between the depth of DAGCN and its accuracy. Note that the datasets used in the experiments is Cora and the results are obtained on a Nvidia GTX1080 GPU. As can be seen from Fig. 2, the training accuracy of DAGCN increases from 72.3% to 85.2% when we increase the number of hops from 3 to 21, respectively. Therefore, to achieve a higher accuracy for deep GCNs, it's required to necessary the depth of the networks.

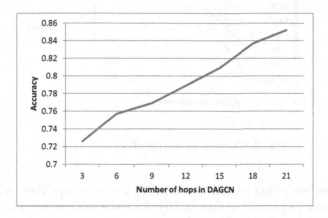

Fig. 2. The relationship between DAGCN depth and accuracy

Although we can improve the training accuracy of DAGCN by increasing the network's depth, the computational complexity and storage overhead of DAGCN will increase accordingly. We further carried out experiments to evaluate the computation and storage overhead of the DAGCN network. As shown in Fig. 3, when the number of graph convolution layers is increased from 3 to 12, the operations increased threefold, while the storage space required is increased by nearly 1.5 times. According to [10], the state-of-the-art Xilinx XVCU9P FPGA contains 6840 DSPs and 77.8 MB of on-chip storage. When the number of hops is three per chip, the storage requirements are in the tens of gigabytes. It is obvious that a single chip cannot bear the acceleration of a deep network; accordingly, we focus on accelerating deep DAGCN on a multi-FPGA platform.

3.3 Partitioning Scheme for Large-Scale Sparse MV

Geng et al. [3] revealed that the adjacent matrix and input features for the first layer in GCNs have a high degree of sparsity (over 90%). A similar conclusion is drawn for deep GCN (or DAGCN) in the present work. As we can be seen in Table 2, the data sparsity of GCN common networks is extremely large; in particular, for the adjacent matrix A, the sparsity exceeds 99%, which indicates that there are a large number of zero-multiplication operations in the hardware computation.

Similar to the general GCN algorithm, the DAGCN computing model is a large-scale sparse matrix-matrix multiplication. For large-scale problems, as noted above, the storage and computation costs arising from a large-scale matrix

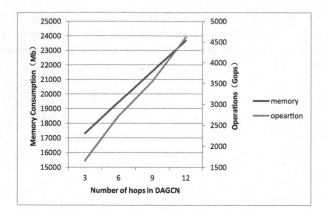

Fig. 3. Changes in computation and storage

multiplication algorithm cannot be borne by a single chip; thus, we adopt the block algorithm mentioned in Dou [2] to reduce the storage resource pressure on the chip. Regarding the sparsity problem, moreover, the adjacency matrix in the convolutional layer is not only large in size but also of high sparsity. On the one hand, this results in a high storage cost; on the other hand, many unnecessary computations are introduced into the matrix multiplication algorithm. Therefore, it is to reduce the number of zero value in the matrix in order to accelerate the matrix multiplication and balance the load.

Table 2. Sparsity and dimensions of matrices in a two-layer GCN for the three most widely evaluated GCN graph datasets.

		CORA	CITESEER	PUBMED
Dense	A	0.18%	0.12%	0.028%
	W	100%	100%	100%
	X_1	1.26%	0.86%	10.0%
	X_2	77.8%	89.3%	77.6%
Dim	Node	2708	3327	19717
	Edge	5429	4732	44338
	Class	7	6	3

3.4 Blocking Compression Scheduling Algorithm

As for the matrix sparsity, we apply the Sparstition [8] algorithm to preprocess the matrix. As shown in Fig. 4, this algorithm first divides the matrix needing to

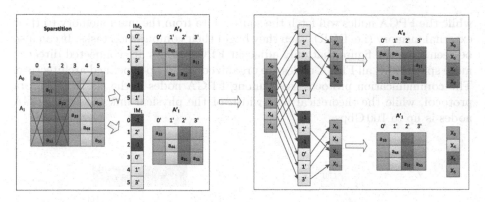

Fig. 4. Block compression scheduling algorithm

be preprocessed into blocks, then builds a vector for each sub-block, where each component of the vector is the order in which the "1" of the sub-block appears in the corresponding row. The "1" in the first line is row zero; all-zero rows are labeled -1. Next, referring to this vector, we obtain the matrix A', this matrix is reordered by rows and all zero rows are removed, which greatly reduces the algorithm calculation and storage overhead.

Because the matrix multiplication algorithm requires corresponding elements in A's row and B by accumulator, the first step disrupts the order of the elements in A's row, so the algorithm of the second step is to arrange the elements in the matrix B. The rearrangement process is carried out in accordance with the IM recorded in the value; as shown in Fig. 4, the algorithm first deletes the x element in the $x3$ and $x4$ elements. Finally, all the blocks of matrix A are multiplied by their corresponding B vectors following transformation to obtain the final result.

4   Deep-Pipelined Acceleration Scheme for DAGCN

In this section, we will introduce the details of our proposed mapping scheme for mapping the entire DAGCN onto a CPU-multi-FPGA platform. This includes the system architecture, the computation engine for the graph convolutional layer, and the dataflow among the entire system.

4.1   System Design

An overview of the hybrid CPU-multi-FPGA system used in this work is presents in Fig. 5. It can be seen from the figure our system compares a host CPU, a main memory and multiple FPGA nodes. Both the CPU and FPGA nodes can access the main memory via the PCI-E bus. Due to resource limitations on the FPGA nodes, the source graph data and final results are stored in the main memory, while the FPGA nodes will fetch the source data from the main memory to their external memory (i.e. DDR) when they begin their computation tasks. It can also be seen from the figure that only adjacent FPGA nodes are connected directly; more specifically, all FPGA nodes are organized as a deep-pipelined architecture. The communication protocol used among FPGA nodes is the classical Aurora protocol, while the theoretical bandwidth of the physical link between FPGA nodes is up to 100 Gbps.

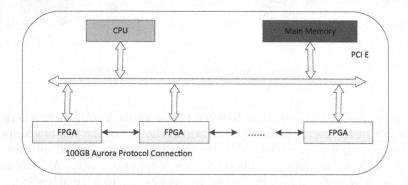

Fig. 5. Overview of the CPU-multi-FPGA system

4.2 Neural Network Mapping Scheme

First of all, we roughly divide the DAGCN network layers into specific groups according to their coarseness; these layers include the pooling layer, the fully-connected layer, and the large calculated amount layer, AGC layer. In order to accelerate the computation of the AGC layer and increase the parallelism, the CPU is used to process the fully connected layer and the pooling layer, the preprocess of the adjacent matrix A, and the computation of the pooling. Since the computation amount is relatively small and there is no good parallelism, CPU is a highly appropriate option for processing the layers, as it can ensure the normal operation of the pipeline.

The process of pooling and matrix preprocessing has been described above. In this section, we will focus on the rules of mapping AGC layers on to FPGA nodes. The most direct mapping scheme involves evenly distributing the calculated load of K layers of graph convolution to FPGA nodes; i.e., each FPGA node is allocated K/N graph convolutional layers, which ensures that the processing time of each FPGA node is relatively balanced. In the next step, we only need to ensure that the computation time of each FPGA node is basically the same, as this will ensure the normal operation of the pipeline.

4.3   GCN Accelerator

As shown in Fig. 6, we organize a graph convolutional neural accelerator on each of the FPGA nodes. Its basic components include: storage controller, several graph convolutional neural accelerators (GCNAR), data cache, data reorganization and network interface module. The storage controller is mainly used to load/write back input/output data onto/off the chip. The graph convolutional neural computing engine is responsible for the acceleration of a GCN layer. The number of graph convolutional layer computation engines is determined by the network mapping scheme. The data cache is responsible for storing data between hops of the network. Moreover, the data reorganization module will reorganize the inter-layer data and provide feature data for the next convolutional layer computation engine, while the network interface module will be used to send the intermediate network layer data generated by the local FPGA node to the neighbor node. We will discuss the role of each module in detail in the following sections.

We prefetch the value of the adjacent matrix A into the FPGA node, while the value of the feature matrix X will transmit with the direction of the data stream. First, we use all the computation engines to calculate the XW of the first hop; this is mainly because the XW of the first hop has a large amount of computation and will become a warp level. After the computation is complete, all nodes will pass the result back to the off-chip, with this results being regarded as the input of the first hop to conduct the pipeline computation.

The PE chain of each computation engine calculates XW, then performs the computation after dividing X into blocks. The computation result should be stored in the Buffer, as the computation result of this step is required in the later process of A(XW). We use adjustable PE, so that in the two-step computation process, dense and sparse matrix computations can be performed to speed up the computation. Subsequently, the block computation results, A(XW) of each step are passed to the next computation engine for computation.

We will discuss the role of each module in more detail in the following sections.

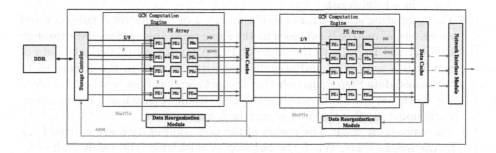

Fig. 6. The structure of the proposed GCN accelerator

GCN Computation Engine. The GCN computation engine is the core functional unit of our GCN accelerator. As shown in Fig. 6, it is composed of several 1-D systolic arrays, each of which contains several basic processing units (PE); thus, we refer to it as the PE link.

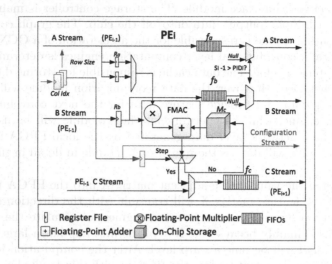

Fig. 7. Block diagram of the PE module

As shown in Fig. 7, PE stores input data from two sets of data registers, while three FIFOs (first-in first-out) are used to transfer data (FMAC) between PEs. The configuration stream is implemented so that the results calculated in the first step can be used directly in the second step. All PEs are configurable, so that they can support the computation of sparse and dense matrix-matrix multiplications. When used to perform dense matrix-matrix multiplication, the workflow can be described as follows:

(1) The prefetch stage: the block of matrix A is prefetched to PE, and then stored in a local register Ra.
(2) The computing stage: PE reads column $K + 1$ of the matrix A subblock and row K of the matrix B subblock at the same time, then passes data through FIFO_A and FIFO_B to the adjacent PE. After Rb is updated, the data of matrix B is multiplied and accumulated with the data in Ra. According to the passed data Step, judge whether the last step has been reached. If it does not finish, return Mc to wait for the next result to continue accumulation. Ra is updated after the data of matrix A is passed in.
(3) The write-back stage: The results of the new computation are written back to the memory Mc. When the last hop is calculated, the result is directly written to the on-chip cache.

When sparse matrix multiplication is performed, since the matrix A has been processed on the host side, we send the non-zero data to all PEs in the data prefetch stage, and each PE receives the corresponding amount of block A data according to the value in the register. In this way, the data flow of the matrix A subblock is eliminated, and only the prefetch process of the reconstituted matrix B subblock is available. At the end of the prefetch process, the state control needs to be carried out according to the number of data in the register. Once the threshold value is reached, this indicates that the non-zero element has been calculated and the computation result can be returned.

Data Reorganization Module. During the computation, the adjacency matrix A and the feature matrix X are characterized by great sparsity. Therefore, we compress them before computation; according to our data compression format, A and XW need to be compressed and reorganized before computation. The processing of the A matrix is done in the CPU, while the XW matrix results are generated during the computation process; thus we need to reorganize the XW matrix. The function of the data reorganization module is to reorganize the data before XW passes back to the GCN computing engine in order to prepare for the next step of matrix product computation with A.

Network Interface Module. The main function of the network interface module is to help the FPGA nodes to achieve high bandwidth transmission. As illustrated in Fig. 6, the network interface module is connected to the last GCN computing engine and immediately passes to the next FPGA node after receiving the sub-block data of the feature matrix. To simplify the design, we used the Aurora IP core provided by Xilinx as the basis for the network interface module design.

4.4 Data Flow

The matrix multiplication data flow is presented in Fig. 8. It introduces the flow process of computation; each computation engine contains 8 PE links. Xij represents the j-th matrix X block data of the i-th convolution layer, while Ak represents the k-th block of the adjacency matrix A. First of all, due to the large calculating amount of X1W, it becomes a bottleneck and seriously lengthens the time delay of the pipeline. Therefore, we first distribute X1 and W to all FPGA nodes for block computation. After the results are returned to the off-chip and regrouping, they become the first input node. The result of XW is transferred to the Rb of the PE of the first GCNAR by row, while the column components of the adjacency matrix A are multiplied and accumulated when they are transferred to Ra. The result of each block is passed as a block of X2 to the second GCNAR, multiplied by W. In this way, provided that there are enough A blocks, the establishment time of the assembly line can be ignored. The parallel between the first node and the second node is realized.

After each block result is generated, it can be transmitted directly to the next GCNAR to calculate the corresponding block, which enables the pipeline processing of the entire computation process to be achieved. Subsequently, the result of XW will be directly cached to the chips, and the result of A (XW) will be transmitted to the next hop computation.

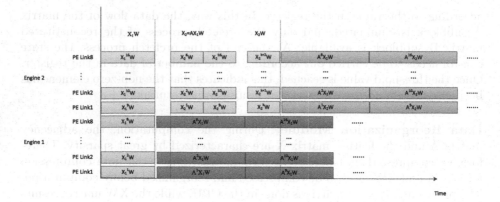

Fig. 8. Accelerator pipeline structure diagram

5   Performance and Resource Modeling

In this chapter, we will introduce the mapping relationship of the model to the FPGA. Firstly, we parameterize the model, summarize the DAG (Database Availability Group) of the data in the experiment; next, we introduce the mapping relationship between parameters and FPGA, ascertain how to determine each of the parameter indicators, and finally, build models featuring constraint conditions such as bandwidth, computing power, and storage capacity.

In order to find better resource utilization parameters, we build a model that estimates the effect of selecting each parameter on the FPGA. Due to the limitations of on-chip storage resources, we store the initial data on DDR (double data rate), and further need to consider factors such as computing time, computing power, and inter-chip bandwidth.

The computation time of the i-th hop is shown in Eq. (2), where freq is the accelerator operating frequency,

$$T_{Ex} = \frac{L \times S \times \{(K + 1) \times Sj + Stage_f\}}{freq} \tag{2}$$

The overall time required for computation is slightly longer than the time outlined in Eq. (2); since on-chip resources are limited, we need to obtain the data off-chip. However, we can calculate while taking data, so we can temporarily ignore the time for taking data.

In the next part, the computing power of the FPGA will be considered. We will constrain the setting of our parameters with reference to the DSP (digital signal processing) usage limits and average handling capacity.

The DSP required for each FPGA computation is outlined in Eq. (3).

$$DSP = Cor \times L \times S \times x \tag{3}$$

Increasing the average handling capacity is also one of the significant criteria for improving computing performance. The average handling capacity of the distributed system is presents in Eq. (4),

$$TP = \frac{\text{hop} \times (OP_{XW} + OP_{A(XW)})}{(T_{DE} + (\text{Stage} - 1) \times T_{Stage}) \times 10^9} \qquad (4)$$

Here, the hop is the iteration times of the entire network, Op denotes the operation times of XW and AXW, TDE is the delay time, and T is the bottleneck time. In order to make the model optimal, we need to maximize the average handling capacity.

BRAM is organized into distributed memory blocks. The memory capacity of each block of our target FPGA board is 18 Kb. The input, output, and weight buffers are made up of various BRAM blocks, which store tiled input/output feature maps and weights. A single BRAM block has two ports (a read port and a write port), supports concurrent read and write, and can stored 512 32-bit or 1024 16-bit words. The usage of BRAM is expressed as Eq. (5), which represents the BRAM used by the storage controller.

$$BRAM = \left\lceil \frac{2 \times K \times S_j}{1024} \right\rceil + 3 \times \text{Cor} \times L \times S + BRAM_{CM} \qquad (5)$$

After calculating these conditions, we will establish the constraints of the model. Firstly, in terms of hardware resources, the DSP is less than 6840 and BRAM is limited to 4320. Secondly, the total time should be greater than the computation time, while also less than the sum of the computation time and the transmission time.

$$T_{EX} < T_{Tatol} < T_{EX} + T_{Trans} \qquad (6)$$

Equation (7) represents the bandwidth limitation; the bandwidth of the engine needs to be greater than the data transmission, and BW represents the maximum bandwidth between nodes.

$$BW > \frac{L \times S \times \{(K + 1) \times Sj + Stage_f\}}{T_{Trans}} \qquad (7)$$

According to the above constraints, we build a model and calculate the optimal case of PE chain number and PE number of each chain in the engine. The specific situation will be explained in the experiment section.

6 Experiments

We evaluate our model and compare it to other GCN models with the same GCN networks, including Cora, Citeseer, and Pubmed [7], on CPU and GPU.

6.1 Experiment Settings

The experimental platform comprises a server, a PCI-E hub and six VCU118 FPGA development boards. We implement the accelerator solution on the Xilinx VCU118 development board. The pipelined accelerator is described in verilog language. Finally, we use Xilinx Vivado (2017.4) for system synthesis, implementation and binary code generation purposes. The chip uses Dual 4×28 Gbps QSFP28 cages for communication. VCU118 FPGA can provide 6840 x usable with the DSPs and 76 Mb of BRAMs.

In order to contrast with the FPGA platform, we conduct experiments on the CPU and GPU platforms. We implement the GCN networks in Pytorch on CPU Intel Xeon ES-2660 and GPU GTX1080. We use 32-bit floating-point data precision for the input/output data and weights. The datasets used for evaluation are Cora, Citeseer and Pubmed, which are the most popular networks in GCN training.

6.2 Experiment Evaluation

Resource Utilization. In the experiment, we test the usage of DSPs, BRAMs and other hardware resources on different datasets with different hop numbers and compare them with the operation UWB-GCN.

Table 3. Resource utilization on an FPGA node

Resource	DSP	BRAM	LUT	Flip-Flop
Available	6840	4320	1182K	2364K
Usage	5386	65.6%	330K	756K
Percentage	78.7%	73.7%	28%	32%

In the test, we utilized the Cora data set. In the network, we set the hop number to 12, and the PE chain number to 8, while each PE chain uses 84 PE. From Table 3, we can see that when the utilization rate of DSP exceeds 78%, the utilization rate of BRAM will approach 75%. The average number of hops assigned to each node will be used as the number of computing engines in the node; that is, each engine calculates a hop operation. The length of the PE chain of each engine is determined by the number of engines at each node.

Scalability. We tested the scalability of the mapping scheme by changing the number of FPGA nodes. Since we have only six VCU118 FPGA boards, we only vary the number of FPGA nodes from 2 to 6 in the experiments. In addition, Pubmed is used as the target datasets, and the number of hops of DAGCN is fixed to 12.

Table 4. Hardware setup under different number of FPGA nodes

Number of FPGA nodes	2	3	4	5	6	
Number of GCNAR	6	4	3	2	3	2
Number of PE links per GCNAR	4	6	7	8	7	8
Number of PE per PE link	56	56	64	84	64	84

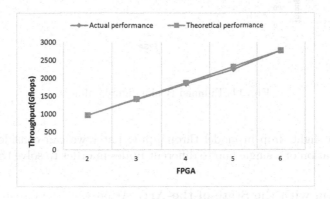

Fig. 9. Cora (Color figure online)

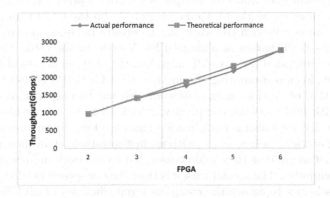

Fig. 10. Citeseer (Color figure online)

As shown in Table 4, we integrated the hardware resources on the chip, then determined the PE chain number and the most reasonable PE number on a single chain through calculation.

Figure 9, 10 and 11 present the sustained (blue dot line) and theoretical system throughput (red dot line) for Cora, Citeseer and Pubmed respectively under different numbers of FPGA nodes. It can be seen that the sustained system throughput for all datasets is very close to the theoretical throughput, which demonstrates that our system has good scalability. When the number of nodes is 5, there is a certain gap in the calculation amount between nodes, which

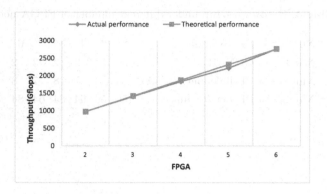

Fig. 11. Pubmed (Color figure online)

results in a slight drop in model throughput. Later we can consider assigning the computation of a single hop to different nodes in order to solve this problem.

Comparison with the State-of-the-Art. At present, the research into neural network FPGA accelerators is still in the initial stage; thus, we selected only two representative works for comparison. Table 5 presents the results of the comparison. Since the work [13] focuses primarily on the GCN training process, we mainly compare it with the work [3]. To ensure fairness, moreover, we only calculated the performance of a single FPGA node in the table. On the same platform, we integrated the more PE unit, but the utilization rate of PE (93% to 90%) and single chip computing throughput (524 GFLOPS of 460.8 GFLOPS) were below that of [3]. The main reason for this has two aspects: one is that we are faced with high network complexity, which leads to our design than their high complexity, establish a performance than we long linear accelerator, and our working accelerator frequency (without optimization) than their job is much lower, it is not as good as their main reason led to the performance of the single chip microcomputer. The second reason is that they proposed two load-balancing scheduling schemes to ensure the computational efficiency of all PEs. However, our work did not optimize the load balance between PE, with the result that load imbalance exists between PEs in the same PE chain. Therefore, as far as the PE utilization rate is concerned, our accelerators are not effective and need to be further improved.

If the influence of the accelerator frequency is not considered, our accelerator performs better than that under the GFLOP/freq measurement [3]. Considering that we are dealing with a more complex depth-mapped convolutional neural network, and that our solution is deployed on a more complex CPU+FPGA distributed platform, we believe that our work is somewhat advanced.

Table 5. Calculation throughput comparison

Work	[3]	[13]	Ours
Platform	Xilinx VCU118	Xilinx Alveo U200	Xilinx VCU118x6
Frequency (MHz)	275	200	200
Number of PE	1024	576	1280 * 6
PE utilization	93%	~	90%
Throughput (GFLOPS)	524	~	460.8 (single)

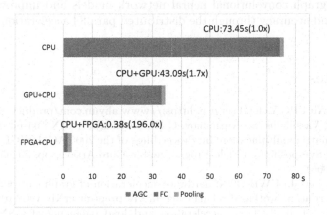

Fig. 12. The breakdown of processing latency of DAGCN on different platforms

Cross-Platform Comparison. As shown in Figure 12, we tested the Cora data set. When the hop number is set to 12, the time occupied by AGC, FC and the pooling layers in the three platforms. We can conclude that, in the CPU+FPGA, CPU+GPU and CPU platforms, the AGC layer accounts for the proportion of time in the whole DAGCN. The CPU platform takes 75.8 s, while the AGC layer takes 74.85 s. The computing time of the AGC layer on GPU+CPU platform was 43.9 s, while the computation time of the AGC layer of FPGA+CPU platform was 0.38 s. It is not difficult to observe that we mainly accelerate the AGC layer, while the acceleration of GCNAR shortens the time for CPU by 196 times and that of CPU+GPU by 115 times.

7 Conclusion and Future Work

In this paper, we implement an FPGA-based parallel acceleration program designing for deep graphic convolutional neural networks. The acceleration system is composed of distributed FPGA, which can realize DAGCN in parallel.

In order to meet the requirements of hardware storage and computing power, we compress the matrix by the Sparstition algorithm, and use the blocking method to streamline the entire computation process, saving hardware storage and computation overhead. And we can dynamically divide the hardware resources according to the user's hardware resources and computation accuracy requirements. According to the number of user's FPGA nodes, the hop number, FPGA resources can be dynamically adjusted. For the application of graph convolutional neural network, this paper is the first article for FPGA hardware acceleration of complex and multi-chip connection. In future study, we will study more deep graph convolutional neural network models and improve the training speed and accuracy through the distributed parallel acceleration system and parallelism.

References

1. Aliyun: Ali FPGA cloud service. https://www.aliyun.com/product/ecs/fpga
2. Dou, Y., Vassiliadis, S., Kuzmanov, G.K., Gaydadjiev, G.N.: 64-bit floating-point FPGA matrix multiplication. In: Proceedings of the 2005 ACM/SIGDA 13th International Symposium on Field-Programmable Gate Arrays, pp. 86–95. ACM, New York (2005)
3. Geng, T., et al.: UWB-GCN: hardware acceleration of graph-convolution-network through runtime workload rebalancing. arXiv preprint arXiv:1908.10834 (2019)
4. Geng, T., et al.: FPDeep: acceleration and load balancing of CNN training on FPGA clusters. In: 2018 IEEE 26th Annual International Symposium on Field-Programmable Custom Computing Machines (FCCM), pp. 81–84. IEEE (2018)
5. Guo, K., et al.: Compressed CNN training with FPGA-based accelerator. In: Proceedings of the 2019 ACM/SIGDA International Symposium on Field-Programmable Gate Arrays, pp. 189–189. ACM (2019)
6. Nakahara, H., Jinguji, A., Shimoda, M., Sato, S.: An FPGA-based fine tuning accelerator for a sparse CNN. In: Proceedings of the 2019 ACM/SIGDA International Symposium on Field-Programmable Gate Arrays, pp. 186–186. ACM (2019)
7. Prithviraj, S., Galileo, N., Mustafa, B., Lise, G.: Collective classification in network data. Technical report CS-TR-4905 and UMIACS-TR-2008-04, University of Maryland, College Park, Washington, USA (2008)
8. Sigurbergsson, B., Hogervorst, T., Qiu, T.D., Nane, R.: Sparstition: a partitioning scheme for large-scale sparse matrix vector multiplication on FPGA. In: 2019 IEEE 30th International Conference on Application-Specific Systems, Architectures and Processors (ASAP), vol. 2160, pp. 51–58. IEEE (2019)
9. Venkataramanaiah, S.K., et al.: Automatic compiler based FPGA accelerator for CNN training. In: 2019 29th International Conference on Field Programmable Logic and Applications (FPL), pp. 166–172. IEEE (2019)
10. XILINX: Xilinx virtex ultrascale+ FPGA VCU118 evaluation kit. https://www.xilinx.com/products/boards-and-kits/vcu118.html#hardware

11. Yan, M., et al.: HyGCN: a GCN accelerator with hybrid architecture. In: 2020 IEEE International Symposium on High Performance Computer Architecture (HPCA), pp. 15–29. IEEE (2020)
12. Yang, L., Chen, Z., Gu, J., Guo, Y.: Dual self-paced graph convolutional network: towards reducing attribute distortions induced by topology. In: Proceedings of the Twenty-Eighth International Joint Conference on Artificial Intelligence (IJCAI-2019), pp. 4062–4069 (2019)
13. Zeng, H., Prasanna, V.: GraphACT: accelerating GCN training on CPU-FPGA heterogeneous platforms. In: The 2020 ACM/SIGDA International Symposium on Field-Programmable Gate Arrays, pp. 255–265 (2020)
14. Zhao, W., et al.: F-CNN: an FPGA-based framework for training convolutional neural networks. In: 2016 IEEE 27th International Conference on Application-Specific Systems, Architectures and Processors (ASAP), pp. 107–114 (2016)

Linear Scalability from Sharding and PoS

Chenlong Yang[1], Xiangxue Li[1,2,3($\boxtimes$)], Jingjing Li[1,4], and Haifeng Qian[1($\boxtimes$)]

[1] School of Software Engineering, East China Normal University, Shanghai, China
{xxli,hfqian}@cs.ecnu.edu.cn
[2] Shanghai Key Laboratory of Trustworthy Computing, Shanghai, China
[3] Westone Cryptologic Research Center, Beijing, China
[4] WanXiang Blockchain Lab, Shanghai, China

Abstract. Scalability is one of the most important problems in blockchain and has been the focus of both industry practitioners and academic researchers since Bitcoin was born. The blockchain has insufficient ability in handling large-scale concurrent transactions. The more transactions that are processed in the network, the more scalability problems appear in the network. Compared to the transaction throughput achieved by the electronic payment channels of mature development, the limitation is magnified. In this paper, we propose a novel consensus mechanism from sharding and Proof-of-Stake (PoS)—a scalable blockchain model that supports high concurrency, and achieve the linear expansion of transaction processing scale while ensuring security in order to prove the feasibility of the proposal. The proposal views the blocks in the network as two levels, namely, intermediate transition blocks (i.e., middle blocks) and final confirmation blocks (i.e., final blocks), and takes epoch as the basic unit of the consensus mechanism operation cycle. An epoch is a recursive process for each cycle of the consensus mechanism operation. Each epoch is equipped with four types of interactions, namely, node sharding, transaction sharding, internal consensus, and final block generation, and thereby determines the network status via main chain and shard chains. Given n nodes in the network, we construct one Validity group and $p - 1$ Regular groups (each one contains n/p nodes). The regular groups create transition blocks according to the transaction pool; the validity group extracts information from the transition blocks created by the regular groups, generates and sends final confirmation blocks to the main chain. PoS consensus is exploited to ensure that adversaries are not able to launch attacks on specific shards (neither transaction nor node shards). We also describe how to re-group the nodes in and add new node to the network. We provide the security analysis under several standard attack models.

Keywords: Blockchain · Linear scalability · Proof-of-stake · Sharding · Regular group · Validity group · Leader node

© Springer Nature Switzerland AG 2020
M. Qiu (Ed.): ICA3PP 2020, LNCS 12452, pp. 548–562, 2020.
https://doi.org/10.1007/978-3-030-60245-1_37

1   Introduction

Blockchain technology, originated from the Bitcoin system proposed by Satoshi in 2008 [1], truly implements a digital payment system that does not rely on a trusted third party in an open peer-to-peer network. The decentralized characteristic is greatly different from existing payment systems, and improves the security and trust model in the existing systems [4,31]. In the trust model of Bitcoin, the trust relationship between users ordinates from the entire system, instead of a trusted third party. The bitcoin network maintains a public ledger that aims to reach a consensus on transactions confirmed by miners heretofore on the network. Provided that the attacker does not control too large a fraction of the computational resources on the network (i.e., the majority of the computing power on the network is honest), he can not undo or alter any transaction already accepted by the system (with a high-probability guarantee). This makes the system able to prevent double-spending (i.e., the risk that a digital currency token may be copied and spent more than once).

There are several methods of reaching consensus on blockchain networks including the proof-of-work algorithm (PoW), the proof-of-stake algorithm (PoS), etc. The best-known method of reaching consensus in a blockchain is the PoW scheme, which is used by Bitcoin. PoW uses a hash function to create conditions under which one single participant is permitted to announce the proof about the submitted information. The proof can then be independently verified by all other system participants .

Consider the Bitcoin system more specifically. The participant that verifies publicly the information on behalf of the network is in turn rewarded with newly created bitcoins (which is extremely overpriced in terms of energy cost and computing resources in practice) for its participation . The process of searching for valid nonce (solutions to the hash function) is called mining. Yet one has to concede the fact that it is really tremendous waste of energy when participants mine blocks. To circumvent the restriction of PoW, the community presents PoS algorithm in an effort to offering a more efficient and environment-friendly alternative. In contrast to PoW, PoS leverages virtual resources (such as the stake of a node) to perform leader election and maintain consensus on the network. Since the mining resources are virtual, PoS-based consensus process is instant and gives rise to a minuscule amount of costs.

One of the prominent issues facing the application of blockchain technology in practical scenarios (such as Internet of Things [2,3], supply chains [5], etc.) is the scalability of blockchain. As the most popular blockchain projects, Bitcoin and Ethereum also face the shortcoming of throughput. For instance, the max throughput of Bitcoin in theory is 6 transactions per second, when it comes to Ethereum, the throughput is 10. Compared to traditional payment channels such as VISA and MasterCard [6], the cryptocurrencies basing on blockchain technology still have shortcomings. As one can see, blockchain scalability becomes an essential research area.

Obviously, the change of some key values of blocks may work. For instance, if we reduce the transaction verification time by reducing the difficulty of proof-of-work, or maybe more simply, reducing the data size of transactions, we may increase the transaction throughput to a limited extent. However, the possibility of forks and vicious attacks cannot be ignored. Consequently, we need to surpass the underlying structure of the blockchain to achieve the purpose of improving transaction throughput. Currently, there are various blockchain scalability solutions in academia, e.g., sharding [30], directed acyclic graph [26], lightning network [27], sidechain [28,29], etc. In addition, there are improved solutions based on consensus mechanisms that are recognized by researchers (for example, Bitcoin-NG [7]).

In this paper, we propose a novel consensus mechanism from blockchain sharding and PoS—a scalable blockchain model that supports high concurrency, and achieve the linear expansion of transaction processing scale while ensuring security in order to prove the feasible of the proposal. The proposal views the blocks in the network as two levels, namely intermediate transition blocks (i.e., middle blocks) and final confirmation blocks (i.e., final blocks), and takes Epoch as the basic unit of the consensus mechanism operation cycle. Given n nodes in the network, we construct one Validity group and $p-1$ Regular groups (each one contains n/p nodes). The regular groups create transition blocks according to the transaction pool; the validity group extracts information from the transition blocks created by the regular groups, generate and sends a confirmation block to the main chain. An epoch is a recursive process for each cycle of the consensus mechanism operation. Each epoch is equipped with four types of interactions, namely, node sharding, transaction sharding, internal consensus, and final block generation. PoS consensus is exploited to ensure that adversaries are not able to launch attacks on specific shards (neither transaction nor node shards). We also describe how to re-group the nodes in and add new node to the network. We provide the security analysis under several standard attack models.

2   Preliminaries

2.1   Proof-of-Stake

Proof-of-stake (PoS) is a widely-used consensus mechanism [8]. In 2012, the developer of Peercoin [9] invented the consensus protocol of PoS. This encrypted currency uses the PoS mechanism to issue new coins and maintain network security. Peercoin is the first application of the proof-of-stake mechanism in the area of cryptocurrencies. The core idea of PoS is to avoid competition for computing power by miners (in the speed of obtaining the value of *Nonce*). Different from proof-of-work, the allocation of block generation rights in the proof of equity is determined by the number of *Coin ages* owned by the participants [10]. The higher the age of the coins, the greater the probability of winning the competition. As the name suggests, the coin age is calculated as the product of the amount of cryptocurrency owned by a node and the time it is owned (for example, if a node has owned 100 coins for 100 days, its coin age is 10,000). The coin age of nodes determines the probability of successfully obtaining the accounting right under the PoS mechanism. In other words, in a PoS competition, the coin age (as a weight) affects the probability of winning the right of accouting transactions.

The most representative application of the PoS mechanism is Ethereum. The consensus protocol of Ethereum is a mix of PoW and PoS. In Ethereum, the coin age of nodes can be converted into computing power to a certain extent. Though the miners still have to solve the hash puzzle by calculating, the higher the coin ages they own, the larger the target value of the PoW puzzle that is needed to calculate; and the lower the cost of the calculation time actually required. Therefore, the larger the probability of successful block generation has. Also, it can be seen that the *Target* value corresponding to each node is different, depending on the coin age. In contrast, in the pure PoW consensus protocol represented by Bitcoin, all the nodes need to face an identical Target value.

In addition, PoS has the following advantages [11].

1. It is not necessary to consume extra resources (such as mining machines and electricity) to ensure the security of the main chain in PoS. If a node has a tendency to launch an attack, the node needs to have sufficient equity on the chain at first to ensure that the probability of winning the bookkeeping right becomes larger; however, the high cost of preparing a attack and the low profit of a successful attack ensures that miners do not have the motive for evil interests;
2. The PoS can be used to ensure that cryptocurrencies will retain their market value. Since PoS avoids high consumption of real economic costs, the system does not need to frequently "mint" coins (compared to Bitcoin) in order to maintain the enthusiasm of participants in the network;
3. The PoS can reduce the crisis of centralization [12]. In the PoW, the miners are assembled into node clusters under the driving of interests, it is extremely difficult for individual node mining to obtain block rights. Miners participate in the mining pool and then make dividends according to the computing power provided by themselves. This makes the growing mining pool violate the core idea of the Bitcoin blockchain design-the original intention of decentralization. However, in PoS, the concept of the *mining cluster* is abandoned, and Ethereum returned the mining right to each individual miner, which is a huge improvement of the decentralized system.

Compared with PoW, PoS cannot be simply used in reaching consensus among nodes in the network. In fact, PoS needs improvements to make it more applicable to the actual situation [13]. In fact, there exist some explict limitations in PoS including the following.

1. PoS does not fully fulfill the disadvantages of traditional consensus protocols [14]. For instance, the nodes in Ethereum still need to solve the hash problem by PoW mechanism (though the difficulty of finding the nonce is lower than that in Bitcoin) [15];
2. The overmuch usage of PoS may cause the centralization in some extreme cases. As described above, the nodes with more coin ages are more probably to get the right of generating blocks; therefore, they may have a greater right to speak for the entire network; in contrast, nodes with less coin ages cannot pose a threat to the overall network.

3. In addition, the longer the PoS protocol runs, the easier it is for nodes with more coin ages to get rewards, thereby increasing the gap between the rich and the poor nodes [16].

In this paper, we propose a novel consensus protocol that solves the problems of pure PoS by introducing the sharding method. Moreover, we increase the average throughput of transactions of the proposal.

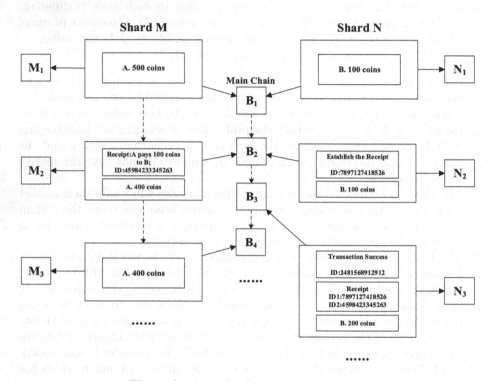

Fig. 1. An example of sharding blockchains

2.2 Blockchain Sharding

The concept of **sharding** technology was proposed much earlier than the blockchain, and is mainly used to improve the storage of databases, especially the access performance optimization of enterprise-level large databases. The ordinary distributed ledger is a single chain, and all the nodes in the peer-to-peer network have to issue transactions and generate blocks on a specific chain (i.e., the main chain) [17]. Meanwhile, the probability of the appearance of forks is high, which is not conducive to improving the speed of block processing and transaction verification. When the idea of sharding is applied to the blockchain technology, the blockchain network with a large number of nodes can be divided into different sub-networks—each sub-network (that is, the sharding of network) [18] owns a part of nodes, and each shard handles corresponding transaction sub-pool.

The core idea of sharding blockchain is divided into three steps:

1. **Network Sharding**: method for dividing all nodes in the entire network into different shards;
2. **Transaction Sharding**: method for dividing all transactions in the entire transaction pool into different shards;
3. **Status Sharding**: method for the nodes only being responsible for verifying transactions within their respective shards, without affecting the security and efficiency of the entire blockchain system.

In the model of sharding blockchains, a shard chain is responsible for transaction accounting in the corresponding shard transaction pool, and the main chain is responsible for packaging the sub-blocks sent by shard chains [19]. Different from traditional blockchains, the miners are replaced by **collators**. As the computing power of the entire network linearly increases (i.e., the number of nodes dynamically increases), the transaction throughput of blockchain also increases, which improves the scalability of blockchain system [20]. A typical example of sharding blockchain is shown in Fig. 1. The following describes the typical scenario of computations and interactions in the model.

1. There is a node A creates a pre-payment transaction, which is packaged in the collating block M_1 in shard M. The transaction is verified by the collator and sent to the main chain (the same below);
2. Node B also creates a pre-payment transaction at the same time, which is packaged in the segment K into the proofreading block N_1;
3. Node A continues to create a receipt transaction with the content of "A pays 100 coins to B" and generates a receipt ID. The transaction is packaged into the collating block M_2, and at this time, the balance of node A is changed;
4. Node B then creates a consumption receipt and generates another receipt ID, which is packaged into the collating block N_2.

Similar to traditional blockchains, there also exists the problem of forks in the fragmented multi-chain structure. If a fork occurs, we compare the length of the main chain at first, and then the length of the shard chains. The longer fork is recognized by the entire network as the winner of main chain [21].

3   The Proposed Framework

In this section, we describe the proposed scalable sharding platform, including the consensus protocol and the reward mechanism. The core idea of the proposal is to simplify the generation process of blocks and emphasize the role of proof-of-stake. Similar to the sharding blockchain model proposed above, the blocks in the proposal are also divided into two levels, namely intermediate transition blocks (i.e., middle blocks) and final confirmation blocks (i.e., final blocks), and the basic unit of the consensus mechanism operation cycle is called the Epoch.

3.1   Consensus Protocol

.We assume that there is a peer-to-peer network with a total number of n nodes. All the nodes are randomly divided into p groups (i.e., the shards), and there are n/p nodes in each group; a specific group c_i is selected as the **Validity Group**, and the remaining $p-1$ groups are named as the **Regular Groups**. The regular groups create transition blocks according to the transaction pool allocated to each group (represented as b); the validity group extracts information from the transition blocks created by the regular groups, generates a confirmation block (denoted as B), and sends it to the main chain. An epoch is a recursive process for each cycle of the consensus mechanism operation. Each epoch is divided into the following steps.

3.1.1   Node Sharding

Let the first ρ bit of the public key of a node θ be converted into a decimal number γ, and the node traverses γ blocks forward according to the value of γ in the final confirmation block where the last intermediate transition block created by itself is located to get the block B_γ (if the node has not generated a block, B_γ is the genesis block on the main chain). After that, the node θ needs to perform a single-layer $SHA256$ PoW calculation, as shown below:

$$H(B\gamma) = SHA256(TimeStamp(n),$$
$$MerkleRoot(B\gamma), H(B_{\gamma-1}), Nonce). \tag{1}$$

Herein, the notations are listed below:

- $H(B_\gamma)$ denotes the hash value of the block;
- $TimeStamp(n)$ is the timestamp of the current calculation time;
- $MerkleRoot(B_\gamma)$ is the Merkle tree root of B_γ;
- $H(B_{\gamma-1})$ denotes the hash value of the previous parent block;
- $Nonce$ is a random number, which satisfies

$$H(B_\gamma) < Target,$$

where $Target$ is the PoW target value, and the system can change the value of $Target$ during each Epoch.

It should be noted that, all the collator nodes need to complete the calculation process mentioned above. After the calculation is completed, the nodes sharding process is implemented. According to the different $Nonce$ values, the system divides the nodes into different shards (depending on the preamble of $Nonce$), and randomly selects a node group as the validity group and the rest are regular groups. Then, each node group randomly selects a **leader node**, and the rest nodes in the group send their public keys to the leader node; after that, the leader node is responsible for collecting and generating the identity list of all nodes in the group, and broadcasting the list to the leaders of other groups. This reduces the complexity of the main chain communication from $O(n^2)$ to $O(p)$, and the complexity of the shard chain communication is $O(n/p)$.

3.1.2   Transaction Sharding

The sharding process of transactions sorts by the timestamp of them. Let the set of mining pool shards be

$$T = \{T_1, T_2...T_n\},$$

where T_n belongs to the validity group, the rest of shards are regular groups. The transactions in the entire mining pool are sorted according to the timestamp sequence, and let the set be

$$Tx = \{Tx_1, Tx_2...Tx_r\}.$$

Each transaction is mapped to the corresponding shard in turn, and the set of mapping results is

$$U_1 = \{Tx_1 \rightarrow T_1, Tx_2 \rightarrow T_2...Tx_n \rightarrow T_n\}.$$

When the process encounters the validity group, it needs to skip to the next group and continue to allocate. After the cycle has been executed once, the leader node of T_n broadcasts the set U_1 to the leader nodes of other groups. After that, T_n will continue to randomly select the next leader node and set the result of the next round of mapping

$$U_2 = \{Tx_{n-1} \rightarrow T_1, Tx_n \rightarrow T_2...Tx_r \rightarrow T_{r-n}\}$$

broadcasts to the leader nodes of each shard. If the node sharding process has ended to T_{n-1}, return to T_1 to continue mapping. The purpose of this design is to enable the nodes of each shard to receive transactions of newly added shards in time, and to avoid verifying transactions that are not related to their own node group.

3.1.3   Internal Consensus

Each regular node group is randomly assigned to a shard of the total transaction pool (Transaction Pool Shard), and PoS consensus is run within the group. Take T_1 as an example, and the corresponding mining pool shard is set to TP_1 as follows:

1. The leader node of T_1 receives the mapping set U_t broadcast by the leader node of T_n, and identifies the transactions corresponding to this group, then adds it to the mining pool shard;
2. All the nodes of T_1 can package transactions in TP_1 into blocks and complete a two-layer SHA256 calculation:

$$H(b_\omega) = SHA256(SHA256(TimeStamp(n), \\ MerkleRoot(b_\omega), H(b_\omega - 1), Nonce)) \tag{2}$$

which should satisfy the following requirement:

$$H(b_\omega) < Target \times min[CoinAge(\omega), MaxCoinAge] \tag{3}$$

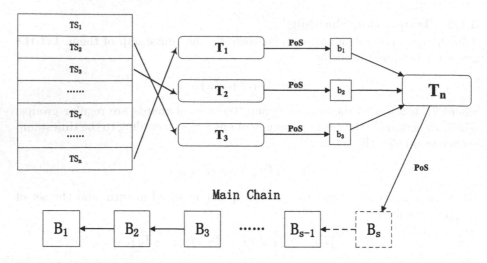

Fig. 2. An example of an epoch in the proposal

where $CoinAge(\omega)$ is the coin age held by the node ω that created the block, and $MaxCoinAge$ is the maximum available coin age specified by the system.

The current coin ages of the nodes are distributed as a transcript, and all the nodes in the group will store a transcript of the current coin age of the group. After the block created by ω (denoted by B_1^1)has been broadcast to the group, the coin age of ω is changed as follows (let $NewCoinAge(\omega)$ be the node's new coin age):

$$NewCoinAge(\omega) =$$
$$\begin{cases} CoinAge(\omega) - MaxCoinAge, \text{ if } CoinAge(\omega) > MaxCoinAge, \\ \qquad\qquad 0, \qquad\qquad\qquad\qquad\qquad \text{else.} \end{cases}$$

In other words, the node actually consumes a certain amount of coin age, but keeps the unused coin age.

3. The leader of T_1 is responsible for sending B_1^1 to the leader of T_n. Meanwhile, it can receive extra mapping sets at the same time.
4. During the period of this epoch, the nodes in T_1 can also repeat the process of generating blocks such as B_1^2, B_1^3 and so on till the epoch ends. However, each time a leader of a regular group receives a new mapping set, the node can only send one block to the leader of T_n.

3.1.4   Final Blocks

The leader node of T_n collects all the transition blocks sent by each shard in the epoch, and broadcasts the blocks to the nodes in T_n. The nodes in T_n should complete the verification below:

1. the transactions in the transition block should correspond to the right node group;

2. there do not exist double-spending problems in the transactions.

After that, T_n also runs the consensus within the group to generate the final block. The nodes in this group need to store the hash values of all transition blocks as leaf nodes of a Merkle tree, and calculate the value of the Merkle tree root:

$$H(B_\sigma) = SHA256(SHA256(TimeStamp(n), \atop MerkleRoot(B_\sigma), Nonce)) \tag{4}$$

Similarly, the node σ needs to continuously increase the value of $Nonce$ to satisfy the condition:

$$H(B_\sigma) < Target \times min[CoinAge(\sigma), MaxCoinAge] \tag{5}$$

The node σ that has successfully wined the final block generation right broadcasts the block to the entire network and adds it to the main chain, and its coin age will be consumed in the same way mentioned above.

3.1.5 Regrouping

After the protocol runs for r cycles, the system regroups all nodes, the grouping rules remain unchanged, and the process mentioned above continues to cycle. r is a variable, and its value depends on the system's efficiency in verifying transactions. Before regrouping, the leader of T_n firstly sends a $ShardRequest$ message to the leaders of other groups with the format

$$< ShardRequest, EpochID, Timestamp >,$$

where $EpochID$ is the serial number of the current epoch. The purpose of design is to help the receiver to confirm that the message corresponds to the current epoch, preventing the messages from being received late due to network delays. After the leader of a regular group receives the message, it sends a $ShardPrepare$ message to the leader of the validity group in the format

$$< ShardPrepare, ShardID, Timestamp, CoinAgeList >,$$

where $CoinAgeList$ refers to the coin age list of nodes in the shard. The leader of validity group merges the coin age lists and broadcasts them to all the nodes in the entire network for the usage in the next round of grouping. An example of an epoch is shown in Fig. 2.

When a new node applies for joining the network, it should firstly broadcast an application message to the entire network in the format of

$$< Apply, PubKey, Timestamp, Hash >,$$

where $PubKey$ is the public key of the node, and $Hash$ is the hash value of message calculated by $SHA256$. Then, before the start of the next round of epoch, the node needs to perform PoW calculation to obtain identity access. Since the node has not generated a block, the block that originally needs to be traversed can be a genesis block, and then according to $Nonce$ for subsequent sharding of nodes.

3.2   Reward Mechanism

In the traditional PoW mechanism, miners add *Coinbase* transactions to the block and set the receiver's address as their own address, so that each mining of blocks can get a certain number of mining rewards, as well as additional transactions **Handling Fee** [22]. In the design of our proposal, we need to realize the reward for collators through **interest**. When a collator generates a block and the block is approved by the validity group, the coin age of the collator will be consumed by a certain amount. Therefore, we consider that if

- a node holds ε coin age,
- every time α coin age is cleared,
- β coin interest will be obtained in the block,

then the interest that can be obtained is

$$\varepsilon \times \beta / \alpha$$

coins. The value of ε, α and β depends on the market value of the token. Obviously, the nodes that pay more coin ages, the higher the rewards obtained through the protocol, that is, relying on holding tokens and converting them into computing power to make a profit [23].

4   Security Analysis

In this section, we provide the security analysis of the proposal. In this model, we mainly focus on two standard attack methods: **Sybil Attack** and **51% Attack**.

The Sybil Attack refers to a type of malicious actions that the attackers create a huge number of *pseudonymous identities* and use them to gain a disproportionately large influence. In peer-to-peer networks, the identities are used as the abstractions, in order that a remote entity can be aware of identities without necessarily finding the relationship between identities and entities in reality. In theory, we think that each identity corresponds to a specific entities in reality; however in fact, different identities may correspond to the same entity in reality. The malicious nodes may create pseudonymous identities that are controlled by themselves. Those identities (obviously dishonest) may launch the attack by changing the result of votes.

51% attack (also called the Double-spending Attack) [24] is also widely discussed in Bitcoin communities. When a group of miners control more than 50% of the hash power of entire network (computing power), a 51% attack may occur. The malicious attackers may launch the attack by generating new blocks without verification and confirmation, revoke the completed transactions on the current block, or initiate a large number of double-spending transactions on the network.

4.1 Sybil Attack

As is mentioned above, the consensus protocol designed by this paper operates on the Epoch as the basic period cycle, before the beginning of each cycle, all the nodes in the entire network need to complete a simple PoW calculation [25]. According to the value of the first few digits of the node's public key converted into a decimal number, the nodes calculates a hash value that satisfies the target value (according to the current timestamp), and traverses several blocks toward the final confirmation block where the last intermediate transition block created by itself. Therefore, the sharding of nodes can be performed by a random number that meets the conditions. In this design, it is not feasible for a malicious attacker to launch the attack by controlling a certain shard. Due to the characteristics of hash functions, a malicious attacker cannot predict which node shard a malicious node will be mapped to, nor can it predict which part of a transaction will be verified by a specific node shard. Meanwhile, at the end of each epoch, the node still needs to complete a single PoW calculation as the identity admission for the next epoch period. Even if the attacker intends to advance the PoW process by controlling its intentional delay in creating blocks, it cannot control the time and order of the honest nodes to send the blocks in the main chain before. Therefore, it is computationally infeasible for attackers to control a certain node shard by creating a large number of malicious nodes or to create a large number of empty transactions to reduce the verification efficiency of the entire network. Therefore, the proposal can resist Sybil attacks.

4.2 51% Attack

A PoW-based 51% attack requires the attacker to control at least "50%+1" of the computing power of the entire network. In the PoS mechanism, launching a 51% attack requires the attacker to occupy at least 50%+1 of the coin age of the entire network. Therefore, the attacker needs to hold a large amount of tokens for a long time before the attack. Compared with PoW, mining coins can theoretically control 50%+1 of the computing power of the entire network. However, since there is no coinage mechanism in traditional PoS, the total amount of cryptocurrencies is fully preset at the beginning of developing. Stakes can only be acquired from existing users and cannot be invested outside the system. Therefore, the cost of launching a 51% attack on a PoS-based blockchain is equal to the cost of purchasing stake from the market. Thus there are no economic benefits for the attackers.

The solution of the proposal is setting a maximum limit value when performing PoS coin age statistics, and the coin ages that exceed the limit are not counted (that is, as mentioned above, a threshold $MaxCoinAge$ is set so that all the nodes have an upper limit of difficulty when doing calculations). Even if the attacker's coin age reaches more than 51% of the entire network, it does not mean that the probability of the attacker winning in the PoS operation reaches more than 51%, that is, there will be no absolute advantage, and when the size of the shard is large enough, the attacker launches a 51% attack Probability is also negligible. In addition, suppose that an attacker launches a 51% attack in order to cause a malicious fork, and hopes that the fork will continue to exist so that the main chain will be replaced. However, members of the validity group are constantly changing with the cycle of epochs, and the coin age of the collators who have win the right of generating blocks in the main chain will be consumed or cleared. Therefore, the attacker needs to control a large number of sequences with high coin ages, and to ensure that the validity group of each epoch has the ability to launch malicious attacks, and the malicious attacks will win in each round of PoS competition. The probability of them can be negligible.

In summary, the 51% attack launched by malicious attackers has no economic benefits, and even if the adversary abandons the economic benefits, it is not feasible to simply launch the attack, hence there is no motive for 51% attacks.

5   Conclusion and Future Work

We mainly focus on how to ensure the that the malicious nodes will not launch the attack on specific node shards. In this paper, we propose a novel scalable blockchain framework based on the sharding technology and Proos-of-Stake concensus protocol. The model has the advantages of dynamic random grouping (both nodes and transactions), independent and internal consensus, and linear scalability. Improved dynamic random grouping based on proof-of-work calculation can ensure that each node cannot predict its sharding situation, so that malicious attackers cannot control specific shards to launch a Sybil attack or a 51% attack.

In future work, we will proceed blockchain scalability and analyze the applications of blockchain projects, including cross-chain communication of side chains, path finding algorithms for off-chain transaction networks, and incentive mechanisms for blockchain sharding. Blockchain scalability provides a credible data mechanism for dealing with high-traffic transactions in decentralized scenarios. However, there are still many problems and challenges, i.e., how to put research results into well-operated landing projects.

Acknowledgement. The work was supported by the National Natural Science Foundation of China (Grant Nos. 61971192, 6191101004) and the National Cryptography Development Fund (Grant No. MMJJ20180106).

References

1. Nakamoto, S., Bitcoin, A.: A peer-to-peer electronic cash system (2008)
2. Fernández-Caramés, T.M., Fraga-Lamas, P.: A review on the use of blockchain for the Internet of things. IEEE Access **6**, 32979–33001 (2018)
3. Shao, Z., Xue, C., Zhuge, Q., Qiu, M., Xiao, B., Sha, E.H.-M.: Security protection and checking for embedded system integration against buffer overflow attacks via hardware/software. IEEE Trans. Comput. **55**(4), 443–453 (2006)
4. Tian, Z., Li, M., Qiu, M., Sun, Y., Shen, Su: Block-DEF: a secure digital evidence framework using blockchain. Inf. Sci. **491**, 151–165 (2019)

5. Bocek, T., Rodrigues, B.B., Strasser, T., Stiller, B.: Blockchains everywhere–a use-case of blockchains in the pharma supply-chain. In: Proceeding of IFIP/IEEE Symposium on Integrated Network Service Manage. (IM), pp. 772–777, May 2017
6. Murdoch, J.S., Anderson, J.: Verified by visa and mastercard securecode: or, how not to design authentication. In: Financial Cryptography, pp. 336–342 (2010)
7. Eyal, I., Gencer, A.E., Renesse, R.V.: Bitcoin-NG: a scalable blockchain protocol. In: Usenix Conference on Networked Systems Design & Implementation, pp. 45–59 (2016)
8. Kiayias, A., Russell, A., David, B., Oliynykov, R.: Ouroboros: a provably secure proof-of-stake blockchain protocol. In: Katz, J., Shacham, H. (eds.) CRYPTO 2017. LNCS, vol. 10401, pp. 357–388. Springer, Cham (2017). https://doi.org/10.1007/978-3-319-63688-7_12
9. King, S., Nadal, S.: PPCoin: peer-to-peer crypto-currency with proof-of-stake. Self–published Paper, 19 August 2012
10. Tosh, D., et al.: CloudPoS: a proof-of-stake consensus design for blockchain integrated cloud. In: IEEE 11th International Conference on Cloud Computing (CLOUD), pp. 302–309 (2018)
11. Bentov, I., Gabizon, A., Mizrahi, A.: Cryptocurrencies without proof of work. In: Clark, J., Meiklejohn, S., Ryan, P.Y.A., Wallach, D., Brenner, M., Rohloff, K. (eds.) FC 2016. LNCS, vol. 9604, pp. 142–157. Springer, Heidelberg (2016). https://doi.org/10.1007/978-3-662-53357-4_10
12. Peifang, N., Hongda, L., Meng, X., Pan, D.: UniqueChain: a fast, provably secure proof-of-stake based blockchain protocol in the open setting. IACR Cryptol. ePrint Arch. **2019**, 456 (2019)
13. Bartoletti, M., Lande, S., Podda, A.S.: A proof-of-stake protocol for consensus on bitcoin subchains. In: Brenner, M., et al. (eds.) FC 2017. LNCS, vol. 10323, pp. 568–584. Springer, Cham (2017). https://doi.org/10.1007/978-3-319-70278-0_36
14. Chepurnoy, A., Duong, T., Fan, L., Zhou, H.: TwinsCoin: a cryptocurrency via proof-of-work and proof-of-stake. IACR Cryptol. ePrint Arch. **2017**, 232 (2017)
15. Badertscher, C., Gazi, P., Kiayias, A., Russell, A., Zikas, V., Ouroboros genesis: composable proof-of-stake blockchains with dynamic availability. In: ACM Conference on Computer and Communications Security, pp. 913–930 (2018)
16. Gazi, P., Kiayias, A., Russell, A.: Stake-bleeding attacks on proof-of-stake blockchains. In: CVCBT, pp. 85–92 (2018)
17. Luu, L., et al.: A secure sharding protocol for open blockchains. In: ACM Sigsac Conference on Computer & Communications Security, pp. 17–30 (2016)
18. Fidelman, Z.: A generic sharding scheme for blockchain protocols. CoRR abs/1909.01162 (2019)
19. Dang, H., Dinh, T., Loghin, D., Chang, E., Lin, Q., Ooi, B.: Towards scaling blockchain systems via sharding. In: SIGMOD Conference, pp. 123–140 (2019)
20. Tong, W., Dong, X., Shen, Y., Jiang, X.: A hierarchical sharding protocol for multi-domain IoT blockchains. In: ICC, pp. 1–6 (2019)
21. Croman, K., et al.: On scaling decentralized blockchains. In: Clark, J., Meiklejohn, S., Ryan, P.Y.A., Wallach, D., Brenner, M., Rohloff, K. (eds.) FC 2016. LNCS, vol. 9604, pp. 106–125. Springer, Heidelberg (2016). https://doi.org/10.1007/978-3-662-53357-4_8
22. Vaidya, K.: Decoding the enigma of Bitcoin Mining - Part I: Mechanism. https://medium.com/all-things-ledger/decoding-the-enigma-of-bitcoin-mining-f8b2697bc4e2/
23. Buterin, V.: A Proof of Stake Design Philosophy. https://medium.com/@VitalikButerin/a-proof-of-stake-design-philosophy-506585978d51/

24. Lam, W.: Attack-prevention and damage-control investments in cybersecurity. Inf. Econ. Policy **37**, 42–51 (2016)
25. Asfia, U., Kamuni, V., Sutavani, S., Sheikh, A., Wagh, S., Singh, N.M.: A blockchain construct for energy trading against sybil attacks. In: MED, pp. 422–427 (2019)
26. Andy, W.: Directed Acyclic Graph: the Future of Blockchain Development. https://www.coinspeaker.com/directed-acyclic-graph-blockchain/
27. Khan, N., State, R.: International Congress on Blockchain and Applications. Lightning Network: A Comparative Review of Transaction Fees and Data Analysis, pp. 11–18. Springer, Berlin (2019)
28. Guo, J., Gai, K., Zhu, L., Zhang, Z.: An approach of secure two-way-pegged multi-sidechain. In: Wen, S., Zomaya, A., Yang, L.T. (eds.) ICA3PP 2019. LNCS, vol. 11945, pp. 551–564. Springer, Cham (2020). https://doi.org/10.1007/978-3-030-38961-1_47
29. Back, A., et al.: Enabling blockchain innovations with pegged sidechains. http://www.opensciencereview.com/papers/123/enablingblockchain-innovations-with-pegged-sidechains
30. Luu, L., et al.: A secure sharding protocol for open blockchains. In: ACM Conference on Computer and Communications Security, pp. 17–30 (2016)
31. Qiu, H., Noura, H., Qiu, M., Ming, Z., Memmi, G.: A user-centric data protection method for cloud storage based on invertible DWT. IEEE Trans. Cloud Comput. 1–12 (2019). https://doi.org/10.1109/TCC.2019.2911679

Tree2tree Structural Language Modeling for Compiler Fuzzing

Haoran Xu[1], Shuhui Fan[1], Yongjun Wang[1(✉)], Zhijian Huang[2], Hongzuo Xu[1], and Peidai Xie[1]

[1] College of Computer, National University of Defense Technology, Changsha, China
{xuhaoran12,fanshuhui18,wangyongjun,xuhongzuo13}@nudt.edu.cn,
xpd2002@126.com
[2] National Key Laboratory of Science and Technology on Information System Security, Institute of System Engineering, Chinese Academy of Military Science, Beijing, China
zjhuang@nudt.edu.cn

Abstract. Compiler fuzzing requires well-formed test cases. Only syntactically correct programs can pass the parsing stage of a compiler. Recently, advanced compiler fuzzers produce test cases by learning a generative language model of regular programs. They treat programs as natural language texts without leveraging any syntactic structure, making them hard to produce syntactically correct programs when programs get long. In this paper, we propose a novel tree-to-tree (tree2tree) model to leverage the structural information for a robust test case generation. We adopt an encoder-decoder architecture to map a partial abstract syntax tree (AST) to a complete AST. We introduce a C compiler fuzzing framework, named TSmith. Specifically, TSmith employs a tree-based encoder to encode the input partial AST to capture the hierarchical structure information. It then adopts a tree decoder to generate a complete AST by expanding the non-terminals in a top-down manner. Finally, the output ASTs are converted into corresponding test programs. Experimental results show that our generation strategies have a maximum parsing pass rate of 83%, which is about 21% higher than sequential models. Besides, TSmith significantly improves the code coverage of the compiler. Benefiting from the high pass rate and broad code coverage, TSmith has found 14 new bugs in GCC.

Keywords: Fuzzing · Compiler · Tree2tree · Neural network · Generation

1 Introduction

Compilers are among the most fundamental tools in software development. Various techniques have been used to improve the correctness and reliability of compilers. Fuzzing has been proven to be a highly successful technique to uncover

M. Qiu (Ed.): ICA3PP 2020, LNCS 12452, pp. 563–578, 2020.
https://doi.org/10.1007/978-3-030-60245-1_38

bugs in compilers by feeding a large number of test programs to the target compiler. Effective compiler fuzzing requires well-formed test cases. Only programs with correct lexicons and syntax could pass the parsing stage of a compiler, after which the most complex and vulnerable functionalities can be tested. Traditional grammar-based fuzzing methods [19] produce test cases based on the given grammars and specifications. However, researchers point out that providing such specifications is a time-consuming and laborious task.

Deep neural networks have been introduced to tackle this problem in recent studies. Recent works show that approaches in natural language processing can be applied to test program generation. In these methods, programs are treated as natural language texts. The state-of-the-art methods in this field adopt the sequence-to-sequence (seq2seq) model [10] which builds a generative language model for code sequence generation. However, programming languages have strict grammars and are not tolerant to syntax errors. Besides, the generation approaches in sequential models are based on prefix sequences. Therefore, it will lose global information which is out of the scope of conditional sequences. It has been demonstrated that it is hard for the seq2seq model to produce syntactically correct programs when the lengths grow large [9].

In this work, we observe that programs naturally contain rich and explicit structural information, which means that program generation is a structured prediction problem. For program generation problems, a useful property is that each program corresponds to a parse tree. Rather than modeling the program as a sequence of chars or tokens, we consider the problem as generating parse trees. Recently, tree-based methods have shown successes on various semantic parsing [5,13] and software engineering tasks [3,20]. Following previous works, we propose a novel tree2tree model to generate syntactically correct test cases. The key idea of our work is to leverage the syntactic structures and global information in source code to learn a structural code model.

We focus on generating test programs in a context in this paper. Specifically, we generate new test programs by completing the input partial programs. Our model employs the Abstract Syntax Tree (AST). An AST is a tree representation of the abstract syntactic structure of source code. Compared to a concrete syntax tree which represents every syntax detail in a program, an AST is more refined because it just contains structural or content-related details. Most programming languages are accompanied with a well-developed parser, which could be used for the conversion between ASTs and programs. Recent studies [1,17,18] have shown that modeling source code with AST can obtain better representation than traditional natural language processing methods. We combine AST and a tree encoder to capture both the lexical (i.e., leaf nodes such as identifiers) and syntactical (i.e., non-leaf nodes like grammar construct `Decl`) information in programs. ASTs are used for both the input and output of our model. Given a partial AST, our goal is to complete it to get a new AST. The generated AST can then be deterministically converted into the corresponding program.

To the best of our knowledge, it is the first work to employ AST-based deep neural networks for fuzzing input generation. We implement a prototype of the proposed methods in a tool called TSmith. Firstly, TSmith employs a tree-based encoder to encode the input AST in order to capture the hierarchical structure information and global context of the partial program. Secondly, it adopts a tree decoder to generate a complete AST by expanding the non-terminals in a top-down manner. Rather than only accessing a fixed encoding, we use an attention-based representation of the input AST. The flexibility of the AST-based model makes it readily applicable for completion tasks. Compared with sequential models, it is easier to design diverse generation strategies for a tree model. To make better use of the learned model, we have designed five generation strategies to produce various new test programs.

To demonstrate the effectiveness of TSmith, we adopt TSmith for C compiler fuzzing. Extensive experiments show that the parsing pass rate of programs generated by TSmith is higher than the previous methods. The highest pass rate of our generation strategies reaches 83%, which is about 21% higher than the sequential model. Even the average pass rate of our five generation strategies has 7% improvements against the highest pass rate of the sequential model. In addition, combined with sampling methods, TSmith significantly improves the code coverage of target compilers.

In summary, in this paper we make the following contributions:

- We propose a tree2tree model to leverage the structural information for compiler fuzzing input generation. AST is leveraged to capture both the lexical and syntactical information in programs. Experiments show that programs generated by our model are more well-formed than previous methods. Further, tree-based structural language modeling supports more generation strategies to produce diverse test programs.
- We present TSmith, a novel compiler fuzzing framework. We design five generation strategies in TSmith to produce various test cases. TSmith has been applied to the fuzz testing of GCC on all currently supported versions. We have found and reported 14 bugs to the development team.

2   Problem Statement

Given a seed program P_s, our task is to generate new test programs. We attack this problem by constructing a new AST T_t given the context information of a partial AST T_s. This simplifies the test program generation problem to a AST completion problem. P_s can be converted into an AST T. Based on the normal AST, we can construct the partial AST T_s by removing nodes or inserting nodes. We define a structural probabilistic grammar model of generating a complete AST T_t given T_s: $p(T_t|T_s)$. The most likely target AST T_t^* is then given by

$$T_t^* = \arg\max_{T_t} p(T_t|T_s) \tag{1}$$

T_t^* can then be converted into the corresponding program P_t.

3   TSmith

In this section, we provide an overview on the overall design and implementation details of TSmith.

3.1   Design Overview

We show the workflow of TSmith in Fig. 1. There are three main components in TSmith: learning module, generation module and fuzzing module. Learning module learns the tree2tree model from a corpus of code samples. Programs have been converted into ASTs before being sent to the model. After training, the generation module produces new test programs according to the five proposed generation strategies. We use code samples from dataset as seed programs during applying generation strategies. Finally, the fuzzing module checks whether new test cases trigger unexpected behaviors (e.g., a crash, a freeze) in compilers.

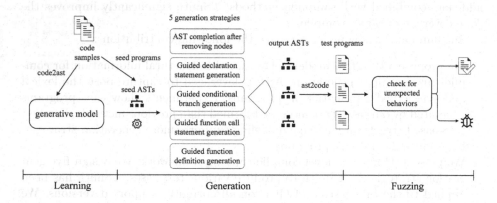

Fig. 1. Workflow of TSmith

3.2   Tree2tree Neural Network

To tackle the test program generation task presented in previous section, we design the tree2tree network. Our model follows an encoder-decoder architecture. The encoder encodes the input partial AST to an embedding. Then the decoder decodes the embedding into a complete AST. Rather than only accessing a fixed encoding, the decoder uses an attention mechanism to locate the corresponding source sub-tree when expanding a node. The structure of an AST is consistent with the syntax structure of the program which can facilitate the learning of the grammar.

When training the tree2tree model, we remove nodes from a normal AST and use the remaining partial tree as input T_s. The normal AST is used as target T_t. Any node in the ASTs except the root can be removed. When removing a node, we remove the entire sub-tree rooted at the node. The removed sub-tree may correspond to a single token(e.g., identifiers), an expression(e.g., grammar construct like `BinaryOp` node), or a full statement(e.g., grammar construct like `Decl` node), etc. It has been demonstrated that encoder and decoder for binary trees could be more effective [3]. Thus, we convert both T_s and T_t into binary trees. For this conversion, we use the Left-Child Right-Sibling representation.

The binary tree encoder employs a Tree-LSTM [15] to encode the input tree from the bottom up. Specifically, for a node N with value v, there are two children N_L and N_R, which denotes its left child and right child respectively. The LSTM state of N is computed as

$$(h, c) = LSTM(([h_L; h_R], [c_L; c_R]), x) \tag{2}$$

where (h_L, c_L) and (h_R, c_R) denote the LSTM state of N_L and N_R respectively, and x denotes the embedding of v. By encoding the partial tree, we can capture the global context of the seed program.

After encoding, the decoder uses the LSTM state of the root node and all the encoder outputs to produce the target tree with a top-down manner. To leverage the global context of source tree, we use the hidden state of the root node of source tree as the state of starting root in the target tree. The decoder maintains a queue of nodes to be expanded. During decoding process, new nodes are generated and appended to the queue. The decoder expands nodes in the queue one by one until the queue is empty.

The decoder first predict the value of the expanding node. It computes the embedding of the expanding node N with attention. When the depth of input partial AST grows large, it is hard for the decoder to complete it only depending on a fixed-length vector representation. Attention mechanism is used to locate the corresponding sub-tree in the input AST when expanding a non-terminal. The decoder uses this information to guide the node expansion. We calculate the alignment scores using the decoder hidden state and encoder hidden states:

$$score(\overline{h}_s, h_t) = \overline{h}_s{}^\top W_a h_t \tag{3}$$

where $\overline{h}_s$ denotes each encoder hidden state, and h_t denotes the decoder hidden state. We then compute the probability of that N_s is sub-tree in the source AST corresponding to N_t in the target AST. We calculate it as:

$$P_t(s) \propto exp(score(\overline{h}_s, h_t)) \tag{4}$$

After that, We calculate the expectation of the hidden state value across all N_s conditioned on N_t.

$$e_s = E[h_s | N_t] = \sum_{s'} h_{s'} P_t(s') \tag{5}$$

We use this embedding vector and the decoder hidden state to produce an attentional embedding:

$$\widetilde{h}_t = tanh(W_b e_s + W_c h_t) \tag{6}$$

Lastly, the attentional embedding is fed into a softmax network to predict the node value.

$$v = argmax\ softmax(W\widetilde{h}_t) \tag{7}$$

The value of a node could be a non-terminal, a terminal, or a EOS token. All leaf nodes are EOS nodes. The decoder only generates child nodes for terminals and non-terminals. We employ two LSTMs [8] for the two child nodes. The LSTM states of child nodes are computed as:

$$(h_l, c_l) = LSTM_L(([h, c], Ev)) \tag{8}$$

$$(h_r, c_r) = LSTM_R(([h, c], Ev)) \tag{9}$$

where E denotes a word embedding matrix.

In experiments, we set the probability of removing each node to one-twentieth. The number of layers of the Tree-LSTM and LSTMs is 1, and the hidden size and embedding size are set to 256. Batch size is set to 20. We apply gradient clipping to prevent gradients from becoming too large. We choose Adam as our optimizer, and the initial learning rate is set to 0.001.

3.3   Generating New Test Programs

RNN-based models generate code sequence with the given conditional token sequences. However, without grammar rules and structural information, it is a complex task for sequential models to select conditional inputs from seed programs and construct new programs from the generated token sequences. Previous works mainly use the methods of cutting and inserting to construct new test programs. Generally the units of cutting and inserting are code lines.

In contrast, the basic unit of AST is node. It is more flexible to construct a new tree at the node level. With the learned structural code model, TSmith can map a partial AST to a complete one. We send partial ASTs to the model and ask the model to produce complete ones. The generated ASTs are then converted into surface programs. During this process, how to prepare the partial ASTs and how to sample from the prediction results are crucial.

Generation Strategy. Our goal is to produce diverse test programs. To make better use of the learned model, we propose five generation strategies: AST completion after removing nodes, guided declaration statement generation, guided conditional branch generation, guided function call statement generation, and guided function definition generation.

AST Completion After Removing Nodes. We randomly remove nodes from the seed ASTs with a certain probability, and take the remaining part of the AST as the input of our model. Nodes that may be removed include terminals and non-terminals. If the removed node is a leaf node of the seed AST, the newly generated code in the completion process may account for a low proportion of the entire program. In this case, the generated AST is more likely to conform to the grammar specification, and thus passes through the parsing stage of compilers. If the removed node is a non-leaf node, the newly generated code may account for a significant proportion of the entire program. On the one hand, the generated programs are prone to syntax errors. On the other hand, it gives the model a large room for creativity.

Guided Declaration Statement Generation. This strategy refers inserting Decl node to the seed ASTs. Decl node is a non-terminal. This insertion guide the model to complete sub-tree rooted at the Decl node. Many important AST nodes derive from Decl (e.g., TypeDecl and FuncDecl). It is worth noting that this way of inserting non-terminals to induce the model to generate relevant code is not supported by sequential models.

Guided Conditional Branch Generation. We insert If node to the seed ASTs. If node is a non-terminal. This insertion guide the model to complete sub-tree rooted at the If node. This will guide the model to produce a conditional statement, thereby introducing new conditional branches and scopes for the program.

Guided Function Call Statement Generation. We insert FuncCall node to the seed ASTs. This insertion leads the model to generate a statement for the function call.

Guided Function Definition Generation. We insert FuncDef node to the seed ASTs. This will guide the model to generate a function definition, thereby introducing new scopes for the program.

Benefiting from the structural language model, we can insert various types of nodes into the seed AST to induce the generation of specific statements or expressions. In this work, we mainly test the five aforementioned strategies.

Sampling Method. We sample the node value from the probability distribution of decoder outputs with two sampling methods. The first one is greedy sampling, which always selects the node value with the greatest possibility. Programs produced by this approach is well-formed. However, the generated programs lack variety. The second one is sampling from a multinomial distribution. Compared with completely random sampling which randomly selects a node value based on the probability distribution, stochastic sampling from a multinomial distribution is a more reasonable way. When sampling from the multinomial distribution, the output with the greatest probability can be selected with high probability. Moreover, there is a certain probability to explore other outputs. We use temperature to control the randomness of the stochastic sampling process. A lower temperature value controls the sampling process to be more conservative. Therefore the generated programs are more likely to be well-formed. A high temperature value makes the sampling more aggressive. The outputs are more error-prone, but it will also bring more variety.

4 Experiments and Evaluation

In this section we evaluate TSmith via empirical analysis. Then we demonstrate its effectiveness in fuzzing real-world compilers.

4.1   Dataset and Preprocessing

We collect well-formed programs from the test suites in compiler source releases as the original dataset. We remove all the comments, and expand all macros. Considering that the inclusion of header files will introduce a number of duplicate standard library headers, we remove programs that contain inclusion of header files.

Due to the flexibility of identifier naming, source code vocabulary can be significantly large. Large vocabularies severely affect code models, degrading their performance and rendering them unable to scale. To facilitate the model to learn the syntax of C programming language, we further process the code samples in the dataset, trying to reduce the size of the vocabulary. First, we rename all the user-declared identifiers in each program using an consistent pattern. For example, based on the order of declaration, we rename function names in each program according to the pattern of {sub_0, sub_1, sub_2, . . .}. For variable names, we rename them according to the pattern of {var_0, var_1, var_2, . . .}. Renamed identifiers also include the parameter name, structure name, etc. This naming convention will not modify the program behavior. Second, we unify the value of literals. We rename all the character literals to a same value of "C", string literals to a same value of "STRING", and floating point literals to a fixed number of 1.0. Changing the value of literals may change the control flow and data flow of the original program. We think it is a trade off to reduce the vocabulary.

4.2   Evaluation Metrics

We evaluate TSmith via three metrics: pass rate of the generated test programs, the code coverage improvements of the target compiler, and the number of bugs found in main stream compilers.

The pass rate of generated programs indicates how well our model learns the syntax of the programming languages. It reflects the proportion of the syntactically correct programs in all the generated programs. It is critical to produce syntactically correct inputs when fuzzing compilers. Only syntactically correct programs could pass the legitimacy check (i.e., lexical analysis and syntax analysis) in compiler parsing stage. After that, the most complex and vulnerable parts of compilers (i.e., middle-end and back-end) could be tested [19].

We select the code coverage improvements as a metric. Code coverage information is widely used by fuzzers to find interesting inputs. Coverage-guided fuzzers [2,7,11,14,21] mutate the same input file to increase the code coverage and thus get closer to new paths where a crash could happen. While TSmith is not coverage-guided, coverage information can helps us analyse the fuzzing process. We use gcov[1] to collect coverage information.

We apply TSmith to the fuzzing of main stream C compilers, and use the number of found bugs as one indicator of the effectiveness of our methods. Compilers like GCC and LLVM have defined a special kind of error called "internal

---

[1] https://gcc.gnu.org/onlinedocs/gcc/Gcov.html.

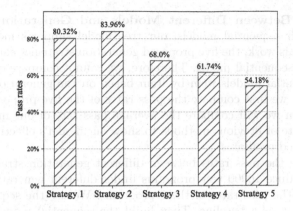

Fig. 2. Pass rate of different strategies in TSmith

compiler error". An internal compiler error is an error that occurs not due to erroneous source code, but rather due to a bug in the compiler itself.

4.3 Pass Rate

In this part, we will analyze how the pass rate varies with different generation strategies, and how the sampling methods affect the compilation pass rate of generated programs. As long as no error occur during compilation, we assume that the test program has pass the parsing stage.

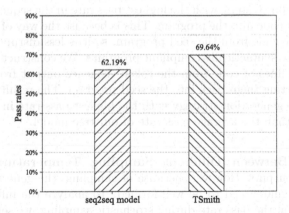

Fig. 3. Pass rate between models. The pass rate of seq2seq model is the highest pass rate of generation strategies in that model. The pass rate of TSmith is the average pass rate of five generation strategies in TSmith.

Comparison Between Different Models and Generation Strategies.
Compared with sequential models, our tree model supports more generation
strategies. In this work, the five proposed generation strategies cannot be implemented in the sequential model. Therefore, we cannot compare our tree model
with the sequential models item by item based on the generation strategy. In
our evaluation, we first compare the pass rates of the five proposed generation
strategies. Then we will compare the average pass rate of our model with the
highest pass rate of previous methods to show our model's effectiveness to generate well-formed programs.

To compare the pass rates between different generation strategies, we use
TSmith to produce 10000 test programs under different generation strategies,
and compared the pass rates of these programs. We use the seq2seq approach
in DeepFuzz [10] as a baseline. They build the sequential model at character
level which requires a lot of effort in dealing with the token-level syntax. In
order to further enhance the performance of the model and make a meaningful
comparison, we implement the token-level seq2seq model as a baseline based
on the same dataset and preprocessing method we proposed. We use greedy
sampling to perform this comparison.

The results are shown in the Fig. 2. Overall, the test program generated by
strategy 2 has the highest pass rate. This means that it is easier to insert well-
formed declaration statements into a program. The test program generated by
strategy 1 also has a high pass rate. This means that our model makes good use
of global information to expand non-terminals, thus completing the program.
The test program generated by strategy 5 has the lowest pass rate. One possible
explanation is that syntax errors are more likely to occur because more new code
is generated when adding new function definitions.

The generation strategy with the highest pass rate in the sequential models is
to insert lines of code into the program. This is because the way of code insertion
does not remove code from the seed program, so it is less disruptive, making it
easier to produce syntactically compliant programs. We construct new programs
by intercepting lines of code from the token sequence output from the seq2seq
model and inserting them back into the seed program. The results are shown in
Fig. 3. Even the generation strategy with the highest pass rate in the sequential
model is lower than the average pass rate of our tree model.

Comparison Between Different Sampling Temperatures. Sampling
brings diverse outputs. However, because of sampling, the generated programs
are not always guaranteed to be well-formed. To analyze the influence of temperature value on the pass rate during stochastic sampling, we select 4 different
temperature values of 0.5, 0.75, 1.0, and 1.25, to compare the pass rate of the
programs generated by TSmith with each generation strategy.

The results are shown in Fig. 4. In summary, for all generation strategies, the
pass rate continues to decline as the sampling temperature increases. It is because
the generative model becomes more aggressive as temperature rise. This may
bring unexpected but useful test inputs to trigger compiler bugs. For strategy

1, when the temperature gets 0.5, the pass rate is reduced by 5% compared to greedy sampling. When the temperature gets 1.0, the pass rate is reduced by 17%.

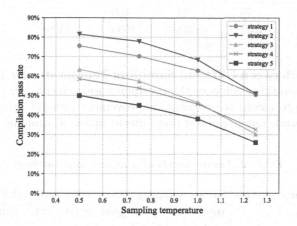

Fig. 4. Pass rate of different sampling temperatures

4.4 Code Coverage

In this part, we evaluate TSmith's impact on the code coverage of the target compiler. We select 5000 programs as seed programs and use the code coverage for compiling these programs as baseline. Then TSmith is used to produce 5000 new programs with each generation strategy based on these seed programs. For large software like compilers, the compilation options at build time will affect the total amount of code of the compiler. Besides, the code coverage will also be influenced by the compilation options when using the compiler to compile test programs. To make a meaningful comparison, we collect the increment of three kinds of coverage information during our analysis: the number of covered lines, the number of covered functions, and the number of covered branches.

Comparison Between Generation Strategies. We use GCC-7 for coverage collection, with the compilation option -S, which means compile only, do not assemble or link. We choose greedy sampling for this comparison. The coverage improvements with each generation strategy are shown in Table 1.

Overall, with each generation strategy, TSmith significantly improves the code coverage. Strategy 3 achieves the best effect in all the 3 kinds of coverage metrics. It brings 1973-line improvement for line coverage, 100 new functions in function coverage, and 1636 new branches in branch coverage.

Table 1. Improvements of compiler code coverage

Improvements	Strategy 1	Strategy 2	Strategy 3	Strategy 4	Strategy 5
Line	1627	1620	1973	1737	1595
Function	75	81	100	75	66
Branch	1522	1278	1636	1409	1331

Comparison Between Different Sampling Temperatures. To analyze how sampling temperatures influence the coverage improvements, we take strategy 2 as an example to record the coverage improvement brought by generating and compiling the same number of test programs at different sampling temperatures.

Table 2. Improvements of compiler code coverage

Improvements	Greedy sampling	0.5	0.75	1.0	1.25
Line	1620	2606	2500	2741	3198
Function	81	155	154	166	177
Branch	1278	1808	1831	2263	2515

Table 2 demonstrates the impact of different sampling temperatures on code coverage improvements. As the sampling temperature rises, our model becomes more aggressive. At the same time, code coverage is gradually increasing. Compared with greedy sampling, when the sampling temperature reaches 1.25, the increased coverage almost doubles. This experiment shows the usefulness of sampling from model outputs. The results are also shown in Fig. 5.

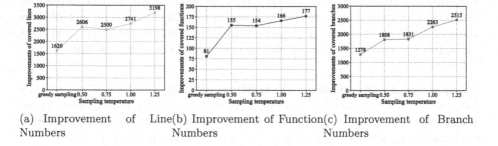

(a) Improvement of Line (b) Improvement of Function (c) Improvement of Branch
Numbers   Numbers   Numbers

Fig. 5. Improvements of compiler code coverage

4.5 New Bugs

TSmith is applied to the fuzz testing of GCC. We test three GCC versions that are still supported by the GCC development team. They are GCC-8, GCC-9 and

GCC-10. In addition, we test GCC-11, which is currently under development and will be released soon. In our experiments, TSmith has found 14 new bugs. The bug numbers and the affected versions are shown in Table 3. All bugs information has been submitted to the GCC development team.

Table 3. New bugs.

ID	Affected versions
bug-95161	10/11
bug-95145	10/11
bug-95124	10
bug-94968	10/11
bug-94966	10
bug-94903	8/9
bug-94902	8/9/10
bug-94842	8/9
bug-94780	8/9
bug-94755	9/10/11
bug-94731	8/9/10/11
bug-94730	8/9/10/11
bug-94726	10/11
bug-94705	10

We elaborate on three test programs which successfully triggered bugs in GCC with more details to further illustrate the effectiveness of TSmith.

Bug-94755. The test case is generated by strategy 1. Raw seed program does not trigger any bug. When cropping the seed AST, we remove a FuncCall node. The value of its original ID child node is __builtin_infl. Then the partial AST is sent to the tree2tree model. The model generates a new FuncCall node, and the value of its ID child node is __builtin_speculation_safe_value. This new test program finally triggers the internal compiler error, and it has been added to the latest test suite in GCC (pr94755.c) after our reporting. This example shows the effectiveness of strategy 1.

```
1    extern void foo ( void ) ;
2    void bar ( double x )
3    {
4  -      if ( x == __builtin_infl() )
5  +      if ( x == __builtin_speculation_safe_value () )
6        foo ( ) ;
7    }
```

Bug-94730. This bug exists in the code that deals with types determined from array initializers. The test input is generated by strategy 2. Raw seed program does not trigger any bug. We insert a `Decl` node to the AST of the seed program. The model then complete the sub-tree rooted at the `Decl`, and produce line 12. The introduction of the array declaration leads to the crash. The bug affected all GCC supported versions, including GCC-11. This example proves valuable for strategy 2.

```
1    int x , y ;
2
3    int foo ( )
4    {
5        x = 0 ;
6        y = 0 ;
7        if (x != 0)
8            y ++ ;
9        else
10           x ++ ;
11   }
12 + int x [] = { 0 } ;
```

Bug-94968. This test case is generated by strategy 3. We add a `If` node to the function `foo`. This guides the model to expand this non-terminal. After generation, a new conditional branch is produced. The addition of these two new lines eventually lead to an internal compiler error. This example demonstrates the usefulness of strategy 3.

```
1    extern void abort (void);
2
3    int foo() {
4        if (__SLASS (0x112233, 4) != 0x1122330)
5            abort ( ) ;
6
7 +      if (__builtin_speculation_safe_value ( 1 , x ))
8 +          abort ( ) ;
9    }
```

5   Related Work

Generation-Based Fuzzing. Fuzzing has become a standard method for software testing and bug finding. For complex input formats like programming languages, generation-based approaches prevent generated inputs from being rejected immediately by the target. CSmith [19] randomly generates new programs guided by a probabilistic grammar which covers a subset of the C programming language. It randomly selects an allowable rule from the grammar to generate C programs avoiding undefined and unspecified behaviors. However, several works have mentioned the difficulties of preparing grammar specifications of programming languages for fuzzers [16,19].

Learning Grammars for Generation-Based Fuzzing. Various efforts have been invested to learn grammars from existing examples. Patra et al. [12] propose to combine learned probabilistic language models of structured data with fuzz testing. Though TreeFuzz does not require prior knowledge of the input format, complex model extractors have been designed to infer the generative model from a given corpus of examples.

More recently, deep learning has been successfully applied to fuzzing input generation. Compared with traditional methods, deep learning requires less human engineering and is capable of capturing complex features. Learn&Fuzz [6] presents learning a generative language over the set of PDF object characters with a seq2seq network. The learned model is then used for PDF parser's fuzzing. Learn&Fuzz generate new PDF object as a sequence of chars. DeepSmith [4] learns a generative language model of OpenCL programming language with LSTM neural network. The learned model is used for OpenCL compiler testing. However, these works treat fuzzing inputs as a sequence of chars or tokens without leveraging any structural information.

Compiler fuzzing requires well-formed input programs. In this work, we focus on representing code as a tree to further learn the code model with structural information. AST has been used in various code generation tasks to leverage the available syntactic structure information and constrain the search space, ensuring generation of well-formed code. Rabinovich et al. [13] introduce abstract syntax networks utilizing a modular decoder whose submodels are composed to generate ASTs in a top-down manner. Their model outperforms previous seq2seq methods in code generation and semantic parsing tasks. To parse natural language descriptions into source code, Yin et al. [20] define a model that transduces an natural language description into an AST for the target programming language.

6   Conclusion

In this paper, we present TSmith, a prototype for effective compiler fuzzing. The core idea is to learn code models with the syntactic structure information. We implement a tree2tree neural network which maps a partial AST to a complete one. Based on the tree2tree model, TSmith supports an extensible set of generation strategies, enabling it to produce diverse test cases. Our evaluation shows that TSmith significantly improves the parsing pass rate of generated programs. After applying TSmith to the testing of the currently supported versions of GCC, we have found and reported 14 bugs.

Acknowledgments. This work is supported by National Key Research and Development Program of China (No. 2018YFB0204301), and the National Natural Science Foundation of China (No. 61472439).

References

1. Chakraborty, S., Allamanis, M., Ray, B.: Tree2tree neural translation model for learning source code changes. arXiv preprint arXiv:1810.00314 (2018)
2. Chen, P., Chen, H.: Angora: efficient fuzzing by principled search. In: 2018 IEEE Symposium on Security and Privacy (SP), pp. 711–725. IEEE (2018)
3. Chen, X., Liu, C., Song, D.: Tree-to-tree neural networks for program translation. In: Advances in Neural Information Processing Systems, pp. 2547–2557 (2018)
4. Cummins, C., Petoumenos, P., Murray, A., Leather, H.: Compiler fuzzing through deep learning. In: Proceedings of the 27th ACM SIGSOFT International Symposium on Software Testing and Analysis, pp. 95–105. ACM (2018)
5. Dong, L., Lapata, M.: Language to logical form with neural attention. arXiv preprint arXiv:1601.01280 (2016)
6. Godefroid, P., Peleg, H., Singh, R.: Learn&fuzz: machine learning for input fuzzing. In: Proceedings of the 32nd IEEE/ACM International Conference on Automated Software Engineering, pp. 50–59. IEEE Press (2017)
7. Google: Honggfuzz (2016). https://github.com/google/honggfuzz
8. Hochreiter, S., Schmidhuber, J.: Long short-term memory. Neural Comput. **9**(8), 1735–1780 (1997)
9. Karpathy, A., Johnson, J., Fei-Fei, L.: Visualizing and understanding recurrent networks. arXiv preprint arXiv:1506.02078 (2015)
10. Liu, X., Li, X., Prajapati, R., Wu, D.: Deepfuzz: automatic generation of syntax valid c programs for fuzz testing. In: Proceedings of the AAAI Conference on Artificial Intelligence (2019)
11. LLVM: libfuzzer: a library for coverage-guided fuzz testing (2017). https://llvm.org/docs/LibFuzzer.html
12. Patra, J., Pradel, M.: Learning to fuzz: Application-independent fuzz testing with probabilistic, generative models of input data. TU Darmstadt, Department of Computer Science, Technical report, TUD-CS-2016-14664 (2016)
13. Rabinovich, M., Stern, M., Klein, D.: Abstract syntax networks for code generation and semantic parsing. arXiv preprint arXiv:1704.07535 (2017)
14. Rawat, S., Jain, V., Kumar, A., Cojocar, L., Giuffrida, C., Bos, H.: Vuzzer: application-aware evolutionary fuzzing. In: NDSS, vol. 17, pp. 1–14 (2017)
15. Tai, K.S., Socher, R., Manning, C.D.: Improved semantic representations from tree-structured long short-term memory networks. arXiv preprint arXiv:1503.00075 (2015)
16. Wang, J., Chen, B., Wei, L., Liu, Y.: Skyfire: data-driven seed generation for fuzzing. In: 2017 IEEE Symposium on Security and Privacy (SP), pp. 579–594. IEEE (2017)
17. Wei, H., Li, M.: Supervised deep features for software functional clone detection by exploiting lexical and syntactical information in source code. In: IJCAI, pp. 3034–3040 (2017)
18. White, M., Tufano, M., Vendome, C., Poshyvanyk, D.: Deep learning code fragments for code clone detection. In: 2016 31st IEEE/ACM International Conference on Automated Software Engineering (ASE), pp. 87–98. IEEE (2016)
19. Yang, X., Chen, Y., Eide, E., Regehr, J.: Finding and understanding bugs in c compilers. In: Proceedings of the 32nd ACM SIGPLAN Conference on Programming Language Design and Implementation, pp. 283–294 (2011)
20. Yin, P., Neubig, G.: A syntactic neural model for general-purpose code generation. arXiv preprint arXiv:1704.01696 (2017)
21. Zalewski, M.: American fuzzy lop (2017). http://lcamtuf.coredump.cx/afl/

Research and Design of Distribution Equipment Health Early Warning System

Lei Chen[1], Huihua Yu[2], Li Tong[3], Peipei Jin[4(✉)], Weiyan Zheng[5], Xu Huai[4], and Yu Huang[4(✉)]

[1] State Grid Zhejiang Electric Power Co., LTE., Hangzhou, China
chenlei3909@163.com
[2] State Grid Zhejiang Hangzhou Fuyang District Power Supply Co., Ltd., Hangzhou, China
yuhuihua@126.com
[3] State Grid Zhejiang Electric Power Research Institute, Hangzhou, China
tonyhust@126.com
[4] Peking University, Beijing, China
{ppj,xubad,hy}@pku.edu.cn
[5] State Grid Hangzhou Power Supply Company, Hangzhou, China
zhweiyan@foxmail.com

Abstract. Load forecasting and health early warning of distribution equipment is a very active research field and an important aspect, which can guarantee the system stability under large disturbances and optimize the distribution of energy resources in the smart grid. Therefore, it is of great significance to design and develop a distribution equipment health early warning system based on the demand of staff for monitoring and early warning of distribution equipment. In this paper, the analysis and design of distribution transformer health early warning system are carried out, and the goal of early warning system is defined. The whole frame and deployment architecture of system are presented. Moreover, the design flow of the system core function modules and the design pattern and the framework for system development are given. The system monitors and forewarns the operation status of the distribution network equipment, and sends the abnormal situation of the early warning result to the staff, which can save the manpower and material resources wasted due to manual troubleshooting.

Keywords: Smart grid · Distribution equipment · Health early warning · Load forecasting · System framework

1 Introduction

In recent years, the frequent occurrence of extreme natural weather such as rainstorm and hurricane has posed a great threat to the safe operation of power grid [1, 2]. As

This work is supported by the project of State Grid (No. 5400-201919144A-0-0-00): Research and Application on Key Technologies of Virtual Agent for Power Supply Service Command Based on Artificial Intelligence.

M. Qiu (Ed.): ICA3PP 2020, LNCS 12452, pp. 579–590, 2020.
https://doi.org/10.1007/978-3-030-60245-1_39

an important public infrastructure, electric power equipment is an important foundation to ensure people's livelihood and promote economic and social development [3]. The health status of distribution equipment directly determines the stable operation of power. The real-time monitoring and early warning of distribution equipment health status will have a very positive significance for the health detection and maintenance of distribution network, and also provide a strong guidance and guarantee for the good operation of the whole power system [4–6]. Distribution transformer is the important equipment to transfer AC energy by exchanging voltage and current according to electromagnetic induction law in distribution network, which can meet the different requirements of users by changing the AC voltage. In addition, the distribution transformer is a power transformer that supplies power directly to users, and its health status directly determines the stable operation of power. Thus, it is very important to analyze and warn the health status of the distribution transformer in time [7–9]. Scientific, timely and accurate analysis of the potential risks in the distribution transformer and the corresponding risk response plan will provide strong support for the safe and stable operation of the power grid system, which is essential to improve the reliability and stability of the whole power grid operation [10, 11].

In order to realize the comprehensive early warning and decision support of power grid, researchers in different countries are actively carrying out relevant research work. The research on condition monitoring and fault diagnosis technology of transformer equipment started earlier abroad [12]. J.Fuhr and other scholars think that transformer condition monitoring, fault early warning and health management technology involve the integration of computer technology, detection technology, power technology and online diagnosis, the core of which is online monitoring, analysis diagnosis and early warning technology [13, 14]. The American Electric Power Research Institute has invested in the research and development of on-line monitoring system of oil immersed power transformer for a long time. The transformer on-line monitoring technology proposed by GE company is recognized as the earliest power transformer fault monitoring method by IEEE standard [15]. With the rapid development of artificial intelligence and other new technologies, power grid health early warning technology has been widely used. In 1980, Smeets R.P.P. scholars proposed the fault diagnosis technology of power transformer based on artificial neural network [16]. In 1986, the expert system was first used in the TOGA system published by Riese. Then, S.I.gallant took the lead in combining expert system with artificial neural network to form an intelligent system [17]. In addition, Japan, Europe and other developed countries and regions have also invested a lot of human and material resources, focusing on the transformer online monitoring system and its application to the actual power system.

In recent years, a series of researches have been carried out on the monitoring and early warning system of distribution network in China, and the relevant systems have been gradually applied. Data protection and forecasting become an important issue in today's Cloud environments [18, 19]. Reference [20] proposes a fault early-warning method of smart meter based on data mining technology and builds a fault early-warning system through VS 2016 platform. The improvement plan of intelligent and hierarchical alarm is proposed and the necessity of setting up intelligent alarm system of power grid dispatch in power system is expounded in [21]. Online early warning methods based on historical data probability distribution is designed in [22], which effectively eliminates the influence of the external environment on the transformer. A power transformer status evaluation and warning model is established based on the fuzzy comprehensive evaluation and Bayes discrimination in [23]. A design scheme of system of on-line monitoring, fault diagnosis and early warning for insulation of mine-used dry-type transformer is also proposed, which introduced general structure of the system and fault diagnosis method, and built a fault type library of turn-to-turn insulation of dry-type transformer. What's more, a new heterogeneous scratchpad memory architecture and secure digital evidence framework using blockchain are proposed in [24, 25] and the secure call instructions is designed in [26], which is used to save the operation data of distribution network equipment and send warning signal.

Since distribution transformer is the important equipment in smart distribution network, the early warning of its operating status is of great significance.

The structure of this paper is organized as follows: we first introduce the whole design scheme of the distribution equipment health early warning system and give the whole frame diagram and deployment architecture diagram (Sect. 2). Then we design the flow of the system core function module including data preprocessing module, model prediction module, early warning module and model update module (Sect. 3). Finally, the system development technology is given in Sect. 4.

2 Design of Distribution Equipment Health Early Waring System

2.1 System Requirements Analysis

The distribution transformer health early warning system designed in the paper can monitor and accurately judge the transformer in real time, based on the operation status of distribution transformer. The early warning system receives the load power and additional information sent from the remote end of each distribution transformer in real time. Through the processing, prediction and analysis of the load power and additional information, it can predict the future health status of each distribution transformer, so as to provide reference for the further adjustment of staff. The early warning system helps dispatchers to judge the operation status of distribution transformer, coordinate all aspects of personnel to make scheduling and planning, and also provides information for maintenance personnel to make timely preventive measures.

Firstly, the system stores and preprocesses the real-time data sent by each device terminal to form the input sequence required by the model. Then, the corresponding future load prediction value, hot spot temperature prediction value and life loss prediction value are obtained through model prediction. Finally, through the analysis of distribution transformer health status to make a judgment, to make an early warning of abnormal situation in time. Finally, through the analysis of distribution transformer health status, we can make a timely warning of abnormal conditions.

2.2 System Architecture Description

The system supports the receiving of real-time data, the analysis, monitoring, early warning and the visual display of distribution transformer health status. The system architecture diagram is given (see Fig. 1).

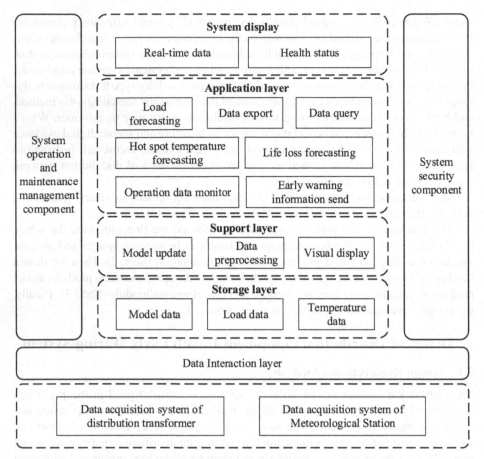

Fig. 1. The health early warning system architecture diagram.

The real-time data obtained by the system includes the real-time power value of distribution transformer and the real-time temperature value of meteorological station. The data interface integrates the real-time information and delivers it to the system storage layer. The storage layer of the system completes the standardized storage of real-time load power data, temperature data and model parameters, which is used for data prediction. The support layer provides model updating, data preprocessing, and visualization of the prediction results to support the upper functions. The application layer mainly includes load forecasting, hot spot temperature forecasting, life loss forecasting and other major health status analysis modules. At the same time, it also realizes the functions of operation data monitoring, data query and early warning information export. The display layer mainly displays the real-time running data and health status analysis results. What's more, the whole system is guaranteed by system operation and maintenance management components and system security components.

The advantages of the framework in the paper are the characteristics of loose coupling of each layer, which is convenient for the planning and adjustment of each layer and the separation of system display layer and application layer, which provides great convenience for system maintenance.

2.3 System Deployment Architecture

In order to ensure data security, data storage, data preprocessing, model prediction, health status analysis, model update, data query and export, running data monitoring are deployed on the intranet server. First, the real-time data is stored in intranet database after passing through firewall through Internet. Then, the health status that is analyzed by application server is fed back to the workbench of the dispatching center in real time. If there is an early warning signal, it will be sent to the front-line staffs' smart phones in time through the SMS service center. The system deployment architecture diagram is as follows (see Fig. 2).

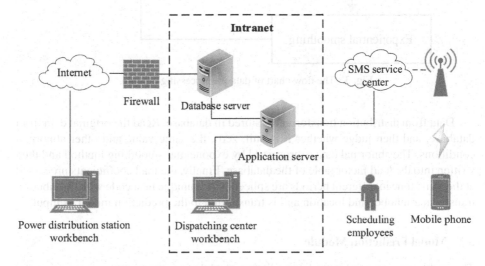

Fig. 2. The system deployment architecture diagram.

3 Design of the System Core Function Module

In this paper, the modular system structure design idea is used to analyze the system from top to bottom, which is conducive to the overall grasp of the system design process, greatly reduces the development difficulty and has the advantages of easy maintenance and high reliability. The design process of the closely related modules of the distribution equipment health warning system is as follows.

3.1 Data Preprocessing Module

The function of the data preprocessing module is to standardize the data received in real time and organize it into the input form required by the prediction model. The flow chart of data preprocessing module is shown in Fig. 3.

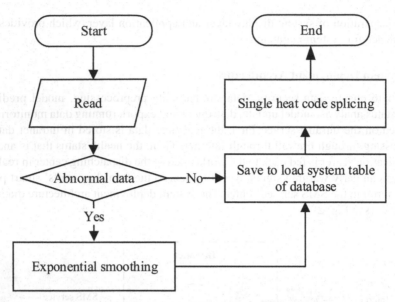

Fig. 3. The flow chart of data preprocessing module.

Data from distribution transformer is stored in database. Read the original data from database, and then judge whether it is null, zero, the same value and other abnormal conditions. The abnormal data is processed by exponential smoothing method and then written into the load factor table of the database. Finally, the load coefficients processed at the same time in different periods are spliced with a unique heat code with distribution transformer number and location and is transmitted to the prediction model as input.

3.2 Model Prediction Module

The main function of the model prediction module is to receive the input from the preprocessing module and then complete the load power, winding hot spot temperature and life loss prediction of each distribution transformer at the next moment. The flow chart of model prediction module is shown in Fig. 4.

Firstly, receive the data processed by the data prediction module and carry out load prediction. And then, set the variable value in the prediction formula of hot spot temperature and life loss according to the time information of the data on the day. Finally, the remaining life of distribution transformer is obtained by using the prediction value of hot spot temperature and life loss.

3.3 Early Warning Module

The main function of the early warning module is to analyze the prediction results and send the early warning information to relevant personnel in time. At the same time, according to the specific threshold value, the estimated load value, hot spot temperature value and remaining life value are divided into different security levels, and the specific

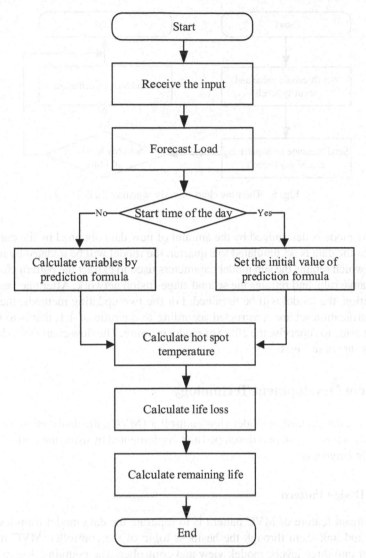

Fig. 4. The flow chart of model prediction module.

value is fed back to the dispatching center workbench. If there is a possible alarm, the early warning information is sent to the front-line maintenance personnel in time. The flow chart of early warning module is shown in Fig. 5.

3.4 Model Update Module

The main function of the model update module is to automatically update the model, so as to further improve the accuracy of prediction. The selection of system update time

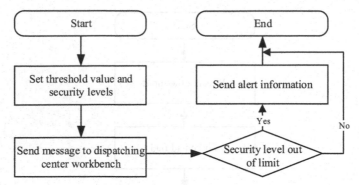

Fig. 5. The flow chart of early warning module.

and update mode is determined by the amount of new data obtained by the current system. When the data is accumulated to a quarter, the model will be updated by migration learning, which retains the sub model parameters used to extract the pattern characteristics of change rule, and retrains the second stage fusion network. After one year of data accumulation, the model will be retrained. For the two updating methods, the training set and verification set are constructed according to the ratio of 4:1, that is to say, 20% data is set aside to supervise the effect of model training. The flow chart of model update module is shown in Fig. 6.

4   System Development Technology

Based on the design idea of model view controller (MVC), the distribution transformer health early warning system is developed and implemented by using the flask framework and python language.

4.1   VC Design Pattern

The significant feature of MVC pattern is to separate the data model from the display interface and link them through the business logic of the controller. MVC mode can be divided into three levels: model, view and controller. The coupling degree between layers is low, which is very conducive to the parallel development of applications and greatly improves the development efficiency.

The model layer mainly implements the data logic of the system, including data access and processing operations, which encapsulates the data processing functions necessary for the realization of business logic to maximize code reuse. The view layer mainly realizes the display function of the system, and provides the interface for the interaction between the system and the user, which not only transmit the user request information to the controller, but also receive the state response data sent by the model and the view selection information sent by the controller. Therefore, it realizes the information display to the user. The controller mainly realizes the business logic of the system and completes the data transfer between the model and the view. After receiving

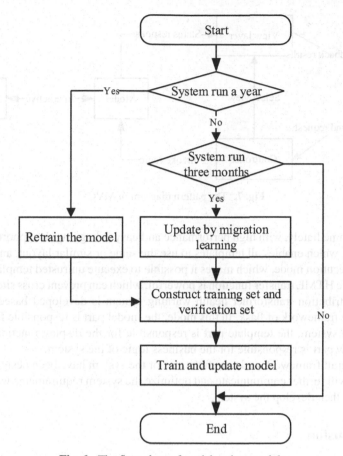

Fig. 6. The flow chart of model update module.

the user's request information, the controller processes the model to realize the business logic of the system, and sends the view selection information to the view. The pattern diagram of MVC is shown in Fig. 7.

4.2 Flask Framework

Flask is a portable web development micro framework, the development language of which is python. The core components of flask are Werkzeug and Jinja2, which are respectively responsible for web routing and template rendering. The Werkzeug routing system can match the URL and the endpoint, generate the URL for the endpoint and capture the variables in the URL, and its response object can wrap other WSGI applications and process the flow data. HTTP utilities can handle entity tags, cache control, user agents and cookies, and test clients can impersonate HTTP requests without running the server.

Jinja2 is one of the most commonly used Python template engines with full Unicode support, which has a wide range of functions. It can convert template code into Python

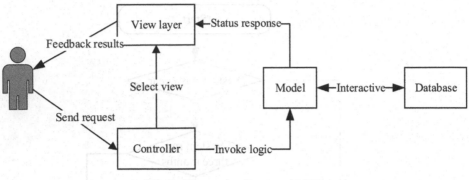

Fig. 7. The pattern diagram of MVC.

bytecode immediately, with high performance and easy debugging. It supports template inheritance, which enables all templates to use the same or similar layout, and supports sandbox execution mode, which makes it possible to execute untrusted templates safely. Its automatic HTML transfer function is powerful, which can prevent cross site scripting.

The distribution transformer health warning system is developed based on MVC mode in the framework of flask. In this mode, the model part is responsible for the data logic of the system, the template part is responsible for the display function of users, and the view part is responsible for the business logic of the system.

The overall framework and core functions of the system have been designed. In the future, we will further communicate and optimize the system requirements with domain experts and then develop the system.

5   Conclusion

This paper focuses on one of the difficult problems in distribution network, which is how to effectively monitor the health status of distribution transformer in real time and give early warning in time. On the basis of monitoring and early warning demand analysis of distribution network equipment, the whole design scheme of the distribution equipment health early warning system is proposed in this paper. First of all, the goal of the early warning system is defined. Then, the whole frame structure of the system is determined based on the core business process and functional requirements. What's more, the core function module design including the data preprocessing module, model prediction module, early warning module and model update module is completed. The health warning system of distribution transformer obtains the load condition, hot spot temperature and remaining life of distribution transformer in time, which provides a reference for the maintenance of staff and saves the manpower and material resources wasted due to manual troubleshooting.

References

1. Peng, K., Chen, X.Y., Li, B., Liao, Y.C., Yu, K.: Analysis of the impact factors of meteorological environment on power load. Power Demand Side Manage. **18**(1), 8–13 (2016)

2. Rahman, A., Liu, X., Kong, F.: A survey on geographic load balancing based data center power management in the smart grid environment. IEEE Commun. Surv. Tutorial **16**(1), 214–233 (2014)
3. Liu, P.C., Li, X.L.: Fault- section location of distribution network containing distributed generation based on the multiple- population genetic algorithm. Power Syst. Prot. Control **44**(2), 36–41 (2016)
4. Liu, S., Feng, J.Q., Chen, Y.H.: Two dimensional simulation analysis of winding temperature field of oil-immersed power transformer. Transformer **46**(9), 35–38 (2001)
5. Su, W., Eichi, H., Zeng, W., et al.: A survey on the electrification of transportation in a smart grid environment. IEEE Trans. Ind. Inform. **8**(1), 1–10 (2012)
6. Liu, G., Jin, Y.J., Ma, Y.Q., Sun, L.P., Chi, C.: Numerical analysis of fluid field and temperature field of oil-immersed transformer. Transformer **54**(5), 22–26 (2017)
7. Huang, J.H., Quan, L.S.: Current status and development of condition-based maintenance of high-voltage electric power equipment in substation. Autom. Electr. Power Syst. **16**, 56–61 (2001)
8. Wang, H.B., Du, S.Y., Dai, J.Z., Ren, M., Dong, M.: Condition pre-warning method of power transformer based on load time series model. High Voltage Apparatus **53**(8), 204–210 (2017)
9. Li, C.Y., Liu, L., Zhang, L.Y.: Research on fault diagnosis of transformer in 10 kV substation based on infrared thermal image. Northeast. Electr. Power Technol. **39**(2), 43–46 (2018)
10. Qingbo, Z.: State grid contributes to clean energy development. Electr. **21**(3), 48–50 (2010)
11. Yimin, W., Jianhui, W., Xuzhu, D., Pengwei, D., et al.: Guest editorial smart grid technologies and development in China. IEEE Trans. Smart Grid. **7**(1), 379–380 (2016)
12. Wang, L., Wang, P.: On-line monitoring and condition-based maintenance technology of electric power equipment. Mod. Electr. Power **50**, 40–45 (2002)
13. Chiang, J.H., Yuan, J.: Optimal maintenance policy for a Markovian system under periodic inspection. Reliab. Eng. Syst. Saf. **71**(2), 165–172 (2001)
14. Fuhr, J., Aschwanden, T.: Experience with diagnostic tools for condition assessment of large power transformers. In: IEEE International Symposium on Electrical Insulation, pp. 508–511 (2004)
15. Dual, M.: New techniques for dissolved gas-oil analysis. IEEE Electr. Insul. Mag. **19**(2), 6–15 (2003)
16. Smeets, R.P.P., Kertesz, V.: Evaluation of high-voltage circuit breaker performance with validated are model. IEEE Proc. Gener. Transm. Distrib. **147**(2), 121–125 (2000)
17. Wang, M., Vandermaar, A.J., Srivas, K.D.: Review of condition assessment of power transformers in service. IEEE Electr. Insul. Mag. **18**(6), 122–125 (2002)
18. Qiu, H., Noura, H., Qiu, M., Ming, Z., Memmi, G.: A user-centric data protection method for cloud storage based on invertible DWT. In: IEEE Transactions on Cloud Computing (2019)
19. Qiu, H., Qiu, M., Lu, Z., Memmi, G.: An efficient key distribution system for data fusion in V2X heterogeneous networks. Inf. Fusion **50**, 212–220 (2019)
20. Zhang, Y., Fan, Y.F., Liu, Q.J.: Design and development of fault early-warning system for smart meter. Electrical Measurement & Instrumentation, pp. 1–7 (2020)
21. Liang, T.: Discussion on intelligent alarm of power grid dispatching automation system. Guangdong Electr. Power **18**(5), 41–44 (2005)
22. Zang, L.C., Tian, P., Li, B.: Research on Online Warning Based on Probability Distribution of Historical Data. Yunnan Electric power **43**(6), 48–50 (2015)
23. Shi, S.W., Wang, K., Chen, L., Yang, F., Liu, P., Tao, D., Lu, G.F.: Power transformer status evaluation and warning based on fuzzy comprehensive evaluation and Bayes discrimination. Electr. Power Autom. Equipment **36**(9), 60–66 (2016)
24. Qiu, M., Chen, Z., Niu, J., et al.: Data allocation for hybrid memory with genetic algorithm. IEEE Trans. Emerg. Top. Comput. **3**(4), 544–555 (2015)

25. Tian, Z., Li, M., Qiu, M., et al.: Block-DEF: a secure digital evidence framework using blockchain. Inf. Sci. **491**, 151–165 (2019)
26. Shao, Z., Xue, C., Zhuge, Q., Qiu, M., Xiao, B., Sha, E.M.: Security protection and checking for embedded system integration against buffer overflow attacks via hardware/software. IEEE Trans. Comput. **55**(4), 443–453 (2006)

Parallel Processing Algorithms for the Vehicle Routing Problem and Its Variants: A Literature Review with a Look into the Future

Bochra Rabbouch[1,2], Hana Rabbouch[2], and Foued Saâdaoui[2,3]($\boxtimes$)

[1] LEONI Wiring Systems, Zone Industrielle Manzel Hayet, 5033 Monastir, Tunisia
bochra.rabbouch@gmail.com
[2] Lab: LR18ES15 Algèbre, Théorie de Nombres et Analyse Non-linéaire,
Faculté des Sciences, University of Monastir, 5019 Monastir, Tunisia
hana.rabbouch@gmail.com
[3] Department of Statistics, Faculty of Sciences, King Abdulaziz University,
P.O BOX 80203, Jeddah 21589, Saudi Arabia
foued.saadaoui@isgs.rnu.tn

Abstract. Vehicle Routing Problems (VRPs) are well-know combinatorial optimization problems used to design an optimal route for a fleet of vehicles to service a set of customers under a number of constraints. Due to their NP-hard complexity, a number of purely computational techniques have been proposed in recent years in order to solve them. Among these techniques, nature-inspired algorithms have proven their effectiveness in terms of accuracy and convergence speed. Some of these methods are also designed in such a way to decompose the basic problem into a number of sub-problems which are subsequently solved in parallel computing environments. It is therefore the purpose of this paper to review the fresh corpus of the literature dealing with the main approaches proposed over the past few years to solve combinatorial optimization problems in general and, in particular, the VRP and its different variants. Bibliometric and review studies are conducted with a special attention paid to metaheuristic strategies involving procedures with parallel architectures. The obtained results show an expansion of the use of parallel algorithms for solving various VRPs. Nevertheless, the regression in the number of citations in this framework proves that the interest of the research community has declined somewhat in recent years. This decline may be explained by the lack of rigorous mathematical results and practical interfaces under famous calculation softwares.

Keywords: Combinatorial optimization · Vehicle Routing Problem · Metaheuristics · Nature-inspired algorithms · Parallel computing

1 Introduction

Determining optimal solutions subject to a number of imposed constraints is the main objective of a mathematical optimization problem. This objective is often

© Springer Nature Switzerland AG 2020
M. Qiu (Ed.): ICA3PP 2020, LNCS 12452, pp. 591–605, 2020.
https://doi.org/10.1007/978-3-030-60245-1_40

exactly intractable for many complicated optimization problems. In practice, especially for large-scaled problems, operation researchers are commonly satisfied with approximate solutions found by nature-inspired algorithms. Among these algorithms, meta-heuristics [31–34] continue to draw attention as a promising techniques for approximating the global optimum of a given function in a fairly reasonable computation time. They especially include evolutionary algorithms, greedy search procedures, and the simulated annealing. However, in order to make metaheuristic methods more efficient, many extensions involving parallel computing strategies have appeared in the last few years. The main principle of parallel processing is to accelerate computation by decomposing the main task among numerous processors. From an algorithmic side, pure parallel computing methodologies employ the partial order of algorithms (i.e., the groups of operations that may be run concurrently in time while keeping the same solving method and the final obtained solution) and thus correspond to the natural parallelism present in the algorithm. The partial order of algorithms provides two main sources of parallelism: data and functional parallelism [9].

One of the most studied combinatorial optimization problems is the Vehicle Routing Problem (VRP). A VRP addresses the problem of delivering goods ordered optimally by visiting customers at their locations and using a fleet of vehicles with respecting required constraints. Different parameters are needed to successfully ensure this operation. Parameters that characterize VRPs are first the road network, which describes the connectivity between customers and depot. Then, the vehicles that transport the delivery among customers, depot in the road network, and the customers that request services, place orders and receive goods. Auxiliary constraints may possibly be added to those four main parameters added to the obtained solution and the objective function to be optimized [31]. Nowadays, in various sectors of the real world, these models are used to optimize problems whose structure can be similar to that of a graph, such as electrical networks, roads, sewers, etc. However, the nature of the problem as well as its size can be decisive factors in the complexity of the problem to be solved. Therefore, the proper design and treatment of the problem with appropriate resolution tools is a very important fact.

In the literature, multitude of methods and strategies have been proposed for solving the VRP and its different variants, with varying degrees of success. Except exact methods, which have proven their limits for tackling this kind of problems, several nature-inspired methods have shown their effectiveness in this area. This survey paper aims to elucidate the complexity of the available approaches, with a special focus on the pros and cons of the individual methods. A short technical background on the VRP with the review of its main drawbacks and reasons to extend it into a parallel version are put forth in Sect. 2. In an attempt to define the popularity of the VRP and the various parallel computing approaches used for solving it, in Sect. 3, we provide an outline of the existing literature including a bibliometric study based on the Web of Science, Scopus and Google Scholar databases, and a brief summary of the review/survey publications on this topic. Then, in Sect. 4, we look back over the last three years of VRP, combinatorial optimization and parallel meta-heuristics, in an attempt to better understand the rapidly growing literature. In particular, we propose a comparisons between the different studies, stress the importance of parallel processing for VRPs, and highlight some recent trends. Finally, in the last section, we look ahead and indicate directions VRP, and combinatorial optimization in general, will or should take in the next few years.

2   General Background

The VRP is a widespread studied combinatorial optimization problem and has become, since its appearance by Dantzig and Ramser [13] in 1959, one of the most investigated research topics. VRPs deal with how to deliver goods ordered optimally by visiting customers at their locations and using a fleet of vehicles with respecting required constraints. The traveling salesman problem (TSP) formulation of Danzig et al. [12] has been generalized to form the two index flow for the VRP:

$$\min \sum_{i \in A} \sum_{j \in A} \alpha_{ij} x_{ij} \tag{1}$$

subject to:

$$\sum_{i \in A} x_{ij} = 1 \quad \forall j \in A \backslash \{0\} \tag{2}$$

$$\sum_{j \in A} x_{ij} = 1 \quad \forall i \in A \backslash \{0\} \tag{3}$$

$$\sum_{i \in A} x_{i0} = K \tag{4}$$

$$\sum_{j \in A} x_{0j} = K \tag{5}$$

$$\sum_{i \notin S} \sum_{j \in S} x_{ij} \geq \gamma(S) \quad \forall S \subseteq A \backslash \{0\}, S \neq \emptyset \tag{6}$$

$$x_{ij} \in \{0, 1\} \quad \forall i, j \in A \tag{7}$$

In this system, α_{ij} represents the cost of moving from i to j, x_{ij} is a binary variable which returns 1 if the arc going from i to j is part of the solution, and 0 otherwise, K is the size of the fleet and $\gamma(S)$ corresponds to the minimum number of vehicles required to serve a customer set S. Constraints (2) and (3) mean that exactly an arc enters and exactly one leaves each vertex associated with a customer, respectively. Constraints (4) and (5) indicate that the number of vehicles leaving the depot and the number of vehicles entering are equal. Constraint (6) is the capacity cut-off constraint, which requires that the routes must be connected and that the demand on each route must not exceed the capacity of the vehicle. Finally, Constraint (7) defines the domain of constants. It is therefore on this formulation that all VRP variants arise.

Nevertheless, finding optimal solutions for the VRP is a hard task. Even mathematical algorithms and optimization models for solving large-scale instances under VRP are still limited. Therefore, several metaheuristic optimization algorithms have been applied with varying degrees of success. Although metaheuristics provide fairly effective strategies for finding approximate solutions to combinatorial optimization problems, the computation time associated with exploring the solution space can be very significant. With the proliferation of parallel computers, powerful workstations and fast communication networks, parallel implementations of metaheuristics quite naturally appear as an alternative to accelerate the search for approximating solutions [6,11,17,25,39]. They also allow to find more accurate solutions in comparison with their sequential counterparts, due to the partitioning of the research space and more possibilities for the intensification and diversification of the research. Therefore, parallelism is a way not only to reduce the operating time of metaheuristics, but also to improve their efficiency and robustness. These are likely to be the most important contribution of parallelism to metaheuristics [19]. The last few years have seen the emergence of a large number of parallel IT architectures. Almost all desktop or laptop computers, and even mobile phones, have multiple integrated processor cores.

Parallel and distributed computing can be used in the design and the implementation of metaheuristics for the main following reasons: **1.** Accelerating the research procedure: One of the main objectives of parallelizing metaheuristics is to reduce the research time. This allows to design interactive and real-time optimization methods. **2.** Improving robustness: Parallel metaheuristics can be more robust when solving different optimization problems and different instances of a given problem. Robustness can also be measured in terms of the sensitivity of metaheuristics to their parameters. **3.** Improving the quality of obtained solutions: The exchange of information between cooperative metaheuristics will modify their behavior in terms of research in the space associated with the problem. Better convergence and reduced search time may occur. Noteworthy that a parallel model for metaheuristics can be more efficient than sequential models even on a single processor. **4.** Solve large-scale problems: Parallel metaheuristics are used to solve large-scale instances for complex optimization problems. The challenge is then to solve very large instances that cannot be resolved by a sequential machine. Another similar challenge is to solve more precise mathematical models associated with different optimization problems. Improving the accuracy of mathematical models generally increases the size of the problems associated with solving. In addition, some optimization issues require the handling of huge databases such as data mining problems.

3   Bibliometric Analysis of VRP and Parallel Metaheuristics

In any research area, the best-known method to learn about a new field is to perform a literature query using one or some acknowledged databases, such as

Web of Science, Scopus, etc., and find the 'hottest' topics, highly cited papers, and new publishing trends. To help a new reader to the field of VRP and parallel metaheuristics, we have firstly performed a short bibliometric analysis. The statistics reported below are essentially based on the three following databases: Google Scholar, Web of Science and Scopus. The objective of this section is mainly, to better know the field of VRP, in particular, where is this field positioned between the different disciplines, in which regions of the world has it been the most developed in recent years, and which are the main publishers where VRPs have grown the most. In this same context, we also focus on parallel-computing-based meta-heuristic approaches that have been developed in recent years for solving these problems.

In this part, we firstly focus on the VRP relative bibliometrics over the past three years. Preliminary research on the Web of Science database reveals that the total number of documents dealing with the problem and published over the same period is 1,469. Table 1 reports the distribution of the number (and percentage) of journal articles according to their discipline. A total of 10 research areas is chosen, which represents the research areas where VRP is most present. We can, for example, notice the absence of important sciences such as economics and mathematics among these fields, which is somewhat unexpected. On the other hand, it is clear and obvious that fields like operations research (OR) and computer science are the two disciplines where publications on VRPs are the most abundant. In Table 2, the distribution of publications indexed by Web of Science is rather made according to the country of origin of the research team. Taking into account the demographic factor, we note the dominance of the countries of Latin Europe and Canada. It is also surprising to notice that a considerable part of the total number of publications is held by a developing country like Iran.

In Fig. 1, we present a bibliometric analysis of the main scientific resources dealing with the VRP topic on both theoretical and applied levels. We can clearly notice that the Dutch-British giant dominates the fields of publication on VRPs with more than 1,700 documents published between 2018 and 2020, followed by Springer Publishers. Noteworthy also that journal articles represent the majority of documents dealing with the subject, followed by conference papers. At the bottom of Fig. 1, we outline the list of the 15 Scopus-indexed journals most contributing to the VRP area. As mentioned above, the majority of these journals belong to Elsevier, with the prestigious European Journal of Operations Research (EJOR) at the top of the list. Below, in Fig. 2 we focus instead on the contribution of parallel computing methods to the field of combinatorial optimization and VRP. The bibliometric analysis here is rather based on the Google Scholar database, where the number of documents dealing with each research topic between 2010 and 2019 is reported as a time series. This allows to speculate evolution of each theme in the future. The first remark to make is about the growing and monotonous trend during the last ten years of work on meta-heuristic methods in combinatorial optimization as in VRP. This is however not the case for parallel metaheuristics, which have grown for a few years but have known a certain stationarity for the past few years. This may be due to the complexity of generalization and implementation of such mechanisms. On the other hand, if we read the results of Fig. 3, describing the number of Scholar Google citations since 2009, we will in fact understand that the interest of the research community to the VRP has decreased in recent years. This decrease was a little less serious for parallel calculation methods. The interpretation of such statistics is that, rather non-researcher practitioners who continue to be most interested to these two techniques. The cause of this recession may also be due to the lack of pure mathematical results concerning the theory of VRPs. In the near future, it would therefore be interesting to better orient research activities towards new theoretical axes. Besides, to better familiarize users with parallel approaches, it is recommended to better explain the different interfaces for solving this type of problems and make them easier to use, especially in the real world.

Table 1. Distribution of the number of journal articles dealing with VRP published on Web of Science between 2018 and 2020 according to their categories. The total number of documents published over the same period is 1,469.

Web of science category	Record count	Percentage of 1,496
Operations Research Management Science	535	35.762%
Computer Science Artificial Intelligence	236	15.775%
Engineering Industrial	217	14.505%
Computer Science Interdisciplinary Applications	210	14.037%
Transportation Science Technology	193	12.901 %
Engineering Electrical Electronic	174	11.631%
Management	164	10.963%
Computer Science Theory Methods	150	10.027%
Transportation	126	8.422%
Computer Science Information Systems	110	7.353%

4   Major Review and Survey Publications

There are many advantages of employing parallel processing for nature-inspired optimization algorithms. Implementing them into numerical or metaheuristic algorithms can accelerate the process of optimal solution finding. Parallel processing can also contribute to improving the quality of the obtained solutions, robustness, and dissolution of large-scale problems [39]. The objective of this section is to review the literature of recent research works having exploited parallel computing techniques for the improvement of the different schemes used for

Table 2. Distribution of the number of journal articles dealing with VRP published on Web of Science between 2018 and 2020 according to the country/region. The total number of documents published over the same period is 1,469.

Country	Record count	Percentage of 1,496
China	395	26.404%
USA	180	12.032%
France	107	7.152%
Italy	96	6.417%
Iran	86	5.749%
Canada	83	5.548%
Germany	83	5.548%
Turkey	68	4.545%
Spain	67	4.479%
India	56	3.743%

solving the VRP or its variants. The survey also includes references dealing with some parallel processing techniques which have been successfully applied to other combinatorial optimization problems, but can also be implemented for different VRP models. As it can be noted below, the criticism that can be made to the literature is that the majority of the evaluations of new proposed approaches were based on benchmarking instances and simulations. The lack of publicly accessible interface devoted to practitioners and real-world applications is also a major drawback. This may explain the decline of the influence of this area in recent years.

Let us start with some prominent early work on the implementation of parallel computing in the field of optimization in general, and VRP in particular. Crainic et al. [8] have shown that cooperative parallelization allows to obtain solutions of quality equivalent to sequential algorithms for similar computational efforts, but in reduced computational time. It also allows greater robustness in comparison with situations when different processes use different solving methods, different parameters, and different starting solutions. Several published reviews also exist on parallel optimization for particular technologies, problems, applications, methodologies and research disciplines [37]. Reviews of parallelization for certain optimization problems were especially performed for vehicle routing problems (VRPs) [10], non-linear optimization [24], mixed integer programming [28] and multi-objective optimization [27]. Many reviews have, for instance, focused on parallel optimization schemes applied to several well-known methodologies such us the simulated annealing (SA) [2], variable neighborhood search (VNS) [30], swarm intelligence algorithms [40], particle swarm optimization algorithms [41], and different types of evolutionary algorithms, including genetic algorithms (GAs) and ant colony optimization algorithms [27,29].

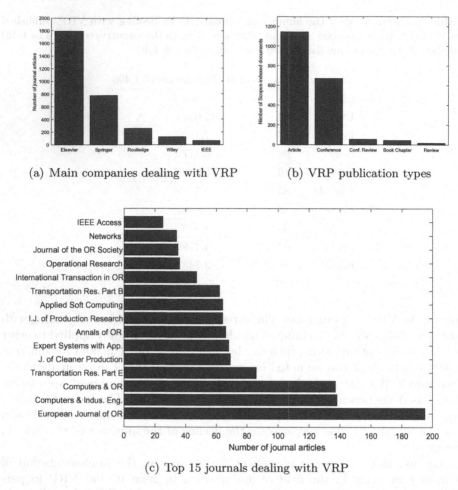

(a) Main companies dealing with VRP

(b) VRP publication types

(c) Top 15 journals dealing with VRP

Fig. 1. Bibliometric analysis of VRP journal articles between 2018 and 2020 using two well-established and generally acknowledged databases: Web of Science and Scopus.

In recent years, VRP models have also evolved to be more flexible, which led to the growth of their computational complexity. For example, optimizing route networks for transportation companies not only amounts to performing complex computations, but also doing that in the shortest possible amounts of time. It has therefore become essential to find the appropriate means to simplify the calculation tasks. It was then proposed to parallelize this type of problem given that Graphics Processing Units (GPUs) afford huge computation when needed operations are correctly parallelized. Cordeau et al. [7] offered a parallel iterated tabu search heuristic for solving four different routing problems: classical VRP, multi-depot VRP, periodic VRP, and site-dependent VRP. Their approach have been also proved valuable for time-window constrained variant of these problems. Jagiello et al. [21] underlined the value of using parallel Tabu

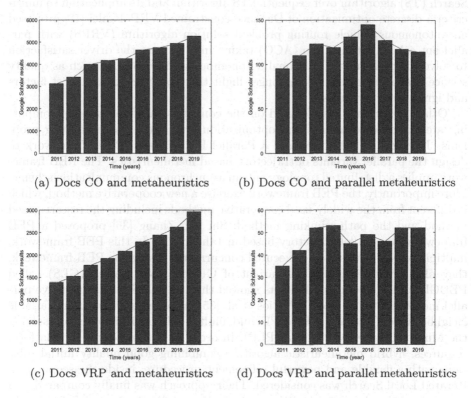

(a) Docs CO and metaheuristics   (b) Docs CO and parallel metaheuristics

(c) Docs VRP and metaheuristics   (d) Docs VRP and parallel metaheuristics

Fig. 2. Google Scholar bibliometric analysis (# documents) of parallel processing methods applied to Combinatorial Optimization (CO) and vehicle routing problems between 2010 and 2019.

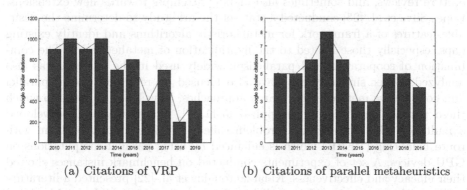

(a) Citations of VRP   (b) Citations of parallel metaheuristics

Fig. 3. Google Scholar bibliometric analysis (# citations) of VRP and parallel processing methods between 2009 and 2019.

Search (TS) algorithm over sequential TS algorithm and its application to multi-criteria discrete optimization of Distance-constrained VRP. Fanlei [18] proposed an autonomous vehicle routing problem solution algorithm (VRPS) with parallel ant colony optimization (ACO) taking into account the driver satisfaction to solve the problem of traditional autonomous vehicle routing, such as quality service is affected by the pilot's longer flight time to overemphasize cost factors and ignore delivery times.

Other works have also focused on the contribution of the parallel computing area to other combinatorial optimization and integer programming problems. Jialong et al. [20] proposed a Parallel Elite Biased (PEB) framework to design the parallel variants of trajectory-based metaheuristics. The PEB framework applies a distributed topology and an asynchronous communication scheme. More importantly, the PEB framework exercise a new cooperative method, which is different from the widely-used cooperative methods including the restart-based method and the path-relinking method. Shi and Zhang [36] proposed a PEB framework for parallel trajectory-based metaheuristics. In this PEB framework, multiple search processes are executed concurrently. Using the PEB framework, they also developed a parallel variant of Guided Local Search (GLS) called PEBGLS. The experimental results showed that PEBGLS is a competitive parallel metaheuristic for the TSP. Rios et al. [35] proposed a Distributed Variable Neighborhood Descent under CPU and multi-GPU environments for solving the Minimum Latency Problem (PLP). In order to better exploit the platform resources, hybrid metaheuristic algorithm combining good quality initial solutions with a flexible and powerful refinement procedure, inside the scope of an Iterated Local Search was considered. Their approach was finally compared to a set of classic local search algorithms, producing results that outperformed well-known state-of-the-art sequential algorithms. Nevertheless, despite their great interest as innovations in the area of combinatorial optimization, these works unfortunately have very little use in real fields.

Other recent contributions were rather in the context of state-of-the-art literature reviews, and sometimes also making attempts towards new extensions. Lopes Silva et al. [25] conducted a state-of-the-art study to determine the desirable features of a framework for metaheuristic algorithms and identify existing gaps, especially those related to the hybridization of metaheuristics. The combination of cooperation and parallelism, widely used in the context, was also analyzed. Other similar works have also focused on reviewing recent parallel and cooperative metaheuristic search approaches, as well as the areas to which these parallel versions can be applied [6,11,17]. Zhang et al. [42] developed a multi-objective optimization problem called a vehicle routing problem with route balancing (VERB) which is enhanced by using parallel computations on GPU devices. A set of experiments conducted on benchmark instances showed their efficacy and effectiveness. Kalatzantonakis et al. [22] presented a literature review with recent successful parallel implementations of variable neighborhood search applied to different variants of VRP. They also proposed three parallelization strategies that coordinate the communication of the multiple processors. Obtained results constituted a first proof-of-concept for the benefits of this new self-adaptive parameterized cooperative approach, especially for computationally hard instances. In their survey paper, Dokeroglu et al. [14] distinguish fourteen innovative metaheuristics that appeared over the two last decades. The metaheuristics are selected according to their efficient performance, high number of citations, and other efficiency criteria. Recent trends, open problems, advances in parallel metaheuristics and new research opportunities were also investigated in their study. For solving the vehicle routing problem with time windows, Marinakis et al. [26] proposed a multi-adaptive particle swarm optimization with a number of adaptive strategies to find the most effective set of parameters. Liu et al. [23] proposed a fast decomposition and reconstruction framework to solve the pickup and delivery problem with time windows run under different processors of multi-core CPU in parallel.

The most recent methodological contributions on combinatorial optimization and parallel processing are also the few references described below, which in fact, such as those above-mentioned, were often limited to evaluations based upon benchmark instances and simulations. Bach et al. [4] developed a GPU parallelization-based Adaptive Large Neighborhood Search (ALNS) for the much studied Distance Constrained Capacitated Vehicle Routing Problem (DCVRP) and compared it with a state-of-the-art methods. While it proved hard to implement certain commonly used mechanisms efficiently on the GPU, experiments showed that their approach yields highly competitive performance. In a literature paper, Schryen [37] suggested a universally applicable framework for parallel optimization in operations research, which can be used by academia to systematically describe their parallelization studies and position these in the landscape of parallel optimization. Abdelhafez et al. [1] presented two expanded studies to the solution goodness, execution time and energy consumption for a number of different metaheuristics (SA, GA and VNS) and their distributed counterparts. They also examined the effectiveness of parallel execution of the metaheuristics when implemented under new computing environments. To outperform the issues of premature convergence, stagnation prevention, and other numerical problems, Dokeroglu et al. [15] introduced hybrid artificial bee colony algorithms combined with teaching learning-based optimization metaheuristic and investigated their performances for real-parameter optimization functions. They also designed island parallel versions of their hybrid algorithms and showed that they can give better results than that of their sequential versions. Blocho [5] reviewed the most important parallelization techniques and cooperative search strategies for solving different variants of the Vehicle Routing Problems. Starzec et al. [38] designed a parallel and distributed ant colony optimization to be capable of efficiently leveraging the available HPC infrastructure. Their approach could be also exploited to solve various VRP models. Azzoug and Boukra [3] redraw the historic development of all vehicular ad-hoc network routing problems that concern either directly related routing tasks or targeting a set of diverse routing-related techniques with the aid of the nature-inspired algorithms including parallelized ones. Finally, a recapitulation of the main references published between 2018 and 2020 and cited above is set out in Table 3.

Table 3. Recapitulative table outlining the main references published between 2018 and 2020 on combinatorial optimization and parallel processing.

Reference	Year	Contribution
[6]	2018	Review
[18]	2018	Parallel Ant Colony Optimization
[20]	2018	Parallel Elite Biased (PEB) Framework
[25]	2018	Review
[35]	2018	Parallel Multi-GPU Local Search
[36]	2018	PEB Framework for Guided Local Search
[4]	2019	GPU-based Adaptive Large Neighborhood Search
[11]	2019	Review
[17]	2019	Review
[16]	2019	Review
[22]	2019	Parallel Variable Neighborhood Search Strategies
[23]	2019	Fast Decomposition and Reconstruction Framework
[26]	2019	Multi-Adaptive Particle Swarm Optimization Algorithm
[42]	2019	GPU-based Multiobjective Memetic Algorithms
[1]	2020	Parallel Optimization Algorithm with Multi-Core System
[3]	2020	Review
[5]	2020	Review
[15]	2020	Parallel Hybrid Artificial Bee Colony Algorithm
[37]	2020	Review
[38]	2020	Distributed Ant Colony Optimization Algorithm

5 Conclusion

Optimization algorithms have emerged in many fields, where complicated problems need to be solved. In these situations, exact search techniques cannot give good solutions for most of the real-life problems in a reasonable time. Metaheuristics have proved to be efficient alternatives for such problems. Despite this, however, these techniques may require large time to reach their optimum solutions. Parallelization is a hopeful approach for overcoming time consumption values of these process. Although recent approaches are running metaheuristics in parallel, the organization still lacks for novel investigation comparing the optimization techniques being running in parallel. On the other hand, the vehicle routing problem is a combinatorial optimization problem which continues to draw attention of practitioners and academia due to its importance especially in transportation, logistics, supply chain management, and commerce. This model, which is very widespread in the real world of management, has also benefited from the principle of parallel calculation. In this paper, a look over recent publications dealing with various parallel-processing meta-heuristics proposed to tackle the VRP and its main variants is carried out. For this aim, we have conducted both a bibliometric analysis and a critical review of the state-of-the-art/survey publications that are out there since 2018. Studies have shown that interest in these methods has not stopped growing, with applications in various fields of the contemporary world. The extrapolation of obtained results in the future shows an upward dynamics regarding the continuation of exploitation and development of parallel computing tools for solving various VRP approaches. On the other hand, the reduction in the number of citations on works dealing with VRP and parallel metaheuristics shows that the interest of the research community has somewhat declined in the last few years. This may be due not only to the deficiency of rigorous mathematical results, but also to the lack of easy interfaces under famous calculus softwares, such as Matlab and R. It would therefore be advisable in the future to push for more fundamental research paths and practical software packages.

References

1. Abdelhafez, A., Luque, G., Alba, E.: Parallel execution combinatorics with meta-heuristics: comparative study. Swarm Evol. Comput. **55**, 100692 (2020)
2. Aydin, M.E., Yigit, V.: Parallel simulated annealing Chapter 12, pp. 267–287. Wiley, Hoboken (2005)
3. Azzoug, Y., Boukra, A.: Bio-inspired VANET routing optimization: an overview. Artif. Intell. Rev., 1–58 (2020). https://doi.org/10.1007/s10462-020-09868-9
4. Bach, L., Hasle, G., Schulz, C.: Adaptive large neighborhood search on the graphics processing unit. Eur. J. Oper. Res. **275**(1), 53–66 (2019)
5. Blocho, M.: Parallel algorithms for solving rich vehicle routing problems. In: Smart Delivery Systems, pp. 185–201 (2020)
6. Codognet, P., Munera, D., Diaz, D., Abreu, S.: Parallel Local Search. In: Hamadi, Y., Sais, L. (eds.) Handbook of Parallel Constraint Reasoning, pp. 381–417. Springer, Cham (2018). https://doi.org/10.1007/978-3-319-63516-3_10
7. Cordeau, J.F., Maischberger, M.: A parallel iterated tabu search heuristic for vehicle routing problems. Comput. Oper. Res. **39**(9), 2033–2050 (2012)
8. Crainic, T.G., Toulouse, M., Gendreau, M.: Towards a taxonomy of parallel tabu search algorithms. INFORMS J. Comput. **9**(1), 61–72 (1997)
9. Crainic, T.G., Toulouse, M.: Parallel strategies for meta-heuristics. In: Glover, F., Kochenberger, G.A. (eds.) Handbook of Metaheuristics. ISOR, vol. 57, pp. 475–513. Springer, Boston (2003). https://doi.org/10.1007/0-306-48056-5_17
10. Crainic, T.G.: Parallel solution methods for vehicle routing problems. In: Golden, B., Raghavan, S., Wasil, E. (eds.) The Vehicle Routing Problem: Latest Advances and New Challenges. ORCS, vol. 43, pp. 171–198. Springer, Boston (2008). https://doi.org/10.1007/978-0-387-77778-8_8

11. Crainic, T.: Parallel metaheuristics and cooperative search. In: Gendreau, M., Potvin, J.-Y. (eds.) Handbook of Metaheuristics. ISORMS, vol. 272, pp. 419–451. Springer, Cham (2019). https://doi.org/10.1007/978-3-319-91086-4_13
12. Dantzig, G.B., Fulkerson, R., Johnson, S.M.: Solution of a large-scale traveling salesman problem. Oper. Res. **2**(4), 393–410 (1954)
13. Dantzig, G., Ramser, J.: The truck dispatching problem. Manag. Sci. **6**(1), 80–91 (1959)
14. Dokeroglu, T., Sevinc, E., Kucukyilmaz, T., Cosar, A.: A survey on new generation metaheuristic algorithms. Comput. Ind. Eng. **137**, 106040 (2019)
15. Dokeroglu, T., Pehlivan, S., Avenoglu, B.: Robust parallel hybrid artificial bee colony algorithms for the multi-dimensional numerical optimization. J. Supercomput. **76**(9), 7026–7046 (2020). https://doi.org/10.1007/s11227-019-03127-7
16. Eskandarpour, M., Ouelhadj, D., Fletcher, G.: Decision making using metaheuristic optimization methods in sustainable transportation. In: Sustainable Transportation & Smart Logistics, pp. 285–304 (2019)
17. Essaid, M., Idoumghar, L., Lepagnot, J., Brévilliers, M.: GPU parallelization strategies for metaheuristics: a survey. Int. J. Parallel Emergent Distrib. Syst. **34**(5), 497–522 (2019)
18. Fanlei, Y.: Autonomous vehicle routing problem solution based on artificial potential field with parallel ant colony optimization (ACO) algorithm. Pattern Recogn. Lett. **116**, 195–199 (2018)
19. Grid, M.: Bee life Parallèle sur GPU pour résoudre le problème dynamique de tournées de véhicules avec une contrainte de capacité. Université Mohamed Khider Biskra, Algérie, Thèse de Docorat (2018)
20. Jialong, S., Qingfu, Z.: A new cooperative framework for parallel trajectory-based metaheuristics. Appl. Soft Comput. **65**, 374–386 (2018)
21. Jagiello, S., Zelazny, D.: Solving multi-criteria vehicle routing problem by parallel tabu search on GPU. Procedia Comput. Sci. **18**, 2529–2532 (2013)
22. Kalatzantonakis, P., Sifaleras, A., Samaras, N.: Cooperative versus non-cooperative parallel variable neighborhood search strategies: a case study on the capacitated vehicle routing problem. J. Global Optim. (9), 1–22 (2019). https://doi.org/10.1007/s10898-019-00866-y
23. Liu, F., Gui, M., Yi, C., Lan, Y.: A fast decomposition and reconstruction framework for the pickup and delivery problem with time windows and LIFO loading. IEEE Access **7**, 71813–71826 (2019)
24. Lootsma, F.A., Ragsdell, K.M.: State-of-the-art in parallel nonlinear optimization. Parallel Comput. **6**, 133–155 (1988)
25. Lopes Silva, M.A., De Souza, S.R., Freitas Souza, M.J., De Franca Filho, M.F.: Hybrid metaheuristics and multi-agent systems for solving optimization problems: a review of frameworks and a comparative analysis. Appl. Soft Comput. **71**, 433–459 (2018)
26. Marinakis, Y., Marinaki, M., Migdalas, A.: A multi-adaptive particle swarm optimization for the vehicle routing problem with time windows. Inf. Sci. **481**, 311–329 (2019)
27. Nebro, A.J., Luna, F., Talbi, E.G., Alba, E.: Parallel multiobjective optimization Chapter 16, pp. 371–394. Wiley, Hoboken (2005)
28. Nwana, V., Mitra, V.: Parallel mixed integer programming: a status review. Technical report, Department of Mathematical Sciences, Brunel University (2000)
29. Pedemonte, M., Nesmachnow, S., Cancela, H.: A survey on parallel ant colony optimization. Appl. Soft Comput. **11**, 5181–5197 (2011)

30. Prez, J.A.M., Hansen, P., Mladenovi, N.: Parallel variable neighborhood search Chapter 11, pp. 247–266. Wiley, Hoboken (2005)
31. Rabbouch, B., Mraihi, R., Saâdaoui, F.: A recent brief survey for the multi depot heterogenous vehicle routing problem with time windows. In: Abraham, A., Muhuri, P.K., Muda, A.K., Gandhi, N. (eds.) HIS 2017. AISC, vol. 734, pp. 147–157. Springer, Cham (2018). https://doi.org/10.1007/978-3-319-76351-4_15
32. Rabbouch, B., Saâdaoui, F., Mraihi, R.: Empirical-type simulated annealing for solving the capacitated vehicle routing problem. J. Exp. Theor. Artif. Intell. **32**(3), 437–452 (2020)
33. Rabbouch, B., Saâdaoui, F., Mraihi, R.: Efficient implementation of the genetic algorithm to solve rich vehicle routing problems. Oper. Res.: Int. J. https://doi.org/10.1007/s12351-019-00521-0
34. Rabbouch, B., Saâdaoui, F., Mraihi, R.: Constraint programming based algorithm for solving large-scale vehicle routing problems. In: Pérez García, H., Sánchez González, L., Castejón Limas, M., Quintián Pardo, H., Corchado Rodríguez, E. (eds.) HAIS 2019. LNCS (LNAI), vol. 11734, pp. 526–539. Springer, Cham (2019). https://doi.org/10.1007/978-3-030-29859-3_45
35. Rios, E., Ochi, L.S., Boeres, C., Coelho, V.N., Coelho, I.M., Farias, R.: Exploring parallel multi-GPU local search strategies in a metaheuristic framework. J. Parallel Distrib. Comput. **111**, 39–55 (2018)
36. Shi, J., Zhang, Q.: A new cooperative framework for parallel trajectory-based meta-heuristics. Appl. Soft Comput. **65**, 374–386 (2018)
37. Schryen, G.: Parallel computational optimization in operations research: a new integrative framework, literature review and research directions. Eur. J. Oper. Res. **287**(1), 1–18 (2020)
38. Starzec, M., Starzec, G., Byrski, A., Turek, W., Piętak, K.: Desynchronization in distributed ant colony optimization in HPC environment. Future Gener. Comput. Syst. **109**, 125–133 (2020)
39. Talbi, E.G.: Metaheuristics: From Design to Implementation. Wiley, Hoboken (2009)
40. Tan, Y., Ding, K.: A survey on GPU-based implementation of swarm intelligence algorithms. IEEE Trans. Cybern. **46**, 2028–2041 (2016)
41. Zhang, Y., Wang, S., Ji, G.: A comprehensive survey on particle swarm optimization algorithm and its applications. Math. Problems Eng. **2015**, 1–38 (2015)
42. Zhang, Z., Sun, Y., Xie, H., Teng, Y., Wang, J.: GMMA: GPU-based multiobjective memetic algorithms for vehicle routing problem with route balancing. Appl. Intell. **49**(1), 63–78 (2018). https://doi.org/10.1007/s10489-018-1210-6

Multi-scaled Non-local Means Parallel Filters for Medical Image Denoising

Hana Rabbouch[1], Othman Ben Messaoud[1,2], and Foued Saâdaoui[1,3(✉)]

[1] Lab: LR18ES15 Algèbre, Théorie de Nombres et Analyse Non-linéaire,
Faculté des Sciences, 5019 Monastir, Tunisia
hana.rabbouch@gmail.com
[2] Compagnie des Phosphates de Gafsa (CPG), Cité Bayech, 2100 Gafsa, Tunisia
bmo.stat@gmail.com
[3] King Abdulaziz University, P.O BOX 80203, Jeddah 21589, Saudi Arabia
foued.saadaoui@isgs.rnu.tn

Abstract. In recent years, there has been an increased interest in denoising techniques that are applicable in various medical imaging fields. The extraordinary development of the denoising area is no doubt due to the ever expanding and successful computing technology, but also to the emergence of the multi-resolution analysis (MRA) on both mathematical and algorithmic levels. However, many denoising techniques still remain ineffective in dealing with certain types of noise. Other methods can be too expensive, given their nested and complicated structure. Therefore, in this paper, A new multi-scale parallel denoising paradigm is defined and tested. A comparative study is conducted between the two best-known MRA-based decomposition techniques: the Empirical Mode Decomposition (EMD) and the Discrete Wavelet Transform (DWT). The comparison is carried out in this framework of multi-scaled parallel denoising, where a Non-Local Means (NLM) filter is implemented and adjusted scale-by-scale to a sample of X-ray benchmark images. Some state-of-the-art denoising methods were also used in the evaluation. The numerical results proved the effectiveness of the multi-scaled parallel denoising in terms of accuracy and speed of convergence, especially when the NLM filtering is coupled with the EMD. This shows a bright future for their medical use in the next few years.

Keywords: Medical imaging · Non-local means · Multi-scaled denoising · Empirical mode decomposition · Wavelets

1 Introduction

Image de-noising is a fundamental task in signal processing, which still plays a vital role in the field of medical imaging, especially in ultrasound imaging, X-ray imaging, Computer Tomography (CT) and Magnetic Reasoning Imaging (MRI) [1–4]. The main aim of image de-noising is to reduce the noise as well as preserve the important features such as edges, textures, boundaries, and sharp structures.

© Springer Nature Switzerland AG 2020
M. Qiu (Ed.): ICA3PP 2020, LNCS 12452, pp. 606–613, 2020.
https://doi.org/10.1007/978-3-030-60245-1_41

This field has seen much research and encouraging advance the last few decades. Despite this, image de-noising is still drawing attention of researchers as image de-noising can cause blurring and introduce artifacts. That is why some adaptive multi-resolution decomposition approaches have recently been used to perform a joint space-spatial frequency de-noising [2]. The most common examples of this kind of image processing tools are the Empirical Mode Decomposition (EMD) [5] and the wavelet transform [6], which gained great interest in biomedical images de-noising [3,7,8].

The main idea behind the principle of multi-scaled NLM parallel denoising is that it firstly establishes a Multi-Resolution Analysis (MRA) -based decomposition of the image. Then, coarser components (approximation coefficients) are slightly NLM-filtered in order to preserve the gray-scale intensity of the image. High-frequency subband components are also filtered but more aggressively, before being joined to processed approximations. The last step consists in recombining smoothed approximation and detail coefficient by performing an adequate inverse transform. The aim of this paper is therefore to define and formalize a new family of adaptive parallel filters which are inspired by the principle of multi-scaled denoising. The two main variants addressed in this framework are those coupling two among the most practical MRA-decomposition methods (EMD and wavelets) with the NLM technique. Although the second category exists in the literature [3], the EMD-based approach is new and has not yet been used. It is also noteworthy that a comparison between EMD and wavelets in this context of parallel denoising has not yet been addressed in the literature. That is why a set of experiments is finally conducted in order to compare the two strategies in a real framework of medical images denoising.

The paper is structured as follows. Section 2 and 3 provide illustrations of algorithmic formulations of the EMD and the wavelet transform. The multi-scaled parallel denoising algorithm is presented in Sect. 4. The experimental setup and numerical results are finally described and discussed in Sect. 5.

2   Empirical Mode Decomposition (EMD)

The Empirical Mode Decomposition (EMD) [5] of an information represents a decomposition technique by which almost any signal $z_{s,t}$, $\forall\, s, t \in \Omega \subset \mathbb{Z}^2$ could be disaggregated into a finite number of mono-component subsignals, called Intrinsic Mode Functions (IMFs), plus a residual. The EMD assumes that the data, depending on its complexity, may have many different coexisting modes of oscillations at the same time. The EMD's IMFs satisfy the following two conditions: (i) Over the entire length of the data (signal or image), the number of local extremums and the number of zero-crossings must either be equal or differ at most by one; (ii) At each pixel of our image (s, t), the average of the polygon connecting local maximums and the polygon connecting local minimums is zero. The IMFs are extracted through a sifting process, whose algorithm can be described in Table 1.

The Cauchy convergence test is commonly used as stopping criterion. This criterion is defined as a normalized squared difference between two successive

Table 1. Empirical Mode Decomposition (EMD).

Algorithm: EMD
1. Using splines, identify extremums of the signal $z_{s,t}$, $\forall\, s,t \in \Omega \subset \mathbb{Z}^2$
2. Interpolate maxima (minima), ending up with a polygon $e_{\max}(s,t)$ $(e_{\min}(s,t))$
3. Calculate the midpoints $m_{s,t} = [e_{\max}(s,t) + e_{\min}(s,t)]/2$
4. Extract the detail $h_{s,t} = z_{s,t} - m_{s,t}$. $h_{s,t}$ should satisfy the definition of IMF
5. Treat $h_{s,t}$ as input and repeat steps (1.) to (4.) k times until the polygons are symmetric with respect to zero. After k cycles, if $h_{s,t}^{(k)}$ is IMF, take $h_{s,t}^{(k)}$ as the ith IMF (denoted $\mathrm{IMF}_{s,t}^{(i)}$), and replace $z_{s,t}$ with $r_i(s,t) = z_{s,t} - h_{s,t}^{(k)}$
6. Repeat steps (1.) to (5.) by sifting the residual until it satisfies a predefined stopping criterion. After finishing the whole sifting process, we have a residual and a collection of several IMFs, denoted as $\mathrm{IMF}_{s,t}^{(i)}$ $(i = 1, 2, \ldots, M)$

sifting operations, i.e.: $\sigma_k = \sum_{s=0}^{S} \sum_{t=0}^{T} \{|h_{s,t}^{(k-1)} - h_{s,t}^{(k)}|^2/(h_{s,t}^{k-1})^2\}$, where $h_{s,t}^k$ is the kth extracted signal through the sifting, and k is the iteration number. If σ_k is smaller than the predetermined value ϵ, the sifting process will be stopped. Moreover, the original $z_{s,t}$ can be exactly reconstructed by a simple summation:

$$z_{s,t} = r_M(s,t) + \sum_{i=1}^{M} \mathrm{IMF}_{s,t}^{(i)}, \tag{1}$$

where $r_i(t) = r_{i-1}(t) - \mathrm{IMF}_t^{(i)}$, $r_M(t)$ is the residual component, which is generally a constant or a monotonic function representing the trend of the time series. EMD performs thus a scale-by-scale analysis that is totally data-driven and that can be applied to all oscillatory time series, including non stationary and locally stationary ones and those generated by a nonlinear system. It is noticeable that this last point represents the superiority of the EMD method in comparison with wavelets.

3   Wavelet Transform

Wavelets are oscillatory and compact functions that have zero-mean and limit width in both the time and frequency domains. The minimum requirements imposed on such functions to qualify for being wavelets are that $\Psi : \mathbb{R}^2 \to \mathbb{R}$ with $\Psi \in L^2(\mathbb{R}^2)$[1] and also fulfill a technical condition, usually referred to as the admissibility condition, which reads as $\int_{(\mathbb{R}_+)^2} |\hat{\Psi}(\xi)|^2 / |\xi|^2 \, d\xi < +\infty$, where $\hat{\Psi}(\xi)$ denotes the adequate Fourier transform of $\Psi(X)$ for $X \in \mathbb{R}^2$. Let us consider the ordinary monodimensional wavelet transform with $\psi(s)$ and $\varphi(s)$ its respective mother and father functions [3]. The product of the two functions allow us to define a bi-dimensional father function:

$$\Phi(s,t) = \varphi(s)\varphi(t), \tag{2}$$

---

[1] A function $f(X)$ is square-integrable if it satisfies $\int_{\mathbb{R}^2} f^2(X)dX < \infty$.

whereas the following products: $\psi^h(s,t) = \psi(s)\varphi(t)$, $\psi^v(s,t) = \varphi(s)\psi(t)$ and $\psi^d(s,t) = \psi(s)\psi(t)$ yield the three principal 2-dimensional wavelet details functions, where the superscripts h, v and d respectively correspond to horizontal, vertical and diagonal direction details. In this basis, we can define

$$\Phi_{l,k_1,k_2}(s,t) = 2^{l/2}\Phi(2^l s - k_1, 2^l t - k_2) \tag{3}$$

as the coarse component, whereas

$$\Psi_{l,k_1,k_2}(s,t) = 2^{l/2}\psi^{dir}(2^l s - k_1, 2^l t - k_2), \quad dir = \{h,v,d\} \tag{4}$$

are the details components.

The bi-dimensional discrete wavelet transform (2D-DWT) of a function $z_{(s,t)}$ of dimension $S \times T$ is written as:

$$T_\Phi(l_0, k_1, k_2) = \frac{1}{\sqrt{ST}} \sum_{s=0}^{S-1} \sum_{t=0}^{T-1} z_{(s,t)}\Phi_{l_0,k_1,k_2}(s,t), \tag{5}$$

$$T_\Psi^{dir}(l_0, k_1, k_2) = \frac{1}{\sqrt{ST}} \sum_{s=0}^{S-1} \sum_{t=0}^{T-1} z_{(s,t)}\Psi_{l_0,k_1,k_2}^{dir}(s,t), \tag{6}$$

where $dir = \{h,v,d\}$, with an arbitrary start resolution l_0 and a scale parameter L. We usually choose $l_0 = 0$ and $N = M = 2^L$ to have $l = 0, 1, \ldots, L-1$ and $m, n = 0, 1, \ldots, 2^l - 1$.

A 2D-DWT algorithm is used to decompose I according to the following recursive way:

$$z_{l+1}[m,n] = \sum_{k_1,k_2 \in \mathbb{Z}} h[k_1]h[k_2]z_l[2m - k_1, 2n - k_2] \tag{7}$$

$$W_{l+1}^h[m,n] = \sum_{k_1,k_2 \in \mathbb{Z}} g[k_1]h[k_2]z_l[2m - k_1, 2n - k_2] \tag{8}$$

$$W_{l+1}^v[m,n] = \sum_{k_1,k_2 \in \mathbb{Z}} h[k_1]g[k_2]z_l[2m - k_1, 2n - k_2] \tag{9}$$

$$W_{l+1}^d[m,n] = \sum_{k_1,k_2 \in \mathbb{Z}} g[k_1]g[k_2]z_l[2m - k_1, 2n - k_2] \tag{10}$$

where $h(.)$ and $g(.)$ are respectively, low-pass and high-pass filter coefficients, $\bar{h}[n] = h[-n]$, $n \in \mathbb{Z}$ is the reversed version of h. The signal z_{l+1} is an approximation of z_l at a lower resolution. This approximation is computed from z_l by low-pass filtering and decimating by 2 along its rows and columns. The signals W_{l+1} contain the detail of z_l.

4   Multiscaled NLM-Denoising

The multi-scaled parallel denoising is based upon the simple principle of filtering EMD or wavelet sub-components by the MLM method before reassembling them. Let $p = (x_i, y_j)$ be a pixel of a bi-dimensional sub-band I. This component represents a noisy discrete signal defined on a domain $\Omega \subset \mathbb{Z}^2$. The non-local means (NLM) filter can be formulated as follows:

$$NLM[I](p) = \sum_{q \in \Omega} \omega(p,q)I(q), \tag{11}$$

where the ω's are weights satisfying $0 < \omega(p,q) < 1$ and $\sum_q \omega(p,q) = 1$. Accordingly, an adjusted pixel $NLM[I](p)$ is a weighted average of all pixels in the image. The weights are based on the similarity between the neighborhoods of pixels p and q. In other words, $\omega(p,q)$ depends on a distance $\delta(p,q)$ between observed gray level vectors at pixels p and q.

Given an odd number n and a pixel $p \in \Omega$, we define the patch $I(\mathcal{N}_p)$ of width n centered at p as the n^2-dimensional vector whose coordinates are gray values of the pixels in a square neighborhood of p with side n: $I(\mathcal{N}_p) = I(p+j)_{||j||_\infty \le \frac{n-1}{2}}$. Therefore, we have: $\delta(p,q) = ||I(\mathcal{N}_p) - I(\mathcal{N}_q)||_{2,\sigma}^2$, where $|| \cdot ||_{2,a}$ is an Euclidean norm weighted by $\sigma > 0$, the standard deviation of the Gaussian filter on the neighborhood. This measure is so much more adapted to any additive white noise such that a noise alters the distance between windows in a uniform way. Indeed,

$$\mathbb{E}||I(\mathcal{N}_p) - I(\mathcal{N}_q)||_{2,\sigma}^2 = ||J(\mathcal{N}_p) - J(\mathcal{N}_q)||_{2,\sigma}^2 + 2\varrho, \tag{12}$$

where J and I are, respectively, the original and noisy image components and ϱ is the noise variance. This equality shows that, in expectation, the Euclidean distance preserves the order of similarity between pixels. So the most similar pixels to p in I also are expected to be the most similar pixels to p in J [1,3]. The NLM weights can be expressed as

$$\omega(p,q) = \frac{\Delta(p,q,h)}{\sum_q \Delta(p,q,h)}, \tag{13}$$

where $\Delta(p,q,h) = \exp \frac{-\delta(p,q)}{h^2}$ with h controlling the decay of the exponential function, and therefore the decay of the weights, as a function of the Euclidean distances. Once AMR-subcomponents are appropriately filtered (one by one in parallel), a reconstruction of the image is carried out using the adequate inverse transform. It is noticeable that the choice of filtering parameters at each resolution level is purely empirical. At this stage, well-known deviation criteria, such as the mean square error or the signal-to-noise ratio can be used. A diagram summarizing the proposed approach is given in Fig. 1.

5 Numerical Results

In this numerical study, the proposed approach is compared to a set of similar de-noising techniques. The two benchmark images of Fig. 2 are first noised, then de-noised using each of the compared techniques. The benchmark images represent X-ray scans of coronary arteries and a left-side shoulder (see [4,9]). The images are corrupted by Additive White Gaussian Noise (AWGN) and Multiplicative Speckle Noise (MSN) the standard deviations of which are gradually increased.

The first experiment is conducted on the left-side image (coronary arteries) and its results are reported in Table 2. Generated noises are denoted as follows: $\mathcal{G}(a)$, a AWGN with variance $Var_a = a \times 10^{-3}$, and, $\mathcal{S}(b)$, a MSN with variance $Var_b = b \times 10^{-3}$. The table reports the improvements of the Peak Signal-to-Noise Ratio (PSNR) for the image de-noising. In this experiment, the three used

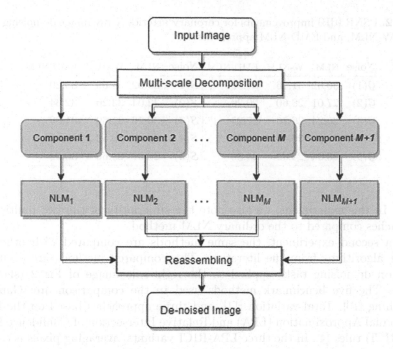

Fig. 1. Multi-scaled NLM parallel de-noising principle.

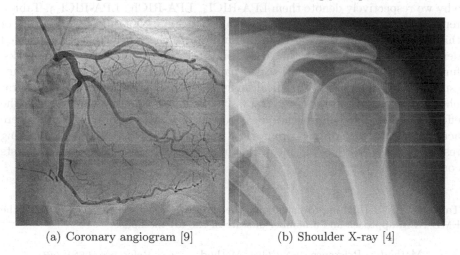

(a) Coronary angiogram [9]   (b) Shoulder X-ray [4]

Fig. 2. Original noise-free X-ray images.

methods are the non-local means (NLM), the wavelet-based NLM (W-NLM), and the EMD-based NLM (EMD-NLM). The results show the effectiveness of the EMD-NLM method, especially when the noise is of a relatively high intensity. On the other hand, the wavelet-based method proves its ability to remove low level

Table 2. PSNR (dB) improvements for coronary arteries X-ray image de-noising using NLM, W-NLM, and EMD-NLM approaches.

Noise	NLM	W-NLM	EMD-NLM	Noise	NLM	W-NLM	EMD-NLM
$\mathcal{G}(1)$	30.02	31.56	31.80	$\mathcal{S}(1)$	35.02	35.09	33.29
$\mathcal{G}(2)$	27.01	28.60	29.20	$\mathcal{S}(2)$	32.04	33.20	32.64
$\mathcal{G}(3)$	25.24	26.76	27.27	$\mathcal{S}(3)$	30.26	31.69	31.80
$\mathcal{G}(4)$	23.99	25.49	25.84	$\mathcal{S}(4)$	29.03	30.53	30.86
$\mathcal{G}(5)$	23.04	24.49	24.80	$\mathcal{S}(5)$	28.07	29.60	30.02

noises. In this experiment, we also note the superiority of the two multiscaled approaches compared to the ordinary NLM method.

In a second experiment, the same methods are compared with other de-noising algorithms from the literature. The compared methods are evaluated based on de-noising tasks applied to the right-side image of Fig. 2 (shoulder X-ray). The five benchmark methods used in the comparison are: Gaussian Smoothing [10], Total variation [11], and three approaches based on the Local Polynomial Approximation (LPA) and Relative Intersection of Confidence Intervals (RICI) rules [4]. In the three LPA-RICI variants, averaging pixels is carried out on quadrilateral regions, octagonal regions, and Hexadecagonal regions, that why we respectively denote them LPA-RICI$_4$, LPA-RICI$_8$, LPA-RICI$_{16}$. Table 3 reports PSNR results, as well as the main benchmark methods. The results of this second experiment prove again the superiority of the principle of multiscaled denoising and in particular, that based on the EMD technique. In future works, further evaluation of the multi-scaled parallel denoising should be carried out, especially in other areas of medical imaging, such as those related to tomography [3] and magnetic resonance [12,13]. It would also be interesting to study the effectiveness of these approaches for X-ray chest images of patients with severe acute respiratory syndrome corona virus of type 2 (SARS-CoV-2). Encouraging results in this direction will mean that rapid and effective scanner-based tests could be easily developed.

Table 3. PSNR (dB) improvements for the shoulder X-ray image denoised using the EMD-NLM approach and a number of state-of-the-art methods.

Method	Reference	PSNR (dB)	Method	Reference	PSNR (dB)
EMD-NLM		39.52	LPA-RICI$_8$	[4]	39.37
W-NLM	[3]	35.56	LPA-RICI$_{16}$	[4]	33.61
NLM	[1]	33.94	Gauss. Smooth	[10]	34.69
LPA-RICI$_4$	[4]	38.73	Tot. Variation	[11]	35.02

6   Conclusion

The results of numerical experiments carried out on the basis of medical images from X-ray scanners prove the interest of the multi-scaled de-noising, especially when it is based on NLM filters. In this framework, the comparison between EMD and wavelet decompositions shows the superiority of EMD in terms of accuracy and speed of convergence. Nevertheless, the EMD-based denoising approach deserves to be tested in other areas of medical imaging, such as those related to tomography [3], magnetic resonance [13], and chest X-ray in patients with COVID-19 [14]. The various algorithms also deserve to be tested on 3D images, where cost-effectiveness criteria are used for comparison.

References

1. Buades, A., Coll, B., Morel, J.-M.: A non-local algorithm for image denoising. IEEE CVPR **2**, 60–65 (2005)
2. Lahmiri, S.: Denoising techniques in adaptive multi-resolution domains with applications to biomedical images. Healthc. Technol. Lett. **4**(1), 25–29 (2017)
3. Rabbouch, H., Saâdaoui, F.: A wavelet-assisted subband denoising for tomographic image reconstruction. J. Vis. Commun. Image R. **55**, 115–130 (2018)
4. Mandić, I., Peić, H., Lerga, J., Štajduhar, I.: Denoising of X-ray images using the adaptive algorithm based on the LPA-RICI algorithm. J. Imaging **4**(2), 34 (2018)
5. Huang, N.E., Shen, Z., Long, S.R., Wu, M.C., Shih, H.H., Zheng, Q.: The empirical mode decomposition and the Hilbert spectrum for nonlinear and nonstationary time series analysis. Proc. R. Soc. Lond. A Math. Phys. Eng. Sci. **454**(1971), pp. 903–995 (1998)
6. Fowler, J.: The redundant discrete wavelet transform and additive noise. IEEE Signal Process. Lett. **12**(9), 629–632 (2005)
7. Arfia, F.B., Sabri, A., Messaoud, M.B., Abid, M.: The bidimensional empirical mode decomposition with 2D-DWT for gaussian image denoising, In: Proceedings of 17th International Conference on DSP, pp. 1–5 (2011)
8. Flandrin, P., Goncalves, P., Rilling, G.: Detrending and denoising with empirical mode decomposition. In: Proceedings of the 12th EUSIPCO, pp. 1581–1584 (2004)
9. Kočka, V.: The coronary angiography-an old-timer in great shape. Cor et Vasa, **57**(6), e419–e424 (2015)
10. Kopparapu, S.K., Satish, M.: Identifying optimal Gaussian filter for Gaussian noise removal. 3rd NCVPRIPG, pp. 126–129 (2011)
11. Rudin, L.I., Osher, S., Fatemi, E.: Nonlinear total variation based noise removal algorithms. Phys. D **60**(1–4), 259–268 (1992)
12. Benameur, N., et al.: Left ventricular MRI wall motion assessment by monogenic signal amplitude image computation. Magn. Reson. Imaging **54**, 109–118 (2018)
13. Chang, L., Bang, G.C., Xi, Y.: A MRI denoising method based on 3D nonlocal means and multidimensional PCA. Comput. Math. Method. M., Art. No. 232389, p. 11 (2015)
14. Jacobi, A., Chung, M., Bernheim, A., Eber, C.: Portable chest X-ray in coronavirus disease-19 (COVID-19): a pictorial review. Clin. Imag. **64**, 35–42 (2020)

Optimized HybridSketch: More Efficient with Analysis and Algorithm

Xiaolei Zhao[1(✉)], Mei Wen[1], Minjin Tang[1], Qun Huang[2],
and Chunyuan zhang[1]

[1] School of Computer Science, National University of Defense Technology,
Changsha, China
zhaoxiaolei14@nudt.edu.cn
[2] Institute of Computing Technology, Chinese Academy of Sciences, Beijing, China

Abstract. The sketch structure is widely applied in the network measurement field due to its limited memory usage and simple operation. However, When the less memory space the system occupied, the accuracy decreases. However, as the flow rate rapidly increases, the on-chip memory will become a system bottleneck. HybridSketch provides a methods to save memory and maintain the accuracy of the measurement system. We apply analysis and new algorithms to make it more efficient. We analyze the error bound of the system and observed that the sketch part of the system will lose the precision of the information along with less memory inevitably. So we propose the data augmentation algorithm based on our analysis. We apply it and propose the optimized HybridSketch. We evaluate the performance and present a comparison with the origin algorithm. The results show that optimized HybridSketch provides an 80% precision rate compared to the original one which occupied 10× the memory size.

Keywords: Sketch · Network measurement · Algorithm

1 Introduction

Network measurement is deployed in data center networks widely to obtain useful information of the traffic as regards Heavy Hitter detection [18], Flow Frequency Estimation [19], Entropy [20], etc. Structure and algorithms based on sketch are preferred by many researchers due to its limited memory usage and simple operation. As a results of the rapid development of network bandwidth, precise measurement methods are unsuitable for deployment in high-speed networks because of the high memory usage and bandwidth utilization required. Accordingly, some sketch algorithms [1–3,7,9] have been proposed. However, with the development of sketch algorithm, more structures expended more memory space on some auxiliary process which aimed to achieve better results. And the consumption caused that the sketch algorithm needs more extra memory space to fulfill a measurement task.

© Springer Nature Switzerland AG 2020
M. Qiu (Ed.): ICA3PP 2020, LNCS 12452, pp. 614–626, 2020.
https://doi.org/10.1007/978-3-030-60245-1_42

An algorithm focus on memory usage is proposed, HybridSketch, which focuses on the memory and precision of the system; this is accomplished through the mixing of two measurement methods by quantitatively analyzing, modeling and allocating appropriate memory space to each method and achieve better results. But we found that the algorithm does not consider the boundary of the data plane. And lower memory usage for the sketch part inevitably leads to the increase of fitting deviation of the distribution formed by the data in sketch structure. Less data in the sketch part influences the distribution of record keys, which are used to recovery the results of some applications.

Accordingly, the contributions of this paper are as follows:

- We analyze the boundary of the top-k part and sketch part of HybridSketch, certify the error rate of two parts of the system.
- We propose the Data Augmentation Algorithm and optimized HybridSketch algorithm, solve the fitting deviation of the distribution result from less memory.
- We evaluate the performance and present a comparison with the origin algorithm. The results show that optimal HybridSketch provides an 80% precision rate compared to the original with 10× the memory size.

2   Background and Motivations

HybridSketch is designed by mixing the top-k and sketch algorithms. Through allocating memory precisely to each algorithm, the system can achieve better results with less memory space.

It combines the advantages of the both algorithms, and saves more memory usage to achieve the same effect. In the top-k part, it applies the Hashpipe [5] algorithm to extract the heavy hitters, then the flows expelled by the top-k part will be enter into the sketch part. It applies SketchLearn [1] algorithm to maintain the bit level counter groups.

It extracts heavy hitters through top-k part, record the bit-level distribution through sketch part, overcomes the shortage that some large flows cannot be extracted by the sketch part due to less counters.

But top-k part occupies the $\frac{1}{4}$ of all the memory of data plane. Less memory for the sketch part means less counters to record the flows. When the counters is not enough, the accuracy of bit level distribution of the flowkey will decrease.

By focusing on this, we analyze the bound of the system and propose the optimal algorithm for overcoming the recorded data shortages.

One of the challenges of our work is how to analyze the two part of the system. Because the structures are different, we cannot use one way to analyze the two part. We use two different methods to analyze the error bound of the two part, and get a specific quantitative range; The other challenge is how to do data expansion and data recovery to overcome the drawback that caused by data shortage. We propose a data augmentation algorithm based on the statistical regularity, supply 10× data size, maintain the accuracy of bit level distribution.

3   Boundary Analysis

In this section, we analyze the top-k part bound and sketch part bound of HybridSketch, which certifies the superiority of the system. Through this proving process, we can clearly identify the theoretical support for the design. We can prove that the top-k part of the system can extract at least 48.99% of the traffic, and the flow which occupies more than 0.1% of the traffic would be identified as a large flow. Moreover, we can obtain the error rate of the sketch part based on the features of the data set.

3.1   Top-K Part Bound Analysis

In order to prove the upper bound, we introduce Zipf'law, which is close to the distribution of the most flows. It is an empirical law formulated using mathematical statistics. It is notable that many types of data studied in the physical and social sciences can be approximated with a Zipfian distribution, which is one of a family of related discrete power law probability distributions. Formally, let:

- N: the number of elements;
- k: their rank;
- s: the value of the exponent characterizing the distribution.

Zipf's law then predicts that out of a population of N elements, the frequency of elements of rank k, $f(k; s, N)$, is as follows:

$$f(k; s, N) = \frac{1/k^s}{\sum_{n=1}^{N}(1/n^s)} \tag{1}$$

For most flows, the exponent characterizing the distribution s is between $1 \sim 2$.

Next, we introduce the Riemann Zeta Function to compute the denominator of $f(k; s, N)$;

$$\zeta(s) = \sum_{n=1}^{\infty}(1/n^s) \tag{2}$$

When $s = 1$, $\zeta(1) = \infty$. We next introduce the finite partial sums of the diverging harmonic series H_n:

$$H_n(N) = \sum_{n=1}^{N}(1/n) \tag{3}$$

We then have:

$$\sum_{n=1}^{N} \frac{1}{n} = \ln N + \gamma + \varepsilon_N \tag{4}$$

where γ is the Euler–Mascheroni constant and $\varepsilon_N \sim \frac{1}{2N}$ which approaches 0 as N goes to infinity.

We next focus on $f(k; 1, N)$:

$$f(k; 1, N) = \frac{1}{k(\ln N + \gamma + \varepsilon_N)} \tag{5}$$

Generally, we set $k = 200$, $N > 200000$, $f(200; 1, 200000) < 0.001$. This means that the flow that ranks 200th occupies less than 0.1% of all traffic. Moreover, the larger the value of s, the smaller the function value.

Next, we calculate the percentage of the top K flows:

$$\begin{aligned} F(k; 1, N) &= \sum_{k=1}^{K} f(k; 1, N) \\ &= [\sum_{k=1}^{K}(\frac{1}{k})] \frac{1}{(\ln N + \gamma + \varepsilon_N)} \\ &= \frac{(\ln K + \gamma + \varepsilon_K)}{(\ln N + \gamma + \varepsilon_N)} \end{aligned} \tag{6}$$

We can thus determine that the top 200 flows occupy 48.99% of the traffic.

When $s = 2$, $\zeta(2) = \pi^2/6$:

$$f(k; 2, N) \geqslant \frac{6}{\pi^2 k^2} \tag{7}$$

We use the same parameters, $f(200; 2, 200000) < 1.6 \times 10^{-5}$; the flow behind is smaller compared with the situation where $s = 1$. Under these circumstances, we can tell that top 200 flows are larger.

Accordingly, the top-k part in the system can extract flows that amount to more than 48.99% of the traffic.

3.2 Sketch Part Bound Analysis

Before analyzing the bound of the sketch part, we first need to declare the initial condition. The sketch part obtains the final results by recovering the data using the information of the distribution which generated by the data in the sketch. What we want to know is how much data is required to form an acceptable normal distribution.

The sketch part maps every flow to a specific space through the use of a hash algorithm. For the space of every bit, the data located there forms an approximate normal distribution, while later operations area based on the features of the distribution. The amount of data apparently has a significant impact on the results.

For the sketch part, we analysis its bound using statistics. Since the sketch structure is in fact a kind of statistical structure, we analyze the data and compute its distribution; subsequently, we obtain the information we want by computing its features. Accordingly, in order to determine the effect that can be obtained by the sketch part, we introduce the minimum sample size calculation method. Let:

- T: The number of the sample;
- σ^2: Sample variance;
- E: Sampling error;
- $Z_{\alpha/2}$: Confidence of the sample.

$$T \approx \frac{(Z_{\alpha/2}^2 \sigma^2)}{E^2} \tag{8}$$

It is worth noting that the equation is designed based on the standard normal distribution. That means $\sigma^2 < 1$. When $Z = 1.64, E = 10\%, \sigma^2 = 0.25$, we can get $T = 67$. The parameter c in the sketch part is same as the number of the sample T. Thus, the sketch part can obtain the error E with the equation above.

In order to compute the error rate, we need to normalize the original distribution to a normal distribution. Moreover, we also need to that we have 104 bits for flowkey, and each of these bits is related to a normal distribution. When these error rates overlap, we need to decrease the every error rate to a specific level to obtain an acceptable result. However, when the error rate we require is less than 1%, we need more than $100 \times T$, which is calculated when $E = 10\%$. We then need to expand the sketch part, which is the parameter c. Thus, we need that allows us to obtain enough data when space is lacking.

4   Data Augmentation Algorithm

In this section, we introduce the data augmentation algorithm that is implemented in order to save resources and maintain the approximate results with the original system in the control plane.

We find that the distribution of flows in the sketch part is approximate to the *Normal Distribution*. On this basis, we decide to use a small scale of data to construct a larger scale of data through using the *Data Augmentation Algorithm*.

There are many solutions for expanding the scale of a dataset, and many papers have proposed efficient new algorithms to solve this challenge in the machine learning area. These methods deal with multidimensional data before the training set is trained; however, what we are dealing with is simple one-dimensional data, where the only feature is the statistical parameters. We do not need such complex methods because expanding the dataset based on the distribution is precisely our focus.

4.1   Ideal Preliminary Algorithm

We consider the situation in which the data in the sketch part strictly follows the normal distribution. We use the standard normal distribution to reconstruct the data. First, we build a set that follows standard normal distribution $N(0, 1)$, $S_0 = \{X | X \sim N(0, 1)\}$. Second, we can obtain the set of the small-scale data in the sketch part, $S_1 = \{X | X \sim N(\mu, \sigma)\}$. Moreover, in the sketch part, the parameter c is related to the standard deviation of the data. When the data scale expands, the standard deviation is inversely proportional to it. Accordingly, we use τ to denote the times that we expand. Let $S_2 = \phi$. Next, for every $X \in S_0$, $X' = (\sigma/\tau)X + \mu/\tau$, $S_2 = S_2 \cup \{X'\}$. We posit that S_2 is a Reasonable Normal Expansion of S_1. What we simulate is the traffic being hashed to the τc boxes while keeping the distribution, as the system really provides τc boxes for every bit recorded.

However, if our input data set does not form an approximate normal distribution, some flows will influence the feature and cause bumps and depressions on the curve of the data. This method would not extract them, but will rather the errors and disperse them to a larger scale of the dataset; this makes it difficult to eliminate the effects of errors. In light of this, we need to preprocess the dataset and make sure that the set can form a smooth normal curve.

Algorithm 1. Ideal preliminary algorithm.

Require: The set of origin data S_1, The scale of data sum;
Ensure: The set of expanded data, S_2;
  1: Generate a set S_0 that has sum elements and follows standard normal distribution and an empty set S_2;
  2: Calculate the amount of data sum', average value μ and standard variance σ of set S_1;
  3: For $\forall x \in S_0$, $x' = (sum'\sigma/sum)x + sum'\mu/sum$, $S_2 = S_2 \cup \{x'\}$;
  4: **return** S_2.

4.2   Preprocessing of the Dataset

The purpose of preprocessing the dataset is to decrease the adverse effects caused by a few dissonant flows. The target is a dataset can form a smooth distribution curve. One of the parameters of the data that deviates from the distribution is the *Standard Deviation*. We first obtain the mean and standard deviation through fitting the origin data, then check the departure of data values from the fitting curve. Next, we decrease this record to the normal level by setting the threshold λ. At the same time, we note the reduced value in another residual sketch, which we call the *excluded sketch*. After that, we conduct the fitting again, repeating the above steps until the data meets the threshold requirement.

Finally, we obtain a dataset that can form a relatively smooth distribution and an excluded sketch. The set will be input into the **Algorithm 1** for expansion. We next deal with the excluded sketch to recover the information that we cut.

For the excluded sketch, most of the space is filled by zeroes. Because of the characteristic of SketchLearn structure, we can recover the flowkey by checking the same location of all bits: if the location of the specific bit is zero, we set this bit to 0, and set it to 1 otherwise, after which we obtain a flowkey f. We then hash it and check whether the value fits the location: if it fits, we extract the flow with the smallest of all the l records; otherwise, we set the bit related to the smallest 0 and repeat the operations, until the excluded sketch is filled completely with zeroes.

Algorithm 2. Preprocessing of data set.
___
Require: The residual sketch $r[l][c]$, the threshold λ, the range of definition R ;
Ensure: The optimal residual sketch, $r'[l][c]$, The excluded flow list L_e;
    for $0 \leqslant i < l$, compute the mean value $\mu[i]$ and standard deviation $\sigma[i]$ of $r[i]$;
    for $0 \leqslant j < c$; predict the probability of the value for that location $p[i][j]$;
    for $0 \leqslant j < c$; if the quantity of the data in the scale $(r[i][j] - R) - (r[i][j] + R)$ which is $Q[i][j] > \lambda p[i][j] \times c$, remove it to the excluded sketch the same location $E[i][j]$.
    extract the flows from the excluded sketch and merge them into the excluded flow list L_e;
    return $r'[l][c]$, L_e.

---

5   Implementation of Optimized HybridSketch

We implement the data plane of Optimized HybridSketch in P4 code, the control plane in C code. In the data plane, we combine the top-k part and sketch part using Hashpipe and SketchLearn. In the control plane, we collect the results of the two parts in order to produce the large flow list and the flow distribution. As for the distribution, we deal with it using the data augmentation algorithm. We then try to run applications to evaluate the system.

We set a bit of status for every packet: the bit equals 0 when the packet enters into the system, while if the packet is expelled by the top-k part, we set it to 1, and let the packet enter into the sketch part. This process is illustrated in **Algorithm 3**.

Algorithm 3. The data plane framework of Optimized HybridSketch.
___
Require: The set of packets, P_n;
Ensure: The set of large flows, L_n ; Bit-level counter distributions $N\{p, \sigma\}$;
1: For $P_i \in P_n$, insert P_i into the first stage of the top-k part;
2: Track the minimum flows and expel f_{min};
3: Insert f_{min} into the sketch part;
4: Hash for f_{min} and update counters according to the flowID of f_{min};
5: Collect the large flow in the top-k part f_i, $L_n = f_i \cup L_n$;
6: Extract large flow f_j and maintain the normal distribution in the sketch part;
7: $L_n = f_j \cup L_n$ and compute bit-level counter distributions $N\{p, \sigma\}$
8: **return** L_n and $N\{p, \sigma\}$.

For the control plane, moreover, we combine the large flows in the top-k part with those extracted by the sketch part. The flow contribution is computed by the data in the residual sketch. Then we process the data by data augmentation algorithm to obtain a lager scale of data to analysis the feature which the applications need. This process is illustrated in **Algorithm 4**.

Algorithm 4. The control plane framework of Optimized HybridSketch.

Require: The residual sketch $r[l][c]$, the threshold λ, the range of definition R, the expansion rate β;

Ensure: The optimal residual sketch, $r'[l][\beta c]$, the excluded flow list L_e;

1: Preprocessing of data set $r[l][c]$, get the optimal residual sketch, $r'[l][c]$ and the excluded flow list L_e;

2: for $0 \leqslant i < l$, expand the dataset $r'[i]$ through the ideal preliminary algorithm, get the final dataset $r'[l][\beta c]$;

3: **return** $r'[l][\beta c]$ and L_e.

---

6 Performance Evaluation of the System

In this section, we present our experimental results on a real-world dataset. We evaluate the *accuracy* of the frequency estimation queries maintained by Optimized HybridSketch. Moreover, we also compared it with the origin simple version, which does not apply the data augmentation algorithm, and SketchLearn, which applies enough memory on the data plane.

6.1 Experimental Setup

Dataset: We use an hour IP-packet trace in **CAIDA 2018**[13] to evaluate the system. The IP-packet has 20.55k flows and 1563570347 IPv4 packets with 1398437622398 bytes. This dataset is similar to a **Zipf** distribution of skew 1.0.

Parameter Settings: We fix the total memory M at 64KB: 16KB for the top-k part M_k and 48KB for the sketch part M_s. We set $k = 200$, $c' = 120$ for $z = 1.0$. For the solutions, we choose SketchLearn in 64KB (SL-64KB), original HybridSketch in 64KB (HS-64KB), original HybridSketch in 640KB (HS-640KB) and optimal HybridSketch in 64KB (HS-optimized-64KB). We also set the threshold $\lambda = 1.2$, the range of definition $R = 8$, the expansion rate $\beta = 10$ in optimal HybridSketch. This allows us to clearly observe the improvements brought about by the optimal algorithm.

Queries: We evaluate the *frequency estimation queries*: in short, given a data item, we estimate its frequency. The queries are different from those in [4], which were obtained by sampling the data items based on their frequencies; instead, we estimate every flow once in the specific trace. We also perform an evaluation of the top-k frequent items query: this involves determining the top-k most frequent items, where k is given by the top-k part.

Metrics: We consider the following metrics:

- Recall: the ratio of true instances reported;
- Precision: the ratio of reported true instances;
- Average relative error (ARE): $\frac{1}{n} \sum_{i=0}^{n-1} \frac{|\widehat{v_i}-v_i|}{v_i}$, where v_i is the true value of i and $\widehat{v_i}$ is the estimate of i[1].

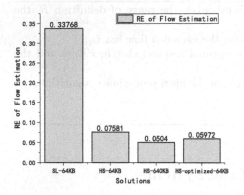

Fig. 1. The RE of FE under the different solutions.

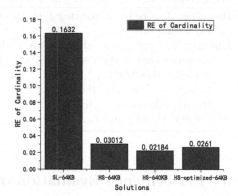

Fig. 2. The RE of Cardinality under the different solutions.

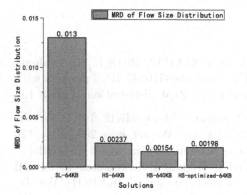

Fig. 3. The MRD of Flow Size Distribution under the different solutions.

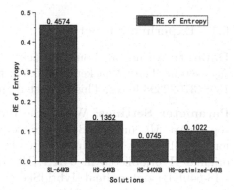

Fig. 4. The RE of Entropy under the different solutions.

6.2   Performance Evaluation of Optimal System

In this part, we compare the performance of the original and optimal systems through the use of applications based on the residual sketch, in order to prove the effect of data augmentation algorithm.

Specifically, we estimate the *FlowEstimate*, *Cardinality* and *Entropy* of the system; this is done in order to complete the applications based on the information in the residual sketch. In addition to those mentioned above, we also consider the *MRD* of the *Flow size distribution*.

Figure 1 presents the ARE of Flow Estimation for the four solutions, HS is more effective than SL under the same memory size conditions, while the HS-optimized-64KB exhibited better performance than HS-64KB, achieving an error rate nearly 1% higher than HS-640KB. In short, it has basically achieved the function of optimization.

Figure 2, 3 and 4 also illustrate that the HS-optimized-64KB achieved better results than HS-64KB, close to HS-640KB.

We can therefore conclude that HS-64KB is more effective than SL-64KB; at the same time, along with the 10× improvement in memory relative to the original HybridSketch (HS-640KB), it provided 30% − 50% promotion. The optimal system (HS-optimized-64KB) achieves about 80% of the effect of HS-640KB. In essence, it has achieved the same effect as original HS using around 10x the memory size.

7   Related Work

SketchVisor [2], Elastic Sketch [3] and ASketch [4] choose two algorithms to implement the measurement functions, making the system running more accurately and efficiently. While they mixed the algorithms for different goals, they did not discuss the method to optimized the system, which is the main goal of our work.

The SketchVisor [2] specially designed a fast path for the quick processing of the trace. Its goal is to improve the throughput of the system. The flow entering the fast path can use the statistics of the fast top-k algorithm to improve the speed. In the normal path, it deploys sketch algorithms. The SketchVisor is a hybrid architecture, but the problem is that not all packets will come into the fast path. Moreover, although the fast path can achieve higher packet rate, it is less precise than the normal path, meaning that accuracy degrades significantly when the majority of packets go to the fast path [6]. This approach has a certain universality, but does not propose a systematic hybrid method and resource allocation scheme, and also does not put forward certain requirements on the system; instead, it only adds extra 8KB for the fast path while maintaining the memory size that the normal path deployed [2]. By contrast, our system is hybrid under certain storage conditions to reduce the memory consumption.

The Elastic Sketch [3] system is divided into a light part and a heavy part, the heavy part uses a voting algorithm, while the light part adopts the structure of the CM sketch. The main goal is to improve the bandwidth of the system. From a general point of view, this is a combination of the two algorithms; however, the paper did not mention the ideas of mixing the methods, but instead simply uses the voting mechanism to filter out flows, making it more like an auxiliary method. Furthermore, CM sketch is a kind of structure that consumes more memory. The method of compression was specifically mentioned in the paper [3], and was detailed in the proof. However, the compression operation will affect the quantitative precision, which is inevitable; in addition, when the elephant flows conflicts in the voting process, there will be a large loss of precision. By contrast, our system does not suffer from problems like compression and stream conflict.

ASketch [4] adds a filter on the basis of the CM sketch that filters the elephant flows and conducts flow updates in the CM sketch. It focus on the precision of the system. The filter is a similar top-k algorithm, so it is also a kind of the combination of top-k and the sketch; however, although these authors' key point was to improve the throughput and accuracy of the whole system, they did not come up with a specific limit for storage requirements. Moreover, the basis of ASketch is the CM sketch, and as the precision of the CM sketch is not particularly outstanding, the ASketch only improved the accuracy on the basis of the CM sketch. However, the accuracy of SketchLearn used in our system is higher than most of the sketch structures, and the improvement effect will be better on this basis.

In summary, these solutions mixed measurement algorithms for different goals, they did not propose the idea on the memory space. Therefore, they are all not memory-saved.

8   Conclusion

In this paper, we present Optimized HybridSketch, which is an optimized algorithm to apply in flow measurement. By combining the top-k and sketch methods, analyzing the characteristics of the trace, modeling and formalizing the allocation in question and designing the method of allocation for the two parts, HybridSketch can achieve better results than similar systems (e.g. [2–4]) while using less memory and supporting more applications (e.g. entropy and flow distributions). We present an algorithm that is optimized for achieving better results than the original one, attaining performance basically equal to that of HS when using 10× the memory size without any optimization. In addition, we also evaluate the system performance; our results demonstrate that, compared with other methods, Optimized HybridSketch is more precise and more memory-saved. The analysis and algorithms we propose make the system more efficient.

References

1. Huang, Q., Lee, P. P., Bao, Y.: Sketchlearn: relieving user burdens in approximate measurement with automated statistical inference. In Proceedings of the 2018 Conference of the ACM Special Interest Group on Data Communication (SIGCOMM '18). ACM, New York, NY, USA, pp. 576–590. ACM (2018)
2. Huang, Q., et al.: SketchVisor: robust network measurement for software packet processing. In Proceedings of the Conference of the ACM Special Interest Group on Data Communication (SIGCOMM '17). ACM, New York, NY, USA, pp. 113–126. ACM (2017)

3. Yang, T., et al.: Elastic sketch: adaptive and fast network-wide measurements. In Proceedings of the 2018 Conference of the ACM Special Interest Group on Data Communication (SIGCOMM '18). ACM, New York, NY, USA, pp. 561–575. ACM (2018)

4. Roy, P., Khan, A., Alonso, G.: Augmented sketch: faster and more accurate stream processing. In Proceedings of the 2016 International Conference on Management of Data (SIGMOD '16). ACM, New York, NY, USA, 1449–1463. ACM (2016)

5. Sivaraman, V., Narayana, S., Rottenstreich, O., Muthukrishnan, S., Rexford, J.: Heavy-hitter detection entirely in the data plane. In Proceedings of the Symposium on SDN Research (SOSR '17). ACM, New York, NY, USA, 164–176. ACM (2017)

6. Liu, Z., et al.: Nitrosketch: robust and general sketch-based monitoring in software switches. In Proceedings of the ACM Special Interest Group on Data Communication (SIGCOMM '19). Association for Computing Machinery, New York, NY, USA, pp. 334–350. ACM (2019)

7. Li, Y., Miao, R., Kim, C., Yu, M.: Flowradar: a better netflow for data centers. In Proceedings of the 13th Usenix Conference on Networked Systems Design and Implementation (NSDI'16). USENIX Association, Berkeley, CA, USA, pp. 311–324. ACM (2016.)

8. Cormode, G., Muthukrishnan, S.: An improved data stream summary: the count-min sketch and its applications. J. Algorithms. **55**(1), 58–75 (2005)

9. Zhang, Y., Zhu, H., Bao, N., Zhang, L.: Comparative analysis of different sketch methods in practical use. In 2018 Sixth International Conference on Advanced Cloud and Big Data (CBD), Lanzhou, pp. 124–129 (2018)

10. Yang, T., Gao, S., Sun, Z., Wang, Y., Shen, Y., Li, X.: Diamond sketch: accurate per-flow measurement for big streaming data. IEEE Trans. Parallel Distrib. Syst. **30**(12), 2650–2662 (2019)

11. Liu, Z., Manousis, A., Vorsanger, G., Sekar, V., Braverman, V.: One sketch to rule them all: rethinking network flow monitoring with univmon. In: Proceedings of the 2016 ACM SIGCOMM Conference (SIGCOMM '16). ACM, New York, NY, USA, pp. 101–114. ACM (2016)

12. Yang, T., et al.: HeavyKeeper: an accurate algorithm for finding top-k elephant flows. IEEE/ACM Trans. Netw. **27**(5), 1845–1858 (2019)

13. Caida Anonymized Internet Traces 2018 Dataset, (2019). http://www.caida.org/data/passive/passive_dataset.xml

14. Canini, M., Fay, D., Miller, D.J., Moore, A.W., Bolla, R.: Per flow packet sampling for high-speed network monitoring: first international communication systems and networks and workshops. Bangalore **2009**, 1–10 (2009)

15. Kandula, S., Mahajan, R.: Sampling biases in network path measurements and what to do about it. In: Proceedings of the 9th ACM SIGCOMM conference on Internet measurement (IMC '09). ACM, New York, NY, USA, pp. 156–169. ACM (2009)

16. Ben Basat, R., Einziger, G., Friedman, R., Luizelli, M.C., Waisbard, E.: Constant time updates in hierarchical heavy hitters. In: Proceedings of the Conference of the ACM Special Interest Group on Data Communication (SIGCOMM '17). ACM, New York, NY, USA, pp. 127–140. ACM (2017)

17. Metwally, A., Agrawal, D., El Abbadi, A.: Efficient computation of frequent and top-k elements in data streams. In: Eiter, T., Libkin, L. (eds.) ICDT 2005. LNCS, vol. 3363, pp. 398–412. Springer, Heidelberg (2004). https://doi.org/10.1007/978-3-540-30570-5_27

18. Manku, G.S., Motwani, R.: Approximate frequency counts over data streams. In: Proceedings of the 28th international conference on Very Large Data Bases (VLDB '02), VLDB Endowment, pp. 346–357 (2002)
19. Kumar, A., Sung, M., Xu, J., Wang, J.: Data streaming algorithms for efficient and accurate estimation of flow size distribution. ACM SIGMETRICS Perform. Eval. Rev. 32(1), 177–188 (2004)
20. Harvey, N.J., Nelson, J., Onak, K.: Sketching and streaming entropy via approximation theory. In: 2008 49th Annual IEEE Symposium on Foundations of Computer Science, pp. 489–498. IEEE (2008)

An Overlapping Community Detection Algorithm Based on Triangle Reduction Weighted for Large-Scale Complex Network

Hanning Zhang[1,2,5], Bo Dong[3,4,5](✉), Boqin Feng[1,5], and Haiyu Wu[1,5]

[1] School of Computer Science and Technology, Xi'an Jiaotong University,
Xi'an 710049, China
zhanghn@stu.xjtu.edu.cn

[2] Shaanxi Province Key Laboratory of Satellite and Terrestrial Network Technology
Research and Development, Xi'an Jiaotong University, Xi'an 710049, China

[3] School of Continuing Education, Xi'an Jiaotong University, Xi'an 710049, China
dong.bo@xjtu.edu.cn

[4] National Engineering Lab for Big Data Analytics Xi'an Jiaotong University,
Xi'an 710049, China

[5] Xi'an Network Computing Data Technology Co., Ltd., Xi'an 710049, China

Abstract. In this digital age, dramatically developed internet makes the data of complex networks appear an explosive growth, which aggrandizes the importance of multilevel community detection used in large-scale complex networks. Nowadays, there are so many community detection algorithms that could perform well on accuracy. However, none of them has an expected function to handle the time increasing problem which is caused by the inflated network scale. Hence, we propose An Overlapping Community Detection Algorithm based on Triangle Reduction Weighted for Large-scale Complex Network (TRWLPA). It consists of two main steps: 1) Transforming the original network to a small-scale triangle reduction network. This network could not only dramatically reduce the running time of community detection, but recover the original network structure by the inverse transformation. Moreover, the scale of the triangle reduction network could be controlled by setting the iteration times. 2) Doing the multi-label propagation on the reduced networks where the weight of each node is the number of initial nodes it contains. The experiments illustrate that the TRWLPA algorithm significantly reduces the running time of community detection on Youtube, DBLP, and LiveJournal datasets. Particularly, comparing with the MOSES algorithm, it achieves 98.1% running time reduction on the Youtube dataset. Furthermore, our algorithm performs well on both modularity and Normalized Mutual Information measure (NMI).

Keywords: Triangle reduction · Multi-label propagation · Complex networks · Community detection

© Springer Nature Switzerland AG 2020
M. Qiu (Ed.): ICA3PP 2020, LNCS 12452, pp. 627–644, 2020.
https://doi.org/10.1007/978-3-030-60245-1_43

1 Introduction

With the fast development of the internet, the data of complex networks increases rapidly. Community detection in large-scale complex networks has become a requisite process during network data mining. Therefore, how to efficiently discover overlapping communities in the large-scale complex network has become the current research hotspot.

As the scale of networks inflated, the increased running time of the algorithm cannot fit the time requirement on application. Due to this issue, we propose the TRWLPA algorithm. Firstly, the triangle reduction network is proposed. On the premise of guaranteeing the quality of the original network community structure, the reduction of network scale is carried out. Then the multi-label overlapping communities with weight are found on the reduced network. Finally, the inverse reduction is carried out to restore the community structure to the original network, Fig. 1 illustrates the overall processes.

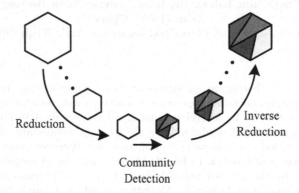

Fig. 1. The structure of TRWLPA algorithm, where white hexagons represent the processes of reduction and colored hexagons stand for the processes of inverse reduction. Moreover, community detection is implemented when the network reaches the defined sizes.

The experimental results show that TRWLPA could keep a higher accuracy while using less running time.

The structure of this paper is: Sect. 1 introduces the research background; Sect. 2 introduces related algorithms and metrics; Sect. 3 introduces the theory, algorithm, and analyses of triangle reduction network; Sect. 4 introduces the details of TRWLPA, which includes setting the weights of nodes, the processes of the algorithm, and analyses of the complexity of algorithm; Sect. 5 introduces the datasets and platform of our experiment, and analyses of the experimental results; Sect. 6 conclusions.

2   Related Work

2.1   Overlapping Community Detection

Generally, each node is attributed only to one community. However, nodes lying at the boundary between communities may make it be difficult to assign to one community or another, based on their connection with other nodes. In this situation, it is rational to consider these nodes as belonging to more communities, so-called overlapping communities. Detecting these overlapped communities could give us more characteristics of the network. Furthermore, it could help us to predict the evolution of the community.

2.2   Modularity and NMI

Modularity [12] measures the goodness of the division of the network with respect to the whole network. Higher modularity indicates better partitioning of the network. The function of moudularity Q_{ov} is given as:

$$Q_{ov} = \frac{1}{2m} \sum_{ij} (A_{ij} - \frac{d_i d_j}{2m}) \delta(C_i, C_j) \tag{1}$$

where A is the adjacency matrix, m is the total number of edges, d_i is the degree of the node i, and $\delta(C_i, C_j) = 1$ if node i and node j belong to the same community, otherwise $\delta(C_i, C_j) = 0$.

NMI is a measure of similarity between the partitions based on the information theory.

$$\text{NMI}(A, B) = \frac{-2 \sum_{i=1}^{C_A} \sum_{j=1}^{C_B} N_{ij} log(\frac{N_{ij} N}{N_{i_s} N_{j_s}})}{\sum_{i=1}^{C_A} N_{i_s} log(\frac{N_{i_s}}{N}) + \sum_{j=1}^{C_B} N_{j_s} log(\frac{N_{j_s}}{N})} \tag{2}$$

where N is the confusion matrix in which rows correspond to the 'real' communities and columns correspond to the 'observed' communities. C_A and C_B represent the number of real communities and observed communities, respectively. N_{ij} is the number of nodes in the real community i that appear in the observed community j. N_{i_s} and N_{j_s} represent the sum over the ith row of matrix N and jth column of the matrix N, respectively. The value of NMI is from 0 to 1. Specifically, NMI equals to 1 if the estimated community matches the reference community and 0 in the opposite case.

2.3   Label Propagation Algorithm

Depend upon the main idea of different community detection algorithms, there are five strategies: graph partition [6,11], agglomerative clustering [2,12,16], genetic algorithm [9,13,14], label propagation, and semantic partition [1,3,24]. In these, label propagation is one of the most widely used because of its near-linear time complexity and high-accuracy community classification. In the beginning, it adds a community label for each node in the network. In the process of propagation, the node obtains the most frequent label from its neighbor nodes, and finally divides the nodes with the same label into the same community. [15] first proposed Label Propagation Algorithm (LPA). In the beginning, each vertex is initialized with a unique label. Subsequently, the algorithm progresses from one iteration to another and at every iteration, each node updates its current label according to a given protocol until no nodes change any more. The time complexity of each iteration is O(m), where m is the number of edges.

[19] extended LPA to the field of overlapping community detection by introducing the mechanism of multiple labels, and proposed Speaker-listener Label Propagation Algorithm (SLPA). The time complexity of SLPA is O(Tn), where T represents the number of label propagation iterations set in advance by the user, and n represents the total number of nodes. Based on SLPA, [5] proposed Weighted Label Propagation Algorithm.

[4] improved the LPA algorithm and the proposed Community Overlap Propagation Algorithm (COPRA). [20] algorithm uses the Markov clustering algorithm to improve the LPA algorithm. The time complexity of the improved algorithm is $O(m)$, where m represents the number of edges. [18] improved the Labelrank algorithm which can not only deal with the weighted graph but also the directed graph. At the same time, this algorithm can solve the community detection problem of dynamic complex networks. [17] proposed a balanced multi-label propagation algorithm to find overlapping communities in complex networks. The time complexity of each iteration is $O(nlogn)$, where n is the number of nodes.

3   Triangle Reduction Network

The main idea of the triangle reduction network is that, in an undirected network, three nodes connected with each other constitute a triangle, and we fuse the three nodes forming a triangle into a composite node. Through repeated iterative fusion, a large-scale network will be transformed into a small-scale triangle reduction network.

3.1   Triangle Reduction Network Structure

We find that triangle topology has strong community nature; Most networks are rich in triangle topology; the Triangle reduction process is sustainable. These three factors are the basis for constructing a triangle reduction network structure.

3.2   Triangle Topology Has Strong Community Nature

The three nodes in the triangle topological structure can reach each other and are a fully connected structure. Therefore, it is in line with the requirements of community definition to partition them into the same community. The SCP algorithm proposed by [7]. When k is equal to 3, the formed k-cliques and k-clique communities are the triangular topological structure, so the triangular topological structure has strong community nature, and the fusion of three nodes in the triangular topological structure can maintain the structure of the initial network.

3.3 Most Networks Are Rich in Triangle Topology

[22] studied 230 complex networks in the real world, including social networks, collaborative networks, and information networks, etc. in this paper, a triangle participation ratio (TPR) index is proposed to measure the quality of community structure. Its calculation formula is as follows:

$$f(S) = \frac{|u : u \in S, (v,w) : v,w \in S, (u,v) \in E, (u,w) \in E, (v,w) \in E \neq \varnothing|}{n_s} \quad (3)$$

where S represents a set of nodes, representing the nodes of the community which needs to be evaluated; E represents the set of all edges in the network; $n_s = |S|$ indicates the number of nodes in set S; $f(S)$ indicates the proportion of nodes in triangle topology is set S to the total nodes. This research shows that the TPR index accounts for a high proportion of the community structure of the real-world network, which fully shows that there are a lot of triangle topological structures in the community structure of the network. [23] directly pointed out that in more than 85% of social information networks, the number of triangle topologies is at least equal to the number of edges.

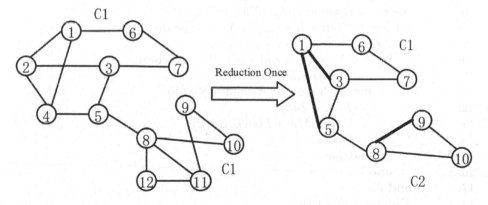

Fig. 2. Result of using triangle reduction once on polygon network, where C1s are the edges before reduction, C2 is the edge after reduction process.

3.4 Triangle Reduction Process Is Sustainable

In the process of triangle reduction, with the increase of the number of iterations, the polygons in the initial network will gradually move towards the triangle topology, providing a new reduction source for the subsequent iterative reduction, so the triangle reduction process is sustainable, which also ensures that the large-scale network can be transformed into a smaller network. An example of triangle reduction network is shown in Fig. 2. The bold edges in the figure represent the newly generated edges after reduction. In community 1 before reduction, nodes 1, 2, 3, 6 and 7 form a Pentagon. When nodes 1, 2 and 4 form a triangle topology structure and are reduced to node 1, two sides of the original Pentagon become a new one, and the Pentagon becomes a quadrilateral. Similarly, the quadrilateral formed by nodes 8, 9, 10 and 11 in community 2 becomes a triangle.

3.5   Triangle Reduction Algorithm

Triangle reduction is an iterative process. In each iteration, it traverses the triangle topology of the network, and merges the three nodes in each triangle into a composite node. The reduced network is the input of the next iteration, and its reduction algorithm is shown as follows:

Algorithm 1. The algorithm of the triangle reduction

1: **Input:** $G=(V,E)$:large-scale network; T_r: the iteration number of reduction
2: **Begin**
3: **for** $t = 1 : T_r$ **do**
4:     NodeSort = Nodes.sortDegree();
5:     **for** $i = 1 : n$ and $F(i) == true$; **do**
6:         $Nebrs_i = NodeSort(i).getNbs()$;
7:         **for** $j = 1 : Nebrs_i.len$ and $F(j) == true$ **do**
8:             $Nebrs_j = NodeSort(j).getNbs()$;
9:             **if** $\exists k \in Nebrs_i.len$ and $F(k) == true$ **then**
10:                 $s = Min(i, j, k)$
11:                 **for** each $u \in \{i, j, k\}$ and $u \neq s$ **do**
12:                     $F(u) = false$
13:                     $UpdateMergeList(L, i, j, k)$;
14:                     **break**;
15:                 **end for**
16:             **end if**
17:         **end for**
18:         $Update(NodeSort)$;
19:     **end for**
20:     $Update(Nodes)$;
21: **end for**

---

where $F[n]$ represents the presence flag bit of n nodes, $F(i)$ represents the flag bit of node i, and the flag bit of each node is initialized to true, indicating that the node exists in the current network and has not been fused. In triangle reduction, once a node is fused into other nodes, its flag position is false, which means that the node has been removed from the current network. The function of setting flag bit is to distinguish nodes in different states during reduction. L is a list array, and each list in the array represents a collection of nodes fused into a composite node; $SortDegree()$ indicates that nodes in the network are sorted in ascending order according to their degrees, so that nodes appearing in fewer triangle topologies can be fused in priority; $getnbs()$ indicates the collection of neighbor nodes of corresponding nodes; $Updatemergelist(L, i, j, k)$ indicates that the information fused by i, j and k is updated to the list array L; $Update(Nodes)$ indicates that the network represented by Nodes is updated according to the information of flag bit $F[n]$, and the main function is to transfer the neighbor node of the disappeared node to the neighbor node of its corresponding composite node.

3.6   Complexity Analysis of Triangle Reduction

Spatial Complexity Analysis. Triangle reduction process uses adjacency matrix storage network, where the spatial complexity is $O(nd_{avg})$ represents the average degree of nodes; The space occupied by list array L is $O(n)$; The spatial complexity of other variables involved in the process is $O(1)$, thus the spatial complexity of triangle reduction process is $O(nd_{avg})$.

Time Complexity Analysis. In the initial stage of triangle reduction, the time complexity is $O(1)$; in the reduction stage, the outer loop iterates T_r times. The inner loop traverses all nodes every time, and the operation on each node is a constant time complexity, so the time complexity of the reduction stage is $O(nT_r)$. The time complexity of triangle reduction process is $O(nT_r)$.

4   TRWLPA Algorithm for Reduction Network

Based on the triangle reduction network proposed in Sect. 2, we use the weighted multi-label propagation method to detect overlapping communities, and propose the TRWLPA community detection algorithm. The algorithm includes two steps: node weight calculation and community detection.

4.1   Node Weight Calculation

The reason why node weight is introduced is that there are composite nodes with different fusion degrees in the reduction network. For example, there are three nodes u, v and w in the network, where u is not a composite node, v is a composite node with three nodes, w is a composite node with five nodes. The higher the degree of integration of nodes, the larger the scale of the community which represents the internal nodes meaning that the greater the influence of this node in the process of tag propagation. Therefore, the labels of nodes with high fusion degree should have a large weight in the process of label propagation. In this paper, the weights of nodes in the reduction network are defined as follows:

$$w(v) = \begin{cases} 1 & v \text{ is not a composite node} \\ x & otherwise \end{cases} \tag{4}$$

where $w(v)$ represents the weight of node v and x is initial number of network nodes included in v. We can get:

$$\sum_{v=1}^{N'} w(v) = N \tag{5}$$

where N stands for the total number of nodes in the initial network, N' is the total number of nodes in the reduction network.

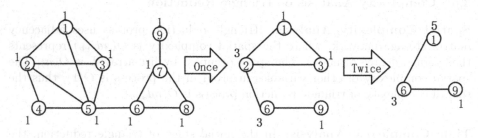

Fig. 3. In the initial stage of the figure, the weights of all nodes are 1. After a triangle reduction, nodes 2 and 6 become composite nodes, and the weights become 3. After the second triangle reduction, nodes 1, 2 and 3 are fused into composite node 1, and the weight becomes 5.

4.2 Community Detection

The specific steps of TRWLPA algorithm are shown in Algorithm 2, where Step 3.2 could be explained as follows: Define A as the community label set of all neighboring nodes received by the current node, that is, $A = \{a1, a2, L, a_k\}$. N_j is defined as the set of neighbor nodes of all a_j labels, that is, $N_j = \{i|c_i = a_j\}$, where c_i represents the community labels from neighbor nodes i. Define function $f(j)$ as follows:

$$f(j) = \sum_{i \in N_j} w_i \tag{6}$$

where w_i represents the weight of node i. We could get the maximum value J of $f(j)$ by using:

$$J = argmax(f(j)) \tag{7}$$

So that in this iteration, the community label selected by the current node is a_j.

Algorithm 2. The algorithm of the triangle reduction

step 1 Calculate the weights of nodes in the reduction network according to formula 4

step 2 Initialize a unique community label for every node in the reduction network

step 3 Traverse all nodes, and manipulate them as follows:

  step 3.1 Receive the most frequent community labels in the community label list from each neighboring node. If there are multiple cases, select one randomly

  step 3.2 For all received community labels, calculating the weight sum of the adjacent nodes corresponding to each label, then storing the weight and the largest community label in the community label list of this node. If there are multiple cases, select on as the label this iteration randomly.

step 4 If arrive the setting propagation iterations T_p, step into Step 5, otherwise Step 3

step 5 Traverse the community label list of each node, and remove the labels whose frequency is less than the label filtering threshold

step 6 According to the fusion node list array generated by triangle reduction, the community label is assigned to each node in the initial network

Step 2, in the above steps, is to initialize the community label. Step 3 and Step 4 constitute the community label propagation process. When the number of label propagation iterations is reached, label filtering is performed, that is, Step 5. Step 6 is an inverse reduction process.

TRWLPA algorithm is shown in Algorithm 3. Moreover, in order to make it easier to understand, the explanations of labels and methods are: $W[n]$ represents the weight of n nodes, $W(v)$ represents the weight of node v; $minNode()$ method represents the node with the smallest number in the node list; map stores the tuple composed of community label and node weight; $getNbs()$ method represents the set of neighbor nodes; $mostlabel()$ method represents the label with the most times in the node label set; $map.find(lm)$ indicates whether there is a tuple labeled lm in the map; $map.add(lm, W(n))$ means to add the tuple consisting of the weight of label lm and node n to the map; $map.labelWithMaxWeight()$ is the community label with the largest weight in the map; $Mem.add()$ respresents the community label that is added to the label set of the corresponding node.

Algorithm 3 MLPA Algorithm

1:    **Input:** G_r: reduced network of G; L: merged list of triangle reduction;

        T_p: the iteration number of propagation; α: label filtering threshold

    Begin

2: $[n, Nodes] = loadnetwork()$;

3: Stage 1: weight initialization

4: $W[n] = 1$;

5: **for** $i = 1 : L.len$ **do**

6:     $W(L(i).minNode()) = L(i).len$;

7: **end for**

8: Stage 2: label initialization

9: **for** $i = 1 : n$ **do**

10:    $Nodes(i).Mem = i$;

11: **end for**

12: Stage 3: evolution

13: **for** $T = 1 : T_p$ **do**

14:    $Nodes.ShuffleOrder()$;

15:    $Map < Label, Weight > map = \varnothing$;

16:    **for** $i = 1 : n$ **do**

17:       $Nebrs = Nodes(i).getNbs()$;

18:       **for** $j = 1 : Nebrs.len$ **do**

19:          $lm = Nebrs(j).mostLabel()$;

20:          **if** $map.find(lm) == true$ **then**

21:             $map(lm)+ = W(Nebrs(j))$;

22:          **else**

23:             $map.add(lm, W(Nebrs(j)))$;

24:          **end if**

25:          $lt = map.labelWithMaxWeight()$;

26:          $Nodes(i).Mem.add(lt)$;

27:       **end for**

28:     **end for**

29: **end for**

30: Stage 4: post-processing

31: **for** $i = 1 : n$ **do**

32:    remove $Nodes(i)$ labels seen with probility $< \alpha$;

33: **end for**

34: Stage 5: inversing reduction

35: **for** $i = 1 : L.len$ **do**

36:    $m = L(i).minNode()$;

37:    **for** $j = 1 : L(i).len$ **do**

38:       L(i)(j).label = m.label;

39:    **end for**

40: **end for**

41: **End**

42: **Output:** $Label_i (i = 1, 2, \ldots, n)$: label of all nodes

---

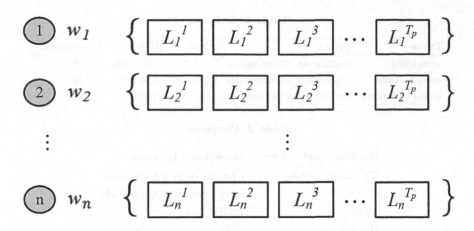

Fig. 4. It shows the way to store the labels during running the MLPA algorithm, where ws are the weights of each node, Ls are the labels stored in each node.

4.3   Complexity Analyses

Spatial Complexity Analysis The label storage structure of TRWLPA is shown in Fig. 4. Because of the need to store the community labels of the T_p iteration process, the spatial complexity of TRWLPA is $O(nT_p)$.

Time Complexity Analysis. In the stage of weight initialization, we need to traverse the merged list L generated by triangle reduction, so the time complexity is $O(L.len)$, where $L.len$ is less than n; In the stage of label initialization, we need to traverse all nodes in the reduction network, so the time complexity is $O(n)$; In the stage of label propagation, the outer loop needs to iterateT_p The time complexity is $O(nT_p)$; In the label filtering stage, it is necessary to traverse each label on each node, so the time complexity is $O(nT_p)$; In the inverse reduction stage, we need to traverse every node in the merge list L, and the worst-case time complexity is $O(n)$. The time complexity of the TRWLPA algorithm for reduction network is $O(n)$.

5   Experimental Analyses

5.1   Datasets and Platform

This section will introduce the datasets and experimental platform of algorithm experiment. As shown in Table 1, three SNAP datasets are used in the experiment. Among them, com-DBLP is the academic cooperation network of internal medicine researchers in the field of computer science. Com-YouTube is a social network established through video sharing based on YouTube website, and com-LiveJournal is the social network data based on a comprehensive social network

Table 1. Datasets

Dataset	Type	#Node	#Edge	#Community
com-DBLP	Undirected, Communities	317,080	1,049,866	13,477
com-Youtube	Undirected, Communities	1,134,890	2,987,624	8,385
com-LiveJournal	Undirected, Communities	3,997,962	34,681,189	287,512

Table 2. Plateform

Hardware and Software	Version and Information
Operating system	Ubuntu 16.04.2 LTS 64bit
CPU	Intel Xeon X5650 2.67GHz
Memories	40GB
Hard drive	1TB
g++	5.4.0
JDK	1.8.0_161

in Russia. All three datasets are undirected networks, and the community structure after deleting the communities with less than three nodes is given.

The type of dataset files in Table 1 are .txt. Each row is the edge represented by the source node and the target node. The source node and the target node are separated by the tab key. The hardware and software configuration of the experimental platform in this section is shown in Table 2.

5.2 Comparative Experiment and Parameter Setting

For the TRWLPA algorithm, we select SLPA [19], COPRA [4], MOSES [10] and GCE [8] to do comparative experiments. SLPA and CPPRA algorithms are based on the idea of label propagation. The difference is that, in COPRA algorithm, the maximum overlap should be preset, that is, a node belongs to several communities at most. Furthermore, each label on the node has a subordination coefficient, which is used to indicate how likely this node belongs to this community. SLPA algorithm does not limit the maximum overlap of nodes but selects nodes through the label filtering stage. MOSES algorithm is based on the idea of modularity optimization. It selects the seed node first, and then optimizes the global objective function by trying to add other nodes to the current community until the global objective function can no longer be optimized. GCE algorithm selects the largest group as the seed node, and partition overlapping communities by maximizing fitness function. In this paper, the module degree of partition result, NMI value of partition result, and actual community structure and algorithm running time are selected for comparative analysis.

Due to the different values of various parameters in the experiment have a great influence on the experimental results, this paper sets the parameters of the comparison algorithm according to the recommendations of relevant literature [4,8,21], and the specific parameters are shown in Table 1.

Among the algorithms involved in this experiment, TRWLPA, SLPA, GCE and MOSES are all implemented in the C++ language, and COPRA is implemented in Java language. There are three parameters to be set in the TRWLPA algorithm, which are triangle reduction iterations T_r, label propagation iterations T_p and label filtering threshold α. In addition, the number of iterations of label propagation T_p and the threshold value of label filtering α is set the same as that of the SLPA algorithm, taking 30 and 0.30 respectively. In this paper, the influence of parameter T_r on the modular Q_{ov} and NMI of TRWLPA algorithm is analyzed through experiments, so as to determine T_r. The experimental results on three datasets are shown in Fig. 5.

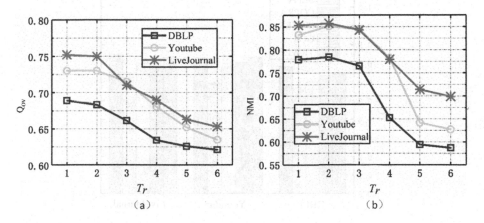

Fig. 5. It shows the influence of the different number of iterations (T_r) on the effect of the algorithm a) The value of modularity on three datasets as the T_r increased b) The value of NMI on three datasets as the T_r increased

As shown in Fig. 5, when $T_r = 2$, Q_{ov} and NMI are relatively large on all three data sets. With the increase of T_r, the reduction degree of the network becomes higher, the community structure in the initial network will be covered, and Q_{ov} and NMI will also decline. Therefore, in the following comparative experiment, the parameter T_r of the TRWLPA algorithm is 2.

5.3   Analyses of Experimental Results

In this section, modularity, NMI and running time are selected to compare the performance of the algorithm. The experimental results are as follows (Table 3).

Modularity. Modularity is an important indicator to measure community partition results. In this section, the TRWLPA algorithm and comparison algorithms are tested on three datasets. The experimental results are shown in Fig. 6.

Table 3. Setting the value of parameters

Algorithm	Parameter	Value
SLPA	Iteration number	30
	Label filter parameter	0.30
COPRA	Max number of communities per node	8
	Max number of iteration	30
GCE	Minimum clique size	3
	Minimum community distance	0.6
	Scaling parameter	1.0

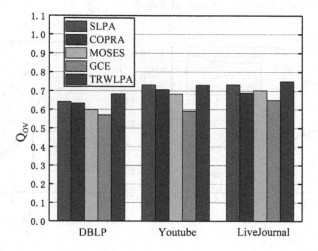

Fig. 6. This chart illustrates the experimental results of the modularity (Q_{ov}) of SLPA, COPRA, MOSES, GCE, and our algorithm running on DBLP, Youtube, and LiveJournal datasets

According to Fig. 6, regarding the results of the SLPA algorithm as the baseline, the modularity of the TRWLPA algorithm decreases by 0.16%, increases by 6.62%, and increases by 2.45% on three datasets respectively. On the whole, when TRWLPA algorithm is used for community discovery, the partition result has better modularity than the comparison algorithm.

NMI. This section uses NMI to measure the similarity between the actual community structure and the divided community structure, so as to measure the accuracy of the division results. The experimental results are shown in Fig. 7.

From Fig. 7, the results illustrate the NMI value of TRWLPA algorithm is increased by 1.59%, increased by 3.91% and decreased by 0.67% on three datasets compared with the results of SLAP algorithm, that is to say, the communities divided by TRWLPA algorithm are closer to the actual situation.

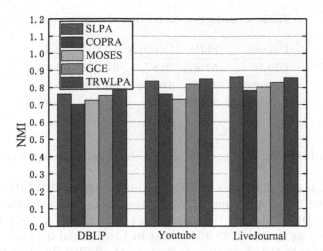

Fig. 7. This chart illustrates the experimental results of the NMI of five different algorithms running on three datasets

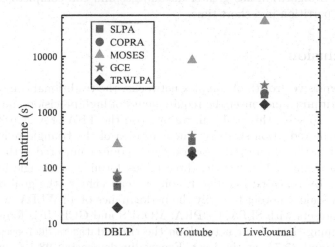

Fig. 8. This chart illustrates the experimental results of the running time of five different algorithms running on three datasets

Running Time. The running time of the community partition algorithm can reflect the running speed of the algorithm. The experimental results about the running time of the algorithm in this section are shown in Fig. 8.

In order to show the difference between the data clearly, the log 10 is used as the vertical coordinate in Fig. 8. It can be seen from the figure that the running time of the MOSES algorithm is the longest on three datasets, and that of other algorithms is relatively close. The running time of the TRWLPA algorithm proposed in this paper is the shortest on three datasets. The decreasing

Table 4. Reduction proportion of running time

Dataset	SLPA	COPRA	MOSES	GCE
DBLP	**57.8%**	70.8%	**92.7%**	75.6%
Youtube	44.3%	**16.3%**	**98.1%**	55.8%
LiveJournal	**32.9%**	36.3%	**96.8%**	54.8%

percentages of running time compared with the other four algorithms are shown in Table 4.

From Table 4, it illustrates that the running time of TRWLPA algorithm is reduced by 92.7% at most and 57.8% at least compared with the comparison algorithm in DBLP dataset; 98.1% at most and 16.3% at least in YouTube dataset; 96.8% at most and 32.9% at least in LiveJournal dataset.

Combined with the above three indicators, the TRWLPA algorithm is better than other algorithms in terms of modularity and NMI. In terms of running time, the TRWLPA algorithm has greater advantages and can complete overlapping community partition in a short time.

6   Conclusion

With the explosive growth of complex networks, the traditional community discovery algorithm cannot meet the requirements of high precision and short execution time. To solve this problem, we propose the TRWLPA algorithm based on the triangle reduction strategy. The core idea of the triangle reduction network is that in the undirected network, three nodes connected with each other constitute a triangle. We fuse the three nodes forming a triangle into a composite node. By repeated iterative fusion, we can achieve the goal of reducing the network scale. In order to verify the performance of TRWLPA, we test it on the same datasets with SLPA, COPRA, MOSES and GCE. It is found that the average running time of our network on the three datasets decreases by 95.9% at the most and 35.7% at the least. Especially decreasing 98.1% on YouTube compared with the MOSES algorithm. Moreover, our algorithm outperforms other networks in modularity and NMI. In general, compared with the traditional algorithm, TRWLPA perfectly meets the requirements of high precision and short execution time. Looking forward to better performance of TRWLPA in the application.

References

1. Blei, D.M., Ng, A.Y., Jordan, M.I.: Latent dirichlet allocation. J. Mach. Learn. Res. **3**(Jan), 993–1022 (2003)
2. Chen, J., Yuan, B.: Detecting functional modules in the yeast protein-protein interaction network. Bioinform. **22**(18), 2283–2290 (2006)

3. Erétéo, G., Gandon, F., Buffa, M.: Semtagp: semantic community detection in folksonomies. In: 2011 IEEE/WIC/ACM International Conferences on Web Intelligence and Intelligent Agent Technology. 1, pp. 324–331. IEEE (2011)
4. Gregory, S.: Finding overlapping communities in networks by label propagation. New J. Phys. 12(10), 103018 (2010)
5. Hu, W.: Finding statistically significant communities in networks with weighted label propagation (2013)
6. Kernighan, B.W., Lin, S.: An efficient heuristic procedure for partitioning graphs. Bell Syst. Tech. J. 49(2), 291–307 (1970)
7. Kumpula, J.M., Kivelä, M., Kaski, K., Saramäki, J.: Sequential algorithm for fast clique percolation. Phys. Rev. E 78(2), 026109 (2008)
8. Lee, C., Reid, F., McDaid, A., Hurley, N.: Detecting highly overlapping community structure by greedy clique expansion. arXiv preprint arXiv:1002.1827 (2010)
9. Liu, X., Li, D., Wang, S., Tao, Z.: Effective algorithm for detecting community structure in complex networks based on GA and clustering. In: Shi, Y., van Albada, G.D., Dongarra, J., Sloot, P.M.A. (eds.) ICCS 2007. LNCS, vol. 4488, pp. 657–664. Springer, Heidelberg (2007). https://doi.org/10.1007/978-3-540-72586-2_95
10. McDaid, A., Hurley, N.: Detecting highly overlapping communities with model-based overlapping seed expansion. In: 2010 International Conference on Advances in Social Networks Analysis and Mining, pp. 112–119. IEEE (2010)
11. Newman, M.E.: Fast algorithm for detecting community structure in networks. Phys. Rev. E 69(6), 066133 (2004)
12. Newman, M.E., Girvan, M.: Finding and evaluating community structure in networks. Phys. Rev. E 69(2), 026113 (2004)
13. Pizzuti, C.: GA-Net: a genetic algorithm for community detection in social networks. In: Rudolph, G., Jansen, T., Beume, N., Lucas, S., Poloni, C. (eds.) PPSN 2008. LNCS, vol. 5199, pp. 1081–1090. Springer, Heidelberg (2008). https://doi.org/10.1007/978-3-540-87700-4_107
14. Pizzuti, C.: A multiobjective genetic algorithm to find communities in complex networks. IEEE Trans. Evol. Comput. 16(3), 418–430 (2011)
15. Raghavan, U.N., Albert, R., Kumara, S.: Near linear time algorithm to detect community structures in large-scale networks. Phys. Rev. E 76(3), 036106 (2007)
16. Rattigan, M.J., Maier, M., Jensen, D.: Graph clustering with network structure indices. In: Proceedings of the 24th international conference on Machine learning, pp. 783–790 (2007)
17. Wu, Z.H., Lin, Y.F., Gregory, S., Wan, H.Y., Tian, S.F.: Balanced multi-label propagation for overlapping community detection in social networks. J. Comput. Sci. Technol. 27(3), 468–479 (2012)
18. Xie, J., Chen, M., Szymanski, B.K.: Labelrankt: incremental community detection in dynamic networks via label propagation. In: Proceedings of the Workshop on Dynamic Networks Management and Mining, pp. 25–32 (2013)
19. Xie, J., Szymanski, B.K.: Towards linear time overlapping community detection in social networks. In: Tan, P.N., Chawla, S., Ho, C.K., Bailey, J. (eds.) PAKDD 2012. LNCS (LNAI), vol. 7302, pp. 25–36. Springer, Heidelberg (2012). https://doi.org/10.1007/978-3-642-30220-6_3
20. Xie, J., Szymanski, B.K.: Labelrank: A stabilized label propagation algorithm for community detection in networks. In: 2013 IEEE 2nd Network Science Workshop (NSW), pp. 138–143. IEEE (2013)
21. Xie, J., Szymanski, B.K., Liu, X.: Slpa: uncovering overlapping communities in social networks via a speaker-listener interaction dynamic process. In: 2011 ieee 11th international conference on data mining workshops, pp. 344–349. IEEE (2011)

22. Yang, J., Leskovec, J.: Defining and evaluating network communities based on ground-truth. Knowl. Inf. Syst. **42**(1), 181–213 (2013). https://doi.org/10.1007/s10115-013-0693-z
23. Ying, K., Gu, X., Bo, Y., et al.: A multilevel community detection algorithm for large-scale social information networks. Chin. J. Comput. **1**, 169–182 (2016)
24. Yu, X., Yang, J., Xie, Z.Q.: A semantic overlapping community detection algorithm based on field sampling. Expert Syst. Appl. **42**(1), 366–375 (2015)

Parallel Belief Propagation Optimized by Coloring on GPUs

Junteng Hou[1,2]([✉]), Chengxiang Si[3], Shupeng Wang[1]([✉]), Guangjun Wu[1], and Lei Zhang[1]

[1] Institute of Information Engineering, Chinese Academy of Sciences, Beijing, China
{houjunteng,wangshupeng,wuguangjun,zhanglei1}@iie.ac.cn
[2] School of Cyber Security, University of Chinese Academy of Sciences, Beijing, China
[3] National Computer Network Emergency Response Technical Team/Coordination Center of China, Beijing, China
sichengxiang@cert.org.cn

Abstract. Belief propagation (BP) is a message passing algorithm that infers over probabilistic graphical models. Its main computational workload, messages update, is suitable for GPU's massively parallel architecture. However, the efficiency of fully parallel BP is low, and traditional algorithms implemented on GPUs occupy amount of computing and memory resources. In this paper, we propose several GPU-friendly BP algorithms optimized by coloring. Color Wave (CW) algorithm performs multi-step coloring on residuals of non-convergent vertices to quickly obtain multiple disjoint partitions and vertices with the largest residuals in each partition, and then updates batches of messages in a fixed order. These operations are all suitable for parallelization and require little additional memory. To save time in each iteration, the Color Extract (CE) algorithm only update messages on edges with the largest residuals among all adjacent edges. The Random Drop (RD) algorithm steadily increases the convergence degree by progressively reducing the messages update ratio of non-convergent edges. The experiments on different GPUs show that our algorithms perform well throughout the calculation process. Compared with state-of-the-art algorithms, CW algorithm converged most of the messages in previous iterations. The convergence degree of CE is higher than all other algorithms in most calculation processes. RD converges fast and always has a high degree of convergence.

Keywords: Belief propagation · GPUs · Parallel optimization · Coloring operations · Progressively reducing update ratio

1 Introduction

Belief propagation (BP) is a message-passing algorithm that approximatively infers over probabilistic graphical models (PGMs). It is a basic algorithm in

This work was supported by the National Natural Science Foundation of China (No. 61931019).

many important fields, including polar codes [1], Low-Density Parity-Check (LDPC) codes [2,3], computer vision [4], and protein-folding [5]. In computer vision, many "pixel-labeling" applications such as stereo matching and image denoising can be mapped to maximum a posteriori (MAP) problems [16], which are successfully solved by Max-Product belief propagation (BP). For example, stereo matching sweeps to find matching pixels from a pair of stereo images and statistically infer depth based on their pixel-wise horizontal displacement, which is inversely proportional to depth [17]. BP decoding performances iterative message-passing over the encoding factor graph in polar codes. The factor graph has multiple stages including basic processing elements(PEs). Each node within one PE is associated with two categories of log-likelihood ratio messages. The decoding result is obtained through continuous updating of this messages. General-purpose Graphical Processing Unit (GPU) has been widely used in parallel graph computing [6–8] in recent years. Messages update calculation account for most of the workload in BP algorithm, which is suitable for massive parallel architecture of GPUs.

In early researches, serial BP algorithms are the main implementation methods, which optimize algorithm efficiency by adjusting update order of messages. Residual belief propagation (RBP) [9] only updates messages with the largest residuals in each iteration. Compared with randomly updating messages, RBP efficiently speeds up convergence. Parallel BP algorithms gradually increase with the development of multi-core CPUs. Loopy BP (LBP) [10,13] is the most straightforward parallel scheme, which updates all non-convergent messages in each iteration. But Joseph et al. [11] demonstrated that the natural, fully synchronous parallelization of BP is highly inefficient. They generalized the laws found in chain graphical models to ordinary graphical models, and proposed Residual Splash (RS) algorithm that achieved high efficiency on CPUs. However, new characters may appear in GPU-based calculation of BP algorithm. By analyzing RBP and RS algorithms implemented on GPUs, Mark et al. [12] concluded that BP can obtain less convergence but faster speed as algorithm parallelism increases. According to this, they propose Random Belief Propagation (RnBP) algorithm, which randomly updates a part of non-convergent messages to tradeoff between algorithm speed and convergence. We analyze multiple parallel BP algorithms including RS, RnBP, and LBP. It is proved that the message update sequence in RS can also accelerate GPU-based BP algorithm, but the vertex sorting and stack operations in RS are not suitable for parallel architecture of GPUs. At the same time, like the GPU-based RS algorithm implemented by Mark [12], it requires a large amount of memory to record the vertices traversed in each step. We propose multiple BP algorithms according to hardware architecture characteristics of GPUs and algorithm characteristics of belief propagation. These algorithms perform well in the entire calculation process, and are superior to state-of-the-art algorithms in their applicable calculation phases.

(1) We propose Color Wave (CW) algorithm that divides non-convergent vertices into multiple partitions using coloring operations and takes a vertex with the largest residual in each partition as the pivot. It updates messages on edges from the farthest vertices to pivot and from pivot to the farthest vertices in each partition. This method can quickly converge messages as many as possible.

(2) We propose Color Extract (CE) algorithm that directly colorizes all edges in PGMs. It efficiently reduces the calculation workload by only updating messages on edges with the largest residuals among their adjacent edges, so that CE algorithm maintains a high convergence speed in each calculation stage.

(3) It is found that whenever the update proportion of non-convergent messages reduces, the convergence degree will have a clear upward trend. Random Drop (RD) algorithm gradually reduces the update proportion of non-convergent messages to maintain a high convergence speed and have a high convergence degree when stabilizing. RD algorithm also avoids the convergence and speed being affected by preset low parallelism parameters mentioned in RnBP.

We have organized the remainder of this paper as follows: Sect. 2 provides the definition of belief propagation and related knowledge. Section 3 shows detailed implementation of our proposed BP algorithms. In Sect. 4, we evaluate our methods and compare them with other algorithms. Section 5 concludes this work.

2   Preliminaries

Belief propagation is a message-passing algorithm that approximatively infers over probabilistic graphical models (PGMs). Markov random fields (MRFs) are the most representative PGMs. In this paper, we mainly research BP algorithm on discrete pairwise MRFs. Among various of BP algorithms, we take Sum-Product BP as a model, and expect that the Sum-Product BP over discrete paired MRFs can be extended to other BP algorithms over various PGMs. A discrete pairwise MRF is an undirected graph based on many discrete random variables. Suppose there are n random variables included in set $\mathcal{X} = \{X_1, X_2, ..., X_n\}$ taking on values $X_i \in A_i$, where A_i is a label set. MRF can be expressed as $G = (V, E)$ where the vertex $v_i \in V$ takes a discrete random variable corresponding to $X_i \in \mathcal{X}$, and each edge $(v_i, v_j) \in E$ corresponds to the probabilistic relations between variables on vertices v_i and v_j. for a certain random variable $x_i \in X_i$, its corresponding unary potential functions is $\{\psi_i : A_i \to \mathbb{R}^+ | i \in V\}$, so the possible relationship between variables can be expressed as $\{\psi_{i,j} : A_i \times A_j \to \mathbb{R}^+ | (i, j) \in E\}$, which is the set of binary potential functions for each edge. And the joint distribution over X on MRFs is shown in formula (1). The ultimate goal of belief propagation is to obtain the "belief" of vertices, which can be calculated by marginal distribution $P(x_i)$ as shown in formula (2).

$$P(x_1, x_2, ..., x_n) \propto \prod_{i \in V} \psi_i(x_i) \prod_{\{i,j\} \in E} \psi_{i,j}(x_i, x_j) \tag{1}$$

$$P(X_i = x_i) \approx b_i(x_i) \propto \psi_i(x_i) \prod_{k \in \Gamma_i} m_{k \to i}(x_i) \tag{2}$$

where Γ_i indicates the neighbors of v_i, $m_{k \to i}$ represents a message from vertex v_k to vertex v_i. It can be seen that the marginal distribution of a certain vertex will converge if the messages no longer changes on all edges that end with this vertex. It often takes many iterations to ensure that all messages are invariable for PGMs in tree structure. And for PGMs containing loops, there are always some changing messages, but the number of such messages gradually stabilizes after many iterations. According to the different starting vertices, there are two messages in each edge. The message $m_{i \to j}$ on edge $(i, j) \in E$ can be calculated as follows:

$$m_{i \to j} \propto \sum_{x_i \in A_i} \psi_{i,j}(x_i, x_j)\psi_i(x_i) \prod_{k \in \Gamma_i \backslash j} m_{k \to i}(x_i) \tag{3}$$

Assuming that the message $m_{i \to j}$ becomes $m_{i \to j}^t$ after t iterations of calculation, then the message difference on edge (i, j) of the t-th BP calculation is shown in formula (4). In literatures [9,11,12], $r(m_{i \to j}^t)$ is defined as the residual of message $m_{i \to j}$, and it is the basis for judging algorithm convergence.

$$r(m_{i \to j}^t) = ||m_{i \to j}^t - m_{i \to j}^{t-1}|| \tag{4}$$

In this paper, we define coloring operation as: assuming that the color value of vertices on edge (V_i, V_j) are $C(V_i)$ and $C(V_j)$, if V_i colors V_j along edge (V_i, V_j), the value of $C(V_j)$ is changed to $C(V_i)$.

3   The BP Implementation Based on GPUs

The main processes of BP algorithm include updating messages on edges and updating marginal distributions on vertices. BP always requires multiple iterative calculations because updating a certain message will cause their adjacent messages to be recalculated. We can update messages multiple times until they no longer change, then calculate marginal distributions according to formula (3). So messages updating is the main calculation workload. Although updating messages on edges will affect message calculations of their adjacent edges, as long as all messages are buffered and the messages of their adjacent edges are read from the buffer memory, all non-convergent messages can be easily updated in parallel. But it is proved that the natural, fully synchronous parallelization of belief propagation is highly inefficient whether on GPUs [12] or CPUs [11]. The update order of edges or vertices corresponding to non-convergent messages directly affects convergence speed and convergence degree. We have designed different BP algorithm implementations according to different algorithm requirements. The main difference between these algorithms is characteristic and update order of messages in each iteration, which makes the algorithms show optimal performance at different algorithm stages.

3.1   The Color Wave Algorithm

Gonzalez et al. [11] took the chain MRFs with n vertices as an example to explain the inefficiency of updating all messages simultaneously. They explained that using two processors in parallel can achieve the same running time $n - 1$ as using $2(n - 1)$ processors. The two processors processed messages following the order of forward $(m_{1 \to 2}, ..., m_{n-1 \to n})$ and backward $(m_{n \to n-1}, ..., m_{2 \to 1})$. They proposed Residual Splash (RS) algorithm extending this theory to ordinary MGFs. In RS algorithm, it presets a step parameter h. For each processor, it selects a vertex with the largest residual to perform breadth-first search (BFS) h times, and records the traversed order of each vertex. Then messages are updated in the order of their corresponding vertices traversed from large to small and from small to large. This algorithm converges well on CPUs in parallel. However, it is not suitable for parallel processing on GPUs. The first reason is that stack and sorting operations used in RS algorithm are not suitable for the large-scale parallel architecture of GPUs. Secondly, like the GPU-based RS algorithm implemented by Mark et al. [12], it requires a frontier with the same size as vertices to record the order of traversing each vertex in each forward traversal. As step h increases, it takes lots of memory. Although Mark et al. performed some optimizations when implementing RS algorithm on GPUs [12] according to existing methods [14,15], its efficiency is much lower than RnBP algorithm.

Although the vertex update sequence in [11] is inspirational, there are many differences between the parallel architecture of GPUs and CPUs. Color Wave (CW) algorithm is a BP algorithm specially designed for GPUs as shown in Algorithm 1. It consists of two main processes: color process and wave process. The color process quickly detects multiple partitions on PGMs, and selects a pivot in each partition whose residual is greater than other vertices. The wave process updates messages according to the order of wave value from large to small and from small to large.

Algorithm 1. Color Wave belief propagation Algorithm

Input: $G(V, E)$: the PGMs with vertex set V and edge set E. h: color steps, a preset parameter. ξ: residual error threshold. $Vcolor$: the color value of each vertex. $Vwave$: the wave value of each vertex. $Ewave$: the wave value of each eage. M: the message on each edge. R: the message residual on each vertex.
Output: B: the marginal distribution of each vertex.
1: $Vcolor, Vwave, Ewave \leftarrow$ Initialize($Vcolor, Vwave, Ewave$)
2: $R \leftarrow$ Residual_calculate(G, M)
3: **while** $\exists r_x > \xi, r_x \in R$ **do**
4:     /* Color */
5:     **for** $i \in [1, ..., h]$ **do**
6:         $Vwave, Ewave \leftarrow$ Color($G, R, Vcolor, Vwave, Ewave, \xi$)
7:     **end for**
8:     $Ewave \leftarrow$ Clean_color($Ewave, Vcolor$)
9:     /* Wave */
10:     **for** $step_i \in [h, ..., 1]$ **do**
11:         $M \leftarrow$ Wave($G, R, Ewave, step_i$)
12:     **end for**
13:     **for** $step_i \in [1, ..., h]$ **do**
14:         $M \leftarrow$ WaveR($G, R, Ewave, step_i$)
15:     **end for**
16:     $R \leftarrow$ Residual_calculate(G, M)
17:     $Vcolor, Vwave, Ewave \leftarrow$ Initialize($Vcolor, Vwave, Ewave$)
18: **end while**
19: $B \leftarrow$ Belief_calculate(G, M)

In the detailed processes of CW algorithm, all operations are parallelizable to fully utilize the massive parallel architecture of GPUs. In order to reduce memory consumption, it only applies for three additional arrays to record the color or wave values of vertices and edges. In the input data, $G(V, E)$ represents the graph structure and probabilistic graphical model, where the vertex set V contains different labels of variables and their corresponding potential functions, and the edge set E contains connections between all vertices and potential functions of transitions between different labels. h is a preset parameter recording the execution times of color process, which is also the distance from pivot to furthest vertices in the same partition. ξ is residual error threshold. $Vcolor$ records color value of each vertex. Vertices with the same color value belong to the same partition, and the color value is the ID of pivot. $Vwave$ and $Ewave$ record wave values of vertices and edges, respectively. They represent the distance between current vertices or edges and their pivot. M records messages calculated in each iteration. R is the message residual on each vertex. It outputs marginal distributions on B.

CW algorithm initializes color and wave values at the beginning. **Initialize** initializes the color values of vertices to their ID, and initializes the wave values to 0. **Residual_calculate** calculates residuals of vertices. It calculates messages in parallel according to formula (3), and the difference of messages between two consecutive calculations is the message residual on edge. The vertex residual in R is the maximum value of message residuals on edges that end with this vertex. As long as there is a residual in R greater than ξ, the algorithm will continue to perform color and wave operations.

Color performs color operation on PGMs to obtain wave values. The operation process of **Color** is shown in Algorithm 2. For a certain non-convergent vertex v_x, if there is a vertex v_y in its adjacent vertices, whose residual is greater than v_x's residual, v_x will be colored with the color value of $Ncolor(v_y)$. Therefore, if v_x and v_y belong to the same partition colored by the vertex with ID of $Ncolor(v_y)$, and the distance from v_y to vertex $Ncolor(v_y)$ is $Nwave(v_y)$, then the distance from vertex v_x to vertex $Ncolor(v_y)$ is $Nwave(v_y) + 1$. Because messages are updated on edges, in order to facilitate wave operations, the wave values are recorded to $Ewave(v_y, v_x)$. **Clean_color** is used to reduce unnecessary operations in wave phase. After multiple color operations, if vertex $Ncolor(v_x)$ is colored by other vertex, v_x belongs to an incomplete partition. Then the wave values of edges connected to vertex v_x will be set to 0.

Algorithm 2. Color process of CW Algorithm

Input: $G(V, E)$: the PGMs with vertex set V and edge set E. $Vcolor$: the color value of each vertex. ξ: residual error threshold. $Vwave$: the wave value of each vertex. $Ewave$: the wave value of each eage. R: the message residual on each vertex.

Output: $Vwave, Ewave$.

1: **for** $v_x \in V$ **do**
2:    **if** $R(v_x) > \xi$ **then**
3:        **while** $(v_y, v_x) \in$ adjacent eages of v_x **do**
4:            **if** $R(v_y) > R(v_x)$ **then**
5:                $Vcolor(v_x) \leftarrow Vcolor(v_y)$
6:                $Vwave(v_x) \leftarrow Vwave(v_y) + 1$
7:                $Ewave(v_y, v_x) \leftarrow Vwave(v_x)$
8:            **end if**
9:        **end while**
10:    **end if**
11: **end for**

In wave phase, CW algorithm calculates messages in a specific order. **Wave** only updates messages on edges whose wave values are $step\_i$. And **WaveR** updates messages on edges opposite to **Wave**, that is, if **Wave** updates the message on edge (v_a, v_b), **WaveR** will update the message on edge (v_b, v_a). After wave operations, **Residual_calculate** recalculates message residuals on vertices. **Initialize** initializes color and wave values. The operations continue iteratively. Finally, **Belief_calculates** calculates marginal distributions according to formula (2).

In CW algorithm, there is no stack and sorting operations that are not suitable for GPUs' architecture. Color operations guarantee that the message residual on selected pivot is greater than other vertices in the same partition, and the distance between pivot and other vertices does not exceed h, which allows both forward and reverse wave operations to be completed in h iterations. There is no intersection of different partitions, so that it will not conflict to update messages in different partitions. Therefore, it doesn't need to apply for memory for "fronter" of each coloring, and *Ewave* can record the update order of all messages. CW algorithm converge most vertices in a short time.

3.2   The Color Extract Algorithm

CW algorithm has a high convergence degree at the beginning, which is benefited from updating messages in a strict sequence centered on vertices with the largest residuals. But as calculation continues, its convergence speed is not fast. Because CW algorithm needs color operations to determine the update order of edges in each iteration, and performs wave operations $2 * h$ times. Although such a message update sequence is conducive to messages convergence, it is so time consuming that it reduces the convergence speed. We propose Color Extract (CE) algorithm that quickly select the messages to be updated in each iteration. It is mentioned in [9,11] that updating messages with the largest residuals in each iteration can effectively reduce the computational complexity. But sorting operation is not suitable for parallel architecture of GPUs. For parallel messages updating, we only need to determine which messages need to be updated, and don't need to know the accurate sorting results of all messages. Therefore, we still use color operations to select messages with the largest residuals in small partitions. Color Extract algorithm is simplified from Color Wave algorithm, which can ensure that messages with the largest residuals within local ranges are updated quickly in each iteration.

Algorithm 3. Color Extract belief propagation Algorithm

Input: $G(V, E)$: the PGMs with vertex set V and edge set E. ξ: residual error threshold. M: the message on each edge. R: the message residual on each edge. $Ecolor$: the color value of each edge.

Output: B: the marginal distribution of each vertex.

1:    $Ecolor \leftarrow$ Initialize($Ecolor$)
2:    $R \leftarrow$ Residual_calculate(G, M)
3:    **while** $\exists r_x > \xi, r_x \in R$ **do**
4:      /* Color */
5:      **for** $e_i(v_a, v_b) \in E$ **do**
6:        **if** $R(e_i) > \xi$ **then**
7:          **for** $e_j \in$ EdgeSetStarting(v_b)   *or* $e_j \in$ EdgeSetEnding(v_a) **do**
8:            **if** $R(e_j) > R(e_i)$ **then**
9:              $Ecolor(e_i) \leftarrow Ecolor(e_j)$
10:            **end if**
11:          **end for**
12:        **end if**
13:      **end for**
14:      /* Calculate */
15:      **for** $e_i \in E$ **do**
16:        **if** $Ecolor(e_i) = e_i$ **then**
17:          $M \leftarrow$ Message_calculate(G, M, e_i)
18:        **end if**
19:      **end for**
20:      $R \leftarrow$ Residual_calculate(G, M)
21:      $Ecolor \leftarrow$ Initialize($Ecolor$)
22: **end while**
23: $B \leftarrow$ Belief_calculate(G, M)

Most of input parameters in Algorithm 3 are the same as Algorithm 1, except that $Ecolor$ and R record color values and message residuals on edges, respectively. At the beginning of CE algorithm, **Initialize** initializes color values of edges to their ID, and **Residual_calculation** calculates message residual on each edge like Algorithm 1. CE algorithm also consists of two main phases. In color phase, $e_i(v_a, v_b)$ represents a edge e_i starting with vertex v_a and ending with v_b, $EdgeSetStarting(v_b)$ and $EdgeSetEnding(v_a)$ represent edge sets formed by edges starting with v_b and ending with v_a, respectively. If there is an edge e_j with residual greater than e_i in these sets, e_j is used to color e_i. In calculation phase, only the messages on edges whose color values are the same as their own ID are updated. **Message_calculate**(G, M, e_i) updates message on edge e_i. After recalculating residuals and initializing color values, the algorithm iteratively calculates until all messages converge. Finally, **Belief_calculate** calculates marginal distributions according to formula (2). Although CE algorithm does not strictly restrict the update order of messages as CW algorithm to converge messages as many as possible in each iteration, it can quickly select messages with the largest residuals in local partitions and quickly update messages in each iteration. Therefore, CE converges quickly during the entire calculation process.

3.3 The Random Drop Algorithm

Randomized Belief Propagation (RnBP) algorithm [12] sets different parallelism parameters to tradeoff algorithm speed and convergence. The parallelism parameter is the proportion of messages updated in each iteration to all non-convergent messages. It is found that the algorithm with high parallelism parameter converges quickly, but its convergence degree is low, and low parallelism is the opposite. Therefore, in RnBP algorithm, all non-convergent messages are updated if messages converge rapidly, otherwise only a part of non-convergent messages are updated according to the preset low parallelism parameter. When it updates all non-convergent messages, the algorithm has the highest parallelism, which can quickly converge messages that are close to convergence. However, as shown in formula (3), the calculation of each message is related to its adjacent messages. It will not converge if the adjacent messages of a certain message change greatly in each iteration. If there is only a part of non-convergent messages updated and the rest remains unchanged, there will be more convergent messages. So the algorithm with low parallelism parameter has high convergence degree. An extreme example is that it only updates one message at a time in serial belief propagation algorithm. Figure 1 shows the variation of convergence degree with different parallelism parameters in 100 iterations, where Ising($N * N$) represents an Ising graph with $N * N$ vertices. In Fig. 1(a), all non-convergent messages are updated, but ten percent of them are updated after 20 iterations. On the contrary, in Fig. 1(b), ten percent of non-convergent messages are updated, but all non-convergent messages are updated after 20 iterations. It can be seen that when parallelism decreases, convergence degree increase noticeably, while if parallelism increases, convergence degree decreases significantly before increasing. According to these analyses, we propose Random Drop (RD) algorithm based on RnBP algorithm. RD algorithm gradually reduces the parallelism, so that it further improves convergence speed on the basis of original convergence speed in each iteration. And when the number of non-convergent messages is stable, RD algorithm has a particularly high degree of convergence.

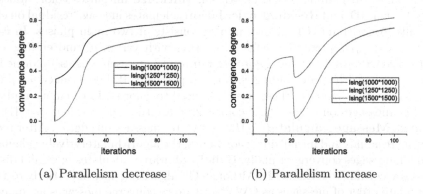

(a) Parallelism decrease   (b) Parallelism increase

Fig. 1. The variation of convergence degree in different parallelism parameters

4   Experimental Methodology

4.1   Experimental Setup

Datasets: we use Ising dataset as our experiment PGMs, which is a standard benchmark for message propagation used in BP algorithms [9,12]. The graph model of Ising is a regular $N * N$ grids of binary variables. The potential function ψ_i of variable on vertices is a random value selected in $[0, 1]$. The potential function $\psi_{i,j}$ is set to $e^{\lambda C}$ when $x_i = x_j$ and $e^{-\lambda C}$ otherwise. λ is sampled from $[-0.5, 0.5]$ to make potential function $\psi_{i,j}$ take random values within a certain range. By setting a larger C, the range of potential function values will be larger, and inference will be more difficult.

In each analysis experiment, we verify the universal performance of algorithms by performing belief propagation on different hardwares over Ising graph model with different parameters. The experiment graph is named as Ising_N_C to represent an Ising graph with $N * N$ vertices and parameter C, and it is given in figure titles.

Baselines and Hardware Platforms: we compare six implementations including: (1) LBP: Murphy's BP algorithm [10] for parallel updating all non-convergent messages implemented on the GPU. (2) RnBP1: Der's algorithm [12] that updates only a part of non-convergent messages on GPUs. (3) RnBP2: Der's algorithm [12] that optimizes RnBP1 by changing parallelism. (4) CW: Our Color Wave algorithm. (5) CE: Our Color Extract algorithm. (6) RD: Our Random Drop algorithm. We use gcc and nvcc with the $-O3$ optimization option for compilation along with $-arch = sm\_37$ on two NVIDIA GPUs including Tesla K80 and GeForce 1050ti(notebooks).

4.2   Performance Analysis of Color Wave Algorithm

The following sections analyze experimental performance of different algorithms. Figure 2(a) shows the change in the number of non-convergent messages with iterations in 100 s. The step parameter h is in parenthesis. It is obvious that more than half of messages converge in previous iterations, and as h increases, the number of messages converged in previous iterations significantly increases, except that the convergence degree of CW algorithm with a step of 3 is surpassed by CW algorithm with a step of 2 after the 21st iteration. Therefore, it is the main advantage of CW algorithm that most messages converge in a short time, and the CW algorithm with a larger step converges more messages in each iteration. But there is a disadvantage that each iteration of calculation takes plenty of time with the increase of h. Within a 100-second runtime, CW algorithm with a step of 2 can perform 133 iterations of calculation, while CW algorithm with a step of 8 only performs 45 iterations. This can also be seen from the change in the number of non-convergent messages with time in Fig. 2(b). The RnBP1 algorithm only updates a part of non-convergent messages, and the updated percentage is in parentheses. LBP updates all non-convergent messages. As shown in Fig. 2(b),

CW algorithm has obvious advantages in the first 20 s, but its convergence degree in subsequent processes is lower than other algorithms, and the convergence degree is similar to others when it is stable.

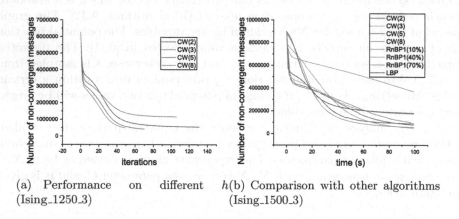

(a) Performance on different (b) Comparison with other algorithms
(Ising_1250_3)   (Ising_1500_3)

Fig. 2. Performance analysis of CW algorithm

4.3 Performance Analysis of Color Extract Algorithm

In view of the disadvantages of CW algorithm, we propose CE algorithm with performance shown in Fig. 3. Figure 3(a) compares the performance of CE algorithm and several CW algorithms with different steps in the first 50 iterations of calculation. CW algorithm strictly stipulates the update order of non-convergent messages, so there is more messages convergence in each iteration. In the first 50 iterations, the convergence degree of CE algorithm can hardly surpass various CW algorithms, except for the CW algorithm with a step of 3 after the 42nd iteration. However, the purpose of designing CE algorithm is reducing calculation time in each iteration to guarantee higher convergence speed in the entire process. From the change in the number of non-convergent messages with time in Fig. 3(b), it can be seen that although CE algorithm has a much lower degree of convergence in the first 17 s than CW algorithm, it has a high convergence speed since then. And the convergence degree is far better than RnBP1 and LBP algorithms before stabilization. There is a similiar phenomenon in Fig. 3(c). It means that CE algorithm has a higher convergence speed and convergence degree in the entire process.

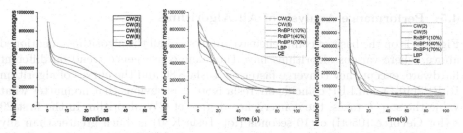

(a) Comparison with CW (Ising_1500_3)

(b) Comparison with other algorithms (Ising_1500_3)

(c) Comparison with other algorithms (Ising_1000_3)

Fig. 3. Performance analysis of CE algorithm

4.4 Performance Analysis of Random Drop Algorithm

The experiment result of random algorithms with different parallelism is shown in Fig. 4(a). Although there is a opposite phenomenon in the first 25 iterations, the algorithms' most iterative processes are in accordance with Der et al.'s proposition [12]. In Fig. 4(b)(c), we analyze the performance comparison between RD algorithm and the algorithms of changing parallelism proposed by Der et al. (RnBP2) and the algorithm that only randomly updates a part of non-convergent messages (RnBP1). In the first 90 iterations of RD algorithm, the proportion of randomly updated non-convergent messages gradually decreases from 100% to 10% and then it stabilizes at 10%. According to Fig. 4 (b)(c), the result curve of RnBP2 almost coincides with RnBP1, so the acceleration of algorithm using two parallel degrees is not obvious. Our algorithm surpasses all other algorithms except in previous iterations. Therefore, RD algorithm keeps a fast convergence speed in almost the entire calculation process, and it has a high convergence degree when stable.

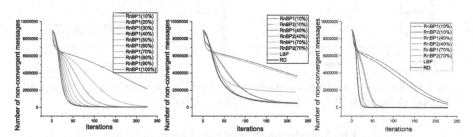

(a) Performance for different update ratios (Ising_1500_2.5)

(b) Compasion with other algorithms (Ising_1500_3)

(c) Comparison with other algorithms (Ising_1500_2)

Fig. 4. Performance analysis of PD algorithm

4.5   Performance Analysis of All Algorithms

Figure 5 is on the basis of comprehensive analysis of all the algorithms mentioned above, where various algorithms have the same convergence trend on different hardware, except they converge faster on Tesla K80 and the curves of algorithm RnBP2(10%) and LBP coincide on Tesla K80. It is obvious that no matter what value the step h takes, the convergence degree of CW algorithm in the first 20 s (for GeForce 1050ti) or 10 seconds (for Tesla K80) is much greater than any other algorithms, and when the algorithms are stable, the convergence degree of CW algorithm is also close to others. Therefore, CW algorithm is very suitable for the requirement of converging more messages in a short time. Before CE algorithm stabilizes, its convergence degree is greater than other algorithms except CW algorithm, and within a period after 20 (GeForce 1050ti) or 10 (Tesla K80) s, the convergence degree of CE algorithm also exceeds that of CW algorithm. Therefore, CE algorithm is suitable for the requirement of high convergence degree before stabilization. For RD algorithm, although its convergence degree is less than that of CW and CE algorithms in most of the time, it has a higher degree of convergence than algorithms that randomly update a part of non-convergent messages and update all non-convergent messages during the entire running time. And RD algorithm has a high convergence degree when it is stable. Therefore, RD algorithm is suitable for the needs of always having high convergence speed and convergence degree. Moreover, RD algorithm avoids the algorithm performance being affected by preset parallel parameters.

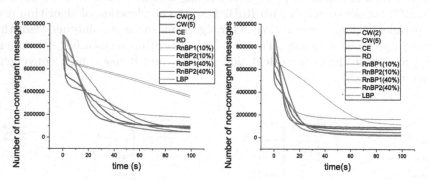

(a) Performance analysis on GeForce 1050ti (Ising_1500_3)

(b) Performance analysis on Tesla K80 (Ising_1500_3)

Fig. 5. Performance analysis of all algorithms

5   Conclusion

Belief propagation is a message passing algorithm over probabilistic graphical models, which is widely used in many fields including machine learning and error-correcting. The main operations of BP algorithm are very suitable for computing on the massive parallel architecture of GPUs. But updating all non-convergent messages in parallel does not always perform well. we implement different algorithms to optimize different algorithm stages. CW algorithm can quickly divide non-convergent vertices into multiple partitions, and select vertices with the largest residuals as pivots in all partitions. It updates messages in a fixed order to ensure that most vertices converge in a short time. CE algorithm selects edges with the largest residuals in local partitions by edge-centered color operations. It reduces calculation complexity of each iteration by only updating messages on the selected edges. RD algorithm steadily increases the convergence degree by progressively reducing the update ratio of non-convergent messages. These algorithms show high performance throughout the entire calculation process, and are superior to state-of-the-art algorithms in their applicable calculation stages. In the future, we will explore more common BP algorithms on multiple types of PGMs.

References

1. Liu, Z., Liu, R., Yan, Z., Zhao, L.: GPU-based implementation of belief propagation decoding for polar codes. In: International Conference on Acoustics Speech and Signal Processing (ICASSP 2019), pp. 1513–1517. IEEE, Brighton (2019)
2. Shan, B., Fang, Y.: GPU accelerated parallel algorithm of sliding-window belief propagation for LDPC codes. Int. J. Parallel Prog. **48**(3), 566–579 (2020)
3. Romero, D.L., Chang, N.B.: Sequential decoding of non-binary LDPC codes on graphics processing units. In: Asilomar Conference on Signals, Systems and Computers (ACSCC 2012), pp. 1267–1271. IEEE, Pacific Grove (2012)
4. Felzenszwalb, P.F., Huttenlocher, D.P.: Efficient belief propagation for early vision. In: IEEE Conference on Computer Vision and Pattern Recognition (CVPR 2004), pp. 261–268. IEEE, Washington (2004)
5. Yanover, C., Weiss, Y.: Approximate inference and protein-folding. In: Neural Information Processing Systems (NIPS 2002), pp. 1481–1488. Vancouver (2002)
6. Wang, H., Geng, L., Lee, R., Hou, K., Zhang, Y., Zhang, X.: SEP-graph: finding shortest execution paths for graph processing under a hybrid framework on GPU. In: ACM Sigplan Symposium on Principles and Practice of Parallel Programming (PPoPP 2019), pp. 38–52. ACM, Washington (2019)
7. Segura, A., Arnau, J.-M., González, A.: SCU: a GPU stream compaction unit for graph processing. In: International Symposium on Computer Architecture (ISCA 2019), pp. 424–435. IEEE, Phoenix (2019)
8. Alabandi, G., Powers, E., Burtscher, M.: Increasing the parallelism of graph coloring via shortcutting. In: ACM Sigplan Symposium on Principles and Practice of Parallel Programming (PPoPP 2020), pp. 262–275. ACM, San Diego (2020)
9. Elidan, G., McGraw, I., Koller, D.: Residual belief Propagation: informed scheduling for asynchronous message passing. In: Uncertainty in Artificial Intelligence (UAI 2006), Cambridge, pp. 165–173 (2006)
10. Murphy, K., Weiss, Y., Jordan, M.I.: Loopy belief propagation for approximate inference: an empirical study. In: Uncertainty in Artificial Intelligence (UAI 1999), Stockholm, pp. 467–475 (1999)
11. Gonzalez, J., Low, Y., Guestrin, C.: Residual splash for optimally parallelizing belief propagation. In: International Conference on Artificial Intelligence and Statistics (AISTATS 2009), Clearwater Beach, pp. 177–184 (2009)

12. Der Merwe, M.V., Joseph, V., Gopalakrishnan, G.: Message scheduling for performant, many-core belief propagation. In: IEEE High Performance Extreme Computing Conference (HPEC 2019), pp. 1–7. IEEE, Waltham (2019)
13. Mooij, J.M., Kappen, H.J.: Sufficient conditions for convergence of the sum-product algorithm. IEEE Trans. Inf. Theory **53**(12), 4422–4437 (2007)
14. Wang, Y., et al.: Gunrock: GPU graph analytics. ACM Trans. Parallel Comput. **4**(1), 1–49 (2017)
15. Pingali, K., et al.: The tao of parallelism in algorithms. In: SIGPLAN Conference on Programming Language Design and Implementation (PLDI 2011), pp. 12–25. ACM, San Jose (2011)
16. Szeliski, R., et al.: A comparative study of energy minimization methods for markov random fields with smoothness-based priors. IEEE Trans. Pattern Anal. Mach. Intell. **30**(6), 1068–1080 (2008)
17. Choi, J., Patil, A.D., Rutenbar, R.A., Shanbhag, N.R.: Analysis of error resiliency of belief propagation in computer vision. In: International Conference on Acoustics, Speech, and Signal Processing (ICASSP 2016), pp. 1060–1064. IEEE, Shanghai (2016)

A Multiplatform Parallel Approach
for Lattice Sieving Algorithms

Michal Andrzejczak[1]($\boxtimes$) (iD) and Kris Gaj[2]

[1] Military University of Technology, Warsaw, Poland
michal.andrzejczak@wat.edu.pl
[2] George Mason University, Fairfax, VA, USA
kgaj@gmu.edu

Abstract. Lattice sieving is currently the leading class of algorithms for solving the shortest vector problem over lattices. The computational difficulty of this problem is the basis for constructing secure post-quantum public-key cryptosystems based on lattices. In this paper, we present a novel massively parallel approach for solving the shortest vector problem using lattice sieving and hardware acceleration. We combine previously reported algorithms with a proper caching strategy and develop hardware architecture. The main advantage of the proposed approach is eliminating the overhead of the data transfer between a CPU and a hardware accelerator. The authors believe that this is the first such architecture reported in the literature to date and predict to achieve up to 8 times higher throughput when compared to a multi-core high-performance CPU. Presented methods can be adapted for other sieving algorithms hard to implement in FPGAs due to the communication and memory bottleneck.

Keywords: Lattice sieving · Hardware acceleration · Cryptography · Multi-platform parallel approach · Parallel algorithms

1 Introduction

Over the last decade, post-quantum cryptography (PQC) has emerged as one of the most important topics in the area of theoretical and applied cryptography. This new branch of cryptology is considered an answer to the threat of quantum computers. A full-scale quantum computer will be able to break popular public-key cryptosystems, such as RSA and ECDSA, using Shor's algorithm.

In 2016, the United States National Institute of Standards and Technology (NIST) announced the Post-Quantum Cryptography Standardization Process (NIST PQC), aimed at developing new cryptographic standards resistant to attacks involving quantum computers. In January 2019, 26 of these candidates (including results of a few mergers) advanced to Round 2.

The biggest group of submissions were lattice-based algorithms. The difficulty of breaking these cryptosystems relies on the complexity of some well-known and

© Springer Nature Switzerland AG 2020
M. Qiu (Ed.): ICA3PP 2020, LNCS 12452, pp. 661–680, 2020.
https://doi.org/10.1007/978-3-030-60245-1_45

computationally-hard problems regarding mathematical objects called lattices. One of these problems is the Shortest Vector Problem (SVP). Lattice sieving, which is a subject of this paper, is a family of algorithms that can be used to solve SVP (at least for relatively small to medium dimensions of lattices).

Recently, significant progress in lattice sieving has been made, especially due to Albrecht et al. [4]. Although multiple types of lattice sieving algorithms emerged in recent years, all of them share a single fundamental operation called vector reduction. As a result, an efficient acceleration of vector reduction is likely to work with most of the sieves and give them a significant boost in performance.

Sieving is a popular technique in cryptography. It was used previously, for example, for factoring large integers. However, it is a memory-intense method, so there exists a data transfer bottleneck disrupting any potential hardware acceleration, which is the biggest problem.

1.1 Contribution

To take full advantage of modern hardware parallel capabilities, we propose a modified approach to lattice sieving algorithms and present a massively parallel FPGA sieving accelerator. In the modified sieving algorithm, due to proper caching techniques on the hardware side, there is no data transfer bottleneck. Thus, the accelerator works with full performance, and a significant speed-up is achieved. In the end, the cost comparison for solving SVP instances with Amazon AWS is presented.

2 Mathematical Background

A *lattice* $\mathcal{L}$ is a discrete additive group generated by n linearly independent vectors $\mathbf{b_1}, \mathbf{b_2}, \ldots, \mathbf{b_n} \in \mathbb{R}^m$

$$\mathcal{L}(\mathbf{b_1}, \mathbf{b_2}, \ldots, \mathbf{b_n}) = \left\{ \sum x_i \mathbf{b}_i \mid x_i \in \mathbb{Z} \right\} \tag{1}$$

The vectors $\mathbf{b_1}, \ldots, \mathbf{b_n}$ are called a *basis* of the lattice, and we define $\mathbf{B}$ as a $m \times n$ matrix consisting of basis vectors as columns. In this paper, we consider only the case of $m = n$.

The lattice, understood as a set of vectors, can be written as

$$\mathcal{L}(\mathbf{B}) = \{ \mathbf{x}\mathbf{B} \mid \mathbf{x} \in \mathbb{Z}^n \} \tag{2}$$

We define the length of a vector as the Euclidean norm $\|\mathbf{x}\| = \sqrt{\sum \mathbf{x}_i^2}$. The Shortest Vector Problem (SVP) aims at finding a linear combination of basis vectors with the shortest length possible. For a given basis $\mathbf{B} \in \mathbb{Z}^{n \times m}$, the shortest vector $\mathbf{v} \in \mathcal{L}(\mathbf{B})$ is a vector for which

$$\forall \mathbf{x} \in \mathbb{Z}^n \ \|\mathbf{v}\| \le \|\mathbf{x}\mathbf{B}\| \tag{3}$$

The shortest vector in a lattice is also called the first successive minimum and is denoted $\lambda_1(\mathcal{L})$. There are known estimates on boundaries of the length of the shortest vector in a given lattice. For the most well-known SVP Challenge [1], a found vector $\mathbf{v}$ should be shorter than

$$\|\mathbf{v}\| \leq 1.05 \cdot \frac{\left(\Gamma(n/2+1)\right)^{1/n}}{\sqrt{\pi}} \cdot (\det \mathcal{L})^{1/n}, \tag{4}$$

where Γ is Euler's gamma function, and det is determinant of the basis $\mathbf{B}$ generating the lattice $\mathcal{L}$.

If two different vectors $\mathbf{v}, \mathbf{u} \in \mathcal{L}(\mathbf{B})$ satisfy $\|\mathbf{v} \pm \mathbf{u}\| \geq \max(\|\mathbf{v}\|, \|\mathbf{u}\|)$, then $\mathbf{v}, \mathbf{u}$ are called *Gauss-reduced*. If every pair of two vectors $(\mathbf{v}, \mathbf{u})$ from the set $A \in \mathcal{L}(\mathbf{B})$ is Gauss-reduced, then the set A is called *pairwise-reduced*.

In this paper, we denote vectors as bold lowercase letters. Matrices are denoted as bold uppercase letters. Lattice points and vectors are used alternatively.

3   Lattice Sieving

SVP is one of the best-known problems involving lattices. Due to its computational complexity, it can be used as a basis for the security of lattice-based cryptosystems. Lattice sieving is one of several approaches to solve SVP. It is not a single algorithm, but rather a class of algorithms. These algorithms are similar to one another and rely on a similar basic operation, but differ in terms of their time and space complexity.

The term "lattice sieving" was proposed in the pioneering work of Ajtai–Kumar–Sivakumar [2,3]. In 2001, these authors introduced a new randomized way for finding the shortest vector in an n-dimensional lattice by sieving sampled vectors.

The main idea was to sample a long list $L = \{\mathbf{w}_1, \ldots, \mathbf{w}_N\}$ of random lattice points, and compute all possible differences among points from this list L. As the algorithm progresses, during reduction, shorter and shorter vectors are discovered. By repeating this step many times, the shortest vector in the lattice is being found as a result of subtracting two vectors $\mathbf{v}_i - \mathbf{v}_j$.

The method proposed by Ajtai et al. is the main element of lattice sieving algorithms. Other algorithms differ mostly in the way of handling lattice vectors, grouping them, or using some techniques of prediction. However, the main idea is still to sample new random vectors and reduce them using those already accumulated.

3.1   The GaussSieve

In 2010, Micciancio and Voulgaris [11] presented two new algorithms: *List-Sieve* with the time complexity $2^{3.199n}$ and the space complexity $2^{1.325n}$, and *GaussSieve*, able to find a solution in the running time $2^{0.52n}$, using memory space in the range of $2^{0.2n}$. The *GaussSieve* is shown below as Algorithm 1. The key idea is taken from Ataji's work and is based mostly on pairwise vector reduction. The GaussSieve starts with an empty list of lattice points L and an

---

Algorithm 1: GaussSieve(**B**) — algorithm that can compute the shortest vector. The *.pop()* operation returns the first vector from a given queue. *KleinSampler()* is a method for random sampling of new vectors. *GaussReduce* reduces vector by other vectors from the set L.

Data: B - lattice basis, c - maximum number of collisions, $\lambda_1(\mathbf{B})$- targeted norm

Result: $\mathbf{t} : \mathbf{t} \in \mathcal{L}(\mathbf{B}) \wedge \|\mathbf{t}\| \leq \lambda_1(\mathbf{B})$

```
 1 begin
 2 │   L ⟵ ∅, S ⟵ ∅, i ⟵ 0, t ⟵ KleinSampler(B)
 3 │   while i < c and ‖t‖ > λ₁(B) do
 4 │   │   if S ≠ ∅ then
 5 │   │   │   v_new ⟵ S.pop()
 6 │   │   else
 7 │   │   │   v_new ⟵ KleinSampler(B)
 8 │   │   end if
 9 │   │   v_new ⟵ GaussReduce(v_new, L, S)
10 │   │   if ‖v_new‖ = 0 then
11 │   │   │   i ⟵ i + 1
12 │   │   else
13 │   │   │   L ⟵ L ∪ {v_new}
14 │   │   │   if ‖v_new‖ < ‖t‖ then
15 │   │   │   │   t ⟵ v_new
16 │   │   │   end if
17 │   │   end if
18 │   end while
19 │   return t
20 end
```

---

empty stack S. The stack is the first source of points to be processed in the next iteration of reduction. In the case of an empty stack, a new point is sampled using Klein's method for sampling random lattice points [9], with modifications and extensions from [7].

Next, a sampled lattice point $\mathbf{v}$ is pairwise reduced by every vector from the list L. The reduction method called *GaussReduce* returns vectors $\mathbf{u}, \mathbf{v}$ satisfying $max(\|\mathbf{u}\|, \|\mathbf{v}\|) \leq \|\mathbf{u} \pm \mathbf{v}\|$. This method is shown below as Algorithm 2. Thus, the list L is always Gauss reduced, so in the case of reducing a vector already on the list, the vector is moved to the stack. If the vector $\mathbf{v}$ is non-zero after reducing by the whole list, it is added to L. Otherwise, the number of collisions i is incremented. A collision occurs when the point is reduced to zero, which means that the same point has been sampled before. The algorithm stops when the number of collisions exceeds the given boundary c, or the shortest vector already found is at least as short as the targeted estimate.

By analyzing Algorithm 1 and Algorithm 2, the number of *Reduce()* calls can be approximated as k^2 for the case of k vectors.

Algorithm 2: GaussReduce(p, L, S)

Data: p - lattice vector, L - list of already reduced vectors, S - list of vectors to reduce

Result: p - reduced vector

```
1  begin
2  |   for vᵢ ∈ L do
3  |   |   Reduce(p, vᵢ)
4  |   end for
5  |   for vᵢ ∈ L do
6  |   |   if Reduce(vᵢ, p) then
7  |   |   |   L ← L\{vᵢ}
8  |   |   |   S.push(vᵢ − p)
9  |   |   end if
10 |   end for
11 |   return p
12 end
```

There have been many papers improving the complexity of presented algorithm and proposing modifications that speed up the computations by several orders of magnitude by applying additional techniques. However, the *GaussSieve* is still a part of newer methods and the vector reduction step is crucial for every lattice sieving algorithm.

3.2 Parallel Sieves

There have been several papers devoted to developing a parallel version of a lattice sieve. In addition to the *gsieve* and *fplll* libraries, Milde and Schneider [12] proposed a parallel version of the GaussSieve algorithm. Their main idea was to use several instances of the algorithm connected in a circular fashion. Each instance has its own queue Q, list L, and stack S. When a vector is reduced by one instance, it is moved to another one. The stacks contain vectors that were reduced in a given instance and need to pass the instance's list once more. During the algorithm, vectors in the instances lists are not always Gauss reduced.

Ishiguro et al. [8] modified the idea of the parallel execution of the *GaussSieve* algorithm. The stack is only one, global for all instances (threads). The execution of the algorithm is divided into three parts. In the first part, sampled vectors in the set V (new or from the stack) are reduced by vectors in the local instance lists. After reduction, the reduced vectors are compared. If any vector is different than before the reduction step, it is moved to the global stack. In the next step, sampled vectors are reduced by themselves. In the last step, vectors from the local lists are reduced by the sampled vectors. The procedure ends with moving vectors from the set V to local lists in instances. A new batch of vectors is sampled, and the procedure starts from the beginning. The advantage of this approach for parallel execution is that vectors in local lists are always pairwise reduced.

In [5], Bos et al. combined ideas from [12] and [8]. As in Milde and Schneider, each node maintains its own local list, but the rounds are synchronized between nodes. During synchronization, the vectors are ensured to be pairwise reduced as in the Ishiguro et al. approach.

Yang et al. [13] proposed a parallel architecture for *GaussSieve* on GPUs. A single GPU executes a parallel approach proposed by Ishiguro et al. Communication and data flow between multiple GPUs is performed by adopting ideas from Bos et al.

Every paper listed above targeted a multi-thread or multi-device implementation, but due to FPGAs structure, some ideas might be also adapted to hardware.

As for FPGAs, there is no publicly available paper about hardware implementation of lattice sieving. FPGAs have been used for solving SVP, but using a class of enumeration algorithms. In 2010, Detrey et al. proposed an FPGA accelerator for the Kannan-Fincke-Pohst enumeration algorithm (KFP) [6]. For a 64-dimensional lattice, they obtained an implementation faster by a factor of 2.12, compared to a multi-core CPU, using an FPGA device with a comparable cost (Intel Core 2 Quad Q9550 vs. Xilinx Virtex-5 SXT 35). For a software implementation, the fplll library was used.

4   Hardware Acceleration of Vector Reduction

Lattices used in cryptography are usually high-dimensional. The hardest problem solved in the SVP Challenge, as of April 2020, is for a 157-dimensional lattice [1]. This dimension is still significantly smaller than dimensions of lattices used in the post-quantum cryptography public-key encryption schemes submitted to the NIST Standardization Process. However, it is still a challenge, similar to RSA-Challenge, to solve an as big problem as possible. Thus, a hardware acceleration might help to find solutions for higher dimensions in a shorter time.

Algorithm 3 describes a common way of implementing the *Reduce* function in software. This method is dependant on the lattice dimension, which affects the dot product computation and the update of the vector's value. The number of multiplications increases proportionally to the dimension. A standard modern CPU requires more time to perform the computations as the lattice dimension increases. However, both affected operations are highly parallelizable. Almost all multiplications can be performed concurrently by utilizing low-level parallelism. Thus, specialized hardware can be competitive even for modern CPUs capable of performing vectorized instructions and can offer a higher level of parallelism.

In this section, we present a new hardware architecture for lattice vector reduction. The analysis is performed step by step, line by line, and the corresponding hardware is proposed.

4.1   Computing a Vector Product

Computing the product of two vectors and an update of the vector's value are the two most time-consuming operations during reduction, but there is a chance for

---

Algorithm 3: Reduce($\mathbf{v}$, $\mathbf{u}$) – vector reduction. The return value is *true* or *false*, depending on whether reduction occurs or not.

Data: $\mathbf{v}, \mathbf{u}$ - lattice vectors
Result: *true* or *false*

1 **begin**
2 $\quad$ $dot = \sum v_i \cdot u_i$
3 $\quad$ **if** $2 \cdot |dot| \leq \|\mathbf{u}\|^2$ **then**
4 $\quad\quad$ | **return** *false*
5 $\quad$ **else**
6 $\quad\quad$ $q = \left\lfloor \frac{dot}{\|\mathbf{u}\|^2} \right\rceil$
7 $\quad\quad$ **for** $i = 0; i < n; i++$ **do**
8 $\quad\quad\quad$ | $v_i -= q \cdot u_i$
9 $\quad\quad$ **end for**
10 $\quad\quad$ $\|\mathbf{v}\|^2 += q^2 \cdot \|\mathbf{u}\|^2 - 2 \cdot q \cdot dot$
11 $\quad\quad$ **return** *true*
12 $\quad$ **end if**
13 **end**

FPGAs to accelerate these computations with massive parallelism. The proposed hardware circuit for obtaining a vector product is shown in Fig. 1. The first step is a multiplication of corresponding coefficients. This multiplication is performed in one clock cycle, even for a very large lattice. After executing the multiplication step, the results are moved to an addition tree, consisting of $\lceil \log_2(n) \rceil$ addition layers. For the majority of FPGAs, the critical path for addition is shorter than for multiplication. Thus, it is possible to perform more than one addition in a single clock cycle without negatively affecting the maximum clock frequency. Let β denote the number of additions performed in one clock cycle, with a shorter latency path than multiplication. This parameter depends on an FPGA vendor and device family. For our target device, $\beta = 4$ addition layers are executed in one clock cycle, and the latency of the addition tree for an n-dimensional vector is $\lceil \log_2(n)/\beta \rceil$ clock cycles.

The proposed design also offers an option for the pipelined execution. It is possible to feed new vectors to registers $\mathbf{v}$ and $\mathbf{u}$ in each clock cycle, reaching the highest possible performance for a given set of vectors. The total latency required for computing the vector product is $1 + \lceil \log_2(n)/\beta \rceil$ cycles. Using this approach, the maximum level of parallelism is achieved.

4.2 Division with Rounding to the Nearest Integer

The next operation performed in the proposed accelerator for the *Reduce* function is a division with rounding to the nearest integer. The division involves the computed vector product, dot, and the square of the norm of the second vector $\|\mathbf{u}\|^2$. Instead of performing normal division, we take advantage of the fact that the result of the division is rounded to the nearest integer in a limited range,

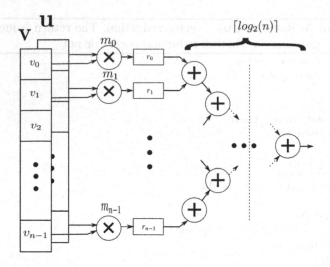

Fig. 1. Hardware module for the pipelined vector product computation. $\mathbf{v}$ and $\mathbf{u}$ are input vectors stored in registers.

so several conditions can be checked instead of performing a real division. The comparisons being made are listed in Table 1 and are easily executed in hardware by using simple shifts, additions, and subtractions.

Table 1. Conditions checked for the rounded division with the restricted result range

Division result	Value	Condition
$<0; 0.5)$	0	$2 \cdot |dot| < \|\mathbf{u}\|^2$
$<0.5; 1.5)$	1	$\|\mathbf{u}\|^2 \le 2 \cdot |dot| < 3 \cdot \|\mathbf{u}\|^2$
$<1.5; 2.5)$	2	$3 \cdot \|\mathbf{u}\|^2 \le 2 \cdot |dot| < 5 \cdot \|\mathbf{u}\|^2$
$<2.5; 3.5)$	3	$5 \cdot \|\mathbf{u}\|^2 \le 2 \cdot |dot| < 7 \cdot \|\mathbf{u}\|^2$
$<3.5; 4.5)$	4	$7 \cdot \|\mathbf{u}\|^2 \le 2 \cdot |dot| < 9 \cdot \|\mathbf{u}\|^2$

The full range of possible results is not covered. The selection of the results range is based on statistical data and the chosen assumption. Assuming that sampled vectors provided to the sieving algorithm are no longer than x times the approximate shortest vector, the rounded division will never generate a result bigger than x. Based on experiments, $x = 4$ is sufficient to accept all sampled vectors. The division in the selected range is necessary to make the comparison with CPU implementations more accurate and avoid rare events when the vector is required to be reduced again, which may lead to data desynchronization in the accelerator.

4.3   Update of Vector Values

Having all the necessary values, it is possible to update the reduced vector element and its norm. These two operations can be performed in parallel.

The element update function simply subtracts the product $q \cdot \mathbf{u}$ from $\mathbf{v}$. This operation can also be executed in parallel. In the first step, the products $q \cdot u_i$ are calculated. They are then subtracted from v_i in the second step. Each step is executed in a separate clock cycle to decrease the critical path's length and obtain a higher maximum clock frequency.

In hardware, the norm update function is executed in three steps, taking one clock cycle each. At first, $2 \cdot dot$ and $q \cdot \|\mathbf{u}\|$ are computed. Secondly, the multiplication by q is applied to both partial results. At the end, the subtraction and addition operations are performed.

4.4   Reduce Module

The described above parts were used to develop the entire *Reduce* algorithm. The hardware block diagram combining previously discussed steps is shown in Fig. 2.

With additional shift registers required to store data for further steps of the algorithm, it is possible to start computations for a new vector pair in each clock cycle, utilizing pipeline properties of used building blocks, and increasing the total performance.

The latency for one pair of vectors depends only on the dimension of a lattice. For an n-dimensional lattice, the latency $f_{cl}(n)$ equals exactly

$$f_{cl}(n) = \left\lceil \frac{\log_2(n)}{\beta} \right\rceil + 5 \tag{5}$$

clock cycles. This is also the number of pairs of vectors being processed in the module concurrently. Therefore, for 200 MHz clock frequency, the pipelined version can perform up to 200,000,000 vector reductions per second, and the branched version can perform twice as many reductions. This calculation does not include the communication overhead, so the practical performance for a standard approach will be lower.

5   Caching Approach to Lattice Sieving for Multi-platform Environment

In the previous section, the hardware accelerator for the vector reduction was introduced. The next step is to develop a way of using this accelerator to increase the performance of any sieve. Due to the communication overhead, a simple call to the accelerator for every occurrence of the reduce operation will not give any speedup. The data transfer takes almost 90% of the total execution time, and the performance is lower than on standard CPU. Moreover, it is not possible to run the entire algorithm on an FPGA due to its lack of sufficiently large memory to

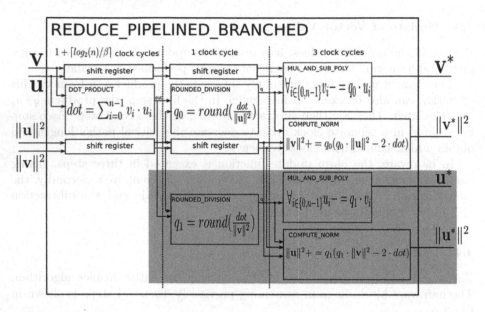

Fig. 2. The architecture of the reduce accelerator with pipelining and branching. The shaded part denotes logic implemented in the branched version and omitted in the standard implementation.

perform standalone sieving on FPGAs. Thus, in this section, a caching approach for lattice sieving algorithms in a multi-platform environment is presented. Our modification allows eliminating the communication delays, omitting the memory limitations, and fully utilizing the proposed parallel architecture for lattice sieving by combining previously reported methods with caching techniques. The proposed techniques will also work for other kinds of sieves.

A software/hardware approach is considered, where only a part of computations is performed in FPGAs, the rest of an algorithm is executed on CPU, and the majority of necessary data is stored on CPU. Currently, there are several approaches to combining CPUs with FPGAs. Thus, the calculations are not focused on any particular solution, but rather on a universal approach, applicable to each practical realization of the system combining both device types.

5.1   Data Transfer Costs

The biggest issue with current algorithms is the data transfer cost. Even the largest FPGAs are not able to store all required data to run sieving standalone for currently attacked dimensions. Thus, only a hybrid solution is considered. However, with only a part of the algorithm being executed on the FPGA side, some data is required to be exchanged between both sides. In lattice sieving, the transferred data will consist mostly of lattice points. For a simple vector expressing the lattice point, its size depends on the dimension of the lattice.

In the presented accelerator, each vector element is stored in 16 bits. It can be extended to 32 bits if needed, but due to our experiments on reduced lattices, 16 bits is sufficient. Additionally, the squared value of a vector length is also stored in another 32 bits. Thus, the number of bits $f_{nb}(n)$ required for a simple n-dimensional vector is expressed as:

$$f_{nb}(n) = n \cdot 16 + 32 = (n + 2) \cdot 16 \tag{6}$$

This number also matches the number of bits required for one vector transfer in any direction between CPU and FPGA. The communication time depends on the size of data and on the width of a data bus. The commonly used data buses are $w = \{32, 64, 128, 256\}$-bits wide and are able to deliver a new data in every FPGA clock cycle. The data transfer latency f_{tl} for one vector is expressed as the number of clock cycles and can be obtained from the equation:

$$f_{tl}(n, w) = \left\lceil \frac{f_{nb}(n)}{w} \right\rceil = \left\lceil \frac{(n + 2) \cdot 16}{w} \right\rceil \tag{7}$$

5.2 Reducing Newly Sampled Vectors by a Set

In large lattice dimensions, the total required memory is significantly larger than the memory available in any FPGA device. For dimensions larger than 85, the accelerator must cooperate with CPU during the reduction of the newly sampled vectors due to memory limitations. The vectors will be processed in smaller sets, and an efficient way to manage the data transfer is required.

In this approach, every new vector is used for reduction at least $2 \cdot |L|$ times. Assuming that the set L is going to be divided into smaller sets L_i, capable of fitting in FPGA memory, the data transfer costs may reduce or even eliminate the acquired acceleration.

In the most basic approach, the newly sampled vector $\mathbf{v}_n$ is reduced by the set L_i that fits FPGA's memory. In the first step, $\mathbf{v}_n$ is reduced sequentially by elements from L_i, while the reduction of elements from L_i by $\mathbf{v}_n$ in the second step is executed in parallel. Elements used in the first step of the reduction are replaced by other elements from L. This approach is not efficient due to the data transfer requirements, and several changes must be made to achieve the best performance.

5.3 On-the-Fly Reduction

It is not necessary to wait until all data is available on the FPGA side. The designed algorithm should take advantage of the fact that reductions may start right after sending the first two vectors. Every new vector will be reduced by those transferred so far, and the communication will happen in the background. This approach will allow to reduce the combined time of computations and data transfers.

The gains from the on-the-fly reduction depend on the approach for sieving. Applying ideas from Bos et al. [5] or Milde and Schneider [12] will require a different data transfer schedule and will be affected differently by the continuous memory transfer. The ideal algorithm should allow avoiding any data transfer costs.

5.4   Maximizing Performance with the Proper Schedule of Operations

To efficiently accelerate any sieving with FPGAs (or any other devices), the aforementioned elements must be included in the algorithm's design.

Taking ideas from literature for parallel sieve, let us divide the *GaussSieve* execution into three parts, as proposed by Ishiguro et al. [8] and extend it to meet our requirements.

The algorithm will operate on a set S of newly sampled vectors, instead of only one vector. The first part is the reduction of the set S by already reduced vectors in the list L. A data transfer latency for one lattice vector depends on the lattice dimensions and the data bus width w, as shown in Eq. 7. Thus, to avoid the data transfer overhead, one reduced lattice vector should be processed during the exact time required for a new one to be transferred. This can be done by extending the size of the set S from one to $k = f_{tl}(n, w)$. Then, taking into account the pipelining capabilities of the reduce function accelerator, during the first reduction, after k clock cycles, the accelerator should be able to start processing a new vector. The algorithm can take advantage of the pipelined execution of instructions due to the lack of any data dependency between vectors from the set S and an already used vector from the list L. The state of registers during the first step of sieving is visualized in Fig. 3, 4 and 5. The number of reductions in the first step is equal to $k \cdot |L|$, and this is also the number of clock cycles spent on computations. The communication cost will include only sending first $k+1$ vectors, where the remaining vectors will be transferred during computations. The FPGA latency will be then:

$$f_{el}^{p1}(n, w) = k^2 + k \cdot |L| + f_{cl}(n) \tag{8}$$

In the second step, elements from the set S are reduced by themselves. The set is already in FPGA memory, so there is no transfer overhead. The accelerator is going to execute the normal *GaussSieve* algorithm. Without the transfer overhead, the latency of computations is

$$f_{el}^{p2} = \frac{k^2}{2} \cdot f_{cl}(n) + \frac{k^2}{2} + f_{cl}(n) \tag{9}$$

During the second stage computations, a new batch S' of sampled vectors can be transferred to FPGA. The number of clock cycles required to transfer a new data consisting of k vectors is expressed as $k \cdot f_{tl}(n, w) = f_{tl}(n, w)^2 = k^2$, and is smaller than the computational latency of the second step.

In the last step, all vectors from the list L are reduced by vectors from the set S, which is already in the FPGA memory. Each vector $\mathbf{v} \in L$ is going to be reduced by k vectors, and reductions may be performed in parallel.

Again, there is no communication overhead. If any reduction occurs, the lattice vector is transferred back to CPU in the background. Otherwise if no reduction happens, there is no need for moving a given vector back to CPU. The latency of the third step is then

$$f_{el}^{p3} = k \cdot |L| + f_{cl}(n) \tag{10}$$

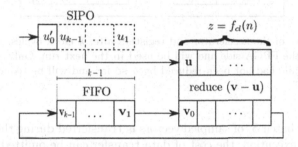

Fig. 3. The state of the accelerator and registers after the first clock cycle. The first part of the next vector $\mathbf{u}'$ is loaded, while the remaining parts of the SIPO unit contain parts of the previously loaded $\mathbf{u}$. The first vector $\mathbf{v}_0$ from the internal set is delivered to the reduce module to be reduced by the vector $\mathbf{u}$. FIFO contains $k-1$ elements and is smaller by one element than the SIPO unit.

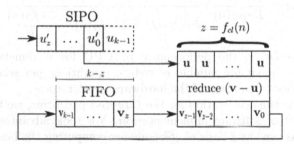

Fig. 4. The state of the accelerator and registers after $z = f_{cl}(n)$ clock cycles. The reduction of the first vector is finished and the vector $\mathbf{v}_0$ is going to be put in the FIFO queue. In every part of the reduce module, the same vector $\mathbf{u}$ is used for reduction. Only z parts from k parts of the new vector $\mathbf{u}'$ had been transferred so far. The FIFO queue contains $k-z$ elements.

By adding all the three steps together, it is possible to compute the latency of adding k new vectors to the list L of already Gauss-reduced vectors. The latency for the first execution will be then

$$f_{el}(n, w) = k^2 + 2 \cdot k \cdot |L| + \frac{k^2}{2} \cdot f_{cl}(n) + k^2 + 3 \cdot f_{cl}(n) \tag{11}$$

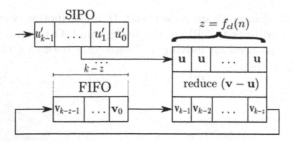

Fig. 5. The state of the accelerator and registers after k clock cycles. The entire new vector $\mathbf{u}'$ is on the FPGA side and will be used in the next run. Only v_i remaining in the **reduce** module had not been reduced by u so far and will be reduced in the next z clock cycles.

As the new batch S' of sampled vectors is transferred during the second step, for every next execution, the cost of data transfer can be omitted and then the final latency becomes:

$$f_{el}(n, w) = k^2 + 2 \cdot k \cdot |L| + \frac{k^2}{2} \cdot f_{cl}(n) + 3 \cdot f_{cl}(n) \tag{12}$$

Compared to CPU, the expected acceleration can be computed as

$$A = \frac{2 \cdot |L| + k}{P_{CPU}(n)} \cdot \frac{H}{k + 2 \cdot |L| + \frac{k}{2} \cdot f_{cl}(n) + \dfrac{3 \cdot f_{cl}(n)}{k}} \tag{13}$$

where the $P_{CPU}(n)$ is the performance of CPU for n dimensional lattice, expressed as the maximum number of reduce operations per second, and H is the maximum clock frequency of the hardware accelerator.

To determine the acceleration for the targeted platforms, we first measured the performance of software implementation. We took advantage of the *fplll* library, that was used as a basis of *g6k* code for computing the best result in the TU Darmstadt SVP Challenge (as of July 2020, dimensions from 153 to 170). Thus, the sieving operations implemented in *fplll* were used for constructing experiments aimed at measuring performance of software sieving.

In the designed experiment, a large set of vectors is sampled and pairwise reduced. Only the reduction time is measured. The size of the set is large enough to exceed the processor's cache memory, which allows us to measure the performance in a real scenario.

For the experiments, an Amazon Web Service `c5n.18xlarge` instance equipped with 72-core, a 3.0 GHz Intel Xeon Platinum processor was used.

In Fig. 6, an expected acceleration for the targeted lattice dimensions is presented, as the combined visualisation of Eq. 13 and obtained experimental data. For dimensions being currently considered in the SVP challenge (dimensions between 155 and 160), the expected acceleration from the proposed FPGA accelerator, compared to one CPU core, is around 30x. As for FPGA, clock frequency was set to 200 MHz, and in our algorithm, there is no visible difference between the considered data bus widths. The number of elements in L was set to 10,000.

The accelerator almost always performs the pipelined vector reduction. Only in a small part of the second stage, the reductions are not pipelined. The communication bottleneck has been completely eliminated. Almost the maximum theoretical performance of the proposed accelerator (i.e., the performance without taking into account the communication overhead) has been achieved with this approach. Moreover, the accelerator can be adapted to other parallel sieves and work with other devices. It is possible to use the proposed accelerator in the parallel implementation of *g6k* as one of the devices performing the basic step of sieving, a vector reduction.

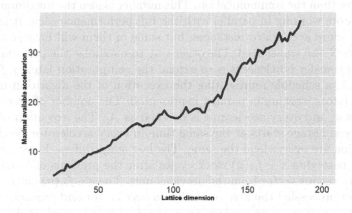

Fig. 6. Expected acceleration offered by one accelerator deployed on FPGA.

Some of the algorithms (e.g., [5]) allow immediately reducing both processed vectors ($\mathbf{v} - \mathbf{u}$ and $\mathbf{u} - \mathbf{v}$) in the next consecutive steps. In that case, a branched version of the accelerator can be used. Then, the third scenario changes, reducing the execution only to the two first steps. In the branched version, the first and the last steps are computed at a time. Thus, the total acceleration can increase by around 1.5 times.

6 Multiple Parallel Instances of the Accelerator in One FPGA

During the second stage, a new batch of sampled vectors is transferred to an FPGA. However, communication requires less time than the computations during the second stage. After all k vectors are transmitted, the data bus waits unused until the third stage of the algorithm. The clock latency for the data transfer $C_T(n)$ is k^2 cycles, whereas the computational latency $C_C(n)$ is $k^2 \cdot f_{cl}(n) + k^2 + f_{cl}(n))$ cycles. Then, the ratio $\frac{C_C(n)}{C_T(n)}$ indicates how many times the computations are longer. A difference between the two times can be used to send several other sets S to other accelerators implemented in the same FPGA. The maximum number of accelerators working in parallel is expressed then as:

$$\frac{C_C(n)}{C_T(n)} = \frac{k^2 \cdot f_{cl}(n) + k^2 + f_{cl}(n)}{k^2}$$

$$= f_{cl}(n) + 1 + \frac{f_{cl}(n)}{k^2} \tag{14}$$

$$\approx f_{cl}(n) + 1$$

The term $\frac{f_{cl}(n)}{k^2}$ for large dimensions is always lower than 1 and can be omitted. Then, for the targeted dimensions, $n > 64$, the computations are $f_{cl}(n) + 1$ times longer than the communication. This number is also the maximum number of accelerators working in parallel with the full performance each. It is possible to connect more accelerator instances, but some of them will have to wait until all new sets S' are transferred. The other way to maximize the performance and avoid data transfer bottlenecks is to extend the computation latency f_{cl}.

In Fig. 7, a schedule representing the execution of the algorithm using several accelerators working in parallel is presented. The number of accelerators is denoted as σ, and every accelerator is denoted as A_i. The execution of the first and the second stage starts at the same time in every accelerator. Vectors used for reduction are everywhere the same. The last stage differs. Every next accelerator starts sieving $k + f_{cl}(n)$ clock cycles after the previous one and takes as an input an output vector from the previous unit. The $k + f_{cl}(n)$ clock cycles are required for processing the first vector by an accelerator and pushing it further. The last accelerator starts working $(\sigma - 1) \cdot (k + f_{cl}(n))$ clock cycles after the first one. This approach is different from the first stage because the reduction $v_i - u_j$ is performed instead of $u_j - v_i$, where vectors v_i are loaded from CPU.

The inbound transmission is divided into two parts. In the first and the last stage, the previously reduced vectors, marked by small rectangles, are transferred to FPGA. In the second stage, a new batch of sampled vectors divided into σ sets, marked as a bigger rectangle, is transferred.

The outbound transmission starts in the third stage and can proceed until the end of the second stage. At first, the Gauss-reduced sets S_i of vectors from the second stage are sent back to CPU. As for results from the third stage, only shortened vectors are pushed back to CPU. It is hard to estimate the number of shortened vectors. To avoid data loses, the output FIFO queue should have a large enough memory available. Fortunately, there is only one FIFO queue, receiving data from the last accelerator.

The performance of multiple accelerators, implemented in one FPGA, scales with their number. For σ accelerators, the performance will be $\approx \sigma$ times higher compared to a single one. The acceleration is not exactly σ times better due to the $(\sigma - 1) \cdot (k + f_{cl}(n))$ clock cycles delay for the last module and proportionally less for other modules in the third step.

After the entire vector is transferred, this vector may be saved to one of the internal FIFO queues, currently not used by the reduce module, or directly

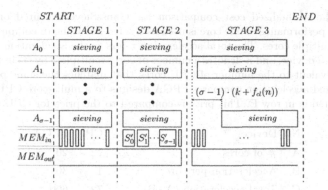

Fig. 7. The activity diagram for multiple accelerators in one FPGA. A_i denotes an i-th instance of the accelerator in FPGA, and σ is the number of accelerators. k is the size of sampled vectors sets S and S'. S is currently used, where S' will be used in the next run. $f_{cl}(n)$ is a function representing the reduction latency for n-dimensional vectors.

provided as one of the input vectors for reduction. Vectors from CPU are saved in queues only in the second stage of the algorithm. When the reduction is applied, there are two options. In the first and the second stage, vectors are always written back to the currently used internal queue. In the second and the third stage, vectors that were shortened during the reduction are placed in the output FIFO queue. In the second step, all vectors are transferred back to CPU, but in the third stage, only the reduced vectors are moved back. If vectors stay the same (i.e., there was no reduction), then they are overwritten in FPGA.

A multi-core version consists of several instances connected into a chain of accelerators. An output from one element is connected to the input of the next element. The last instance is responsible for the data transfer back to CPU.

6.1 Final Results

To measure the highest possible acceleration, we tried to fit as many instances as possible in one FPGA. The final number for 160-dimensional lattices is 20 accelerators working in parallel. This number is bigger than the boundary in Eq. 14, so we extended the f_{cl} latency to 12 to achieve higher clock frequency (150 MHz) and this way mitigated the second stage data transfer bottleneck. The design was described in the VHDL language and verified in simulation. Our code passes all stages of the FPGA design process. However, the actual run would be too long to be attempted with the current equipment and algorithm for a 160-dimensional lattice. Then, we used a proposed in the literature method for comparing cross-platform implementations [10], and we cost-compared our estimated results using two Amazon AWS instances: *f1.2xlarge* equipped with Xilinx FPGAs and *c5.18xlarge* aforementioned Intel Xeon. The results are presented in Table 2. The FPGA-based AWS instance can solve an equivalent problem for only 6% of the CPU-based instance price.

Table 2. The normalized cost comparison for GaussSieve executed on CPU and FPGAs. The performance of one core is used as a reference value to compute the acceleration for multiple cores. The total acceleration refers to the acceleration obtained by fully utilizing a device, and it denotes a number of cores multiplied by their acceleration, which is equivalent to the number of CPU cores that matches the same performance. The normalized acceleration compares FPGA designs to a multi-core CPU. The price per acceleration is in row E. This price is compared to the price for CPU in row F.

No	Device	CPU	FPGA
A	# of cores	72	20
B	Acceleration per core	1	30
C	Total acceleration (A · B)	72	600
D	Normalized acceleration	1	8.32
E	AWS price ($/h)	3.05	1.65
F	Price per acceleration (E/D)	3.05	0.20
G	Compared to CPU (F/F.CPU)	1	0.06

6.2   Comparison to Other Results

It is hard to compare cross-platform implementations. Looking only at the performance, the presented implementation achieves more than 8x speed-up compared to a 72-core CPU for a 160-dimensional lattice, so the implementation has the performance of around 576 CPU cores. The [13] achieved 21.5x acceleration for a 96-dimensional lattice when compared to [8] (2x CPUs with 8 cores), so it has the performance of around 344 cores (in a lower dimension). And the cost of power consumption will be very likely probably lower for FPGA when compared to GPU.

7   Conclusions

This paper introduces a new approach to lattice sieving by using a massively parallel FPGA design to accelerate the most common operation in every lattice sieving algorithm – vector reduction. As an example, the *GaussSieve* algorithm was accelerated. The acceleration is possible only with the proposed modification to parallel versions of sieving algorithms. The modification is devoted to eliminating the communication overhead between the specialized circuit, implemented in FPGA, and the CPU, running the rest of the algorithm, by using a caching strategy. The acceleration depends on the lattice dimension and increases linearly as a function of that dimension. For the targeted 160-dimensional lattice, the proposed solution is estimated to achieve 8.32 better performance compared to CPU. The results were obtained from FPGA simulation and CPU experiments. Comparing the cost of solving the SVP problem in AWS, the presented architecture will require only 6% of the CPU-based costs. Our project is also the first attempt reported to date to accelerate lattice sieving with specialized hardware.

The proposed hardware accelerator can be used directly for almost any lattice sieve performing a vector reduction operation. In this paper, the GaussSieve algorithm was investigated as an example algorithm. The parallel hardware architecture with the proposed caching strategy can be adapted to other GaussSieve modifications reported in the literature [5,8,12], as well as for other lattice sieving algorithms with a better complexity. As a part of future work, the adoption of the presented solution to algorithms other than GaussSieve will be explored. Additionally, an application of the proposed solution to other algorithms hard to implement in FPGAs due to the communication and memory bottleneck will be investigated.

References

1. SVP Challenge. https://www.latticechallenge.org/svp-challenge/
2. Ajtai, M., Kumar, R., Sivakumar, D.: An overview of the sieve algorithm for the shortest lattice vector problem. In: Silverman, J.H. (ed.) CaLC 2001. LNCS, vol. 2146, pp. 1–3. Springer, Heidelberg (2001). https://doi.org/10.1007/3-540-44670-2_1
3. Ajtai, M., Kumar, R., Sivakumar, D.: A sieve algorithm for the shortest lattice vector problem. In: Proceedings of the Thirty-Third Annual ACM Symposium on Theory of Computing, STOC 2001, pp. 601–610. ACM Press (2001). https://doi.org/10.1145/380752.380857. http://portal.acm.org/citation.cfm?doid=380752.380857
4. Albrecht, M.R., Ducas, L., Herold, G., Kirshanova, E., Postlethwaite, E.W., Stevens, M.: The general sieve kernel and new records in lattice reduction. In: Ishai, Y., Rijmen, V. (eds.) EUROCRYPT 2019. LNCS, vol. 11477, pp. 717–746. Springer, Cham (2019). https://doi.org/10.1007/978-3-030-17656-3_25
5. Bos, J.W., Naehrig, M., van de Pol, J.: Sieving for shortest vectors in ideal lattices: a practical perspective. Int. J. Appl. Cryptol. 3(4), 313–329 (2017). https://doi.org/10.1504/IJACT.2017.089353
6. Detrey, J., Hanrot, G., Pujol, X., Stehlé, D.: Accelerating lattice reduction with FPGAs. In: Abdalla, M., Barreto, P.S.L.M. (eds.) LATINCRYPT 2010. LNCS, vol. 6212, pp. 124–143. Springer, Heidelberg (2010). https://doi.org/10.1007/978-3-642-14712-8_8
7. Gentry, C., Peikert, C., Vaikuntanathan, V.: Trapdoors for hard lattices and new cryptographic constructions. In: Proceedings of the Fortieth Annual ACM Symposium on Theory of Computing, STOC 2008, p. 197. ACM Press (2008). https://doi.org/10.1145/1374376.1374407. http://dl.acm.org/citation.cfm?doid=1374376.1374407
8. Ishiguro, T., Kiyomoto, S., Miyake, Y., Takagi, T.: Parallel gauss sieve algorithm: solving the SVP challenge over a 128-dimensional ideal lattice. In: Krawczyk, H. (ed.) PKC 2014. LNCS, vol. 8383, pp. 411–428. Springer, Heidelberg (2014). https://doi.org/10.1007/978-3-642-54631-0_24
9. Klein, P.: Finding the closest lattice vector when it's unusually close. In: Proceedings of the Eleventh Annual ACM-SIAM Symposium on Discrete Algorithms, SODA 2000, pp. 937–941. Society for Industrial and Applied Mathematics, USA (2000)

10. Kuo, P.-C., et al.: Extreme enumeration on GPU and in clouds. In: Preneel, B., Takagi, T. (eds.) CHES 2011. LNCS, vol. 6917, pp. 176–191. Springer, Heidelberg (2011). https://doi.org/10.1007/978-3-642-23951-9_12
11. Micciancio, D., Voulgaris, P.: Faster exponential time algorithms for the shortest vector problem. In: Proceedings of the Twenty-First Annual ACM-SIAM Symposium on Discrete Algorithms, pp. 1468–1480. Society for Industrial and Applied Mathematics. https://epubs.siam.org/doi/10.1137/1.9781611973075.119
12. Milde, B., Schneider, M.: A parallel implementation of GaussSieve for the shortest vector problem in lattices. In: Malyshkin, V. (ed.) PaCT 2011. LNCS, vol. 6873, pp. 452–458. Springer, Heidelberg (2011). https://doi.org/10.1007/978-3-642-23178-0_40
13. Yang, S.-Y., Kuo, P.-C., Yang, B.-Y., Cheng, C.-M.: Gauss Sieve algorithm on GPUs. In: Handschuh, H. (ed.) CT-RSA 2017. LNCS, vol. 10159, pp. 39–57. Springer, Cham (2017). https://doi.org/10.1007/978-3-319-52153-4_3

Effect of Evaporation on Aggregation Kinetics of Clusters: A Monte Carlo Simulation Study

Nongdie Tan[1], Lei Chen[1], Xianglin Ye[1], Hao Zhou[1], and Hailing Xiong[1,2(✉)]

[1] College of Computer and Information Science, Southwest University,
Chongqing 400715, China
xionghl@swu.edu.cn
[2] Business College, Southwest University, Chongqing 402460, China

Abstract. In this paper, an aggregation model suitable for an evaporation system was constructed based on the cluster-cluster aggregation model. Evaporation affects not only the diffusion and aggregation probability of clusters, but also the rate of the whole aggregation process. The analysis of the aggregation model and the evaporation properties in an open system shows that the rate of colloidal aggregation usually depends on the diffusion probability and the aggregation probability in the diffusion process. The diffusion model and aggregation model of the colloid was derived from the electric double-layer theory and the properties of the evaporation system. The evaporation model was constructed to simulate the movement of the colloid in the evaporation process. The cluster distribution was analyzed using the cluster weight average. The influence of monomer aggregation on the aggregation process was also analyzed. The results show that the cluster dominated by Brownian diffusion has a decreasing effect on the aggregation as the simulation progresses. With slow evaporation, the aggregation process is almost the same as that of the basic aggregation model, whereas with fast evaporation, the evaporation accelerates the aggregation process.

Keywords: Sticking model · Collision probability · Aggregation probability · Aggregation kinetics · Monte Carlo simulation

1 Introduction

Evaporation is a common phenomenon in nature and has been widely studied in many fields. In the field of colloidal chemistry, evaporation has been most widely studied in connection with the preparation of thin films. The evaporation system is a typical open system: for example, droplet evaporation [1], thin film growth [2,3], and nanofilm preparation [4,5]. In the drying process of colloidal suspensions such as nanofluids, the particles distributed in the fluid can self-assemble to form various complex structures [1] (e.g., branched structures [6]). Researchers have mostly focused on the self-assembly properties of dispersed particles during

© Springer Nature Switzerland AG 2020
M. Qiu (Ed.): ICA3PP 2020, LNCS 12452, pp. 681–694, 2020.
https://doi.org/10.1007/978-3-030-60245-1_46

evaporation and the branching structure pattern after drying [1]. For example, Yuki et al. [7] used the lattice-gas model to study the preparation of nanofilms by evaporation. The state of the lattice is determined by calculating the energy change of each lattice. Studies have shown that, at high Pclet numbers, dense clusters are easily obtained by evaporation. Zigelman et al. [8] studied the evaporation of droplets with volatile suspensions. They found that the rate of liquid evaporation and the local aggregation rate of particles determine the geometry of the sediment. The evaporation rate affects the stability of the solution in which the particles are located, which in turn affects the aggregation properties of the particles and the final aggregation structure. Semenov et al. [9] studied the process of droplet diffusion and evaporation in the solid mechanism in the case of complete wetting and partial wetting at the same time. The results showed that diffusion and evaporation dominated in different stages of the process. Li et al. [10] studied the evaporation characteristics of immobilized nanofluid droplets. Their results showed that the volume percentage of nanoparticles and surface wettability affects their diffusion behavior.

The influence of evaporation on the motion properties of particles determines the structure of clusters. Previous studies have mostly been based on the molecular-dynamics method, which calculates the interaction potential between particles at each step. In evaporation experiments, advanced measuring instruments or high-precision analysis technology are needed to obtain the microstructure parameters of clusters: for example, particle size, density, and porosity. This method is lengthy and expensive. Moreover, it is difficult to find the law underlying the random phenomenon by ordinary experimental methods, and the method ignores the evolution of particle morphology in the process of aggregation. Monte Carlo simulation is a recently developed computer-simulation method that can be used to study the interaction of colloidal clusters. It transforms the complex forces between particles into corresponding probability models and reduces the amount of calculation required. Moreover, it is suitable for large-scale particle simulation and can be used to simulate the thermodynamic properties, dynamic properties, structural properties, and morphological characteristics of colloids [11]. It is therefore of benefit to summarize the characteristics of evaporation systems and use Monte Carlo simulation to establish an evaporation model that reflects the dynamic process of particle aggregation [12].

Based on the clusterCcluster aggregation (CCA) model, this paper proposes an evaporation model using Monte Carlo simulation to study the diffusion and aggregation of cluster particles in an evaporation system. In the evaporation system, a large number of colloidal particles are distributed in the systems initial state, and water evaporates at a constant rate. As time passes, the amount of solvent in the system decreases, and the stability of the dispersion in the system is affected by the solvent evaporation. solvent evaporations Its influence is mainly reflected in the following aspects: (1) the decrease of water content in the system leads to an increase of electrolyte concentration, a change in the electric double-layer interaction between particles in the colloidal dispersion, and a change in the particle aggregation probability; (2) the decrease of solvent and the constant number of particles lead to an increase in particle concentration and the probability of collision between particles; (3) evaporation causes the liquid level to drop, which reduces the number of clusters on the surface of the body and affects their movement. Evaporation therefore not only affects the diffusion and aggregation probability of clusters, but also affects the rate of the whole aggregation process. The analysis of the open-system aggregation model and evaporation properties shows that the rate of colloidal aggregation usually depends on the collision probability and the aggregation probability in the collision process. This paper defines the collision probability and adhesion probability of colloids according to the characteristics of the evaporation system. The evaporation model is then constructed.

This paper is structured as follow: By referring to the idea of evaporation process of the CCA model, the diffuse model and aggregation model proposed in Sect. 2.1 and 2.2. Section 3.1 demonstrated effect of evaporation rate on aggregation results. Section 3.2 analyzed effect of evaporation rate on weight average of system. Section 3.3 demonstrated Cluster distribution under the conditions of rapid evaporation and slow evaporation. Section 3.4 analyzed Influence of monomer aggregation probability on aggregation process.

2 Construction of Evaporation Model

2.1 Construction of Diffuse Model

For the evaporation system, it is assumed that the cube with height L acts as the container of colloidal dispersion, and a large number of colloidal particles are distributed. All evaporation takes place on the liquid surface, the evaporation is carried out at a stable rate, and the surface area of the liquid remains stable. When the surface water evaporates at a specified rate, the surface of the suspension will drop, and the particles and clusters on the surface of the suspension will drop together. In the process of falling, the particles will collide with the particles below the liquid surface. Similar to the collisions caused by the Brownian motion of particles, these collisions may lead to aggregations. The height $H(T)$ of the suspension can be expressed as:

$$H(t) = L - V_e \tag{1}$$

where V_e is the evaporation rate and t is the current time. The corresponding unit time t can be expressed as [13]:

$$\Delta t = \frac{l_B^2}{6M_1(0)} \tag{2}$$

where l_B is the step length of Brownian motion and $M_1(0)$ is the diffusion coefficient of elementary particles at the initial time ($t = 0$).

According to the Einstein relation, the diffusion coefficient $M_1(t)$ of a single particle can be expressed as [13]:

$$M_1(t) = \frac{k_B T}{6\pi \eta(t) R} \tag{3}$$

where K_B is the Boltzmann constant, T is the absolute temperature, $\eta(t)$ is the solution viscosity, and R is the radius of the colloidal particles.

The solution viscosity $\eta(t)$ is related to the volume fraction $C(t)$ of the dispersion (i.e., the aggregation of the dispersion). Therefore, $\eta(t)$ can be expressed as [14]:

$$\eta(t) = \eta_0 * (1 + 2.5C(t)) \tag{4}$$

where η_0 is the viscosity of the pure solution. The volume fraction $C(t)$ of the dispersion is determined by the number of dispersions and the size of the simulation system:

$$C(t) = N(t)/V_s(t) \tag{5}$$

where $N(t)$ is the number of dispersions in the current system. If aggregation is considered in the open system without considering the occurrence of cluster dispersion, then small clusters will aggregate to form larger clusters, and the number of dispersions in the system will decrease.

The volume of the floating liquid, which decreases with the evaporation, can be expressed as:

$$V_s(t) = L^2 * H(t) \tag{6}$$

Additionally, according to formulas (1) and (6), the expression of $D_i(t)$ is as follows:

$$D_i(t) = \frac{k_B T}{6\pi\eta_0 r}[1 - \frac{2.5N(t)}{L^2(L - v_e t) + 2.5N(t)}] * S^r \tag{7}$$

2.2 Construction of Aggregation Model

The stability of colloids can be quantified by the stability ratio W, which is the ratio of the number of collisions between particles to the effective number of collisions after the collision between particles [15]. The larger the W, the more stable the system is, and the less likely the particles in the system are to aggregate. Therefore, assuming that the particle-bonding probability P_1 is in opposition to the system stability rate W, P_1 can be expressed as:

$$P_1 = 1/W \tag{8}$$

The Derjaguin-Landau-Verwey-Overbeek (DLVO) theory is used to describe the stability of charged colloidal particle aggregation and the interaction between particles, which is applicable to the charged colloidal solution theory without considering chemical adsorption [16]. When the solution in the system is a strong electrolyte solution–that is, the electrolyte can be completely ionized [17]–the stability of the system can be related to the electrolyte concentration C_E by using DLVO theory. When particles move in the electrolyte solution, the particles move close to one another, which can produce several forces. The classical forces are van der Waals attraction and electrostatic repulsion. The relationship between van der Waals gravity U_A, electrostatic repulsion U_R, and the total potential energy U is shown in Fig. 1. Different electrolyte solutions have different U curves. The repulsion or attraction between particles can be judged by the total potential energy U between them.

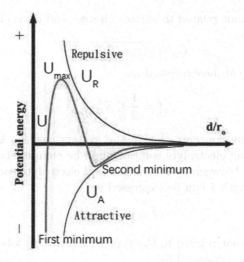

Fig. 1. Variation curves of gravitational potential energy U_A, repulsion potential energy U_R, and comprehensive total potential energy U by distance between particles [18]

As shown in Fig. 1, when a larger barrier peak U_{max} appears on the total potential energy curve, repulsion is dominant. If particles are to aggregate, they must cross this barrier. Therefore, the size of the repulsion barrier is the key to the stability of colloids. It has been shown that W is almost completely dependent on the barrier.

$$W \approx exp(U_{max}/k_B T) \qquad (9)$$

where U_{max} is the maximum barrier and k_B is the Boltzmann constant. The logarithm of W can be expressed as follows:

$$ln(W) \approx U_{max}/k_B T \qquad (10)$$

U_{max} can be expressed as [19]:

$$U_{max} = \frac{C_1}{k} - C_2 \qquad (11)$$

where C_1 and C_2 are the surface charge constants of ions, which depend on the type of solution; and k is the Debye–Huckel change. Its reciprocal is often used to characterize the thickness of the electric double layer. The expression is [19]:

$$k^{-1} = \sqrt{\varepsilon_0 \varepsilon_r T/2000 F^2 I} \qquad (12)$$

where ε_0 and ε_1 denote the relevant dielectric constants, F is the Faraday constant, and I is the ionic strength. Formula (13) can be further simplified as follows:

$$k^{-1} = C_3/\sqrt{I} \qquad (13)$$

where C_3 is a constant related to surface charge, and its value can be expressed as:

$$C_3 = \sqrt{\varepsilon_0 \varepsilon_0 T / 2000 F^2} \tag{14}$$

The ionic strength can be expressed as:

$$I = \frac{1}{2} \sum_B Z_B^2 / 2 \tag{15}$$

where M_B is the ion concentration in the system and Z_B is the ionic valence number. Since strong electrolyte solutions can be completely ionized, there is a positive correlation between ionic strength and electrolyte concentration. Therefore, the ionic strength I can be expressed as:

$$I = C_4 * C_E \tag{16}$$

where C_4 is a constant related to the type of electrolyte solution, which can be expressed as: can be expressed as:

$$C_4 = \sum_B Z_B^2 / 2 \tag{17}$$

C_E is the electrolyte concentration of the system. Combining equations (9) and (17), the aggregation probability can be expressed as:

$$P_1 = e^{\frac{K_1}{\sqrt{C_E}}} / e^{k_2} \tag{18}$$

Taking into account the evaporation of solvent, the C_E concentration of electrolyte varies with time [20]:

$$C_E(t) = \frac{L}{L - v_e t} C_0 \tag{19}$$

where C_0 is the initial electrolyte concentration. The expression of $P_{ij}(T)$ can be obtained by synthesizing formulas (5), (19), and (20):

$$P_{ij}(t) = \frac{e^{K_1 \sqrt{\frac{L - v_e t}{L C_0}}}}{e^{K_2}} * (i * j)^{\sigma} \tag{20}$$

The solution presented above was used to determine the influence of solvent evaporation on the diffusion coefficient and aggregation probability. This was then combined with the CCA model to simulate the evaporation of an open system.

3   Experiment and Analysis

3.1   Effect of Evaporation Rate on Aggregation Results

The simulation was carried out in a three-dimensional lattice simulation system with side lengths of 100 cm. The initial cluster concentration was 0.01, and

1,000 particles were randomly distributed in the system. In this study, the initial aggregation probability P_1 is set to 0.1. The final results of cluster aggregation at different evaporation rates obtained after water evaporation of the system are shown in Fig. 2. The evaporation rates V_E were set to 0.001 $d\Delta t^{-1}$, 0.01 $d\Delta t^{-1}$, 0.1 $d\Delta t^{-1}$, and 1 $d\Delta t^{-1}$, where d is the diameter of the particle. The side length is the length of the lattice unit. To distinguish clusters, the simulation program identifies each cluster using different colors. To distinguish clusters, the simulation program identifies each cluster using different colors.

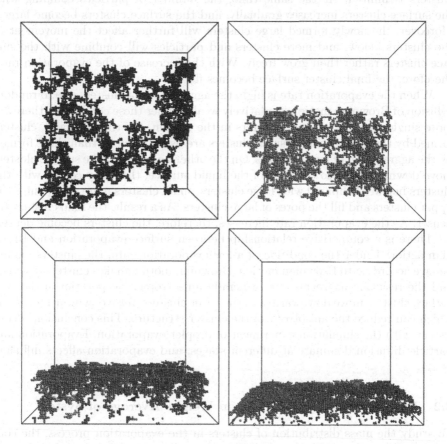

Fig. 2. The final state diagram of cluster aggregation at different evaporation rates

Evaporation mainly affects the water on the surface of the system. With the increase of the evaporation rate, therefore, the clusters on the surface of the system are susceptible to the drop in the liquid level and move continuously downward. In the downward movement, a cluster may collide with other clusters and aggregate, and so the whole process and the final cluster structure are susceptible to evaporation.

When the evaporation rate is low, the particles or clusters in the system are mainly affected by Brownian motion and there is a large probability of random diffusion, which leads to collision between particles and clusters and aggregation growth, resulting in the final open and loose formation of the aggregate. For example, the diffusion-limited aggregation model taking into account only Brownian motion results in the formation of branching clusters centered on the root node [21]. However, with the increase of evaporation rate, the downward movement of surface clusters occurs earlier, and the free diffusion of surface clusters is limited. At the same time, the number of particles colliding with the surface clusters increases gradually, and the surface clusters become larger. Moreover, the newly formed large clusters will further affect the movement of the clusters below, and more clusters and particles will combine with the surface clusters rather than grow freely. With the increase of the evaporation rate, therefore, the final cluster surface becomes flatter.

When the evaporation rate is high, the aggregation of clusters due to random collision of Brownian motion is relatively weak. Under these conditions, there are more single particles or small clusters in the system. On the one hand, clusters formed by the aggregation of small clusters are usually denser than those formed by the aggregation of large clusters. On the other hand, when the surface clusters move downward due to the drop of the liquid surface, they may collide with the clusters below. Compared with large clusters, small clusters more easily enter the upper clusters and fill the pores of large clusters. As a result, the final clusters are denser. As the evaporation rate increases, therefore, the clusters become denser.

There is a competitive relationship between surface evaporation and Brownian action. Under the condition of a low evaporation rate, the clusters mainly behave according to Brownian motion. Brownian motion makes clusters disperse, and the resulting aggregates are generally anisotropic. Evaporation makes the surface clusters move downward, and the lower clusters move downward together, which can reduce the anisotropy of the cluster structure. This conclusion is consistent with the simulation experiment of droplet evaporation. Evaporation and particle diffusion dominate at different stages, and evaporation affects diffusion [9, 22].

3.2   Effect of Evaporation Rate on Weight Average of System

To study the mass distribution of clusters in the evaporation process, the concept of the mass weight average (referred to as weight average) of clusters is introduced [13]. The change in weight average can reflect the growth rate of clusters in the system.

$$S(t) = \Sigma s^2 n_s(t) / \Sigma s n_s(t) \qquad (21)$$

where $n_s(t)$ is the cluster concentration with mass s at time t. In the actual simulation, the number of clusters is used to express the quality of clusters, and so $n_s(t)$ can be understood as the number of clusters with the size s at time t, as shown in Fig. 3.

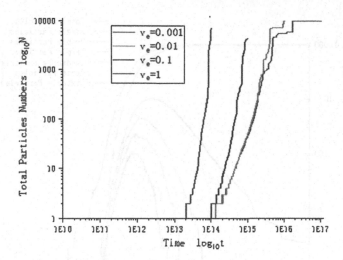

Fig. 3. Time dependence of cluster weight at different evaporation rates

In the simulation process, the cluster weight changes with time under different evaporation rates, as shown in Fig. 3. It was found that the average weight of clusters in the system changes very slowly initially, but after a long period of time, the average weight begins to increase rapidly and reaches the maximum value in a short time. This is because, in the beginning, the system consists of single particles or small clusters, and the cluster motion is mainly affected by Brownian action. As the simulation progresses, a layer of clusters gradually forms on the surface, which continuously sweep the lower clusters during the process of falling, and the clusters collide with one another to form larger clusters. The weight of the system increases rapidly and reaches the maximum value in a short time. In addition, with the increase of the evaporation rate, the liquid level drops faster, and the surface clusters accumulate faster, which makes the weight average increase faster.

3.3 Cluster Distribution Under the Conditions of Rapid Evaporation and Slow Evaporation

Figure 4 shows the time-dependence of clusters of different sizes in the system at evaporation rates of 0.001 $d\Delta t^{-1}$ (slow evaporation) and 1 $d\Delta t^{-1}$ (rapid evaporation). Given that the trend in the concentration change of clusters of sizes other than those marked in the graph is similar to that of six-particle clusters, this paper examines only the changes from single-particle to six-particle clusters and total clusters. It can be seen from Fig. 4 and Fig. 5 that the total number of clusters and the number of single particles decrease as the simulation progresses, and the number of clusters of other sizes increase first and then decrease.

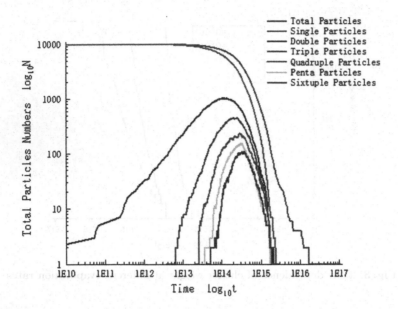

Fig. 4. Time-dependence of cluster numbers in the system at evaporation rate of $0.001d\Delta t^{-1}$

As shown in Fig. 4, when the solvent is evaporated to dryness, only one cluster remains in the final simulation system. Because there is little difference between the aggregation at a low evaporation rate and the aggregation under Brownian motion. The cluster has enough time for random motion, collision and aggregation, and finally aggregation into a cluster. In Fig. 5, when the solvent is evaporated to dryness, many clusters are still left in the system. Due to the aggregation at a high evaporation rate, the frequency of downward movement of clusters is higher. The collision aggregation caused by Brownian motion is frequently interrupted, and the aggregation process is weakened. When the system is evaporated to dryness, some clusters still exist at the bottom. This situation is similar to the cluster state in Fig. 2. In addition, in terms of simulation time, a high evaporation rate takes less time to simulate than a low evaporation rate. This phenomenon is also reflected in the change diagram of cluster weight average. In other words, evaporation accelerates the aggregation process and affects the final cluster structure.

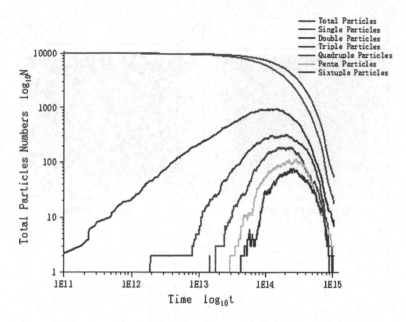

Fig. 5. Time-dependence of cluster numbers in the system at evaporation rate of $0.1d\Delta t^{-1}$

3.4   Influence of Monomer Aggregation Probability on Aggregation Process

According to the evaporation model, in addition to the evaporation rate, monomer aggregation probability is an important factor affecting the evaporation process. According to the different probabilities of particle aggregation, the classical CCA model can be divided into reaction–limited colloid aggregation (RLCA) and diffusion–limited colloid aggregation (DLCA) models. These two models correspond to slow aggregation and fast aggregation, respectively. The particle aggregation probability P_1 is set to 0.01 and 1, respectively, and the evaporation rate is fixed at 0.01 $d\Delta t^{-1}$. The termination condition of the simulation is still the complete evaporation of water from the system. The cluster state diagram is shown in Fig. 5. This paper presents the state diagrams of intermediate clusters with liquid surface heights of 80 D, 40 D, and 0 D under two kinds of aggregation probabilities.

Figure 6 (a), (c), and (e) represent slow aggregation ($P_1 = 0.01$). The aggregation mainly occurs on the surface of the system. The surface cluster aggregation is caused by the drop of the liquid level. The final clusters are dense, which is consistent with the characteristics of RLCA (slow speed and high fractal dimension). Figure 6 (b), (d), and (f) represent fast aggregation. Although there is growth in surface clusters due to the drop of the liquid level in the system, other clusters far from the surface layer have a high probability of aggregating with other clusters before being affected by the surface clusters, which leads to the formation of clusters with many branches. Figure 6 (e), (c), and (f) represent the

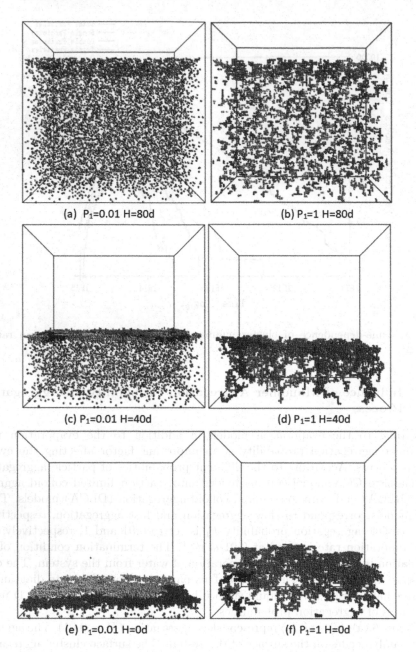

(a) P₁=0.01 H=80d   (b) P₁=1 H=80d

(c) P₁=0.01 H=40d   (d) P₁=1 H=40d

(e) P₁=0.01 H=0d   (f) P₁=1 H=0d

Fig. 6. Cluster aggregation snapshot in the intermediate state of fast aggregation and slow aggregation at evaporation rate of $0.01\mathrm{d}\Delta t^{-1}$

final clusters when the aggregation probability is 0.01, 0.1, and 1, respectively. As the aggregation probability increases, the structure of clusters becomes more open. This is consistent with the conclusion in the classic CCA model [23].

4 Conclusion

In this paper, an evaporation aggregation model was established by analyzing the characteristics of an evaporation system and the CCA aggregation model. This study also examined the particle motion in the cubic lattice simulation space. As there are too many free variables involved in the model, the aggregation probability P_1 is set as a constant. Additionally, the paper discussed the effects of evaporation rate and monomer aggregation probability on the aggregation process, and it analyzed the final state of cluster aggregation, the number of clusters at all levels, and the weight average of clusters. The results show that evaporation leads to the formation of dense clusters on the surface of the system and drives the clusters below to move downward, which accelerates the aggregation process. In addition, to a certain extent, the increase of the evaporation rate limits the free diffusion of clusters, resulting in the formation of denser clusters.

Acknowledgements. This work was financially supported by the National Natural Science Foundation of China (41271292), the Key Project of Chongqing Science and Technology Bureau (cstc2019jscx-gksbX0103), the Fundamental Research Funds for the Central Universities of China (SWU2009107) and the Key Project of Education Department of Anhui Province of China (KJ2019A0864).

References

1. Rabani, E., et al.: Drying-mediated self-assembly of nanoparticles, **14**(4), 1449–1454 (2012)
2. Crivoi, A., Duan, F.: Evaporation-induced formation of fractal-like structures from nanofluids. Physical Chem. Chemical Phys. **426**(6964), 271–274 (2016)
3. Crivoi, A., Duan, F.: Evaporation-induced branched structures from sessile nanofluid droplets. J. Phys. Chem. C. **117**(15), 7835–7843 (2013)
4. Yosef, G., Rabani, E.: Self-assembly of nanoparticles into rings: a lattice-gas model. J. Phys. Chem. B. **110**(42), 20965–20972 (2006)
5. Wasik, P., et al.: Hierarchical surface patterns upon evaporation of a ZnO nanofluid droplet: effect of particle morphology. Langmuir **34**(4), 1645–1645 (2018)
6. Zhang, H., et al.: Modeling the self-assembly of nanoparticles into branched aggregates from a sessile nanofluid droplet. Appl. Therm. Eng. **94**, 650–656 (2016)
7. Kameya, Y.: Kinetic Monte Carlo simulation of nanoparticle film formation viananocolloid drying. J. Nanopart. Res. **19**(6), 214 (2017)
8. Zigelman, A., Manor, O.: Simulations of the dynamic deposition of colloidal particles from a volatile sessile drop. J. Colloid Interface Sci. **525**, 282–290 (2018)
9. Semenov, S., et al.: Simultaneous spreading and evaporation: recent developments. Adv. Colloid Interface Sci. **206**, 382–398 (2014)
10. Li, Y., et al.: Nanoparticle-tuned spreading behavior of nanofluid droplets on the solid substrate. Microfluid. Nanofluid. **18**(1), 111–120 (2015)
11. Xiong, H.-L., Yang, Z.-M., Li, H.: Coupling effects of diffusive model and sticking model on aggregation kinetics of colloidal particlesA Monte Carlo simulation study. Acta Phys. -Chim. Sin. **30**(3), 413–422 (2014)
12. Hai-Ling, X., Yong-Zhi, Y., Hang, L., Hua-Ling, Z., Xian-Jun, J.: Computer simulation of colloidal aggregation induced by directionalism of long range van der waals forces. Acta Phys. -Chim. Sin. **23**(08), 1241–1246 (2007)

13. Leone, R., et al.: Coupled aggregation and sedimentation processes: three-dimensional off-lattice simulations. Eur. Phys. J. E. **7**(2), 153–161 (2002)

14. Yuan, X., et al.: Cluster formation and rheology of photoreactive nanoparticle dispersions. Langmuir **24**(10), 5299–5305 (2008)

15. Mellema, M., van Opheusden, J.H.J., van Vliet, T.: Relating colloidal particle interactions to gel structure using brownian dynamic simulations and the fuchs stability ratio. J. Chem. Phys. **111**(13), 6129 (1999)

16. Vincent, B.: Early (pre-DLVO) studies of particle aggregation. Adv. Colloid Interface Sci. **170**(1), 56–67 (2012)

17. Sun, G., Yi, Z., Ngai, T.: Particle-stabilized interfaces and their interactions at interfaces. Acta Phys. Chim. Sin. **36**(10), 1910005 (2020)

18. Zhen, Z., Li, N.: Molecular Force and Colloidal Stability and Sedimentation. Higher Education Press, Beijing (1995)

19. Kim, S., et al.: Three-dimensional off-lattice Monte Carlo simulations on a direct relation between experimental process parameters and fractal dimension of colloidal aggregates. J. Colloid Interface Sci. **344**(2), 353–361 (2010)

20. Xiong, H., et al.: Application of the ClusterCCluster Aggregation model to an open system. J. Colloid Interface Sci. **344**(1), 37–43 (2010)

21. Alves, S.G., Ferreira, S.C., Martins, M.L.: Strategies for optimize off-lattice aggregate simulations. Braz. J. Phys. **38**(1), 81–86 (2008)

22. Li, Y., et al.: Nanoparticle-tuned spreading behavior of nanofluid droplets on the solid substrate. Microfluid. Nanofluid. **18**(1), 111–120 (2015)

23. Li, C., Xiong, H.: 3D simulation of the ClusterCCluster Aggregation model. Comput. Phys. Commun. **185**(12), 3424–3429 (2014)

Processing in Memory Assisted MEC 3C Resource Allocation for Computation Offloading

Yang Yang[1(✉)], Xiaolin Chang[1], Ziye Jia[2], Zhu Han[3], and Zhen Han[1]

[1] Beijing Key Laboratory of Security and Privacy in Intelligent Transportation,
Beijing Jiaotong University, Beijing, China
{16112082,xlchang,zhan}@bjtu.edu.cn
[2] State Key Laboratory of ISN, Information Science Institute, Xidian University,
Xi'an, China
ziyejia@stu.xidian.edu.cn
[3] Department of Electrical and Computer Engineering, University of Houston,
Houston, TX, USA
hanzhu22@gmail.com

Abstract. The improvement of Internet of Things (IoT) applications has led to a substantial increase in the number of multiple resources of computation, communication, and caching (3C). The fifth generation (5G) and multi-access edge computing (MEC) are promising to enhance the computation offloading of IoT applications with high performance and reliability. According to resource-consuming preferences, IoT applications can be divided into computation-hungry applications and memory-hungry applications. To deal with the computation-hungry applications, Graphics Processing Units (GPUs) are increasingly used to process simple computation tasks. Meanwhile, the running of memory-hungry applications is accompanied by massive data transfers between processing core and memory. These transfers can result in significant energy and performance costs. Processing in memory (PIM) is a computing paradigm that avoids most data movement costs by performing a part of the computations directly in the memory. In this paper, we focus on offloading computation tasks in MEC that require 3C resources with high efficiency and low energy consumption considering latency and resilience constraints in a PIM-assisted multi-core (PAMC) architecture of physical machines (PMs). We formulate an optimization problem to minimize the total weighted resource costs and energy consumption. We also present an algorithm based on the column generation to solve the problem. Simulation results demonstrate that the proposed PAMC architecture can achieve good results in terms of energy consumption and resources utilization in comparison with the traditional PMs' architecture with the same resources.

Keywords: Column generation · Computation offloading · GPU · Multi-access edge computing · Processing in memory

© Springer Nature Switzerland AG 2020
M. Qiu (Ed.): ICA3PP 2020, LNCS 12452, pp. 695–709, 2020.
https://doi.org/10.1007/978-3-030-60245-1_47

1   Introduction

The Internet of Things (IoT) is invaluable due to its ability to connect everything in the physical world with wireless communications [1]. The improvement of the IoT applications has led to a substantial increase in the number of multiple resources of computation, communication, and caching (3C). The fifth generation (5G) is regarded as the most promising technology to provide efficient and flexible support for IoT applications. 5G providers superior performance and quality of experience for applications that consume more resources and are also timing and resilience sensitive [2]. To guarantee the quality of service (QoS), it is critical to design an efficient and reliable computation offloading mechanism due to the limited battery capacity and computing ability of mobile terminal devices (MTDs) [3]. Transferring some compute-intensive processes from an MTD to a multi-access edge computing (MEC) host enables an acceleration of the application on the various types of mobile devices [4]. Besides, electrical energy consumption is one of the major operating costs of MEC. With the unprecedented growth in MEC, reducing energy consumption is a primary challenge for service providers (SPs). Computation offloading is a complex task and it may result in high energy consumption if it is not addressed effectively [5].

According to the resource-consuming preferences, the IoT applications can be divided into two types, computation-hungry applications, and memory-hungry applications, respectively. To deal with the computation-hungry applications, Graphics Processing Unit (GPU) which features high-performance parallel computing is increasingly used to proceed simple computation tasks in conjunction with CPU. IoT workloads such as big data analytic, machine learning, and artificial intelligence (AI), require highly parallel computing that the CPU is not specialized at. A group of GPU cores can execute hundreds of threads in parallel, where cores running the same instructions can have divergent control flows [6]. In cloud computing scenarios, GPUs have been used for AI and high-performance computing (HPC) by IBM [7]. Reference [8] studied the problem of how to construct a runtime target device between GPU and CPU, which also should be considered in the MEC computation offloading.

Data transfer overhead between computing cores and memory hierarchy has been a persistent issue for von Neumann architectures and the problem has only become more challenging with the emergence of multi-core systems [9]. The energy and performance costs to move this data between the memory subsystem and the processing cores now dominate the total costs of computation. Processing in memory (PIM) is a computing paradigm that avoids most data movement costs by doing a part of computations directly in the memory [10]. Both CPU and GPU architectures and applications, where main memory bandwidth is a critical bottleneck can benefit from the use of PIM. Authors in [11] proposed designs that enable PIM in the three major memory technologies. A survey [12] has been studied in which the authors present an analysis of various dynamic random access memory (DRAM) designs and PIM systems to reduce memory access time. In [13], a task-offloading condition and GPU-based PIM configurations are

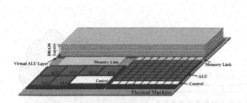
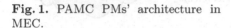
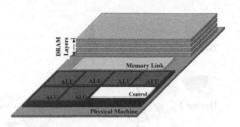

Fig. 1. PAMC PMs' architecture in MEC.

Fig. 2. Traditional PMs' architecture in MEC.

proposed while focusing on energy efficiency. But they didn't apply it in 5G and MEC computation offloading scenario.

The computation offloading problem is usually defined as integer linear programming (ILP) and the total solution set is exponential. General exhaustive searching is impossible with a large number of IoT devices. The column generation is an efficient method for large-scale ILP problems which have been widely used in scheduling problems [14–16].

Driven by the above considerations, this paper aims to explore an effective and efficient 3C resource allocation approach for computation offloading in MEC. The approach can auto-select offloading destinations according to services type to minimize energy consumption (defined later in (2)) and SP's long-term cost (defined later in (3)) considering latency and resilience constraints. We design a PIM-assisted multi-core (PAMC) architecture for physical servers in MEC to help improving the performance of computation offloading. Then we formulate the problem as an ILP and propose a column generation-based computation offloading algorithm which named CGBA to solve it. A comprehensive comparative study is made between the PAMC and traditional topology with the same amount of resources using CGBA. To the best of our knowledge, no performance evaluation of these two architectures has been carried out in the study of 3C computation offloading.

The remainder of this paper is organized as follows. In Sect. 2, the system model is presented. Section 3 presents the proposed CGBA algorithm. Section 4 provides the results for performance evaluation. Finally, we conclude the paper in Sect. 5.

2   System Model

This section presents the system model of the computation offloading problem. Firstly, we introduce the structures of PAMC. Next we describe the architectures of substrate topologies and computation tasks. Finally, we provide the corresponding problem formulations for the computation offloading problem.

Figure 1 and Fig. 2 show the PAMC and the traditional architecture of a physical machine, respectively. The most significant difference from the traditional architecture is the PMAC addition of an virtual arithmetic logic unit

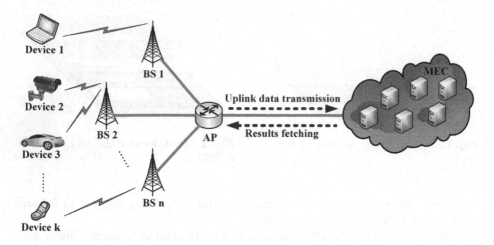

Fig. 3. MTD-BS-AP-MEC system.

(ALU) layer at the lowest level of the memory. A small amount of computation resources is deployed on the virtual ALU layer to support the simple calculations of memory-hungry services. This operation avoids the frequent transferring of data between processing cores and memory. The second difference is adding a set of GPUs which feature parallel computing next to CPU. These GPUs can help to process some auxiliary, simple computation tasks. To this end, an application can be properly partitioned and scheduled to execute on either CPU, GPU or memory according to its service type.

Figure 3 describes the MTD-BS-AP-MEC system. The MTD-BS-AP-MEC system consists of four components: mobile terminal devices (MTDs), base stations (BS), an aggregation point (AP), and MEC. Application mobility is a unique feature of the MEC system. To keep the services active during the computation offloading, multiple BSs are located close together. An AP is located between BSs and MEC to serve several BSs [4]. In this system, MTDs and BSs communicate by orthogonal frequency division multiplexing (OFDM) while BSs and the AP, the AP and MEC are both connected by fiber-optic networks. MTDs generate several computation tasks, such as payment requests, video, chatting, etc. These computation tasks cannot all be processed by MTDs due to the limited battery capacity and computing ability. Computation offloading can help with this situation by distributing computation tasks to MEC.

A tenant can use "virtual requests" as a means of specifying their resource requirements to SP. This paper defines a virtual request to represent a virtual network (including virtual nodes, which represent different types of service, and virtual links) allocated to an MTD. Each component of the virtual requests has its own physical resource demands [5]. Computation offloading can be completed in three steps: uplink data transmission, data processing, and results fetching [17]. Here is the computation offloading procedure of virtual requests. Firstly, the virtual requests generated from MTDs are transmitted to a BS over OFDM.

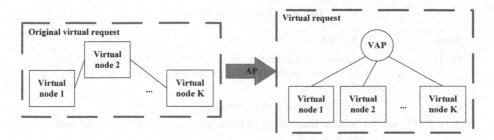

Fig. 4. Virtual requests.

Then, the BS will transmit the complete requests to the AP. In the AP, the virtual nodes will be transmitted to different destinations through physical links according to the proposed offloading strategy. Then, after being processed by offloading destinations, the processing results will be transmitted back to devices through the AP and BS. We only discuss the communication between the AP and MEC in this paper.

To express the computation resources of processing core better and reduce the calculation complexity, we use U to represent a data unit that need to be processed. Since the computing capability of a processing core is related to its frequency, we assume that it takes Q cycles to process a unit (U bit) data. The AP-MEC topology can be modeled as weighted graphs, denoted by $G_P(N^P, E^P)$. The physical nodes in MEC are denoted by N^S while n^{AP} represents the AP. Thus, $n^P \in N^P = n^{AP} \cup N^S$. Each physical node $n^S \in N^S$ in MEC is associated with processing core frequency (including CPU frequency and GPU frequency) and memory. Each link $e^P(n^{AP}, n^S) \in E^P$ between n^{AP} and a physical node n^S is associated with bandwidth capacity. All the physical resources (e.g. computing capability of processing core, memory and bandwidth) in $G_P(N^P, E^P)$ are limited.

Since virtual nodes in an original virtual request generated from MTDs represents different services of an application, we add a virtual aggregation point (VAP) for each virtual request to manage it better. This operation can be done in the AP, which is shown in Fig. 4. The star topology is suitable for hosting many types of applications. By doing this, we can model virtual requests as weighted graphs denoted by $G_V(N^V, E^V)$, $n^V \in N^V$, $e^V \in E^V$. Each virtual node $n^V \in N^V$ has different requirements of workload $w(n^V)$, memory $m(n^V)$ and also, resilience constraint $R(n^V)$. A virtual node with resilience constraint means that it has a higher requirement for the running environment. It should be processed in a safe and secure environment to protect it from being attacked. Referring to some resiliency measures which have already been in commercial use, we try to satisfy the resilience constraints by pre-installing virtual firewalls on some physical servers. Virtual nodes with resilience constraints can only be offloaded to the physical servers on which a virtual firewall have been pre-installed. Virtual links $e^V(n^{AP}, n^V) \in E^V$ between VAP and virtual nodes n^V are used to communicate between different services. Bandwidth requirements

Table 1. Variable definition.

Term	Definitions
$f(u)$	Frequency of physical node u associated with CPU or GPU used to process data according to service types of virtual nodes. $u \in N^S$
$A_m(u)$	Available memory of physical node u. $u \in N^S$
$A_b(e^P(n^{AP}, u))$	Available bandwidth of links between the AP and physical node u. $u \in N^S$
$R(u)$	A binary variable. $u \in N^S$. Denote whether a physical node u has by pre-installed virtual firewall. 1 means installed; otherwise 0
$w(v)$	Workload need to be processed of virtual node v. $v \in N^V$
$m(v)$	Memory request of virtual node v. $v \in N^V$
$b(e^V(n^{VP}, v))$	Bandwidth requirement between the VAP and virtual node v. $v \in N^V$
$R(v)$	A binary variable. $v \in N^V$. Denote whether a virtual node has resilience constraint. 1 means has; otherwise 0
$T_{\max}$	Latency constraint of a virtual request

$b(e^V)$ should to be satisfied. All the bandwidth used by virtual links cannot exceed the total available bandwidth of the physical links. Besides, each virtual request should be completed within given latency constraint $T_{\max}$. Table 1 defines the variables used in the following.

Assume that MTDs generate K active virtual requests in interval T. Equation (1) defines the revenue of serving the kth virtual request G_{Vk}.

$$R(G_{Vk}) = \sum_{v \in N^{Vk}} \alpha_w \cdot w(v) + \sum_{v \in N^{Vk}} \alpha_m \cdot m(v) + \sum_{e^V \in E^{Vk}} \alpha_b b(e^V). \quad (1)$$

Here, α_w, α_m, and α_b are the weighted parameters to determine the relative importance of physical resources.

Assume that the energy consumption of MEC is related to the power required to maintain the operation of the servers. When calculating the total energy consumption, we can only consider the physical servers which are power on. What's more, energy consumption is independent of the number of services running on the server. That means when a virtual node is offloaded to a newly powered physical server, the energy consumption increases. To evaluate the performance of the proposed computation offloading strategy more intuitively, we only consider the energy change on arriving time of virtual requests in interval T. That means the energy consumption changing happens after a computation offloading decision has been made. We define $E(G_{Vk})$ to represent the energy consumption for serving the G_{Vk} as

$$E(G_{Vk}) = \sum_{u \in N^P} \omega \cdot (y_u - x_u). \quad (2)$$

Here, ω represents the power consumption of physical node u in interval T. x_u and y_u are both binary variables to denote whether a physical node is active or not before and after offloading. 1 means active; otherwise 0. Equation (3) defines the cost of serving G_{Vk} in terms of the consumed physical computing resources, memory and bandwidth resources.

$$C(G_{Vk}) = \eta_w \cdot \sum_{v \in N^{Vk}} w(v) + \eta_m \cdot \sum_{v \in N^{Vk}} m(v) + \eta_e \cdot \sum_{e^V \in E^{Vk}} b(e^V). \tag{3}$$

Here η_w, η_m and η_e are weighted parameters.

The total latency $L(G_{Vk})$ of virtual request G_{Vk} consists of three time intervals: uplink transmission time L_t from the AP to offloading destinations, processing time L_p on offloading destinations and results fetching time L_r from offloading destinations to the AP. Since the virtual nodes are transmitted by the same physical link in both uplink transmission and results fetching steps, we assume that L_t and L_r are the same. The total latency of G_{Vk} can be calculated by

$$L(G_{Vk}) = \max_{v \in N^{Vk}} \{ L_t(v) + L_p(v) + L_r(v) \}. \tag{4}$$

Each latency interval can be calculated by (5), (6) and (7), respectively. Here, d_v represents the offloading destination of virtual node v. We have

$$L_t(v) = \frac{Uw(v)}{b(e^P(n^{AP}, d_v))}, \tag{5}$$

$$L_p(v) = \frac{Qw(v)}{f(d_v)}, \tag{6}$$

$$L_r(v) = L_t(v). \tag{7}$$

To satisfy the resilience constraints of virtual nodes, we pack virtual firewalls as virtual machines and pre-install them on several physical servers. A virtual firewall can protect the services running on a physical server from being attacked. If a physical node $u \in N^S$ has been pre-installed a virtual firewall, we set $R(u) = 1$; otherwise, $R(u) = 0$.

3   The Proposed Computation Offloading Approach

In this section, we first formulate the optimization problem as ILP. Then we decompose the problem to a master problem and a set of pricing problems to fit the column generation algorithm. Thirdly, we propose the CGBA algorithm based on column generation and give the algorithm described in detail.

3.1   Overview

The objective of the proposed approach is to minimize energy consumption (defined in (2)) and SP's long-term cost (defined in (3)) under constraints. The objective function can be defined as

Table 2. Variable definition

Term	Definitions
λ_c	A binary decision variable. Its value is 1 if configuration c is used in offloading; otherwise, set to 0. $c \in C$
a_c^v	A binary decision variable. Its value is 1 if configuration c serves virtual node v; otherwise, set to 0. $v \in N^V$. $c \in C$
z_c	The physical node associated with configuration c. $c \in C$

$$
\begin{aligned}
EC(G_V) &= C(G_V) + \rho \cdot E(G_V) \\
&= \eta_w \cdot \sum_{v \in N^V} w(v) + \eta_m \cdot \sum_{v \in N^V} m(v) + \eta_e \cdot \sum_{e^V \in E^V} b(e^V) + \eta_\rho \cdot \sum_{u \in N^P} y_u.
\end{aligned}
\tag{8}
$$

The computation offloading problem is ILP and the total solution set is exponential. General exhaustive searching is impossible with a large number of IoT devices. In this paper, we use a column generation-based algorithm CGBA to solve it. According to the principle of the column generation algorithm, the computation offloading problem can be decomposed into a master problem and a set of pricing problems. In both the master problem and pricing problems, the resource constraints of substrate topology should be guaranteed. We denote computation offloading configurations (COCs) as resource allocation strategies for physical nodes. That means a COC c can provide a resource allocation strategy to demonstrate whether virtual node $v \in n^V$ is offloaded in physical node $u \in n^S$. C is the set of all possible COCs. $c \in C$. We also denote $COST_c$ to calculate the cost of configuration c. Table 2 defines the variables used in the following.

3.2 Master Problem

The purpose of the master problem is to choose a maximum of $|N^S|$ configurations among COCs to minimize the total energy consumption and cost. Since it is hard to consider all the COCs in the calculations, the master problem should be restricted to a restricted master problem in column generation algorithm. By solving the restricted master problem, we can get an relaxed approximate optimal solution of the master problem. Equation (9) defines the objective function of the restricted master problem associated with λ_c. As listed in Table 2, the binary decision variable λ_c is to determine whether COC c is used in the offloading.

$$
\begin{aligned}
f(G_V) &= \sum_{c \in C} COST_c \lambda_c + \rho \sum_{c \in C} \lambda_c \\
&= \sum_{c \in C} (COST_c + \rho) \lambda_c \\
&= \sum_{c \in C} \sum_{v \in N^V} a_c^v (w(v) + m(v) + b(e^V(n^{VA}, v))) \lambda_c + \rho \sum_{c \in C} \lambda_c.
\end{aligned}
\tag{9}
$$

ρ is weighted parameter. The formulations of the restricted master problem are following.

$$\min_{\lambda_c \in \{0,1\}} f(G_V) \tag{10}$$

s.t.

$$\sum_{v \in N^V} \lambda_c a_c^v w(v) Q \leq f(z_c) T_{\max}, \forall c \in C \tag{11}$$

$$\sum_{v \in N^V} \lambda_c a_c^v m(v) \leq A_m(z_c), \forall c \in C \tag{12}$$

$$\sum_{v \in N^V} \lambda_c a_c^v b(e^V(n^{VA}, v)) \leq A_b(e^P(n^{AP}, z_c)), \forall c \in C \tag{13}$$

$$\sum_{c \in C} \lambda_c a_c^v = 1, \forall v \in N^V \tag{14}$$

$$\sum_{c \in C} \lambda_c \leq |N^S|. \tag{15}$$

Constraint set (11) enforces the computing ability of physical nodes. Constraint set (12) enforces the memory bounds of physical nodes. Constraint set (13) enforces the capacity bounds of physical bandwidth. Constraint set (14) makes sure that all the virtual nodes should be offloaded. Constraint set (15) means that each physical node can not be associated with more than one configuration.

3.3 Pricing Problem

As mentioned above, the pricing problem is to generate additional COCs. The newly generated COC should satisfy all the constraints of virtual requests. According to the steps of column generation algorithm, we can get the dual variables associated with constraints (11), (12), (13) (14), and (15) by solving the dual problem of the restricted master problem. We denote them with γ, η, ϕ, φ, and δ, respectively. Then the reduced cost function can be written as

$$\begin{aligned} RC(G_V) = &\sum_{c \in C}(COST_c + \rho) \\ &+ \gamma \cdot \sum_{c \in C} \sum_{v \in N^V} a_c^v w(v) Q + \eta \cdot \sum_{c \in C} \sum_{v \in N^V} a_c^v m(n^V) \\ &+ \phi \cdot \sum_{c \in C} \sum_{v \in N^V} a_c^v b(e^V(n^{VA}, v)) - \psi \cdot \sum_{v \in N^V} a_c^v + \delta. \end{aligned} \tag{16}$$

The formulations of the pricing problem are presented as follows, which is also ILP. The objective function is defined in (17). By solving the pricing problem, we can get a newly generated COC. By checking the minimum value of reduced cost, we can determine whether the newly generated COC can be added to the restricted master problem.

$$\min_{a_c^v \in \{0,1\}} RC(G_V) \tag{17}$$

Table 3. Steps of CGBA

Algorithm	
Input:	(G_P, G_V)
Output:	The offloading solution of G_V
Step 1:	Restrict the master problem to the restricted master problem. Set initial columns of the restricted master problem which make sure the solution of λ_c is feasible
Step 2:	Solve the dual problem of the linear relaxation of the restricted master problem and generate the dual multipliers γ, η, ϕ, φ, and δ
Step 3:	Pass the values of dual multipliers to the reduced cost function (defined in (16)), solve the pricing problem to get the minimum reduced cost value
Step 4:	Check the solution of the pricing problem. If the minimum reduced cost value is negative, this newly generated configuration can be added to the restricted master problem, then go to **Step 2**
Step 5:	Otherwise, stop solving the pricing problem. Solve the linear relaxation of restricted master problem directly
Step 6:	Return solution of the restricted master problem. If the solution is not an integer, we round it up and get the final solution

s.t.

$$\sum_{v \in N^V} a_c^v w(v) Q \le f(z_c) T_{\max}, \tag{18}$$

$$\sum_{v \in N^V} a_c^v m(v) \le A_m(z_c), \tag{19}$$

$$\sum_{v \in N^V} a_c^v b(e^V(n^{VA}, v)) \le A_b(e^P(n^{AP}, z_c)), \tag{20}$$

$$\max_{v \in N^V} \left\{ \frac{2U w(v) a_c^v}{b(e^P(n^{AP}, z_c))} + \frac{Q w(v) a_c^v}{f(z_c)} \right\} \le T_{\max}, \tag{21}$$

$$R(v) a_c^v = R(z_c), \forall v \in N^V \tag{22}$$

$$\sum_{v \in N^V} a_c^v >= 1. \tag{23}$$

Constraint sets (18), (19), and (20) enforce the capacity bounds of physical resources. Constraint sets (21) and (22) guarantee that the latency and resilience constraints are satisfied. Constraint set (23) means that at least one feasible COC should be host at least one virtual node.

3.4 Details of CGBA

To clearly describe the flow of CGBA, we give the algorithm described in detail. Firstly, the master problem needs to be restricted to a restricted master problem which has a smaller scale. By this operation, we can start the iteration in a small searching area to reduce the complexity. To solve the linear relaxation of the restricted master problem, it is necessary to set initial columns that can make sure the solution of λ_c is feasible. Secondly, solving the dual problem of linear relaxation of the restricted master problem to yield the dual multipliers of λ_c. γ, η, ϕ, φ, and δ are dual variables associated with constraints (11), (12), (13), (14), and (15), respectively. Thirdly, passing the values of γ, η, ϕ, φ, and δ to the reduced cost function (defined in (16)). Solving the pricing problem to get a new COC and the minimum reduced cost value of the newly generated COC. Finally, checking the minimum reduced cost value we achieved in the last step. If it is negative, we add the newly generated COC to the restricted master problem. Then the next iteration begins to solve the linear relaxation of restricted master problem. Otherwise, we stop solving the pricing problem and solve the linear relaxation of restricted master problem directly. Notice that, the solution returned by the restricted master problem may not be an integer. We round it up and get the final solution. Table 3 shows the key steps of the CGBA.

4 Performance Evaluation

We carry out experimental assessments using CPLEX. Simulations are conducted by evaluating the performance of different numbers of virtual requests in PAMC and traditional PM's architecture. These two settings have the same amount of resources. Firstly, the scenario parameters used in the simulations are set. Then, we present the simulation results and discussions.

4.1 Simulation Environment Configurations

The physical network topologies and virtual requests are both generated by using the GT-ITM tool. To simplify the expression, we assume that the number of cores of a physical node is equivalent to CPU/GPU frequency. In MEC with the PAMC architecture, the substrate topology contains 20 physical servers with 8 CPU cores, 32 GPU cores, and 256G memory. Bandwidth between the AP and MEC is set to 1Gbps. In MEC with traditional architecture, the substrate topology contains 100 physical servers with 8 CPU cores and 256G memory. Bandwidth between the AP and MEC is set to 1Gbps. Virtual firewalls are pre-installed randomly on these physical nodes. Virtual nodes are defined to represent different services. To simplify the calculation, in this paper we define three types of services. Table 4 shows the details of their resource requirements. The total number of virtual nodes in interval T generated from MTDs is increasing from 3 to 60. Resilience constraints are generated randomly accompanied by the virtual nodes. The latency constraint of a virtual request is 1,000 s. The metrics considered in simulations include:

Table 4. Resource requirements of virtual nodes

Service type	CPU requirements	GPU requirements	Memory requirements
Type 1	4 cores	0	32G
Type 2	0	8 cores	32G
Type 3	0	0	48G

- **Active node number** which measures the energy consumption of the substrate topology.
- **Relaxation gap of active node number** which demonstrates the gap between the real active node number and the rounding up result.
- **Computing resource utilization** and **Memory utilization** which measure the utilization of physical resources.

To compare the metrics under the same network state, we control all virtual nodes to be successfully offloaded in the simulation. That means the acceptance rate of virtual requests is 100 and the offloading revenues in PAMC and traditional architecture are the same. All the weights defined in Section II are set to 1. Simulation runs 10 times in each scenario. The simulations are done in CentOS 5.4 running in VMware workstation virtual machine. The PC configuration is Intel Core i7-10710 CPU and 16 GB memory.

4.2   Simulation Result Analysis

The simulations aim to compare the numbers of physical nodes that are powered on, defined as active node number in this paper, and resource utilization in MEC with different topology. Fig. 5 and Fig. 6 describe the results about active node number. The results in Fig. 5 shows that with the enlarging of virtual nodes, the number of active nodes in PAMC increases significantly slower than that in traditional architecture. This is because GPU can help CPU process some computation-hungry services on the same PM. Meanwhile, PIM shares the pressure of the CPU to process memory-hungry services, so that the CPU can free up computing power to process other computation-hungry services. This strategy improves the throughput of the processing cores and helps reduce the number of active nodes. The results in Fig. 6 shows the relaxation gap between real active node number and round up results. We get the solutions of computation offloading problems using the column generation algorithm by solving the linear relaxation of the restricted master problem. According to the actual physical meaning, we round up the results which are not integers to get the final results. In some cases, the resources of the relaxation gap are idle. The results in Fig. 6 show that the PAMC has a higher resource idle rate than that in traditional architecture, which means the performance can be improved further. Figure 7 and Fig. 8 describe the results of computing resources utilization and memory utilization, respectively. PAMC has higher computing resources utilization and lower memory utilization than that of traditional architecture. That is because

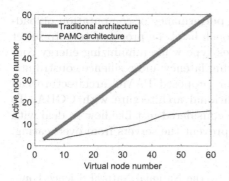

Fig. 5. Results of active node number.

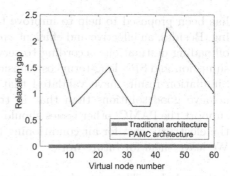

Fig. 6. Relaxation gap between real active node number and round up results.

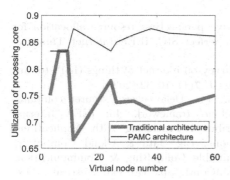

Fig. 7. Computing resource utilization.

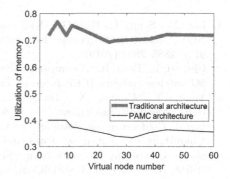

Fig. 8. Memory utilization.

PAMC always keeps a higher processing cores throughput during computation offloading. At the same time, PIM reduces frequent data transferring between processing cores and memory. This operation helps the memory-hungry services to be completed in a shorter time. Memory can be freed up earlier so that the memory utilization is lower.

CGBA used in this paper can get the approximate optimal solutions by considering all kinds of configurations and selecting the most reasonable one among them. This strategy leaves room for the computation offloading of future virtual nodes to improve resource utilization. The scale of MEC is small, we can get the solution in a short time. Large scale networks can highlight the advantages of CGBA which can narrow down the searching fields more quickly than heuristic algorithms.

5 Conclusions

This paper has studied the 3C resource allocation for computation offloading in MEC. A memory assisted multi-core architecture for physical servers in MEC has been proposed to help to improve the performance of computation offloading. Besides, an effective and efficient approach has been proposed to auto-select offloading destinations according to services type while minimizing energy consumption and SP's long-term cost considering latency and resilience constraints. Simulation results have validated that our proposed PAMC architecture can achieve good solutions than that in traditional architecture with CGBA. To support the PAMC, other issues should be considered just like how to deal with the electricity cost for air conditioning to prevent the servers from overheating. We leave them for future work.

Acknowledge. This research was supported by the National Natural Science Foundation of China under Grant U1836105.

References

1. Liu, M., Song, T., Gui, G.: Deep cognitive perspective: resource allocation for NOMA-based heterogeneous IoT with imperfect SIC. IEEE Internet Things J. **6**(2), 2885–2894 (2019)
2. Chettri, L., Bera, R.: A comprehensive survey on internet of things (IoT) toward 5G wireless systems. IEEE Internet Things J. **7**(1), 16–32 (2020)
3. Yang, Y., Chang, X., Han, Z., Li, L.: Delay-aware secure computation offloading mechanism in a fog-cloud framework. In: IEEE International Conference on Parallel & Distributed Processing with Applications, Ubiquitous Computing & Communications, Big Data & Cloud Computing, Social Computing & Networking, Sustainable Computing & Communications (ISPA/IUCC/BDCloud/SocialCom/SustainCom), pp. 346–353. Australia, Dec, Melbourne (2018)
4. Kekki, S., Featherstone, W., Fang, Y., et al.: MEC in 5G networks. ETSI White Paper, No. 28 (2018)
5. Yang, Y., Chang, X., Liu, J., Li, L.: Towards robust green virtual cloud data center provisioning. IEEE Trans. Cloud Comput. **5**(2), 168–181 (2017)
6. Liu, S., Wei, Y., Chi, J., Shezan, F.H., Tian, Y.: Side channel attacks in computation offloading systems with GPU virtualization. In: IEEE Security and Privacy Workshops (SPW), San Francisco, CA, pp. 156–161, May 2019
7. IBM Cloud GPU Solutions for AI and HPC Workloads. https://www.ibm.com/downloads/cas/RDPDBJ3X
8. Chikin, A., Amaral, J.N., Ali, K., Tiotto, E.: Toward an analytical performance model to select between GPU and CPU execution. In: IEEE International Parallel and Distributed Processing Symposium Workshops (IPDPSW), Rio de Janeiro, Brazil, pp. 353–362, May 2019
9. Pattnaik, A., et al.: Opportunistic computing in GPU architectures. In: ACM/IEEE 46th Annual International Symposium on Computer Architecture (ISCA), Phoenix, AZ, pp. 210–223, June 2019
10. Ghose, S., Boroumand, A., Kim, J., G®mez-Luna, J., Mutlu, O.: A Workload and Programming Ease Driven Perspective of Processing-in-Memory, arXiv.org (2019). http://search.proquest.com/docview/2267321794/
11. Gupta, S., Imani, M., Rosing, T.: Exploring processing in-memory for different technologies. In: Proceedings of the 2019 on Great Lakes Symposium on VLSI (GLSVLSI 19), New York, NY, May 2019, pp. 201–206 (2019)

12. Hazarika, A., Poddar, S., Rahaman, H.: Survey on memory management techniques in heterogeneous computing systems. IET Comput. Digit. Tech. **14**(2), 47–60 (2019)
13. Kim, B., Lim, E.C., Rhee, C.E.: Exploration of a PIM design configuration for energy-efficient task offloading. In: IEEE International Symposium on Circuits and Systems (ISCAS). Sapporo, Japan, May 2019
14. Kohni, M., Janacek, J.: Acceleration strategies of the column generation method for the crew scheduling problem. In: IEEE International Conference on Service Operations and Logistics, and Informatics (SOLI), Italy, Bari, pp. 54–57, September 2017
15. Pakpoom, P., Charnsethikul, P.: A column generation approach for personnel scheduling with discrete uncertain requirements. In: 2nd International Conference on Informatics and Computational Sciences (ICICoS), Semarang, Indonesia, October 2018
16. Riazi, S., Wigstrom, O., Bengtsson, K., Lennartson, B.: A column generation-based gossip algorithm for home healthcare routing and scheduling problems. IEEE Trans. Autom. Sci. Eng. **16**(1), 127–137 (2019)
17. Sheng, M., Wang, Y., Wang, X., Li, J.: Energy-efficient multiuser partial computation offloading with collaboration of terminals, radio access network, and edge server. IEEE Trans. Commun. **68**(3), 1524–1537 (2020)

A Greedy Heuristic Based Beacons Selection for Localization

Fuhua Ma[1], Qianqian Ren[1(✉)], and Jun Li[2(✉)]

[1] Department of Computer Science and Technology, Heilongjiang University,
Harbin 150080, China
renqianqian@hlju.edu.cn
[2] China Industrial Control Systems Cyber Emergency Response Team, Beijing, China
lijun@cics-cert.org.cn

Abstract. Wi-Fi based localization technology is a hot issue in recent indoor localization research. Due to the exist of obstacles and signal fluctuation in indoor environment, RSSI measurements from beacons are often noisy. To solve this problem, this paper first proposes a greedy heuristic algorithm to choose optimal beacons involved in localization. During the localization process, the reference points in the area covered by the selected beacons form triangles. The gravity centers of the triangles jointly determine the target's location. Finally, a comprehensive set of simulations are provided to invalidate the performance of the proposed algorithm.

Keywords: Localization · Beacon · Greedy heuristic

1 Introduction

In wireless sensor networks, mobile target tracking and localization is a basic function, which plays an important role in many location service based applications, such as intelligent traffic monitoring, human health monitoring, battlefield monitoring, disaster prediction, etc. [1–3]. However, the limits of sensor networks and environmental factors make the research on localization still face many challenges [4]. In this paper, the indoor localization technology in wireless sensor networks is studied and an effective localization algorithm is proposed.

The presence of obstacles in indoor environment leads to multipath effect and signal fluctuation, which makes the indoor environment complex. Moreover, considering the limits of GPS in indoor environment, many other signals are used for indoor positioning, such as Wi-Fi, Bluetooth, magnetic field, ultrasonic wave, etc. [5–8]. Due to the extensive deployment of wireless local area network (WLAN) and the universal use of mobile devices, Wi-Fi localization technique has attracted attention. In the scheme, the location of the interested target can be characterized by a series of detected signal (e.g., RSSIs from different beacons). However, RSSI is often influenced by obstacles in indoor environment, thus sampled values are often noisy, which reduces the localization accuracy. To address this issue and release the influence of noisy data on localization

M. Qiu (Ed.): ICA3PP 2020, LNCS 12452, pp. 710–718, 2020.
https://doi.org/10.1007/978-3-030-60245-1_48

results, this paper aims to choose optimal beacons involved in localization procedure. The contributions of the paper is as following:

1) A grid based network model is given, which are the basis of the proposed localization algorithm.
2) A greedy heuristic based beacons selection algorithm is introduced to select optimal beacons that participate in localization.
3) An overlapping grids based localization method is presented to determine the area that the target appears. Moreover, the accurate location of the target is specified via calculating gravity centers of triangles cover by selected beacons.
4) An experimental simulation platform is constructed to verify the performance of the proposed algorithm in the paper.

This paper is organized as follows. In Sect. 2, we give some work related to this paper. In Sect. 3, we present the network model; In Sect. 4, a beacon selection algorithm is proposed; In Sect. 5, we propose the localization algorithm; In Sect. 6, the simulation results and analysis are presented. The conclusion is given in the last section. Table 1 gives the notations definition used in this paper.

Table 1. Notations definition

Notations	Definitions
M	Number of beacons
N	Number of reference points
m	Number of beacons that detect the target
BN	The set of beacons
DB	The set of beacons that detect the target
DB'	The sorted DB by income function in ascending order
DB''	The set of chosen beacons
RP	The set of reference points
RP'	The set of reference points that locate within the area covered by beacons in DB''
V_R	Variance vector of RP

2 Related Work

In this section, we will briefly review the work related to this paper, including Wi-Fi based localization.

Xiao et al. propose a regression supported deep learning structure and classification support vector machine (SVM) to directly output the estimated position from the measured fingerprints [9]. Mizmizi et al. present a KNN based localization method [10], which is simple and easy to implement. However it is at the price of high time complex, especially when the number of features is very large. Chen et al. study the influence of environmental changes, devices orientation and different devices response factors on RSSI measurement [11]. Due to the fluctuation of Wi-Fi signal in indoor environment, it is easy to introduce extra localization error. To solve this problem, a triple loss function is designed to measure the rank difference of RSSI distance and position distance [11], which reduces the influence of signal fluctuation on localization efficiently. Hou et al. propose a fingerprint selection and location estimation algorithm [12], which filters those access points affected by environmental changes. Peng et al. introduce a BPNN (back propagation neural network) method to carry out three-dimensional indoor positioning [13], which can improve the positioning accuracy effectively. Wang et al. propose the APIT algorithm [14], its principle is to select three beacons randomly around the unknown node to form a triangular region, and use RSSI values to judge whether the unknown node is within it. Several overlapping regions of triangles are used to estimate the area of the target. Our localization algorithm is proposed based on this idea.

Inspired by the above research methods, considering that RSSI values are easily influenced by the environment, this paper proposes a beacons selection algorithm based on greedy heuristic. According to the fluctuation of RSSI values, a subset of optimal beacons are selected to locate the target accurately.

3   Network Model

We assume that the monitoring area of the sensor network is a rectangular area of $L_1 \times L_2$, which is divided into $N = a \times b$ grids. M beacons are deployed uniformly in the area, which is denoted as $BN = \{b_1, b_2, \cdots, b_M\}$. We further assume that the sensing area and transmitting area of each beacon is a disk with radius of l. There is one reference node at the center of each grid. We use $RP = \{rp_1, rp_2, \cdots, rp_N\}$ to denote the reference points set, the number of reference points is N. Figure 1 shows an example of the network model.

4   Beacons Selection Algorithm

In real applications, RSSI is often influenced by obstacles in the environment, thus sampled values are often noisy, which reduces the localization accuracy. To address this issue and release the influence of noisy data on localization results, we aim to choose the beacons with lower interference and better quality participating in localization.

When the target appears in the monitoring area, it can be detected by multiple beacons. These beacons collect RSSI values. Each beacon samples RSSI measurements s times, the measurements set of db_i is denoted as $R^i = \{r_1^i, r_2^i, \cdots, r_s^i\}$, and the average value is denoted as $\bar{R}^i = \frac{\sum_{j=1}^s r_j^i}{s}$. The variance of RSSI measurements from db_i at a reference point is denoted as $v_r^i = \frac{\sum_{j=1}^s \left(r_j^i - \bar{R}^i\right)^2}{s}$, and the variance set is represented as $V_R = \{v_r^1, v_r^2, \ldots, v_r^m\}$, m is the number of beacons that detect the target. We further denote the RSSI vector sampled by the target as $T = \{\bar{T}^1, \bar{T}^2, \ldots, \bar{T}^m\}$, and its corresponding variance is denoted as $V_T = \{v_t^1, v_t^2, \ldots, v_t^m\}$.

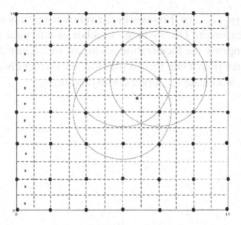

Fig. 1. An example of the network model. Black nodes are beacons, brown nodes are reference points and blue star is the target to be located. (Color figure online)

In this section, we give a greedy heuristic based beacons selection algorithm, which chooses beacons incrementally from DB. We define an income function g, the greedy heuristic algorithm selects the beacons that maximize the income function g each time. Given beacons set $BN = \{b_1, b_2, ..., b_M\}$, for the problems solved in this section, the income function g is defined as follows:

$$g(b_i) = \frac{\bar{T}^i}{v_t^i} \tag{2}$$

where $\bar{T}^i$ represents the RSSI measurement from b_i sampled by the target, and v_t^i represents variance from b_i. The greedy heuristic based beacon selection algorithm consists of three stages:

(1) Initiation. Given RSSI vector T sampled by the target and BN, calculate $g(b_i)$. For each b_i, if $g(b_i) > 0$, then b_i can detect the target and put b_i into DB.

(2) Sorting. Sort DB by income function in ascending order and get the sorted set $DB' = \{db'_1, db'_2, ..., db'_m\}$.

(3) Iteration. For each time, we choose the beacons from DB' that maximizes the income function and its residue energy is greater than a given threshold. The chosen beacons set is denoted as $DB'' = \{db''_1, db''_2, ..., db''_k\}$. k is the number of chosen beacons, which is decided by the user or the localization algorithm.

5   Target Localization Algorithm

This section describes the target localization algorithm in detail. When the target enters the monitoring area, beacons can detect it. According to the previous beacons selection algorithm, a subset beacons are chosen to participate in localization, the overlapping of sensing disks of these beacons form the area that the target appears, as shown in Fig. 2

(a). Let $RP' = \{rp'_1, rp'_2, \ldots, rp'_v\}$ represent the reference points that locate in the area jointly covered by beacons in DB'', v is the number of reference points. To shrink the appearing area and improve the localization accuracy, we choose three points from RP' each time, if these three points can form a triangle, we further check whether the target is within the triangle and locate the target correspondingly.

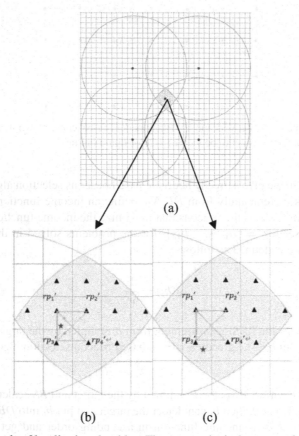

Fig. 2. An example of localization algorithm. The green point is the center of a triangle, and the star is the target to be located. (Color figure online)

The localization algorithm consists of five phases.

(1) We choose three RPs from RP', that's rp_i, rp_j and rp_s, these RPs form a triangle. We say that the target is in the triangle, if the following condition is satisfied:

$$|rp_i, L| + |rp_j, L| + |rp_s, L| < H_{ij} + |rp_i, rp_j| + cons. \tag{3}$$

where $|rp_i, L|$, $|rp_j, L|$ and $|rp_s, L|$ are the distance from the target to rp_i, rp_j and rp_s, respectively. H_{ij} is the perpendicular distance from rp_s to the side formed by rp_i and rp_j. *cons* is constant.

(2) If the target is within the triangle, we calculate the gravity center of the triangle.
(3) Repeat step (1) until all the triangles formed by three *PR*s are found and checked.
(4) The average of all gravity centers is the target's estimation location, as shown in Fig. 2 (b).
(5) If the target is not in any triangle, we use the centriod of k nearest *RP*s to estimate the target, as shown in Fig. 2 (c).

6 Experimental Analysis

In order to verify the feasibility and priority of the proposed algorithm, we use MATLAB to build a simulation platform. In the experiments, the detection area of the sensor network is a 50×50 unit area, and 36 beacons are deployed in the network. We divide the monitoring area into several small grids according to the grid resolution of 1×1 unit, 1.2×1.2 unit and 1.5×1.5 unit, respectively. We investigate the the impact of system parameters, including the number of reference points and grid size on the localization performance. When one parameter is evaluated, we fix other parameters. For each sampled value, we sample it three times and use the average result over three times. The metirc we consider to measure the localization quality is localization error.

6.1 Impact of *RP*s Number and Grid Size on Localization Error

The experimental results study the influence of parameters including involved *RP*s number and grid size on localization accuracy. Figure 3, Fig. 4 and Fig. 5 show localization error when the number of *RP*s is varying from 5 to 12 and the grid size is 1, 1.2 and 1.5, respectively.

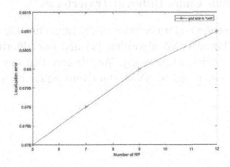

Fig. 3. The influence of *RP* number on localization when grid size $= 1$

In the algorithm, the *RP*s locate in the area covered by beacons in DB'' participate in localization procedure. To further release the computation cost, we try to choose

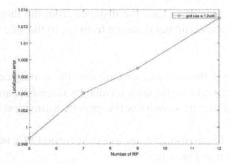

Fig. 4. The influence of *RP* number on localization when grid size = 1.2

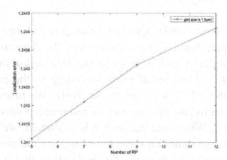

Fig. 5. The influence of *RP* number on localization when grid size = 1.5

partial *RP*s closest to the target involved in work. It can be seen from the figures that the localization error increases with the increasing number of *RP*s. However, the difference is not very obvious. It is also observed that as the raise of grid size, the localization error increases. When grid resolution is 1 × 1unit, the localization error is minimal.

6.2   Localization Results Under Different Trajectories

Figure 6 and Fig. 7 show the real trajectories of the target and the estimated trajectories of support vector machines(SVM) algorithm [9] and our algorithm under linear and parabolic moving trajectories, respectively. We observe that our algorithm can obtain better tracking results compared to SVM algorithm under both linear and parabolic moving trajectories.

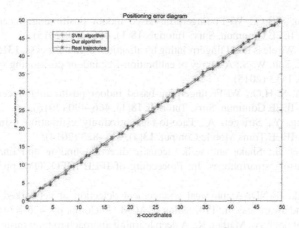

Fig. 6. Linear moving trajectories

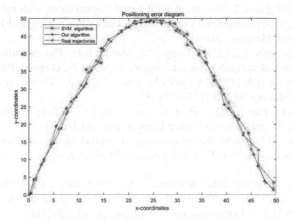

Fig. 7. Parabolic moving trajectories

7   Conclusion

This paper considers the problem of indoor environment factors on localization accuracy and introduce a greedy heuristic algorithm to select beacons for target localization. In the localization procedure, the reference points in the area covered by the selected beacons are investigated for localization. The experimental results show that the beacons selection based localization algorithm we present in this paper not only reduces the redundancy and calculation, but also can obtain excellent localization quality.

References

1. Li, T., Chen, Y., Zhang, R., Zhang, Y., Hedgpeth, T.: Secure crowdsourced indoor positioning systems. In: IEEE INFOCOM 2018 - IEEE Conference on Computer Communications, Honolulu, HI, pp. 1034–1042 (2018)

2. He, S., Chan, S.H.G.: Wi-Fi fingerprint-based indoor positioning: recent advances and comparisons. IEEE Commun. Surv. Tutorials **18**(1), 466–490 (2015)
3. Yiu, S., et al.: Wireless RSSI fingerprinting localization. Signal Process. **131**, 235–244 (2017)
4. Ossain, A.M., Soh, W.-S.: A survey of calibration-free indoor positioning systems. Comput. Commun. **66**, 1–13 (2015)
5. He, S., Chan, S.-H.G.: Wi-Fi fingerprint-based indoor positioning: recent advances and comparisons. IEEE Commun. Surv. Tutorials **18**(1), 466–490 (2016)
6. Liu, S., Jiang, Y., Striegel, A.: Face-to-face proximity estimation using Bluetooth on smartphones. IEEE Trans Mobile Comput. **13**(4), 811–823 (2014)
7. Huang, W., et al.: Shake and walk: acoustic direction finding and fine-grained indoor localization using smartphones. In: Proceeding of IEEE INFOCOM, pp. 370–378, April 2014
8. Sun, Z., et al.: PANDAA: physical arrangement detection of networked devices through ambient-sound awareness. In: Proceeding of ACM UbiComp, pp. 425–434 (2011)
9. Xiao, L., Behboodi, A., Mathar, R.: A deep learning approach to fingerprinting indoor localization solutions. In: 2017 27th International Telecommunication Networks and Applications Conference (ITNAC). IEEE (2017_
10. Mizmizi, M., Reggiani, L.: Design of RSSI based fingerprinting with reduced quantization measures. In: 2016 International Conference on Indoor Positioning and Indoor Navigation (IPIN), Alcala de Henares (2016)
11. Chen, Y., Kleisouris, K., Li, X., Trappe, W., Martin, R.P.: The robustness of localization algorithms to signal strength attacks: a comparative study. In: Gibbons, P.B., Abdelzaher, T., Aspnes, J., Rao, R. (eds.) DCOSS 2006. LNCS, vol. 4026, pp. 546–563. Springer, Heidelberg (2006). https://doi.org/10.1007/11776178_33
12. Hou, Y., Sum, G., Fan, B.: The indoor wireless location technology research based on WiFi. In: International Conference on Natural Computation. IEEE (2014)
13. Peng, L., et al.: 3D indoor localization based on spectral clustering and weighted back propagation neural networks. In: IEEE/CIC International Conference on Communications in China (ICCC), Qingdao (2017)
14. He, T., Huang, C., Blum, B.M., Stankovic, J.A., Abdelzaher, T.F.: Range-free localization schemes for large scale sensor networks. In: Proceedings of the 9th Annual International Conference on Mobile Computing and Networking, pp. 81–95 (2003)

A Periodic Variable Star Observation System with High Accuracy Based on Star Sensors

Chen Jiwei[✉] and Tang Guojian

College of Aerospace Science and Engineering, National University of Defense Technology, Changsha, China
chenjiwei@spacechina.com, gjtang@263.net

Abstract. In order to observe the photodynamic characteristics of periodic variable stars, a periodic variable star ground observation system is designed with high frequency and high precision star sensor. The high precision star sensor is used to determine the attitude information of the system, and the NFOV star sensor is used to detect atmospheric changes and other environmental factors to track the variable stars accurately. The accuracy measurement matrix and the pointing vector of the star sensor are obtained from the attitude matrix and the accuracy measurement transformation matrix of the star sensor. Finally, the pointing and rolling accuracy of the star sensor are obtained by the accuracy evaluation standard. The system is validated for the observation of periodic variable stars and high dynamic detection, which provides the observation basis for the application of periodic variable stars.

Keywords: Periodic variable star · Star sensors · Observation accuracy

1 Introduction

Periodic variable stars refer to special variable stars whose photodynamic characteristics have periodic variation laws. Periodic variable stars have a longer optical variation period [1]. Their photodynamic characteristics can be obtained through full-period phase measurement. Moreover, the observable spectral range of periodic variable stars is widely distributed, which is convenient for ground-based observation and ground-based applications and has a wide range of applications.

Much that we know about stars and the universe came from studying variable stars. There are over 200,000 variable stars catalogued with many more suspected. With modest equipment any amateur can make observations that are valuable scientific contributions [2]. The minimum required equipment for this award is a pair of binoculars, but any size telescope or go-to telescope can be used.

Variable stars are objecting whose light is not constant. The observer's goal is to determine the brightness of the star when compared to stars of fixed brightness in or near the same field of view. The thrills of variable star observation are many finding the right star in the patterns in the field; pushing the limits of your telescope and your observational skills to glimpse that mag 14.5 star; improving your observing methods,

© Springer Nature Switzerland AG 2020
M. Qiu (Ed.): ICA3PP 2020, LNCS 12452, pp. 719–728, 2020.
https://doi.org/10.1007/978-3-030-60245-1_49

equipment, and ability to yield more and more variables per observing session; seeing your favorite stars change in brightness as you watch them from week to week; catching in outburst [3]. You get all this plus you get to provide useful scientific data to scientists and other observers.

Star sensor is a high-precision space attitude measurement device which takes fixed stars as reference and starry sky as working object. So, in this paper, with the development of research on star sensor and the emergence of new materials and new technologies, the accuracy of star sensor is continuously improved, power consumption is reduced, cost is reduced, and new type star sensors with increasingly wide application fields are constantly introduced.

The designed periodic variable star ground observation system is mainly composed of: high-precision star sensor, NFOV (narrow field of view) (1°) star sensor, two-dimensional turntable and related accessories. During observation, the high-precision star sensor is used to match the star map, determine the attitude information of the system, and NFOV star sensor is used to accurately track the variable stars by detecting environmental factors such as atmospheric changes, thus realizing periodic observation and high dynamic detection of the variable stars. The two-dimensional turntable is used to realize 360-degree rotation control in yaw and roll directions, which has various working modes and can accurately control the rotation angle and angular velocity.

2   Related Works

The accuracy of the observation system based on star sensors is relatively high, but the accuracy of star sensors is affected by the atmosphere to some extent. Therefore, it is very important to study the sources of star sensor errors and analysis methods, and then calibrate and compensate the errors. The accuracy of star sensors is the cornerstone of the observation accuracy of the whole system. Only by mastering the accuracy evaluation process can we obtain more accurate photodynamic characteristics of periodic variable stars [4].

When we look up at the night sky it is easy to imagine that the stars are unchanging. Apart from twinkling due to the effects of our atmosphere stars appear fixed and constant to the untrained eye. Careful observations, some even done with the naked eye, show that some stars do in fact appear to change in brightness over time. Some exhibit periodic behavior, brightening quickly then diminishing in brightness slowly only to repeat themselves [5]. With some these changes take place over several days whilst others occur in a matter of hours or many months. Other stars exhibit a once-off dramatic change in brightness by orders of magnitude before fading away to obscurity. All of these are examples of what are termed variable stars. A variable star is simply one whose brightness (or other physical property such as radius or spectral type) changes over time.

At a fundamental level all stars are variable as they evolve and change over time (from a main sequence to a red giant star as in the Sun's case for example). Furthermore, we can infer that all stars are likely to vary their light output to some extent due to variations caused by phenomena such as sunspots. In the section however, we focus on stars with a measurable change in brightness. In order to try and understand variable stars, astronomers have sought to classify them according to observable properties. The diagram below the main types of variable stars.

A supernova is a cataclysmic stage towards the end of a star's life that is characterized by a sudden and dramatic rise in brightness. A typical supernova may see a star become brighter by up to 20 magnitudes to an absolute magnitude of about −15 [6]. This means that a typical supernova may outshine the rest its galaxy for several days or a few weeks.

Supernovae are caused by one of two main mechanisms. The first takes place when accreting material falling onto a white dwarf in a binary system takes it over the mass set by the Chandrasekhar limit [7]. The resulting instability triggers a runaway thermonuclear explosion that destroys the star and releases large amounts of radioactive and heavy elements into space. The second process occurs in very massive stars once all the material in their core has been fused into iron. As fusion cannot occur in elements heavier than iron the drop in outwards radiation pressure means that gravitational collapse overwhelms the core which rapidly implodes. The core material gets crushed to form degenerate neutron-density material whilst the extreme temperature and pressure in the surrounding layers cause rapid (R-process) nuclear reactions that synthesize the heaviest elements [8]. A huge flux of neutrinos is thought to interact with the super-dense material, ripping the star apart. Such core collapse supernovae may result in neutron stars and black holes forming from the remaining core material. More details are given in the later section on star death.

Observationally, supernovae are classified according to their spectra. Type I supernova exhibit no hydrogen lines in spectra taken soon after the supernova event. Those with silicon lines present are further classified as Type Ia and are thought to be due to thermonuclear explosions as in accreting white dwarfs. If no Si lines are present they are Type Ib or Ic depending on the high or low abundance of He lines respectively. These types occur due to core collapse following the outer layers being stripped away in Wolf-Rayet or binary stars [9].

3 Observation Probability and Attitude

3.1 Observation Probability

The NFOV star sensor is mainly used to analyze the atmospheric environment in the sky field when the high-precision star sensor has captured the variable star, so as to realize high-precision tracking of the variable stars. First of all, the variable star observation capability of NFOV star sensor shall be analyzed. The detection capability of star sensors is mainly reflected in the detectable limit magnitude and detection probability. The formula of detectable limit magnitude is:

$$M_V = lg \frac{S}{5\sqrt{S + N_{bkg} + N_{sensor}^2}} \bigg/ lg2.512 \tag{1}$$

Where,

$$S = \int_{\lambda_1}^{\lambda_2} F_0(\lambda)\tau_0(H, \mu, \lambda)\frac{\pi D^2}{4}\tau_{opt}(\lambda)t\frac{1}{W_{ph}}Q_0(\lambda)Kd\lambda \tag{2}$$

$$N_{bkg} = \int_{\lambda_1}^{\lambda_2} I_b(\lambda, \mu, \mu_0, \phi, H) \frac{\pi D^2}{4} \tau_{opt}(\lambda) t \frac{1}{W_{ph}} Q_0(\lambda) \Omega d\lambda \tag{3}$$

S represents the energy of the picture element in the light spot center, and $F_0(\lambda)$ represents the spectral radiant exitance of zero-magnitude stars. The G2 spectrum of the sun has an apparent magnitude of -26.7, the surface temperature of 5800 K, and the radiance on the earth is about 1.4 KW/m^2. Therefore, the energy of the sun measured on the earth is about $2.512^{26.7} = 4.79 \times 10^{10}$ times that of the zero-magnitude stars, and the total radiant exitance of the zero-magnitude stars is about 2.9228×10^{-8} W/m^2. Under the given wavelength and temperature conditions, the energy of the black-body radiation is distributed according to the wavelength:

$$I(\lambda, T) = \frac{2\pi hc^2}{\lambda^5 \left(e^{hc/\lambda k_B T} - 1 \right)} \tag{4}$$

$h = 6.626 \times 10^{-34} Js$, c is the light velocity, $c = 2.997 \times 10^8$ m/s, K_B is the boltzmans constant, $K_B = 1.38 \times 10^{-23}$ J/K, T is the temperature, $T = 5800$ k.

$\tau_0(H, \mu, \lambda)$ is the atmospheric transmittance at the observation height of H and the observation zenith angle of μ, D the optical system aperture, $\tau_{opt}(\lambda)$ the optical system transmittance, $W_{ph} = h \frac{c}{\lambda}$ the energy of a single photon, N_{bkg} the background radiation energy, N_{sensor} the detector noise, and $I_b(\lambda, \mu, \mu_0, \phi, H)$ represents the sky background radiance reaching the detector when the observation height is H, the observation zenith angle is μ, the solar zenith angle is μ_0, and the observation azimuth angle is ϕ. $Q_0(\lambda)$ is the quantum efficiency of the detector, Ω is the solid angle occupied by the picture element, t is the integration time, and λ is the working spectral wavelength. $\Omega = \sigma/f^2$, σ is the area of a single picture element, f is the focal length of the optical system.

According to the function provided by the SKY2000 Star Catalog, the corresponding relationship between the average number of stars and the apparent magnitude:

$$N(M_V) = 6.5 e^{1.107 M_V} \tag{5}$$

Where, $N(M_V)$ represents the number of guide-stars whose magnitude is less than or equal to M_V, and M_V the apparent magnitude.

It is assumed that the average number of observable variable stars is not less than 180, and that stars are evenly distributed on the celestial sphere, the average number of stars in the field of view can be calculated according to the sphericity corresponding to the field of view of the star sensor which satisfies the following relation:

$$N_{FOV} = N(M_V) \frac{\left[sin^2 \left(\frac{\theta_{FOV}}{2} \right) \right]}{4} \tag{6}$$

Where, N_{FOV} represents the number of stars in the field of view, θ_{FOV} the angle of the field of view. If a one-dimensional rotating mechanism and coding system are equipped, the number of stars in the field of view can be expressed as:

$$N_{FOV} = N(M_V) sin \left(\frac{\theta_{FOV}}{2} \right) sin(\delta) \tag{7}$$

When $\delta = 45°$, $N_{FOV} \approx 3.3318$.

The number of stars in any field of view follows the Poisson distribution, and its distribution law is as follows:

$$P(X = k) = e^{-N_{FOV}} \frac{N_{FOV}^k}{k!} \qquad (8)$$

Where, $P(X = k)$ represents the probability of k stars appearing in the field of view, and N_{FOV} represents the average number of stars in the field of view. Here, it is assumed that $N_{FOV} = 3.33$, the probability that the number of stars in the field of view is greater than or equal to 1 can be calculated as follows:

$$P(X \geq 1) = 1 - P(X = 0) = 96.42\% \qquad (9)$$

3.2 Attitude Matrix

Fix the star sensor on the earth with its rolling axis pointing to the zenith, and a star map stored in the star sensor. According to the direction vector of the guide star in the coordinate system of the star sensor and the direction vector ($\vartheta_{CRFJ2000}$) in the J2000.0 rectangular coordinate system, the optimal attitude matrix of the star sensor $q_i = \begin{bmatrix} q_1 & q_2 & q_3 & q_4 \end{bmatrix}$ and the actual shooting time (T + Δt_i) of the corresponding star map are obtained and output.

The optimal attitude matrix $A_q(T + \Delta t_i)$ obtained according to q_i is as follows:

$$A_q(T + \Delta t_i) = \begin{bmatrix} q_1^2 - q_2^2 - q_3^2 + q_4^2 & 2(q_1q_2 + q_3q_4) & 2(q_1q_3 - q_2q_4) \\ 2(q_1q_2 - q_3q_4) & -q_1^2 + q_2^2 - q_3^2 + q_4^2 & 2(q_2q_3 + q_1q_4) \\ 2(q_1q_3 + q_2q_4) & 2(q_2q_3 - q_1q_4) & -q_1^2 + q_2^2 - q_3^2 + q_4^2 \end{bmatrix}$$

$$(10)$$

3.3 Accuracy Measurement Matrix

According to the actual shooting time of the star sensor (T + Δt_i) and the precession, nutation and rotation of the earth, the accuracy test transformation matrix $R_{T+\Delta t_i}$ associated with the star sensor can be obtained.

The transformation matrix $R_{ERF}(-\theta_1)$ conversed from the J2000.0 rectangular coordinate system to the epoch ecliptic coordinate system: Based on the J2000.0 rectangular coordinate system ($X_{CRFJ2000}$, $Y_{CRFJ2000}$, $Z_{CRFJ2000}$), the epoch ecliptic coordinate system (X_{ERF}, Y_{ERF}, Z_{ERF}) can be obtained by rotating the J2000.0 rectangular coordinate system counterclockwise around the X axis of the J2000.0 rectangular coordinate system by $23°26'21''$:

$$(X_{ERF}, Y_{ERF}, Z_{ERF}) = (X_{CRFJ2000}, Y_{CRFJ2000}, Z_{CRFJ2000}) \cdot R_X \left(-23°26'21''\right) \quad (11)$$

Then, $R_{ERF}(-\theta_1) = R_X \left(-23°26'21''\right)$, where R_X is the coordinate transformation basis. Obtaining a transformation matrix $R_{CRFT}(-\theta_2)$ for converting the epoch ecliptic coordinate system into the celestial coordinate system at the current time T;

Transforming the epoch ecliptic coordinate system $(X_{ERF}, Y_{ERF}, Z_{ERF})$ into the celestial coordinate system $(X_{CRFT}, Y_{CRFT}, Z_{CRFT})$ at the current time T is obtained by the following steps:

(1) Rotate the epoch ecliptic coordinate system $(X_{ERF}, Y_{ERF}, Z_{ERF})$ clockwise $50.29'' \times T$ around its Z axis;
(2) Then rotate clockwise $-23°26'21''$ around the X axis of the coordinate system after the first rotation;
(3) Then rotate counterclockwise ε_A around the X axis of the coordinate system after the second rotation;
(4) Then rotate clockwise Δ_φ around the Z axis of the coordinate system after the third rotation;
(5) Then rotate clockwise $\varepsilon_A + \Delta_\varepsilon$ around the X axis of the coordinate system after the fourth rotation to obtain the celestial coordinate system $(X_{CRFT}, Y_{CRFT}, Z_{CRFT})$ of the current time T containing nutation term, where, Δ_φ is celestial longitude nutation, and Δ_ε is and oblique nutation.

The celestial coordinate system structure $(X_{CRFT}, Y_{CRFT}, Z_{CRFT})$ is obtained by the following formula:

$$(X_{CRFT}, Y_{CRFT}, Z_{CRFT}) = (X_{ERF}, Y_{ERF}, Z_{ERF}) \cdot R_Z\left(50.29'' \times T\right) \cdot R_X\left(23°26'21''\right) \cdot R_X(-\varepsilon_A) \cdot R_Z(\Delta_\varphi) \cdot R_X(\varepsilon_A + \Delta_\varepsilon) \quad (12)$$

Where, R_x and R_z are the base of coordinate transformation, so

$$R_{CRFT}(-\theta_2) = R_Z\left(50.29'' \times T\right) \cdot R_X\left(23°26'21''\right) \cdot R_X(-\varepsilon_A) \cdot R_Z(\Delta_\varphi) \cdot R_X(\varepsilon_A + \Delta_\varepsilon) \quad (13)$$

According to the IAU2000B nutation model, ε_A, celestial longitude nutation Δ_φ, and oblique nutations Δ_ε are respectively as below:

$$\varepsilon_A = \varepsilon_0 - 46.84024'' t - 0.00059''t^2 + 0.001813''t^3 \quad (14)$$

$$\Delta_\varphi = \Delta_{\varphi p} + \sum_{i=1}^{77}\left[(Q_{i1} + Q_{i2}t)\sin\gamma_i + Q_{i3}\cos\gamma_i\right] \quad (15)$$

$$\Delta_\varepsilon = \Delta_{\varepsilon p} + \sum_{i=1}^{77}\left[(Q_{i4} + Q_{i5}t)\sin\gamma_i + Q_{i6}\cos\gamma_i\right] \quad (16)$$

Where, $\Delta_{\varphi p} = -0.000135''$, $\Delta_{\varepsilon p} = -0.000388''$, $\varepsilon_0 = 84381.488''$, t is the number of Julian centuries starting from J2000.0 and obtained based on the current time T;

The breadth angle γ_i is a linear combination of the angles as follows, n_{ik} is an integer and F_k is the Delaunay angle related to the position of the sun and moon. The value of "n_{ik} and $Q_{i1} - Q_{i6}$" in the nutation expression can be found on the service website of the International Earth Rotation and Reference System Service.

$$\gamma_i = \sum_{k=1}^{5} n_k F_k = n_{i1}l + n_{i2}l' + n_{i3}F + n_{i4}D + n_{i5}\Omega \quad (17)$$

The conversion from celestial coordinate system $(X_{CRFT}, Y_{CRFT}, Z_{CRFT})$ at the current time T to the ground-fixed coordinate system $(X_{TRF}, Y_{TRF}, Z_{TRF})$ at the actual shooting time $(T + \Delta t_i)$ is obtained by rotating the celestial coordinate system $(X_{CRFT}, Y_{CRFT}, Z_{CRFT})$ around the Z axis of the celestial coordinate system counterclockwise at $\Omega = 7.292115 \times 10^{-5}$ rad/s:

$$(X_{TRF}, Y_{TRF}, Z_{TRF}) = (X_{CRFT}, Y_{CRFT}, Z_{CRFT}) \cdot R_Z(-\Omega \Delta t) \qquad (18)$$

Therefore, the transformation matrix $R_{TRF}(-\theta_3) = R_Z(-\Omega \Delta t)$ of the ground-fixed coordinate system at the actual shooting time $(T + \Delta t_i)$. Get the transformation matrix of the star sensor accuracy test $R_{T+\Delta t_i}$:

$$
\begin{aligned}
R_{T+\Delta t_i} &= R_{ERF}(-\theta_1) \cdot R_{CRFT}(-\theta_2) \cdot R_{TRF}(-\theta_3) = R_{ERF}(\theta_1)^{-1} \cdot R_{CRFT}(\theta_2)^{-1} \cdot R_{TRF}(\theta_3)^{-1} \\
&= (R_{TRF}(\theta_3) \cdot R_{CRFT}(\theta_2) \cdot R_{ERF}(\theta_1))^{-1}
\end{aligned}
\qquad (19)
$$

Using the optimal attitude matrix $A_q(T + \Delta t_i)$ and the accuracy test transformation matrix $R_{T+\Delta t_i}$, the accuracy test matrix is obtained as $A_{test}(T + \Delta t_i) = A_q(T + \Delta t_i) \cdot R_{T+\Delta t_i}$.

4 Experiment Results

4.1 Experiment Setup

The variable star observation system is fixed on the earth through a two-dimensional turntable. In order to minimize the influence of the atmosphere and the like, the star sensor is directed to the zenith as far as possible, and the star sensor can output corresponding attitude information along with the movement of the earth. The precision testing method comprises the following steps:

(1) acquiring the direction vector of the guide star at the time T under the J2000.0 Rectangular Coordinate System;
(2) deducing the optimal attitude matrix of the star sensor $A_q(T + \Delta t_i)$;
(3) acquiring an accuracy measurement transformation matrix associated with the star sensor according to the actual shooting time of the star sensor and the precession, nutation and rotation of the earth;
(4) obtaining an accuracy measurement matrix according to the attitude matrix and the accuracy measurement transformation matrix of the star sensor;
(5) obtaining the Triaxial directional vector of the star sensor according to the accuracy measurement matrix;
(6) Obtaining the pointing accuracy and rolling accuracy of the star sensor according to the accuracy evaluation standard.

4.2 Accuracy Evaluation

Determine the Triaxial directional vector of the star sensor $P(T + \Delta t_i)$ according to the accuracy test matrix $A_{test}(T + \Delta t_i)$:

$$
P(T + \Delta t_i) = A_{test}(T + \Delta t_i)^T \begin{bmatrix} 1 & 0 & 0 \\ 0 & 1 & 0 \\ 0 & 0 & 1 \end{bmatrix} \qquad (20)
$$

According to the triaxial directional vector of the star sensor $P(T + \Delta t_i)$, the angles $(\alpha_i, \beta_i, \varepsilon_i)$ respectively between the three optimal directing vectors of the star sensor and the X-axis, Y-axis, and Z-axis vectors of the star sensor at the actual shooting time $(T + \Delta t_i)$ are obtained as the following steps:

(1) The obtained triaxial directional vector of the star sensor $P(T + \Delta t_i)$ is expressed as a row vector as follows, and normalized;

$$P(T + \Delta t_i) = \left[Px(T + \Delta t_i), \quad Py(T + \Delta t_i), \quad Pz(T + \Delta t_i) \right] \tag{21}$$

(2) According to the row vector of the triaxial directional vector of the star sensor, the optimal vectors of the X-axis, Y-axis and Z-axis of the star sensor $P_{opt}(T + \Delta t_i)$ are obtained, to make it minimum sum of each of the three row vector of $P_{opt}(T + \Delta t_i)$ $[Px_{opt}(T + \Delta t_i), Py_{opt}(T + \Delta t_i), Pz_{opt}(T + \Delta t_i)]$ and the square of the vector angles at different actual shooting time $(T + \Delta t_i)[Px(T + \Delta t_i), Py(T + \Delta t_i), Pz(T + \Delta t_i)]$, and normalize the three row vectors;

(3) According to the optimal triaxial directional vector of star sensor $P_{opt}(T + \Delta t_i)$ and the triaxial directional vector $P(T + \Delta t_i)$ at different actual shooting time $(T + \Delta t_i)$, the cosine matrix C is obtained:

$$C = \begin{bmatrix} c_{11} & c_{12} & c_{13} \\ c_{21} & c_{22} & c_{23} \\ c_{31} & c_{32} & c_{33} \end{bmatrix} = P_{opt}(T + \Delta t_i)^T \cdot P(T + \Delta t_i) \tag{22}$$

(4) According to the cosine matrix C, the angle $(\alpha_i, \beta_i, \varepsilon_i)$ between the three optimal directional vectors of star sensor and the vectors of X-axis, Y-axis and Z-axis of the star sensor at the actual shooting time $(T + \Delta t_i)$ is further obtained:

$$\begin{bmatrix} \alpha_i \\ \beta_i \\ \varepsilon_i \end{bmatrix} = \begin{bmatrix} \arccos(|c_{11}|) \\ \arccos(|c_{22}|) \\ \arccos(|c_{33}|) \end{bmatrix} \tag{23}$$

Where, All of $(\alpha_i, \beta_i, \varepsilon_i)$ are in the range of $[0, \frac{\pi}{2}]$.

According to the statistical law, the angle between the optimal vector of the rotation axis of the star sensor and the angle of the vector of the rotation axis can be obtained through the three-axis vector of the star sensor, η_i, which is conforms to mean value of zero, and the variance is normal distribution of σ^2. But due to the measurement error, the angle must be positive. So its probability density function is slightly different from the probability density function of normal distribution. The probability density function is expressed as:

$$p(\eta_i) = \begin{cases} 2f(\eta_i) & \eta_i \geq 0 \\ 0 & \eta_i < 0 \end{cases} \tag{24}$$

$$f(\eta_i) = \frac{1}{\sqrt{2\pi}\sigma} e^{-\frac{\eta_i^2}{2\sigma^2}} \tag{25}$$

Where, σ is expressed by the following formula:

$$\sigma = \sqrt{\frac{\sum_0^n \eta_i^2}{n-1}} \tag{26}$$

$\alpha_i, \beta_i, \varepsilon_i$ are expressed with η_i in a unified way. $\sigma_X, \sigma_Y, \sigma_Z$ can be obtained by substituting them into the above formula η_i respectively. n represents the total sampling times of the star sensor.

The schematic diagram is shown in Fig. 1:

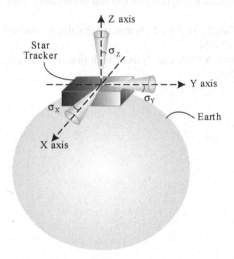

Fig. 1. Schematic diagram of direction accuracy and rolling accuracy of the star sensor

$\alpha_i, \beta_i, \varepsilon_i$ can reflect the minor changes of the three axes of the star sensor due to errors, which is used as the accuracy evaluation standard of the star sensor. The roll accuracy of the star sensor can be obtained, $3\sigma_X (99.7\%)$ or $3\sigma_Y(99.7\%)$, and the direction accuracy is $3\sigma_Z(99.7\%)$.

5 Conclusions

This paper uses the ground observation system based on star sensor to obtain the attitude information of the system and the observation probability of periodic variable stars. It deduces the direction accuracy and rolling accuracy of the star sensor, which provides theoretical support for the application of subsequent observation system and provides application system for the observation of periodic variable stars.

References

1. George, J., Terejanu, G., Singla, P.: Spacecraft attitude estimation using adaptive gaussian sum filter. J. Astronaut. Sci. **57**(1), 31–45 (2009)

2. Godsill, S.J., Doucet, A., West, M.: Monte Carlo smoothing for nonlinear time series. J. Am. Stat. Assoc. **99**(465), 156–168 (2004)
3. Lyalikov, A.M.: Revealing macrodefects in periodic structures of the transmission type in white light on the basis of shift of images. Opt. Spectrosc. **98**(3), 477–482 (2005)
4. Mutapcic, A., Boyd, S., Farjadpour, A., Johnson, S.G., Avniel, Y.: Robust design of slow-light tapers in periodic waveguides. Eng. Optim. **41**(4), 365–384 (2009)
5. Wei, X., Zhang, G., Jiang, J.: Study on the method of star subdivision and positioning of star image in the star sensor. J. Beihang Univ. (09) (2003)
6. Pengju, H., Bin, L., Tao, Z., Jun, Y.: Research on calibration technology of star sensor with large field of view. Acta Opt. Sin. **31**(10), 1023001 (2011)
7. Guangpu, Y., et al.: Research progress of CCD star photometry. Prog. Astron. **30**(4), 467–486 (2012)
8. Bin, L., Hailong, Z., Tao, Z., Yuchan, T.: Status and development trend of star sensor technology research. China Opt. (2016)
9. Ting, S., Fei, X., Zheng, Y.: Error analysis of high precision of optical system in star sensor. Acta opt. Sin. (03) (2013)

Author Index

Ai, Zhengpeng III-426
Anand Gopalakrishnan, Atul III-601
Anderson, Scott II-415
Andrzejczak, Michal I-661
Asare, Bismark Tei III-580

Ba, Cheikh I-344
Bader, David A. II-157
Baek, Nakhoon II-723
Bai, Jing II-619
Bao, Yungang II-705
Ben Messaoud, Othman I-606
Bian, Haodong II-111
Borowiec, Damian II-492
Bu, Deqing I-47

Cai, Meng-nan II-677
Cai, Shubin III-155
Cai, Wei II-200, II-215
Cai, Wenzheng III-703
Cao, Jian II-587
Cao, Weipeng I-219, I-448, II-352, II-538
Cao, Zhengjia II-215
Cérin, Christophe III-381
Chang, Peng III-494
Chang, Xiaolin I-695, II-619
Chao, Pingfu I-190
Chen, Juan I-92
Chen, Lei I-579, I-681
Chen, Lin III-676
Chen, Shuang II-97
Chen, Xuxin III-537
Chen, Zhijun III-441
Chen, Zhuang II-230
Cheng, Dongxu III-144
Cheng, Qixuan I-528
Cheng, Yongyang II-142, II-646
Chi, Yuanfang II-200
Christopher Victor, Ashish III-601
Chu, Zhaole I-15
Cong, Peijin III-564

Cooke, Jake II-415
Cui, Jiahe II-259

Dai, Chuangchuang III-654
Dai, Hua III-218
Dai, Jialu II-383
Dai, Wenhao III-475
De Giusti, Armando I-262
Deng, Anyuan III-81
Deng, Xiaoge I-495
Ding, Jia III-396
Ding, Jiafeng I-190
Ding, Kai III-3
Ding, Yepeng I-480
Ding, Zhenquan II-398
Dong, Bo I-627
Dong, Changkun II-142
Dong, Runting II-111
Dong, Xiaoyun II-259
Dong, Yong I-92, II-82, II-97
Dowling, Anthony I-3
Dricot, Jean-Michel III-309
Du, Xinlu I-465
Du, Zhihui II-157
Duan, Haihan II-200, II-215
Duan, Huifeng III-367
Duan, Lijuan III-367
Dun, Ming I-232

Eliassen, Frank II-230
Ellinidou, Soultana III-309
Engel, Fabian Herbert I-3

Faiz, Sami II-3
Fan, Dongrui I-61, II-14
Fan, Jianxi II-47
Fan, Kai III-676
Fan, Shuhui I-563
Fan, Weibei II-47
Fan, Xiaopeng I-174
Fang, Junhua I-190

Fang, Xuqi   III-537
Fang, Zhengkang   III-19
Fei, Haiqiang   II-398
Feng, Boqin   I-627
Feng, Yujing   I-61
Friday, Adrian   II-492
Fu, Shaojing   III-520
Fu, Xianghua   II-432, II-477, II-523, III-340
Fu, Zongkai   I-247, II-259

Gai, Keke   II-200, III-3, III-19, III-35,
     III-110, III-282
Gaj, Kris   I-661
Gan, Yu   I-159, II-82
Gao, Hang   III-184
Garraghan, Peter   II-492
Ge, Shuxin   II-306
Gogniat, Guy   III-309
Groß, Julian   I-369
Gu, Xiaozhuo   III-475
Guan, Hongtao   III-609
Guan, Jianbo   III-126
Guan, Zhenyu   III-144
Gueye, Abdoulaye   I-344
Gunturi, Venkata M. V.   I-125
Guo, Guibing   III-426
Guo, Qiang   I-275
Guo, Xiaobing   III-654
Guo, Yi   I-330
Guo, Yuluo   II-111
Guojian, Tang   I-719

Hamdi, Wael   II-3
Han, Ru   III-639
Han, Yetong   II-383
Han, Yuejuan   II-47
Han, Zhen   I-695
Han, Zhu   I-695
Hao, Long   II-321
Hao, Zhiyu   II-398
Harper, Richard   II-492
He, Fubao   III-297
He, Jianwei   II-477, II-523
He, Kai   III-65, III-81
He, Yulin   II-509
Honan, Reid   II-415
Hong, Xiaoguang   II-690
Hou, Aiqin   III-197

Hou, Biao   II-633
Hou, Junteng   I-31, I-645
Hu, Jiale   III-144
Hu, Jingkun   II-157
Hu, Lin   II-449
Hu, Mengtao   I-386
Hu, Wei   I-159, I-330, II-82, II-97
Hu, Wenhui   II-603
Hu, Xinrong   III-65
Hu, Yuekun   I-465
Hu, Ziyue   I-174, III-110
Huai, Xu   I-579
Huang, Chunxiao   III-65
Huang, Feng   I-495
Huang, Han   I-401
Huang, Hua   II-463, III-297, III-355, III-549
Huang, Jiahao   III-270
Huang, Jianqiang   II-111
Huang, Joshua Zhexue   II-509
Huang, Kaixin   I-433
Huang, Linpeng   I-433
Huang, Qun   I-614
Huang, Xiaofu   I-465
Huang, Xiaoping   II-575
Huang, Xin   II-184
Huang, Xinli   III-537, III-564
Huang, Yu   I-579, II-603
Huang, Zhijian   I-563
Hui, Zhao   II-142, II-646

Jayanthi, Akhilarka   III-601
Ji, Jiawei   I-143
Ji, Yimu   II-449
Jia, Ziye   I-695
Jiang, Congfeng   III-381
Jiang, Feng   II-142, II-646
Jiang, Jian-guo   II-677
Jiang, Linying   III-426
Jiang, Peng   III-35
Jiang, Shang   II-677
Jiang, Shenghong   I-143
Jiang, Zoe L.   II-126
Jiao, Qiang   II-82, II-97
Jin, Hao   III-93
Jin, Honghe   I-512
Jin, Peipei   I-579
Jin, Peiquan   I-15, III-623
Jin, Shuyuan   III-459

Jin, Xiaolong   III-50
Jing, Junchang   I-112
Jiwei, Chen   I-719

Kandoor, Lakshmi   II-184
Kay, Savio   II-184
Khedimi, Amina   III-381
Köster, Marcel   I-369
Krüger, Antonio   I-369
Kuang, Xiaoyun   III-676

Lagwankar, Ishaan   III-592
Lan, Dapeng   II-230
Lei, Yongmei   I-143
Lewis, Trent W.   II-415
Li, Chao   III-297
Li, Dawei   III-409
Li, Fei   I-416
Li, Geng   III-324
Li, Haochen   III-35
Li, Huiyong   I-247
Li, Jianchuan   III-623
Li, Jiawei   III-218
Li, Jinbao   I-305, III-509
Li, Jingjing   I-548
Li, Jun   I-710
Li, Ling   I-47
Li, Lun   II-398
Li, Ningwei   III-184
Li, Peilong   III-93
Li, Peng   II-290
Li, Sukun   III-663
Li, Wei   I-355
Li, Wenming   I-61, II-14
Li, Xiangxiang   II-32
Li, Xiangxue   I-548
Li, Xin   I-314
Li, Yi   II-14
Li, Yunchun   I-232
Li, Yuwen   III-282
Li, Yuxiang   I-78, I-112
Li, Zhenhao   II-82
Li, Zhong   III-50
Liao, Chenyang   III-270
Liao, Xiaojian   I-416
Lin, Chen   II-32
Lin, Mufeng   II-32
Lin, Yang   II-463, III-270
Lin, Zhen   III-520

Liu, Bin   I-78
Liu, Fanghan   III-703
Liu, Feng   I-495
Liu, Hongli   III-367
Liu, Jianwei   III-144, III-324, III-409
Liu, Kaihang   II-449
Liu, Lei   II-230
Liu, Li   I-386
Liu, Liang   III-184
Liu, Meiqin   III-549
Liu, Qiang   II-449
Liu, Shangdong   II-449
Liu, Wuji   III-197
Liu, Xin   II-463
Liu, Xinxin   III-549
Liu, Xuefeng   I-247, III-687
Liu, Xueyang   II-603
Liu, Yanlan   II-449
Liu, Yao   I-386
Liu, Ye   II-383
Liu, Yiyang   III-218
Liu, Yizhong   III-324, III-409
Liu, Yu   I-3, III-297
Liu, Yuan   III-426
Liu, Yufei   III-537, III-564
Liu, Zhe   I-416
Liu, Zihao   III-170
Long, Hao   I-448
Lu, Fengyuan   III-537, III-564
Lu, Jintian   III-459
Lu, Ming   III-170
Lu, Youyou   I-416
Lu, ZhongHua   I-290
Lu, Zhonghua   III-654
Luan, Hua   I-401
Luan, Zerong   I-232
Luo, Yan   III-93
Luo, Yongping   I-15
Luo, Yuchuan   III-520
Lv, Huahui   III-676
Lv, Xiangyu   I-159
Lv, Xingfeng   III-509
Lyu, Xukang   I-205

Ma, Bingnan   I-31
Ma, Dongchao   I-465
Ma, Fuhua   I-710
Ma, Kun   III-703
Ma, Qiangfei   III-355

Ma, Xingkong   III-609
Mao, Yupeng   III-170
Markowitch, Olivier   III-309
Maxwell, Thomas   II-184
Memmi, Gerard   II-274
Meng, Lianxiao   II-552
Menouer, Tarek   III-381
Miao, Weikai   III-231
Ming, Zhong   I-219, II-352, II-538, III-155
Mishra, Abhishek   I-125
Mišić, Jelena   II-619
Mišić, Vojislav   II-619
Mu, Lin   III-623
Musariri, Manasah   I-47

Nagpal, Rahul   III-592, III-601
Naiouf, Marcelo   I-262
Nana, Laurent   III-580
Nie, Feiping   II-337
Nie, Yu   III-297
Niu, Beifang   III-654
Niu, DanMei   I-112
Niu, Jianwei   I-247, II-259, III-687

Ou, Yan   I-61, II-14
Ou, Zhixin   I-92
Ouyang, Zhenchao   I-247, II-259

Pan, Yu   I-305
Park, Seung-Jong   II-723
Pei, Songwen   II-173
Peng, Jianfei   III-184
Peng, Yaqiong   II-398
Peng, Zhaohui   II-690
Pousa, Adrián   I-262

Qian, Haifeng   I-548
Qian, Shiyou   II-587
Qiao, Yuanhua   III-367
Qin, Xiaolin   I-314
Qiu, Han   II-274
Qiu, Meikang   I-159, II-173, II-274, II-463,
      III-3, III-35, III-297, III-355, III-549,
      III-654, III-663
Quist-Aphetsi, Kester   III-580

Rabbouch, Bochra   I-591
Rabbouch, Hana   I-591, I-606
Rajapakshe, Chamara   II-184

Rajashekar, Vishwas   III-592
Ramnath, Sarnath   I-125
Ren, Qianqian   I-305, I-710
Ren, Shuangyin   II-552
Ren, Yi   III-126
Rong, Guoping   II-32

S N, Durga Prasad   III-592
Saâdaoui, Foued   I-591, I-606
Sang, Yafei   III-494
Sanz, Victoria   I-262
Sato, Hiroyuki   I-480
Shao, Pengpeng   III-170
Sharma, Gaurav   III-309
Shen, Chongfei   I-61
Shen, Junzhong   I-528
Shen, Siqi   III-609
Shen, Wei   III-197
Shen, Xiaoxian   II-245
Shi, Jiakang   III-282
Shi, Jiaoli   III-65, III-81
Shi, Quanfu   II-337
Shi, Yimin   II-200
Shu, Jiwu   I-416
Si, Chengxiang   I-645
Sugizaki, Yukimasa   II-365
Sun, Jie   III-218
Sun, Qingxiao   I-232
Sun, Shibo   II-14
Sun, Tao   I-495
Sun, Xiaoxiao   I-512
Sun, Xudong   II-523
Sun, Zhi   III-50
Suo, Siliang   III-676
Swiatowicz, Frank   I-3

Taherkordi, Amir   II-230
Takahashi, Daisuke   II-365
Tan, Nongdie   I-681
Tan, Yusong   III-126
Tang, Gaigai   II-552
Tang, Minjin   I-614
Tang, Shuning   II-449
Tang, Xudong   III-231
Teng, Meiyan   I-314
Teng, Yajun   III-475
Tian, Mao   III-494
Tian, Mengmeng   III-426
Tian, Ze   III-639

Tolnai, Alexander John I-3
Tong, Li I-579
Tu, Yaofeng I-433

Valiullin, Timur I-448, II-321

Wan, Shouhong I-15
Wang, Changming III-367
Wang, Danghui II-563, III-639
Wang, Deguang I-528
Wang, Jianwu II-184
Wang, Jihe II-563, II-575
Wang, Jikui II-337
Wang, Li I-275
Wang, Qiang I-448, III-231
Wang, Ruili III-110
Wang, Sa II-705
Wang, Shiyu II-575
Wang, Shupeng I-31, I-645
Wang, Shuxin II-432, II-477, II-523, III-340
Wang, Si-ye II-677
Wang, Tianbo III-251
Wang, Wei I-386
Wang, Weixu II-306
Wang, Xiaoying II-111
Wang, Xinyi I-47
Wang, Xiwen III-170
Wang, Xizhao II-352
Wang, Yan II-47
Wang, Yang I-174
Wang, Yaobin I-47
Wang, Yonghao I-159, I-330, II-97
Wang, Yongjun I-563
Wang, Yu II-603
Wang, Zhe III-126
Wei, Chenghao I-448, II-321
Wei, Guoshuai II-662
Wei, Xiaohui II-245
Wei, Yanzhi II-432, II-477, II-523
Wei, Yihang III-282
Wen, Mei I-528, I-614
Wen, Yuan I-159, II-82
Wen, Yujuan III-564
Wu, Chase Q. I-205, III-197
Wu, Guangjun I-31, I-645
Wu, Haiyu I-627
Wu, Jie I-314
Wu, Jing I-330
Wu, Jiyan III-687

Wu, Kaishun II-383
Wu, Qianhong III-324, III-409
Wu, Quanwang II-662
Wu, Xinxin I-61, II-14
Wu, Yuhao II-538
Wu, Yusheng II-173

Xia, Chunhe III-251
Xia, Jingjing II-47
Xia, Yuanqing I-219
Xiao, Ailing I-465
Xiao, Bowen II-215
Xiao, Wan II-449
Xie, Peidai I-563
Xie, Tao III-520
Xie, Wenhao II-432, II-477, III-340
Xie, Zhongwu II-352
Xiong, Hailing I-681
Xu, Aidong III-676
Xu, Chen III-93
Xu, Chengzhong I-174
Xu, Fang III-81
Xu, Haoran I-563
Xu, Hongzuo I-563
Xu, Jiajie I-190
Xu, Jianqiu III-218
Xu, Jiawei II-587
Xu, Lei III-19
Xu, Liwen I-512
Xu, Wenxing II-14
Xu, Yuan II-705
Xu, Zhengyang II-449
Xu, Zhiwu II-538, III-396
Xue, Guangtao II-587
Xue, Wei I-386

Yan, Ruibo II-142, II-646
Yang, Chenlong I-548
Yang, Geng III-218
Yang, Guang II-690
Yang, Hailong I-232
Yang, Hang III-676
Yang, Jingying III-340
Yang, Lei II-215
Yang, Lin II-552
Yang, NingSheng III-155
Yang, Renyu II-492
Yang, Saqing III-126
Yang, Shaokang III-687

Yang, Wu   II-552
Yang, Xueying   III-654
Yang, Yang   I-47, I-695, II-619
Yang, Yifan   II-32
Yang, Zhe   I-416
Yang, Zhengguo   II-337
Ye, Feng   III-170
Ye, Qianwen   III-197
Ye, Xianglin   I-681
Ye, Xiaochun   I-61
Ye, Xuan   II-509
Yeung, Gingfung   II-492
Yin, Hao   III-282
Yin, Jiayuan   III-324
Ying, Yaoyao   I-433
Yiu, Siu Ming   II-126
You, Xin   I-232
You, Ziqi   III-126
Yu, Dunhui   III-441
Yu, Hui   III-324, III-409
Yu, Huihua   I-579
Yu, Shucheng   II-62
Yuan, Baojie   II-383
Yuan, Yuan   I-92
Yue, Chen   III-639
Yue, Hengshan   II-245

Zeng, Xiao   I-159
Zeng, Yi   II-274
Zeng, Yunhui   I-275
Zhan, Ke   I-290
Zhang, Chaokun   II-306
Zhang, Chunyuan   I-528, I-614
Zhang, Hanning   I-627
Zhang, Jie   I-275
Zhang, Jiyong   I-448, II-538
Zhang, Jun   II-126
Zhang, Junxing   II-633
Zhang, Lei   I-31, I-645
Zhang, Lu   III-639
Zhang, Mengjie   III-441
Zhang, Minghui   II-603
Zhang, Sen   II-157
Zhang, Shengbing   II-575
Zhang, Tao   II-290
Zhang, Tianwei   II-705
Zhang, Wentao   III-355
Zhang, Xingsheng   III-441

Zhang, Yan-fang   II-677
Zhang, Yang   I-305
Zhang, Yanlin   II-290
Zhang, Yongzheng   III-494
Zhang, Yue   III-3
Zhang, Yunfang   I-92
Zhang, YunQuan   I-290
Zhang, Yunyi   III-459
Zhang, Zhibo   II-184
Zhang, Zhiyong   I-78, I-112
Zhang, Zijian   III-110
Zhao, Bo-bai   II-677
Zhao, Da-ming   II-62
Zhao, Hui   II-603
Zhao, Jiaxiang   II-563
Zhao, Jie   III-623
Zhao, Lei   I-190
Zhao, PengPeng   I-190
Zhao, Qinglin   II-306
Zhao, Shuyuan   III-494
Zhao, Wenqian   II-32
Zhao, Xiaolei   I-614
Zhao, Yonglin   II-432, II-477, II-523
Zheng, Hongqiang   I-330
Zheng, Jianyu   II-184
Zheng, Shengan   I-433
Zheng, Weiyan   I-579
Zhou, Changbao   II-245
Zhou, Cong   II-538
Zhou, Fangkai   III-270
Zhou, Hao   I-681
Zhou, Huaifeng   III-155
Zhou, Jian-tao   II-62
Zhou, Jingren   III-623
Zhou, Xiaobo   II-306
Zhu, Guanghui   I-275
Zhu, Haoran   II-619
Zhu, Junwei   II-587
Zhu, Liehuang   III-19, III-35, III-110, III-282
Zhu, Qingting   I-386
Zhu, Weidong   II-587
Zhu, Yanmin   II-587
Zhu, Zongyao   II-587
Zhuang, Chuanzhi   III-50
Zhuang, Yuan   I-275
Zou, Weidong   I-219
Zou, Xingshun   I-355
Zou, Yongpan   II-383